# Recent Advances in Synthesis, Characterization and Applications of Innovative Materials in Removal of Water Contaminants

# Recent Advances in Synthesis, Characterization and Applications of Innovative Materials in Removal of Water Contaminants

Editor

**Thomas Dippong**

MDPI • Basel • Beijing • Wuhan • Barcelona • Belgrade • Manchester • Tokyo • Cluj • Tianjin

*Editor*
Thomas Dippong
Chemistry and Biology
Technical University of
Cluj Napoca
Baia Mare
Romania

*Editorial Office*
MDPI
St. Alban-Anlage 66
4052 Basel, Switzerland

This is a reprint of articles from the Special Issue published online in the open access journal *International Journal of Molecular Sciences* (ISSN 1422-0067) (available at: www.mdpi.com/journal/ijms/special_issues/water_contaminants).

For citation purposes, cite each article independently as indicated on the article page online and as indicated below:

LastName, A.A.; LastName, B.B.; LastName, C.C. Article Title. *Journal Name* **Year**, *Volume Number*, Page Range.

**ISBN 978-3-0365-6357-2 (Hbk)**
**ISBN 978-3-0365-6356-5 (PDF)**

© 2023 by the authors. Articles in this book are Open Access and distributed under the Creative Commons Attribution (CC BY) license, which allows users to download, copy and build upon published articles, as long as the author and publisher are properly credited, which ensures maximum dissemination and a wider impact of our publications.

The book as a whole is distributed by MDPI under the terms and conditions of the Creative Commons license CC BY-NC-ND.

# Contents

**About the Editor** . . . . . . . . . . . . . . . . . . . . . . . . . . . . . . . . . . . . . . . . . . . . . . . . . . . . . ix

**Thomas Dippong**
Recent Advances in Synthesis, Characterization and Applications of Innovative Materials in Removal of Water Contaminants
Reprinted from: *Int. J. Mol. Sci.* **2022**, 24, 330, doi:10.3390/ijms24010330 . . . . . . . . . . . . . . . 1

**Ze Liao, Yang Zi, Chunyan Zhou, Wenqian Zeng, Wenwen Luo and Hui Zeng et al.**
Recent Advances in the Synthesis, Characterization, and Application of Carbon Nanomaterials for the Removal of Endocrine-Disrupting Chemicals: A Review
Reprinted from: *Int. J. Mol. Sci.* **2022**, 23, 13148, doi:10.3390/ijms232113148 . . . . . . . . . . . . . 9

**Thomas Dippong, Erika Andrea Levei, Iosif Grigore Deac, Ioan Petean and Oana Cadar**
Dependence of Structural, Morphological and Magnetic Properties of Manganese Ferrite on Ni-Mn Substitution
Reprinted from: *Int. J. Mol. Sci.* **2022**, 23, 3097, doi:10.3390/ijms23063097 . . . . . . . . . . . . . . 29

**Thomas Dippong, Erika Andrea Levei and Oana Cadar**
Investigation of Structural, Morphological and Magnetic Properties of $MFe_2O_4$ (M = Co, Ni, Zn, Cu, Mn) Obtained by Thermal Decomposition
Reprinted from: *Int. J. Mol. Sci.* **2022**, 23, 8483, doi:10.3390/ijms23158483 . . . . . . . . . . . . . . 45

**Thomas Dippong, Oana Cadar, Firuta Goga, Dana Toloman and Erika Andrea Levei**
Impact of Ni Content on the Structure and Sonophotocatalytic Activity of Ni-Zn-Co Ferrite Nanoparticles
Reprinted from: *Int. J. Mol. Sci.* **2022**, 23, 14167, doi:10.3390/ijms232214167 . . . . . . . . . . . . . 57

**Ahmed R. Weshahy, Ahmed K. Sakr, Ayman A. Gouda, Bahig M. Atia, H. H. Somaily and Mohamed Y. Hanfi et al.**
Selective Recovery of Cadmium, Cobalt, and Nickel from Spent Ni–Cd Batteries Using Adogen® 464 and Mesoporous Silica Derivatives
Reprinted from: *Int. J. Mol. Sci.* **2022**, 23, 8677, doi:10.3390/ijms23158677 . . . . . . . . . . . . . . 73

**Claudia Mihaela Ninciuleanu, Raluca Ianchiș, Elvira Alexandrescu, Cătălin Ionuț Mihăescu, Sabina Burlacu and Bogdan Trică et al.**
Adjusting Some Properties of Poly(methacrylic acid) (Nano)Composite Hydrogels by Means of Silicon-Containing Inorganic Fillers
Reprinted from: *Int. J. Mol. Sci.* **2022**, 23, 10320, doi:10.3390/ijms231810320 . . . . . . . . . . . . . 95

**Simona Filice, Viviana Scuderi, Sebania Libertino, Massimo Zimbone, Clelia Galati and Natalia Spinella et al.**
Sulfonated Pentablock Copolymer Coating of Polypropylene Filters for Dye and Metal Ions Effective Removal by Integrated Adsorption and Filtration Process
Reprinted from: *Int. J. Mol. Sci.* **2022**, 23, 11777, doi:10.3390/ijms231911777 . . . . . . . . . . . . . 115

**Tae-Hyun Kim, Chihyun Seo, Jaeyoung Seon, Anujin Battulga and Yuhoon Hwang**
Granulation of Bismuth Oxide by Alginate for Efficient Removal of Iodide in Water
Reprinted from: *Int. J. Mol. Sci.* **2022**, 23, 12225, doi:10.3390/ijms232012225 . . . . . . . . . . . . . 129

**Xiaoxin Chen, Jiacheng Lin, Yingjie Su and Shanshan Tang**
One-Step Carbonization Synthesis of Magnetic Biochar with 3D Network Structure and Its Application in Organic Pollutant Control
Reprinted from: *Int. J. Mol. Sci.* **2022**, 23, 12579, doi:10.3390/ijms232012579 . . . . . . . . . . . . . 145

Zhifei Liu, Yaqi Tan, Xuefeng Ruan, Jing Guo, Wei Li and Jiajun Li et al.
Spark Plasma Sintering-Assisted Synthesis of $Bi_2Fe_4O_9/Bi_{25}FeO_{40}$ Heterostructures with Enhanced Photocatalytic Activity for Removal of Antibiotics
Reprinted from: *Int. J. Mol. Sci.* **2022**, 23, 12652, doi:10.3390/ijms232012652 . . . . . . . . . . . . . 169

Jiaming Zhu, Zuohua Liu, Hao Wang, Yue Jian, Dingbiao Long and Shihua Pu
Preparation of a Z-Type $g-C_3N_4/(A-R)TiO_2$ Composite Catalyst and Its Mechanism for Degradation of Gaseous and Liquid Ammonia
Reprinted from: *Int. J. Mol. Sci.* **2022**, 23, 13131, doi:10.3390/ijms232113131 . . . . . . . . . . . . . 181

Fangyuan Zheng, Pedro M. Martins, Joana M. Queirós, Carlos J. Tavares, José Luis Vilas-Vilela and Senentxu Lanceros-Méndez et al.
Size Effect in Hybrid $TiO_2$:Au Nanostars for Photocatalytic Water Remediation Applications
Reprinted from: *Int. J. Mol. Sci.* **2022**, 23, 13741, doi:10.3390/ijms232213741 . . . . . . . . . . . . . 199

Jian Feng, Xia Ran, Li Wang, Bo Xiao, Li Lei and Jinming Zhu et al.
The Synergistic Effect of Adsorption-Photocatalysis for Removal of Organic Pollutants on Mesoporous $Cu_2V_2O_7/Cu_3V_2O_8/g-C_3N_4$ Heterojunction
Reprinted from: *Int. J. Mol. Sci.* **2022**, 23, 14264, doi:10.3390/ijms232214264 . . . . . . . . . . . . . 217

Yingjie Su, Yuqing Shi, Meiyi Jiang and Siji Chen
One-Step Synthesis of Nitrogen-Doped Porous Biochar Based on N-Doping Co-Activation Method and Its Application in Water Pollutants Control
Reprinted from: *Int. J. Mol. Sci.* **2022**, 23, 14618, doi:10.3390/ijms232314618 . . . . . . . . . . . . . 233

Alessandra Piras, Chiara Olla, Gunter Reekmans, An-Sofie Kelchtermans, Dries De Sloovere and Ken Elen et al.
Photocatalytic Performance of Undoped and Al-Doped ZnO Nanoparticles in the Degradation of Rhodamine B under UV-Visible Light: The Role of Defects and Morphology
Reprinted from: *Int. J. Mol. Sci.* **2022**, 23, 15459, doi:10.3390/ijms232415459 . . . . . . . . . . . . . 255

Dorota Babilas, Anna Kowalik-Klimczak and Anna Mielańczyk
Recovery of the *N,N*-Dibutylimidazolium Chloride Ionic Liquid from Aqueous Solutions by Electrodialysis Method
Reprinted from: *Int. J. Mol. Sci.* **2022**, 23, 6472, doi:10.3390/ijms23126472 . . . . . . . . . . . . . 275

Xiao-Xia Zheng, You-Cheng Pan and Wei-Feng Sun
Water-Tree Characteristics and Its Mechanical Mechanism of Crosslinked Polyethylene Grafted with Polar-Group Molecules
Reprinted from: *Int. J. Mol. Sci.* **2022**, 23, 9450, doi:10.3390/ijms23169450 . . . . . . . . . . . . . 287

Xingyu Liu, Yan Liang, Yongtao Peng, Tingting Meng, Liling Xu and Pengcheng Dong
Sensitivity of the Transport of Plastic Nanoparticles to Typical Phosphates Associated with Ionic Strength and Solution pH
Reprinted from: *Int. J. Mol. Sci.* **2022**, 23, 9860, doi:10.3390/ijms23179860 . . . . . . . . . . . . . 301

Hosung Yu, Kang Hoon Lee and Jae-Woo Park
Impact of Acetate in Reduction of Perchlorate by Mixed Microbial Culture under the Influence of Nitrate and Sulfate
Reprinted from: *Int. J. Mol. Sci.* **2022**, 23, 10608, doi:10.3390/ijms231810608 . . . . . . . . . . . . . 315

Xia Ran, Li Wang, Bo Xiao, Li Lei, Jinming Zhu and Zuoji Liu et al.
Effective Removal of Methylene Blue on $EuVO_4/g-C_3N_4$ Mesoporous Nanosheets via Coupling Adsorption and Photocatalysis
Reprinted from: *Int. J. Mol. Sci.* **2022**, 23, 10003, doi:10.3390/ijms231710003 . . . . . . . . . . . . . 327

**Cristina Iosif, Stanca Cuc, Doina Prodan, Marioara Moldovan, Ioan Petean and Mîndra Eugenia Badea et al.**
Effects of Acidic Environments on Dental Structures after Bracket Debonding
Reprinted from: *Int. J. Mol. Sci.* **2022**, *23*, 15583, doi:10.3390/ijms232415583 . . . . . . . . . . . . . **343**

# About the Editor

**Thomas Dippong**

Thomas Dippong (Associate Professor, Doctor at the Technical University of Cluj Napoca) is a chemical engineer with a PhD and hability in chemistry. His current research activities are related to obtained and characterization nanoparticles for various applications, as part of ongoing research in partnership with the Technical University of Cluj-Napoca within the field of ferrite embedded in silica matrix. He is an expert in analytical chemistry, organic/inorganic chemistry, thermal treatment, instrumental analysis, and synthesis of nanomaterials. Dr Dippong has published 131 peer-reviewed publications (75 papers in high ranked scientific ISI-Thomson journals (40 Q1, 12 Q2 and 23 Q3), 56 in other national and international journals), 1309 citations, h-index: 31 (WoS) and has given 34 lectures at international conferences (ICTAC 14 Brazil, ESTAC Brasov, JTACC Budapest, CEEC-TAC5 Roma, etc). He has also published two books at international publishing houses and 15 books at national publishing houses. He has been the contract manager for projects totalling a budget of EUR 4 million and is currently an active member of four other projects. Dr. Dippong has reviewed 400 scientific articles for 75 ISI-Thomson journals. He is the guest editor of seven Special Issues of five prestigious Q1 ISI-Thomson journals.

*Editorial*

# Recent Advances in Synthesis, Characterization and Applications of Innovative Materials in Removal of Water Contaminants

Thomas Dippong

Faculty of Science, Technical University of Cluj-Napoca, 76 Victoriei Street, 430122 Baia Mare, Romania; dippong.thomas@yahoo.ro

**Citation:** Dippong, T. Recent Advances in Synthesis, Characterization and Applications of Innovative Materials in Removal of Water Contaminants. *Int. J. Mol. Sci.* **2023**, *24*, 330. https://doi.org/10.3390/ijms24010330

Received: 15 December 2022
Revised: 19 December 2022
Accepted: 21 December 2022
Published: 25 December 2022

**Copyright:** © 2022 by the author. Licensee MDPI, Basel, Switzerland. This article is an open access article distributed under the terms and conditions of the Creative Commons Attribution (CC BY) license (https://creativecommons.org/licenses/by/4.0/).

Water is a scarce resource with a close and intricate nexus with energy. Water contamination has been reported in almost every region of the world, with a significant impact on human health [1,2]. There are numerous available water decontamination methods, based on various techniques with different efficiencies and operational costs. Engineered nanomaterials with magnetic properties allow the adsorption of contaminants, followed by magnetic separation, while other nanomaterials allow the contaminant's photodegradation [3,4]. The use of these innovative materials, whether for pollutant adsorption or decomposition by photocatalysis, or for constructing low-cost sensors for the detection of contaminants, has gained interest in the preceding decades.

This Special Issue focuses on the (*i*) application of innovative materials in water decontamination, (*ii*) the synthesis and characterization of engineered nanocomposites, (*iii*) water decontamination by photocatalysis, adsorption and other techniques, and (*iv*) computational and theoretical studies of the reaction mechanisms, kinetics and thermodynamics of water depollution processes.

In this Special Issue, Liao et al. [5] reported the large-scale production and characterization of carbon nanomaterials to remove endocrine-disrupting chemicals (EDCs). EDCs are continuously released and widely spread pollutants in natural environments. At low levels, EDC exposure may cause metabolic, sexual development, and reproductive disorders in aquatic animals and humans [5]. The removal of EDCs from wastewater by adsorption to nanomaterials has wide applicability. Carbon-based nanomaterials (carbon nanotubes, graphene, magnetic carbon nanomaterials, carbon membranes, carbon dots, carbon sponges) have been extensively explored for EDCs adsorption because they are eco-friendly, have good chemical stability, structural diversity, low density, and are suitable for large-scale production [5]. The applications of carbon nanomaterials for the removal of different kinds of EDCs and the adsorption mechanism, as well as recent advances in carbon nanocomposite synthesis and characterization, are discussed [5]. The preparation cost of carbon nanomaterials, such as carbon nanotubes and graphene, is relatively high and there are some technical difficulties with their recyclability [5,6]. It is still challenging to develop new, safe, efficient, and lower-cost carbon nanocomposite adsorbents [5].

Magnetic spinel ferrite ($MFe_2O_4$, where M = Zn, Co, Mn, Ni, etc.) nanoparticles are of high interest to researchers in the fields of materials science and nanotechnology [3,4,6]. The sol–gel route is the most popular means of preparing nanosized ferrites due to its simplicity, low cost, and good control over their structure and properties [3,4,7]. The microwave-assisted sol–gel method combines the advantages of microwave and sol–gel methods, constituting a faster, energy-saving procedure for obtaining single-phase nanopowders of high purity, with accurate control of stoichiometry and capability of industrial scale-up [3,4,7]. Embedding ferrites into a silica ($SiO_2$) matrix allows the control of particle growth, minimizes particle agglomeration, and enhances their magnetic guidability and overall biocompatibility [3,4,7].

This Special Issue also includes the study of Dippong et al. [7], an investigation into the structure, morphology and magnetic properties of $MFe_2O_4$ (M = Co, Ni, Zn,

Cu, Mn), obtained by thermal decomposition. Unlike similar ferrites embedded in $SiO_2$ matrices, single crystalline phases were obtained at both temperatures, excepting the presence of $Co_3O_4$ ($CoFe_2O_4$) and $\alpha$-$Fe_2O_3$ ($MnFe_2O_4$) at 700 °C [7]. $CuFe_2O_4$ showed the largest particle size (85 nm), while $MnFe_2O_4$ had the smallest particle size (32 nm) [7]. The crystalline $CoFe_2O_4$, heat treated at 1000 °C, displayed the highest saturation magnetization ($M_s$), coercive field ($H_C$) and anisotropy constant ($K$) values, presenting superparamagnetic behavior. Conversely, the other ferrites exhibited paramagnetic behavior [7].

The dependence of structural, morphological and magnetic properties on Ni-Mn substitution in manganese ferrite, synthesized by the sol–gel method and annealed at different temperatures, was reported by Dippong et al. [8]. A number of features were identified, ranging from the presence of poorly crystalline ferrite and highly crystalline mixed cubic spinel ferrite at low annealing temperatures, accompanied by secondary phases at high annealing temperatures, to the gradual decreases in lattice parameters and increases with rising Ni content in the crystallite size, volume, and X-ray density of $Mn_{1-x}Ni_xFe_2O_4@SiO_2$ NCs [8]. With increasing Ni content, the saturation magnetization, remanent magnetization, squareness, magnetic moment per formula unit, and anisotropy constant all increase as well, while the coercivity decreases [8]. The magnetic properties of the NCs were strongly dependent on chemical composition, cation distribution between tetrahedral (A) and octahedral (B) sites, as well as on surface effects derived from the synthesis methods [8].

The study of Dippong et al. [9], focusing on the impact of Ni content on the structure and sonophotocatalytic activity of Ni-Zn-Co ferrite nanoparticles, was also included in this Special Issue. The obtained results indicated the formation and decomposition of metal succinate precursors in two stages, with distinct formation and decomposition of divalent ($Ni^{2+}$, $Zn^{2+}$, $Co^{2+}$) and trivalent ($Fe^{3+}$) succinates [9]. The XRD analysis revealed the presence of well-crystallized ferrites along two crystalline phases of the $SiO_2$ matrix (cristobalite and tridymite) [7]. In samples with high Zn content, traces of hematite were also identified [9]. Both the agglomeration of particles and the particle size of Ni-Zn-Co ferrites increase with the rising Ni content, the latter growing from 34 nm to 40 nm [9]. All samples showed an excellent optical response in the visible range, the best sonophotocatalytic performance being found for the $Ni_{0.3}Zn_{0.3}Co_{0.4}Fe_2O_4@SiO_2$ sample, most likely due to the equilibrium between Ni-ferrite and Zn-ferrite [9].

In this Special Issue, the selective recovery of cadmium, cobalt, and nickel from spent Ni–Cd batteries was reported by Weshahy et al., who used adogen® 464 and mesoporous silica derivatives [10]. Spent Ni–Cd batteries are now considered an important source for many valuable metals [10]. Adogen 464 was used for $Cd^{2+}$ extraction, followed by precipitation as a yellow CdS product with 0.5% $Na_2S$ solution, by setting the pH at 1.25 and maintaining room temperature conditions. The optimum leaching process was achieved using 20% $H_2SO_4$, solid/liquid (S/L) 1/5 at 80 °C for 6 h [10]. The leaching efficiency of Fe, Cd, and Co was nearly 100%, whereas the leaching efficiency of Ni was 95% [10]. The prepared 1,1′-(4-hydroxy-1,3-phenylene) bis(3-(3-(triethoxysilyl)propyl)urea-bridged mesoporous organosilica (PTU-MS) silica was applied for adsorption of Co(II) ions from aqueous solution, while the desorption process was performed using 0.3 M $H_2SO_4$ [10].

The study of Ninciuleanu et al. [11], focusing on the impact of adjusting some properties of poly (methacrylic acid) (nano) composite hydrogels by means of silicon-containing inorganic fillers, was also included in this Special Issue. The effect of these fillers, in correlation with their characteristics, structure and swelling, as well as the viscoelastic and water decontamination properties of (nano)composite hydrogels, were studied comparatively [11]. The experiments demonstrated that the nanocomposite hydrogel morphology was determined by the way the filler particles were dispersed in water [11]. The structure of poly (methacrylic acid) hydrogels was also affected by the reinforcing agent through the pH of its aqueous dispersion [11]. The two Laponite XLS/XLG clays and montmorillonite led to exfoliated and intercalated nanocomposites, respectively, while pyrogenic silica formed agglomerations of spherical nanoparticles within the hydrogel [11]. The equilibrium swelling degree depended on both the pH of the environment and the filler nature [11]. At approximately constant swelling degree, the filler addition improved the mechanical properties of the (nano)composite hydrogels, while after equilibrium swelling (pH 5.4), the viscoelastic moduli values depended on the filler [11]. The rheological measurements also showed that the strongest hydrogels were obtained in the case of the Laponite XLS/XLG clays reinforcing agent [11]. The synthesized (nano)composite hydrogels displayed a different ability to decontaminate cationic dye-containing waters as a function of the filler included, with the highest absorption rate and absorption capacity being displayed by the Laponite XLS-reinforced hydrogel [11]. The (nano)composite hydrogels discussed here may also find applications in the pharmaceutical field as substances to mediate the controlled release of drugs [11].

In this Special Issue, the effective removal of sulfonated pentablock copolymer coating of polypropylene filters for dye and metal ions by integrated adsorption and filtration process was reported by Filice et al. [12]. The polypropylene (PP) fibrous filters, equipped with sulfonated pentablock copolymer (s-PBC) layers, were tested for their capacity to remove cationic organic dyes, such as methylene blue and heavy metal ions ($Fe^{3+}$ and $Co^{2+}$), from water by adsorption and filtration [12]. Polymer-coated filters showed high efficiency in removing methylene blue from an aqueous solution in both the absorption and filtration processes, with 90% and 80% removal rates, respectively [12]. The coated filters showed a better performance removing heavy metal ions ($Fe^{3+}$ and $Co^{2+}$) during filtration than adsorption [12]. In the adsorption process, controlled interaction times allow the ionic species to interact with the surface of the filters leading to the formation and release of new species in solution [12]. During filtration, the ionic species are easily trapped in the filters, especially those that are UV-modified. A total removal (>99%) via a single filtration process was observed for $Fe^{3+}$ ions [12]. The filtration processes are faster, and, therefore, the interaction time is not sufficient to release reaction byproducts into the solution; the ionic species are easily trapped in the filters, in particular the filters whose surface was modified by UV treatment [12]. It has also been shown that the treatment increases the hydrophilicity of the filters, enhancing their filtration capacity [12]. Although further work is needed to extensively investigate the lifetime and regeneration processes of such filters, the functional polymeric coating of commercial and low-cost filters is a promising strategy for the effective removal of pollutants from water [12].

This Special Issue also includes the study of Kim et al. [13], which highlighted the granulation of bismuth oxide by alginate for iodide removal from water. The granulation of bismuth oxide by alginate were presented, along with the iodide adsorption efficacy of alginate–bismuth oxide for different initial iodide concentrations and contact time values [13]. Bismuth oxide appeared in two forms: $Bi_2O_{2.33}$ and $\gamma$-$Bi_2O_3$, and was successfully granulated with alginate, yielding spherical beams with a diameter of 3 mm. The intraparticle pores in the granule could enhance iodide adsorption [13]. Iodide adsorption by alginate–bismuth oxide gradually increased and did not reach a plateau, even at an initial iodide concentration of 1000 mg/L. The process occurred as monolayer adsorption by the chemical interaction and precipitation between bismuth and iodide, followed by physical multilayer adsorption at a very high concentration of iodide in solution [13]. The surface and cross-section after iodide adsorption indicated that the adsorbed iodide interacted with bismuth oxide in alginate–bismuth oxide through Bi-O-I complexation [13]. These data showed that alginate–bismuth oxide is a promising iodide adsorbent with a high absorption capacity and stability, and it can help to prevent secondary pollution [13].

In this Special Issue, Chen et al. reported on the one-step carbonization synthesis of magnetic biochar (BMFH/$Fe_3O_4$) with a 3D network structure, undertaken by controlling different heating conditions in a high-temperature process, as well as its application in organic pollutant control [14]. The microbial filamentous fungus *Trichoderma reesei* was used as a template, and $Fe^{3+}$ was added to the culture process, which resulted in uniform recombination through the bio-assembly properties of fungal hyphae [14]. The presence of magnetic nanoparticles allows researchers to recover biochar from water and convert it into a Fenton-like catalytic reagent that improves treatment efficiency [14]. The catalytic degradation of organic pollutants reached 99% in 60 min [14]. After 10 cycles, malachite and tetracycline hydrochloride removal by BMFH/Fe$_3$O$_4$ remained above 80% [14].

This Special Issue also includes the study of Liu et al. [15] on a facile and quick preparation of Bi$_2$Fe$_4$O$_9$/Bi$_{25}$FeO$_{40}$ hetero structures with enhanced photocatalytic activity for the removal of antibiotics by the hydrothermal method, combined with spark plasma sintering (SPS). Bismuth ferrite-based heterojunction composites are promising visible light-responsive photocatalysts because of their narrow band gap structure; however, the synthetic methods reported in the literature were usually time-consuming [15]. It was found that the formation of a well-defined heterojunction between Bi$_2$Fe$_4$O$_9$ and Bi$_{25}$FeO$_{40}$ speeds up the transformation and separation of photoinduced carriers, enhancing their photoelectric properties and photocatalytic performance [15]. The possible influence factors of spark plasma sintering on photoelectric and photocatalytic performance of bismuth ferrite-based composites were also discussed [15]. This study provides a simple, feasible and economical method for the facile and quick synthesis of a highly active bismuth ferrite-based visible light-driven photocatalyst for practical applications [15].

The study of Zhu et al. [16] on the preparation of a Z-type g-C$_3$N$_4$/(A-R)TiO$_2$ composite catalyst and its mechanism for the degradation of gaseous and liquid ammonia is also presented in this Special Issue. The g-C$_3$N$_4$/(A-R)TiO$_2$ composite catalyst had a better dispersion, a smaller band gap width, a larger specific surface area, a stronger light absorption capacity, and a stronger photogenerated carrier separation ability than (A-R)TiO$_2$ catalyst [16]. Gaseous and liquid ammonia were used as the target pollutants to investigate the activity of the prepared catalysts, and the results showed that the air wetness and initial concentration of ammonia had a great influence on its degradation [16]. The superoxide anion radical (O$_2^-$) and hydroxyl radical (OH) were the main active components in the photocatalytic reaction process [16]. The photogenerated electrons in the conduction band of (A-R)TiO$_2$ catalyst transferred to the valence band of g-C$_3$N$_4$ and combined with the photogenerated holes in the valence band of g-C$_3$N$_4$, forming a Z-type heterostructure that significantly improved the efficiency of the photogenerated electron–hole migration and separation, thus increasing the reaction rate [16].

This Special Issue also includes the study of Zheng et al. [17], that reported a size effect in hybrid TiO$_2$:Au nanostars as a promising alternative method with which to remove contaminants of emerging concern from wastewaters under sunlight irradiation. TiO$_2$:Au nanostars with different Au component sizes and branching were generated and tested in the degradation of ciprofloxacin. They showed the highest photocatalytic degradation, between 83% and 89% under UV and VIS radiation, together with a threshold in photocatalytic activity in the red region [17]. The large size of the Au-branched nanoparticles extended the light absorption to the visible range, in addition to part of the NIR region, and reduced the bandgap from 3.10 eV to 2.86 eV, respectively [17]. The applicability of TiO$_2$:Au-NSs with lower branching and optimum performance was further explored with their incorporation into a porous matrix, based on PVDF-HFP. The concept behind this was to open the way for a reusable energy cost-effective system in the photo-degradation of emerging contaminants [17]. The membranes were produced successfully and presented high porosity and a well-distributed porous structure [17].

In this Special Issue, Feng et al. reported that the synergistic effect of adsorption–photocatalysis for the removal of organic pollutants on mesoporous Cu$_2$V$_2$O$_7$/Cu$_3$V$_2$O$_8$/g-C$_3$N$_4$ heterojunctions (CVCs) enhanced visible light absorption and improved the separation efficiency of photoinduced charge carriers [18]. CVCs exhibited superior adsorption capacity and photocatalytic performance in comparison with pristine g-C$_3$N$_4$ (CN) CVC-2 (containing 2 wt% of Cu$_2$V$_2$O$_7$/Cu$_3$V$_2$O$_8$) [18]. Good synergistic removal efficiencies were obtained for dyes (96.2% for methylene blue, 97.3% for rhodamine B) and antibiotics (83.0% for ciprofloxacin, 86.0% for tetracycline and 80.5% for oxytetracycline) [18]. The pseudo-first-order rate constants of methylene blue and rhodamine B photocatalytic degradation on CVC-2 were 3 times and 10 times that of pristine g-C$_3$N$_4$. This work provides a reliable reference for wastewater treatment [18].

This Special Issue also includes the study of Su et al. [19] on the one-step synthesis of nitrogen-doped porous biochar, based on the nitrogen-doping co-activation method and its application in the control of water pollutants. Birch bark (BB) was used for the first time to prepare porous biochars via different one-step methods, including direct activation (BBB) and N-doping co-activation (N-BBB) [19]. The specific surface area and total pore volume of BBB and N-BBB were 2502.3 and 2292.7 m$^2$/g, and 1.1389 and 1.0356 cm$^3$/g, respectively, proving the feasibility of N-doping co-activation in pore-forming [19]. The large specific surface area and the high total pore volume played a substantial role in the adsorption process. Both BBB and N-BBB showed excellent capacity to remove methyl orange dye and Cr$^{6+}$. The adsorption capacity of N-BBB remained above 80% after five regenerations, which fully proved the stability of regeneration [19]. Moreover, the excellent adsorption performance of N-BBB may have been influenced by pore filling, π–π interaction, H-bond interaction, and electrostatic attraction, all of which supported the biochars' high performance. This study provided new directions for biomass valorization [19].

In this Special Issue, Piras et al. reported the influence of the defects and morphology of undoped and Al-doped ZnO nanoparticles on the photocatalytic degradation efficiency of Rhodamine B under UV-Visible [20]. The undoped ZnO nanopowder, annealed at 400 °C, resulted in the highest degradation efficiency of ca. 81% after 4 h under green light irradiation (525 nm) in the presence of 5 mg of catalyst [20]. Photoluminescence showed that the insertion of a dopant increases the oxygen vacancies, reducing the peroxide-like species responsible for photocatalysis [20]. The annealing temperature helps to increase the number of red-emitting centers up to 400 °C, while at 550 °C the photocatalytic performance drops due to the aggregation tendency [20]. These results suggest the interconnection between defects, synthesis and post-synthesis routes, particle size and photocatalytic activity [20]. The high amount of defect that is not absorbed in the green region, and the reduction of $O_2^{2-}$ species, possibly deactivate the photocatalytic process. This has the effect of rendering the Al-doped ZnO catalysts inactive for the photodegradation of the Rhodamine B dye in an aqueous solution under green light [20].

This Special Issue also includes the study of Babilas et al. [21] on the recovery of the N,N-Dibutylimidazolium chloride ionic liquid from aqueous solutions by an electrodialysis method. The fact that the recovery ratio, the [C$_4$C$_4$IM]Cl molar flux, and the electric current efficiency increase with a rising concentration of [C$_4$C$_4$IM]Cl in the feed solution was reported [21]. The energy consumption also highly depends on the initial [C$_4$C$_4$IM]Cl content and increases linearly with the increase in the [C$_4$C$_4$IM]Cl concentration in the initial solution. The ionic liquids concentration in the feed solution influences the solution conductivity, electrical resistance, and concentration polarization, as well as the electrodialysis efficiency [21] by a reduction in electrical resistance of the initial dilute with the increasing ionic liquids concentration and the acceleration of ions transport across membranes [21].

In this Special Issue, Zheng et al. reported water tree characteristics and its mechanism of crosslinking polyethylene grafted with polar group molecules. [22]. The researchers aimed to restrain the electric stress impacts of water micro-droplets on insulation defects under electric fields in crosslinked polyethylene material. To do so, chemical graft modifications were performed by introducing chloroacetic acid allyl ester and maleic anhydride individually, as two specific polar group molecules, into polyethylene material via a peroxide melting approach [22]. Combined with Monte Carlo molecular simulations, it is verified that water tree resistance can be significantly promoted by grafting polar group molecules [22]. The grafted polar groups can enhance Van der Waals' forces between polyethylene molecules and are available as heterogeneous nucleation centers for polyethylene crystallization, both of which lead to the increased densities of spherulites with reduced-volume and increased-tenacity amorphous regions between lamellae. These developments account for the considerable improvement in water tree resistance [22]. The crosslinked polyethylene materials improve the insulation performances of polyethylene insulation materials in high-voltage cable manufacturing by grafting on polar group molecules [22].

This Special Issue also includes the study of Liu et al. [23] on the sensitivity of plastic nanoparticle transportation to typical phosphates associated with ionic strength and solution pH. The influence of phosphates on the transport of plastic particles in porous media is environmentally relevant due to their ubiquitous coexistence in the environment [23]. The trends of plastic nanoparticles transport vary with increasing concentrations of $NaH_2PO_4$ and $Na_2HPO_4$ due to the coupled effects of increased electrostatic repulsion, the competition for retention sites, and the double layer compression [23]. Hydrogen bonding from two phosphates that act as proton donors contributes to variations in the interactions of plastic nanoparticles and porous media and thus influences plastic nanoparticles transport [23]. High pH values tend to increase the rate of plastic nanoparticles transportation due to their enhanced deprotonation of surfaces [23]. The presence of physicochemical heterogeneities on solid surfaces can reduce rates of plastic nanoparticles transport and increase the sensitivity of the transport to ionic strength [23]. This study highlights the sensitivity of plastic nanoparticles transport to phosphates and contributes to the better understanding of the fate of plastic nanoparticles and other colloidal contaminants in the environment [23].

In this Special Issue, Yu et al. reported the impact of acetate on the reduction of perchlorate by mixed microbial culture under the influence of nitrate and sulfate, the kinetic parameters of the Monod equation and the optimal ratio of acetate to perchlorate for the perchlorate-reducing bacterial consortium. [24]. The biological reduction of contaminants such as perchlorate ($ClO_4^-$) is considered to be a promising water treatment technology: the process is based on the ability of a specific mixed microbial culture to use perchlorate as an electron acceptor in the absence of oxygen. Both fixed optimal hydraulic retention times and the effect of nitrate on perchlorate reduction were investigated with various concentrations of the electron donor [24]. The presence of sulfate in wastewater did not affect the perchlorate reduction. However, perchlorate reduction was inhibited in the presence of nitrate during exposure to a mixed microbial culture [24].

This Special Issue also includes the study of Ran et al. [25] on the removal of methylene blue by $EuVO_4/g-C_3N_4$ mesoporous nanosheets via coupling adsorption and photocatalysis. The ultrathin and porous structure of the $EuVO_4/g-C_3N_4$ increased the specific surface area and active reaction sites, and the formation of the heterostructure extended visible light absorption and accelerated the separation of charge carriers [25]. These two factors were advantageous in promoting the synergistic effect of adsorption and photocatalysis, and ultimately enhanced the methylene blue's adsorption capability and photocatalytic removal efficiency [25]. The methylene blue adsorption on $EuVO_4/g-C_3N_4$ followed the pseudo second-order kinetics model, and the adsorption isotherm data complied with the Langmuir isotherm model [25]. The photocatalytic degradation data of methylene blue on $EuVO_4/g-C_3N_4$ obeyed the zero-order kinetics equation for 0–10 min and the first-order kinetics equation for 10–30 min [25]. This study provided promising $EuVO_4/g-C_3N_4$ heterojunctions, with superior synergetic effects of adsorption and photocatalysis, for potential application in wastewater treatment [25].

The study of Iosif et al. [26] is focused on the orthodontic bracket's interaction with an acidic environment, induced by regular consumption of soft drinks and energy drinks, and the enamel quality after debonding. It was found that the adhesive layer has the most important role in properly binding the brackets to the enamel surface [26]. Two categories of adhesives were investigated: resin-modified glass ionomer and resin composite adhesives [26]. It was observed that the resin-modified glass ionomer is more resistant in an acidic environment, assuring a better bond strength than the resin composite because of better insulation of the filler particles, preventing their direct contact with acids in soft drinks [26]. The SEM and AFM complex investigations of the enamel surface, after bracket debonding and exposure to acidic environments, reveal that it is more affected by soft drinks containing phosphoric acid after resin composite debonding and by energy drinks with citric acid content after resin-modified glass ionomer debonding [26]. Additionally, it was found that proper polishing of the enamel surface after debonding assures good protection against further erosion [26].

I am grateful to all the authors for their contributions, covering the most recent progress and new developments in the field of metal–ferrite nanocomposites. I hope that the published studies will pave the way for novel real-world synthesis, characterization, and applications of innovative materials.

**Acknowledgments:** I am grateful to all the authors for submitting their studies to the present Special Issue, as well as to all reviewers for their helpful suggestions, which improved the manuscripts. I would also like to thank for their excellent support during the development and publication of the Special Issue.

**Conflicts of Interest:** The author declares no conflict of interest.

# References

1. Dippong, T.; Hoaghia, M.A.; Mihali, C.; Cical, E.; Calugaru, M. Human health risk assessment of some bottled waters from Romania. *Environ. Poll.* **2020**, *267*, 115409. [CrossRef]
2. Dippong, T.; Mihali, C.; Năsui, D.; Berinde, Z.; Butean, C. Assessment of Water Physicochemical Parameters in the Strîmtori-Firiza Reservoir in Northwest Romania. *Water Environ. Res.* **2018**, *90*, 220–233. [CrossRef]
3. Dippong, T.; Levei, E.A.; Deac, I.G.; Petean, I.; Cadar, O. Dependence of Structural, Morphological and Magnetic Properties of Manganese Ferrite on Ni-Mn Substitution. *Int. J. Mol. Sci.* **2022**, *23*, 3097. [CrossRef]
4. Dippong, T.; Levei, E.A.; Cadar, O. Formation, structure and magnetic properties of $MFe_2O_4@SiO_2$ (M = Co, Mn, Zn, Ni, Cu) nanocomposites. *Materials* **2021**, *14*, 1139. [CrossRef]
5. Liao, Z.; Zi, Y.; Zhou, C.; Zeng, W.; Luo, W.; Zeng, H.; Xia, M.; Luo, Z. Recent Advances in the Synthesis, Characterization, and Application of Carbon Nanomaterials for the Removal of Endocrine-Disrupting Chemicals: A Review. *Int. J. Mol. Sci.* **2022**, *23*, 13148. [CrossRef]
6. Li, H.; Xing, J.; Xia, Z.; Chen, J. Preparation of Coaxial Heterogeneous Graphene Quantum Dot-Sensitized $TiO_2$ Nanotube Arrays via Linker Molecule Binding and Electrophoretic Deposition. *Carbon* **2015**, *81*, 474–487. [CrossRef]
7. Dippong, T.; Levei, E.A.; Cadar, O. Investigation of Structural, Morphological and Magnetic Properties of $MFe_2O_4$ (M = Co, Ni, Zn, Cu, Mn) Obtained by Thermal Decomposition. *Int. J. Mol. Sci.* **2022**, *23*, 8483. [CrossRef]
8. Dippong, T.; Levei, E.A.; Cadar, O.; Deac, I.G.; Lazar, M.; Borodi, G.; Petean, I. Effect of amorphous $SiO_2$ matrix on structural and magnetic properties of $Cu_{0.6}Co_{0.4}Fe_2O_4/SiO_2$ nanocomposites. *J. Alloy. Comp.* **2020**, *849*, 156695. [CrossRef]
9. Dippong, T.; Cadar, O.; Goga, F.; Toloman, D.; Levei, E.A. Impact of Ni Content on the Structure and Sonophotocatalytic Activity of Ni-Zn-Co Ferrite Nanoparticles. *Int. J. Mol. Sci.* **2022**, *23*, 14167. [CrossRef]
10. Weshahy, A.R.; Sakr, A.K.; Gouda, A.A.; Atia, B.M.; Somaily, H.H.; Hanfi, M.Y.; Sayyed, M.I.; El Sheikh, R.; El-Sheikh, E.M.; Radwan, H.A.; et al. Selective Recovery of Cadmium, Cobalt, and Nickel from Spent Ni–Cd Batteries Using Adogen® 464 and Mesoporous Silica Derivatives. *Int. J. Mol. Sci.* **2022**, *23*, 8677. [CrossRef]
11. Ninciuleanu, C.M.; Ianchis, R.; Alexandrescu, E.; Mihaescu, C.I.; Burlacu, S.; Trica, B.; Nistor, C.L.; Preda, S.; Scomoroscenco, C.; Gîfu, C.; et al. Adjusting Some Properties of Poly(methacrylic acid) (Nano)Composite Hydrogels by Means of Silicon-Containing Inorganic Fillers. *Int. J. Mol. Sci.* **2022**, *23*, 10320. [CrossRef]
12. Filice, S.; Scuderi, V.; Libertino, S.; Zimbone, M.; Galati, C.; Spinella, N.; Gradon, L.; Falqui, L.; Scalese, S. Sulfonated pentablock copolymer coating of polypropylene filters for dye and metal ions effective removal by integrated adsorption and filtration process. *Int. J. Mol. Sci.* **2022**, *23*, 11777. [CrossRef]
13. Kim, T.H.; Seo, C.; Seon, J.; Battulga, A.; Hwang, Y. Granulation of bismuth oxide by alginate for efficient removal of iodide in water. *Int. J. Mol. Sci.* **2022**, *23*, 12225. [CrossRef]
14. Chen, X.; Lin, J.; Su, Y.; Tang, S. One-Step Carbonization Synthesis of Magnetic Biochar with 3D Network Structure and Its Application in Organic Pollutant Control. *Int. J. Mol. Sci.* **2022**, *23*, 12579. [CrossRef]
15. Liu, Z.; Tan, Y.; Ruan, X.R.; Guo, J.; Li, W.; Li, J.; Ma, H.; Xiong, R.; Wei, J. Spark Plasma Sintering-Assisted Synthesis of $Bi_2Fe_4O_9/Bi_{25}FeO_{40}$ Heterostructures with Enhanced Photocatalytic Activity for Removal of Antibiotics. *Int. J. Mol. Sci.* **2022**, *23*, 12652. [CrossRef]
16. Zhu, J.; Liu, Z.; Wang, H.; Jian, Y.; Long, D.; Pu, S. Preparation of a Z-Type $g-C_3N_4/(A-R)TiO_2$ Composite Catalyst and Its Mechanism for Degradation of Gaseous and Liquid Ammonia. *Int. J. Mol. Sci.* **2022**, *23*, 13131. [CrossRef]
17. Zheng, F.; Martins, P.M.; Queirós, J.M.; Tavares, C.J.; Vilas-Vilela, J.L.; Lanceros-Méndez, S.; Reguera, J. Size Effect in Hybrid $TiO_2$: Au Nanostars for Photocatalytic Water Remediation Applications. *Int. J. Mol. Sci.* **2022**, *23*, 13741. [CrossRef]
18. Feng, J.; Ran, X.; Wang, L.; Xiao, B.; Lei, L.; Zhu, J.; Liu, Z.; Xi, X.; Feng, G.; Dai, Z.; et al. The Synergistic Effect of Adsorption-Photocatalysis for Removal of Organic Pollutants on Mesoporous $Cu_2V_2O_7/Cu_3V_2O_8/g-C_3N_4$ Heterojunction. *Int. J. Mol. Sci.* **2022**, *23*, 14264. [CrossRef]
19. Su, Y.; Shi, Y.; Jiang, M.; Chen, S. One-Step Synthesis of Nitrogen-Doped Porous Biochar Based on N-Doping Co-Activation Method and Its Application in Water Pollutants Control. *Int. J. Mol. Sci.* **2022**, *23*, 14618. [CrossRef]

20. Piras, A.; Olla, C.; Reekmans, G.; Kelchtermans, A.-S.; De Sloovere, D.; Elen, K.; Carbonaro, C.M.; Fusaro, L.; Adriaensens, P.; Hardy, A.; et al. Photocatalytic Performance of Undoped and Al-Doped ZnO Nanoparticles in the Degradation of Rhodamine B under UV-Visible Light:The Role of Defects and Morphology. *Int. J. Mol. Sci.* **2022**, *23*, 15459. [CrossRef]
21. Babilas, D.; Kowalik-Klimczak, A.; Mielanczyk, A. Recovery of the N,N-Dibutylimidazolium Chloride Ionic Liquid from Aqueous Solutions by Electrodialysis Method. *Int. J. Mol. Sci.* **2022**, *23*, 6472. [CrossRef]
22. Zheng, X.-X.; Peng, Y.C.; Sun, W.F. Water-Tree Characteristics and Its Mechanical Mechanism of Crosslinked Polyethylene Grafted with Polar-Group Molecules. *Int. J. Mol. Sci.* **2022**, *23*, 9450. [CrossRef]
23. Liu, X.; Liang, Y.; Peng, Y.; Meng, T.; Xu, L.; Dong, P. Sensitivity of the Transport of Plastic Nanoparticles to Typical Phosphates Associated with Ionic Strength and Solution pH. *Int. J. Mol. Sci.* **2022**, *23*, 9860. [CrossRef]
24. Yu, H.; Lee, K.H.; Park, J.-W. Impact of Acetate in Reduction of Perchlorate by Mixed Microbial Culture under the Influence of Nitrate and Sulfate. *Int. J. Mol. Sci.* **2022**, *23*, 10608. [CrossRef]
25. Ran, X.; Wang, L.; Xiao, B.; Lei, L.; Zhu, J.; Liu, Z.; Xi, X.; Feng, G.; Li, R.; Feng, J. Effective Removal of Methylene Blue on EuVO$_4$/g-C$_3$N$_4$ Mesoporous Nanosheets via Coupling Adsorption and Photocatalysis. *Int. J. Mol. Sci.* **2022**, *23*, 10003. [CrossRef] [PubMed]
26. Iosif, C.; Cuc, S.; Prodan, D.; Moldovan, M.; Petean, I.; Badea, M.E.; Sava, S.; Tonea, A.; Chifor, R. Effects of Acidic Environments on Dental Structures after Bracket Debonding. *Int. J. Mol. Sci.* **2022**, *23*, 15583. [CrossRef] [PubMed]

**Disclaimer/Publisher's Note:** The statements, opinions and data contained in all publications are solely those of the individual author(s) and contributor(s) and not of MDPI and/or the editor(s). MDPI and/or the editor(s) disclaim responsibility for any injury to people or property resulting from any ideas, methods, instructions or products referred to in the content.

*Review*

# Recent Advances in the Synthesis, Characterization, and Application of Carbon Nanomaterials for the Removal of Endocrine-Disrupting Chemicals: A Review

Ze Liao [1], Yang Zi [1], Chunyan Zhou [1], Wenqian Zeng [1], Wenwen Luo [1], Hui Zeng [1], Muqing Xia [1] and Zhoufei Luo [1,2,*]

[1] College of Bioscience and Biotechnology, Hunan Agricultural University, Changsha 410128, China
[2] Hunan Provincial Key Laboratory of Phytohormones and Growth Development, Hunan Agricultural University, Changsha 410128, China
* Correspondence: zhoufeiluo@hunau.edu.cn

**Abstract:** The large-scale production and frequent use of endocrine-disrupting chemicals (EDCs) have led to the continuous release and wide distribution of these pollutions in the natural environment. At low levels, EDC exposure may cause metabolic disorders, sexual development, and reproductive disorders in aquatic animals and humans. Adsorption treatment, particularly using nanocomposites, may represent a promising and sustainable method for EDC removal from wastewater. EDCs could be effectively removed from wastewater using various carbon-based nanomaterials, such as carbon nanofiber, carbon nanotubes, graphene, magnetic carbon nanomaterials, carbon membranes, carbon dots, carbon sponges, etc. Important applications of carbon nanocomposites for the removal of different kinds of EDCs and the theory of adsorption are discussed, as well as recent advances in carbon nanocomposite synthesis technology and characterization technology. Furthermore, the factors affecting the use of carbon nanocomposites and comparisons with other adsorbents for EDC removal are reviewed. This review is significant because it helps to promote the development of nanocomposites for the decontamination of wastewater.

**Keywords:** endocrine-disrupting chemicals; carbon nanomaterials; synthesis; adsorption; application

## 1. Introduction

### 1.1. Endocrine-Disrupting Chemicals and Their Damaging Effects

With the rapid development of the global economy, the release of endocrine-disrupting chemicals (EDCs) into the environment has become continuous. The International Security Program (IPCS) of the World Health Organization (WHO) has defined EDCs as exogenous substances or mixtures that alter the function(s) of the endocrine system and consequently adversely affect the health of populations or subpopulations of intact organisms or their progeny [1,2]. EDCs bind to receptors by imitating natural hormones and thereby interfere with the metabolism of the organism. There are two categories of EDCs: natural and synthetic. Natural EDCs are found in animals and humans and include genistein, coumarins, estradiol, and adrenocortical hormones [3,4]. Synthetic EDCs are classified into a variety of categories based on their application, chemical, and structural characteristics, such as pharmaceutical, organochlorine pesticides, detergents and surfactants, heavy metals, and dyes. The primary types and sources of synthetic EDCs are shown in Figure 1. Therefore, regardless of their class, substances exhibiting estrogenic activity in organisms can be considered as EDCs.

The public concern about the impacts of EDCs on both human health and the environment has grown particularly, owning to their wide occurrence, high persistence, bioaccumulation capabilities, and high toxicity. The main sources of synthetic EDCs are derived from the anthropogenic activity and industrial consumption of insecticides, plastic products, food packaging materials, electronics and construction materials, personal care items, medical tubing, antibacterial agents, detergents, cosmetics, pesticides, fabrics, and clothes [5]. In polar regions, EDCs were deposited by the "global distillation effect" and "the grasshopper effect" due to their semi-volatile nature and degradation resistance, leading to a global spread of contamination. The concurrence of EDCs was found in the Arctic and Antarctic regions and alpine regions [6,7]. The altitude dependence of polychlorinated biphenyls (PCBs) and polybrominated diphenyl ethers (PBDEs) was increased along a high-altitude aquatic food chain in the Tibetan Plateau, China [8]. The bioaccumulation of EDCs in aquatic organisms, especially fish, mussels, and daphnia, is an important criterion in risk assessment. The majority of organochlorine pesticides (OCPs) including DDTs could be enriched from the surrounding media into aquatic organisms through bioamplification of the food chain [9]. EDCs and their intermediate compounds are toxic and have harmful effects on the environment and human health. After an organism is exposed to estrogen, unstable intermediates with strong reactivity are generated by the biological metabolic enzymes and signalling transfer process [10]. Some of the metal ions of EDCs are covalently combined with cellular polymer components such as proteins and nucleic acids, resulting in irreversible chemical modifications [11]. Bisphenol A (BPA) is a ubiquitous EDC that has recently been associated with adverse effects on human health, precocious puberty, and sexual dysfunction [12,13]. Examples of synthetic EDCs and their negative impacts were show in Table 1. These organic molecules will enter the water cycle through rivers, lakes, or groundwater infiltration; the river's organic chemicals flow into the ocean, while these pollutant EDCs enter the ocean via seawater intrusion. These EDCs are widely considered to be the major alarming sources of EDCs of great concern for the aquatic ecosystem.

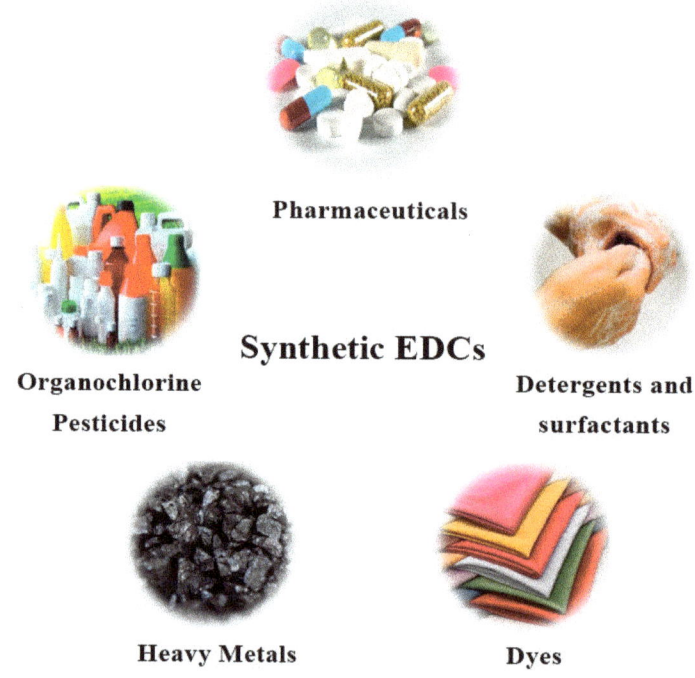

**Figure 1.** Different sources of synthetic EDCs.

Table 1. Examples of synthetic EDCs and their negative impacts.

| EDC Class | Example | Negative Impact | Refs |
|---|---|---|---|
| Pesticides | Dichlorodiphenyltrichloro-ethane (DDT)<br>2,4-dichlorophenoxyacetic acid(2,4-D)<br>Polychlorinated biphenyl (PCBs)<br>Methoxychlor (MXC)<br>Cyhalothrin<br>Organophosphate pesticide | Disturb hormonal balance in women, harm to the liver, Parkinson's, preterm birth (PTB) metabolic syndrome, infertility, testosterone concentration reduction. | [14–17] |
| Steroidal pharmaceuticals | Estradiol<br>Ethynylestradiol<br>Testosterone<br>Androstenedione<br>Corticosteroids<br>Progesterones | Osteonecrosis of the jaw, interferes with the thyroid endocrine system, placenta previa, lung cancer, stimulates lung adenocarcinoma cell production. | [18–22] |
| Detergents and Surfactants | Alkylphenols (APs)<br>Bisphenol A<br>Diethylstilbestrol (DES)<br>Nonyphenol<br>Alkylphenol Ethoxylates (APEO)<br>Perfluorooctanoic acid (PFOA)<br>Perfluorooctane sulphonate (PFOS)<br>Alkylphenol carboxylates | Cardiovascular risk factors, ovarian, uterine, pituitary, testicular cancers, diabetes, low sperm count, recurrent miscarriages. | [23–28] |
| Personal care products | Benzophenone<br>Oxybenzone<br>Chlorophene<br>N,N-Diethyl-m-toluamide (DEET)<br>tris(2-chloroethyl)phosphate (TCEP) | DNA damage, obesity, hepatic steatosis. | [29–31] |
| Nonsteroid pharmaceuticals | Paracetamol<br>Indomethacin<br>Aspirin<br>Ibuprofen<br>Tetracycline | Wheeze and asthma risk in children, hepatocyte senescence, risk of severe intraventricular hemorrhage, interference with the endocrine activity of the thyroid gland. | [32–36] |
| Dyes | Methylene blue (MB)<br>Aniline yellow<br>Rhodamine B (RB)<br>Thiazine | Reduce soil fertility and the photosynthetic activity of aquatic plants, potentially promoting toxicity, mutagenicity, and carcinogenicity. | [37–39] |
| Disinfection by-products | Haloacetamides<br>Bromoacetonitriles<br>Cyanoformaldehyde<br>Bromoaldehydes | Induced liver and kidney injury, mammalian cell cytotoxicity and genotoxicity. | [40–42] |
| Heavy Metals | $Ni^{2+}$, $Pb^{2+}$, $As^{3+}$, $Cr^{3+}$, $Hg^{2+}$, $Cd^{2+}$ | Prostate cancer, hepatotoxicity, nephrotoxicity, skeletal toxicity. | [43–47] |
| Industrial additives and agents | Polybrominateddiphenyl ethers (PBDEs)<br>2,3,7,8-tetrachlorodibenzo-p-dioxin (TCDD)<br>Polyfluoroalkyl substances (PFAS)<br>Phthalate Esters (PAEs)<br>Polyfluoroalkyl | Risk of diabetes, alteration of cognitive functions, risk of atherosclerosis, various cancers, elevated cholesterol levels, decreased immune and liver functionalities, birth defects. | [48–50] |

EDCs are discharged into aquatic ecosystems via surface runoff, natural circulation, and the food chain. In particular, EDCs accumulate as they travel up the food chain, finally becoming harmful to human health. Low levels of EDCs cause neuroendocrine disorders, developmental malformations, and reproductive disorders [51,52]. Organisms such as aquatic plants, aquatic animals, and humans are at risk of long-term exposure to EDCs in the ecological environment. The accumulation of bisphenol A inhibits the growth of aquatic rice seedlings [53]. After exposure to some steroid EDCs, the expression of the VTG-A gene was disrupted in freshwater fish, leading to reproductive and developmental disfunctions, as well as decreased hatchability and reduced vitellogenin levels [54]. EDCs may lead to the feminization of some aquatic organisms, as well as metabolic diseases, cardiovascular risks, and reproductive issues in humans [55].

*1.2. Distribution and Regularity of EDC Pollution*

With the worldwide growth of manufacturing industries, EDC pollution in aquatic ecosystems is becoming more serious. In a 2008–2010 study of 46 representative active EDCs in 182 rivers in the United States by the U.S. Environmental Protection Agency, the approximate concentrations of steroid EDCs in surface waters reached 63 ng/L [56]. Aquatic plants such as algae absorb alkylphenols from polluted soil or water through their roots; EDC accumulation in algae was as high as 53 μg/L in the Bohai Sea region of China [57]. Eleven phthalate esters (PAEs) were measured in urine and serum samples collected from males diagnosed with infertility in Tianjin, China, and the median levels of mPAEs in the serum (n.d. to 3.63 ng/mL) were 1–2 orders of magnitude lower than the median levels of mPAEs in the urine (n.d. to 192 ng/mL) [58]. The bioaccumulation of EDCs harms the growth and development of animals and plants and endangers human health.

*1.3. Treatment of EDC Pollution*

Owing to the strong degradation resistance and high toxicity of EDCs, it is challenging to remove EDCs from wastewater. EDC removal methods mainly include chemical/electrochemical treatments, biological treatments, and physical treatments. Chemical/electrochemical methods mainly include chlorination, ozonation, photocatalysis, electrocoagulation, and anodic oxidation. Chemical/electrochemical treatments do not reduce the generation of chlorine residues or several other harmful intermediate products [59]. Biological treatments are widely applied in the secondary processes of sewage treatment plants and include aerobic and anaerobic treatments. However, these methods may produce degradation pollutants which increase the ecological risks of sewage treatments. Gadd et al. [60] reported that anaerobic treatments increased estrogenic content and estrogenic activity in cowshed wastewater. Compared with the former two technologies, physical adsorption is considered a more environmentally friendly wastewater treatment. The physical adsorption method is effective, efficient, and economical, and different adsorbents can be designed for use in this treatment method. Nanocomposites may represent a useful new wastewater treatment method; this treatment is advantageous due to the large specific surface area and high adsorption capacity.

## 2. Development of Carbon Nanomaterials

In recent years, carbon has become one of the most multilateral elements present in the periodic table due to its strength and ability to form bonds with other elements. Carbon-based nanomaterials and associated modified composites have excellent adsorption properties due to their environmentally friendly, extremely high specific surface area, large pore volume, uniform microporosity, and adjustable surface chemical properties [61,62]. Biomass carbon nanocomposites are a low-cost, readily available, widely distributed, and renewable nanomaterial. Several promising reports have emerged in recent years on the synthesis of carbon nanomaterials from cost-effective, rich, and renewable biomaterial resources such as saw dust, crab shells, bagasse, olive stone waste, and activated carbon cloth [63,64]. Activated carbon is widely used for the control of synthetic and naturally occurring EDCs in drinking water [65]. Carbon nanotubes (CNTs), which were discovered back in the year 1991, have been extensively adapted to study the adsorption capability in water treatment [66]. Graphene, a new material for which the Noble Prize was won, has received increasing attention due to its unique physico-chemical properties for removing EDC pollutants from wastewater [67]. Carbon dots with abundant functional groups (-OH, -COOH, -C=O) on their surface were specially designed to enhance the adsorption capacity [68]. Highly porous carbon sponges always contain some functional groups, which could enhance the surface sensitivity and selectivity of EDC pollutants [69]. Many carbon nanocomposites have been synthesized and used as adsorbents for the removal of EDC pollutants from wastewater. There are many industries such as mining, battery manufacturers, the pharmaceutical industry, the cultivation industry, galvanization, and metal finishing which generate wastewater containing EDCs and emit it directly or indirectly into the nearest water resources [5]. Carbon nanomaterials play an important role in nanoadsorbents. A schematic diagram of carbon nanoadsorbents which are used in EDC pollution treatments is illustrated in Figure 2. Carbon nanocomposites can be classified into seven categories including carbon nanofibers, carbon nanotubes, graphene family, magnetic carbon nanomaterials, carbon nanomembranes, carbon dots, and carbon sponges depending on their composition, structure, and characteristics. The seven categories of carbon nanocomposites according to their specific characteristics are depicted in Table 2.

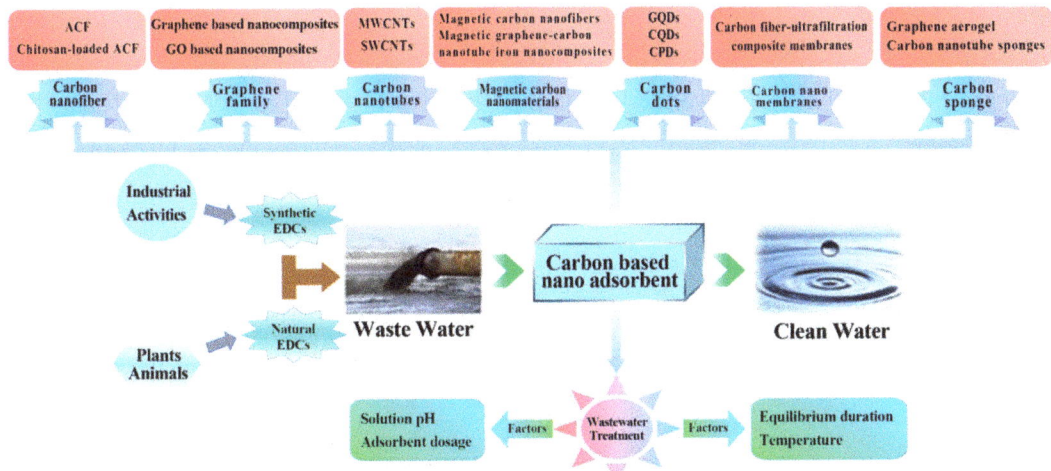

**Figure 2.** Schematic diagram of carbon nanoadsorbents for EDC adsorption for wastewater treatment, the categories of carbon nanoadsorbents, and nanoadsorbents in applications for the decontamination of wastewater.

**Table 2.** Different categories of carbon nanoadsorbents.

| Classification | Category | Type of EDCs | Refs |
|---|---|---|---|
| Carbon nanofiber | Activated carbon fiber (ACF) | Organochlorine pesticides | [70] |
|  | Chitosan-loaded activated carbon fiber | $Pb^{2+}$-EDTA complex | [71] |
|  | Activated carbon fiber supported/modified nanotubes | 2-Chlorophenol | [72] |
| Carbon nanotubes | Multiwalled carbo nanotubes (MWCNTs) | Triton X-100, 17β-Estradiol, Sodium dodecylbenzene, $Hg^{2+}$, $Cr^{3+}$, $As^{3+}$, Bisphenol A | [73–75] |
|  | Single-wall carbon nanotubes (SWCNTs) | $Hg^{2+}$, $Cr^{3+}$, $As^{3+}$, Hexachlorocyclohexane, Dichloro-diphenyl-trichloroethane | [76] |
| Graphene family | Graphene oxide (GO) and other nanocomposites | Perfluoroalkyl substances $Pb^{2+}$, $Hg^{2+}$, $As^{5+}$, $Cr^{6+}$, Methylene blue | [77–79] |
| Magnetic carbon-nanomaterials | Magnetic carbon nanofibers | Phenol, Rhodamine B | [80] |
|  | Magnetite/porous graphene-base nanocomposites | $Cr^{6+}$, $Pb^{2+}$, $Cd^{2+}$, $As^{3+}$, Polychlorinated biphenyl, Dye | [81] |
|  | Magnetic graphene–carbon nanotubes iron nanocomposites | $Pb^{2+}$, $Cd^{2+}$ | [82] |
| Carbon nanomembranes | Carbon fiber ultrafiltration composite membranes | Steroid hormones | [83] |
| Carbon dots | Graphene quantum dots (GQDs) Carbon quantum dots (CQDs) Carbonized polymer dots (CPDs) | carbamate pesticide oxamyl, $Cd^{2+}$, $Hg^{2+}$, $Pb^{2+}$, tetracycline, Carbamazepine | [84–88] |
| Carbon sponge | Graphene sponges | Methylene blue, $Pb^{2+}$ | [89] |
|  | Carbon nanotube sponges | Polychlorinated biphenyl | [69] |

### 2.1. Carbon Nanofiber

Carbon nanofiber (CN), also known as nano-activated carbon, is widely used for the treatment of organic wastewater and removal of EDC substances. In addition to the high adsorption capability, the regeneration of ACF could be carried out under a lower temperature than granular activated carbon [90]. The adsorption efficiency of ACF is significantly higher than that of granular or powdered activated carbon. Murayama et al. used ACF to recover organic chlorine pesticides (OCPs) from rainwater, river water, and seawater samples and confirmed that ACF adsorbed sub-ng/L level OCPs from environmental water samples [70].

### 2.2. Carbon Nanotubes (CNTs)

CNTs are a kind of carbon allotrope with an aromatic surface rolled up to form a cylindrical structure; the length of CNTs varies from 10 s of nm to 10 s of mm [76]. They can be divided into single-walled carbon nanotubes (SWCNTs) and multiwalled carbon nanotubes (MWCNTs), both of which have high a mechanical strength and elasticity as well as chemical and structural stability. SWCNTs and MWCNTs are widely used in removing EDCs from wastewater after being modified by different functional groups (polydopamine, magnetic particles, and molecular imprinting polymers e.g.,) [91]. Mashkoor et al. [92] summarized the adsorption capacities of the original and modified CNTs for different dyes; the adsorption capacity of glycine-$\beta$-cyclodextrin MWCNTs for methylene blue was up to 90.90 mg/g, while the adsorption capacities of modified MWCNTs for $Hg^{2+}$, $Cr^{4+}$, and $As^{3+}$ were 123.45 mg/g, 146.5 mg/g, and 133.33 mg/g, respectively [74]. The large specific surface area, large pore volume, and average pore diameter contribute to the excellent adsorption capacity of CNTs [93].

### 2.3. Graphene Family

Graphene and graphene oxide (GO) have also been widely used for the adsorption of EDCs because of their special properties such as extraordinary quantum hall effects, high mobility, excellent electronic and mechanical properties, specific magnetism, and high thermal conductivity [94]. The excellent adsorption capacity of GO is improved by the electrostatic interaction between the oxidation groups and the adsorbates, and GO is mainly used for the adsorption of metal ions, anionic dyes, and cationic dyes. Modified GO has a good adsorption capacity for different heavy metals (60.2–1076.65 mg/g) [78]. GO also has a specific adsorption capacity for steroids: the removal efficiency of 17$\alpha$-ethinylestradiol by magnetic GO can reach more than 25 mg/g [95].

### 2.4. Magnetic Carbon Nanocomposites

Various magnetic nanocomposites, including nanomaterials such as magnetic carbon nanotubes, magnetic graphene oxide, and magnetic metal–organic frameworks, exhibit excellent adsorption capabilities [96]. The advantage of magnetic nanomaterials is that their magnetic properties allow them to be separated from solutions and regenerated using an external magnetic field. Recently, magnetic nanocomposites have been extensively applied to the adsorption of EDC pollutants such as metals, steroid hormones, alkylphenols, and dyestuff from water [80,81]. Nanocomposites are primarily magnetized using ferriferous oxide ($Fe_3O_4$) and manganese ferrite ($MnFe_2O_4$). Different modified magnetic nanomaterials have different adsorption properties for specific anions, cations, and macromolecular dyes [97]. After the modification of PS-EDTA resin with magnetic ferric oxide nanoparticles, the Cr adsorption rate reached 99.3% [98]. An amino-functionalized mesoporous silica-magnetic GO nanocomposite, which was synthesized using a unique magnetic nanomaterial, removed oxytetracycline more effectively than the original magnetic graphene oxide (mGO) [99].

## 2.5. Carbon Membranes

Carbon membranes have become one of the most important materials employed in wastewater treatment. Carbon membranes have a low production cost and are highly permeable, selective, and stable [100]. Carbon membrane techniques are used for adsorption and filtration in a variety of wastewater treatment plants. The retention rates of nanoparticle-modified polyamide membranes for bisphenol A reached 99.4% [101]. The adsorption capacity of biopolymer carbon membranes modified with lignin, oat, soybean protein, sodium alginate, and chitosan for $Pb^{2+}$ was 35 mg/g [102]. In addition, the adsorption capacity of an electrospun lignin–carbon membrane for methyl blue was 10 times greater than that of active carbon [103].

## 2.6. Carbon Dots

Carbon dots are single-layer or multilayer graphite structures or polymer aggregate carbon particles with a diameter less than 10 nm. Carbon dots have been acknowledged as discrete, quasispherical, fluorescent carbon particles, most of which are $sp^2$ or $sp^3$ hybrid carbon structures [104]. According to their structure and composition characteristics, carbon dots could be divided into graphene quantum dots (GQDs), carbon quantum dots (CQDs), and carbonized polymer dots (CPDs). The synthesized carbon dots show water solubility, good chemical stability, low toxicity, and good biocompatibility. In the adsorption treatment of water pollution, GQDs have received increasing attention as excellent adsorbents. It was reported that GQDs have good adsorption properties in removing pesticides, dyes, heavy metals, and drugs from water [68,105]. Modifying GQDs improves the adsorption effect of nanocomposites by increasing the specific surface area for removing anionic and cationic dyes. Meanwhile, the oxygen-containing functional groups on GQDs also enhance the electrostatic interaction between dyes and nanocomposites. It is the contribution of the formation of hydrogen bonds to the surface between the adsorbate and GQDs. The removal efficiency of $Hg^{2+}$ and $Pb^{2+}$ by GQD adsorption were 98.6 and 99.7%, respectively [86]. The other kinds of carbon dot nanocomposites also have an excellent EDC removal ability. For example, the maximum adsorption capacities of carbon dot nanocomposites for carbamazepine and tetracycline could reach up to 65 mg/g and 591.72 mg/g, respectively [87]. GQDs present significant opportunities for the adsorption of EDCs and pose challenges for future work in environmental application fields.

## 2.7. Carbon Sponges

Carbon sponges are a spongy nanomaterial with a high temperature resistance, excellent elasticity, fast adsorption rate, and low cost [106,107]. Graphene-based and carbon nanotube-based materials with aerogel structures were developed in recent years for various adsorption applications. Graphene aerogels and carbon nanotube aerogels are macroscopic porous materials with a unique isotropic structure. Dubey et al. [108] synthesized the graphene aerogel, which is the lightest material with a density of 0.16 g/L, in 2013. Graphene aerogels have a high adsorption efficiency for EDC pollutants such as dyes. For example, Li et al. [89] reported a new graphene sponge by using polyvinyl alcohol cross-linked GO for adsorbing methylene blue. The results showed that the excellent adsorption performance for the GO/polyvinyl alcohol sponge captured methylene blue in the flow state. In addition, carbon nanotube sponges are widely used to treat water pollution. Wang et al. [106] used carbon nanotube sponges as adsorbents to enrich trace polychlorinated biphenyls (PCBs) in water samples; the recovery rates of this analysis method for carbon nanotube sponges for PCBs are 81.1% to 119.1%, which have a good application prospect.

## 3. Innovative Methods of Carbon Nanocomposite Synthesis

Carbon nanocomposite synthesis methods can be divided into four categories according to the reaction conditions and formation processes used: in situ polymerization, direct compounding, solvothermal synthesis, and electrospinning.

## 3.1. In Situ Polymerization

In situ polymerization is the most common method of nanocomposite synthesis. In situ polymerization prevents the agglomeration of nanocomposites and results in uniform distributions and structures [109]. Youssef et al. [110] added prepared polyvinyl alcohol (PVA) to a solution of MWCNT materials, stirring for 3 h at room temperature, and dripping in 0.01 g of citric acid over 30 min to obtain an MWCNT/PVA hybrid polymer nanocomposite film. The addition of 6% MWCNT to PVA resulted in a remarkable increase in tensile strength (to 11.45 MPa) prior to the formation of the nanocomposite film. The removal rates of $Cr^{2+}$, $Cd^{2+}$, and $Pb^{2+}$ by this composite material were more than 90%. In addition, the adsorption performance of carbon nanomaterials was improved after modification using in situ polymerization, and this method may thus play an important role in environmental applications.

## 3.2. Direct Compounding

Direct compounding is a synthesis method that does not require an activator or special reaction conditions. Direct compounding is widely used because of its simple steps, relatively low cost, and ease of expansion. For example, nanocomposites can be synthesized directly by mixing the polymer matrix and the nanomaterials under mechanical forces [80]. Yang et al. [111] synthesized an anionic polyacrylamide-functionalized GO composite by pouring an anionic polyacrylamide solution into a GO solution with slight mechanical stirring. The maximum adsorption capacity of this composite for basic fuchsin was up to 1034.3 mg/g, which indicates that anionic polyacrylamide/GO aerogels are promising adsorbents for the removal of dye pollutants from aqueous solutions.

## 3.3. Solvothermal Synthesis

Solvent thermal synthesis, for which water or organic solvents are the reaction medium, is usually used in a closed environment to create a critical reaction state (i.e., high temperature and high pressure) to generate homogeneous nanomaterials. Because water molecules hydrolyze at high temperatures, solvothermal synthesis has the advantage of forming nanoparticles of ideal shape, uniform size, and specific surface area. In solvent thermosynthesis, a molecule can easily be functionalized by simply replacing the original organic ligand with functional groups. Solvothermal synthesis can be finished in one step without adding a catalyst, such as the green synthesis of functional GO. Ammonia-modified GO was synthesized by solvothermal reaction at 180 °C for 10 h [112]. Guo et al. [113] synthesized $Fe_3O_4$–GS composites by solvothermal reaction by mixing $FeCl_3 \cdot 6H_2O$, ethylenediamine, and GO at 200 °C for 8 h.

## 3.4. Electrospinning

Electrospinning is a useful fabrication technique that produces NF membranes. Electrospinning relies on the electrostatic repulsion between surface charges to absorb nanofibers in a viscoelastic fluid. For example, Sun et al. [114] prepared a super-hydrophobic carbon fiber membrane by electrospinning. Carbon nanofibers were prepared by electrospinning for their special phase morphology, crystal structure, and surface geometry. Electrospinning provides a simple and versatile method for generating ultrathin fibers from a rich variety of materials that include polymers, composites, and ceramics. Electrospinning synthesis technology can greatly improve the production efficiency of carbonyl nanomaterials.

## 4. Characterization Techniques

### 4.1. Morphological and Microstructural Analysis

Adsorption rates are related to nanocomposite porosity, particle size, and other factors. The surface morphology of a nanocomposite can be observed using scattering electron microscopy (SEM), transmission electron microscopy (TEM), and atomic force microscopy (AFM) [115,116]. SEM and AFM more accurately reflect grain size and other surface properties as compared to other available techniques, but TEM is commonly preferred due to its high resolution. TEM provides two-dimensional (2D) projected images of three-dimensional (3D) objects, directly revealing the size and shape of nanocomposites. Zhang et al. [117] analysed the specific shape and micromorphology of carbon/SiO$_2$ core-sheath nanofibers by combining TEM and SEM technologies. The physical microstructure and compositional elements of the nanomaterial surface can be observed using various techniques, including SEM-energy dispersive X-ray spectroscopy (EDX), TEM-EDX, SEM-elemental mapping, ultraviolet–visible spectroscopy (UV-VIS), X-ray diffraction (XRD), Fourier transform infrared (FT-IR), and X-ray photoelectron spectroscopy (XPS). Wang et al. [118] investigated the surface morphologies and chemical compositions of carbon nanofibrous membranes using SEM-EDX. EDX detectors are equipped with ultra-thin element light windows that detect elements. XRD can be used to identify the structures of diffraction patterns based on various indexes such as element type. The functional groups of nanocomposites were identified using FT-IR spectroscopy. Wu et al. [119] analyzed the formation of Ag nanoparticle decorating in carbon nanotube sponges by FT-IR characterization. Surface area and morphology are important characteristics that affect the adsorption of organic pollutants. Raman spectroscopy is a vibrational spectroscopic technique that can provide detailed information about the structure, chemical composition, and morphology of nanocomposites. Raman spectroscopy is an important method of GO composite characterization. The methods used to synthesize and characterize some carbon nanocomposites are presented in Table 3.

**Table 3.** Synthesis and characterization of some carbon nanoadsorbents.

| Classification | Category | Characterizations | Synthesis | References |
|---|---|---|---|---|
| Carbon nano tubes (CNTs) | Activated carbon fiber-supported/modified titanate nanotubes | TEM, FE-SEM, XRD, FTIR, XPS, UV-vis | Hydrothermal method | [72] |
| Graphene | Graphene oxide (GO) | XPS, FTIR, AFM, SEM, TEM | Chemical reduction Hummer's method | [77,78] |
|  | Magnetic chitosan and graphene oxide (MCGO) | TGA, FTIR, TG, SEM, DSC | Hummer's method Ultrasonic dispersion | [79] |
| Magnetic carbon-nanomaterials | Magnetic carbon nanotube iron nanocomposites | BET, XRD, XPS, TEM, SEM, AFM, SAED, μXRF | Solvothermal synthesis Microwave irradiation | [81] |
|  | Magnetite/porous graphene-based nanocomposites | FE-SEM, EDX, XRD, FTIR, RAM, VSM, UV-vis, XPS, SSA | Sonication Solvothermal synthesis | [82] |
|  | Electrospun magnetic carbon nanofibers | SQUID, AGM, VSM, MOKE | Electrospinning Field-assisted electrospinning Solvothermal technique | [80] |
| Carbon membranes | Carbon fiber ultrafiltration composite membranes | FE-SEM, BET, FTIR | Filtration process Sonication mixing | [83] |
| Carbon dots | Graphene quantum dots (GQDs) Carbon quantum dots (CQDs) Carbonized polymer dots (CPDs) | XRD, AFM, TEM, UV-vis, XPS, FTIR | Solvothermal technique Hydrothermal method | [120–122] |
| Carbon sponge | Graphene aerogel Carbon nanotube sponges | XRD, XPS, FTIR, FE-SEM | Solvothermal technique | [119] |

Abbreviations: FE-SEM, field emission scanning electron microscope; XRD, X-ray diffractometer; TEM, transmission electron microscopy; FTTR, Fourier transform infrared; XPS, X-ray photoemission spectroscopy; UV-vis, ultraviolet–visible spectroscopy; AFM, atomic force microscopy; SEM, scanning electron microscopy; DSC, differential scanning calorimetry; VSM, vibrating sample magnetometer; BET, Brunauer–Emmett–Teller; RAM, Raman spectra; SQUID, superconducting quantum interference device; AGM, alternate gradient magnetometer; MOKE, magneto-optical Kerr effect; μXRF, micro X-ray fluorescence; SSA, specific surface area analysis.

## 4.2. Magnetic Properties

Magnetism is also useful for adsorbent separation. Adsorbents can be easily separated from solutions without secondary contamination using an external magnetic field. Superconducting quantum interference devices (SQUIDs), alternating gradient magnetometers (AGMs), and vibrating sample magnetometers (VSMs) are commonly used to measure the magnetization and coercivity of magnetic nanocomposites [82]. SQUID measurements were used to investigate the magnetic properties of pure and Mn-doped electrospun barium titanate nanofibers. VSM plays a very important role in the measurement of the magnetization of magnetic nanocomposites. VSM measurements were used to investigate magnetic nanofibers containing ferromagnetic Ni nanoparticles. The significant hysteresis loops in the S-shaped curves indicate the ferromagnetic behaviors of the nanocomposites. In addition, VSM differentiates between measurements parallel and perpendicular to the axes of the aligned nanofibers. VSM has also been used to measure the effects of magnetic material doping on magnetization, the squareness of hysteresis loops, and the straightening force of materials such as samarium–cobalt nanofibers [80].

## 5. Application in Wastewater

Several recent studies have investigated the real-world application of nanoadsorbent materials in wastewater treatment. Material regeneration and reuse in real-word applications and comparisons of nanoadsorbents with other common adsorbent materials are summarized below.

### 5.1. Regeneration and Reuse

Due to the economic benefits of EDC pollution treatments, it is necessary to summarize the regeneration and reuse efficiency of various nanoadsorbents.

A variety of adsorbent regeneration technologies have been developed, including thermal, steam, chemical, microwave-assisted, electrochemical, and biological regeneration methods. It was shown that thermally regenerated MWCNT (incubated at 300 °C for 2 h) maintained its adsorption efficiency for cyclophosphamide, ifosfamide, and 5-fluorouracil in water. These results indicate that, when regenerating an adsorbent, a suitable regeneration technique should be selected according to the chemical and physical properties of the adsorbent itself [123]. The adsorption rates of neutral and $Pb^{2+}$ by the $MnFe_2O_4$/GO nanocomposite were 94% and 98.8%, respectively, and these magnetic nanomaterials could be recycled five times with good stability [124].

Generally, the adsorption efficiency of a recycled carbon nanoadsorbent will decrease as the number of use-cycles increases. Carbon-based nanocomposites have a better regeneration ability. In studies on the adsorption removal of rhodamine B dye using GO-based nano-nickel composite materials, the adsorption efficiency decreased from 90% to 85% after the first recovery and decreased to 60% after the fifth recovery [125].

### 5.2. Carbon Nanoadsorbent Patents

Patented carbon nanocomposites were reflected upon to understand the development and applicability of carbon nanoadsorbents in wastewater treatment. Dai et al. prepared a novel GO/chitosan composite adsorbent (Dai et al., patent CN113318710A. 31 August 2021). The adsorption capacity of the GO/chitosan composite adsorbent for $Cr^{6+}$ was significantly enhanced. Guo et al. reported a novel method to synthesize a polyamidine/carbon nanomaterial that could be used to remove some anionic dyes in wastewater (Guo et al., patent CN109174035B. 15 June 2021). A novel modified carbon nanotube was prepared by Huang et al. for the adsorption of $Pb^{2+}$. The sulfhydryl functional groups on the surface of carbon nanotubes could strongly chelate with $Pb^{2+}$. The adsorption rate of $Pb^{2+}$ can reach 92.89% (Huang et al., patent CN110449132B. 25 March 2022). Carbon sponge blending by nanocellulose, polyvinyl alcohol, and polyvinylpyrrolidone could efficiently remove $Cr^{6+}$ (9.3948 mg/g) and organic dyes from water (Ma et al., patent CN105597681B. 14 November 2017). The adsorbing capacity of a nitrogen-doped graphene quantum dot hybrid membrane for $Pb^{2+}$ removal is 9 mg/g (Liu et al., patent CN112723346B. 24 June 2022).

### 5.3. Comparison with Other Adsorbents

In Table 4, the EDC adsorbent efficacy among carbon nanocomposites and other materials based on previous studies was compared. Carbon nanocomposites show great promise owing to their many advantages: First, compared with other adsorbents, carbon nanocomposites have a larger surface area and better adsorption capacity [126]. For example, the adsorption capacity of most activated carbons for methyl blue is 100–500 mg/g, while the nanoadsorbent capacity for MB is often higher than 500 mg/g [127]. ACF derived from japonica seed hair fibers has an adsorption capacity for MB of 943.372 mg/g [128]. Second, the application of nanocomposites can greatly reduce the cost of water treatment. Beck et al. [103] reported that the use of carbon nanofiber membranes can reduce energy consumption by 87% during wastewater treatment processes. Third, some biodegradable polymer/multiwalled carbon nanotubes can effectively reduce environmental pollution and have good adsorption properties for EDCs [129]. These carbon nanomaterials can be biodegraded after adsorption and the nanomaterials can be recycled.

Table 4. Comparison with other materials of EDC adsorbents.

| Adsorbents | Advantages | Disadvantages | Refs |
|---|---|---|---|
| Carbon nanomaterials | Large specific surface, surface multi-functionality, regenerative capabilities, eco–friendly, high adsorption capacity, biodegradability | High cost to realize large-scale production and application | [130–132] |
| Biomass | Abundant in nature, available in large quantities, inexpensive, have potential as complexing materials | At the laboratory stage | [133,134] |
| Natural zeolites | Easily available and relatively cheap | Low permeability, continuous need for amendments (pH adjustment) | [135,136] |
| Conventional ion exchange resins | Effectiveness, ease of operation, large available exchange capacities, small footprint, regenerative capabilities | Mostly operated only in low pH ranges | [137,138] |
| Biodegradable polymers | Biodegradability, high local availability, low cost, high surface area, high chemical stability, remarkable flexibility | Risk of contamination, possible release of methane, costly regeneration | [129,139] |

## 6. Factors Affecting the Adsorption of Pollutants on Carbon Nanocomposites

*6.1. Effects of Solution pH*

The solution pH value is an important factor affecting carbon nanocomposite adsorption capacity. The relationships among solution pH, nanocomposite pHpzc, and EDC pollutant pKa should be comprehensively analyzed to determine the effects on adsorption efficiency [97]. When the pH of the solution is less than the $pH_{pzc}$ of the nanocomposite, the surface of the adsorbent is positively charged due to the protonation of the surface functional groups. Conversely, when the surface of the adsorbent is negatively charged, the electrostatic interactions between the adsorbent and the EDC pollutants are altered, which then affects the adsorption capacity of the nanoadsorbent on the EDCs [140]. For example, the removal efficiencies for MB by functionalized MWCNTs at a pH of 2, 4, and 6 were 65%, 85%, and 95%, respectively [141]. Kumar et al. [142] showed that phenolic compounds usually exist in a neutral form when the solution pH is less than the pKa of phenolic EDCs; however, the functional groups of the phenolic compounds dissociate and generate phenoxide groups when the solution pH is greater than the pKa of the EDC pollutant; thus, the phenolic EDCs are repelled by each other and the negatively charged adsorbent, reducing the removal efficiency of the adsorbent. These results show that the removal efficiency of the carbon nanoadsorbent for phenolic EDCs is maximized in solutions with a pH less than the pKa (~6).

*6.2. Effects of Adsorbent Dosage and the Initial Concentration of EDC Pollutants*

Adsorbent dosage and initial EDC concentration are usually considered the most important factors influencing liquid-phase adsorption. Adsorbent dosage is correlated with the adsorption site and capacity. Baharum et al. [143] investigated the effects of adsorbent dosage on the removal of polycyclic aromatic hydrocarbons and found that the adsorption capacity of diazinon decreased with increasing adsorbent dosage. This may have resulted from the aggregation of the active site due to the violent collisions among the excess bioadsorbent particles.

The initial EDC concentration strongly influences the adsorption capacity of the nanocomposite. At low EDC concentrations, the number of EDC molecules initially available to the adsorption sites is low. At high EDC concentrations, the ratio is again low due to the lack of available sites. Thus, the removal rate is affected by the initial EDC concentration. Hameed et al. [144] reported that, within a certain range, increases in the initial pesticide concentration increased the adsorption rate of activated carbon.

*6.3. Effects of Adsorption Equilibrium Duration*

The duration of the adsorption equilibrium between the nanoadsorbent and the EDC adsorbate is a vital parameter for evaluating the adsorption efficiency and time cost. In the initial state, the removal efficiency of the adsorbent usually increases with contact time until the adsorption process reaches equilibrium. The length of the period before the adsorption equilibrium is reached varies by nanocomposite, which greatly impacts the cost of sewage treatment. For example, the equilibrium time for $Pb^{2+}$ removal by porous carbon nanofibers is 60 min, and the removal rate is up to 80% [145]. The equilibrium time for the removal of tetracycline by a copper alginate-carbon nanotube membrane is 2000 min, and the equilibrium adsorption capacity is 120 mg/g [146].

*6.4. Effects of Temperature*

Temperature is an important factor affecting the performance of the nanoadsorbent. In general, when the adsorption process is endothermic, the adsorption capacity increases with temperature; when the process is exothermic, the adsorption capacity decreases as the temperature increases [144]. For example, Yadav et al. [147] adsorbed CAC-500 onto phenol EDCs and found that the adsorption capacity of phenol decreased from 1.65 mg/g to 1.54 mg/g as the temperature increased from 25 °C to 55 °C. Because this adsorption process is exothermic, increasing the temperature reduced the adsorption capacity of the adsorbent.

## 7. Conclusions and Perspectives

EDCs are increasingly threatening human and aquatic life; therefore, treatment of these pollutants is of utmost importance. Adsorption has a wide applicability for the removal of EDC pollutants from wastewater. Carbon-based nanomaterials have been extensively explored for EDC adsorption applications because they are eco-friendly, their good chemical stability, structural diversity, low density, and suitability for large-scale production. This review is a timely summary of recent advances in the treatment of various types of EDC-polluted wastewater via carbon nanocomposite adsorption. This review details the synthesis methods, characterization, and explores the latest developments in carbon nanocomposites as adsorbents for EDC wastewater treatment. The comparisons of maximum adsorption capacities and costs of carbon nanocomposites are reviewed in this work with various adsorbents previously studied. While considering the aforementioned requirements and constraints, the present review critically affirms the capabilities of carbon nanomaterials in adsorbing EDC contaminants from wastewater.

Outlooks and challenges have discussed to inspire more exciting developments in the application of carbon nanocomposites for the remediation of wastewater. The properties of carbon nanoadsorbents increase their potential applicability, and carbon nanoadsorbents may be more useful and beneficial in several fields than other adsorbents. However, it should be pointed out that carbon nanomaterials as adsorbents still face some limitations. The preparation cost of some carbon nanomaterials is relatively high, such as carbon nanotubes and graphene. On the other hand, there are some technical problems with the recyclability of some carbon nanomaterials for large-scale wastewater treatment [148,149]. It is still a challenge to develop new, safe, efficient, and lower-cost carbon nanocomposite adsorbents. Moreover, future studies should further explore novel aspects of carbon nanomaterials, such as chemical stabilization and surface adaptations, to improve their applicability to the treatment of water and wastewater. The challenges involved in the development of these novel nanoadsorbents for the decontamination of wastewaters have also been examined to help identify future directions for this emerging field to continue to grow.

**Author Contributions:** Writing original draft preparation, Z.L. (Ze Liao); investigation and curation, Y.Z., C.Z., W.Z., W.L., H.Z. and M.X.; writing—review and editing, Z.L. (Zhoufei Luo); funding acquisition and supervision, Z.L. (Zhoufei Luo) All authors have read and agreed to the published version of the manuscript.

**Funding:** This research was funded by the Scientific Research Fund of Hunan Provincial Natural Science Foundation (2021JJ40243) and the Scientific Research Project of Hunan Provincial Department of Education (19B279 and 21B0204).

**Institutional Review Board Statement:** Not applicable.

**Informed Consent Statement:** Not applicable.

**Data Availability Statement:** The data presented in this study are available on request from the corresponding author.

**Conflicts of Interest:** The authors declare no conflict of interest.

# References

1. Jones, L.; Regan, F. Endocrine Disrupting Chemicals. *Encycl. Anal. Sci.* **2019**, *3*, 31–38. [CrossRef]
2. Green, M.P.; Harvey, A.J.; Finger, B.J.; Tarulli, G.A. Endocrine Disrupting Chemicals: Impacts on Human Fertility and Fecundity during the Peri-conception Period. *Environ. Res.* **2021**, *194*, 110694. [CrossRef]
3. Campbell, C.E.; Mezher, A.F.; Tyszka, J.M.; Nagel, B.J.; Eckel, S.P.; Herting, M.M. Associations Between Testosterone, Estradiol, and Androgen Receptor Genotype with Amygdala Subregions in Adolescents. *Psychoneuroendocrinology* **2022**, *137*, 105604. [CrossRef]
4. Folley, S.J.; Greenbaum, A.L. Effects of Adrenalectomy and of Treatment with Adrenal Cortex Hormones on the Arginase and Phosphatase Levels of Lactating Rats. *Biochem. J.* **1946**, *40*, 46–51. [CrossRef]
5. Ismanto, A.; Hadibarata, T.; Kristanti, R.A.; Maslukah, L.; Safinatunnajah, N.; Kusumastuti, W. Endocrine Disrupting Chemicals (EDCs) in Environmental Matrices: Occurrence, Fate, Health Impact, Physio-chemical and Bioremediation Technology. *Environ. Pollut.* **2022**, *302*, 119061. [CrossRef] [PubMed]
6. Routti, H.; Harju, M.; Lühmann, K.; Aars, J.; Ask, A.; Goksøyr, A.; Kovacs, K.M.; Lydersen, C. Concentrations and Endocrine Disruptive Potential of Phthalates in Marine Mammals from the Norwegian Arctic. *Environ. Int.* **2021**, *152*, 106458. [CrossRef] [PubMed]
7. Esteban, S.; Moreno-Merino, L.; Matellanes, R.; Catalá, M.; Gorga, M.; Petrovic, M.; López de Alda, M.; Barceló, D.; Silva, A.; Durán, J.J.; et al. Presence of Endocrine Disruptors in Freshwater in the Northern Antarctic Peninsula Region. *Environ. Res.* **2016**, *147*, 179–192. [CrossRef]
8. Ren, J.; Wang, X.; Wang, C.; Gong, P.; Wang, X.; Yao, T. Biomagnification of Persistent Organic Pollutants along a High-Altitude Aquatic Food Chain in the Tibetan Plateau: Processes and Mechanisms. *Environ. Pollut.* **2017**, *220*, 636–643. [CrossRef] [PubMed]
9. Sun, R.X.; Sun, Y.; Xie, X.D.; Yang, B.Z.; Cao, L.Y.; Luo, S.; Wang, Y.Y.; Mai, B.X. Bioaccumulation and Human Health Risk Assessment of DDT and Its Metabolites (DDTs) in Yellowfin Tuna (*Thunnus Albacares*) and Their Prey from the South China Sea. *Mar. Pollut. Bull.* **2020**, *158*, 111396. [CrossRef]
10. Xiang, D.; Yang, J.; Xu, Y.; Lan, L.; Li, G.; Zhang, C.; Liu, D. Estrogen Cholestasis Induces Gut and Liver Injury in Rats Involving in Activating PI3K/Akt and MAPK Signaling Pathways. *Life Sci.* **2021**, *276*, 119367. [CrossRef]
11. Charlet, L.; Chapron, Y.; Faller, P.; Kirsch, R.; Stone, A.T.; Baveye, P.C. Neurodegenerative Diseases and Exposure to the Environmental Metals Mn, Pb, and Hg. *Coord. Chem. Rev.* **2012**, *256*, 2147–2163. [CrossRef]
12. Vichit, S.; Chutima, J.; Wichit, N.; Sopon, P.; Suttipong, W.; Olle, S. Increased Levels of Bisphenol A (BPA) in Thai Girls with Precocious Puberty. *J. Pediatr. Endocrinol. Metab.* **2016**, *29*, 1233–1239. [CrossRef]

13. Matuszczak, E.; Komarowska, M.D.; Debek, W.; Hermanowicz, A. The Impact of Bisphenol A on Fertility, Reproductive System, and Development: A Review of the Literature. *Int. J. Endocrinol.* **2019**, *2019*, 4068717. [CrossRef]
14. Rosenbaum, P.F.; Weinstock, R.S.; Silverstone, A.E.; Sjodin, A.; Pavuk, M. Metabolic Syndrome is Associated with Exposure to Organochlorine Pesticides in Anniston, AL, United States. *Environ. Int.* **2017**, *108*, 11–21. [CrossRef] [PubMed]
15. Zaynab, M.; Fatima, M.; Sharif, Y.; Sughra, K.; Sajid, M.; Khan, K.A.; Sneharani, A.H.; Li, S. Health and Environmental Effects of Silent Killers Organochlorine Pesticides and Polychlorinated Biphenyl. *J. King Saud Univ. Sci.* **2021**, *33*, 101511. [CrossRef]
16. Zeng, J.Y.; Miao, Y.; Liu, C.; Deng, Y.L.; Chen, P.P.; Zhang, M.; Cui, F.P.; Shi, T.; Lu, T.T.; Liu, C.J.; et al. Serum Multiple Organochlorine Pesticides in Relation to Testosterone Concentrations among Chinese Men from an Infertility Clinic. *Chemosphere* **2022**, *299*, 134469. [CrossRef]
17. Dwivedi, N.; Mahdi, A.A.; Deo, S.; Ahmad, M.K.; Kumar, D. Assessment of Genotoxicity and Oxidative Stress in Pregnant Women Contaminated to Organochlorine Pesticides and Its Correlation with Pregnancy Outcome. *Environ. Res.* **2022**, *204*, 112010. [CrossRef]
18. Patel, B.; Priefer, R. Impact of Chronic Obstructive Pulmonary Disease, Lung Infection, and/or Inhaled Corticosteroids Use on Potential Risk of Lung Cancer. *Life Sci.* **2022**, *294*, 120374. [CrossRef] [PubMed]
19. Sun, C.H.; Chou, J.C.; Chao, K.P.; Chang, H.C.; Lieu, F.K.; Wang, P.S. 17α-Ethynylestradiol and 4-Nonylphenol Stimulate Lung Adenocarcinoma Cell Production in Xenoestrogenic Way. *Chemosphere* **2019**, *218*, 793–798. [CrossRef] [PubMed]
20. Herman, H.G.; Volodarsky-Perel, A.; Nu, T.N.T.; Machado-Gedeon, A.; Cui, Y.; Shaul, J.; Dahan, M.H. The Effect of Higher Estradiol Levels during Stimulation on Pregnancy Complications and Placental Histology. *Placenta* **2022**, *126*, 114–118. [CrossRef] [PubMed]
21. Omar, T.F.T.; Ahmad, A.; Aris, A.Z.; Yusoff, F.M. Endocrine Disrupting Compounds (EDCs) in Environmental Matrices: Review of Analytical Strategies for Pharmaceuticals, Estrogenic Hormones, and Alkylphenol Compounds. *TrAC Trends Anal. Chem.* **2016**, *85*, 241–259. [CrossRef]
22. Kim, R.; Kim, S.W.; Kim, H.; Ku, S.Y. The Impact of Sex Steroids on Osteonecrosis of the Jaw. *Osteoporos. Sarcopenia* **2022**, *8*, 58–67. [CrossRef]
23. Palmer, J.R.; Herbst, A.L.; Noller, K.L.; Boggs, D.A.; Troisi, R.; Titus-Ernstoff, L.; Hatch, E.E.; Wise, L.A.; Strohsnitter, W.C.; Hoover, R.N. Urogenital Abnormalities in Men Exposed to Diethylstilbestrol in Utero: A Cohort Study. *Environ. Health* **2009**, *8*, 37–42. [CrossRef]
24. Fénichel, P.; Chevalier, N. Environmental Endocrine Disruptors: New Diabetogens? *C. R. Biol.* **2017**, *340*, 446–452. [CrossRef]
25. Lin, C.Y.; Hwang, Y.T.; Chen, P.C.; Sung, F.C.; Su, T.C. Association of Serum Levels of 4-Tertiary-Octylphenol with Cardiovascular Risk Factors and Carotid Intima-Media Thickness in Adolescents and Young Adults. *Environ. Pollut.* **2019**, *246*, 107–113. [CrossRef] [PubMed]
26. Rubin, B.S. Bisphenol A: An Endocrine Disruptor with Widespread Exposure and Multiple Effects. *J. Steroid Biochem. Mol. Biol.* **2011**, *127*, 27–34. [CrossRef]
27. Wen, H.J.; Chang, T.C.; Ding, W.H.; Tsai, S.F.; Wang, S.L. Exposure to Endocrine Disruptor Alkylphenols and the Occurrence of Endometrial Cancer. *Environ. Pollut.* **2020**, *267*, 115475. [CrossRef]
28. Bhandari, G.; Bagheri, A.R.; Bhatt, P.; Bilal, M. Occurrence, Potential Ecological Risks, and Degradation of Endocrine Disrupter, Nonylphenol, from the Aqueous Environment. *Chemosphere* **2021**, *275*, 130013. [CrossRef] [PubMed]
29. Yan, S.; Wang, J.; Zheng, Z.; Ji, F.; Yan, L.; Yang, L.; Zha, J. Environmentally Relevant Concentrations of Benzophenones Triggered DNA Damage and Apoptosis in Male Chinese Rare Minnows (*Gobiocypris Rarus*). *Environ. Int.* **2022**, *164*, 107260. [CrossRef] [PubMed]
30. Cui, Q.; Zhu, X.; Guan, G.; Hui, R.; Zhu, L.; Wang, J. Association of N,N-Diethyl-m-Toluamide (DEET) with Obesity among Adult Participants: Results from NHANES 2007–2016. *Chemosphere* **2022**, *307*, 135669. [CrossRef] [PubMed]
31. Yang, D.; Wei, X.; Zhang, Z.; Chen, X.; Zhu, R.; Oh, Y.; Gu, N. Tris (2-Chloroethyl) Phosphate (TCEP) Induces Obesity and Hepatic Steatosis via FXR-Mediated Lipid Accumulation in Mice: Long-Term Exposure As a Potential Risk for Metabolic Diseases. *Chem.-Biol. Interact.* **2022**, *363*, 110027. [CrossRef] [PubMed]
32. Sordillo, J.E.; Scirica, C.V.; Rifas-Shiman, S.L.; Gillman, M.W.; Bunyavanich, S.; Camargo, C.A.; Weiss, S.T.; Gold, D.R.; Litonjua, A.A. Prenatal and Infant Exposure to Acetaminophen and Ibuprofen and the Risk for Wheeze and Asthma in Children. *J. Allergy Clin. Immunol.* **2015**, *135*, 441–448. [CrossRef] [PubMed]
33. Orrell, K.A.; Cices, A.D.; Guido, N.; Majewski, S.; Ibler, E.; Huynh, T.; Rangel, S.M.; Laumann, A.E.; Martini, M.C.; Rademaker, A.W.; et al. Malignant Melanoma Associated with Chronic Once-Daily Aspirin Exposure in Males: A Large, Single-Center, Urban, US Patient Population Cohort Study from the "Research on Adverse Drug events And Report" (RADAR) Project. *J. Am. Acad. Dermatol.* **2018**, *79*, 762–764. [CrossRef] [PubMed]
34. Meseguer-Ripolles, J.; Lucendo-Villarin, B.; Tucker, C.; Ferreira-Gonzalez, S.; Homer, N.; Wang, Y.; Lewis, P.J.S.; Toledo, E.M.; Mellado-Gomez, E.; Simpson, J.; et al. Dimethyl Fumarate Reduces Hepatocyte Senescence Following Paracetamol Exposure. *iScience* **2021**, *24*, 102552. [CrossRef] [PubMed]
35. Hammers, A.L.; Sanchez-Ramos, L.; Kaunitz, A.M. Antenatal Exposure to Indomethacin Increases the Risk of Severe Intraventricular Hemorrhage, Necrotizing Enterocolitis, and Periventricular Leukomalacia: A Systematic Review with Metaanalysis. *Am. J. Obstet. Gynecol.* **2015**, *212*, 505.e1–505.e13. [CrossRef]

36. Geng, M.; Gao, H.; Wang, B.; Huang, K.; Wu, X.; Liang, C.; Yan, S.; Han, Y.; Ding, P.; Wang, W.; et al. Urinary Tetracycline Antibiotics Exposure During Pregnancy and Maternal Thyroid Hormone Parameters: A Repeated Measures Study. *Sci. Total Environ.* **2022**, *838*, 156146. [CrossRef] [PubMed]
37. Adenan, N.H.; Lim, Y.Y.; Ting, A.S.Y. Removal of Triphenylmethane Dyes by Streptomyces Bacillaris: A Study on Decolorization, Enzymatic Reactions and Toxicity of Treated Dye Solutions. *J. Environ. Manag.* **2022**, *318*, 115520. [CrossRef] [PubMed]
38. Patil, R.; Zahid, M.; Govindwar, S.; Khandare, R.; Vyavahare, G.; Gurav, R.; Desai, N.; Pandit, S.; Jadhav, J. Constructed Wetland: A Promising Technology for the Treatment of Hazardous Textile Dyes and Effluent. In *Development in Wastewater Treatment Research and Processes*; Elsevier: Amsterdam, The Netherlands, 2022; pp. 173–198. [CrossRef]
39. Al-Tohamy, R.; Ali, S.S.; Li, F.; Okasha, K.M.; Mahmoud, Y.A.G.; Elsamahy, T.; Jiao, H.; Fu, Y.; Sun, J. A Critical Review on the Treatment of Dye-Containing Wastewater: Ecotoxicological and Health Concerns of Textile Dyes and Possible Remediation Approaches for Environmental Safety. *Ecotoxicol. Environ. Saf.* **2022**, *231*, 113160. [CrossRef]
40. Deng, Y.; Zhang, Y.; Lu, Y.; Lu, K.; Bai, H.; Ren, H. Metabolomics Evaluation of the in Vivo Toxicity of Bromoacetonitriles: One Class of High-Risk Nitrogenous Disinfection Byproducts. *Sci. Total Environ.* **2017**, *579*, 107–114. [CrossRef]
41. Bond, T.; Huang, J.; Templeton, M.R.; Graham, N. Occurrence and Control of Nitrogenous Disinfection By-Products in Drinking Water—A Review. *Water Res.* **2011**, *45*, 4341–4354. [CrossRef]
42. Plewa, M.J.; Wagner, E.D. Risks of Disinfection Byproducts in Drinking Water: Comparative Mammalian Cell Cytotoxicity and Genotoxicity. *Encycl. Environ. Health* **2019**, 559–566. [CrossRef]
43. Soriano, J.J.; Mathieu-Denoncourt, J.; Norman, G.; Solla, S.R.; Langlois, V.S. Toxicity of the Azo Dyes Acid Red 97 and Bismarck Brown Y to Western Clawed Frog (*Silurana Tropicalis*). *Environ. Sci. Pollut. Res.* **2013**, *21*, 3582–3591. [CrossRef] [PubMed]
44. Koyu, A.; Gokcimen, A.; Ozguner, F.; Bayram, D.S.; Kocak, A. Evaluation of the Effects of Cadmium on Rat Liver. *Mol. Cell. Biochem.* **2006**, *284*, 81–85. [CrossRef] [PubMed]
45. Wang, X.; Cui, W.; Wang, M.; Liang, Y.; Chen, X. The Association Between Life-Time Dietary Cadmium Intake from Rice and Chronic Kidney Disease. *Ecotoxicol. Environ. Saf.* **2021**, *211*, 111933. [CrossRef] [PubMed]
46. Staessen, J.A.; Roels, H.A.; Emelianov, D.; Kuznetsova, T.; Thijs, L.; Vangronsveld, J.; Fagard, R. Environmental Exposure to Cadmium, Forearm Bone Density, and Risk of Fractures: Prospective Population Study. *Lancet* **1999**, *353*, 1140–1144. [CrossRef]
47. Liang, Y.; Yi, L.; Deng, P.; Wang, L.; Yue, Y.; Wang, H.; Tian, L.; Xie, J.; Chen, M.; Luo, Y.; et al. Rapamycin Antagonizes Cadmium-Induced Breast Cancer Cell Proliferation and Metastasis Through Directly Modulating ACSS2. *Ecotoxicol. Environ. Saf.* **2021**, *224*, 112626. [CrossRef]
48. Zhang, Y.; Jiao, Y.; Li, Z.; Tao, Y.; Yang, Y. Hazards of Phthalates (PAEs) Exposure: A Review of Aquatic Animal Toxicology Studies. *Sci. Total Environ.* **2021**, *771*, 145418. [CrossRef] [PubMed]
49. Bey, L.; Coumoul, X.; Kim, M.J. TCDD Aggravates the Formation of the Atherosclerotic Plaque in ApoE KO Mice with a Sexual Dimorphic Pattern. *Biochimie* **2022**, *195*, 54–58. [CrossRef] [PubMed]
50. Dickman, R.A.; Aga, D.S. A Review of Recent Studies on Toxicity, Sequestration, and Degradation of Per- and Polyfluoroalkyl Substances (PFAS). *J. Hazard. Mater.* **2022**, *436*, 129120. [CrossRef] [PubMed]
51. Pinson, A.; Franssen, D.; Gérard, A.; Parent, A.S.; Bourguignon, J.P. Neuroendocrine Disruption without Direct Endocrine Mode of Action: Polychloro-Biphenyls (PCBs) and Bisphenol A (BPA) As Case Studies. *C. R. Biol.* **2017**, *340*, 432–438. [CrossRef]
52. Patisaul, H.B. Endocrine Disrupting Chemicals (EDCs) and the Neuroendocrine System: Beyond Estrogen, Androgen, and Thyroid. *Adv. Pharmacol.* **2021**, *92*, 101–150. [CrossRef]
53. Ali, I.; Liu, B.; Farooq, M.A.; Islam, F.; Gan, Y. Toxicological Effects of Bisphenol A on Growth and Antioxidant Defense System in *Oryza Sativa* As Revealed by Ultrastructure Analysis. *Ecotoxicol. Environ. Saf.* **2015**, *124*, 277–284. [CrossRef]
54. Tolussi, C.E.; Gomes, A.D.O.; Kumar, A.; Ribeiro, C.S.; Nostro, F.L.L.; Bain, P.A.; Souza, G.B.; Cuña, R.D.; Honji, R.M.; Moreira, R.G. Environmental Pollution Affects Molecular and Biochemical Responses during Gonadal Maturation of *Astyanax fasciatus* (Teleostei: Characiformes: Characidae). *Ecotoxicol. Environ. Saf.* **2018**, *147*, 926–934. [CrossRef] [PubMed]
55. Wu, Y.; Wang, J.; Wei, Y.; Chen, J.; Kang, L.; Long, C.; Wu, S.; Shen, L.; Wei, G. Maternal Exposure to Endocrine Disrupting Chemicals (EDCs) and Preterm Birth: A Systematic Review, Meta-Analysis, and Meta-Regression Analysis. *Environ. Pollut.* **2022**, *292*, 118264. [CrossRef] [PubMed]
56. Batt, A.L.; Kincaid, T.M.; Kostich, M.S.; Lazorchak, J.M.; Olsen, A.R. Evaluating the Extent of Pharmaceuticals in Surface Waters of the United States Using a National-Scale Rivers and Streams Assessment Survey. *Environ. Toxicol. Chem.* **2016**, *35*, 874–881. [CrossRef] [PubMed]
57. Zhang, M.; Shi, Y.J.; Lu, Y.L.; Johnson, A.C.; Sarvajayakesavalu, S.; Liu, Z.Y.; Su, C.; Zhang, Y.Q.; Juergens, M.D.; Jin, X.W. The Relative Risk and its Distribution of Endocrine Disrupting Chemicals, Pharmaceuticals and Personal Care Products to Freshwater Organisms in the Bohai Rim, China. *Sci. Total Environ.* **2017**, *590*, 633–642. [CrossRef]
58. Wang, B.; Qin, X.; Xiao, N.; Yao, Y.; Duan, Y.; Cui, X.; Zhang, S.; Luo, H.; Sun, H. Phthalate Exposure and Semen Quality in Infertile Male Population from Tianjin, China: Associations and Potential Mediation by Reproductive Hormones. *Sci. Total Environ.* **2020**, *744*, 140673. [CrossRef] [PubMed]
59. Azizi, D.; Arif, A.; Blair, D.; Dionne, J.; Filion, Y.; Ouarda, Y.; Pazmino, A.G.; Pulicharla, R.; Rilstone, V.; Tiwari, B.; et al. A Comprehensive Review on Current Technologies for Removal of Endocrine Disrupting Chemicals from Wastewaters. *Environ. Res.* **2021**, *207*, 112196. [CrossRef]

60. Gadd, J.B.; Northcott, G.L.; Tremblay, L.A. Passive Secondary Biological Treatment Systems Reduce Estrogens in Dairy Shed Effluent. *Environ. Sci. Technol.* **2010**, *44*, 7601–7606. [CrossRef]
61. Srivastava, A.; Singh, M.; Karsauliya, K.; Mondal, D.P.; Khare, P.; Singh, S.; Singh, S.P. Effective Elimination of Endocrine Disrupting Bisphenol A and S from Drinking Water Using Phenolic Resin-Based Activated Carbon Fiber: Adsorption, Thermodynamic and Kinetic Studies. *Environ. Nanotechnol. Monit. Manag.* **2020**, *14*, 100316. [CrossRef]
62. Singh, N.B.; Garima, N.; Sonal, A. Rachna Water Purification by Using Adsorbents: A Review. *Environ. Technol. Innov.* **2018**, *11*, 187–240. [CrossRef]
63. Goswami, A.D.; Trivedi, D.H.; Jadhav, N.L.; Pinjari, D.V. Sustainable and Green Synthesis of Carbon Nanomaterials: A Review. *J. Environ. Chem. Eng.* **2021**, *9*, 106118. [CrossRef]
64. Tiwari, S.K.; Bystrzejewski, M.; De Adhikari, A.; Huczko, A.; Wang, N. Methods for the Conversion of Biomass Waste into Value-Added Carbon Nanomaterials: Recent Progress and Applications. *Prog. Energy Combust. Sci.* **2022**, *92*, 101023. [CrossRef]
65. Srivastava, A.; Gupta, B.; Majumder, A.; Gupta, A.K.; Nimbhorkar, S.K. A Comprehensive Review on the Synthesis, Performance, Modifications, and Regeneration of Activated Carbon for the Adsorptive Removal of Various Water Pollutants. *J. Environ. Chem. Eng.* **2021**, *9*, 106177. [CrossRef]
66. Iijima, S. Helical Microtubules of Graphitic Carbon. *Nature* **1991**, *354*, 56–58. [CrossRef]
67. Abu-Nada, A.; Abdala, A.; McKay, G. Removal of Phenols and Dyes from Aqueous Solutions Using Graphene and Graphene Composite Adsorption: A Review. *J. Environ. Chem. Eng.* **2021**, *9*, 105858. [CrossRef]
68. Zhu, S.; Song, Y.; Zhao, X.; Shao, J.; Zhang, J.; Yang, B. The Photoluminescence Mechanism in Carbon Dots (Graphene Quantum Dots, Carbon Nanodots, and Polymer Dots): Current State and Future Perspective. *Nano Res.* **2015**, *8*, 355–381. [CrossRef]
69. Li, H.; Gui, X.; Zhang, L.; Wang, S.; Ji, C.; Wei, J.; Wang, K.; Zhu, H.; Wu, D.; Cao, A. Carbon Nanotube Sponge Filters for Trapping Nanoparticles and Dye Molecules from Water. *Chem. Commun.* **2010**, *46*, 7966–7968. [CrossRef]
70. Murayama, H.; Moriyama, N.; Mitobe, H.; Mukai, H.; Takase, Y.; Shimizu, K.; Kitayama, Y. Evaluation of Activated Carbon Fiber Filter for Sampling of Organochlorine Pesticides in Environmental Water Samples. *Chemosphere* **2003**, *52*, 825–833. [CrossRef]
71. Teng, Y.; Zhu, J.; Xiao, S.; Ma, Z.; Huang, T.; Liu, Z.; Xu, Y. Exploring Chitosan-loaded Activated Carbon Fiber for the Enhanced Adsorption of Pb(II)-EDTA Complex from Electroplating Wastewater in Batch and Continuous Processes. *Sep. Purif. Technol.* **2022**, *299*, 121659. [CrossRef]
72. Duan, J.; Ji, H.; Xu, T.; Pan, F.; Liu, X.; Liu, W.; Zhao, D. Simultaneous Adsorption of Uranium(VI) and 2-Chlorophenol by Activated Carbon Fiber Supported/Modified Titanate Nanotubes (TNTs/ACF): Effectiveness and Synergistic Effects. *Chem. Eng. J.* **2021**, *406*, 126752. [CrossRef]
73. Flor, M.P.; Camarillo, R.; Martínez, F.; Jiménez, C.; Quiles, R.; Rincón, J. Synthesis and Characterization of Bimetallic $TiO_2$/CNT/Pd-Cu for Efficient Remediation of Endocrine Disruptors under Solar Light. *J. Environ. Chem. Eng.* **2022**, *10*, 107245. [CrossRef]
74. Gao, Q.; Chen, W.; Chen, Y.; Werner, D.; Cornelissen, G.; Xing, B.; Tao, S.; Wang, X. Surfactant Removal with Multiwalled Carbon Nanotubes. *Water Res.* **2016**, *106*, 531–538. [CrossRef] [PubMed]
75. Mohammadi, A.A.; Dehghani, M.H.; Mesdaghinia, A.; Yaghmaian, K.; Es'haghi, Z. Adsorptive Removal of Endocrine Disrupting Compounds from Aqueous Solutions Using Magnetic Multi-wall Carbon Nanotubes Modified with Chitosan Biopolymer Based on Response Surface Methodology: Functionalization, Kinetics, and Isotherms Studies. *Int. J. Biol. Macromol.* **2019**, *155*, 1019–1029. [CrossRef]
76. Bassyouni, M.; Mansi, A.E.; Elgabry, A.; Ibrahim, B.A.; Kassem, O.A.; Alhebeshy, R. Utilization of Carbon Nanotubes in Removal of Heavy Metals from Wastewater: A Review of the CNTs' Potential and Current Challenges. *Appl. Phys. A* **2019**, *126*, 1–33. [CrossRef]
77. Khaliha, S.; Bianchi, A.; Kovtun, A.; Tunioli, F.; Boschi, A.; Zambianchi, M.; Paci, D.; Bocchi, L.; Valsecchi, S.; Polesello, S.; et al. Graphene Oxide Nanosheets for Drinking Water Purification by Tandem Adsorption and Microfiltration. *Sep. Purif. Technol.* **2022**, *300*, 121826. [CrossRef]
78. Peng, W.; Li, H.; Liu, Y.; Song, S. A Review on Heavy Metal Ions Adsorption from Water by Graphene Oxide and Its Composites. *J. Mol. Liq.* **2017**, *230*, 496–504. [CrossRef]
79. Fan, L.; Luo, C.; Li, X.; Lu, F.; Qiu, H.; Sun, M. Fabrication of Novel Magnetic Chitosan Grafted with Graphene Oxide to Enhance Adsorption Properties for Methyl Blue. *J. Hazard. Mater.* **2012**, *215–216*, 272–279. [CrossRef]
80. Blachowicz, T.; Ehrmann, A. Most Recent Developments in Electrospun Magnetic Nanofibers: A Review. *J. Eng. Fibers Fabr.* **2020**, *15*, 1–14. [CrossRef]
81. Sharma, V.K.; McDonald, T.J.; Kim, H.; Garg, V.K. Magnetic Graphene-Carbon Nanotube Iron Nanocomposites As Adsorbents and Antibacterial Agents for Water Purification. *Adv. Colloid Interface Sci.* **2015**, *225*, 229–240. [CrossRef]
82. Bharath, G.; Alhseinat, E.; Ponpandian, N.; Khan, M.A.; Alsharaeh, E.H. Development of Adsorption and Electrosorption Techniques for Removal of Organic and Inorganic Pollutants from Wastewater Using Novel Magnetite/Porous Graphene-Based Nanocomposites. *Sep. Purif. Technol.* **2017**, *188*, 206–218. [CrossRef]
83. Zhang, J.; Nguyen, M.N.; Li, Y.; Yang, C.; Schäfer, A.I. Steroid Hormone Micropollutant Removal from Water with Activated Carbon Fiber-Ultrafiltration Composite Membranes. *J. Hazard. Mater.* **2020**, *391*, 122020. [CrossRef] [PubMed]

34. Agarwal, S.; Sadeghi, N.; Tyagi, I.; Gupta, V.K.; Fakhri, A. Adsorption of Toxic Carbamate Pesticide Oxamyl from Liquid Phase by Newly Synthesized and Characterized Graphene Quantum Dots Nanomaterials. *J. Colloid Interface Sci.* **2016**, *478*, 430–438. [CrossRef] [PubMed]
35. Nuengmatcha, P.; Mahachai, R.; Chanthai, S. Removal of Hg(II) from Aqueous Solution Using Graphene Oxide as Highly Potential Adsorbent. *Asian J. Chem.* **2014**, *26*, S85–S88. [CrossRef]
36. Mohammad-Rezaei, R.; Jaymand, M. Graphene Quantum Dots Coated on Quartz Sand as Efficient and Low-Cost Adsorbent for Removal of $Hg^{2+}$ and $Pb^{2+}$ from Aqueous Solutions. *Environ. Prog. Sustain. Energy* **2018**, *38*, S24–S31. [CrossRef]
37. Deng, Y.; Ok, Y.S.; Mohan, D.; Pittman, C.U.; Dou, X. Carbamazepine Removal from Water by Carbon Dot-Modified Magnetic Carbon Nanotubes. *Environ. Res.* **2019**, *169*, 434–444. [CrossRef]
38. Liu, F.; Zhang, W.; Chen, W.; Wang, J.; Yang, Q.; Zhu, W.; Wang, J. One-Pot Synthesis of $NiFe_2O_4$ Integrated with EDTA-Derived Carbon Dots for Enhanced Removal of Tetracycline. *Chem. Eng. J.* **2017**, *310*, 187–196. [CrossRef]
39. Li, X.; Liu, T.; Wang, D.; Li, Q.; Liu, Z.; Li, N.; Zhang, Y.; Xiao, C.; Feng, X. Superlight Adsorbent Sponges Based on Graphene Oxide Cross-Linked with Poly(vinyl alcohol) for Continuous Flow Adsorption. *ACS Appl. Mater. Interfaces* **2018**, *10*, 21672–21680. [CrossRef]
90. Yang, J.; Juan, P.; Shen, Z.; Guo, R.; Jia, J.; Fang, H.; Wang, Y. Removal of Carbon Disulfide (CS2) from Water via Adsorption on Active Carbon Fiber (ACF). *Carbon* **2006**, *44*, 1367–1375. [CrossRef]
91. Ghasemi, S.S.; Hadavifar, M.; Maleki, B.; Mohammadnia, E. Adsorption of Mercury Ions from Synthetic Aqueous Solution Using Polydopamine Decorated SWCNTs. *J. Water Process Eng.* **2019**, *32*, 100965. [CrossRef]
92. Mashkoor, F.; Nasar, A. Inamuddin Carbon Nanotube-Based Adsorbents for the Removal of Dyes from Waters: A Review. *Environ. Chem. Lett.* **2020**, *18*, 605–629. [CrossRef]
93. Hua, S.; Gong, J.L.; Zeng, G.M.; Yao, F.B.; Guo, M.; Ou, X.M. Remediation of Organochlorine Pesticides Contaminated Lake Sediment Using Activated Carbon and Carbon Nanotubes. *Chemosphere* **2017**, *177*, 65–76. [CrossRef] [PubMed]
94. Yu, J.G.; Yu, L.Y.; Yang, H.; Liu, Q.; Chen, X.H.; Jiang, X.Y.; Chen, X.Q.; Jiao, F.P. Graphene Nanosheets as Novel Adsorbents in Adsorption, Preconcentration and Removal of Gases, Organic Compounds and Metal Ions. *Sci. Total Environ.* **2015**, *502*, 70–79. [CrossRef] [PubMed]
95. Luo, Z.F.; Li, H.P.; Yang, Y.; Lin, H.J.; Yang, Z.G. Adsorption of 17α-Ethinylestradiol from Aqueous Solution onto a Reduced Graphene Oxide-Magnetic Composite. *J. Taiwan Inst. Chem. Eng.* **2017**, *80*, 797–804. [CrossRef]
96. Zhang, X.; Lv, L.; Qin, Y.; Xu, M.; Jia, X.; Chen, Z. Removal of Aqueous Cr(VI) by a Magnetic Biochar Derived from *Melia Azedarach* Wood. *Bioresour. Technol.* **2018**, *256*, 1–10. [CrossRef]
97. Mehta, D.; Mazumdar, S.; Singh, S.K. Magnetic Adsorbents for the Treatment of Water/Wastewater—A Review. *J. Water Process Eng.* **2015**, *7*, 244–265. [CrossRef]
98. Mao, N.; Yang, L.; Zhao, G.; Li, X.; Li, Y. Adsorption Performance and Mechanism of Cr(VI) Using Magnetic PS-EDTA Resin from Micro-polluted Waters. *Chem. Eng. J.* **2012**, *200–202*, 480–490. [CrossRef]
99. Panida, P.; Parnuch, H.; Patiparn, P. Amino-Functionalized Mesoporous Silica-Magnetic Graphene Oxide Nanocomposites As Water-Dispersible Adsorbents for the Removal of the Oxytetracycline Antibiotic from Aqueous Solutions: Adsorption Performance, Effects of Coexisting Ions, and Natural Organic Matter. *Environ. Sci. Pollut. Res. Int.* **2020**, *27*, 6560–6576. [CrossRef]
100. Ji, Y.L.; Yin, M.J.; An, Q.F.; Gao, C.J. Recent Developments in Polymeric Nano-based Separation Membranes. *Fundam. Res.* **2021**, *2*, 254–267. [CrossRef]
101. Ji, Y.L.; Lu, H.H.; Gu, B.X.; Ye, R.F.; Gao, C.J. Tailoring the Asymmetric Structure of Polyamide Reverse Osmosis Membrane with Self-assembled Aromatic Nanoparticles for High-Efficient Removal of Organic Micropollutants. *Chem. Eng. J.* **2021**, *416*, 129080. [CrossRef]
102. Kolbasov, A.; Sinha-Ray, S.; Yarin, A.L.; Pourdeyhimi, B. Heavy Metal Adsorption on Solution-Blown Biopolymer Nanofiber Membranes. *J. Membr. Sci.* **2017**, *530*, 250–263. [CrossRef]
103. Beck, R.J.; Zhao, Y.; Fong, H.; Menkhaus, T.J. Electrospun Lignin Carbon Nanofiber Membranes with Large Pores for Highly Efficient Adsorptive Water Treatment Applications. *J. Water Process Eng.* **2017**, *16*, 240–248. [CrossRef]
104. Kappen, J.; Aravind, M.K.; Varalakshmi, P.; Ashokkumar, B.; John, S.A. Quantitative Removal of Hg(II) as Hg(0) Using Carbon Cloths Coated Graphene Quantum Dots and Their Silver Nanoparticles Composite and Application of Hg(0) for the Sensitive Determination of Nitrobenzene. *Colloids Surf. A Physicochem. Eng. Asp.* **2022**, *641*, 128542. [CrossRef]
105. Tshangana, C.S.; Muleja, A.A.; Kuvarega, A.T.; Malefetse, T.J.; Mamba, B.B. The Applications of Graphene Oxide Quantum Dots in the Removal of Emerging Pollutants in Water: An Overview. *J. Water Process Eng.* **2021**, *43*, 102249. [CrossRef]
106. Wang, L.; Wang, X.; Zhou, J.B.; Zhao, R.S. Carbon Nanotube Sponges as a Solid-Phase Extraction Adsorbent for the Enrichment and Determination of Polychlorinated Biphenyls at Trace Levels in Environmental Water Samples. *Talanta* **2016**, *160*, 79–85. [CrossRef] [PubMed]
107. Jang, Y.; Bang, J.; Seon, Y.S.; You, D.W.; Oh, J.S.; Jung, K.W. Carbon Nanotube Sponges as an Enrichment Material for Aromatic Volatile Organic Compounds. *J. Chromatogr. A* **2020**, *1617*, 460840. [CrossRef] [PubMed]
108. Dubey, S.P.; Dwivedi, A.D.; Kim, I.; Sillanpaa, M.; Kwon, Y.N.; Lee, C.H. Synthesis of Graphene-Carbon Sphere Hybrid Aerogel with Silver Nanoparticles and its Catalytic and Adsorption Applications. *Chem. Eng. J.* **2014**, *244*, 160–167. [CrossRef]
109. Mao, H.N.; Wang, X.G. Use of In-Situ Polymerization in the Preparation of Graphene/Polymer Nanocomposites. *New Carbon Mater.* **2020**, *35*, 336–343. [CrossRef]

110. Youssef, A.M.; El-Naggar, M.E.; Malhat, F.M.; Sharkawi, H.E. Efficient Removal of Pesticides and Heavy Metals from Wastewater and the Antimicrobial Activity of f-MWCNTs/PVA Nanocomposite Film. *J. Clean. Prod.* **2019**, *206*, 315–325. [CrossRef]
111. Yang, X.; Li, Y.; Du, Q.; Sun, J.; Chen, L.; Hu, S.; Wang, Z.; Xia, Y.; Xia, L. Highly Effective Removal of Basic Fuchsin from Aqueous Solutions by Anionic Polyacrylamide/Graphene Oxide Aerogels. *J. Colloid Interface Sci.* **2015**, *453*, 107–114. [CrossRef]
112. Lai, L.; Chen, L.; Zhan, D.; Sun, L.; Liu, J.; Lim, S.H.; Poh, C.K.; Shen, Z.; Lin, J. One-Step Synthesis of $NH_2$-Graphene from in Situ Graphene-Oxide Reduction and Its Improved Electrochemical Properties. *Carbon* **2011**, *49*, 3250–3257. [CrossRef]
113. Guo, X.; Du, B.; Wei, Q.; Yang, J.; Hu, L.; Yan, L.; Xu, W. Synthesis of Amino Functionalized Magnetic Graphenes Composite Material and Its Application to Remove Cr(VI), Pb(II), Hg(II), Cd(II) and Ni(II) from Contaminated Water. *J. Hazard. Mater.* **2014**, *278*, 211–220. [CrossRef] [PubMed]
114. Sun, X.; Bai, L.; Li, J.; Huang, L.; Sun, H.; Gao, X. Robust Preparation of Flexibly Super-hydrophobic Carbon Fiber Membrane by Electrospinning for Efficient Oil-Water Separation in Harsh Environments. *Carbon* **2021**, *182*, 11–22. [CrossRef]
115. Li, L.; Hu, J.; Shi, X.; Fan, M.; Luo, J.; Wei, X. Nanoscale Zero-Valent Metals: A Review of Synthesis, Characterization, and Applications to Environmental Remediation. *Environ. Sci. Pollut. Res.* **2016**, *23*, 17880–17900. [CrossRef] [PubMed]
116. Feng, C.; Khulbe, K.C.; Matsuura, T.; Tabe, S.; Ismail, A.F. Preparation and Characterization of Electro-spun Nanofiber Membranes and Their Possible Applications in Water Treatment. *Sep. Purif. Technol.* **2013**, *102*, 118–135. [CrossRef]
117. Zhang, Z.; Zhao, Y.; Li, Z.; Zhang, L.; Liu, Z.; Long, Z.; Li, Y.; Liu, Y.; Fan, R.; Sun, K.; et al. Synthesis of Carbon/$SiO_2$ Core-Sheath Nanofibers with Co-Fe Nanoparticles Embedded in via Electrospinning for High-Performance Microwave Absorption. *Adv. Compos. Hybrid Mater.* **2021**, *5*, 513–524. [CrossRef]
118. Wang, J.; Lu, X.; Ng, P.F.; Lee, K.I.; Fei, B.; Xin, J.H.; Wu, J. Polyethylenimine Coated Bacterial Cellulose Nanofiber Membrane and Application As Adsorbent and Catalyst. *J. Colloid Interface Sci.* **2015**, *440*, 32–38. [CrossRef]
119. Wu, J.; Zhang, J.; Zhou, S.; Yang, Z.; Zhang, X. Ag Nanoparticle-Decorated Carbon Nanotube Sponges for Removal of Methylene Blue from Aqueous Solution. *New J. Chem.* **2020**, *44*, 7096–7104. [CrossRef]
120. Wang, B.; Chen, P.Y.; Zhao, R.X.; Zhang, L.; Chen, Y.; Yu, L.P. Carbon-Dot Modified Polyacrylonitrile Fibers: Recyclable Materials Capable of Selectively and Reversibly Adsorbing Small-Sized Anionic Dyes. *Chem. Eng. J.* **2020**, *391*, 123484. [CrossRef]
121. Pirhaji, J.Z.; Moeinpour, F.; Dehabadi, A.M.; Ardakani, S.A.Y. Synthesis and Characterization of Halloysite/Graphene Quantum Dots Magnetic Nanocomposite as a New Adsorbent for Pb(II) Removal from Water. *J. Mol. Liq.* **2020**, *300*, 112345. [CrossRef]
122. Maldonado-Orozco, M.C.; Ochoa-Lara, M.T.; Sosa-Márquez, J.E.; Talamantes-Soto, R.P.; Hurtado-Macías, A.; Antón, R.L.; González, J.A.; Holguín-Momaca, J.T.; Olive-Méndez, S.F.; Espinosa-Magaña, F. Absence of Ferromagnetism in Ferroelectric Mn-Doped $BaTiO_3$ Nanofibers. *J. Am. Ceram. Soc.* **2019**, *102*, 2800–2809. [CrossRef]
123. Toński, M.; Paszkiewicz, M.; Doonek, J.; Flejszar, M.; Biak-Bielińska, A. Regeneration and Reuse of the Carbon Nanotubes for the Adsorption of Selected Anticancer Drugs from Water Matrices. *Colloids Surf. A* **2021**, *618*, 126355. [CrossRef]
124. Katubi, K.M.M.; Alsaiari, N.S.; Alzahrani, F.M.; Siddeeg, S.M.; Tahoon, M.A. Synthesis of Manganese Ferrite/Graphene Oxide Magnetic Nanocomposite for Pollutants Removal from Water. *Processes* **2021**, *9*, 589. [CrossRef]
125. Jinendra, U.; Bilehal, D.; Nagabhushana, B.M.; Kumar, A.P. Adsorptive Removal of Rhodamine B Dye from Aqueous Solution by Using Graphene–Based Nickel Nanocomposite. *Heliyon* **2021**, *7*, e06851. [CrossRef]
126. Jiang, Y.; Liu, Z.; Zeng, G.; Liu, Y.; Shao, B.; Li, Z.; Liu, Y.; Zhang, W.; He, Q. Polyaniline-Based Adsorbents for Removal of Hexavalent Chromium from Aqueous Solution: A Mini Review. *Environ. Sci. Pollut. Res.* **2018**, *25*, 6158–6174. [CrossRef]
127. Souza, C.C.; Souza, L.Z.M.; Yılmaz, M.; Oliveira, M.A.; Bezerra, A.C.S.; Silva, E.F.; Dumont, M.R.; Machado, A.R.T. Activated Carbon of Coriandrum Sativum for Adsorption of Methylene Blue: Equilibrium and Kinetic Modeling. *Clean. Mater.* **2022**, *3*, 100052. [CrossRef]
128. Wang, D.; Wang, Z.; Zheng, X.; Tian, M. Activated Carbon Fiber Derived from the Seed Hair Fibers of Metaplexis Japonica: Novel Efficient Adsorbent for Methylene Blue. *Ind. Crop. Prod.* **2020**, *148*, 112319. [CrossRef]
129. Sharabati, M.A.; Abokwiek, R.; Al-Othman, A.; Tawalbeh, M.; Karaman, C.; Orooji, Y.; Karimi, F. Biodegradable Polymers and Their Nano-composites for the Removal of Endocrine-Disrupting Chemicals (EDCs) from Wastewater: A Review. *Environ. Res.* **2021**, *202*, 111694. [CrossRef]
130. Ussia, M.; Di Mauro, A.; Mecca, T.; Cunsolo, F.; Cerruti, P.; Nicotra, G.; Spinella, C.; Impellizzeri, G.; Privitera, V.; Carroccio, S.C. ZnO-pHEMA Nanocomposites: An Eco-Friendly and Reusable Material for Water Remediation. *ACS Appl. Mater. Interfaces* **2018**, *10*, 40100–40110. [CrossRef]
131. El-Sayed, M.E.A. Nanoadsorbents for Water and Wastewater Remediation. *Sci. Total Environ.* **2020**, *739*, 139903. [CrossRef]
132. Pillay, K.; Cukrowska, E.M.; Coville, N.J. Multi-walled Carbon Nanotubes As Adsorbents for the Removal of Parts per Billion Levels of Hexavalent Chromium from Aqueous Solution. *J. Hazard. Mater.* **2009**, *166*, 1067–1075. [CrossRef] [PubMed]
133. Gadd, G.M. Biosorption: Critical Review of Scientific Rationale, Environmental Importance and Significance for Pollution Treatment. *J. Chem. Technol. Biotechnol.* **2010**, *84*, 13–28. [CrossRef]
134. Crini, G.; Lichtfouse, E.; Wilson, L.D.; Morin-Crini, N. Conventional and Non-conventional Adsorbents for Wastewater Treatment. *Environ. Chem. Lett.* **2019**, *17*, 195–213. [CrossRef]
135. Yagub, M.T.; Sen, T.K.; Afroze, S.; Ang, H.M. Dye and Its Removal from Aqueous Solution by Adsorption: A Review. *Adv. Colloid Interface Sci.* **2014**, *209*, 172–184. [CrossRef]
136. Delkash, M.; Bakhshayesh, B.E.; Kazemian, H. Using Zeolitic Adsorbents to Cleanup Special Wastewater Streams: A Review. *Microporous Mesoporous Mater.* **2015**, *214*, 224–241. [CrossRef]

137. Dixit, F.; Dutta, R.; Barbeau, B.; Berube, P.; Mohseni, M. PFAS Removal by Ion Exchange Resins: A Review. *Chemosphere* **2021**, *272*, 129777. [CrossRef]
138. Caltran, I.; Heijman, S.G.J.; Shorney-Darby, H.L.; Rietveld, L.C. Impact of Removal of Natural Organic Matter from Surface Water by Ion Exchange: A Case Study of Pilots in Belgium, United Kingdom and the Netherlands. *Sep. Purif. Technol.* **2020**, *247*, 116974. [CrossRef]
139. Berber, M.R. Current Advances of Polymer Composites for Water Treatment and Desalination. *J. Chem.* **2020**, *2020*, 1–19. [CrossRef]
140. Akpomie, K.G.; Conradie, J. Efficient Synthesis of Magnetic Nanoparticle-Musa Acuminata Peel Composite for the Adsorption of Anionic Dye. *Arab. J. Chem.* **2020**, *13*, 7115–7131. [CrossRef]
141. Saxena, M.; Lochab, A.; Saxena, R. Asparagine Functionalized MWCNTs for Adsorptive Removal of Hazardous Cationic Dyes: Exploring Kinetics, Isotherm and Mechanism. *Surf. Interface* **2021**, *25*, 101187. [CrossRef]
142. Kumar, K.A.; Venkata Mohan, V. Removal of Natural and Synthetic Endocrine Disrupting Estrogens by Multi-walled Carbon Nanotubes (MWCNT) as Adsorbent: Kinetic and Mechanistic Evaluation. *Sep. Purif. Technol.* **2012**, *87*, 22–30. [CrossRef]
143. Baharum, N.A.; Nasir, H.M.; Ishak, M.Y.; Isa, N.M.; Hassan, M.A.; Aris, A.Z. Highly Efficient Removal of Diazinon Pesticide from Aqueous Solutions by Using Coconut Shell-Modified Biochar. *Arab. J. Chem.* **2020**, *13*, 6106–6121. [CrossRef]
144. Hameed, B.H.; Salman, J.M.; Ahmad, A.L. Adsorption Isotherm and Kinetic Modeling of 2,4-D Pesticide on Activated Carbon Derived from Date Stones. *J. Hazard. Mater.* **2009**, *163*, 121–126. [CrossRef] [PubMed]
145. Mahar, F.K.; He, L.; Wei, K.; Mehdi, M.; Zhu, M.; Gu, J.; Zhang, K.; Khatri, Z.; Kim, I. Rapid Adsorption of Lead Ions Using Porous Carbon Nanofibers. *Chemosphere* **2019**, *225*, 360–367. [CrossRef]
146. Zhang, Y.; Li, Y.; Xu, W.; Cui, M.; Wang, M.; Chen, B.; Sun, Y.; Chen, K.; Li, L.; Du, Q.; et al. Filtration and Adsorption of Tetracycline in Aqueous Solution by Copper Alginate-Carbon Nanotubes Membrane Which Has the Muscle-Skeleton Structure. *Chem. Eng. Res. Des.* **2022**, *183*, 424–438. [CrossRef]
147. Yadav, N.; Maddheshiaya, D.N.; Rawat, S.; Singh, J. Adsorption and Equilibrium Studies of Phenol and Para-nitrophenol by Magnetic Activated Carbon Synthesised from Cauliflower Waste. *Environ. Eng. Res.* **2019**, *25*, 742–752. [CrossRef]
148. Li, F.; Jiang, X.; Zhao, J.; Zhang, S. Graphene Oxide: A Promising Nanomaterial for Energy and Environmental Applications. *Nano Energy* **2015**, *16*, 488–515. [CrossRef]
149. Li, H.; Xing, J.; Xia, Z.; Chen, J. Preparation of Coaxial Heterogeneous Graphene Quantum Dot-Sensitized $TiO_2$ Nanotube Arrays via Linker Molecule Binding and Electrophoretic Deposition. *Carbon* **2015**, *81*, 474–487. [CrossRef]

Article

# Dependence of Structural, Morphological and Magnetic Properties of Manganese Ferrite on Ni-Mn Substitution

Thomas Dippong [1,*], Erika Andrea Levei [2], Iosif Grigore Deac [3], Ioan Petean [4] and Oana Cadar [2]

1. Faculty of Science, Technical University of Cluj-Napoca, 76 Victoriei Street, 430122 Baia Mare, Romania
2. INCDO-INOE 2000, Research Institute for Analytical Instrumentation, 67 Donath Street, 400293 Cluj-Napoca, Romania; erika.levei@icia.ro (E.A.L.); oana.cadar@icia.ro (O.C.)
3. Faculty of Physics, Babes-Bolyai University, 1 Kogalniceanu Street, 400084 Cluj-Napoca, Romania; iosif.deac@phys.ubbcluj.ro
4. Faculty of Chemistry and Chemical Engineering, Babes-Bolyai University, 11 Arany Janos Street, 400028 Cluj-Napoca, Romania; petean.ioan@gmail.com
* Correspondence: dippong.thomas@yahoo.ro

**Abstract:** This paper presents the influence of $Mn^{2+}$ substitution by $Ni^{2+}$ on the structural, morphological and magnetic properties of $Mn_{1-x}Ni_xFe_2O_4@SiO_2$ (x = 0, 0.25, 0.50, 0.75, 1.00) nanocomposites (NCs) obtained by a modified sol-gel method. The Fourier transform infrared spectra confirm the formation of a $SiO_2$ matrix and ferrite, while the X-ray diffraction patterns show the presence of poorly crystalline ferrite at low annealing temperatures and highly crystalline mixed cubic spinel ferrite accompanied by secondary phases at high annealing temperatures. The lattice parameters gradually decrease, while the crystallite size, volume, and X-ray density of $Mn_{1-x}Ni_xFe_2O_4@SiO_2$ NCs increase with increasing Ni content and follow Vegard's law. The saturation magnetization, remanent magnetization, squareness, magnetic moment per formula unit, and anisotropy constant increase, while the coercivity decreases with increasing Ni content. These parameters are larger for the samples with the same chemical formula, annealed at higher temperatures. The NCs with high Ni content show superparamagnetic-like behavior, while the NCs with high Mn content display paramagnetic behavior.

**Keywords:** manganese ferrite; nanocomposites; amorphous silica matrix; annealing temperature; magnetic properties

## 1. Introduction

Nanoscale materials have remarkable optical, magnetic, electrical, and catalytic properties [1–6]. The structure and composition of spinel nano-ferrites control the functional properties of magnetic nanosized materials [4,5]. Nanocomposites (NCs) are mixtures of different components at the nanometer scale, with properties that depend on the contribution of each component in the mixture [5].

Magnetic spinel ferrite ($MFe_2O_4$, where M = Zn, Co, Mn, Ni, etc.) nanoparticles are of high interest for materials science and nanotechnology, due to their high reactivity, chemical stability, and reusability [7–10]. The structure, magnetic, and electrical properties of nanosized ferrites depend upon the synthesis method, annealing temperature, as well as on the concentration, nature, and distribution of the cations between the tetrahedral (A)- and octahedral (B) sites [2]. Thus, by selecting the suitable synthesis parameters it is possible to design ferrites with the expected properties [7,11]. Particle size, shape, and enhanced surface-to-volume ratio also influence the magnetic characteristics of the nanomaterials [12]. Accordingly, the magnetization parameters are enhanced by the surface spins and spin canting [11]. Below the critical single domain, the nanomaterials have a single domain blocked state and exhibit optimum magnetic properties. In such single-domain systems, the magnetic anisotropy determines the spin alignment along the easy axis of magnetization, while the thermal fluctuations cause these spins to undergo Brownian motion along their axes [12–14]. As the magnetic field allows the control of the shape-memory effect, new types of microstructures may be produced by applying an external magnetic field [6]. The coercivity ($H_C$), remanent magnetization ($M_R$), saturation magnetization ($M_S$), and anisotropy constant ($K$) are the main magnetic properties that determine the spinel ferrites applications [14].

The nickel ferrite, $NiFe_2O_4$, has remarkable magnetic and electrical characteristics such as high $M_S$, permeability, resistivity, Curie temperature, and low eddy current loss [1,2,7–9]. $NiFe_2O_4$ has an inverse spinel structure with the $Fe^{3+}$ ions placed equally in tetrahedral (A) and octahedral (B) sites and the $Ni^{2+}$ ions placed in octahedral (B) sites [2,8]. Moreover, due to the magnetic moments of antiparallel spins between $Fe^{3+}$ ions at the tetrahedral (A) sites and $Ni^{2+}$ ions at the octahedral (B) sites of the spinel structure, $NiFe_2O_4$ displays ferromagnetic behavior [8]. The partial substitution of $NiFe_2O_4$ with magnetic divalent transition metal ions ($Mn^{2+}$, $Cu^{2+}$, $Zn^{2+}$, $Cd^{2+}$, $Mn^{2+}$, etc.) results in exceptional properties.

The manganese ferrite, $MnFe_2O_4$, is a soft ferrite characterized by high magnetic permeability and low hysteresis losses [2]. $MnFe_2O_4$ has a spinel crystal structure with the $Mn^{2+}$ ions occupying only the tetrahedral (A) sites, while the $Fe^{3+}$ ions populate the octahedral (B) sites. The substitution of $Ni^{2+}$ ions in $MnFe_2O_4$ changes its structure, magnetic, electrical, and dielectric properties [8]. When the $Mn^{2+}$ ions are substituted by $Ni^{2+}$ ions, $Ni^{2+}$ ions are expected to occupy the octahedral (B) sites, while $Mn^{2+}$ ions are randomly distributed between tetrahedral (A) and octahedral (B) sites [12,14]. Moreover, when substituting one $Mn^{2+}$ ion with one $Ni^{2+}$ ion, the atomic magnetic moment increases from 2 $\mu_B$ to 5 $\mu_B$ [9]. Mixed Ni-Mn ferrites present attractive magnetic properties with applications as soft and hard magnets due to their high electrical resistivity, $M_S$ and permeability, and low dielectric losses [1,9,12].

For the development of new applications, it is important to tailor the magneto-optic properties of spinel ferrites. The main routes that allow the properties tailoring for a specific application are the optimization of the synthesis parameters and selection of the optimum spinel ferrite composition [13]. Thus, the development of new ways to control the properties, especially the particle size and shape of spinel ferrites by the preparation route become of great interest [13]. The large-scale applications of nanosized spinel ferrites promoted the development of various chemical preparation methods as alternatives to solid-state reactions which produce large agglomerated particles with limited homogeneity and low sinterability [2]. Generally, the chemical methods produce fine-grained particles, but the poor crystallinity and wide particle size distribution can alter the expected properties. Moreover, the use of long reaction time and post-synthesis thermal treatment is needed [7,15–18]. Spinel ferrites are usually prepared by a standard ceramic technique that uses high temperatures and produces particles with a low specific surface [2]. Therefore, in order to obtain nanosized ferrites with high specific surface and homogeneity, alternative methods such as co-precipitation, polymeric gel, hydrothermal, micro-emulsion, heterogeneous precipitation, sono-chemistry, combustion, and sol-gel methods are used. These methods require expensive equipment, energy overriding, and high processing temperature as well as long reaction time [2,7]. The sol-gel route is the most popular way to prepare nanosized ferrites due to its simplicity, low cost, and good control over the structure and properties [10]. The microwave-assisted sol-gel method combines the advantages of microwave and sol-gel methods, being a faster, energy-saving procedure for obtaining single-phase nanopowders of high purity with accurate control of stoichiometry and capability of industrial scale-up [15–20]. The homogenous dispersion of the ferrite particles into an organic matrix also allows the production of composite materials with highly dispersed fine magnetic particles [10]. Embedding ferrites into silica ($SiO_2$) matrix allows the control of the particle growth, minimizes the particle agglomeration, and enhances the magnetic guidability and biocompatibility [19].

The objective of the study was to investigate the effect of Ni content and annealing temperature on the structure, morphology, and magnetic behavior of $Mn_{1-x}Ni_xFe_2O_4@SiO_2$

($x$ = 0, 0.25, 0.50, 0.75, 1.00) NCs. The formation of ferrite and SiO$_2$ matrix was investigated by Fourier transform infrared (FT-IR) spectroscopy, the formation of crystalline phases was studied by X-ray diffraction (XRD), while the shape, morphology, size, and rugosity of nanoparticles were investigated by atomic force microscopy (AFM). The variation of magnetization saturation ($M_S$) vs. the coercive field ($H_C$) of Mn$_{1-x}$Ni$_x$Fe$_2$O$_4$@SiO$_2$ NCs was studied by magnetic measurements.

## 2. Results and Discussion

### 2.1. X-ray Diffraction

The XRD patterns of Mn$_x$Ni$_{1-x}$Fe$_2$O$_4$@SiO$_2$ NCs ($x$ = 0, 0.25, 0.50, 0.75, 1.00) annealed at 400, 800, and 1200 °C are presented in Figure 1. At 400 °C, the baseline noise and the amorphous halo between 10 and 30° (2θ) indicate the formation of poorly crystalized ferrite, while at higher annealing temperatures, the formation of highly crystalline mixed spinel ferrites is confirmed by the sharp diffraction peaks. At higher annealing temperatures, the presence of other crystalline secondary phases is also remarked. The variation of the relative intensities and signal-to-noise ratio indicates distinct crystallinity degrees or different crystallite sizes [7].

In the case of Mn$_x$Ni$_{1-x}$Fe$_2$O$_4$@SiO$_2$ ($x$ = 0.00), the poorly crystallized MnFe$_2$O$_4$ (JCPDS card no 74-2403 [21]) is accompanied by α-Fe$_2$O$_3$ (JCPDS card no. 87-1164 [21]), cristobalite (JCPDS card no. 89-3434 [21], quartz (JCPDS card 85-0457 [21]) and Fe$_2$SiO$_4$ (JCPDS card no.87-0315 [21]) at 800 °C, and α-Fe$_2$O$_3$, cristobalite and quartz at 1200 °C. The diffraction peaks matching with the MnFe$_2$O$_4$ reflection planes (2 2 0), (3 1 1), (2 2 2), (4 0 0), (4 2 2), (5 1 1), and (4 4 0) confirm the cubic spinel structure corresponding to the space group $Fd$-$3m$ [22]. The formation of α-Fe$_2$O$_3$ might be explained by the partial embedding of ferrite in the SiO$_2$ matrix and the unsatisfactory annealing temperature or time needed to produce pure crystalline MnFe$_2$O$_4$ phase [7–9,19]. The formation of Fe$_2$SiO$_4$ could be a consequence of the reducing conditions produced by the decomposition of carboxylate precursors in the matrix pores that partially reduce the Fe$^{3+}$ ions into Fe$^{2+}$ ions, which react with SiO$_2$ leading to the formation of Fe$_2$SiO$_4$ [7–10].

In the case of Mn$_x$Ni$_{1-x}$Fe$_2$O$_4$@SiO$_2$ ($x$ = 1.00), NiFe$_2$O$_4$ (JCPDS card no. 10-0325 [21]) is conveyed by α-Fe$_2$O$_3$, cristobalite, quartz and Fe$_2$SiO$_4$ at 800 °C, and cristobalite, quartz, and Fe$_2$SiO$_4$ at 1200 °C. The distinct formation of secondary phases of α-Fe$_2$O$_3$ and SiO$_2$ could be attributed to the instability of Mn$^{2+}$ ions [23–25]. The SiO$_2$ matrix avoids the aggregation of nanoparticles through steric repulsion [24,25]. The possible oxidation-reduction reactions are also determined by the oxygen partial pressure and the presence of air during the annealing process [23,25].

In the case of Mn$_x$Ni$_{1-x}$Fe$_2$O$_4$@SiO$_2$ ($x$ = 0.25–0.75), at 800 °C, the ferrite is accompanied by cristobalite and quartz ($x$ = 0.25) and cristobalite, quartz, α-Fe$_2$O$_3$, and Fe$_2$SiO$_4$ ($x$ = 0.50 and $x$ = 0.75). At 1200 °C, the crystalline phase of mixed Mn-Ni ferrite is accompanied by cristobalite and quartz ($x$ = 0.25 and $x$ = 0.50), and cristobalite, quartz, and Fe$_2$SiO$_4$ ($x$ = 0.75). Some possible explanations for the formation of secondary phases could be the higher mobility of cations and the strain variation induced by the annealing process which also causes a small shift in 2θ positions and peak broadening, concomitantly with the increase of crystallite sizes [3,26].

The XRD parameters are presented in Table 1. The average crystallite size was estimated using the full width at half-maximum ($w_{hkl}$) of the most intense (311) peak via Scherrer's equation [11]. For the cubic structure, the lattice parameter (a) can be calculated from Miller indices (h, k, l) and inter-planar spacing (d) using the equation $a = d(h^2 + k^2 + l^2)^{1/2}$ and Bragg's law [11]. Larger crystallite sizes were obtained at high annealing temperatures since the small nanoparticles join and form larger nanoparticles during the annealing process [12].

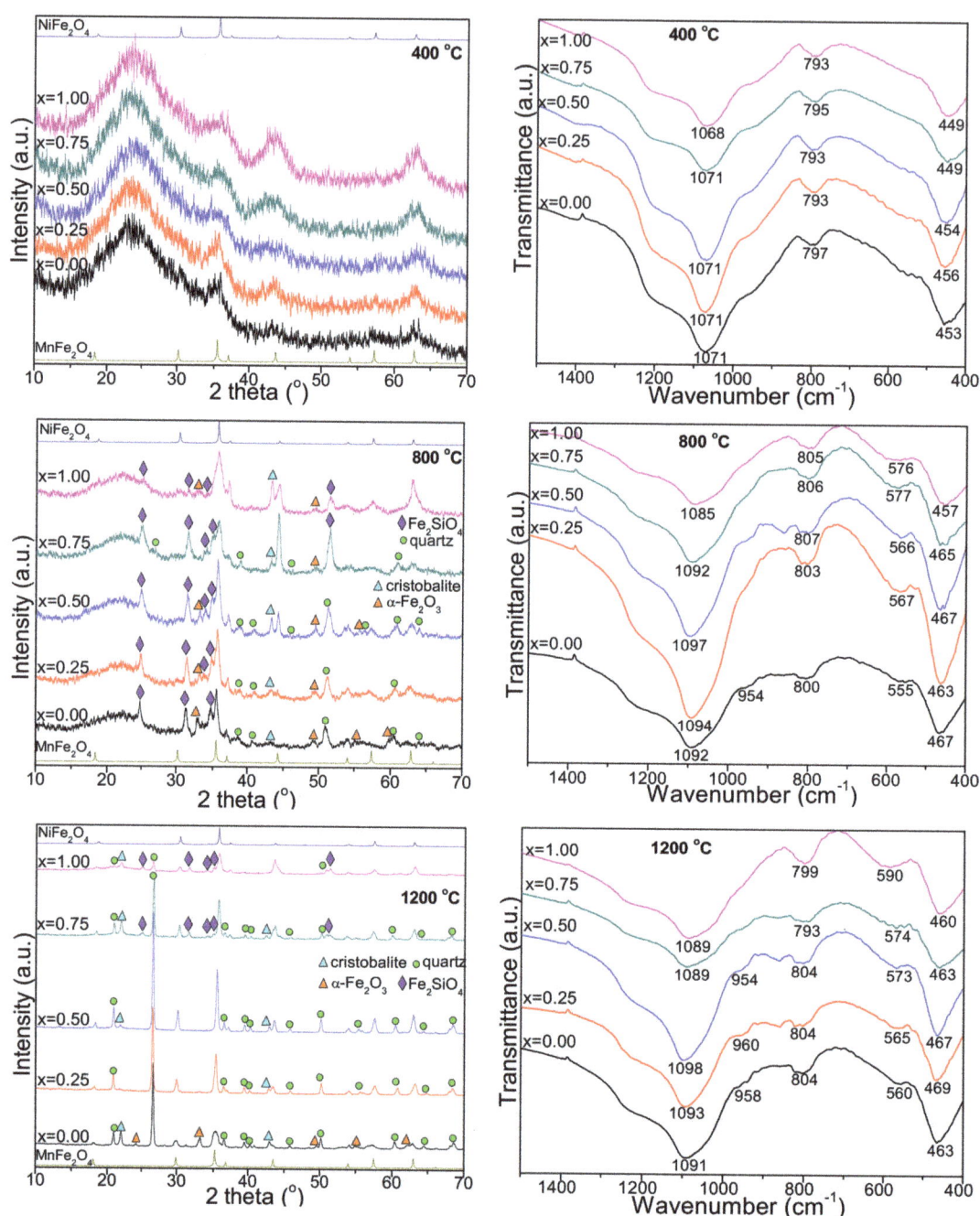

**Figure 1.** X-ray diffraction patterns and FT-IR spectra of Mn$_{1-x}$Ni$_x$Fe$_2$O$_4$@SiO$_2$ NCs annealed at 400, 800, and 1200 °C.

Table 1. XRD parameters of $Mn_{1-x}Ni_xFe_2O_4@SiO_2$ annealed at 400, 800, and 1200 °C.

| NC | Temperature, °C | Crystallite Size, nm | Lattice Parameter, Å | Volume, Å$^3$ | X-ray Density, g·cm$^{-3}$ | Hopping Length in A, Å | Hopping Length in B, Å |
|---|---|---|---|---|---|---|---|
| x = 0.00 | 400 | 6 | 8.465 | 606.6 | 5.050 | 3.665 | 2.993 |
|  | 800 | 10 | 8.472 | 608.1 | 5.037 | 3.668 | 2.995 |
|  | 1200 | 23 | 8.485 | 610.9 | 5.014 | 3.674 | 2.999 |
| x = 0.25 | 400 | 8 | 8.441 | 601.4 | 5.114 | 3.655 | 2.984 |
|  | 800 | 13 | 8.448 | 602.9 | 5.101 | 3.658 | 2.987 |
|  | 1200 | 27 | 8.459 | 605.3 | 5.081 | 3.663 | 2.991 |
| x = 0.50 | 400 | 10 | 8.432 | 599.5 | 5.151 | 3.651 | 2.981 |
|  | 800 | 17 | 8.437 | 600.6 | 5.142 | 3.653 | 2.983 |
|  | 1200 | 30 | 8.443 | 601.9 | 5.131 | 3.656 | 2.985 |
| x = 0.75 | 400 | 12 | 8.416 | 596.1 | 5.202 | 3.644 | 2.976 |
|  | 800 | 21 | 8.423 | 597.6 | 5.189 | 3.647 | 2.978 |
|  | 1200 | 37 | 8.427 | 598.4 | 5.182 | 3.649 | 2.979 |
| x = 1.00 | 400 | 14 | 8.402 | 593.1 | 5.249 | 3.638 | 2.971 |
|  | 800 | 26 | 8.409 | 594.6 | 5.236 | 3.641 | 2.973 |
|  | 1200 | 46 | 8.412 | 595.2 | 5.230 | 3.643 | 2.974 |

In NCs with a low Ni content (x = 0.25–0.50), the expansion of crystallite size is delayed, while at high Ni content (x = 0.75–1.00), the growth of crystallite size at the nucleation centers is preferred [7]. The metal ions are distributed between the tetrahedral (A) and octahedral (B) sites with oxygen as the nearest neighbor [12]. The increase of the lattice parameters at a low Ni content can be ascribed to the replacement of the smaller ionic radii $Ni^{2+}$ (tetrahedral: 0.55Å; octahedral 0.69 Å) by the larger ionic radii $Mn^{2+}$ (tetrahedral: 0.655 Å; octahedral: 0.80 Å). The replacement of $Ni^{2+}$ ions by $Mn^{2+}$ ions causes an increase of the interatomic space and, consequently, the lattice constant increase in accordance with Vegard's law [11,22,26]. The variation of the lattice constant generates internal stress and suppresses additional grain growth during the annealing process. The difference between the theoretical and experimental values can be accredited to the approximation which considers the ions as spheres distributed in a rigid manner [1]. The obtained results are in good agreement with previous studies [8]. The crystallites are more compact in the case of NC (x = 1.00), as a $Ni^{2+}$ ion is smaller and dissolves more easily in the spinel lattice. The decrease of unit cell volume is also observed with the introduction of smaller-sized $Ni^{2+}$ ions in the crystal lattice [1]. There is no significant difference between the molecular weight of the obtained NCs, thus, the decrease of the unit cell volume with the increase of Ni content leads to the increase of X-ray density [1]. The X-ray density also increases with the increase of Ni content and annealing temperature. The variation of X-ray density as a consequence of small fluctuations of the lattice constant is attributed to the variation of the distribution of cations within tetrahedral (A) and octahedral (B) sites [1]. The substitution of $Mn^{2+}$ ion generates an increase in the porosity of grains due to its greater ionic radius, the grains becoming less compact and causing an increase in particle size [26]. The hopping length ($L_A$ and $L_B$) between the magnetic ions in the tetrahedral (A)- and octahedral (B)- sites increases with the increase of annealing temperature and decreases with the Ni content, probably due to the higher ionic radius of $Mn^{2+}$ in comparison to that of $Ni^{2+}$ [1]. Furthermore, the $Mn^{2+}$ and $Ni^{2+}$ ions have a very low tendency for tetrahedral (A) site occupancy, while $Fe^{3+}$ ions are unevenly divided between tetrahedral (A) and octahedral (B) sites, depending on the Ni content in the sample [1].

*2.2. Fourier-Transform Infrared Spectroscopy*

At all annealing temperatures, the FT-IR spectra show the characteristic peaks for ferrite and $SiO_2$ matrix in the range of 1500–400 cm$^{-1}$, while outside this range only the specific bands of adsorbed water are remarked (Figure 1). The specific bands of the $SiO_2$ matrix appear at 1068–1098 cm$^{-1}$ with a shoulder around 954–960 cm$^{-1}$ attributed to the stretching and bending vibration of Si-O-Si chains, 793–807 cm$^{-1}$ attributed to the symmetric and asymmetric vibrations of $SiO_4$ tetrahedron, and 449–469 cm$^{-1}$ attributed to the vibration of the Si-O bond that overlaps the vibration band of the Fe-O bond [19,20,27]. The high intensity of these bands suggests a low polycondensation degree of the $SiO_2$ network. Additionally, to the specific bands of $SiO_2$, the vibration of tetrahedral Zn-O and Ni-O bonds (555–590 cm$^{-1}$) and the octahedral Fe-O bonds (449–469 cm$^{-1}$) are observed [2,19,27]. These bands confirm the formation of cubic spinel structure and are in good agreement with XRD analysis [12]. The vibration band at 555–590 cm$^{-1}$ was not observed at 400 °C, but appears at 800 and 1200 °C and increases with the increasing of annealing temperature, most probably due to the increase of the ferrite crystallization degree [19,27]. The shift of the vibration bands (555–590 cm$^{-1}$) towards lower wavenumbers observed in samples with high Mn content is a consequence of the displacement of Fe, Mn, and Ni ions in the octahedral (B) and tetrahedral (A) sites that further leads to changes of the $Fe^{3+}$–$O^{2-}$ ($M^{3+}$–$O^{2-}$) and $M^{2+}$–$O^{2-}$ distances, respectively. This shift indicates a lower degree of occupancy of tetrahedral sites with $Fe^{3+}$ ions [2].

*2.3. Atomic Force Microscopy*

As the powder samples are slightly agglomerated, the aqueous dispersion facilitates the release of free nanoparticles that are transferred onto a solid substrate as thin films prior to AFM scanning [26,28,29]. AFM images of tailored nanostructures obtained via liquid dispersion of ferrite nanoparticles were previously reported [30–32]. Liquid dispersion of ferrite nanoparticles allows the production of tailored nanostructures with possible application in magnetic resonance imaging [33] and 3D inkjet printing to produce ferrite nanomaterial thin films for magneto-optical devices [34].

AFM images reveal that the annealing process has a high influence on the particle size, smaller size particles being obtained at 400 °C. The particle size increases considerably with the annealing temperature, as follows: 18 nm at 400 °C, 24 nm at 800 °C, and 30 nm at 1200 °C. The smallest nanoparticles were obtained at 400 °C and the largest at 1200 °C. Mn ferrite nanoparticles are smaller than Ni ferrite nanoparticles, while the particle size of mixed Ni-Mn composition increases with the increasing of Ni content. The obtained particle sizes are slightly higher than those reported for Mn ferrite (10 nm) obtained by thermal decomposition [35] and lower than those reported for Mn ferrite annealed at high temperatures [34,36].

The obtained $NiFe_2O_4$ particle size (Figure 2m–o) of 22 nm at 400 °C, 30 nm at 800 °C, and 58 nm at 1200 °C are in good agreement with the data reported by Tong et al. [37] and Ashiq et al. [38] for nanoparticles obtained by reverse micelle technique. The progressive replacement of the $Mn^{2+}$ ions by $Ni^{2+}$ ions has direct consequences on the particle size (Figure 2d–l), which progressively increases with the increase of the Ni content and the annealing temperature. In this regard, the finest nanoparticles were obtained for $Mn_{0.75}Ni_{0.25}Fe_2O_4$ annealed at 400 °C, while the bigger particles for $Mn_{0.25}Ni_{0.75}Fe_2O_4$ were annealed at 1200 °C. The obtained results confirmed that the modified sol-gel method resulted in very fine, highly dense, homogenous, and single-phase ferrite nanoparticles.

The nanoparticles size is slightly higher than the ferrite crystallite size estimated by the Scherrer equation, most probably due to the presence of secondary phases at high annealing temperatures. In all cases, round shape particles with a marked tendency to be adsorbed in uniform layers onto a solid substrate are remarked (Figure 3). The short deposition time allows the optimal arrangement of the particles onto the substrate and prevents their overlapping and agglomeration. Thus, the deposed thin film roughness (Table 2), depends mainly on the nanoparticle's size.

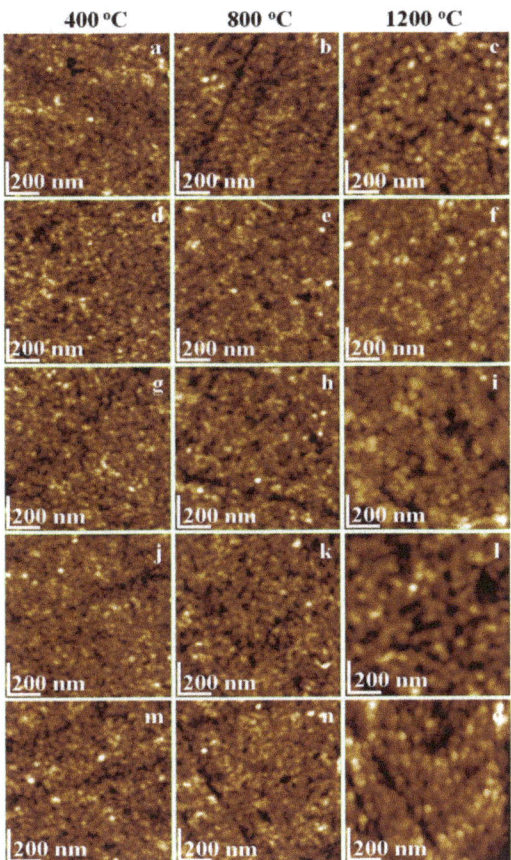

**Figure 2.** Topographical AFM images of $Mn_{1-x}Ni_xFe_2O_4@SiO_2$ NCs annealed at 400, 800, and 1200 °C (x = 0 (**a–c**), x = 0.25 (**d–f**), x = 0.50 (**g–i**), x = 0.75 (**j–l**) and x = 1.0 (**m–o**)).

**Table 2.** AFM parameters of $Mn_{1-x}Ni_xFe_2O_4@SiO_2$ NCs.

| NC's | Temperature, °C | Height, nm | Rq Roughness, nm | Average Particle Size, nm |
|---|---|---|---|---|
| x = 0.00 | 400 | 19 | 1.15 | 18 |
| | 800 | 16 | 1.44 | 24 |
| | 1200 | 9 | 0.87 | 30 |
| x = 0.25 | 400 | 16 | 1.16 | 16 |
| | 800 | 18 | 1.19 | 20 |
| | 1200 | 15 | 1.24 | 35 |
| x = 0.50 | 400 | 14 | 1.08 | 18 |
| | 800 | 12 | 0.93 | 25 |
| | 1200 | 39 | 4.17 | 40 |
| x = 0.75 | 400 | 12 | 1.05 | 20 |
| | 800 | 12 | 1.09 | 30 |
| | 1200 | 19 | 1.97 | 60 |
| x = 1.00 | 400 | 11 | 1.08 | 22 |
| | 800 | 16 | 1.07 | 30 |
| | 1200 | 9 | 0.92 | 58 |

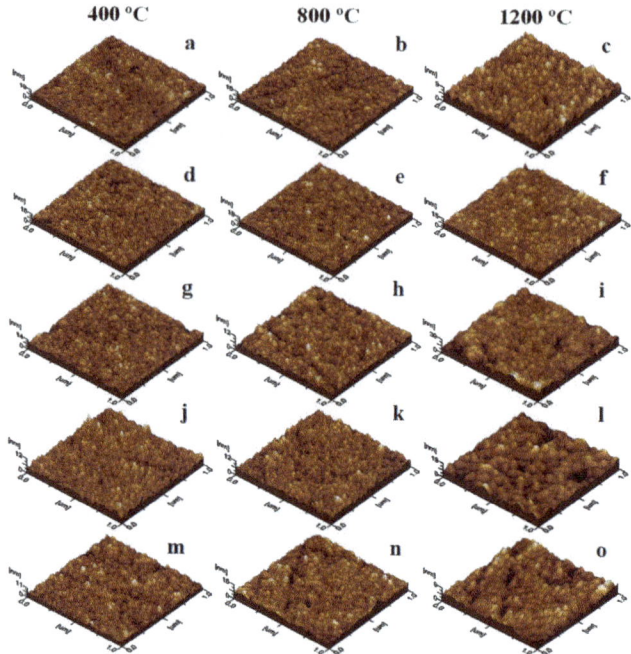

**Figure 3.** AFM 3D images of $Mn_{1-x}Ni_xFe_2O_4@SiO_2$ NCs annealed at 400, 800, and 1200 °C (x = 0 (**a–c**), x = 0.25 (**d–f**), x = 0.50 (**g–i**), x = 0.75 (**j–l**) and x = 1.0 (**m–o**)).

The smoothest thin film was obtained for the powders annealed at 400 °C, while the rougher films result from the powders annealed at 1200 °C. The tridimensional aspect is almost clogged for the ferrites with high Ni content annealed at 1200 °C due to their relatively higher size and agglomeration tendency.

*2.4. Magnetic Properties*

Generally, the magnetic properties of the ferrites are affected by the chemical formula, by the cation distribution between the tetrahedral (A) and octahedral (B) sites of the lattice, as well as by the particle sizes and their distribution [22]. The main magnetic properties of $Mn_{1-x}Ni_xFe_2O_4@SiO_2$ NCs annealed at 800 and 1200 °C are displayed in Figure 4. In all cases, the hysteresis loops have a typical shape for ferrimagnetic materials. The main magnetic parameters, namely saturation magnetization ($M_S$), remanent magnetization ($M_R$), squareness (S), coercivity ($H_C$), the magnetic moment per formula unit ($n_B$) expressed in numbers of Bohr magnetons and anisotropy constant (K) extracted from the hysteresis loops are presented in Table 3.

The magnetic parameters $M_S$, $M_R$, $n_B$, and K increase, while $H_C$ decreases with increasing Ni content. All the magnetic parameters are larger for the NCs with the same Ni content, annealed at higher temperatures. This behavior is different from that reported for bulk ferrites with the same chemical formula, for which an increase of the saturation magnetization was found with increasing Mn content [39]. The difference between the two systems consists in the presence of $SiO_2$ coating in our samples. The largest $M_S$ value was recorded for the samples with x = 1.00 ($NiFe_2O_4@SiO_2$) with the largest particle sizes, for both annealing temperatures. $M_S$ increases almost linearly with increasing Ni content for both annealing temperatures. The $M_S$ is strongly affected by the so-called "surface spin effect" which is a result of the defects and broken chemical bonds which disrupt the parallel alignment of the magnetic moments and give rise to spin canting and spin disorder in the layer from the surface of the particles. The smaller the size of the particle, the larger the surface-to-volume ratio. Increasing the fraction of this layer will make dominant the magnetic behavior of the shell over that from the interior, and the magnetization of the smaller size particles will be reduced. In addition to this size effects, the XRD analysis also showed an increase of the hematite (which is known to have low magnetic properties) and quartz content with increasing Mn content in the samples and this can contribute additionally to the decrease of the $M_S$ value The $M_S$ value is also affected by the cation's distribution between the tetrahedral (A) and octahedral (B) sites [22].

**Table 3.** Saturation magnetization ($M_S$), remanent magnetization ($M_R$), coercivity ($H_C$), squareness (S), magnetic moment per formula unit ($n_B$), and anisotropy constant (K) of $Mn_{1-x}Ni_xFe_2O_4@SiO_2$ NCs.

| NC | Temperature, °C | $M_S$, emu/g | $M_R$, emu/g | $H_C$, Oe | S | $n_B$ | K, erg/dm$^3$ |
|---|---|---|---|---|---|---|---|
| x = 0.00 | 800 | 4.7 | 0.3 | 200 | 0.064 | 0.194 | 0.590 |
|  | 1200 | 16.4 | 4.2 | 260 | 0.246 | 0.677 | 2.678 |
| x = 0.25 | 800 | 6.8 | 1.1 | 190 | 0.162 | 0.282 | 0.811 |
|  | 1200 | 22.4 | 5.8 | 250 | 0.259 | 0.929 | 3.517 |
| x = 0.50 | 800 | 7.8 | 1.7 | 183 | 0.218 | 0.325 | 0.896 |
|  | 1200 | 29.6 | 13.5 | 240 | 0.456 | 1.232 | 4.461 |
| x = 0.75 | 800 | 9.1 | 2.8 | 175 | 0.308 | 0.380 | 1.000 |
|  | 1200 | 37.5 | 14.6 | 220 | 0.389 | 1.567 | 5.181 |
| x = 1.00 | 800 | 10.2 | 3.4 | 166 | 0.333 | 0.428 | 1.063 |
|  | 1200 | 45.7 | 16.1 | 185 | 0.357 | 1.918 | 5.310 |

The increase of the $M_S$ with increasing Ni content can also suggest that in the octahedral (B) sites $Fe^{3+}$ ions (5 $\mu_B$) were replaced by $Ni^{2+}$ (2 $\mu_B$) ions with a smaller magnetic moment which force the $Fe^{3+}$ ions to migrate in the tetrahedral (A) sites. This results in an inverse spinel structure since the $Fe^{3+}$ ions are rearranged in both tetrahedral (A) and octahedral (B) sites and the antiferromagnetic interaction becomes weaker, while the ferromagnetic super-exchange interaction increases. Therefore, the normal spinel Mn ferrite is converted to a dominant inverse spinel ferrite as a result of the $Mn^{2+}$ ions substitution by $Ni^{2+}$ ions. This may be a consequence of the coating of $Mn_{1-x}Ni_xFe_2O_4$ nanoparticles by the $SiO_2$ matrix. Comparative $Ms$ values were reported by Airimioaei et al. [2] for Ni-Mn ferrites obtained by combustion reaction, while Jessudoss et al. [22] reported higher $M_S$ and $M_R$ values for $Ni_{1-x}Mn_xFe_2O_4$ obtained by a microwave combustion reaction route. Köseoğlu reported $M_S$ between 31 and 56 emu/g at room temperature and 41–70 emu/g at 10 K [8]. Opposite to our results, the higher $M_S$ value of undoped $MnFe_2O_4$ of 66.93 emu/g steadily decreased from 64.68 emu/g (x = 0.2) to 35.43 emu/g (x= 1.0) with increasing Ni content, with $NiFe_2O_4$ showing the lower value; the linear decrease of $Ms$ values of the samples could be mainly due to the difference in the magnetic moments of $Mn^{2+}$ and $Ni^{2+}$ ions [7].

The coercive field ($H_C$) slightly decreases with increasing Ni content for both the samples annealed at 800 and 1200 °C (Table 3). This behavior can be attributed to the well-known dependence of the $H_C$ on the sizes of the nanoparticles in the magnetic multidomain range. In this region, the size of the nanoparticles causes them to be composed of many magnetic domains which allow an easy domain wall motion and magnetization reversal, reducing the value of the coercive field by lowering the value of the domain wall energy [1,19,28]. The $H_C$ is a measure of the magneto-crystalline anisotropy of a sample. By increasing the Ni content, the nanoparticle and the crystallite sizes also increase, leading to a decrease of the magnetocrystalline anisotropy. The $H_C$ is strongly affected by the particle's sizes and their shape as well as by their distribution, crystallinity and magnetic domain sizes, and micro-strains induced by the $SiO_2$ matrix [1]. Similarly, Airimioaei et al. reported that the $H_C$ slightly increases with increasing the amount of Mn from 37.4 Oe to 53.7 Oe for x = 0–0.5, respectively [2]. In accordance with Mathubala [7], the $H_C$ and the $M_R$ values decrease with the increase of the Ni content in $MnFe_2O_4$ lattice.

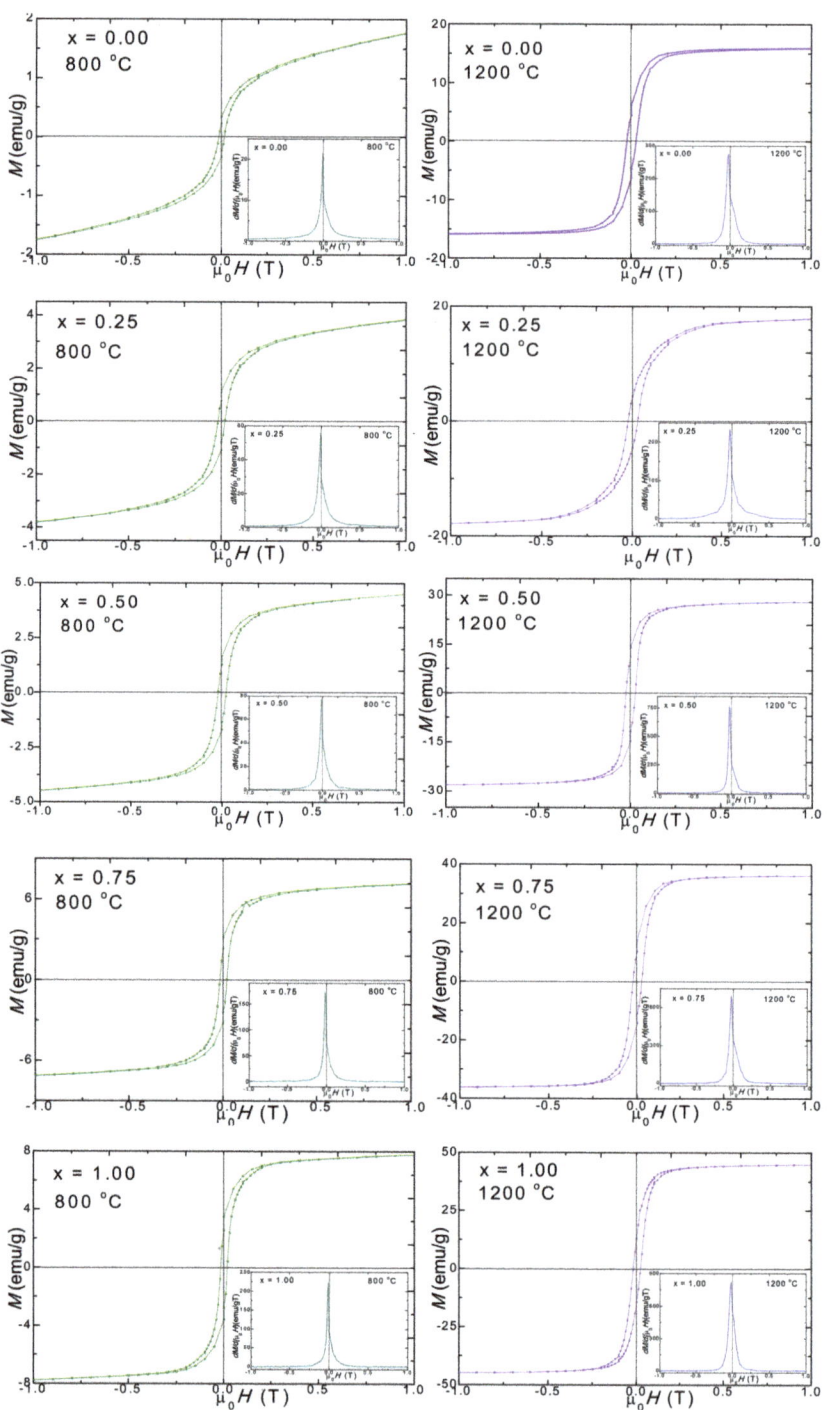

**Figure 4.** Magnetic hysteresis loops and magnetization derivative (in insets) for $Mn_{1-x}Ni_xFe_2O_4@SiO_2$ NCs heat-treated at 800 and 1200 °C.

The magnetic moment per formula unit increases with the increasing of Ni content for the same reasons used to explain the behavior of the $M_S$ since the ratio of $M_S/n_B$ is nearly constant. The increase of remanent magnetization ($M_R$) with the increase of Ni content also needs to be correlated with the variation of the particle's size and the related surface effects as the presence of defects and of secondary phases may act as a pinning center for the magnetic domain walls.

The anisotropy constant ($K$) reveals the energy required to rotate the magnetic moment inside the particle. The $K$ increases with increasing Ni content with a factor of 1.8 for the NCs annealed at 800 °C and with a factor of 1.98 for the samples annealed at 1200 °C. A possible explanation could be the increase of magneto-crystalline anisotropy which originates in spin-orbital contribution since for $MnFe_2O_3$ the orbital quantum number is $L = 0$. Another explanation could be the presence of the spin disorder in the surface layer of the nanoparticles which needs a higher magnetic field for $M_S$, a field that depends on the size of the particles and their distribution within the samples [8].

The squareness ratio ($S = M_R/M_S$) is a measure of how square the hysteresis loop is. A theoretical value of $M_R/M_S$ lower than 0.5 indicates the presence of non-interacting uniaxial single domain particles with the easy axis being randomly oriented [1]. The $S$ increases from 0.064 to 0.333 for the NCs annealed at 800 °C and from 0.246 to 0.357 for the NCs annealed at 1200 °C. Generally, the derivatives of the hysteresis loops exhibited small and broad single peaks indicating partially crystalline samples with a main magnetic phase in the presence of crystal defects. The presence of a high magnetic purity phase is indicated by the sharp peaks. The broad peaks correspond to large particle size distributions and wide coercive fields distributions. The horizontal shifts of peaks from the origin are rather small for all the samples, suggesting that the coercivities distributions are not large as a result of the magnetic interaction between the particles.

### 2.5. Potential Applications

Magnetic nanoparticles that are small enough to remain in circulation after injection and are able to pass through the capillary systems of various organs are non-toxic, well dispersed, and biocompatible and are potential candidates for biomedical applications such as cancer therapy, drug delivery, magnetic resonance imaging, or magnetic hyperthermia [19,40]. The magnetic parameters of nanosized $Mn_{1-x}Ni_xFe_2O_4$ ferrites are related to the synthesis route. The synthesized Mn-Ni ferrite nanoparticles are prospective candidates for biomedicine due to their easy synthesis process, controllable structure and size, stoichiometry control, high magnetization value, and superparamagnetic nature. Moreover, their embedding into mesoporous $SiO_2$ enhances their biocompatibility and reduces their agglomeration and degradation. $SiO_2$ matrix is an excellent non-toxic coating material that can create cross-linking, giving rise to an inert outer shield, avoiding the acute toxicity by ferrite inoculation [24]. However, their biocompatibility, cytotoxicity, pharmacokinetics, and other potential side effects are still underexplored. Although there are several in vitro studies that indicate the cytotoxicity through suppression of proliferation and apoptosis induction of different nanosized magnetic ferrites against human colon cancer (HT29), breast cancer (MCF7), and liver cancer (HepG2) cells, data on the Mn-Ni ferrite cytotoxicity is limited [41]. The dose-dependent cytotoxic effects of $Mn_{1-x}Ni_xFe_2O_4$ nanoparticles against J774 E murine macrophages and U2OS human osteosarcoma was reported [42]. The effective light-to-heat conversion upon exposure of Mn-Ni ferrites to near-infrared irradiation could also be attractive for different bio-applications [42]. In order to increase their chemical stability in biological systems and for enhancing their magnetic properties different doping elements (Ni, Co, Mn, Zn, Mg, etc.) might be added [43]. Intravenous inoculation of ferrite nanosized particles can be useful as contrast agents for magnetic resonance imaging (MRI) [44]. Mn ferrite was found as a very effective MRI contrast agent in comparison with magnetite since it has large $M_S$ and high crystalline anisotropy resulting in a slower magnetic moment of relaxation [45].

## 3. Materials and Methods

### 3.1. Synthesis of NCs

Ferric nitrate nonahydrate (Fe(NO$_3$)$_3$·9H$_2$O) of 99.6% purity, nickel nitrate hexahydrate (Ni(NO$_3$)$_2$·6H$_2$O) of 99.8% purity, manganese nitrate trihydrate (Mn(NO$_3$)$_2$·3H$_2$O) of 100.0% purity, 1,4-butanediol (1,4BD) of 99.9% purity, tetraethyl orthosilicate (TEOS) of 100.0% purity, and ethanol of 99.9% purity were purchased from Merck (Germany) and used for synthesis without additional purifications.

The Mn$_{1-x}$Ni$_x$Fe$_2$O$_4$@SiO$_2$ NCs (60% wt. ferrite, 40% wt. SiO$_2$) were prepared by sol-gel route using Mn:Ni:Fe molar ratios of 0:1:2 ($x = 0.00$), 0.25:0.75:2 ($x = 0.25$), 0.50:0.50:2 ($x = 0.50$), 0.75:0.25:2 ($x = 0.75$) and 1:0:2 ($x = 1.00$). The sols were obtained by mixing the nitrate mixture with 1,4BD, TEOS and ethanol. After 4 weeks at room temperature, the gelation takes place by the formation of an SiO$_2$ matrix that contains the nitrates and 1,4BD. The gels were annealed at 400, 800, and 1200 °C for 4h in air using a LT9 muffle furnace (Nabertherm, Lilienthal, Germany).

### 3.2. Characterization of NCs

The crystallinity and structure of the ferrite were investigated by X-ray diffraction recorded at room temperature, using a D8 Advance (Bruker, Germany) diffractometer, operating at 40 kV and 40 mA with CuKα radiation ($\lambda = 1.54060$ Å).

The formation of ferrite and SiO$_2$ matrix was monitored using a Spectrum BX II (Perkin Elmer, Waltham, MA, USA) Fourier transform infrared spectrometer on pellets containing 1% sample in KBr.

AFM was performed by a JSPM 4210 (JEOL, Tokio, Japan) scanning probe microscope using NSC15 cantilevers (diamond-coated silicon nitride tips) with a resonant frequency of 325 kHz and a force constant of 40 N/m in tapping mode. The samples were dispersed into ultrapure water and transferred on glass slides by vertical adsorption for 30 s, followed by natural drying. Areas of 2.5 µm × 2.5 µm to 1 µm × 1 µm of dried glass slides were scanned for three different macroscopic sites.

A cryogen-free VSM magnetometer (Cryogenic Ltd., London, UK) was used for the magnetic measurements. The $M_S$ was measured in a high magnetic field up to 10 T, while the magnetic hysteresis loops were performed between −2 and 2 T, at 300 K on samples incorporated in epoxy resin.

## 4. Conclusions

The microstructure, morphology, particle size, phase composition, and magnetic properties of Mn$_{1-x}$Ni$_x$Fe$_2$O$_4$ ($x = 0, 0.25, 0.50, 0.75, 1.00$) NCs were investigated. The crystallite size (6–46 nm), X-ray density (5.050–5.249 g/cm$^3$), lattice parameter (8.402–8.485 Å), and volume (593.1–606.6 Å$^3$) of Mn$_{1-x}$Ni$_x$Fe$_2$O$_4$ increase with the increase of Ni content. The XRD patterns showed poor crystalized Mn$_{1-x}$Ni$_x$Fe$_2$O$_4$ at 400 °C and highly crystalline Mn$_{1-x}$Ni$_x$Fe$_2$O$_4$, accompanied by secondary phases of Fe$_2$SiO$_4$, α-Fe$_2$O$_3$, cristobalite, and quartz at 800 and 1200 °C. FT-IR spectroscopy confirmed the formation of the oxidic phases and SiO$_2$ matrix. AFM investigations revealed round-shaped nanoparticles with sizes depending on the annealing temperature and Ni content. The magnetic properties of the NCs were strongly dependent on the chemical composition, cation distribution between tetrahedral (A) and octahedral (B) sites, as well as on the surface effects derived from the synthesis methods. For the NCs annealed at 1200 °C, $M_S$ (16.4–45.7 emu/g), $M_R$, magnetic moment per formula unit and K (2.678–5.310 erg/dm$^3$) increased, while $H_C$ (260–185 Oe) slightly decreased with the increase of Ni content. The small S values indicated the presence of non-interacting single domain uniaxial particles. The magnetic parameters displayed similar behavior for the NCs annealed at 800 °C, but their variations are smaller. All the magnetic parameters increased with the annealing temperature. The obtained magnetic Mn-Ni ferrite nanoparticles are potential candidates for biomedical applications such are cancer therapy, drug delivery, magnetic resonance imaging, or magnetic hydrothermia.

Despite the promising results, further studies of the biocompatibility, differential toxicity, and pharmacokinetics of nanosized Mn-Ni ferrites are required.

**Author Contributions:** T.D., conceptualization, methodology, writing—original draft, writing—review and editing, visualization, supervision; E.A.L., I.G.D., I.P. and O.C., methodology, investigation, writing—original draft, writing—review and editing. All authors have read and agreed to the published version of the manuscript.

**Funding:** This research received no external funding.

**Acknowledgments:** The authors acknowledge the Research Centre in Physical Chemistry "CECHIF" of Babes Bolyai University for AFM assistance.

**Conflicts of Interest:** The authors declare no conflict of interest.

# References

1. Suresh, J.; Trinadh, B.; Babu, B.V.; Reddy, P.V.S.S.S.N.; Mohan, B.S.; Krishna, A.R.; Samatha, K. Evaluation of micro-structural and magnetic properties of nickel nano-ferrite and $Mn^{2+}$ substituted nickel nano-ferrite. *Phys. B Condens. Matter* **2021**, *620*, 413264. [CrossRef]
2. Airimioaei, M.; Ciomaga, C.E.; Apostolescu, A.; Leonite, L.; Iordan, A.R.; Mitoseriu, L.; Palamaru, M.N. Synthesis and functional properties of the $Ni_{1-x}Mn_xFe_2O_4$ ferrites. *J. Alloys Compd.* **2011**, *509*, 8065–8072. [CrossRef]
3. Bonanni, B.; Cannistraro, S. Gold nanoparticles on modified glass surface as height calibration standard for atomic force microscopy operating in contact and tapping mode. *J. Nanotechnol.* **2005**, *1*, 1–14.
4. Mohamed, W.S.; Hadia, N.M.A.; Al Bakheet, B.; Alzaid, M.; Abu-Dief, A.M. Impact of $Cu^{2+}$ cations substitution on structural, morphological, optical and magnetic properties of $Co_{1-x}Cu_xFe_2O_4$ nanoparticles synthesized by a facile hydrothermal approach. *Solid State Sci.* **2022**, *125*, 106841. [CrossRef]
5. Abu-Dief, A.M.; Essawy, A.A.; Diab, A.K.; Mohamed, W.S. Facile synthesis and characterization of novel $Gd_2O_3$–CdO binary mixed oxide nanocomposites of highly photocatalytic activity for wastewater remediation under solar illumination. *J. Phys. Chem. Solids* **2021**, *148*, 109666. [CrossRef]
6. Salaheldeen, M.; Abu-Dief, A.M.; Martinez-Goyeneche, L.; Alzahrani, S.O.; Alkhatib, F.; Alvarez-Alonso, P.; Blanco, J.A. Dependence of the magnetization process on the thickness of $Fe_{70}Pd_{30}$ nanostructured thin film. *Materials* **2020**, *13*, 5788. [CrossRef] [PubMed]
7. Mathubala, G.; Manikandan, A.; Arul Antony, S.; Ramar, P. Photocatalytic degradation of methylene blue dye and magnetooptical studies of magnetically recyclable spinel $Ni_xMn_{1-x}Fe_2O_4$ (x = 0.0-1.0) nanoparticles. *J. Mol. Struct.* **2016**, *113*, 79–87. [CrossRef]
8. Köseoğlu, Y. Structural, magnetic, electrical and dielectric properties of $Mn_xNi_{1-x}Fe_2O_4$ spinel nanoferrites prepared by PEG assisted hydrothermal method. *Ceram. Int.* **2013**, *39*, 4221–4230. [CrossRef]
9. Marinca, T.F.; Chicinaș, I.; Isnard, O.; Neamțu, B.V. Nanocrystalline/nanosized manganese substituted nickel ferrites—$Ni_{1-x}Mn_xFe_2O_4$ obtained by ceramic-mechanical milling route. *Ceram. Int.* **2016**, *42*, 4754–4763. [CrossRef]
10. Shobana, M.K.; Sankar, S. Structural, thermal and magnetic properties of $Ni_{1-x}Mn_xFe_2O_4$ nanoferrites. *J. Magn. Magn. Mater.* **2009**, *321*, 2125–2128. [CrossRef]
11. Abdallah, H.M.I.; Moyo, T. Superparamagnetic behavior of $Mn_xNi_{1-x}Fe_2O_4$ spinel nanoferrites. *J. Magn. Magn. Mater.* **2014**, *361*, 170–174. [CrossRef]
12. Maaz, K.; Duan, J.L.; Karim, S.; Chen, Y.H.; Zhai, P.F.; Xu, L.J.; Yao, H.J.; Liu, J. Fabrication and size dependent magnetic studies of $Ni_xMn_{1-x}Fe_2O_4$ (x = 02) cubic nanoplates. *J. Alloys Compd.* **2016**, *684*, 656–662. [CrossRef]
13. Mohamed, W.S.; Abu-Dief, A.M. Impact of rare earth europium (RE-$Eu^{3+}$) ions substitution on microstructural, optical and magnetic properties of $CoFe_{2-x}Eu_xO_4$ nanosystems. *Ceram. Int.* **2020**, *46*, 16196–16209. [CrossRef]
14. Mohamed, W.S.; Alzaid, M.; Abdelbaky, S.M.; Amghouz, Z.; Garcia-Granda, S.; Abu-Dief, A.M. Impact of $Co^{2+}$ substitution on microstructure and magnetic properties of $Co_xZn_{1-x}Fe_2O_4$ nanoparticles. *Nanomaterials* **2019**, *9*, 1602. [CrossRef]
15. Sudakshina, B.; Suneesh, M.V.; Arun, B.; Chandrasekhar, K.; Vasundhara, M. Effects of Cr, Co, Ni substitution at Mn-site on structural, magnetic properties and critical behaviour in $Nd_{0.67}Ba_{0.33}MnO_3$ mixed-valent manganite. *J. Magn. Magn. Mater.* **2022**, *548*, 168980. [CrossRef]
16. Siakavelas, G.I.; Charisiou, N.D.; AlKhoori, A.; Sebastian, V.; Hinder, S.J.; Baker, M.A.; Yentekakis, I.V.; Polychronopoulou, K.; Goula, M.A. Cerium oxide catalysts for oxidative coupling of methane reaction: Effect of lithium, samarium and lanthanum dopants. *J. Environ. Chem. Eng.* **2022**, *10*, 107259. [CrossRef]
17. Yang, Y.; Li, J.; Zhang, H.; Li, J.; Xu, F.; Wang, G.; Gao, F.; Su, H. $Nb^{5+}$ ion substitution assisted the magnetic and gyromagnetic properties of NiCuZn ferrite for high frequency LTCC devices. *Ceram. Int.* **2022**, in press. [CrossRef]
18. Junaid, M.; Oazafi, I.A.; Khan, M.A.; Gulbadan, S.; Ilyas, S.Z.; Somaily, H.H.; Attia, M.S.; Amin, M.A.; Noor, H.M.; Asghar, H.M.N.H.K. The influence of Zr and Ni co-substitution on structural, dielectric and magnetic traits of lithium spinel ferrites. *Ceram. Int.* **2022**, in press. [CrossRef]

19. Dippong, T.; Levei, E.A.; Cadar, O. Recent advances in synthesis and applications of $MFe_2O_4$ (M= Co, Cu, Mn, Ni, Zn) nanoparticles. *Nanomaterials* **2021**, *11*, 1560. [CrossRef] [PubMed]
20. Barvinschi, P.; Stefanescu, O.; Dippong, T.; Sorescu, S.; Stefanescu, M. $CoFe_2O_4/SiO_2$ nanocomposites by thermal decomposition of some complex combinations embedded in hybrid silica gels. *J. Therm. Anal. Calorim.* **2013**, *112*, 447–453. [CrossRef]
21. Joint Committee on Powder Diffraction Standards. *Powder Diffraction File*; International Center for Diffraction Data: Swarthmore, PA, USA, 1999.
22. Jesudoss, S.K.; Judith Vijaya, J.; John Kennedy, L.; Iyyappa Rajana, P.; Al-Lohedan, A.H.; Jothi Ramalingam, R.; Kaviyarasu, K.; Bououdina, M. Studies on the efficient dual performance of $Mn_{1-x}Ni_xFe_2O_4$ spinel nanoparticles in photodegradation and antibacterial activity. *J. Photochem. Photobiol. B Biol.* **2016**, *165*, 121–132. [CrossRef]
23. Nizam, M.M.N.; Khan, S. Structural, electrical and optical properties of sol-gel synthesized cobalt substituted $MnFe_2O_4$ nanoparticles. *Phys. B* **2017**, *520*, 21–27.
24. Aparna, M.L.; Nirmala Grace, A.; Sathyanarayanan, P.; Sahu, N.K. A comparative study on the supercapacitive behaviour of solvothermally prepared metal ferrite ($MFe_2O_4$, M=Fe, Co, Ni, Mn, Cu, Zn) nanoassemblies. *J. Alloys Compd.* **2018**, *745*, 385–395. [CrossRef]
25. El Mendili, Y.; Bardeau, J.F.; Randrianantoandro, N.; Greneche, J.M.; Grasset, F. Structural behavior of laser-irradiated $\gamma$-$Fe_2O_3$ nanocrystals dispersed in porous silica matrix: $\gamma$-$Fe_2O_3$ to $\alpha$-$Fe_2O_3$ phase transition and formation of $\varepsilon$-$Fe_2O_3$. *Sci. Technol. Adv. Mater.* **2016**, *17*, 597–609. [CrossRef] [PubMed]
26. Hussain, A.; Abbas, T.; Niazi, S.B. Preparation of $Ni_{1-x}Mn_xFe_2O_4$ ferrites by sol–gel method and study of their cation distribution. *Ceram. Int.* **2013**, *39*, 1221–1225. [CrossRef]
27. Dippong, T.; Deac, I.G.; Cadar, O.; Levei, E.A.; Petean, I. Impact of $Cu^{2+}$ substitution by $Co^{2+}$ on the structural and magnetic properties of $CuFe_2O_4$ synthesized by sol-gel route. *Mater. Charact.* **2020**, *163*, 110248. [CrossRef]
28. Fanga, P.P.; Buriez, O.; Labbé, E.; Tian, Z.Q.; Amatore, C. Electrochemistry at gold nanoparticles deposited on dendrimers assemblies adsorbed onto gold and platinum surfaces. *J. Electroanal. Chem.* **2011**, *659*, 76–82. [CrossRef]
29. Ur. Rahman Awana, F.; Keshavarz, A.; Azhar, M.R.; Akhondzadeh, H.; Ali, M.; Al-Yaseri, A.; Abid, H.R.; Iglauer, S. Adsorption of nanoparticles on glass bead surface for enhancing proppant performance: A systematic experimental study. *J. Mol. Liq.* **2021**, *328*, 115398. [CrossRef]
30. Fokin, N.; Grothe, T.; Mamun, A.; Trabelsi, M.; Klöcker, M.; Sabantina, L.; Döpke, C.; Blachowicz, T.; Hütten, A.; Ehrmann, A. Magnetic properties of electrospun magnetic nanofiber mats after stabilization and carbonization. *Materials* **2020**, *13*, 1552. [CrossRef]
31. Anchieta, C.G.; Cancelier, A.; Mazutti, M.A.; Jahn, S.L.; Kuhn, R.C.; Gündel, A.; Chiavone-Filho, O.; Foletto, E.L. Effects of solvent diols on the synthesis of $ZnFe_2O_4$ particles and their use as heterogeneous photo-fenton catalysts. *Materials* **2014**, *7*, 6281–6290. [CrossRef] [PubMed]
32. Januskevicius, J.; Stankeviciute, Z.; Baltrunas, D.; Mažeika, K.; Beganskiene, A.; Kareiva, A. Aqueous sol-gel synthesis of different iron ferrites: From 3D to 2D. *Materials* **2021**, *14*, 1554. [CrossRef]
33. Van Cutsem, E.; Verheul, H.M.W.; Flamen, P.; Rougier, P.; Beets-Tan, R.; Glynne-Jones, R.; Seufferlein, T. Imaging in colorectal cancer: Progress and challenges for the clinicians. *Cancers* **2016**, *31*, 81. [CrossRef]
34. Enuka, E.; Monne, M.A.; Lan, X.; Gambin, V.; Koltun, R.; Chen, M.Y. 3D inkjet printing of ferrite nanomaterial thin films for magneto-optical devices. In *Quantum Sensing and Nano Electronics and Photonics XVII*; International Society for Optics and Photonics: Bellingham, WA, USA, 2020. [CrossRef]
35. Díez-Villares, S.; Ramos-Docampo, M.A.; da Silva-Candal, A.; Hervella, P.; Vázquez-Ríos, A.J.; Dávila-Ibáñez, A.B.; López-López, R.; Iglesias-Rey, V.; Salgueiriño, M.; Manganese, M. Ferrite nanoparticles encapsulated into vitamin e/sphingomyelin nanoemulsions as contrast agents for high-sensitive magnetic resonance imaging. *Adv. Healthc. Mater.* **2021**, *10*, 2101019. [CrossRef] [PubMed]
36. Yang, L.; Ma, L.; Xin, J.; Li, A.; Sun, C.; Wei, R.; Ren, B.W.; Chen, Z.; Lin, H.; Gao, J. Composition tunable manganese ferrite nanoparticles for optimized T2 contrast ability. *Chem. Mater.* **2017**, *29*, 3038–3047. [CrossRef]
37. Tong, S.-K.; Chi, P.-W.; Kung, S.H.; Wei, D.H. Tuning bandgap and surface wettability of $NiFe_2O_4$ driven by phase transition. *Sci. Rep.* **2018**, *8*, 1338. [CrossRef]
38. Ashiq, N.M.; Ehsan, M.F.; Iqbal, M.J.; Gul, I.H. Synthesis, structural and electrical characterization of $Sb^{3+}$ substituted spinel nickel ferrite ($NiSb_xFe_{2-x}O_4$) nanoparticles by reverse micelle technique. *J. Alloys Compd.* **2011**, *509*, 5119–5126. [CrossRef]
39. Hu, J.; Qin, H.; Wang, Y.; Wang, Z.; Zhang, S. Magnetic properties and magnetoresistance effect of $Ni_{1-x}Mn_xFe_2O_4$ sintered ferrites. *Solid State Commun.* **2000**, *115*, 233–235. [CrossRef]
40. Sánchez, J.; Cortés-Hernández, D.A.; Escobedo-Bocardo, J.C.; Jasso-Terán, R.A.; Zugasti-Cruz, A. Bioactive magnetic nanoparticles of Fe–Ga synthesized by sol–gel for their potential use in hyperthermia treatment. *J. Mater. Sci. Mater. Med.* **2014**, *25*, 2237–2242. [CrossRef] [PubMed]
41. Al-Qubaisi, M.S.; Rasedee, A.; Flaifel, M.H.; Ahmad, S.H.; Hussein-Al-Ali, S.; Hussein, M.Z.; Eid, E.E.; Zainal, Z.; Saeed, M.; Ilowefah, M.; et al. Cytotoxicity of nickel zinc ferrite nanoparticles on cancer cells of epithelial origin. *Int. J. Nanomed.* **2013**, *8*, 2497–2508. [CrossRef] [PubMed]
42. Pazik, R.; Zachanowicz, E.; Pozniak, B.; Małecka, M.; Ziecina, A.; Marciniak, L. Non-contact $Mn_{1-x}Ni_xFe_2O_4$ ferrite nano-heaters for biological applications–heat energy generated by NIR irradiation. *RSC Adv.* **2017**, *7*, 18162. [CrossRef]

43. Pu, Y.; Tao, X.; Zeng, X.; Le, Y.; Chen, J.F. Synthesis of Co–Cu–Zn doped $Fe_3O_4$ nanoparticles with tunable morphology and magnetic properties. *J. Magn. Magn. Mater.* **2020**, *322*, 1985–1990. [CrossRef]
44. Bacon, B.R.; Stark, D.D.; Park, C.H.; Saini, S.; Groman, E.V.; Hahn, P.F.; Compton, C.C.; Ferrucci, J.T., Jr. Ferrite particles: A new magnetic resonance imaging contrast agent. Lack of acute or chronic hepatotoxicity after intravenous administration. *J. Lab. Clin. Med.* **1987**, *110*, 164–171.
45. Ravichandran, M.; Velumani, S. Manganese ferrite nanocubes as an MRI contrast agent. *Mater. Res. Express* **2020**, *7*, 016107. [CrossRef]

Article

# Investigation of Structural, Morphological and Magnetic Properties of MFe$_2$O$_4$ (M = Co, Ni, Zn, Cu, Mn) Obtained by Thermal Decomposition

Thomas Dippong [1], Erika Andrea Levei [2] and Oana Cadar [2,*]

1. Faculty of Science, Technical University of Cluj-Napoca, 76 Victoriei Street, 430122 Baia Mare, Romania; dippong.thomas@yahoo.ro
2. INCDO-INOE 2000, Research Institute for Analytical Instrumentation, 67 Donath Street, 400293 Cluj-Napoca, Romania; erika.levei@icia.ro
* Correspondence: oana.cadar@icia.ro

**Abstract:** The structural, morphological and magnetic properties of MFe$_2$O$_4$ (M = Co, Ni, Zn, Cu, Mn) type ferrites produced by thermal decomposition at 700 and 1000 °C were studied. The thermal analysis revealed that the ferrites are formed at up to 350 °C. After heat treatment at 1000 °C, single-phase ferrite nanoparticles were attained, while after heat treatment at 700 °C, the CoFe$_2$O$_4$ was accompanied by Co$_3$O$_4$ and the MnFe$_2$O$_4$ by α-Fe$_2$O$_3$. The particle size of the spherical shape in the nanoscale region was confirmed by transmission electron microscopy. The specific surface area below 0.5 m$^2$/g suggested a non-porous structure with particle agglomeration that limits nitrogen absorption. By heat treatment at 1000 °C, superparamagnetic CoFe$_2$O$_4$ nanoparticles and paramagnetic NiFe$_2$O$_4$, MnFe$_2$O$_4$, CuFe$_2$O$_4$ and ZnFe$_2$O$_4$ nanoparticles were obtained.

**Keywords:** ferrite; thermal decomposition; heating temperature; crystalline phase; specific surface; magnetic behavior

Citation: Dippong, T.; Levei, E.A.; Cadar, O. Investigation of Structural, Morphological and Magnetic Properties of MFe$_2$O$_4$ (M = Co, Ni, Zn, Cu, Mn) Obtained by Thermal Decomposition. *Int. J. Mol. Sci.* **2022**, 23, 8483. https://doi.org/10.3390/ijms23158483

Academic Editor: Daniel Arcos

Received: 11 July 2022
Accepted: 28 July 2022
Published: 30 July 2022

**Publisher's Note:** MDPI stays neutral with regard to jurisdictional claims in published maps and institutional affiliations.

**Copyright:** © 2022 by the authors. Licensee MDPI, Basel, Switzerland. This article is an open access article distributed under the terms and conditions of the Creative Commons Attribution (CC BY) license (https://creativecommons.org/licenses/by/4.0/).

## 1. Introduction

Spinel ferrites of MFe$_2$O$_4$ (M = Co, Ni, Zn, Cu, Mn) type have a cubic, closely packed arrangement of oxygen atoms with M$^{2+}$ and Fe$^{3+}$ ions occupying the tetrahedral (A) and octahedral (B) sites [1]. The spinel structure determines excellent magnetic and electrical properties, high chemical stability and low production costs [1–10]. These interesting properties enable the use of nanostructured materials in a wide range of novel applications in the field of science and technology [9–13].

Cobalt ferrite (CoFe$_2$O$_4$), nickel ferrite (NiFe$_2$O$_4$) and copper ferrite (CuFe$_2$O$_4$) have inverse spinel structures with M$^{2+}$ (M$^{2+}$ = Co$^{2+}$, Ni$^{2+}$ or Cu$^{2+}$) ions occupying the octahedral (B) sites and Fe$^{3+}$ ions equally distributed between the tetrahedral (A) and octahedral (B) sites [5,14]. Zinc ferrite (ZnFe$_2$O$_4$) has a normal spinel ferrite with Zn$^{2+}$ ions in tetrahedral (A) and Fe$^{3+}$ ions in octahedral (B) sites, while manganese ferrite (MnFe$_2$O$_4$) has a partially inverse spinel structure, in which only 20% of divalent Mn$^{2+}$ ions are located at octahedral (B) sites and the remainder of them are positioned at tetrahedral (A) sites [6].

CoFe$_2$O$_4$ is a ferromagnetic material with unique characteristics, such as large coercivity, magnetocrystalline anisotropy, Curie temperature and electrical resistance, remarkable thermal stability, moderate saturation magnetization, good chemical and mechanical stability, low eddy current loss and production cost [2,3,11,15]. These properties make it a promising candidate for various kind of applications such as drug delivery, magnetic resonance imaging, magnetic storage devices, catalysts and adsorption of toxic metals [15–19]. CoFe$_2$O$_4$ with unique architectures, including nanoparticles, hollow nanospheres, mesoporous nanospheres, nanorods and three-dimensional ordered macroporous structures, have been produced in the last years [3].

$NiFe_2O_4$ may display paramagnetic, superparamagnetic or ferrimagnetic behavior depending on the particle size and shape [20]. Due to its high magnetocrystalline anisotropy, high-saturation magnetization and unique magnetic structure combined with high Curie temperature, low coercivity, low eddy current loss, low price and high electrochemical stability [12,20–22] it is one of the most suitable candidates for applications in biosensors, corrosion protection, drug delivery, ceramics, medical diagnostics, microwave absorbers, transformer cores, magnetic liquids, magnetic refrigeration and high-density magnetic recording media, water-oxidation processes, dye removal by magnetic separation, etc. [5,21–23].

$CuFe_2O_4$ is a soft material with low coercivity, low-saturation magnetization, high electrical resistance, low eddy current losses, great resistance to corrosion, thermal stability, excellent catalytic properties and environmental benignity, and it is not readily demagnetized [12,14,24,25]. $CuFe_2O_4$ shows ferromagnetic behavior with a single-domain state and is widely used in magnetic storage, catalysis, photocatalysis, pollutant removal from wastewater, color imaging, magnetic refrigeration, magnetic drug delivery and high-density information storage [8,12,24–26].

$ZnFe_2O_4$ possess exceptional structural, optical, magnetic, electrical and dielectric properties at nanoscale, besides low toxicity, chemical and thermal stability [5,7,9,11,15]. $ZnFe_2O_4$ is antiferromagnetic at temperatures below the Neel temperature, but when the size of $ZnFe_2O_4$ approaches the nanometer range, it transforms into a diamagnetic, superparamagnetic or ferromagnetic substance [12,18]. Consequently, it has a wide potential to be used in microwave absorption, energy storage, drug delivery, magnetic resonance imaging, gas sensors, absorbent material for hot-gas desulphurization, high-performance electrode materials, photocatalysts and pigments [5,9,11,12,27]. Additionally, $ZnFe_2O_4$ is a promising semiconductor that can sensitize and activate under visible light other photocatalysts due to its small band gap [28].

$MnFe_2O_4$ have controllable grain size, high magnetization value, superparamagnetic nature, ability to be monitored by an external magnetic field, an easy synthesis process, surface manipulation ability, greater biocompatibility, thermal stability, non–toxicity, non-corrosion and environmentally friendly ability. Its properties have attracted potential consideration in biomedicine, in ceramic and paint industry as black pigment, and in high-frequency magnetostrictive and electromagnetic applications [6,29,30].

Synthesis methods with low toxicity that are also economical in terms of energy consumption, allowing for the production of fine, nanosized, highly pure, single-phase nanocrystalline ferrites have received considerable interest [31]. Spinel ferrites are commonly synthesized using the ceramic technique, which infers high temperatures and produces particles with small specific surface. In order to achieve ferrites with large specific surface and high degree of homogeneity, different synthesis methods, namely coprecipitation, polymeric gel, hydrothermal, microemulsion, heterogeneous precipitation, sonochemistry, combustion, sol–gel methods, etc. were used [31]. Despite the resulting fine-grained microstructure, the chemical methods have some disadvantages such as necessity of complex apparatus, long reaction time and post–synthesis thermal treatment to complete the formation and crystallization of final products, poor crystallinity and broad particle size distribution, which may negatively influence the related properties (shape, surface area and porosity) [32]. The sol–gel method has been used to prepare fine, homogenous, highly dense and single-phase ferrite nanoparticles. Compared to other conventional methods, the sol–gel method provides a good stoichiometric control and produces ferrites at relatively low temperatures. Furthermore, it allows for the embedding of ferrites into silica ($SiO_2$) matrix to prevent particle growth and particle agglomeration and to improve the magnetic properties [16,33]. However, despite its noticeable advantages, its main disadvantage consists of limited efficiency and long processing time [34]. The thermal decomposition method is a very efficient synthesis strategy, based on the heating of metallic precursors at different temperatures. Additionally, this method is simple and environmentally friendly, has a relatively low cost, requires a low reaction temperature and provides small particle size, narrow size distribution and no toxic by-products [35].

This paper focuses on the structural and morphological characteristics as well as the magnetic properties of nanosized $CoFe_2O_4$, $NiFe_2O_4$, $ZnFe_2O_4$, $CuFe_2O_4$ and $MnFe_2O_4$, obtained by thermal decomposition at 700 and 1000 °C. To the best of our knowledge, this is the first work that investigates the structural, morphological and magnetic properties of nanoferrites obtained by thermal decomposition of nitrates and compares them with those of correspondent nanoferrites embedded in $SiO_2$ matrix obtained by sol–gel method. The reaction progress was monitored by thermal (TG/DTA) analysis, while the nanoferrite composition was investigated by inductively coupled plasma optical emission spectrometry (ICP-OES). The crystalline phases and crystallite size were investigated by X-ray diffraction (XRD), while the particle properties such as shape, size and agglomeration were studied by transmission electron microscopy (TEM). The influence of crystallite size and divalent ions on the magnetic properties and the variation of saturation magnetization, remanent magnetization, coercivity and anisotropy of nanoferrites were also studied.

## 2. Results

Figure 1 presents the thermal decomposition diagrams (thermogravimetric—TG and differential thermal—DTA) of $MFe_2O_4$ systems. On the DTA diagram, the formation of $CoFe_2O_4$ is indicated by three endothermic effects at 63, 143 and 195 °C and two exothermic effects at 253 and 303 °C, respectively. The total mass loss is 65%. $NiFe_2O_4$, $ZnFe_2O_4$ and $CuFe_2O_4$ show two endothermic effects at 96 and 210 °C, 69 and 213 °C, 83 and 204 °C and an exothermic effect at 279, 267 and 289 °C, respectively. The total mass loss is 63% for $NiFe_2O_4$, 57% for $ZnFe_2O_4$ and 62% for $CuFe_2O_4$, respectively. The formation of $MnFe_2O_4$ is showed by three endothermic effects at 69, 132 and 201 °C, and an exothermic effect at 270 °C. The total mass loss shown on the TG diagram is 60%.

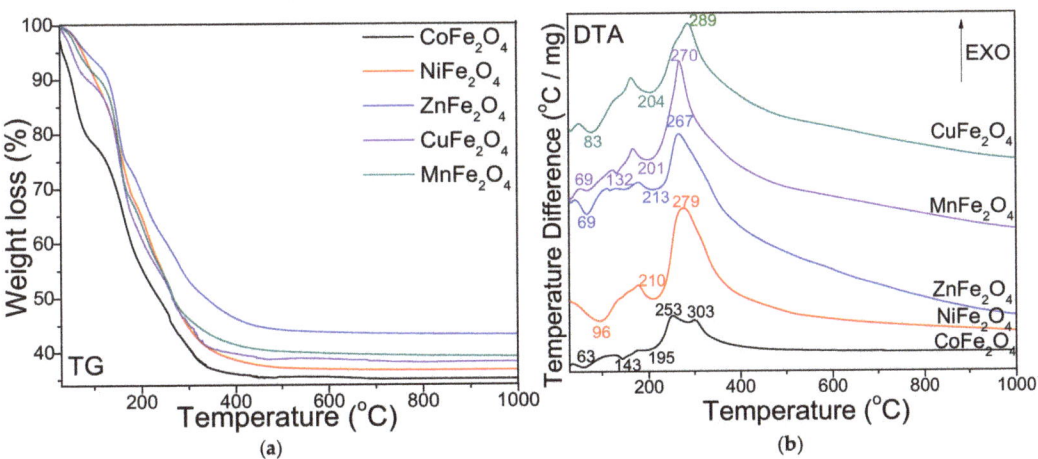

**Figure 1.** Thermogravimetric (TG) (**a**) and differential thermal analysis (DTA) (**b**) diagrams for $MFe_2O_4$ (M = Co, Ni, Zn, Cu, Mn).

The crystalline phases after heat treatment at 700 and 1000 °C are presented in Figure 2. The XRD pattern of $CoFe_2O_4$ exhibits a single-phase cubic spinel $CoFe_2O_4$ (JCPDS card no. 22-1086, [36]), belonging to *Fd3m* space group at 1000 °C, while at 700 °C, the presence of $Co_3O_4$ (JCPDS card no. 80-1451 [36]) is also remarked. In case of $NiFe_2O_4$, $ZnFe_2O_4$ and $CuFe_2O_4$ at both temperatures, single phase crystalline $NiFe_2O_4$ (JCPDS card no. 89-4927 [36]), $ZnFe_2O_4$ (JCPDS card no. 16-6205 [36]) and $CuFe_2O_4$ (JCPDS card no. 25-0283 [36]) are remarked. The presence of ZnO or CuO identified in case of ferrites embedded in $SiO_2$ matrix was not observed [33]. Single-phase crystalline $MnFe_2O_4$ (JCPDS card no.

74-2403 [36]) is obtained at 1000 °C, while at 700 °C, the MnFe$_2$O$_4$ is accompanied by α-Fe$_2$O$_3$ (JCPDS card no. 87-1164 [36]).

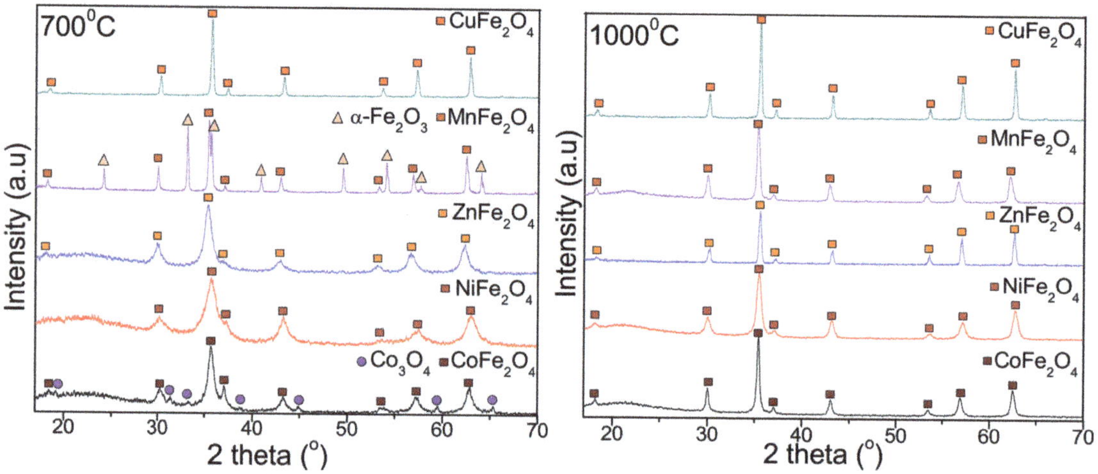

**Figure 2.** X-ray diffraction pattern of MFe$_2$O$_4$ (M = Co, Ni, Zn, Mn, Cu) heat treated at 700 °C (**a**) and 1000 °C (**b**).

The average crystallite size was estimated using the most intense diffraction (311) peak from the Debye–Scherrer formula [7,23,25] (Table 1). The crystallite size increases with the heating temperature, with the largest crystallite size being observed for CuFe$_2$O$_4$ at 1000 °C (81 nm), while the smallest crystallite size was observed for ZnFe$_2$O$_4$ at 700 °C (13 nm). The lattice parameter (a) also increases with the annealing temperature, with the highest value being observed for NiFe$_2$O$_4$ at 1000 °C (8.365 Å), while the lowest value observed was for CuFe$_2$O$_4$ at 700 °C (8.207 Å). The M/Fe molar ratio calculated based on Co, Mn, Zn, Ni and Fe concentrations measured by ICP-OES confirms the theoretical elemental composition of the obtained nanoferrites (Table 1). In all cases, the best fit of experimental and theoretical data is remarked for samples annealed at 1000 °C. In case of CoFe$_2$O$_4$ and MnFe$_2$O$_4$ annealed at 700 °C, the M/Fe molar ratio could not be calculated due to the presence of Co$_3$O$_4$ and α-Fe$_2$O$_3$ as secondary phases.

**Table 1.** Average particle size (D$_{PS}$), average crystallite size (D$_{CS}$), lattice parameter (a) and M/Fe molar ratio for MFe$_2$O$_4$ (M = Co, Ni, Zn, Mn, Cu) heat treated at 700 and 1000 °C.

|  | Temperature (°C) | CoFe$_2$O$_4$ | NiFe$_2$O$_4$ | ZnFe$_2$O$_4$ | MnFe$_2$O$_4$ | CuFe$_2$O$_4$ |
|---|---|---|---|---|---|---|
| D$_{PS}$ (nm) | 1000 | 78 | 52 | 68 | 32 | 85 |
| D$_{CS}$ (nm) | 700 | 23 | 18 | 13 | 14 | 27 |
|  | 1000 | 69 | 49 | 57 | 29 | 81 |
| a (Å) | 700 | 8.275 | 8.258 | 8.278 | 8.269 | 8.207 |
|  | 1000 | 8.334 | 8.365 | 8.342 | 8.318 | 8.302 |
| M/Fe molar ratio | 700 | - | 0.98/2.04 | 0.97/2.01 | - | 0.98/2.03 |
|  | 1000 | 0.99/2.00 | 0.99/2.01 | 0.99/2.00 | 0.98/1.99 | 1.00/2.01 |

Due to the low amount of adsorbed/desorbed nitrogen, the determination of porosity and specific surface area (SSA) for samples heat treated at 700 and 1000 °C was not possible. The SSA below the method detection limit (0.5 m$^2$/g), suggests that all ferrites have a non–porous structure, probably as a consequence of particle agglomeration that limits the absorption of nitrogen.

According to TEM images, the nanoparticles have spherical shape. The particle sizes estimated by XRD and TEM are comparable, the low differences appearing probably due to some large-size nanoparticles. $CuFe_2O_4$ displays the largest particle size (85 nm), while $MnFe_2O_4$ has the smallest particle size (32 nm) (Table 1 and Figure 3).

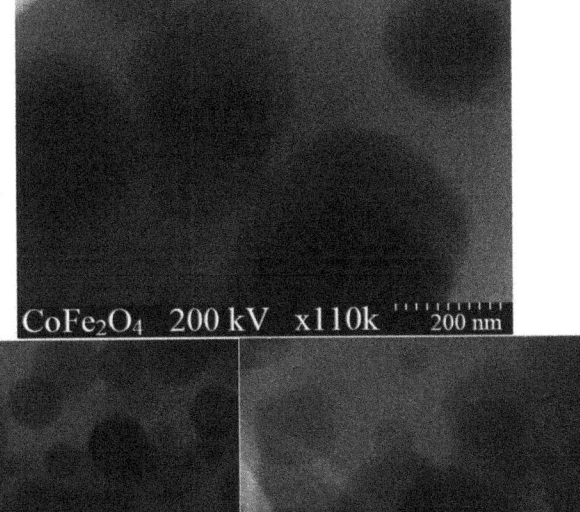

**Figure 3.** TEM images for $MFe_2O_4$ (M = Co, Zn, Ni, Cu, Mn) heat treated at 1000 °C.

The hysteresis loops (Figure 4) have an S-shape at low magnetic fields and are linear at higher fields, indicating the presence of small-sized magnetic particles with superparamagnetic behavior [37]. The spin rotation energy for particles smaller than the critical diameter is lower than the thermal energy. Thus, in the absence of an applied magnetic field, the random orientation of the magnetic moments results in zero average global magnetic moment [36].

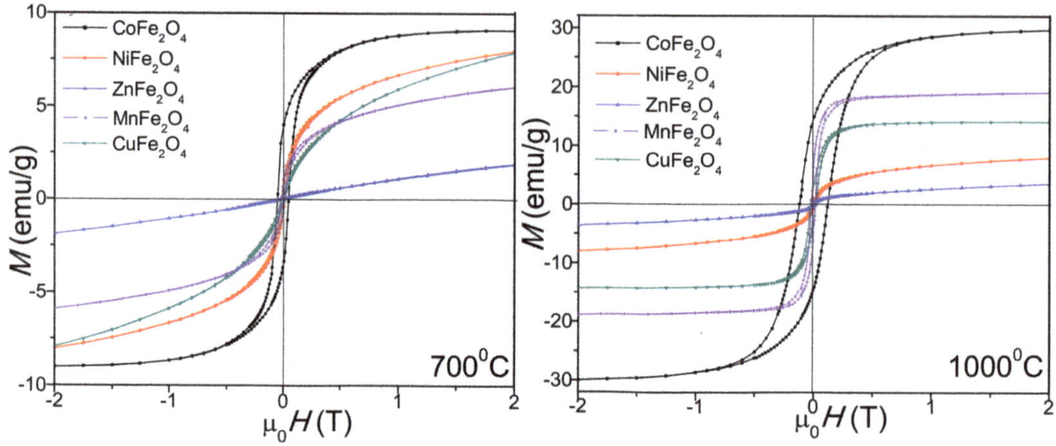

**Figure 4.** Magnetic hysteresis loops of $MFe_2O_4$ (M = Co, Zn, Ni, Cu, Mn) heat treated at 700 °C (**a**) and 1000 °C (**b**).

The saturation magnetization ($M_S$) and remanent magnetization ($M_R$) increase with the increase of heating temperature, with the highest values being measured for $CoFe_2O_4$ (29.7 emu/g) and the lowest for $ZnFe_2O_4$ (2.45 emu/g), as shown in Table 2. The coercivity ($H_C$) increases (from 49.5 Oe to 131 Oe in case of $CoFe_2O_4$) with the increase of heating temperature, indicating that the magnetic moment arrangement is highly disordered at high heating temperatures [27].

**Table 2.** Saturation magnetization ($M_S$), remanent magnetization ($M_R$), coercivity ($H_C$), squareness (S) and magnetic anisotropy constant (K) of $MFe_2O_4$ (M = Co, Zn, Ni, Cu, Mn) heat treated at 700 and 1000 °C.

| | Temperature (°C) | $CoFe_2O_4$ | $NiFe_2O_4$ | $ZnFe_2O_4$ | $CuFe_2O_4$ | $MnFe_2O_4$ |
|---|---|---|---|---|---|---|
| $M_S$ (emu/g) | 700 | 9.20 | 8.08 | 1.89 | 7.91 | 6.08 |
|  | 1000 | 29.7 | 11.2 | 2.45 | 14.2 | 19.1 |
| $M_R$ (emu/g) | 700 | 3.98 | 1.55 | 0.142 | 1.11 | 0.98 |
|  | 1000 | 14.1 | 8.14 | 0.695 | 9.82 | 1.55 |
| $H_C$ (Oe) | 700 | 49.5 | 4.62 | 1.81 | 10.4 | 20.8 |
|  | 1000 | 131 | 12.8 | 10.1 | 14.7 | 51.6 |
| S | 700 | 0.437 | 0.192 | 0.075 | 0.143 | 0.161 |
|  | 1000 | 0.475 | 0.727 | 0.284 | 0.692 | 0.081 |
| $K \times 10^3$ (erg/cm$^3$) | 700 | 0.286 | 0.023 | 0.002 | 0.052 | 0.079 |
|  | 1000 | 2.44 | 0.090 | 0.015 | 0.131 | 0.619 |

In all cases, the squareness ratio (S, 0.075 for $ZnFe_2O_4$ at 700 °C—0.727 for $NiFe_2O_4$ at 1000 °C), the anisotropy constant (K, $0.002 \times 10^3$ erg/cm$^3$ for $ZnFe_2O_4$ at 700 °C—$2.44 \times 10^3$ erg/cm$^3$ for $CoFe_2O_4$ at 1000 °C) also increase with the increase of heating temperature. Compared to the same ferrites embedded in $SiO_2$ matrix, the K value is much lower [33].

## 3. Discussion

The thermal behavior of $Fe^{III}(NO_3)_3$–$M^{II}(NO_3)_2$–1,3-propanediol solutions was studied by DTA. The endothermic effect at 70–100 °C is attributed to the loss of moisture and of crystallization water from the metallic nitrates used in the synthesis. The endothermic effect at 132–213 °C on the DTA diagram is attributed to formation of metal-malonate precursor. The formation of metal-malonate precursor for $CoFe_2O_4$ and $ZnFe_2O_4$ takes place in two stages as indicated by the two endothermic effects. The first endothermic effect was assigned to the divalent metal-malonate formation (143 °C for Co-malonate and 132 °C for Mn-malonate), while the second endothermic effect (around 201 °C) was assigned to the formation of Fe-malonate. In case of the other synthesis, the divalent metal (Ni, Zn, Cu) malonates and the trivalent Fe-malonate formation takes place in a single stage (204–213 °C). The formation of ferrites by decomposition of malonate precursors is indicated on the DTA diagram by a single exothermic effect at 250–350 °C, except for $CoFe_2O_4$, where two exothermic effects appear at 253 and 303 °C. The two-stage formation of $CoFe_2O_4$ in the metal nitrates–diol mixture could be explained by the fact that the aqua cation $[Fe(H_2O)_6]^{3+}$ is a stronger acid than the aqua cation $[Co(H_2O)_6]^{2+}$ [17,36]. The highest mass loss shown on the TG diagram is attributed to $CoFe_2O_4$ (65%), while the lowest mass loss is attributed to $ZnFe_2O_4$ (57%), probably due to the fact that $ZnFe_2O_4$ is quantitatively obtained at lower temperatures compared to other ferrites accompanied by other crystalline or amorphous secondary phases.

The increase of heating temperature from 700 to 1000 °C did not affect the crystal structure of the studied ferrites but improved the phase purity [10]. Moreover, by increasing the heating temperature, the diffraction peaks become sharper and narrower, indicating the formation of larger particles due to grain growth [25,38]. After heat treatment at 1000 °C, an important agglomeration of the particles takes place without consequent recrystallization, supporting the formation of single crystals rather than polycrystals [1,2,6]. Oppositely, at 700 °C, the surface dipole–dipole interactions, high surface energy and tension, as well as the change of cation distribution within the nanocrystallite, induces lattice shrinking, which further inhibits grain growth [1–6]. Generally, the size of the crystallite is higher than the size of the corresponding ferrites embedded in $SiO_2$ matrix, produced by sol–gel method [15–19]. These findings indicate that the heating temperature and the synthesis route plays a key role in determining the crystallite size [8]. Crystallite size has a significant effect on the magnetic and optical properties of the material, especially when the grain size is approaching the crystallite size [4–6,38].

The different particle size of the produced ferrites may be attributed to different kinetics of metal oxides formation, different particle growth rate or presence of structural disorder and strain in the lattice caused by different ionic radii [39] The different particle arrangement is attributed to the formation of well-delimited particles that generate solid boundaries. Moreover, interfacial surface tensions appear most likely due to the agglomeration tendency of small particles, weak surface interaction due to Van der Waals forces and magnetic interactions [39].

The magnetic properties of nanoferrites are strongly influenced by the cation distribution between the tetrahedral (A) and octahedral (B) sites, as well as by the interactions between the magnetic ions [37,40]. Different size and morphology of the nanoparticles at different heating temperatures results in different surface spin disorder, pinned magnetic moment, different surface spin canting and consequently different cation inversion in the spinel structure and magnetic features [8]. The lower $M_S$ values of ferrites heat treated at 700 °C compared to those at 1000 °C result from the lower crystallinity at 700 °C, presence of vacancies, interatomic spacing, low coordination number and surface spin disorder [37].

At 1000 °C, the $CoFe_2O_4$ has superparamagnetic behavior, while the other ferrites display paramagnetic behavior. The superparamagnetic behavior is attributed to the high disorder of the magnetic moment orientation with the increase in the surface-area-to-volume ratio [41]. For $MnFe_2O_4$, the sharp increase in $M_S$ with the increase of heating temperature could be explained by the formation of trace paramagnetic $\alpha$-$Fe_2O_3$ [27,30].

The ZnFe$_2$O$_4$ heat treated at 700 °C is paramagnetic, the low M$_S$ values being attributed to the lattice defects, core–shell interactions, spin canting, disordered cation distribution, A–B super exchange interaction and random spin orientation on the surface of nanoparticles [30]. The M$_S$ values reported for ZnFe$_2$O$_4$ differ from study to study, indicating that M$_S$ strongly depends on the synthesis route and heating temperature [27,41]. The changes of CuFe$_2$O$_4$ magnetic properties following the reduction in bulk grain size of CuFe$_2$O$_4$ nanoparticles by milling was reported by Soufi et al. [12]. The M$_S$ value of NiFe$_2$O$_4$ is lower than that of MnFe$_2$O$_4$ and CoFe$_2$O$_4$, probably due to the increase in surface effects with the decrease in particle size [20]. The influences of surface effect on the magnetic properties may be explained by the different exchange interactions and presence of magnetic defects on the nanoparticles surface [20]. Priyadharsini et al. also reported the M$_S$, M$_R$ and H$_C$ values increasing with the heating temperature [25].

Generally, the H$_C$ of the spinel ferrite nanoparticles is governed by the magnetocrystalline anisotropy, strain, interparticle interaction, grain size and morphology [42]. The H$_C$ also increase with the increase in the surface potential barrier caused by crystalline lattice defects such as the deviation of atoms from the normal positions in the surface layers [43]. The influence of the particle sizes, internal strain, magnetic domain structure, shape and magnetocrystalline anisotropy of the nanoparticles on the H$_C$ value is not fully explained [43]. The low H$_C$ of all ferrites for both heating temperatures indicate an enhanced coalescence of the crystallites that further results in strong magnetic coupling and high magnetization [43]. At both temperatures, the H$_C$ of CoFe$_2$O$_4$ nanoparticles prepared by thermal decomposition increases with the particle-size increase, suggesting the presence of a single magnetic domain [8]. The transition from superparamagnetic to ferromagnetic behavior of CoFe$_2$O$_4$ was noticed after heat treatment at 1000 °C [25]. The different H$_C$ of NiFe$_2$O$_4$ is attributed both to the crystallite size and the presence of shape anisotropy [30].

The increasing squareness ratio (S) at high heating temperatures could be the consequence of the reorientation of grains along the easy axis of magnetization when the field is switched off [25]. The main factors that influence the magnetic anisotropy are the crystallographic directions, surface defects and irregularities [44,45]. The high H$_C$ and K of CoFe$_2$O$_4$ are the consequence of Co$^{2+}$ ions in octahedral (B) sites, which induce frozen orbital angular momentum and strong spin-orbital coupling [8,46].

## 4. Materials and Methods

Fe(NO$_3$)$_3$·9H$_2$O, Co(NO$_3$)$_2$·6H$_2$O, Ni(NO$_3$)$_2$·6H$_2$O, Zn(NO$_3$)$_2$·6H$_2$O, Cu(NO$_3$)$_2$·3H$_2$O, Mn(NO$_3$)$_2$·3H$_2$O and 1,3-propanediol of purity higher than 98% were purchased from Merck (Darmstadt, Germany) and used as received.

MFe$_2$O$_4$ (M = Co, Zn, Ni, Cu, Mn) were synthesized by mixing the metal nitrates in 1M/2Fe molar ratio with 1,3-propanediol in equimolecular ratio of NO$_3^-$/1,3-propanediol. The resulted solutions were heat treated at 700, and 1000 °C (5 h) in air using a LT9 muffle furnace (Nabertherm, Lilienthal, Germany).

The ferrite formation was investigated by thermogravimetry (TG) and differential thermal analysis (DTA) using a Q600 SDT (TA Instruments, New Castle, DE, USA) analyzer, in air, up to 1000 °C, at 10 °C·min$^{-1}$ using alumina standards. The crystalline phases were investigated by X-ray diffraction using a D8 Advance (Bruker, Karlsruhe, Germany), at ambient temperature, with CuKα radiation (λ = 1.54060 Å) and LynxEye detector, operating at 40 kV and 40 mA. The Co/Fe (CoFe$_2$O$_4$), Ni/Fe (NiFe$_2$O$_4$), Zn/Fe (ZnFe$_2$O$_4$), Cu/Fe (CuFe$_2$O$_4$) and Mn/Fe (MnFe$_2$O$_4$) molar ratios were confirmed using an Optima 5300 DV (Perkin Elmer, Norwalk, CT, USA) inductively coupled plasma optical emission spectrometer (ICP-OES), spectrometer, after microwave digestion (Xpert microwave system, Berghof, Eningen, Germany) with aqua regia. Specific surface area (SSA) was calculated using the BET model from N$_2$ adsorption–desorption isotherms recorded at 196 °C on samples degassed for 4 h at 150 °C and 2 Pa pressure using a Sorptomatic 1990 (Thermo Fisher Scientific, Waltham, MA, USA) instrument. The shape and clustering of nanoparticles were studied on samples deposited and dried on carbon-coated copper grids using a transmission electron microscope (TEM, HD-2700, Hitachi, Tokyo, Japan). The magnetic measurements were performed using a 7400 vibrating-sample magnetometer (VSM, LakeShore Cryotronics, Westerville, OH, USA). The hysteresis loops were recorded at room temperature in magnetic fields between −2 to 2 Tesla.

## 5. Conclusions

The structural, morphological and magnetic characteristics of nanosized $CoFe_2O_4$, $NiFe_2O_4$, $ZnFe_2O_4$, $CuFe_2O_4$ and $MnFe_2O_4$ obtained by thermal decomposition were investigated. The formation of ferrites appeared as a single exothermic effect at 250–350 °C, excepting $CoFe_2O_4$ with two exothermic effects. The highest mass loss was attributed to $CoFe_2O_4$ (65%), while the lowest mass loss was assigned to $ZnFe_2O_4$ (55%). Unlike similar ferrites embedded in $SiO_2$ matrix, at both temperatures, single crystalline phases were remarked, excepting the presence of $Co_3O_4$ (in case of $CoFe_2O_4$) and $\alpha$-$Fe_2O_3$ (in case of $MnFe_2O_4$) at 700 °C. The SSA values lower than 0.5 m$^2$/g indicated a non–porous structure due to the particle agglomeration. $CuFe_2O_4$ showed the largest particle size (85 nm), while $MnFe_2O_4$ had the smallest particle size (32 nm). The crystalline $CoFe_2O_4$ heat treated at 1000 °C displayed the highest Ms, $H_C$ and K values, presenting superparamagnetic behavior, while the other ferrites exhibited paramagnetic behavior.

**Author Contributions:** Conceptualization, methodology, investigation, writing—original draft, visualization, supervision, T.D.; methodology, investigation, writing—review, E.A.L. and O.C. All authors have read and agreed to the published version of the manuscript.

**Funding:** The APC was funded by Technical University of Cluj-Napoca Grant Support GC1/2021.

**Institutional Review Board Statement:** Not applicable.

**Informed Consent Statement:** Not applicable.

**Data Availability Statement:** Not applicable.

**Acknowledgments:** This study was supported by the Ministry of Research, Innovation and Digitization through Program 1—Development of the national research & development system, Subprogram 1.2—Institutional performance—Projects that finance the RDI excellence, Contract no. 18PFE/30 December 2021.

**Conflicts of Interest:** The authors declare no conflict of interest. The funders had no role in the design of the study; in the collection, analyses, or interpretation of data; in the writing of the manuscript; or in the decision to publish the results.

## References

1. Mylarappa, M.; Lakshmi Venkata, V.; Mahesh Vishnu, K.R.; Raghavendra, N.; Nagaswarupa, H.P. Cyclic voltammetry, impedance and thermal properties of $CoFe_2O_4$ obtained from waste Li-Ion batteries. *Mater. Today Proc.* **2018**, *5*, 22425–22432. [CrossRef]
2. Xiong, Q.Q.; Tu, J.P.; Shi, S.J.; Liu, X.J.; Wang, X.L.; Gu, C.D. Ascorbic acid-assisted synthesis of cobalt ferrite ($CoFe_2O_4$) hierarchical flower-like microspheres with enhanced lithium storage properties. *J. Power Sources* **2014**, *256*, 153–159. [CrossRef]
3. Shetty, K.; Renuka, L.; Nagaswarupa, H.P.; Nagabhushana, H.; Anantharaju, K.S.; Rangappa, D.; Prashantha, S.C.; Ashwini, K. A comparative study on $CuFe_2O_4$, $ZnFe_2O_4$ and $NiFe_2O_4$: Morphology, impedance and photocatalytic studies. *Mater. Today Proc.* **2017**, *4*, 11806–11815. [CrossRef]
4. Asghar, K.; Qasim, M.; Das, D. Preparation and characterization of mesoporous magnetic $MnFe_2O_4$@$mSiO_2$ nanocomposite for drug delivery application. *Mater. Today Proc.* **2020**, *26*, 87–93. [CrossRef]
5. Chand, P.; Vaish, S.; Kumar, P. Structural, optical and dielectric properties of transition metal ($MFe_2O_4$; M = Co, Ni and Zn) nanoferrites. *Phys. B* **2017**, *524*, 53–63. [CrossRef]
6. Revathi, J.; Abel, M.J.; Archana, V.; Sumithra, T.; Thiruneelakandan, R. Synthesis and characterization of $CoFe_2O_4$ and Ni-doped $CoFe_2O_4$ nanoparticles by chemical Co-precipitation technique for photo-degradation of organic dyestuffs under direct sunlight. *Phys. B* **2020**, *587*, 412136. [CrossRef]
7. Li, X.; Sun, Y.; Zong, Y.; Wei, Y.; Liu, X.; Li, X.; Peng, Y.; Zheng, X. Size-effect induced cation redistribution on the magnetic properties of well-dispersed $CoFe_2O_4$ nanocrystals. *J. Alloy. Compd.* **2020**, *841*, 155710. [CrossRef]
8. Shilpa Amulya, M.A.; Nagaswarupta, H.P.; Anil Kumar, M.R.; Ravikumar, C.R.; Kusuma, K.B.; Prashantha, S.C. Evaluation of bifunctional applications of $CuFe_2O_4$ nanoparticles synthesized by a sonochemical method. *J. Phys. Chem. Solids* **2021**, *148*, 109756. [CrossRef]

9. Peng, S.; Wang, S.; Liu, R.; Wu, J. Controlled oxygen vacancies of $ZnFe_2O_4$ with superior gas sensing properties prepared via a facile one-step self-catalyzed treatment. *Sens. Actuators B Chem.* **2019**, *288*, 649–655. [CrossRef]
10. Cui, K.; Sun, M.; Zhang, J.; Xu, J.; Zhai, Z.; Gong, T.; Hou, L.; Yuan, C. Facile solid-state synthesis of tetragonal $CuFe_2O_4$ spinels with improved infrared radiation performance. *Ceram. Int.* **2022**, *48*, 10555–10561. [CrossRef]
11. Sarkar, K.; Mondal, R.; Dey, S.; Kumar, S. Cation vacancy and magnetic properties of $ZnFe_2O_4$ microspheres. *Phys. B Condens. Matter* **2020**, *583*, 412015. [CrossRef]
12. Soufi, A.; Hajjaoui, H.; Elmoubarki, R.; Abdennouri, M.; Qourzal, S.; Barka, N. Heterogeneous Fenton-like degradation of tartrazine using $CuFe_2O_4$ nanoparticles synthesized by sol-gel combustion. *Appl. Surf. Sci. Adv.* **2022**, *9*, 100251. [CrossRef]
13. Rajini, R.; Ferdinand, A.C. Structural, morphological and magnetic properties of (c-$ZnFe_2O_4$ and t-$CuFe_2O_4$) ferrite nanoparticle synthesized by reactive ball milling. *Chem. Data Collect.* **2022**, *38*, 100825. [CrossRef]
14. Iqbal, M.J.; Yaqub, N.; Sepiol, B.; Ismail, B. A study of the influence of crystallite size on the electrical and magnetic properties of $CuFe_2O_4$. *Mater. Res. Bull.* **2011**, *46*, 1837–1842. [CrossRef]
15. Dey, A.; Saini, B. Effect of various surfactant templates on the physicochemical properties of $CoFe_2O_4$ nanoparticles. *Mater. Today Proc.* **2022**, *61*, 351–355. [CrossRef]
16. Dippong, T.; Levei, E.A.; Lengauer, C.L.; Daniel, A.; Toloman, D.; Cadar, O. Investigation of thermal, structural, morphological and photocatalytic properties of $Cu_xCo_{1-x}Fe_2O_4$ ($0 \leq x \leq 1$) nanoparticles embedded in $SiO_2$ matrix. *Mater. Charact.* **2020**, *163*, 110268. [CrossRef]
17. Stefanescu, M.; Dippong, T.; Stoia, M.; Stefanescu, O. Study on the obtaining of cobalt oxides by thermal decomposition of some complex combinations, undispersed and dispersed in $SiO_2$ matrix. *J. Therm. Anal. Calorim.* **2008**, *94*, 389–393. [CrossRef]
18. Dippong, T.; Levei, E.A.; Goga, F.; Petean, I.; Avram, A.; Cadar, O. The impact of polyol structure on the formation of $Zn_{0.6}Co_{0.4}Fe_2O_4$ spinel-based pigments. *J. Sol-Gel Sci. Technol.* **2019**, *93*, 736–744. [CrossRef]
19. Dippong, T.; Levei, E.A.; Cadar, O.; Mesaros, A.; Borodi, G. Sol-gel synthesis of $CoFe_2O_4$:$SiO_2$ nanocomposites–insights into the thermal decomposition process of precursors. *J. Anal. Appl. Pyrolysis* **2017**, *125*, 169–177. [CrossRef]
20. Sivakumar, A.; Dhas, S.S.J.; Dhas, S.A.M.B. Assessment of crystallographic and magnetic phase stabilities on $MnFe_2O_4$ nano crystalline materials at shocked conditions. *Solid State Sci.* **2020**, *107*, 106340. [CrossRef]
21. Özçelik, B.; Özçelik, S.; Amaveda, H.; Santos, H.; Borrell, C.; de la Fuente, G.F.; Saez-Puche, R.; Angurel, L.A. High speed processing of $NiFe_2O_4$ spinel using laser furnance. *J. Mater.* **2020**, *6*, 661–670. [CrossRef]
22. Hoghoghifard, S.; Moradi, M. Influence of annealing temperature on structural, magnetic, and dielectric properties of $NiFe_2O_4$ nanorods synthesized by simple hydrothermal method. *Ceram. Int.* **2022**, *48*, 17768–17775. [CrossRef]
23. Majid, F.; Rauf, J.; Ata, S.; Bibi, I.; Malik, A.; Ibrahim, S.M.; Ali, A.; Iqbal, M. Synthesis and characterization of $NiFe_2O_4$ ferrite: Sol–gel and hydrothermal synthesis routes effect on magnetic, structural and dielectric characteristics. *Mater. Chem. Phys.* **2021**, *258*, 123888. [CrossRef]
24. Mohanty, D.; Satpathy, S.K.; Behera, B.; Mohapatra, R.K. Dielectric and frequency dependent transport properties in magnesium doped $CuFe_2O_4$ composite. *Mater. Today Proc.* **2020**, *33*, 5226–5231. [CrossRef]
25. Priyadharsini, R.; Das, S.; Venkateshwarlu, M.; Deenadayalan, K.; Manoharan, C. The influence of reaction and annealing temperature on physical and magnetic properties of $CuFe_2O_4$ nanoparticles: Hydrothermal method. *Inorg. Chem. Commun.* **2022**, *140*, 109406. [CrossRef]
26. Xu, P.; Xie, S.; Liu, X.; Wang, L.; Jia, X.; Yang, C. Electrochemical enhanced heterogenous activation of peroxymonosulfate using $CuFe_2O_4$ particle electrodes for the degradation of diclofenac. *Chem. Eng. J.* **2022**, *446*, 136941. [CrossRef]
27. Ge, Y.-C.; Wang, Z.-L.; Yi, M.-Z.; Ran, L.-P. Fabrication and magnetic transformation from paramagnetic to ferrimagnetic of $ZnFe_2O_4$ hollow spheres. *Trans. Nonferrous Met. Soc. China* **2019**, *29*, 1503–1509. [CrossRef]
28. Mohanty, D.; Mallick, P.; Biswall, S.K.; Behera, B.; Mohapatra, R.K.; Behera, A.; Satpathy, S.K. Investigation of structural, dielectric and electric properties of $ZnFe_2O_4$. *Mater. Today Proc.* **2020**, *33*, 4971–4975. [CrossRef]
29. Junlabhut, P.; Nuthongkum, P.; Pechrapa, W. Influences of calcination temperature on structural properties of $MnFe_2O_4$ nanopowders synthesized by co-precipitation method for reusable absorbent materials. *Mater. Today Proc.* **2018**, *5*, 13857–13864. [CrossRef]
30. Sivakumar, P.; Ramesh, R.; Ramanand, A.; Ponnusamy, S.; Muthamizhchelvan, C. Synthesis and characterization of $NiFe_2O_4$ nanoparticles and nanorods. *J. Alloys Compd.* **2013**, *563*, 6–11. [CrossRef]
31. Airimioaei, M.; Ciomaga, C.E.; Apostolescu, A.; Leonite, L.; Iordan, A.R.; Mitoseriu, L.; Palamaru, M.N. Synthesis and functional properties of the $Ni_{1-x}Mn_xFe_2O_4$ ferrites. *J. Alloys Compd.* **2011**, *509*, 8065–8072. [CrossRef]
32. Mathubala, G.; Manikandan, A.; Arul Antony, S.; Ramar, P. Photocatalytic degradation of methylene blue dye and magnetooptical studies of magnetically recyclable spinel $Ni_xMn_{1-x}Fe_2O_4$ (x = 0.0 − 1.0) nanoparticles. *J. Mol. Struct.* **2016**, *113*, 79–87. [CrossRef]
33. Dippong, T.; Levei, E.A.; Cadar, O. Formation, structure and magnetic properties of $MFe_2O_4@SiO_2$ (M = Co, Mn, Zn, Ni, Cu) nanocomposites. *Materials* **2021**, *14*, 1139. [CrossRef]
34. Alarifi, A.; Deraz, N.M.; Shaban, S. Structural, morphological and magnetic properties of $NiFe_2O_4$ nano-particles. *J. Alloy. Compd.* **2009**, *486*, 501–506. [CrossRef]
35. Naseri, M.G.; Bin Saion, E.; Ahangar, H.A.; Hashim, M.; Shaari, A.H. Synthesis and characterization of manganese ferrite nanoparticles by thermal treatment method. *J. Magn. Magn. Mater.* **2011**, *323*, 1745–1749. [CrossRef]
36. Joint Committee on Powder Diffraction Standards. *International Center for Diffraction Data*; ASTM: Philadelphia, PA, USA, 1999.

37. Ati, M.A.; Othaman, Z.; Samavati, A. Influence of cobalt on structural and magnetic properties of nickel ferrite nanoparticles. *J. Mol. Struct.* **2013**, *1052*, 177–182. [CrossRef]
38. Vinosha, P.A.; Xavier, B.; Krishnan, S.; Das, S.J. Investigation on zinc substituted highly porous improved catalytic activity of $NiFe_2O_4$ nanocrystal by co-precipitation method. *Mater. Res. Bull.* **2018**, *101*, 190–198. [CrossRef]
39. Jadhav, J.; Biswas, S.; Yadav, A.K.; Jha, S.N.; Bhattacharyya, D. Structural and magnetic properties of nanocrystalline Ni-Zn ferrites: In the context of cationic distribution. *J. Alloys Compd.* **2017**, *696*, 28–41. [CrossRef]
40. Salunkhe, A.B.; Khot, V.M.; Phadatare, M.R.; Thorat, N.D.; Joshi, R.S.; Yadav, H.M.; Pawar, S.H. Low temperature combustion synthesis and magnetostructural properties of Co-Mn nanoferrites. *J. Magn. Magn. Mater.* **2014**, *352*, 91–98. [CrossRef]
41. Lemine, O.M.; Bououdina, M.; Sajieddine, M.; Al-Saie, A.M.; Shafi, M.; Khatab, A.; Al-Hilali, M.; Henini, M. Synthesis, structural, magnetic and optical properties of nanocrystalline $ZnFe_2O_4$. *Phys. B* **2011**, *406*, 1989–1994. [CrossRef]
42. Yadav, S.P.; Shinde, S.S.; Bhatt, P.; Meena, S.S.; Rajpure, K.Y. Distribution of cations in $Co_{1-x}Mn_xFe_2O_4$ using XRD, magnetization and Mossbauer spectroscopy. *J. Alloys Compd.* **2015**, *646*, 550–556. [CrossRef]
43. Stefanescu, M.; Stoia, M.; Caizer, C.; Dippong, T.; Barvinschi, P. Preparation of $Co_xFe_{3-x}O_4$ nanoparticles by thermal decomposition of some organo-metallic precursors. *J. Therm. Anal. Calorim.* **2009**, *97*, 245–250. [CrossRef]
44. Sontu, U.B.; Yelasani, V.; Musugu, V.R.R. Structural, electrical and magnetic characteristics of nickel substituted cobalt ferrite nanoparticles, synthesized by self combustion method. *J. Magn. Magn. Mater.* **2015**, *374*, 376–380. [CrossRef]
45. Bameri, I.; Saffari, J.; Baniyaghoob, S.; Ekrami-Kakhki, M.S. Synthesis of magnetic nano-$NiFe_2O_4$ with the assistance of ultrasound and its application for photocatalytic degradation of Titan Yellow: Kinetic and isotherm studies. *Colloid Interface Sci. Commun.* **2022**, *48*, 100610. [CrossRef]
46. Belhadj, H.; Messaoudi, Y.; Khelladi, M.R.; Azizi, A. A facile synthesis of metal ferrites ($MFe_2O_4$, M = Co, Ni, Zn, Cu) as effective electrocatalysts toward electrochemical hydrogen evolution reaction. *Int. J. Hydrog. Energy* **2022**, *47*, 20129–20137. [CrossRef]

## Article

# Impact of Ni Content on the Structure and Sonophotocatalytic Activity of Ni-Zn-Co Ferrite Nanoparticles

Thomas Dippong [1], Oana Cadar [2], Firuta Goga [3], Dana Toloman [4] and Erika Andrea Levei [2,*]

[1] Faculty of Science, Technical University of Cluj-Napoca, 76 Victoriei Street, 430122 Baia Mare, Romania
[2] Research Institute for Analytical Instrumentation Subsidiary, National Institute for Research and Development for Optoelectronics INOE 2000, 67 Donath Street, 400293 Cluj-Napoca, Romania
[3] Faculty of Chemistry and Chemical Engineering, Babes-Bolyai University, 11 Arany Janos Street, 400028 Cluj-Napoca, Romania
[4] National Institute for Research and Development of Isotopic and Molecular Technologies, 67-103 Donath Street, 400293 Cluj-Napoca, Romania
* Correspondence: erika.levei@icia.ro

**Abstract:** The structure, morphology, and sonophotocatalytic activity of Ni-Zn-Co ferrite nanoparticles, embedded in a $SiO_2$ matrix and produced by a modified sol-gel method, followed by thermal treatment, were investigated. The thermal analysis confirmed the formation of metal succinate precursors up to 200 °C, their decomposition to metal oxides and the formation of Ni-Zn-Co ferrites up to 500 °C. The crystalline phases, crystallite size and lattice parameter were determined based on X-ray diffraction patterns. Transmission electron microscopy revealed the shape, size, and distribution pattern of the ferrite nanoparticles. The particle sizes ranged between 34 and 40 nm. All the samples showed optical responses in the visible range. The best sonophotocatalytic activity against the rhodamine B solution under visible irradiation was obtained for $Ni_{0.3}Zn_{0.3}Co_{0.4}Fe_2O_4@SiO_2$.

**Keywords:** nickel-zinc-cobalt ferrite; thermal behavior; crystalline phase; sonophotocatalysis

**Citation:** Dippong, T.; Cadar, O.; Goga, F.; Toloman, D.; Levei, E.A. Impact of Ni Content on the Structure and Sonophotocatalytic Activity of Ni-Zn-Co Ferrite Nanoparticles. *Int. J. Mol. Sci.* **2022**, *23*, 14167. https://doi.org/10.3390/ijms232214167

Academic Editor: Shaojun Yuan

Received: 18 October 2022
Accepted: 14 November 2022
Published: 16 November 2022

**Publisher's Note:** MDPI stays neutral with regard to jurisdictional claims in published maps and institutional affiliations.

**Copyright:** © 2022 by the authors. Licensee MDPI, Basel, Switzerland. This article is an open access article distributed under the terms and conditions of the Creative Commons Attribution (CC BY) license (https://creativecommons.org/licenses/by/4.0/).

## 1. Introduction

Despite the measures taken to reduce pollution, industrial effluents containing dyes and pigments used in the textile industry often resurface in the surrounding environment. Dyes are complex organic structures with a high resistance to chemical and biological degradation, high water solubility, and have a negative impact on the environment, particularly aquatic ecosystems [1,2]. Therefore, the efficient treatment of industrial effluents and wastewaters containing dyes is crucial for environmental protection.

The photocatalytic degradation of dyes is a simple, cost-effective, and environmentally friendly approach for wastewater treatment as it allows the decomposition of complex organic structures into $CO_2$ and water [3,4]. In the last few years, the use of sonophotocatalysis for the degradation of a wide range of organic pollutants in aqueous systems has been the topic of several studies [5,6]. Sonophotocatalysis use the synergistic effects of ultrasonic waves, UV-Vis irradiation and photocatalyst to form highly reactive free radicals in an aqueous medium that further react with dyes and lead to their degradation [7]. By providing additional nuclei, the heterogeneous catalyst enhances the formation of cavitation bubbles, which in turn increases the formation of reactive radicals through water pyrolysis [5]. The mechanism of sonophotocatalytic degradation, as well as the main advantages of combined ultrasound and photocatalytic processes, are presented by Abdurahman et al. [5]. The high energy consumption of sonophotocatalysis makes its large-scale application difficult, however, the high costs could be compensated by the low time required for the degradation of organic compounds [7]. Due to their magnetic properties, their recovery using an external magnetic field and their reuse is possible [8,9].

Oxides containing at least two types of metals are potential candidates for photoelectrochemistry and photocatalysis due to their band structure and energy position [10]. The bandgap of several $MFe_2O_4$-type ferrites are presented by Dillert et al. [11]. Nanosized spinel-type ferrites, containing first-row transition metals such are Ni, Co, Zn, Mn are attractive materials in electronics, magnetic storage, ferrofluid technology, gas sensors, catalysis, photocatalysis and biomedicine, including magnet-guided drug carriers, contrast agents and tracers for positive magnetic resonance imaging [12–17]. They are also promising candidates for wastewater treatment as they could act both as adsorbents, due to their high specific surface area, and as photocatalysis, due to their low energy band gap, that allow the conversion of UV or visible light into chemical energy that favors the degradation of dyes [8,17–21]. The strong photodegradation capacity of 3d transition metal ferrites such as $CoFe_2O_4$, $CuFe_2O_4$, $NiFe_2O_4$, and $ZnFe_2O_4$, with magnetic properties, was previously demonstrated for different organic compounds [12,13,21–27]. The surface coating of the ferrite nanoparticles with different materials, especially semiconductor materials such as $TiO_2$ and $SiO_2$, was proven to enhance the photocatalytic activity [28]. $CoFe_2O_4$ nanoparticles coated with $TiO_2$–$SiO_2$ efficiently degraded (up to 98%) the methylene blue dye [2], while Rhodamine B (RhB) degradation by $CoFe_2O_4$, was only 80% [29,30]. The enhancement of the photocatalytic performance of $Ni_xCo_{1-x}Fe_2O_4$, prepared by the coprecipitation method against methylene blue at a high Ni content, was reported by Lassoued and Li [31]. A good photodegradation efficiency (about 80%) of methylene blue under visible light irradiation was also delineated for Co-Zn ferrite with various Co and Zn contents, obtained using the citrate precursor method [32]. The increase of the rate constant with the increasing Co content was also observed [32]. The photocatalytic activity of $Co_{0.6}Zn_{0.4}Ni_xFe_{2-x}O_4$ powders, obtained by the sol-gel method under visible light against methyl orange dye in an aqueous solution, was also reported [33].

The spinel ferrite properties are determined by their composition, structure, particle size, and morphology [34–37]. These characteristics strongly depend on the synthesis route, chemical composition, doping cations, and sintering conditions [36,38]. The change in preparation method and temperature affects the microstructure, cation distribution among tetrahedral (A) and octahedral (B) sites and magnetic properties. Thus, to produce spinel ferrite nanoparticles with a desired stoichiometry, high compositional control, excellent chemical stability, high purity, crystallinity, saturation magnetization, and low coercivity are of interest and the selection of the synthesis route is critical [15,38].

Soft chemical routes such as sol-gel, solid-phase, hydrothermal, coprecipitation, sonochemical, spray pyrolysis, citrate gel, microwave refluxing, flash auto combustion, etc., are currently preferred for the synthesis of nano ferrites [17,37,39,40]. The solid-state reaction is a simple and attractive preparation method that allows large productivity and well-controllable grain sizes [41]. The conventional ceramic method is based on the solid-state reaction of metal oxides/carbonates at high temperatures (>1000 °C) and produces particles in the micrometer regime; however, agglomeration due to slow reaction kinetics is unavoidable [39]. Hence, wet chemical methods have been used intensively to avoid the limitations of conventional ceramic methods and to produce nanoscale materials with improved magnetic properties [35]. Wet chemical routes, such as hydrothermal, sol-gel method, and auto combustion, have been employed to obtain ferrite nanoparticles at low temperatures [34,39,41]. The main drawback of the wet methods is the formation of different oxide impurities, particularly $Fe_2O_3$ [17]. Highly homogenous Ni-Zn nanoferrite powders can be easily produced by a wet chemical route, using low-cost raw materials, and in air atmospheres. Its properties can be adjusted to fit the requirements of different applications by appropriately adjusting the Ni-Zn ratio and the sintering process [37,42]. The properties of Ni-Zn nanoferrites can be further improved by adding low amounts of other divalent ions such as $Co^{2+}$ [37].

This paper investigates the formation, structure, morphology, and sonophotocatalytic activity of mixed Ni-Zn-Co ferrites embedded in $SiO_2$, obtained by a modified sol-gel method and followed by thermal treatment. The reaction progress was investigated through thermal analysis (TG-DTA) and Fourier transform infrared spectroscopy (FT-IR), while the Ni-Zn-Co ferrites composition was investigated by inductively coupled plasma optical emission spectrometry (ICP-OES). The formation of the crystalline phase, crystallite size, and lattice constant were monitored by X-ray diffraction (XRD). The surface (specific surface area and porosity) was investigated using the Brunauer-Emmett-Teller (BET) method. The sonophotocatalytic properties of the samples were evaluated under visible light irradiation, assisted by sonication against RhB.

## 2. Results and Discussion

### 2.1. Thermal Analysis

The TG-DTA curves of all of the samples show three endothermic and two exothermic processes characterized by very close, overlapping peaks (Figure 1).

The endothermic effect at 61–70 °C, accompanied by 3–5% mass loss, is attributed to the loss of crystallization and constitution water. The endothermic effect at 136–144 °C, accompanied by 17–26% mass loss, is assigned to the formation of divalent metal precursors (Co, Ni, and Zn succinates), while the endothermic effect at 187–201 °C accompanied by 9–14% mass loss, is ascribed to the formation of a trivalent metal precursor (Fe succinate). The distinct behavior of Fe succinate, compared to the divalent metal (Co, Ni, Zn) succinates, can be attributed to the redox reaction between $Fe(NO_3)_3$ and 1,4BD, as well as to the stronger acidity of the aqua-cation $[Fe(H_2O)_6]_3$ [43]. The overlapping exothermic effects, at 270–292 °C and 310–325 °C, accompanied by 19–25% mass loss, are attributed to the decomposition of metal succinates to metal oxides, which leads to the formation of ferrites.

The exothermic peak, characteristic of the decomposition of divalent metal succinates, decreases with the increasing Zn content and shifts toward higher temperatures, leading to the increase of the exothermic peak, attributed to the decomposition of the Fe succinates. The $SiO_2$ matrix suffers various transformations during the thermal process, making the demarcation of the processes attributed to the formation and decomposition of succinate precursors difficult [44]. The total mass loss increases in the following order: $Ni_{0.5}Zn_{0.1}Co_{0.4}Fe_2O_4@SiO_2 < Zn_{0.6}Co_{0.4}Fe_2O_4@SiO_2 < Ni_{0.3}Zn_{0.3}Co_{0.4}Fe_2O_4@SiO_2 < Ni_{0.6}Co_{0.4}Fe_2O_4@SiO_2 < Ni_{0.2}Zn_{0.4}Co_{0.4}Fe_2O_4@SiO_2 < Ni_{0.4}Zn_{0.2}Co_{0.4}Fe_2O_4@SiO_2 < Ni_{0.1}Zn_{0.5}Co_{0.4}Fe_2O_4@SiO_2$.

In the case of the samples dried at 40 °C, the intense band at 1379–1389 $cm^{-1}$ is associated with the N-O bonds stretching vibration in metal nitrates. This band disappears in the case of samples dried at 200 °C, indicating the decomposition of nitrates [44]. The bands at 2958–2963 $cm^{-1}$ and 2872–2888 $cm^{-1}$ are specific to the symmetric and asymmetric vibration of C-H bonds in 1,4-BD or succinate precursors. These bands also disappear in samples heat-treated at 200 °C. The bands at 1578–1605 and 3200–3210 $cm^{-1}$ are attributed to the stretching and bending vibrations of O-H in 1,4-BD and adsorbed molecular water [44,45]. In the samples dried at 200 °C, the band at 3200–3210 $cm^{-1}$ is shifted towards higher wavenumbers (3421–3437 $cm^{-1}$), indicating that the metal succinates are hygroscopic [44,45]. The presence of this absorption band could also be due to the O-H stretching vibration and Si-OH deformation vibration caused by the hydrolysis of $-Si(OC_2H_5)_4$ [44,45]. For samples dried at 40 and 200 °C, the bands at 557–568 $cm^{-1}$ are attributed to Ni-O and Zn-O vibrations, while the band at 433–452 $cm^{-1}$ is attributed to the Fe-O vibration in the nitrates [44,45]. In samples at 40 °C, the band at 683–393 $cm^{-1}$ is assigned to the Co-O bond vibration in the cobalt nitrate [44,45]. The formation of the $SiO_2$ matrix in the samples dried at 40 and 200 °C is confirmed by the presence of specific bands of Si-O bond vibration (433–452 $cm^{-1}$), cyclic Si-O-Si bonds vibration (557–568 $cm^{-1}$, more noticeable in samples dried at 40 °C), Si-O symmetric stretching and bending vibration (792–815 $cm^{-1}$), Si-OH bonds (938–943 $cm^{-1}$, well delimited only in case of samples dried at 40 °C) vibration and Si-O-Si bonds stretching vibration (1045 $cm^{-1}$ at 40 °C and 1058–1068 $cm^{-1}$ at 200 °C). The band at 938–943 $cm^{-1}$ is distinguishable for the samples dried at 40 °C, indicating the presence of unreacted TEOS, while the band at 1045 $cm^{-1}$ suggests the formation of amorphous $SiO_2$ [44,45]. Figure 3a shows the FT-IR spectra of NCs thermally treated at 1000 °C. The band at 618–626 $cm^{-1}$ is attributed to the vibration of the M(II)-O (Co-O, Ni-O, Zn-O) bonds, while that at 485–490 $cm^{-1}$ is attributed to the Fe-O bond [44–46].

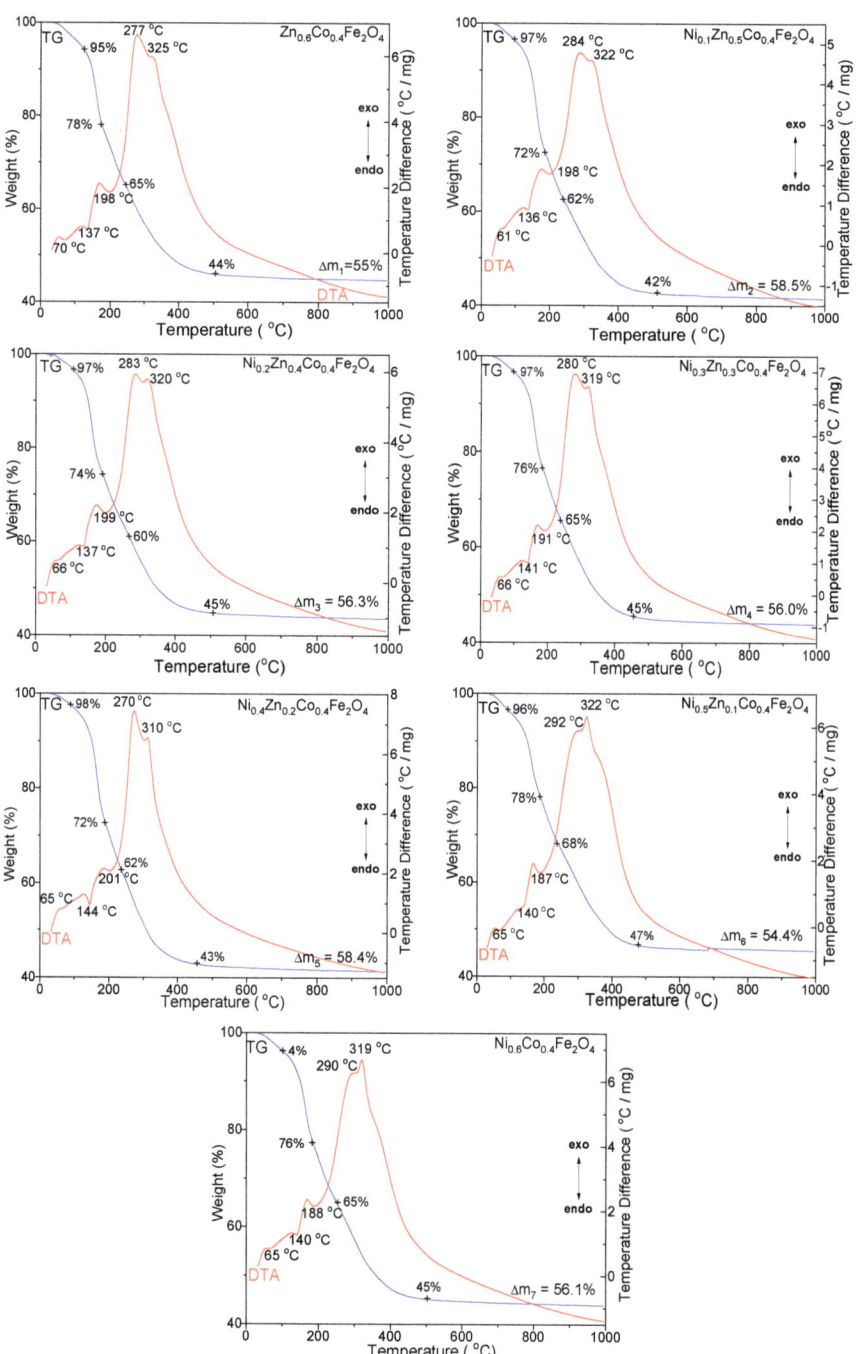

**Figure 1.** TG-DTA diagrams of $Zn_{0.6}Co_{0.4}Fe_2O_4@SiO_2$, $Ni_{0.1}Zn_{0.5}Co_{0.4}Fe_2O_4@SiO_2$, $Ni_{0.2}Zn_{0.4}Co_{0.4}Fe_2O_4@SiO_2$, $Ni_{0.3}Zn_{0.3}Co_{0.4}Fe_2O_4@SiO_2$, $Ni_{0.4}Zn_{0.2}Co_{0.4}Fe_2O_4@SiO_2$, $Ni_{0.5}Zn_{0.1}Co_{0.4}Fe_2O_4@SiO_2$ and $Ni_{0.6}Co_{0.4}Fe_2O_4@SiO_2$ samples.

## 2.2. FT-IR Analysis

As vibrational modes in FT-IR spectroscopy are determined by the bond type, the symmetry of the lattice sites and the elements in the crystal lattice, the monitoring of the ferrite formation process is possible [44]. The FT-IR spectra of the gels dried at 40 and 200 °C are presented in Figure 2.

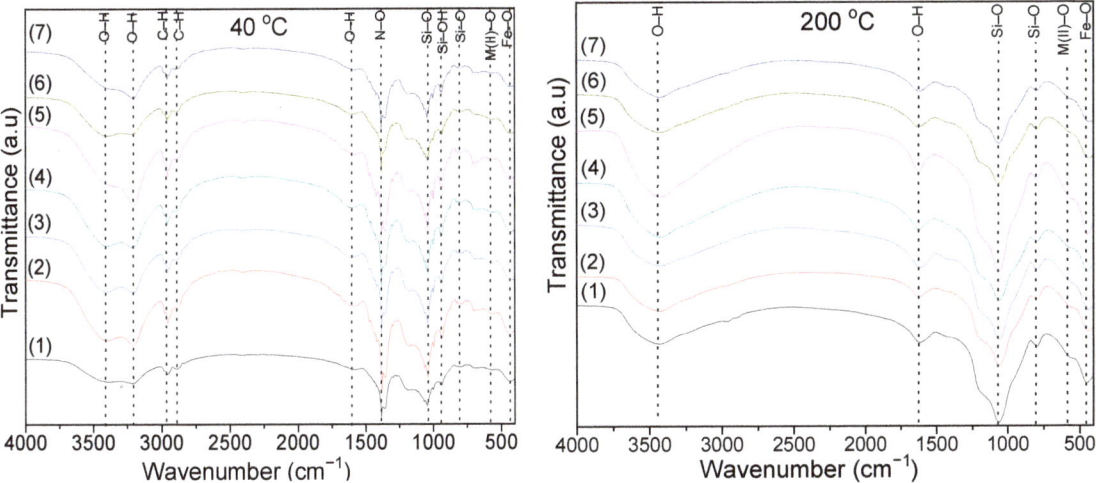

**Figure 2.** FT-IR spectra of $Zn_{0.6}Co_{0.4}Fe_2O_4@SiO_2$ (1) $Ni_{0.1}Zn_{0.5}Co_{0.4}Fe_2O_4@SiO_2$ (2), $Ni_{0.2}Zn_{0.4}Co_{0.4}Fe_2O_4@SiO_2$ (3), $Ni_{0.3}Zn_{0.3}Co_{0.4}Fe_2O_4@SiO_2$ (4) $Ni_{0.4}Zn_{0.2}Co_{0.4}Fe_2O_4@SiO_2$ (5) $Ni_{0.5}Zn_{0.1}Co_{0.4}Fe_2O_4@SiO_2$ (6) and $Ni_{0.6}Co_{0.4}Fe_2O_4@SiO_2$ (7) samples at 40 and 200 °C.

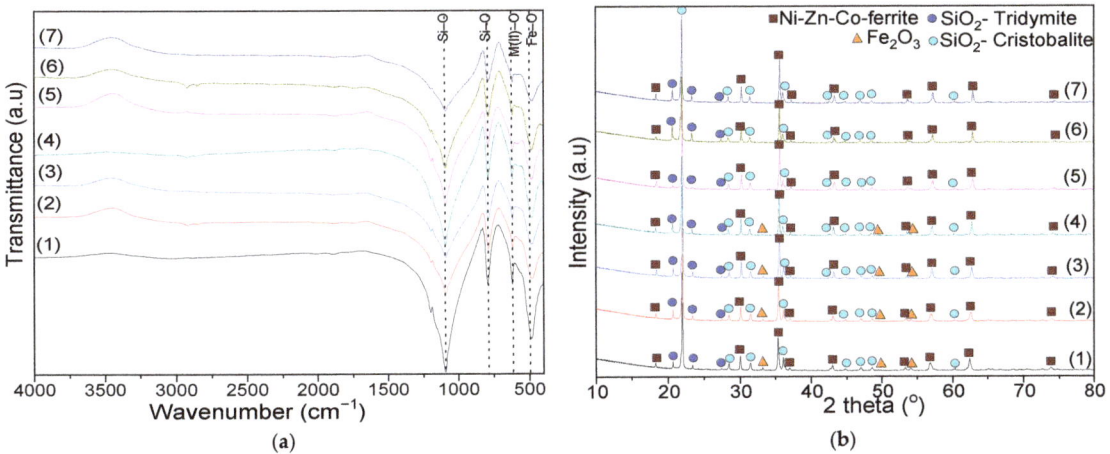

**Figure 3.** FT-IR spectra (**a**) and XRD patterns (**b**) of $Zn_{0.6}Co_{0.4}Fe_2O_4@SiO_2$ (1), $Ni_{0.1}Zn_{0.5}Co_{0.4}Fe_2O_4@SiO_2$ (2), $Ni_{0.2}Zn_{0.4}Co_{0.4}Fe_2O_4@SiO_2$ (3), $Ni_{0.3}Zn_{0.3}Co_{0.4}Fe_2O_4@SiO_2$ (4), $Ni_{0.4}Zn_{0.2}Co_{0.4}Fe_2O_4@SiO_2$ (5), $Ni_{0.5}Zn_{0.1}Co_{0.4}Fe_2O_4@SiO_2$ (6) and $Ni_{0.6}Co_{0.4}Fe_2O_4@SiO_2$ (7) samples at 1000 °C.

In comparison to the samples dried at 40 °C and 200 °C, in the samples thermally treated at 1000 °C, the wavenumbers specific to Co-O bond vibration decrease, and the wavenumbers specific to M(II)-O increase. The Jan-Teller effect, determined by the presence of $Fe^{2+}$ ions in the sublattices, can lead to band splitting, small bands and/or shoulders [47]. The $Fe^{2+}$ ions may result from the hopping process, namely $M^{2+} + Fe^{3+} \leftrightarrow M^{3+} + Fe^{2+}$ (M = Co,

Ni, Zn). The $Co^{3+}$ ion may migrate to tetrahedral (A) sites, while the $Fe^{2+}$ ions remain in their sites [31].

The bands at 790–795 $cm^{-1}$ are characteristic for the vibration of the Si-O bond in $SiO_2$ matrix, while those at 1090–1095 $cm^{-1}$ and 485–490 $cm^{-1}$ are characteristic to the stretching and bending vibration of Si-O-Si chains and show a low degree of polycondensation of the $SiO_2$ network [44,45]. The difference in band position could be attributed to the difference in M-O distance in the tetrahedral and octahedral sites [39].

### 2.3. XRD Analysis

The XRD patterns (Figure 3b) confirm the presence of well-crystallized ferrites in all of the samples, while the positions and intensities of the diffraction lines support the spinel structure [47]. The peaks with 2θ values of 30.07, 35.42, 37.07, 43.05, 53.41, 56.94, and 62.52 correspond to the (220), (311), (222), (400), (422), (511), and (440) planes [41,48]. In all of the samples, the local crystal structure is cubic spinel-type, belonging to the *Fd-3m* space group [21,40,41]. Additionally, two crystalline phases of the $SiO_2$ matrix ($SiO_2$-cristobalite, JCPDS card 39-1425 [48] and $SiO_2$-tridymite, JCPDS card 042-1401 [48]) are identified. In the case of $Zn_{0.6}Co_{0.4}Fe_2O_4@SiO_2$, the well-crystallized spinel is composed by $CoFe_2O_4$ (JCPDS card 22-1086 [48]) and $ZnFe_2O_4$ (JCPDS card 70-6491 [48]). In the case of $Ni_{0.6}Co_{0.4}Fe_2O_4@SiO_2$, the well-crystallized spinel is composed of $CoFe_2O_4$ and $NiFe_2O_4$ (JCPDS card 74-2081 [48]), while in the other samples ($Ni_{0.1}Zn_{0.5}Co_{0.4}Fe_2O_4@SiO_2$, $Ni_{0.2}Zn_{0.4}Co_{0.4}Fe_2O_4@SiO_2$, $Ni_{0.3}Zn_{0.3}Co_{0.4}Fe_2O_4@SiO_2$, $Ni_{0.4}Zn_{0.2}Co_{0.4}Fe_2O_4@SiO_2$, and $Ni_{0.5}Zn_{0.1}Co_{0.4}Fe_2O_4@SiO_2$), the crystalline phase contains $NiFe_2O_4$, $ZnFe_2O_4$ and $CoFe_2O_4$. In ferrites with a high Zn content, the presence of hematite ($Fe_2O_3$, JCPDS card 89-0599 [48]) is also remarked. The presence of $Fe_2O_3$ indicates the decomposition of $Fe(NO_3)_3$ into $Fe_2O_3$, leading to the formation of spinel ferrite [39]. The excess of metal oxides in insoluble secondary phases ($Fe_2O_3$) can contribute to the densification by generating high pore volumes and demagnetizing fields. During synthesis, the homogeneity of the metal oxide particles may result in higher defects and pore volumes in the final products [36,42].

The average crystallites size ($D_{XRD}$) was calculated using the Scherrer equation (Equation (1)).

$$D_{XRD} = \frac{0.9 \cdot \lambda}{\beta \cdot \cos\theta} \quad (1)$$

where λ is the wavelength of the $CuK_\alpha$ radiation (1.5406 Å), β is the full width at half-maximum intensity (FWHM), hkl are the Miller indices and θ is the Bragg angle (°) [43,44,49,50].

The lattice constant (a), was calculated from Bragg's law with Nelson-Riley Equation (2) [44,50].

$$a = \frac{\lambda\sqrt{h^2 + k^2 + l^2}}{2 \cdot \sin\theta} \quad (2)$$

where λ is the wavelength of $CuK_\alpha$ radiation (1.5406 Å), hkl are the Miller indices, θ is the Bragg angle (°) [44,50].

The unit cell volume (V) and the hopping length (L) of magnetic ions for tetrahedral (A) and octahedral (B) sites were calculated using Equations (3)–(5) [44,50].

$$V = a^3 \quad (3)$$

$$L_A = 0.25 \cdot a\sqrt{3} \quad (4)$$

$$L_B = 0.25 \cdot a\sqrt{2} \quad (5)$$

where a is the lattice constant (Å).

The average crystallite size lies in the nanocrystalline range and increases with the increasing Ni content, while the lattice parameter (a) decreases with the increasing Ni content (Table 1).

Table 1. Parameters obtained from TEM, XRD and ICP-OES analysis for samples thermally treated at 1000 °C.

| Nanocomposite | $D_{TEM}$ (nm) | $D_{XRD}$ (nm) | $D_C$ (%) | a (Å) | V (Å$^3$) | $L_A$ (Å) | $L_B$ (Å) | Quantitative Analysis (%) | Ni/Zn/Co |
|---|---|---|---|---|---|---|---|---|---|
| $Zn_{0.6}Co_{0.4}Fe_2O_4@SiO_2$ | 34 | 33.2 | 89.0 | 8.456 | 605 | 14.6 | 12.0 | 12% $Zn_{0.6}Co_{0.4}Fe_2O_4$ / 4% $\alpha$-$Fe_2O_3$ / 84% $SiO_2$ | 0/0.60/0.39 |
| $Ni_{0.1}Zn_{0.5}Co_{0.4}Fe_2O_4@SiO_2$ | 35 | 33.6 | 89.4 | 8.405 | 594 | 14.5 | 11.9 | 14% $Ni_{0.1}Zn_{0.5}Co_{0.4}Fe_2O_4$ / 4% $\alpha$-$Fe_2O_3$ / 82% $SiO_2$ | 0.08/0.49/0.39 |
| $Ni_{0.2}Zn_{0.4}Co_{0.4}Fe_2O_4@SiO_2$ | 36 | 34.1 | 89.8 | 8.375 | 587 | 14.5 | 11.8 | 15% $Ni_{0.2}Zn_{0.4}Co_{0.4}Fe_2O_4$ / 4% $\alpha$-$Fe_2O_3$ / 81% $SiO_2$ | 0.19/0.38/0.41 |
| $Ni_{0.3}Zn_{0.3}Co_{0.4}Fe_2O_4@SiO_2$ | 36 | 34.5 | 90.1 | 8.366 | 586 | 14.5 | 11.8 | 17% $Ni_{0.3}Zn_{0.3}Co_{0.4}Fe_2O_4$ / 2% $\alpha$-$Fe_2O_3$ / 81% $SiO_2$ | 0.28/0.27/0.42 |
| $Ni_{0.4}Zn_{0.2}Co_{0.4}Fe_2O_4@SiO_2$ | 37 | 35.0 | 90.4 | 8.354 | 583 | 14.4 | 11.8 | 13% $Ni_{0.4}Zn_{0.2}Co_{0.4}Fe_2O_4$ / 87% $SiO_2$ | 0.41/0.08/0.38 |
| $Ni_{0.5}Zn_{0.1}Co_{0.4}Fe_2O_4@SiO_2$ | 38 | 36.6 | 90.6 | 8.346 | 581 | 14.4 | 11.8 | 15% $Ni_{0.5}Zn_{0.1}Co_{0.4}Fe_2O_4$ / 85% $SiO_2$ | 0.49/0.08/0.38 |
| $Ni_{0.6}Co_{0.4}Fe_2O_4@SiO_2$ | 40 | 36.9 | 90.8 | 8.335 | 579 | 14.4 | 11.7 | 20% $Ni_{0.6}Co_{0.4}Fe_2O_4$ / 80% $SiO_2$ | 0.61/0.39/0 |

The change in the lattice constant (a) generates internal stress and suppresses additional grain growth during thermal treatment [44,50,51]. The tetrahedral (A) sites have smaller radii (0.52 Å) than the octahedral (B) site (0.81 Å) [9]. The ionic radii of $Ni^{2+}$ (0.69 Å), $Zn^{2+}$ (0.74 Å) and $Co^{2+}$ (0.75 Å) ions are larger than the ionic radius of $Fe^{3+}$ (0.64 Å) [3,38,52]. The amorphous to crystalline phase transformation and the relative content of crystalline phases, after thermal treatment at 1000 °C, were assessed using the relative degree of crystallinity (DC), calculated as the ratio between the area of diffraction peaks and the total area of diffraction peaks and halos. The DC increases with the increase of the crystallite size and Ni content. The Reference Intensity Ratio (RIR) method was used for the quantitative phase analysis of NCs thermally treated at 1000 °C.

*2.4. Elemental Analysis*

The Ni/Zn/Co molar ratios, determined by microwave digestion and combined with inductively coupled plasma optical emission spectrometry, are in good agreement with the theoretical values (Table 1).

*2.5. BET Analysis*

Due to the low amount of adsorbed/desorbed nitrogen, the determination of porosity and specific surface area (SSA) for the samples thermally treated at 1000 °C was not possible. The SSA below the method detection limit (0.5 m$^2$/g) suggests that all ferrites have a non-porous structure, probably due to particle agglomeration that limits nitrogen absorption.

*2.6. TEM Analysis*

The TEM images of the mixed Ni-Zn-Co ferrites following thermal treatment at 1000 °C (Figure 4) reveal spherical, small (high Zn content), or large (high Ni content) nanoparticles that form large spongy aggregates.

The formation of agglomerates with irregular morphology composed of high Zn content ferrite particles and the homogenous dispersion of high Ni content ferrite particles is also remarked. The small grains have a high surface area to volume ratio and allow faster oxygen diffusion than the larger grains, leading to an increase in the stoichiometry of the sample [35]. Although the small particles are closely arranged together, a clear boundary between adjacent particles is still observed. The average particle size is 34–36 nm, the difference being attributed to the grain boundary motion that exerts a dragging force, while the pores delay the force over the grain [41]. Moreover, the driving force increases the grain boundaries over the pores, resulting in lower pore volume and higher density [41]. The average crystallite size estimated by XRD is close to the particle size determined by

TEM, the slight differences being attributed to the amorphous $SiO_2$ matrix and large-size nanoparticles [43,44,50].

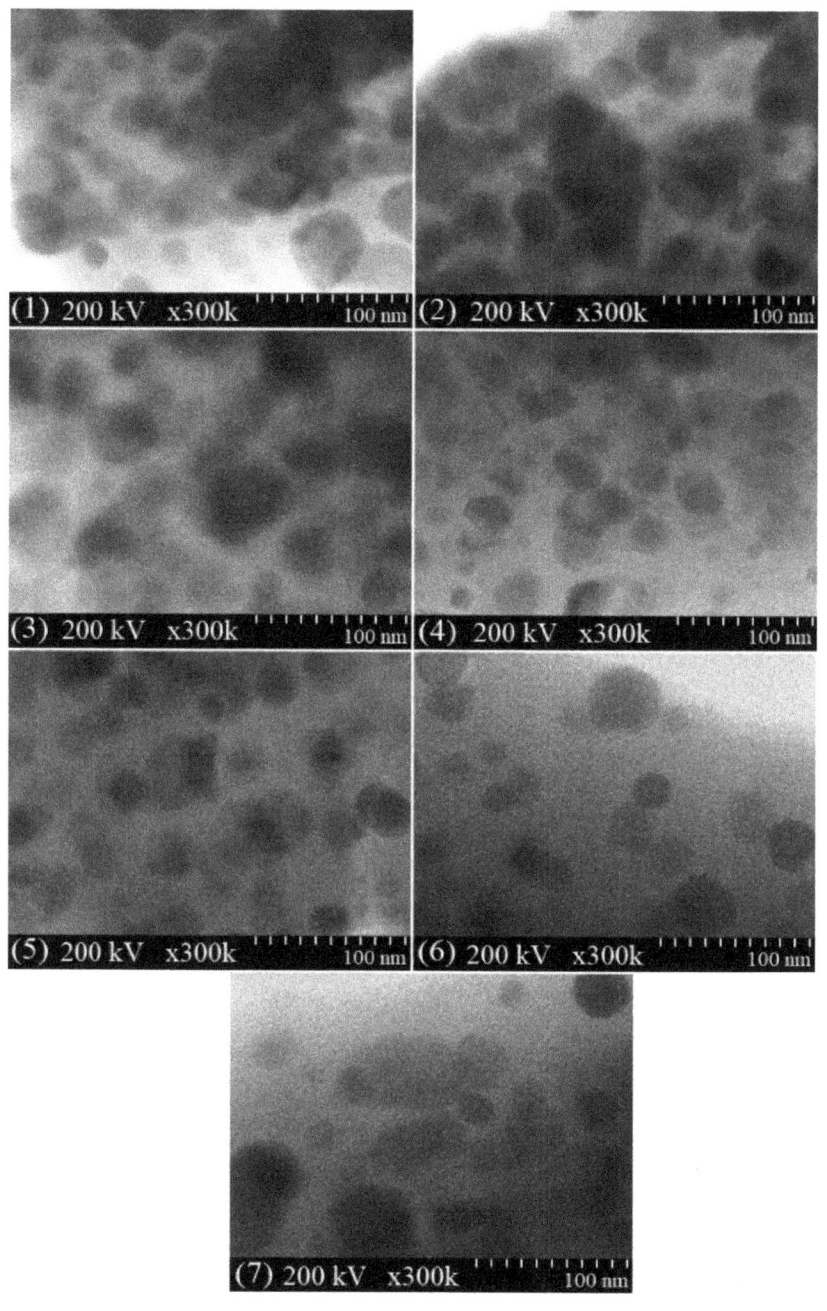

**Figure 4.** TEM images of $Zn_{0.6}Co_{0.4}Fe_2O_4@SiO_2$ (1), $Ni_{0.1}Zn_{0.5}Co_{0.4}Fe_2O_4@SiO_2$ (2), $Ni_{0.2}Zn_{0.4}Co_{0.4}Fe_2O_4@SiO_2$ (3), $Ni_{0.3}Zn_{0.3}Co_{0.4}Fe_2O_4@SiO_2$ (4), $Ni_{0.4}Zn_{0.2}Co_{0.4}Fe_2O_4@SiO_2$ (5), $Ni_{0.5}Zn_{0.1}Co_{0.4}Fe_2O_4@SiO_2$ (6) and $Ni_{0.6}Co_{0.4}Fe_2O_4@SiO_2$ (7) samples at 1000 °C.

## 2.7. UV-VIS Analysis

The optical response of the samples was evaluated by UV-Vis spectroscopy. The UV-Vis absorption (Figure 5a) shows that all the samples have a broad response in the visible range. Based on the absorption spectra and using the Tauc's relation [50], the band gap energy of the samples was evaluated (Figure 5b).

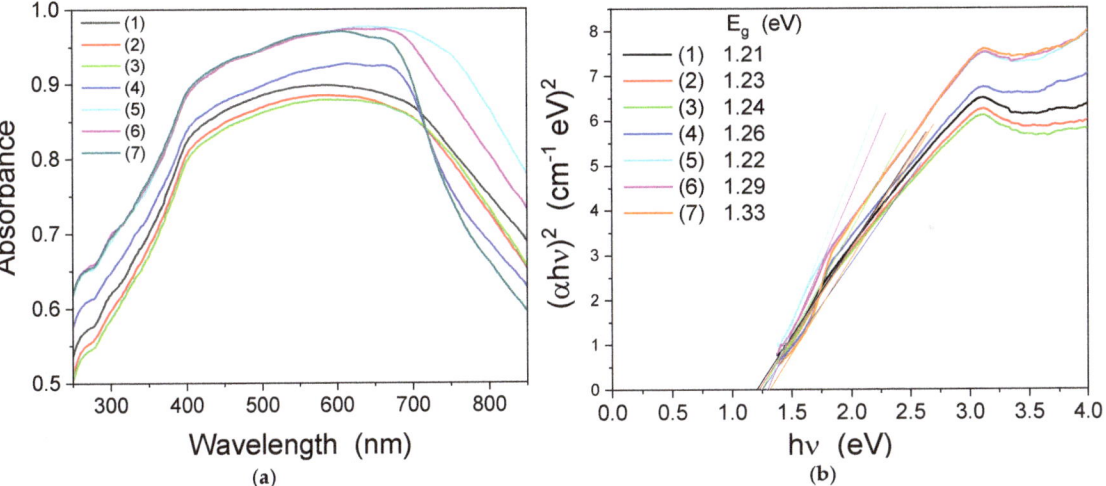

**Figure 5.** (a) UV-Vis absorption spectra and (b) Tauc's plot of the of $Zn_{0.6}Co_{0.4}Fe_2O_4@SiO_2$ (1), $Ni_{0.1}Zn_{0.5}Co_{0.4}Fe_2O_4@SiO_2$ (2), $Ni_{0.2}Zn_{0.4}Co_{0.4}Fe_2O_4@SiO_2$ (3), $Ni_{0.3}Zn_{0.3}Co_{0.4}Fe_2O_4@SiO_2$ (4), $Ni_{0.4}Zn_{0.2}Co_{0.4}Fe_2O_4@SiO_2$ (5), $Ni_{0.5}Zn_{0.1}Co_{0.4}Fe_2O_4@SiO_2$ (6) and $Ni_{0.6}Co_{0.4}Fe_2O_4@SiO_2$ (7) samples at 1000 °C.

The band gap energy values are in the range 1.21–1.49 eV, are lower than that of $NiFe_2O_4$ (2.2 eV), $ZnFe_2O_4$ (1.91 eV) and $CoFe_2O_4$ (2.31 eV), and are comparable to those reported for $CoFe_2O_4$ xerogel calcined at 500 °C (1.5 eV) [33,50]. The band gap of our samples was also lower than those of $Ni_xCo_{1-x}Fe_2O_4$ (1.37–1.78 eV) obtained by coprecipitation [52]. The optical band gap of the samples is due to the d-d transition. The crystal field splits the d level in the $e_g$ and $t_{2g}$ levels and the band gap energy depends on octahedral (B) and tetrahedral (A) sites. The band gap energy, in the case of the octahedral site, is higher than that of the tetrahedral (A) site [53]. The variation of the band gap energy, by replacing $Zn^{2+}$ ions with $Ni^{2+}$, can be explained by the redistribution of $Ni^{2+}$ ions between the octahedral (B) and tetrahedral (A) sites. In the XRD data, the peaks corresponding to the plane (220) and (422) are sensitive to the tetrahedral (A) site, whereas the peak corresponding to the (222) plane is sensitive to the octahedral (B) site [54,55]. The values of the I(220)/I(222) ratio, for the samples annealed at 1000 °C, are 4.29 for $Zn_{0.6}Co_{0.4}Fe_2O_4@SiO_2$, 3.79 for $Ni_{0.1}Zn_{0.5}Co_{0.4}Fe_2O_4@SiO_2$, 4.26 for $Ni_{0.2}Zn_{0.4}Co_{0.4}Fe_2O_4@SiO_2$, 4.10 for $Ni_{0.3}Zn_{0.3}Co_{0.4}Fe_2O_4@SiO_2$, 3.52 for $Ni_{0.4}Zn_{0.2}Co_{0.4}Fe_2O_4@SiO_2$, 3.22 for $Ni_{0.5}Zn_{0.1}Co_{0.4}Fe_2O_4@SiO_2$ and 3.77 for $Ni_{0.6}Co_{0.4}Fe_2O_4@SiO_2$, which indicates that the population at the tetrahedral (A) site tends to decrease with the increase of the $Ni^{2+}$ ions. These findings are correlated with the optical band gap values, which increases with the increase of $Ni^{2+}$ content. The low band gap energy makes our samples suitable for the absorption of visible light. The activation energy of the $Co_{0.6}Zn_{0.4}Ni_xFe_{2-x}O_4$ ferrite nanoparticles obtained by sol-gel route decreased with the increasing Ni content, from 2.71 eV (x = 0.2) to 1.46 eV (x = 1) [33].

## 2.8. Sonophotocatalytic Activity

The sonophotocatalytic activity of the samples was evaluated using an RhB synthetic solution under visible irradiation. Before visible irradiation, the samples were kept in the dark for 1 h to reach the adsorption-desorption equilibrium. The adsorption capacity of the sample varied between 7–28%. The adsorption properties depend on the surface sites and specific surface area. In our case, the samples had almost identical particle sizes; thus, those surface sites were responsible for the adsorption properties. The removal rate (Figure 6) was evaluated after 7 h of visible irradiation and varied between 16 and 75%. Similar removal efficiencies (83.9%) for methylene blue were obtained using Ni-Cu-Zn ferrite@SiO$_2$@TiO$_2$ by Chen et al. [52,55].

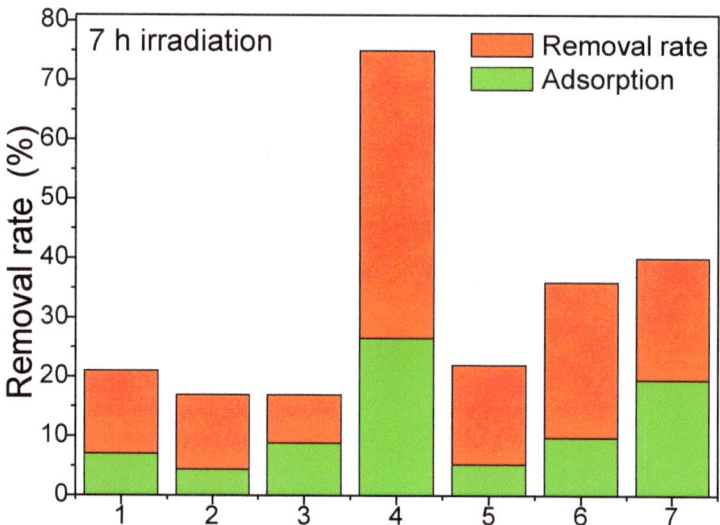

**Figure 6.** Removal rate of Zn$_{0.6}$Co$_{0.4}$Fe$_2$O$_4$@SiO$_2$ (1), Ni$_{0.1}$Zn$_{0.5}$Co$_{0.4}$Fe$_2$O$_4$@SiO$_2$ (2), Ni$_{0.2}$Zn$_{0.4}$Co$_{0.4}$Fe$_2$O$_4$@SiO$_2$ (3), Ni$_{0.3}$Zn$_{0.3}$Co$_{0.4}$Fe$_2$O$_4$@SiO$_2$ (4), Ni$_{0.4}$Zn$_{0.2}$Co$_{0.4}$Fe$_2$O$_4$@SiO$_2$ (5), Ni$_{0.5}$Zn$_{0.1}$Co$_{0.4}$Fe$_2$O$_4$@SiO$_2$ (6) and Ni$_{0.6}$Co$_{0.4}$Fe$_2$O$_4$@SiO$_2$ (7) samples.

The sample with similar Zn$^{2+}$ and Ni$^{2+}$ ions content (Ni$_{0.3}$Zn$_{0.3}$Co$_{0.4}$Fe$_2$O$_4$@SiO$_2$) shows the highest removal capacity, indicating that the equilibrium between Ni-ferrite and Zn-ferrite assures the best photocatalytic performance. In addition, based on the quantitative crystalline phase analysis, this sample contains a lower amount of α-Fe$_2$O$_3$ (2%) compared with samples Zn$_{0.6}$Co$_{0.4}$Fe$_2$O$_4$@SiO$_2$ (1), Ni$_{0.1}$Zn$_{0.5}$Co$_{0.4}$Fe$_2$O$_4$@SiO$_2$ (2), Ni$_{0.2}$Zn$_{0.4}$Co$_{0.4}$Fe$_2$O$_4$@SiO$_2$ (3), which means that in the case of this sample, α-Fe$_2$O$_3$ does not significantly influence photocatalytic activity.

For this sample, the photodegradation kinetic was analyzed with respect to the absorbance of RhB using the pseudo-first order kinetic model (Equation (6)).

$$-\ln\frac{A_t}{A_0^*} = k_i \cdot t \quad (6)$$

where $A_t$ is the absorbance of RhB at time t, $A_0^*$ is the absorbance of RhB at time $t_0$ and $k_i$ is the apparent kinetic constant (min$^{-1}$). A linear relationship with the irradiation time (Figure 7), with a rate constant of $2.79 \times 10^{-3}$ min was obtained.

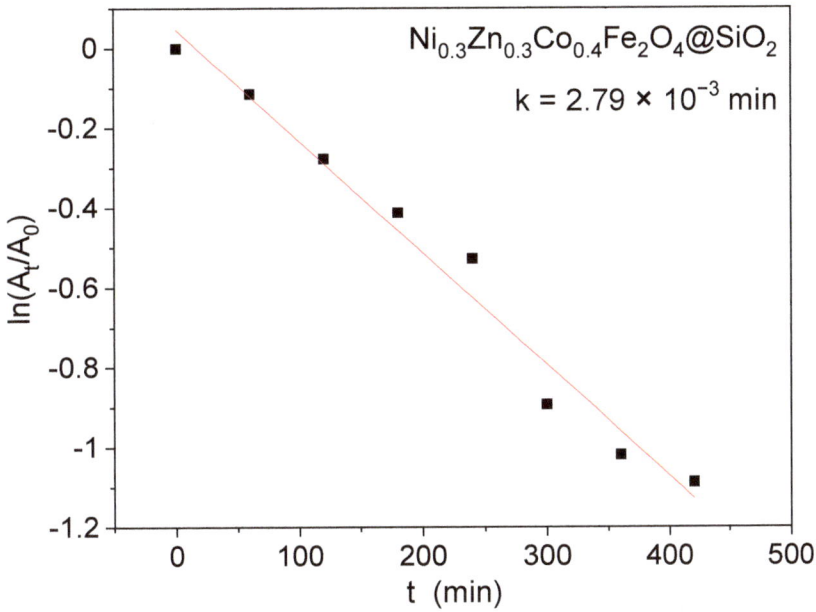

**Figure 7.** Photodegradation kinetics of RhB in the presence of $Ni_{0.3}Zn_{0.3}Co_{0.4}Fe_2O_4@SiO_2$.

## 3. Materials and Methods

### 3.1. Synthesis

The Ni-Zn-Co ferrite embedded in $SiO_2$ matrix (60% wt. ferrite, 40% wt. $SiO_2$) were synthesized by sol-gel method using $Ni(NO_3)_2·6H_2O$, $Zn(NO_3)_2·6H_2O$, $Co(NO_3)_2·6H_2O$, $Fe(NO_3)_3·9H_2O$, 1,4-butanediol (1,4-BD), tetraethyl orthosilicate (TEOS), ethanol and $HNO_3$ 65%, using a Ni:Zn:Co:Fe molar ratio of 0:6:4:20 ($Zn_{0.6}Co_{0.4}Fe_2O_4@SiO_2$), 1:5:4:20 ($Ni_{0.1}Zn_{0.5}Co_{0.4}Fe_2O_4@SiO_2$), 1:2:2:10 ($Ni_{0.2}Zn_{0.4}Co_{0.4}Fe_2O_4@SiO_2$), 3:3:4:20 ($Ni_{0.3}Zn_{0.3}Co_{0.4}Fe_2O_4@SiO_2$), 2:1:2:10 ($Ni_{0.4}Zn_{0.2}Co_{0.4}Fe_2O_4@SiO_2$), 5:1:4:20 ($Ni_{0.5}Zn_{0.1}Co_{0.4}Fe_2O_4@SiO_2$), 6:0:4:20 ($Ni_{0.6}Co_{0.4}Fe_2O_4@SiO_2$) and a nitrate:1,4-BD:TEOS molar ratio of 1:1:0.67. All chemicals were of analytical grade (Merck) and used without further purification. The resulting sols were kept at room temperature until gelation (5 weeks), ground, dried at 40 °C (5 h), and then subjected to thermal treatment 1000 °C.

### 3.2. Characterization

The formation and decomposition of the carboxylate-type precursor were investigated by thermogravimetry (TG) and differential thermal analysis (DTA) using the SDT Q600 thermogravimeter, in air, up to 1000 °C, at 10 °C·min$^{-1}$ heating rate, using alumina standards. The FT-IR spectra were recorded on KBr pellets containing 1% samples using a Perkin Elmer Spectrum BX II spectrometer, while the XRD patterns were recorded at room temperature using a Bruker D8 Advance diffractometer with $CuK_{\alpha 1}$ radiation ($\lambda$ = 1.54060 Å). The Ni/Zn/Co molar ratios were confirmed using Perkin Elmer ICP-OES Optima 5300 DV (Norwalk, CT, USA) after closed-vessel microwave-assisted aqua regia digestion using a Speedwave Xpert system (Berghof, Germany). The specific surface area (SSA) was obtained using the BET model from $N_2$ adsorption-desorption isotherms recorded at −196 °C by a Sorptomatic 1990 (Thermo Fisher Scientific) instrument. The UV–VIS absorption spectra were recorded using a JASCO V570 UV–VIS-NIR spectrophotometer, equipped with a JASCO ARN-475 absolute reflectivity measurement accessory.

*3.3. Sonophotocatalysis*

The sonophotocatalytic activity of the samples was evaluated against RhB solution under visible light irradiation in a Laboratory-Visible-Reactor system using a 400 W halogen lamp (Osram) and an ultrasonic bath. The catalyst (10 mg) was suspended in an aqueous solution of RhB ($1.0 \times 10^{-5}$ mol L$^{-1}$, 20 mL), and the mixture was stirred in the dark to achieve the adsorption equilibrium on the catalyst surface. Each degradation experiment was conducted for 240 min. Samples from a given mixture (3.5 mL) were withdrawn for analysis every 60 min. After separating the catalyst from the suspensions with a permanent magnet, the solution was analyzed using a UV–Vis spectrophotometer by recording the maximum absorbance of RhB at 554 nm. The sonophotocatalytic activity was estimated based on the calculated degradation rate. Before the sonophotodegradation experiments, the RhB adsorption on the surface of the nanoparticles was analyzed. The adsorption was verified in the dark by mixing the photocatalyst into the RhB solution for 60 min until the adsorption-desorption equilibrium was reached.

## 4. Conclusions

Ni-Zn-Co ferrites, with different Ni:Zn:Co ratios ($Zn_{0.6}Co_{0.4}Fe_2O_4$, $Ni_{0.1}Zn_{0.5}Co_{0.4}Fe_2O_4$, $Ni_{0.2}Zn_{0.4}Co_{0.4}Fe_2O$, $Ni_{0.3}Zn_{0.3}Co_{0.4}Fe_2O_4$, $Ni_{0.4}Zn_{0.2}Co_{0.4}Fe_2O_4$, $Ni_{0.5}Zn_{0.1}Co_{0.4}Fe_2O_4$, and $Zn_{0.6}Co_{0.4}Fe_2O_4@SiO_2$), embedded in $SiO_2$ were obtained by sol-gel method, followed by thermal treatment at 1000 °C. The thermal analysis revealed the formation and decomposition of metal succinate precursors in two stages, with distinct formation and decomposition of divalent ($Ni^{2+}$, $Zn^{2+}$, $Co^{2+}$) and trivalent ($Fe^{3+}$) succinates. The shapes of the DTA curves are similar, with the exception of the divalent metal's succinate decomposition stage, where for samples with high Ni content, the intensity of the exothermic peak decreases and shifts to higher temperatures. The total mass losses vary between 54.4–58.5%. The precursor formation and their decomposition into ferrites, as well as the formation of the silica matrix, are also confirmed by the FT-IR spectra. The XRD revealed the presence of well-crystallized ferrites along two crystalline phases of the $SiO_2$ matrix (cristobalite and tridymite). In samples with high Zn content, traces of hematite were also identified. The agglomeration of particles and the particle size of Ni-Zn-Co ferrites increase with the increasing Ni content, from 34 nm to 40 nm. All of the samples show a good optical response in the visible range, the best sonophotocatalytic performance being found for the $Ni_{0.3}Zn_{0.3}Co_{0.4}Fe_2O_4@SiO_2$ sample, most likely due to the equilibrium between Ni-ferrite and Zn-ferrite.

**Author Contributions:** Conceptualization, T.D.; methodology, T.D., O.C. and E.A.L.; formal analysis, T.D., O.C., F.G., D.T. and E.A.L.; investigation, T.D., O.C., F.G., D.T. and E.A.L.; resources, T.D., O.C., F.G., D.T. and E.A.L.; data curation, T.D.; writing—original draft preparation, T.D., O.C. and E.A.L.; writing—review and editing, T.D., O.C. and E.A.L.; visualization, T.D.; supervision, T.D. All authors have read and agreed to the published version of the manuscript.

**Funding:** The APC was funded by the Technical University of Cluj-Napoca Grant Support GC1/2021.

**Institutional Review Board Statement:** Not applicable.

**Informed Consent Statement:** Not applicable.

**Data Availability Statement:** Not applicable.

**Acknowledgments:** This study was supported by the Ministry of Research, Innovation and Digitization through Program 1—Development of the national research and development system, Subprogram 1.2—Institutional performance -Projects that finance the RDI excellence, Contracts no. 18PFE/30.12.2021 (O.C. and E.A.L.) and 37PFE/30.12.2021 (D.T.). The funder had no role in the design of the study; in the collection, analysis and interpretation of data; in the writing of the manuscript; and in the decision to submit the article for publication.

**Conflicts of Interest:** The authors declare no conflict of interest.

## References

1. Kefeni, K.K.; Mamba, B.B. Photocatalytic application of spinel ferrite nanoparticles and nanocomposites in wastewater treatment: Review. *Sustain. Mater. Technol.* **2020**, *23*, e00140. [CrossRef]
2. Sun, M.; Han, X.; Chen, S. Synthesis and photocatalytic activity of nano-cobalt ferrite catalyst for the photo-degradation various dyes under simulated sunlight irradiation. *Mater. Sci. Semicond. Process.* **2019**, *91*, 367–376. [CrossRef]
3. Rajabi, H.R.; Arjmand, H.; Kazemdehdashti, H.; Farsi, M. A comparison investigation on photocatalytic activity performance and adsorption efficiency for the removal of cationic dye: Quantum dots vs. magnetic nanoparticles. *J. Environ. Chem. Eng.* **2016**, *4*, 2830–2840. [CrossRef]
4. AlSalka, Y.; Granone, L.I.; Ramadan, W.; Hakki, A.; Dillert, R.; Bahnemann, D.W. Iron-based photocatalytic and photoelectrocatalytic nano-structures: Facts, perspectives, and expectations. *Appl. Catal. B Environ.* **2019**, *244*, 1065–1095. [CrossRef]
5. Abdurahman, M.H.; Abdullah, A.Z.; Shoparwe, N.F. A comprehensive review on sonocatalytic, photocatalytic, and sonophotocatalytic processes for the degradation of antibiotics in water: Synergistic mechanism and degradation pathway. *Chem. Eng. J.* **2021**, *413*, 127412. [CrossRef]
6. Paustian, D.; Franke, M.; Stelter, M.; Braeutigam, P. Sonophotocatalysis—Limits and possibilities for synergistic effects. *Catalysts* **2022**, *12*, 754. [CrossRef]
7. Joseph, C.G.; Puma, G.L.; Bono, A.; Krishnaiah, D. Sonophotocatalysis in advanced oxidation process: A short review. *Ultrason. Sonochem.* **2009**, *16*, 583–589. [CrossRef]
8. Anjum, M.; Miandad, R.; Waqas, M.; Gehany, F.; Barakat, M.A. Remediation of wastewater using various nanomaterials. *Arab. J. Chem.* **2019**, *12*, 4897–4919. [CrossRef]
9. Munoz, M.; De Pedro, Z.M.; Casas, J.A.; Rodriguez, J.J. Preparation of magnetite-based catalysts and their application in heterogeneous Fenton oxidation—A review. *Appl. Catal. B Environ.* **2015**, *176–177*, 249–265. [CrossRef]
10. Matsumoto, Y. Energy positions of oxide semiconductors and photocatalysis with iron complex oxides. *J. Solid State Chem.* **1996**, *126*, 227–234. [CrossRef]
11. Dillert, R.; Taffa, D.H.; Wark, M.; Bredow, T.; Bahnemann, D.W. Research Update: Photoelectrochemical water splitting and photocatalytic hydrogen production using ferrites ($MFe_2O_4$) under visible light irradiation. *APL Mater.* **2015**, *3*, 104001. [CrossRef]
12. Taffa, D.H.; Dillert, R.; Ulpe, A.C.; Bauerfeind, K.C.L.; Bredow, T.; Bahnemann, D.W.; Wark, D. Photoelectrochemical and theoretical investigations of spinel type ferrites ($M_xFe_{3-x}O_4$) for water splitting: A mini-review. *J. Photon Energy* **2017**, *7*, 012009. [CrossRef]
13. Chakrabarty, S.; Bandyopadhyay, S.; Pal, M.; Dutta, A. Sol-gel derived cobalt containing Ni–Zn ferrite nanoparticles: Dielectric relaxation and enhanced magnetic property study. *Mater. Chem. Phys.* **2021**, *259*, 124193. [CrossRef]
14. Sepelak, V.; Feldhoff, A.; Heitjans, P.; Krumeich, F.; Menzel, D.; Litterst, F.J.; Bergmann, I.; Becker, K.D. nonequilibrium cation distribution, canted spin arrangement, and enhanced magnetization in nanosized $MgFe_2O_4$ prepared by a one-step mechanochemical route. *Chem. Mater.* **2006**, *18*, 3057–3067. [CrossRef]
15. Dolcet, P.; Kirchberg, K.; Antonello, A.; Suchomski, C.; Marschall, R.; Diodati, S.; Muñoz-Espí, R.; Landfester, K.; Gross, S. Exploring wet chemistry approaches to $ZnFe_2O_4$ spinel ferrite nanoparticles with different inversion degrees: A comparative study. *Inorg. Chem. Front.* **2019**, *6*, 1527. [CrossRef]
16. Simon, C.; Zakaria, M.B.; Kurz, H.; Tetzlaff, D.; Blösser, A.; Weiss, M.; Timm, J.; Weber, B.; Apfel, U.P.; Marschall, R. Magnetic $NiFe_2O_4$ nanoparticles prepared via non-aqueous microwave-assisted synthesis for application in electrocatalytic water oxidation. *Chem. Eur. J.* **2021**, *27*, 16990–17001. [CrossRef] [PubMed]
17. Kirchberg, K.; Becker, A.; Bloesser, A.; Weller, T.; Timm, J.; Suchomski, C.; Marschall, R. Stabilization of monodisperse, phase-pure $mgfe_2o_4$ nanoparticles in aqueous and nonaqueous media and their photocatalytic behavior. *J. Phys. Chem. C* **2017**, *121*, 27126–27138. [CrossRef]
18. He, H.Y. Photocatalytic degradations of dyes on magnetically separable $Ni_{1-x}Co_xFe_2O_4$ nanoparticles synthesized by a hydrothermal process. *Part Sci. Technol.* **2016**, *34*, 143–151. [CrossRef]
19. Kefeni, K.K.; Mamba, B.B.; Msagati, T.A.M. Application of spinel ferrite nanoparticles in water and wastewater treatment: A review. *Sep. Purif. Technol.* **2017**, *188*, 399–422. [CrossRef]
20. Casbeer, E.; Sharma, V.K.; Li, X.Z. Synthesis and photocatalytic activity of ferrites under visible light: A review. *Sep. Purif. Technol.* **2012**, *87*, 1–14. [CrossRef]
21. Sundararajan, M.; Sailaja, V.; Kennedy, L.J.; Vijaya, J.J. Photocatalytic degradation of rhodamine B under visible light using nanostructured zinc doped cobalt ferrite: Kinetics and mechanism. *Ceram. Int.* **2017**, *43*, 540–548. [CrossRef]
22. Stergiou, C. Magnetic, dielectric and microwave adsorption properties of rare earth doped Ni-Co and Ni-Co-Zn spinel ferrite. *J. Magn. Magn. Mater.* **2017**, *426*, 629–635. [CrossRef]
23. Shetty, K.; Renuka, L.; Nagaswarupa, H.P.; Nagabhushana, H.; Anantharaju, K.S.; Rangappa, D.; Prashantha, S.C.; Ashwini, K. A comparative study on $CuFe_2O_4$, $ZnFe_2O_4$ and $NiFe_2O_4$: Morphology, impedance and photocatalytic studies. *Mater. Today Proc.* **2017**, *4*, 11806–11815. [CrossRef]
24. Zhu, H.; Jiang, R.; Fu, Y.; Li, R.; Yao, J. Novel multifunctional $NiFe_2O_4$/ZnO hybrids for dye removal by adsorption, photocatalysis and magnetic separation. *Appl. Surf. Sci.* **2016**, *369*, 1–10. [CrossRef]

25. Li, Y.; Zhang, Z.; Pei, L.; Li, X.; Fan, T.; Ji, J.; Shen, J.; Ye, M. Multifunctional photocatalytic performances of recyclable Pd-NiFe2O4/reduced graphene oxide nanocomposites via different co-catalyst strategy. *Appl. Catal. B Environ.* **2016**, *190*, 1–11. [CrossRef]
26. Mahmoodi, N.M. Zinc ferrite nanoparticle as a magnetic catalyst: Synthesis and dye degradation. *Mater. Res. Bull.* **2013**, *48*, 4255–4260. [CrossRef]
27. Revathi, J.; Abel, M.J.; Archana, V.; Sumithra, T.; Thiruneelakandan, R. Synthesis and characterization of $CoFe_2O_4$ and Ni-doped $CoFe_2O_4$ nanoparticles by chemical co-precipitation technique for photo-degradation of organic dyestuffs under direct sunlight. *Phys. B Condens. Matter.* **2020**, *587*, 412136. [CrossRef]
28. Harraz, F.A.; Mohamed, R.M.; Rashad, M.M.; Wang, Y.C.; Sigmund, W. Magnetic nanocomposite based on titania-silica/cobalt ferrite for photocatalytic degradation of methylene blue dye. *Ceram. Int.* **2014**, *40*, 375–384. [CrossRef]
29. Ahmadi Golsefidi, M.; Yazarlou, F.; Naeimi Nezamabad, M.; Naeimi Nezamabad, B.; Karimi, M. Effects of capping agent and surfactant on the morphology and size of $CoFe_2O_4$ nanostructures and photocatalyst properties. *J. Nanostruct.* **2016**, *6*, 121–126.
30. Abbasi, A.; Khojasteh, H.; Hamadanian, M.; Salavati-Niasari, M. Synthesis of $CoFe_2O_4$ nanoparticles and investigation of the temperature, surfactant, capping agent and time effects on the size and magnetic properties. *J. Mater. Sci. Mater. Electron.* **2016**, *27*, 4972–4980. [CrossRef]
31. Lassoued, A.; Li, J.F. Magnetic and photocatalytic properties of Ni–Co ferrites. *Solid State Sci.* **2020**, *104*, 106199. [CrossRef]
32. Chahar, D.; Taneja, S.; Bisht, S.; Kesarwani, S.; Thakur, P.; Thakur, A.; Sharma, P.B. Photocatalytic activity of cobalt substituted zinc ferrite for the degradation of methylene blue dye under visible light irradiation. *J. Alloys Compd.* **2021**, *851*, 156878. [CrossRef]
33. Bhukal, S.; Bansal, S.; Singhal, S. Structural, electrical and magnetic properties of Ni doped Co-Zn nanoferrites and their application in photo-catalytic degradation of methyl orange dye. *Solid State Phenom.* **2015**, *232*, 197–211. [CrossRef]
34. Vatsalya, V.L.S.; Sundari, G.S.; Sridhar, C.S.L.N.; Lakshmi, C.S. Evidence of superparamagnetism in nano phased copper doped nickel zinc ferrites synthesized by hydrothermal method. *Optik* **2021**, *247*, 167874. [CrossRef]
35. Raju, K.; Venkataiah, G.; Yoon, D.H. Effect of Zn substitution on the structural and magnetic properties of Ni–Co ferrites. *Ceram. Int.* **2014**, *40*, 9337–9344. [CrossRef]
36. Huili, H.; Grindi, B.; Viau, G.; Tahar, L.B. Influence of the stoichiometry and grain morphology on the magnetic properties of Co substituted Ni–Zn nanoferrites. *Ceram. Int.* **2016**, *42*, 17594–17604. [CrossRef]
37. Sherstyuk, D.P.; Starikov, A.Y.; Zhivulin, V.E.; Zherebstov, D.A.; Gudkova, S.A.; Perov, N.S.; Alekhina, Y.A.; Astapovich, K.A.; Vinnik, D.A.; Trukhanov, A.V. Effect of Co content on magnetic features and SPIN states in Ni–Zn spinel ferrites. *Ceram. Int.* **2021**, *47*, 12163–12169. [CrossRef]
38. Kumar, R.; Barman, P.B.; Singh, R.R. An innovative direct non-aqueous method for the development of Co doped Ni–Zn ferrite nanoparticles. *Mater. Today Commun.* **2021**, *27*, 102238. [CrossRef]
39. Kaur, M.; Jain, P.; Singh, M. Studies on structural and magnetic properties of ternary cobalt magnesium zinc (CMZ) $Co_{0.6-x}Mg_xZn_{0.4}Fe_2O_4$ (x = 0.0, 0.2, 0.4, 0.6) ferrite nanoparticles. *Mater. Chem. Phys.* **2015**, *162*, 332–339. [CrossRef]
40. Vinnik, D.A.; Sherstyuk, D.P.; Zhivulin, V.E.; Zhivulin, D.E.; Starikov, A.Y.; Gudkova, S.A.; Zherebtsov, D.A.; Pankratov, D.A.; Alekhina, Y.A.; Perov, N.S.; et al. Impact of the Zn–Co content on structural and magnetic characteristics of the Ni spinel ferrites. *Ceram. Int.* **2022**, *48*, 18124–18133. [CrossRef]
41. Mohapatra, P.P.; Singh, H.K.; Kiran, M.S.R.N.; Dobbidi, P. Co substituted Ni–Zn ferrites with tunable dielectric and magnetic response for high-frequency applications. *Ceram. Int.* **2022**, *48*, 29217–29228. [CrossRef]
42. Ramesh, S.; Sekhar, B.C.; Rao, P.S.V.S.; Rao, B.P. Microstructural and magnetic behavior of mixed Ni–Zn–Co and Ni–Zn–Mn ferrites. *Ceram. Int.* **2014**, *40*, 8729–8735. [CrossRef]
43. Stefanescu, M.; Stoia, M.; Dippong, T.; Stefanescu, O.; Barvinschi, P. Preparation of $Co_xFe_{3-x}O_4$ Oxydic System Starting from Metal Nitrates and Propanediol. *Acta Chim. Slov.* **2009**, *56*, 379–385.
44. Dippong, T.; Levei, E.A.; Lengauer, C.L.; Daniel, A.; Toloman, D.; Cadar, O. Investigation of thermal, structural, morphological and photocatalytic properties of $Cu_xCo_{1-x}Fe_2O_4$ ($0 \leq x \leq 1$) nanoparticles embedded in $SiO_2$ matrix. *Mater. Caract.* **2020**, *163*, 110268. [CrossRef]
45. Dippong, T.; Levei, E.A.; Cadar, O.; Goga, F.; Barbu-Tudoran, L.; Borodi, G. Size and shape-controlled synthesis and characterization of $CoFe_2O_4$ nanoparticles embedded in a PVA-$SiO_2$ hybrid matrix. *J. Anal. Appl. Pyrol* **2017**, *128*, 121–130. [CrossRef]
46. Amer, M.A.; Tawfik, A.; Mostafa, A.G.; El-Shora, A.F.; Zaki, S.M. Spectral studies of Co substituted Ni–Zn ferrites. *J. Magn. Magn. Mater.* **2011**, *323*, 1445–1452. [CrossRef]
47. Mallapur, M.M.; Chougule, B.K. Synthesis, characterization and magnetic properties of nanocrystalline Ni–Zn–Co ferrites. *Mater. Lett.* **2010**, *64*, 231–234. [CrossRef]
48. ASTM. *Joint Committee on Powder Diffraction Standards-International Center for Diffraction Data*; ASTM: Philadelphia, PA, USA, 1999.
49. Ghodake, J.S.; Shinde, T.J.; Patil, R.P.; Patil, S.B.; Patil, S.B.; Suryavanshi, S.S. Initial permeability of Zn–Ni–Co ferrite. *J. Magn. Magn. Mater.* **2015**, *378*, 436–439. [CrossRef]
50. Dippong, T.; Levei, E.A.; Cadar, O. Formation, structure and magnetic properties of $MFe_2O_4@SiO_2$ (M = Co, Mn, Zn, Ni, Cu) nanocomposites. *Materials* **2021**, *14*, 1139. [CrossRef]
51. Dalal, M.; Mallick, A.; Mahapatra, A.S.; Mitra, A.; Das, A.; Das, D.; Chakrabarti, P.K. Effect of cation distribution on the magnetic and hyperfine behavior of nanocrystalline Co doped Ni–Zn ferrite ($Ni_{0.4}Zn_{0.4}Co_{0.2}Fe_2O_4$). *Mater. Res.* **2016**, *76*, 389–401.

52. Chen, C.; Jaihindh, D.; Hu, S.; Fu, Y. Magnetic recyclable photocatalysts of Ni-Cu-Zn ferrite@SiO2@TiO2@Ag and their photocatalytic activities. *J. Photochem. Photobiol. A Chem.* **2017**, *334*, 74–85. [CrossRef]
53. Meena, S.; Anantharaju, K.S.; Malini, S.; Dey, A.; Renuka, L.; Prashantha, S.C.; Vidya, Y.S. Impact of temperature-induced oxygen vacancies in polyhedron $MnFe_2O_4$ nanoparticles: As excellent electrochemical sensor, supercapacitor and active photocatalyst. *Ceram. Int.* **2021**, *47*, 14723–14740. [CrossRef]
54. Sharma, D.; Khare, N. Tuning of optical bandgap and magnetization of $CoFe_2O_4$ thin films. *Appl. Phys. Lett.* **2014**, *105*, 032404. [CrossRef]
55. Toloman, D.; Stefan, M.; Pana, O.; Rostas, A.M.; Silipas, T.D.; Pogacean, F.; Leostean, C.; Barbu-Tudoran, L.; Popa, A. Transition metal ions as a tool for controlling the photocatalytic activity of MWCNT-$TiO_2$ nanocomposites. *J. Alloys Compd.* **2022**, *921*, 166095. [CrossRef]

Article

# Selective Recovery of Cadmium, Cobalt, and Nickel from Spent Ni–Cd Batteries Using Adogen® 464 and Mesoporous Silica Derivatives

Ahmed R. Weshahy [1], Ahmed K. Sakr [2,3,*], Ayman A. Gouda [1], Bahig M. Atia [3], H. H. Somaily [4,5], Mohamed Y. Hanfi [3,6], M. I. Sayyed [7], Ragaa El Sheikh [1], Enass M. El-Sheikh [3], Hend A. Radwan [3], Mohamed F. Cheira [3] and Mohamed A. Gado [3,*]

1 Department of Chemistry, Faculty of Science, Zagazig University, Zagazig 44519, Egypt
2 Department of Civil and Environmental Engineering, Wayne State University, 5050 Anthony Wayne Drive, Detroit, MI 48202, USA
3 Nuclear Materials Authority, Cairo P.O. Box 530, Egypt
4 Research Center for Advanced Materials Science (RCAMS), King Khalid University, P.O. Box 9004, Abha 61413, Saudi Arabia
5 Department of Physics, Faculty of Science, King Khalid University, P.O. Box 9004, Abha 61413, Saudi Arabia
6 Institute of Physics and Technology, Ural Federal University, St. Mira, 19, 620002 Yekaterinburg, Russia
7 Department of Physics, Faculty of Science, Isra University, Amman 11622, Jordan
* Correspondence: akhchemist@gmail.com (A.K.S.); mag.nma@yahoo.com (M.A.G.)

**Abstract:** Spent Ni–Cd batteries are now considered an important source for many valuable metals. The recovery of cadmium, cobalt, and nickel from spent Ni–Cd Batteries has been performed in this study. The optimum leaching process was achieved using 20% $H_2SO_4$, solid/liquid (S/L) 1/5 at 80 °C for 6 h. The leaching efficiency of Fe, Cd, and Co was nearly 100%, whereas the leaching efficiency of Ni was 95%. The recovery of the concerned elements was attained using successive different separation techniques. Cd(II) ions were extracted by a solvent, namely, Adogen® 464, and precipitated as CdS with 0.5% $Na_2S$ solution at pH of 1.25 and room temperature. The extraction process corresponded to pseudo-2nd-order. The prepared PTU-MS silica was applied for adsorption of Co(II) ions from aqueous solution, while the desorption process was performed using 0.3 M $H_2SO_4$. Cobalt was precipitated at pH 9.0 as $Co(OH)_2$ using $NH_4OH$. The kinetic and thermodynamic parameters were also investigated. Nickel was directly precipitated at pH 8.25 using a 10% NaOH solution at ambient temperature. FTIR, SEM, and EDX confirm the structure of the products.

**Keywords:** spent Ni–Cd batteries; separation; cadmium; nickel; cobalt

## 1. Introduction

Nowadays, cost-effective and low-cost techniques for collecting important metals from secondary sources have been developed to help meet economic and environmental constraints. In this regard, cadmium is prominent among non-ferrous metals. The principal application of cadmium has moved during the past 40 years from coatings to portable Ni–Cd batteries [1]. Rechargeable batteries (such as Ni–Cd batteries) have the highest quality, when utilized at low temperatures, which makes them a key user priority [2,3]. Thanks to their multiple attributes, i.e., long life spans, easy-of-rechargeing, no maintenance requirements, good shelf life, and reliability. The used Ni–Cd batteries wind up in large amounts in landfill, and cadmium waste pollutes the environment [4]. Spent Ni–Cd batteries are composed mostly of Ni–Cd electrode components corresponding to around 43–49% of battery weight. The remainder comprises the outer case steel parts/fundamental plates (40–49%) and feed- and Ni grids (9%), electrolyte, and 2% plastic constituents [5,6].

The recycling of spent Ni–Cd batteries has been thoroughly established and run commercially at scale [7–11]. Huge amounts of spent Ni–Cd batteries have been recycled internationally with pyrometallurgical procedures, including destructive cadmium distillation. The recycling process was established using demounting, crushing, metallurgic vacuum separation, and magnetic separation to produce environmentally friendly technology [12,13].

Acid leaching is a significant stage in hydrometallurgical improvements of the shredding of separated Ni–Cd battery powder. Cd and Ni efficiently dissolve during processes based on Ni–Cd wasted battery liquidation with hydrochloric acid [14–16]. The hydrochloric acid solution actually has a substantially higher leaching capacity than other acids; yet sulfuric acid has mostly been proposed as the total leaching and regeneration agent [17].

The separation and elimination of Cd from solutions involving various metal ions could be attained by adsorption, precipitation, ion exchange, electrolysis, solvent extraction, etc. [18,19]. With almost equivalent valence configurations, metal species allow co-transportation and selective extraction challenging. Selective extraction of Cd was possible in the presence of Ni, Mn, Zn, Fe, Mg, and Ca. However, Pb and Cu were co-extracted with Cd from processes established for HCl liquidation of Ni–Cd spent batteries [20]. Cd in the same mixture was observed to impair other metals' physiological balance [21].

Solvent extraction is a very adaptable technique for the sometimes difficult extraction, purification, recovery, and separation of the aqueous medium incorporating metal ions [22]. It is a quick and straightforward, economical method [23]. Liquid-liquid extraction is employed for metal separation using immiscible fluids (usually one organic and another, an aqueous phase containing metal cations) in contact with each other. Most investigations employed the distribution ratios and metal separation factors with specified extractants for a particular metal [24,25]. Ionic liquids are task-specific extractants. They have useful and adaptable physicochemical properties, such as heat stability, high polarity, negligible steam pressure, inactivity, and a wide range of miscellaneous effects on other organic solvents, which have been highlighted in various scientific publications [26,27]. Adogen® 464 has been widely investigated in the extraction of metal chloride complexes. It is suitable for cadmium separation from cobalt, copper, zinc, and manganese [28,29].

The present work aims to recycle and separate different valuable constituents of spent Ni–Cd batteries in a pure state using successive different separation techniques (solvent extraction for cadmium, adsorption for cobalt, and precipitation for nickel). Significant physical chemistry factors were investigated to develop the solvent extraction and adsorption approach, solvent, adsorbent concentration, pH, temperature, and extraction time.

## 2. Results and Discussion

### 2.1. Precipitation of Iron

The solution was treated with 1:1 of 32% hydrogen peroxide under agitation for 65 min at 60 °C. A green solution was produced at pH 3.0. The solution pH was increased to pH 4.0 using 10 M NaOH, then raised to pH 5.0 using 32% ammonium solution. After warming for 11 min at 73 °C, iron(III) hydroxide was the obtained product and was identified using EDX analysis [30].

### 2.2. Factors Controlling Adsorption of Cadmium

#### 2.2.1. Effect of pH

Using 25 mL battery liquor, 0.1 M $Na_2SO_4$ was added into deionized water as inert salt to minimize phase separation problems linked to low ionic strength. The samples were equalized at different pH values with 25 mL 0.02 M Adogen 464/kerosene while maintaining the same phase rate because the pH of the solution substantially affects extraction of the metal ions.

The pH was in the range 2.0–5.5. The pH recorded after extraction is 5.0 which was greater than the original pH value. Subsequently, it declined with a rise in the original pH for the ionic fluids. As the pH of the solution increases, $Cd^{2+}$ extraction increases; however, at pH 5, the percentage of cadmium extraction was insufficient, with just 19% extraction (Figure 1a). Increased pH with higher levels of Adogen 464 resulted in a higher percentage of cadmium extraction [31]. This may be because acid competes with $Cd^{2+}$ ions or bisulfate ions at low pH values, hindering $Cd^{2+}$ extraction. As the pH increased, extractable $Cd^{2+}$ complexes were generated, which improved the extraction ratio.

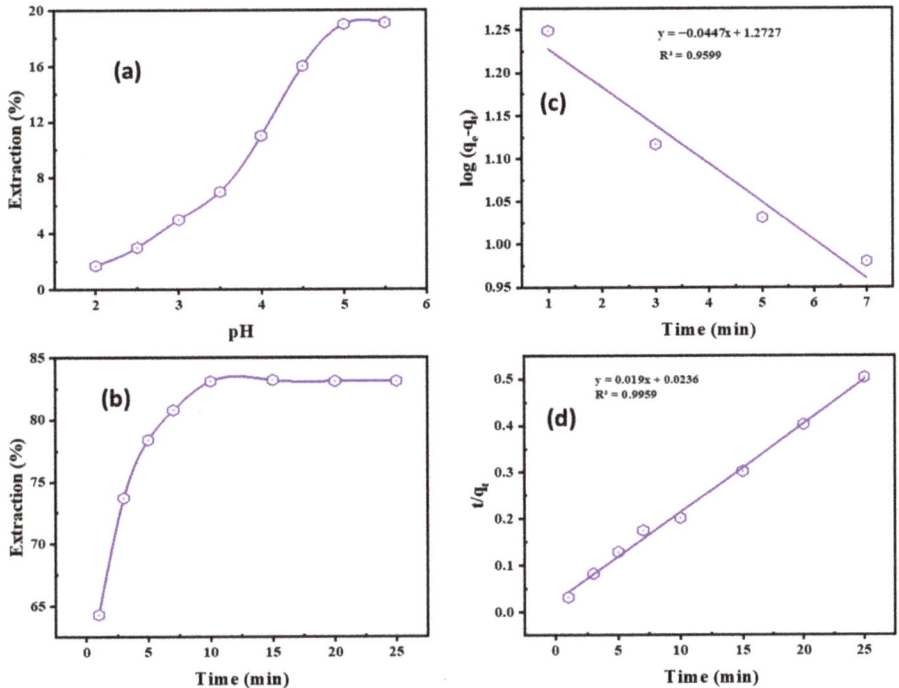

**Figure 1.** (**a**) Impacts of pH (0.1 M $Na_2SO_4$, 0.02 M Adogen 464, 5 min, O:A 1:1, 25 °C); (**b**) equilibration time on $Cd^{2+}$ extraction (0.1 M $Na_2SO_4$, pH 5.0, 0.02 M Adogen 464, O:A 1:1, 25 °C); (**c**) The pseudo-1st-order; (**d**) pseudo-2nd-order modeling of $Cd^{2+}$ extraction by Adogen 464.

2.2.2. Effect of Time

The kinetics of $Cd^{2+}$ removal was tested with the ionic liquid Adogen 464 in kerosene and with a 25 mL solution of leaching liquor and 0.1 M $Na_2SO_4$ in pH 5.0; the influence of balanced cadmium (II) removal time was investigated. As demonstrated in Figure 1b, the extraction efficiency improved gradually with the growth in contacting time until it reached its maximum efficiency at 10 min of contact time, which is sufficient to achieve equilibrium, after which the extraction efficiency became constant with any further increase in the contact time.

The rate of cadmium ion extraction using Adogen 464 is described by its kinetics. The kinetic parameters assist in assessing sorption rates and offer necessary details for extraction approach setup and modelling. The $Cd^{2+}$ extraction process by Adogen 464 and the rate constants were determined using pseudo-1st and 2nd-order modelling. The following mathematical Equation (1) describes the pseudo-1st-order modelling [32–34]:

$$\log(q_e - q_t) = \log q_e - \left(\frac{K_1 t}{2.303}\right) \qquad (1)$$

The following formula describes the quantity of metal generated per unit weight ($q_e$) and the quantity extracted per unit time ($q_t$) increase: ($t$, $min^{-1}$). As Figure 1c illustrates, the extraction rate of the 1st order follows a straight line when plotting log ($q_e - q_t$) vs.

$t$ values. To discover the practical data in Table 1, one can apply the pseudo-1st-order modelling, which matches the results shown in Table 1. A good correlation was established between the calculated value of $q_e$ at a modest extraction rate of ($K_1 = 0.1029$ mg/g), which is around 18.74 mg/g. This is very different from the practical determining capability of 49.712 mg/g.

**Table 1.** The parameters of $Cd^{2+}$ extraction kinetic by Adogen 464.

| Pseudo-1st Order | | | Exp. Capacity $q_e$, mg/g | Pseudo-2nd Order | | |
|---|---|---|---|---|---|---|
| $q_e$ | $K_1$ | $R^2$ | | $q_e$ | $K_2$ | $R^2$ |
| 18.737 | 0.1029 | 0.9599 | 49.712 | 52.632 | 0.0153 | 0.9959 |

It is now apparent that the 1st-order modelling and the empirical data are not alike, and therefore the modelling is not suitable for the system under investigation. Conversely, the pseudo-2nd-order modelling is symbolized by the subsequent Equation (2) [35–38]:

$$\frac{t}{q_t} = \frac{1}{K_2 q_e^2} + \frac{t}{q_e} \qquad (2)$$

The rate constant is given as $K_2$ (g/mg·min). The slope of the linear plotting is $1/q_e$, and the intercept is $1/q_e^2$. From the model in Figure 1d, and the data in Table 2, it is obvious that both the theoretical and experimental uptakes are close, and the correlation coefficient $R^2$ is 0.9959 is higher than the 1st order one. The obtained data establish that the 2nd-order modelling is adapted to the extraction system.

**Table 2.** Thermodynamic parameters acquired from $Cd^{2+}$ extraction temperature investigation.

| $\Delta H°$ (kJ·mol$^{-1}$) | $\Delta S°$ (J·mol$^{-1}$·K$^{-1}$) | $\Delta G°$ (kJ·mol$^{-1}$) | | | | | |
|---|---|---|---|---|---|---|---|
| | | 303 K | 313 K | 323 K | 333 K | 343 K | 353 K |
| 9.981 | 0.0505 | −15.302 | −15.808 | −16.313 | −16.818 | −17.323 | −17.829 |

### 2.2.3. Effect of Adogen 464 Concentration

A total of 25 mL battery leach liquor, and extractant concentrations between 0.001 and 0.1 M, were used in the experiments to examine ionic liquid concentration impact on $Cd^{2+}$ extraction. The results showed that ionic liquid concentration had no marked effect on $Cd^{2+}$ extraction. Figure 2a illustrates that the percent extraction of cadmium rose as the extractant concentration was increased. 0.2 M Adogen 464 was used to extract 99.9% of the cadmium ions from the solution. At pH 5.0, and with other parameters constant, contact time at room temperature was 10 min.

### 2.2.4. Effect of Sulfate Concentration

To explore the influence of $Na_2SO_4$ at concentrations of 0.1–1.0 M upon $Cd^{2+}$ recovery, experiments were conducted at room temperature using 0.2 M Adogen 464. The experiment presented in Figure 2b showed that the percentage of cadmium extraction in the aqueous feed decreased with an enlargement in the concentration of sulfate ions because of the salting-out effect for Adogen 464/kerosene.

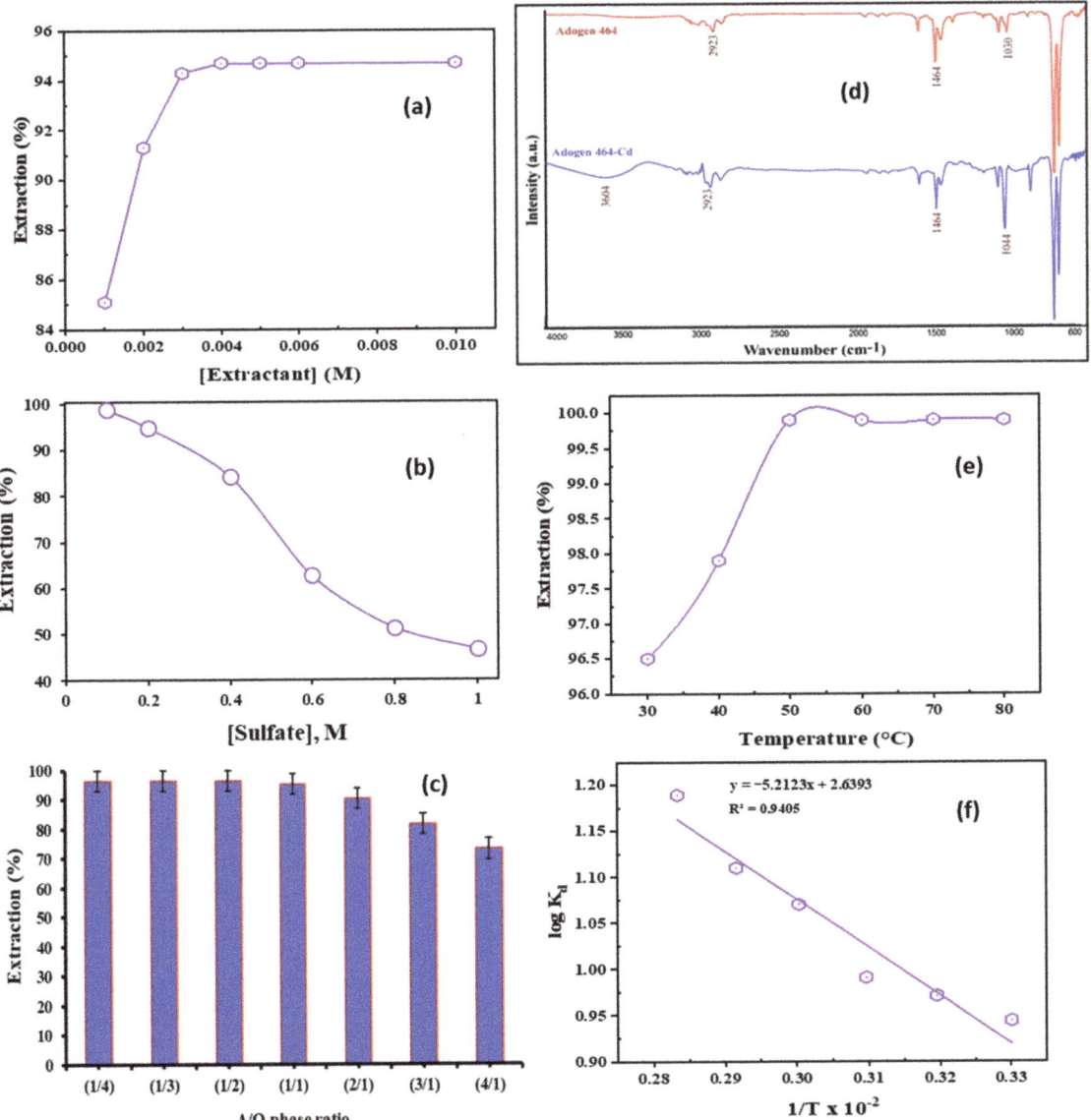

**Figure 2.** (**a**) Impact of Adogen 464 concentration upon $Cd^{2+}$ extraction (0.1 M $Na_2SO_4$, pH 5.0, O:A 1:1, 25 °C, 10 min); (**b**) Impact of sulfate concentration upon $Cd^{2+}$ extraction (0.2 M Adogen 464, pH 5.0, O:A 1:1, 10 min, 25 °C); (**c**) Impact of aqueous to organic ratio upon $Cd^{2+}$ extraction (0.2 M Adogen 464, pH 5.0, 0.1 M $Na_2SO_4$, 10 min, 25 °C); (**d**) FTIR spectra of 0.2 M Adogen 464/toluene and cadmium/Adogen 464/toluene); (**e**) Impact of temp. upon $Cd^{2+}$ extraction (0.2 M Adogen 464, pH 5.0, 0.1 M $Na_2SO_4$, O:A 1:1, 10 min); (**f**) Influence of temp. on $Cd^{2+}$ partition coefficient (0.2 M Adogen 464, pH 5.0, 0.1 M $Na_2SO_4$, O:A 1:1, 10 min).

Nayl (2010) described the relationship between quaternary ammonium extractants, which were employed to extract metal ions, as only occurring with a solvation reaction mechanism. The extraction of the $Cd^{2+}$ mechanism from $SO_4^{2-}$ media by $R_3R'N.SCN$

obeyed the addition mechanism, and the removed dominant species were (R$_3$R′N.CdSO$_4$.SCN) [39].

$$3Cd^{2+}_{(aq)} + 3SO_4^{2-}_{(aq)} + 2H^+ + 3(R_3R'N.SCN)_{(org)} \leftrightarrow 3(R_3R'N.CdSO_4.SCN)_{(org)} + 2H^+$$

where R = C$_{10}$H$_{21}$, C$_9$H$_{19}$, C$_8$H$_{17}$ and R′ = CH$_3$ groups. The extraction ($K_{ex}$) constant for Adogen 464 is gained from Equation (3).

$$K_{ex} = \frac{[(R_3R'N.CdSO_4.SCN)]^3_{(org)}}{[Cd^{2+}]^3_{(aq)}[SO_4^{2-}]^3_{(aq)}[(R_3R'N.SCN)]^3_{(org)}} \quad (3)$$

2.2.5. Effect of Organic to Aqueous Ratio

The ratio between aqueous/organic phases was studied with various ratios or relations of aqueous leach liquid with the fitting organic solvent, generating the balance curve. The influence of varying the aqueous to organic phase ratio (A:O) upon Cd$^{2+}$ extraction was examined using different aqueous/organic phase ratios from 4:1 to 1:4 at 25 °C, 10 min of shaking time, pH 5.0, and 0.2 M Adogen 464/kerosene. According to Figure 2c, the A:O ratio slightly influences the extraction efficiency in the case of increasing the organic volume. In the other case, however, the extraction efficiency decreases gradually with increasing the aqueous phase. Hence, the O:A ratio of 1:1 was selected as the optimal ratio, with 96.4% extraction.

Fourier infrared spectroscopy technique (FTIR) has been employed to detect many characteristic vibrational bands [40]; it can detect the functional groups of Adogen 464 before and after the extraction of cadmium. A study was conducted using Adogen 464, diluted in kerosene and presented in Figure 2d. The peak at 1030 cm$^{-1}$ for Adogen 464 was assigned for –C–N stretching vibration that was transformed to 1044 cm$^{-1}$ for Cd-Adogen 464 complex, after the bonding among Cd$^{2+}$ and Adogen 464 happened. The wide absorption band at 3604 cm$^{-1}$ for Cd-Adogen 464 was assigned for the stretching vibration of –OH of soluble moisture [41]. The distinctive peak of quaternary amine at 1464 cm$^{-1}$ appeared in both spectra and was induced via (CH$_3$)N$^+$. The asymmetric –CH peaks (2923 cm$^{-1}$) for a combination of Adogen 464 prior to and after extraction investigations were equivalent.

2.2.6. Effect of Temperature

The extraction rate may rise or decrease depending on the temperature of the environment. Temperatures ranging from 30 °C to 80 °C were used to scrutinize the temperature influence upon Cd$^{2+}$ extraction via ionic liquid Adogen 464. The aqueous solution used in this experiment was the battery leach liquor with 0.1 M Na$_2$SO$_4$, pH 5.0, and 0.2 M Adogen 464 in kerosene, as shown in Figure 2e. Altogether, the outcomes demonstrated that temperature had a favourable influence on Cd$^{2+}$ extraction. The extraction of Cd$^{2+}$ ions rose as the temperature increased; it reached its maximum E% (99.9%) at 50 °C. From the successive van't Hoff Equation (4) [42–45].

$$\log K_d = \frac{-\Delta H°}{2.303RT} + \frac{\Delta S°}{2.303R} \quad (4)$$

The enthalpy ($\Delta H°$) and entropy ($\Delta S°$) were determined via the plot of log $K_d$ vs. 1/T (Figure 2f). From the relation of $\Delta G° = \Delta H° - T\Delta S°$, the values of $\Delta G°$ were computed and are listed in Table 2.

The positive significance value of the $\Delta H°$ in Table 2 shows that the extraction of cadmium ions through Adogen 464 is an endothermic mechanism, which is confirmed by the temperature impact that affects the E% of Cd$^{2+}$ ions. Negative $\Delta G°$ values show that the suggested extraction contrivance is spontaneous and feasible. The Arrhenius relation is important as the straight-line slope in Figure 2f is applicable for assessing the apparent activation energy ($E_a$) of cadmium ions extraction at different temperatures [46]. Values are calculated from the following Equation (5) [47].

$$\log K_d = \frac{-2.303 E_a}{RT} + \log A \quad (5)$$

The partition coefficient, $K_d$, and $E_a$ (kJ·mol$^{-1}$) denote the extraction activation energy, $R$ (8.314 J·mol$^{-1}$·K$^{-1}$ is the molar gas constant), and $T$ is the absolute temperature in Kelvin. Cadmium ions extraction required 1.882 kJ·mol$^{-1}$ of activation energy, and the Arrhenius constant is 436.4. This signifies that Cd$^{2+}$ ions extraction by Adogen 464 is an endothermic approach, which means that the reaction requires energy to be completed.

### 2.3. Cadmium Stripping from Cadmium/Adogen 464 and Precipitation

The removal of Cd$^{2+}$ ions from the loaded Adogen 464 in toluene was accomplished using different concentrations of H$_2$SO$_4$ and EDTA. Figure 3a demonstrates that raising the acid concentration from 0.1–2.0 M causes an increase in Cd$^{2+}$ stripping from 25% to 83.5%. The use of 2.0 M H$_2$SO$_4$ and/or 0.1 M EDTA in two successive stages at an O:A ratio of 1:2 resulted in a significant increase in Cd$^{2+}$ stripping from 23.4% to 69.5% with an increase in EDTA concentration from 0.01 to 0.1 M, resulting in the complete removal of the entire cadmium content of the solvent, as shown in Figure 3b. As a result, 2.0 M H$_2$SO$_4$ was selected as the most effective stripping agent for this work.

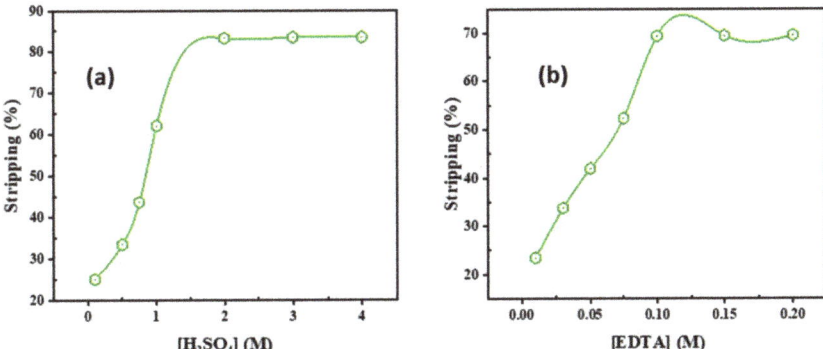

**Figure 3.** Influence of (**a**) H$_2$SO$_4$; (**b**) EDTA concentration upon Cd$^{2+}$ stripping from the loaded Adogen 464.

The sulfate solution containing Cd$^{2+}$ produced from the stripping process was subjected to precipitation using 0.5% Na$_2$S solution. Adjusting the pH to 1.25 at room temperature, a yellow precipitate of CdS was formed; it was separated via filtration and dried in the oven at 70 °C. The obtained product was about 2.0 g. The product has also been subjected to SEM-EDX analysis, as shown in Figure 4, which reveals a maximum purity.

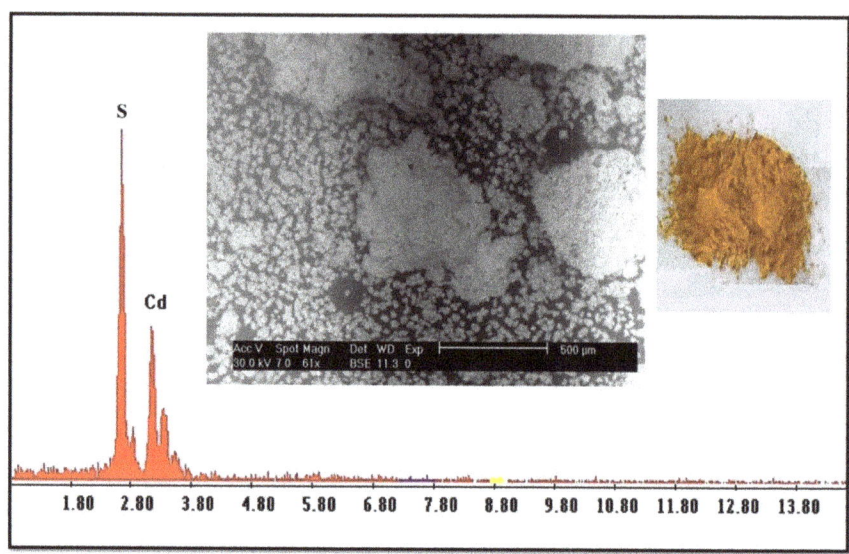

**Figure 4.** SEM-EDX chart for CdS product.

### 2.4. Adsorption of Cobalt Using Prepared Silica Adsorbant

The leachate is now free of cadmium and iron and contains nickel and cobalt. The remaining solution was subjected to the selective separation of $Co^{2+}$ from $Ni^{2+}$ utilizing a new adsorbent, PTU-MSA.

#### 2.4.1. Characterization of PTU-MSA

The morphology of PTU-MS was gained via an electron scanning microscope (SEM-EDX). The matrix mainly consisted of a hexagonal formation and showed a rather homogeneous particle size of about 0.85 mm PTU-MSA adsorbent (mean size) [48]. It was different from PTU-MS (Figure 5a,b). Moreover, the PTU-MS was analyzed using FTIR. The data showed major typical peaks in the range of 2900–3000 cm$^{-1}$ due to aliphatic stretching of C–H. The feature at 1655 cm$^{-1}$ is C=O for urea, that at 3410 cm$^{-1}$ is due to phenolic –OH, and peaks at 1454 cm$^{-1}$ and 1568 cm$^{-1}$ correspond to N–H and N–C, respectively. Three bands are attributed to silica at 1075, 794, and 456 cm$^{-1}$ of Si–O–Si. Moreover, a peak at 949 cm$^{-1}$ is assigned to Si–OH [49–54].

The spectrum of sulphonic acid-modified PTU-MSA shows new bands at 1151 cm$^{-1}$ and a specific peak is recorded at around 584 cm$^{-1}$, resulting from the $SO_3$ group. The peak at 650 cm$^{-1}$ is dispersed due to the stretching vibration of the S–N. The S–N band was predicted to overlap with the Si–O stretching band of the identical area of energy (Figure 5c). The spectrum of FT-IR confirmed that the surface of prepared PTU-MS and PTU-MSA material was successfully functionalized and synthesized [55,56].

The nitrogen adsorption/desorption isotherm curves (BET) of the two samples (PTU-MS and PTU-MSA) were obtained through steep condensing/evaporation capillary stages [57–59]. Recognizable H1 hysteresis loops were shown for the adsorption/desorption of nitrogen in the two samples (PTU-MS and PTU-MSA), which are typical for mesoporous materials with cylindrical mesoporous following the IUPAC classification. The PTU-MSA's specific surface area is decreased due to the sulphonic acid group's incorporation into the PTU-MS framework. The results in Table 3 along with Figure 5d, showed a reduction in the pore diameter and pore volume indicating successful incorporation of the group.

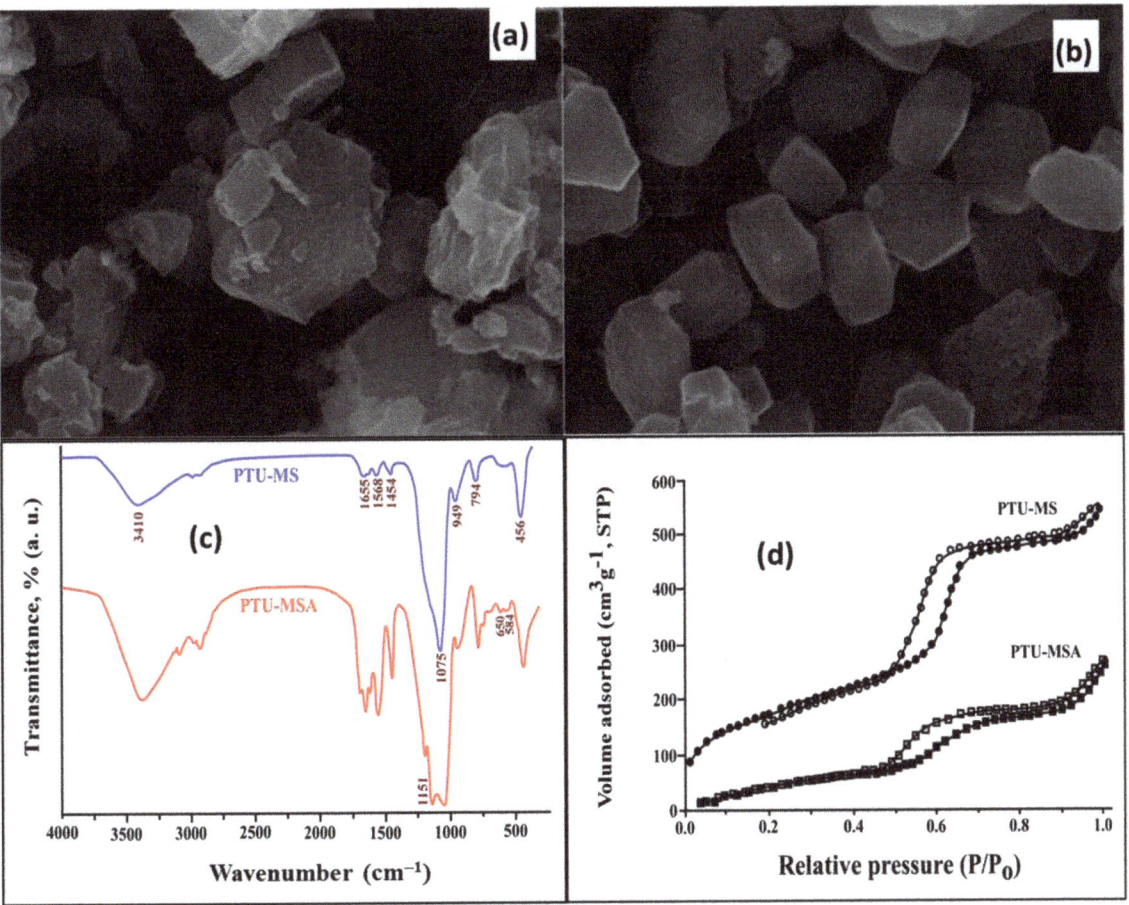

**Figure 5.** (**a**) SEM photograph after sulfonation of PTU-MS; (**b**) SEM photograph after sulfonation of PTU-MSA; (**c**) FT-IR spectra of PTU-MS and PTU-MSA materials; (**d**) N$_2$ adsorption-desorption isotherm curves of PTU-MS and PTU-MSA.

**Table 3.** Physico-chemical properties of PTU-MS material sorption, for the examination of N$_2$ prior and after incorporation of a sulphonic group.

| Materials | Specific Surface Area (m$^2$·g$^{-1}$) | Pore Volume (cm$^3$·g$^{-1}$) | Pore Diameter (Å) |
|---|---|---|---|
| PTU-MS | 634 | 0.71 | 68 |
| PTU-MSA | 357 | 0.42 | 55 |

2.4.2. Factors Controlling the Adsorption Process of Cobalt

The adsorption of cobalt was initially tested using PTU-MSA from a synthetic solution. It was then applied on the leach liquor free of cadmium and iron, containing only cobalt and nickel.

1. Impact of pH

The pH has a vital role in reducing or increasing the adsorption selectivity of efficiently hydrolysable metal ions. As already established, most metal ions are hydrated in water [60]. Figure 6a indicates the pH dependency of Co$^{2+}$ ions on PTU-MSA. The Co$^{2+}$ ions form insoluble aqueous complexes with increased pH, when $Co^{2+}$ ions undergo hydrolysis reactions in water.

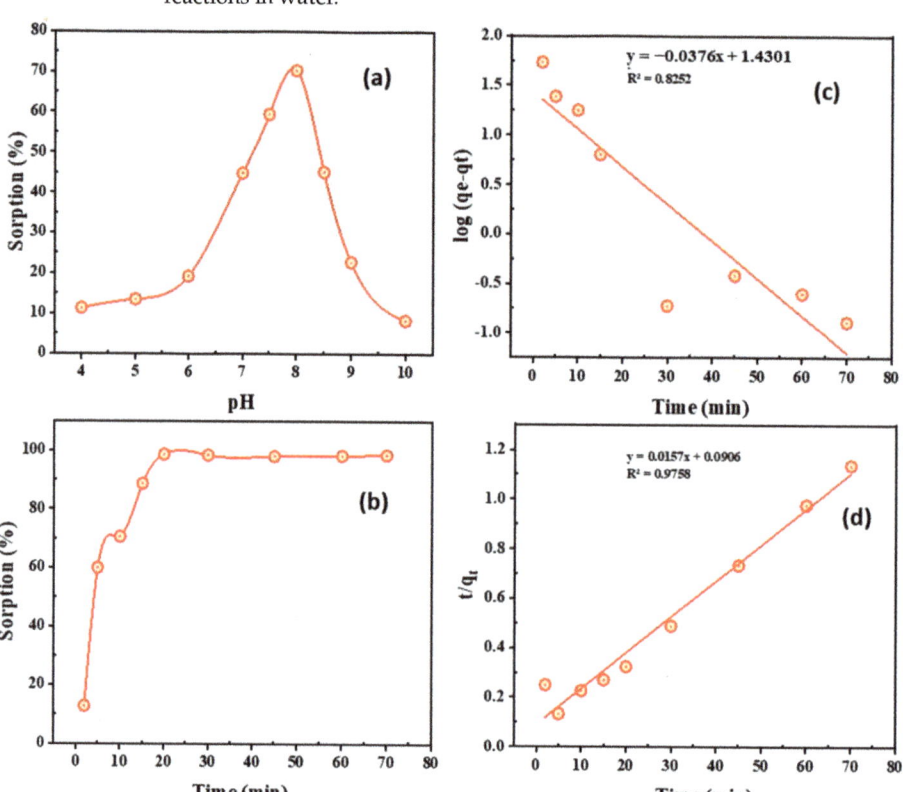

**Figure 6.** (a) impact of pH upon $Co^{2+}$ adsorption (20 mL of 250 mg/L $Co^{2+}$, 10 min, 0.1 g of PTU-MSA, 25 °C); (b) Impact of contacting time upon $Co^{2+}$ adsorption (20 mL of 250 mg/L $Co^{2+}$, pH 8.0, 0.08 g of PTU-MSA, 25 °C); (c) 1st-order kinetic plot; (d) 2nd-order kinetic plot of PTU-MSA's $Co^{2+}$ adsorption (20 mL of 250 mg/L $Co^{2+}$, pH 8.0, 0.08 g of PTU-MSA, 25 °C).

The solution pH influences the speed of surface reactions. The adsorption capacity changes with pH scope mostly due to the impact of pH upon the adsorption aspects of the PTU-MSA surface. Over the pH range 4.0–10.0, $Co^{2+}$ adsorption efficiency increased with higher pH, as demonstrated in Figure 6a. The adsorption conditions were kept constant at $Co^{2+}$ concentration of 250 mg/L, 10 min, and 0.1 g of PTU-MSA at room temperature, while the pH was varied.

The results shown in Figure 6a indicate that the $Co^{2+}$ adsorption on the surface of PTU-MSA steadily rose with the solution pH rising to the highest or maximal at pH 8. The $Co^{2+}$ species exclusively exist in a divalent ionic state at this pH, enhancing the removal of the adsorbent from the solution. The pH effect can be explained by the presence in working solutions with pH values just below 8.0 of various ionic shapes such as $Co^{2+}$, $Co(OH)^+$, $Co(OH)_2$, and $Co(OH)_3^-$, which reduce the overall effectiveness of the $Co^{2+}$ removal [61,62]. PTU-MSA has the maximum adsorption efficiency at pH 8.0. At rising pH levels (pH > 8.0), the $Co^{2+}$ was precipitated as cobalt hydroxide. This outcome displays the pH effect on the adsorbent.

2. Impact of time and kinetics

The foremost important feature of cobalt adsorption was the adsorption time. Sequences of experiments with 0.08 g PTU-MSA adsorbent were examined at differing contact times (5–70 min) while the rest of the experimental conditions were maintained constant. As shown in Figure 6b, $Co^{2+}$ sorption capacity rose by 5 min to 61.69 mg·g$^{-1}$ upon PTU-MSA. However, the $Co^{2+}$ ions uptake was not affected when the contact period extended after 20 min. As a result, the balance was extended to 20 min, and the adsorption conditions of cobalt were then improved.

The kinetics of the adsorption process of $Co^{2+}$ regulates the $Co^{2+}$ uptake rate and, in turn, the adsorbate residence time was controlled by this rate at the solid-solution interface. Lagergren's 1st- and 2nd-order kinetic modellings derived PTU-MSA rate constants were used to quantify $Co^{2+}$ adsorption [63,64]. The correlation coefficient was utilized to measure the consistency of the investigational outcomes by the values estimated through the two models ($R^2$). A higher $R^2$ assessment suggests that $Co^{2+}$ adsorption kinetics were accurately characterized by a particular model.

The pseudo-1st- and pseudo-2nd-order kinetics (see Equations (1) and (2)) were shown in Table 4 and Figure 6c,d. The 2nd-order correlation coefficient is 0.9758 and the theoretical uptake is 61.69 mg·g$^{-1}$, which resembles the investigational finding. The result suggests that $Co^{2+}$ adsorption upon PTU-MSA hinges on the initial $Co^{2+}$ concentration. Thus, pseudo-2nd-order kinetics anticipates the adsorption performance throughout the extent of the whole concentration examined [61].

**Table 4.** Kinetic factors of $Co^{2+}$ adsorption upon PTU-MSA.

| Pseudo-1st Order | | | Exp. Capacity $q_e$ (mg·g$^{-1}$) | Pseudo-2nd Order | | |
| --- | --- | --- | --- | --- | --- | --- |
| $q_e$ | $K_1$ | $R^2$ | 61.69 | $q_e$ | $K_2$ | $R^2$ |
| 26.922 | 0.08659 | 0.8252 | | 63.692 | 0.00272 | 0.9758 |

3. Impact of PTU-MSA dose

The influence of PTU-MSA dosage upon $Co^{2+}$ adsorption was investigated over the range 0.02–0.2 g. As demonstrated in Figure 7a, with augmented PTU-MSA dosage from 0.02 to 0.08 g, the $Co^{2+}$ uptake improved [58,59]. The $Co^{2+}$ adsorption upon PTU-MSA does not differ substantially when the PTU-MSA dosage is more than 0.08 g with the same constant $Co^{2+}$ concentration in the medium. For the subsequent experiments, 0.08 g of PTU-MSA was therefore chosen.

4. Impact of initial $Co^{2+}$ concentration and adsorption isotherms features

For three constant temperatures (298, 323, and 343 K), the effect of the initial $Co^{2+}$ concentration was explored. The $Co^{2+}$ adsorption ability improved as the initial $Co^{2+}$ concentration was increased, according to the results reported in Figure 7b. Increased $Co^{2+}$ ion concentrations enhanced the forces in the aqueous and solid phase that suppress the mass transfer. In addition, a beneficial effect was seen on the uptake of $Co^{2+}$ ions in PTU-MSA which shows the process could have an endothermal origin.

Two isothermal models have been utilized to explore the interaction and distribution mechanism between $Co^{2+}$ and solid interface, the isotherms of Freundlich and Langmuir. The Langmuir isothermal pattern implies that monolayers with homogenous binding sites form on the surface of the PTU-MSA (Figure 7c) [65–67]. The model is usually described by Equation (6):

$$\frac{C_e}{q_e} = \frac{1}{bq_{max}} + \frac{C_e}{q_{max}} \qquad (6)$$

where $C_e$ (mg·L$^{-1}$) is the equilibrium concentration, $q_e$ (mg·g$^{-1}$) is the equilibrating uptake of adsorbed $Co^{2+}$, $q_{max}$ (mg·g$^{-1}$) is maximal sorption capability, and $b$ (L·mg$^{-1}$) is equilibrating adsorbance constant. Freundlich modeling is presumed such that adsorption performance is carried out with variable binding energies in scenarios with a heterogeneous surface (Figure 7d) [68–73]. The Freundlich modeling is expressed as Equation (7):

$$\ln q_e = \ln k_f + \frac{1}{n} \ln C_e \qquad (7)$$

where $k_f$ (mg·g$^{-1}$) is the uptake of the Freundlich continuum and $n$ is the sorption intensity parameter. Table 5 shows that the result agrees more with the Langmuir than Freundlich isothermal for Co$^{2+}$ adsorption. This may be due to the greater correlation coefficient ($R^2 = 0.9992$) and the maximal sorption capacity at all three temperatures, which are closer to the experimental. The outcomes showed that there was sorption upon a homogeneous monolayer surface and that the energy of every metal-binding place is the same [74–76]. In the Freundlich isotherm, the computed value is $1/n$ higher than 1.0 (Table 5), which showed normal adsorption. It may be inferred that the higher the $k_f$, the higher is the adsorption intensity [77,78].

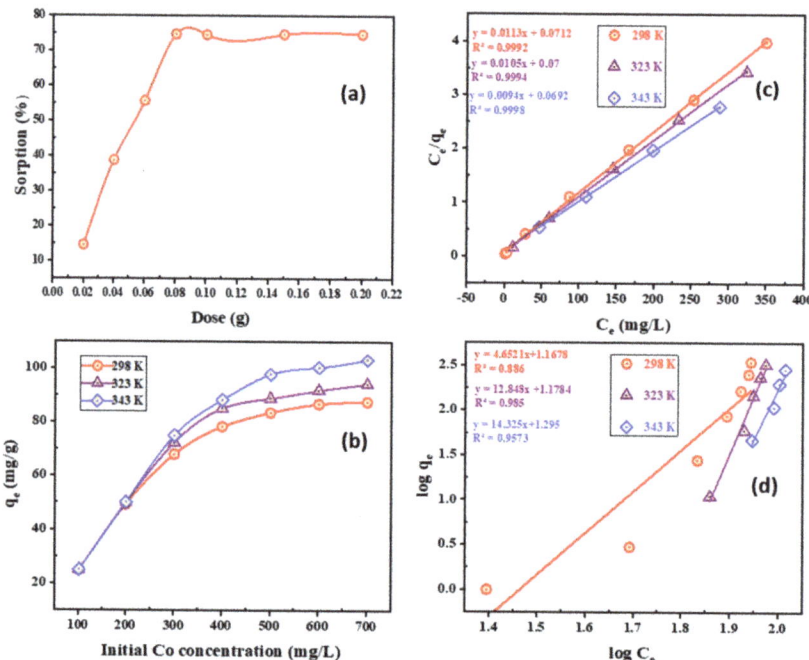

**Figure 7.** (**a**) Impact of PTU-MSA dose upon Co$^{2+}$ adsorption (20 mL of 250 mg/L Co$^{2+}$, pH 8.0, 20 min, 25 °C); (**b**) Impact of initial Co$^{2+}$ concentration upon PTU-MSA uptake (20 mL of 250 mg/L Co$^{2+}$, pH 8.0, 20 min, 0.08 g PTU-MAS); (**c**) Langmuir isotherm model; (**d**) Freundlich isotherm mode for Co$^{2+}$ adsorption upon PTU-MSA (20 mL of 250 mg/L Co$^{2+}$, pH 8.0, 20 min, 0.08 g PTU-MAS).

**Table 5.** Langmuir and Freundlich data for Co$^{2+}$ adsorption by PTU-MSA.

| Temp °C | Langmuir Isotherm | | | $q_e$ (mg·g$^{-1}$) | Freundlich Isotherm | | |
|---|---|---|---|---|---|---|---|
| | $b$ | $q_{max}$ | $R^2$ | | $k_f$ | $1/n$ | $R^2$ |
| 25 | 14.045 | 88.496 | 0.9992 | 87.625 | 14.716 | 4.6521 | 0.886 |
| 50 | 14.288 | 95.24 | 0.9994 | 94.125 | 15.079 | 12.848 | 0.985 |
| 70 | 14.451 | 106.38 | 0.9998 | 103.2 | 19.724 | 14.325 | 0.9573 |

When the values of the separation factor constant ($R_L$) are calculated, the degree of adsorptive capacity of PTU-MSA towards $Co^{2+}$ can be predicted, and this gives an indicator of whether the adsorption development can occur. A suitable environment for Co/PTU-MSA adsorption exists when the $R_L$ values fall within the range of 0 to 1.0. It is found that the relative lightness ($R_L$) ranges between 0.003614 and 0.10626, indicating that the PTU-MSA is capable of adsorbing $Co^{2+}$ from the aqueous phase. The following thermodynamic evaluation factors can prove this Equation (8):

$$R_L = \frac{1}{1 + K_L C_o} \tag{8}$$

5. Thermodynamic parameters for systems of cobalt adsorption

The intercept and slope of the plot of log $K_d$ vs. $1/T$ for the cobalt adsorption system were used to compute the values of $\Delta H°$ and $\Delta S°$, besides the values of $\Delta G°$ from Equation (4) [79]. The thermodynamic parameters are shown in Table 6. Both $\Delta H°$ and $\Delta S°$ are positive, and $T\Delta S°$ is larger than $\Delta H°$.

**Table 6.** Factors controlling the thermodynamics of $Co^{2+}$ adsorption upon PTU-MSA at different temperatures.

| $\Delta H°$ (kJ·mol$^{-1}$) | $\Delta S°$ (J·mol$^{-1}$·K$^{-1}$) | $\Delta G°$ (kJ·mol$^{-1}$) | | |
|---|---|---|---|---|
| | | 298 K | 323 K | 343 K |
| 15.921 | 7.018 | −2.075 | −2.251 | −2.391 |

These findings support the idea that this adsorption mechanism is endothermic. Positive values of $\Delta S°$ indicate that there is a strong affinity between PTU-MSA and $Co^{2+}$ ions. Furthermore, a decline in negative $\Delta G°$ quantities with increasing temperature demonstrates that sorption processes are more effective at more elevated temperatures, most likely because ions are more mobile in the solution [46].

6. Impact of $Ni^{2+}$ ion concentration on PTU-MSA selectivity for $Co^{2+}$ adsorption

In this experiment, we use a mixture of $Co^{2+}$ and $Ni^{2+}$ in the same solution, while changing the concentration of $Ni^{2+}$ to study the influence of $Ni^{2+}$ on the selectivity of PTU-MSA for $Co^{2+}$ adsorption. The data in Figure 8 demonstrate that increasing the concentration of nickel to more than 15-fold the concentration of $Co^{2+}$ ions has no influence on the adsorption of cobalt. There is no affinity of PTU-MSA for adsorption of $Ni^{2+}$ ions. This study covered the application of PTU-MSA to leach liquor that contained only nickel and cobalt ions.

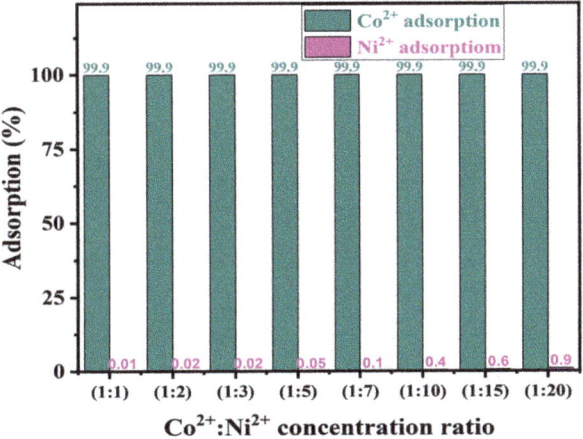

**Figure 8.** Influence of nickel concentration upon the selectivity of PTU-MSA for $Co^{2+}$ ions.

## 2.5. Desorption and Precipitation of Cobalt

The cobalt was eluted from Co/PTU-MSA to a varied phase ratio in the range of 0.01–0.5 M for a 5 min duration using different concentrations of sulfuric acid and ethylene diamine tetraacetic acid (EDTA). Metal values were examined in the aqueous phase. The elution percentage of $Co^{2+}$ vs. $H_2SO_4$ concentration (M) is shown in Figure 9a. The cobalt desorption grew from 10.8 to 99.9% by enhancing the concentration of $H_2SO_4$ from 0.01 to 0.2 M. In case of EDTA, the maximal desorption of $Co^{2+}$ was realized at 0.5 M; the highest elution percentage in $H_2SO_4$ concentration, 0.3 M $H_2SO_4$ was unique as the greatest eluting agent.

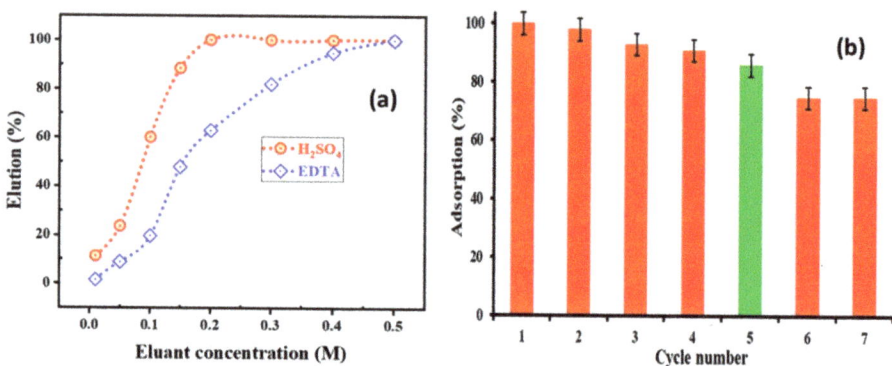

**Figure 9.** (a) Effect of eluents concentration on elution of cobalt loaded on PTU-MSA; (b) Reusability of PTU-MSA.

At this juncture, to identify how often the adsorbent can be used for the adsorption process, and whether its effectiveness changes or not, the regeneration process has been carried out seven times after every adsorption cycle. Figure 9b shows that the adsorption capacity was reduced with higher numbers of cycles. In the first 4 cycles, the positive results were evident. However, the adsorption percentage fell in cycle number 5 compared to the first cycle from around 99 to 84%. Cycles 6 and 7 were extremely close in their adsorption percentages. A decrease in adsorption percentage may be due to poisoning of the active sites or partial leaching.

After desorption of cobalt ions, the precipitation of $Co^{2+}$ was achieved by increasing the pH of the solution pH to 9.0 using $NH_4OH$; a pale pink precipitate of $Co(OH)_2$ was formed. It was filtered and cleaned with distilled $H_2O$ numerous times, after which the precipitate was dried at 60 °C for 2 h. Later, it was calcined at 900 °C to obtain a cobalt (II) oxide (CoO) product. The final product was analyzed by SEM-EDX (Figure 10), which showed that the product is entirely cobalt.

## 2.6. Nickel Recovery

The final solution contains only nickel ions after separation of cadmium and cobalt from the Cd–Ni batteries leach liquor. 200 mL of the sulfate leach liquor contains about 0.823 g of Ni. This has been precipitated at pH 8.25 using 10% NaOH, added to the Ni solution at room temperature. After filtration, the precipitate $Ni(OH)_2$ was washed several times with distilled $H_2O$ to eliminate any impurities and ignited at 650 °C for 1.5 h. The NiO product was recovered and identified using chemical and SEM-EDX techniques, as shown in Figure 11. The recovery of Ni was attained 98% with a purity $\geq$ 97%. In fact, the interference problem from $Co^{2+}$ ions has been overcome by the prior removal of $Co^{2+}$ ions.

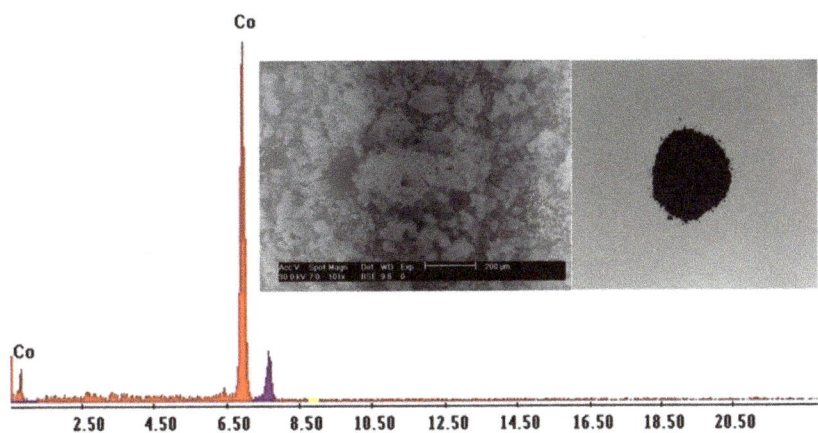

**Figure 10.** EDX and SEM analysis of CoO.

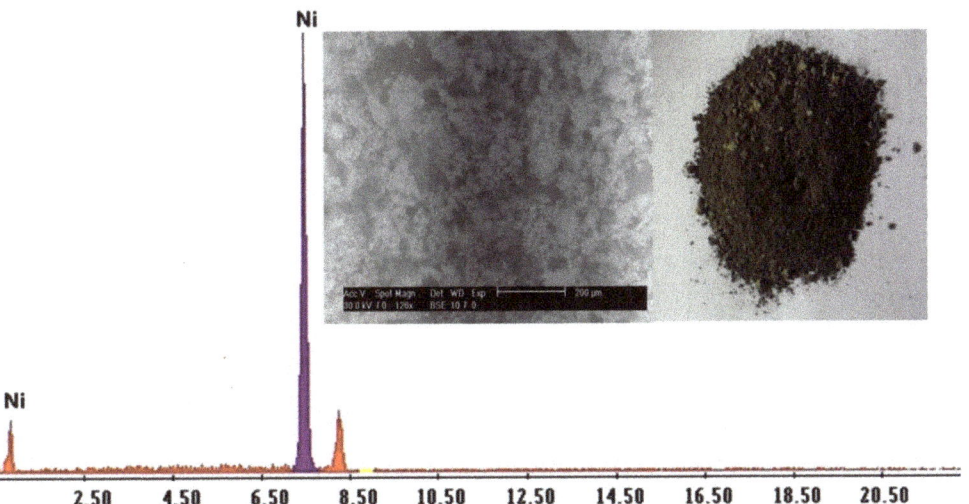

**Figure 11.** EDX and SEM analysis of NiO.

## 3. Materials and Methods

### 3.1. Chemicals and Instruments

Cadmium sulfate, xylenol orange, Adogen 464, KSCN, and sulfuric acid were supplied from Riedel-Dehaen AG and Merck, Merck, Darmstadt, Germany. 3-(Triethoxysilyl) propylisocyanate (95%), poly 2,4-diaminophenol (98%), and ethylene glycol were bought from Sigma-Aldrich chemicals, St. Louis, MO, USA. Tetraethyl orthosilicate, nickel nitrate, block polypropylene glycol-block polyethylene glycol (P123, EO20PO70EO20, MW = 5800 g/mol), chromium nitrate, cobalt nitrate anhydrous, chlorosulphonic acid, triethylamine trihydrochloride and copper nitrate anhydrous were obtained from Merck, Darmstadt, Germany. Distilled water was utilized in all processes. The structure and molecular weight of Adogen 464 are shown in Figure 12. A pH meter (systronics μ pH-System 362) was used to monitor pH for the aqueous phase before and after extraction. An ICP-OES spectrometer (OPTIMA 5300 DV, PerkinElmer, Waltham, MA, USA) was utilized for elemental analysis. A mass balance provided the metal quantities extracted by the solvent extractor. FTIR Affinity-1S (Shimadzu, Kyoto, Japan) was applied to detect the FTIR spectra of samples.

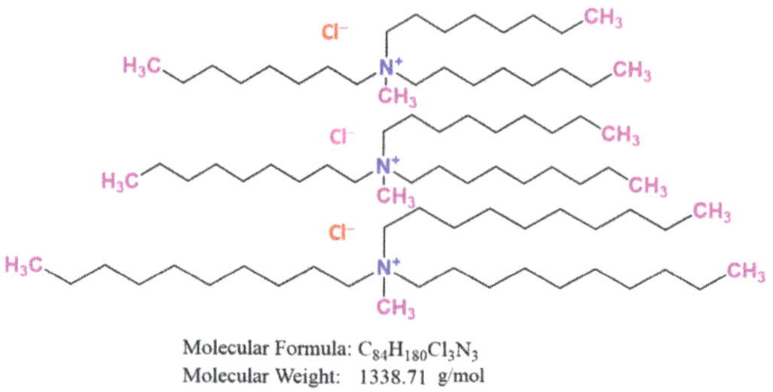

Molecular Formula: $C_{84}H_{180}Cl_3N_3$
Molecular Weight: 1338.71 g/mol

**Figure 12.** Structure of Adogen® 464.

### 3.2. Preparation of Cd–Ni Batteries Leachate

The spent batteries were manually broken into various fractions. Plastic and paper wastes were removed while the electrodes (anode and cathode-active materials) were separated and mixed, ground then washed with distilled water to remove electrolyte (KOH). Figure 13 shows the SEM diagram powder of a deconstructed Ni–Cd battery. After that, the mixture was roasted at 550 °C to oxidize metallic cadmium and decompose cadmium and nickel salts. The net weight of the mixture was about 50 g; the leaching process occurred using 20% $H_2SO_4$, solid/liquid (S/L) of 1/5 at 80 °C for 6 h. Before leaching, a combination of conventional and tooling methods was utilized to scrutinize the chemical formatting of the examined powder (originated from Ni–Cd battery) [80]. Table 7 shows the concentrations of the components that agree with published data before and after the leaching procedure [81–83]. Therefore, the solution contained some iron, which had to be precipitated. The iron-free leaching solution was also employed for selective extraction of $Cd^{2+}$ ions using Adogen 464.

**Figure 13.** SEM image of Ni–Cd battery powder and a digital photograph of a dismantled spent Ni–Cd battery.

**Table 7.** Chemical composition of consumed Ni-Cd battery powder.

| Elements | Chemical Configuration of Powder Ni–Cd Batteries (g/kg) | Elements Concentration in Ni–Cd Batteries Leachate (g/L) | Leaching Efficiency, % |
|---|---|---|---|
| Ni | 21.89 | 20.80 | 94.90 |
| Cd | 13.31 | 13.31 | 99.90 |
| Co | 0.85 | 0.85 | 99.90 |
| Fe | 2.71 | 2.71 | 99.90 |

## 3.3. Preparation of Adogen® 464 for Extraction Process

Adogen® 464 (30%) was converted from chloride to thiocyanate form; this was exposed to 1.5 M KSCN solution, at a 1:1 ratio of water to organic material. It was agitated for 20 min. After two injections of organic solution with a new SCN⁻ solution, there were still traces of thiocyanate in the organic solution. After being in contact with the organic solution, thiocyanate titration was indicated by AgNO₃ concentration [29]. An appropriate volume of Adogen® 464 with commercial-grade kerosene (diluent) was diluted and/or dissolved to generate organic solutions of varied levels unless otherwise indicated. Concentration levels of the ionic fluid were determined.

## 3.4. Extraction Processes and Measurements

A 100 mL separation funnel with the aqueous phase of $Cd^{2+}$ was balanced with organic Adogen 464 solvent at room temperature (except for temperature variation). After the water layer separation, the pH was determined with a pH meter. Each trial was explored three times, with error bars added on each occasion. Following extraction, the amount of a metal in water increases when it moves from the aqueous stage to the organic stage, according to the equation (Equation (8)). $[M]_O$ equals the metal amount in the organic layer, and $[M]_A$ equals the metal amount in the aqueous layer following extraction:

$$D = \frac{[M]_O}{[M]_A} = \frac{[M]_{i,A} - [M]_A}{[M]_A} \quad (9)$$

$$D = \frac{[M]_{i,A} - [M]_A}{[M]_A} \times \frac{V_A}{V_O} \quad (10)$$

$V_A$ and $V_O$ are volumes of aqueous and organic layers. Hence, the extraction efficiency was applied as observed [84]:

$$E\% = \frac{100D}{D + \left(\frac{V_A}{V_O}\right)} \quad (11)$$

## 3.5. Preparation of Precursor, 1,1'-(4-hydroxy-1,3-phenylene)bis(3-(3-(triethoxysilyl)propyl)urea (PTU)

A 2,4-Isocyanate of propyl-di-aminophenol (1.2 g) and 3-(triethoxysilly)propyl isocyanate (5.0 g) was liquefied at the concentration of 1:2 in 85 mL dry acetonitrile (Figure 14). The mix was warmed to 85 °C for 24 h via static reflux. Thin-layer chromatography (TLC) was utilized to assess reaction advancement. The solvent was then evaporated to produce a white precipitate during over 5 h in dry hexane. The puffed white outcome was purified, cleaned with hexane, and dried with a vacuum.

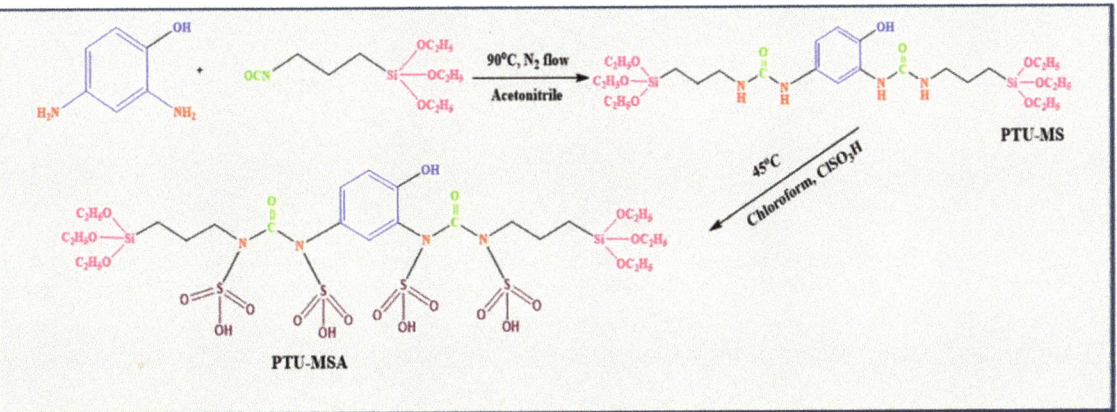

**Figure 14.** Preparation of PTU-MS and PTU-MSA.

*3.6. Synthesis of PTU Bridged Mesoporous Organosilica (PTU-MS)*

N1,1'-(4-hydroxy-1,3-phenylene) bis(3-(3-(triethoxysilyl)propyl)urea, a bridging mesoporous material, was prepared [85]. TEOS as the parent source of silica and the structure-directing agent, PTU as the organic bridging groups, and pluronic P-123 as a sample were used and mixed in a flask. The synthesized materials were designated PTU-MS. Solutions of P123, HCl, and water were mixed with a strong stirring at 35 °C in a typical process. The combination of 2 mixtures was allowed to mix for 24 h at the above temperature until white precipitation was established, and the heterogeneous blend aged with mixing for a further 24 h and 80 °C. The white precipitate was filtered, cleaned with $H_2O$, and dried in air for various times.

Before using the produced adsorbent PTU-MS, it was subjected to washing with HCl (3 mL)–ethanol (150 mL) solution for 12 h at 25 °C to get rid of the surfactant template. This was performed three times until the surfactants had been eliminated. To obtain the final product, PTU-MS was filtered, cleaned with ethanol, and desiccated overnight at 60 °C.

Finally, 2.0 g of PTU-MS was dipped in chloroform, stirred at 45 °C for 1 h and mixed with 0.08 mL of chlorosulphonic acid ($ClSO_3H$), to generate enough triethylamine to eliminate the released hydrochloric acid with 18 h of stirring (Figure 14). The yield was filtered, cleaned with excess chloroform, and vacuum-dried to gain organic-inorganic mesoporous silica improved by a group of sulphonic acids. The end product received the symbol PTU-MSA. The preparation of PTU-MS and PTU-MSA are illustrated in Figure 14.

## 4. Conclusions

The Scarp Ni–Cd batteries (50 g) were treated with 20% $H_2SO_4$, with the 1:5 solid/liquid (S:L) ratio at 80 °C for 6 h. The leaching efficiency of Cd, Fe, and Co was nearly 100%, whereas the leaching efficiency of Ni was 95%. Both newly prepared materials were confirmed via FTIR and SEM techniques. Adogen 464 was used for $Cd^{2+}$ extraction, followed by precipitation as a yellow CdS product with 0.5% $Na_2S$ solution by setting the pH at 1.25, at room temperature. PTU-MS silica adsorbent was used as an ion exchanger for $Co^{2+}$ ions. 0.3 M $H_2SO_4$ was employed as an eluant and the precipitation of cobalt was achieved at up to pH 9.0 as $Co(OH)_2$. The kinetic and thermodynamic parameters were also investigated. Lastly, nickel was directly precipitated at pH 8.25, using a 10% NaOH solution at ambient temperature.

**Author Contributions:** Conceptualization, A.A.G. and A.K.S.; methodology, A.R.W. and M.A.G.; software, A.K.S. and M.A.G.; validation, M.Y.H. and M.I.S.; formal analysis, B.M.A.; investigation, A.R.W.; resources, M.A.G.; data curation, A.A.G.; writing—original draft preparation, E.M.E.-S.; writing—review and editing, A.K.S.; visualization, A.R.W.; supervision, R.E.S. and H.A.R.; project administration, M.F.C.; funding acquisition, H.H.S. All authors have read and agreed to the published version of the manuscript.

**Funding:** This work was supported by King Khalid University through a grant (KKU/RCAMS/22) under the Research Center for Advanced Materials Science (RCAMS) at King Khalid University, Saudi Arabia.

**Institutional Review Board Statement:** Not applicable.

**Informed Consent Statement:** Not applicable.

**Data Availability Statement:** Not applicable.

**Acknowledgments:** This work was supported by King Khalid University through a grant (KKU/RCAMS/22) under the Research Center for Advanced Materials Science (RCAMS) at King Khalid University, Saudi Arabia.

**Conflicts of Interest:** The authors declare no conflict of interest.

## References

1. Plachy, J. *Cadmium Recycling in the United States in 2000 [Electronic Resource] by Jozef Plachy*; U.S. Department of the Interior, U.S. Geological Survey: Reston, VA, USA, 2003.
2. Dehghani-Sanij, A.R.; Tharumalingam, E.; Dusseault, M.B.; Fraser, R. Study of energy storage systems and environmental challenges of batteries. *Renew. Sustain. Energy Rev.* **2019**, *104*, 192–208. [CrossRef]
3. Weshahy, A.R.; Gouda, A.A.; Atia, B.M.; Sakr, A.K.; Al-Otaibi, J.S.; Almuqrin, A.; Hanfi, M.Y.; Sayyed, M.I.; El Sheikh, R.; Radwan, H.A.; et al. Efficient Recovery of Rare Earth Elements and Zinc from Spent Ni–Metal Hydride Batteries: Statistical Studies. *Nanomaterials* **2022**, *12*, 2305. [CrossRef] [PubMed]
4. Mason-Jones, K.; von Blottnitz, H. Flows and fates of nickel–cadmium batteries in the City of Cape Town. *Miner. Eng.* **2010**, *23*, 211–218. [CrossRef]
5. Nogueira, C.A.; Margarido, F. Leaching behaviour of electrode materials of spent nickel–cadmium batteries in sulphuric acid media. *Hydrometallurgy* **2004**, *72*, 111–118. [CrossRef]
6. Hazotte, C.; Leclerc, N.; Diliberto, S.; Meux, E.; Lapicque, F. End-of-life nickel–cadmium accumulators: Characterization of electrode materials and industrial Black Mass. *Environ. Technol.* **2015**, *36*, 796–805. [CrossRef] [PubMed]
7. Morrow, H.; Keating, J. Overview paper on effective recycling of Ni–Cd batteries. In Proceedings of the OECD Workshop on the Effective Collection and Recycling of Nickel–Cadmium Batteries, Lyon, France, 23–25 September 1997; pp. 23–25.
8. Bernardes, A.M.; Espinosa, D.C.R.; Tenório, J.A.S. Recycling of batteries: A review of current processes and technologies. *J. Power Sources* **2004**, *130*, 291–298. [CrossRef]
9. Espinosa, D.C.R.; Tenório, J.A.S. Recycling of nickel–cadmium batteries using coal as reducing agent. *J. Power Sources* **2006**, *157*, 600–604. [CrossRef]
10. Sullivan, J.L.; Gaines, L. Status of life cycle inventories for batteries. *Energy Convers. Manag.* **2012**, *58*, 134–148. [CrossRef]
11. Sohn, J.-S.; Shin, S.; Kang, K.-S.; Choi, M.-J. Trend on the Recycling Technologies for the used Lithium Battery by the Patent Analysis. *J. Korean Inst. Resour. Recycl.* **2007**, *16*, 50–60.
12. Espinosa, D.C.R.; Tenório, J.A.S. Fundamental aspects of recycling of nickel–cadmium batteries through vacuum distillation. *J. Power Sources* **2004**, *135*, 320–326. [CrossRef]
13. Huang, K.; Li, J.; Xu, Z. A Novel Process for Recovering Valuable Metals from Waste Nickel–Cadmium Batteries. *Environ. Sci. Technol.* **2009**, *43*, 8974–8978. [CrossRef]
14. Rudnik, E.; Nikiel, M. Hydrometallurgical recovery of cadmium and nickel from spent Ni–Cd batteries. *Hydrometallurgy* **2007**, *89*, 61–71. [CrossRef]
15. Reddy, B.R.; Priya, D.N. Chloride leaching and solvent extraction of cadmium, cobalt and nickel from spent nickel–cadmium, batteries using Cyanex 923 and 272. *J. Power Sources* **2006**, *161*, 1428–1434. [CrossRef]
16. Fernandes, A.; Afonso, J.C.; Bourdot Dutra, A.J. Hydrometallurgical route to recover nickel, cobalt and cadmium from spent Ni–Cd batteries. *J. Power Sources* **2012**, *220*, 286–291. [CrossRef]
17. Kim, Y.J.; Kim, J.H.; Thi, L.D.; Qureshi, T.I. Recycling of NiCd batteries by hydrometallurgical process on small scale. *J. Chem. Soc. Pak.* **2011**, *33*, 853–857.
18. Jha, M.K.; Kumar, V.; Jeong, J.; Lee, J.-C. Review on solvent extraction of cadmium from various solutions. *Hydrometallurgy* **2012**, *111*, 1–9. [CrossRef]
19. Abdel Geleel, M.; Atwa, S.T.; Sakr, A.K. Removal of Cr (III) from aqueous waste using Spent Activated Clay. *J. Am. Sci.* **2013**, *9*, 256–262. [CrossRef]
20. Gotfryd, L.; Cox, M. The selective recovery of cadmium(II) from sulfate solutions by a counter-current extraction–stripping process using a mixture of diisopropylsalicylic acid and Cyanex® 471X. *Hydrometallurgy* **2006**, *81*, 226–233. [CrossRef]
21. Tanong, K.; Tran, L.-H.; Mercier, G.; Blais, J.-F. Recovery of Zn (II), Mn (II), Cd (II) and Ni (II) from the unsorted spent batteries using solvent extraction, electrodeposition and precipitation methods. *J. Clean. Prod.* **2017**, *148*, 233–244. [CrossRef]
22. Vander Hoogerstraete, T.; Onghena, B.; Binnemans, K. Homogeneous Liquid–Liquid Extraction of Rare Earths with the Betaine—Betainium Bis(trifluoromethylsulfonyl)imide Ionic Liquid System. *Int. J. Mol. Sci.* **2013**, *14*, 21353–21377. [CrossRef]
23. Mahandra, H.; Singh, R.; Gupta, B. Liquid-liquid extraction studies on Zn(II) and Cd(II) using phosphonium ionic liquid (Cyphos IL 104) and recovery of zinc from zinc plating mud. *Sep. Purif. Technol.* **2017**, *177*, 281–292. [CrossRef]
24. Lerum, H.V.; Sand, S.; Eriksen, D.Ø.; Wibetoe, G.; Omtvedt, J.P. Comparison of single-phase and two-phase measurements in extraction, separation and back-extraction of Cd, Zn and Co from a multi-element matrix using Aliquat 336. *J. Radioanal. Nucl. Chem.* **2020**, *324*, 1203–1214. [CrossRef]
25. Zhang, L.; Hessel, V.; Peng, J. Liquid-liquid extraction for the separation of Co(II) from Ni(II) with Cyanex 272 using a pilot scale Re-entrance flow microreactor. *Chem. Eng. J.* **2018**, *332*, 131–139. [CrossRef]
26. Flieger, J.; Feder-Kubis, J.; Tatarczak-Michalewska, M. Chiral Ionic Liquids: Structural Diversity, Properties and Applications in Selected Separation Techniques. *Int. J. Mol. Sci.* **2020**, *21*, 4253. [CrossRef]
27. Park, J.; Jung, Y.; Kusumah, P.; Lee, J.; Kwon, K.; Lee, C.K. Application of Ionic Liquids in Hydrometallurgy. *Int. J. Mol. Sci.* **2014**, *15*, 15320–15343. [CrossRef]
28. Sato, T.; Shimomura, T.; Murakami, S.; Maeda, T.; Nakamura, T. Liquid-liquid extraction of divalent manganese, cobalt, copper, zinc and cadmium from aqueous chloride solutions by tricaprylmethylammonium chloride. *Hydrometallurgy* **1984**, *12*, 245–254. [CrossRef]

29. Daud, H.; Cattrall, R.W. The extraction of Hg(II) from potassium iodide solutions and the extraction of Cu(II), Zn(II) and Cd(II) from hydrochloric acid solutions by aliquat 336 dissolved in chloroform. *J. Inorg. Nucl. Chem.* **1981**, *43*, 779–785. [CrossRef]
30. Zhao, B.; Zhang, Y.; Dou, X.; Yuan, H.; Yang, M. Granular ferric hydroxide adsorbent for phosphate removal: Demonstration preparation and field study. *Water Sci. Technol.* **2015**, *72*, 2179–2186. [CrossRef]
31. Ritcey, G.M.; Ashbrook, A.W. *Solvent Extraction: Principles and Applications to Process Metallurgy*; Elsevier: Amsterdam, The Netherlands, 1984.
32. Atia, B.M.; Khawassek, Y.M.; Hussein, G.M.; Gado, M.A.; El-Sheify, M.A.; Cheira, M.F. One-pot synthesis of pyridine dicarboxamide derivative and its application for uranium separation from acidic medium. *J. Environ. Chem. Eng.* **2021**, *9*, 105726. [CrossRef]
33. Atia, B.M.; Gado, M.A.; Cheira, M.F.; El-Gendy, H.S.; Yousef, M.A.; Hashem, M.D. Direct synthesis of a chelating carboxamide derivative and its application for thorium extraction from Abu Rusheid ore sample, South Eastern Desert, Egypt. *Int. J. Environ. Anal. Chem.* **2021**, 1–24. [CrossRef]
34. Ibrahium, H.A.; Atia, B.M.; Awwad, N.S.; Nayl, A.A.; Radwan, H.A.; Gado, M.A. Efficient preparation of phosphazene chitosan derivatives and its applications for the adsorption of molybdenum from spent hydrodesulfurization catalyst. *J. Dispers. Sci. Technol.* **2022**, 1–16. [CrossRef]
35. Ibrahium, H.A.; Awwad, N.S.; Gado, M.A.; Hassanin, M.A.; Nayl, A.A.; Atia, B.M. Physico-Chemical Aspects on Uranium and Molybdenum Extraction from Aqueous Solution by Synthesized Phosphinimine Derivative Chelating Agent. *J. Inorg. Organomet. Polym. Mater.* **2022**, 1–18. [CrossRef]
36. Alharbi, A.; Gouda, A.A.; Atia, B.M.; Gado, M.A.; Alluhaybi, A.A.; Alkabli, J. The Role of Modified Chelating Graphene Oxide for Vanadium Separation from Its Bearing Samples. *Russ. J. Inorg. Chem.* **2022**, *67*, 560–575. [CrossRef]
37. Ibrahium, H.A.; Gado, M.A.; Elhosiny Ali, H.; Fathy, W.M.; Atia, B.M.; Awwad, N.S. Synthesis of chelating N-hydroxyl amine derivative and its application for vanadium separation from Abu Zeneima ferruginous siltstone ore, Southwestern Sinai, Egypt. *Int. J. Environ. Anal. Chem.* **2021**, 1–23. [CrossRef]
38. Ibrahium, H.A.; Gado, M.A.; Awwad, N.S.; Fathy, W.M. Selective separation of Yttrium and Uranium from Xenotime Concentrate. *Z. Anorg. Allg. Chem.* **2021**, *647*, 1568–1577. [CrossRef]
39. Nayl, A.A. Extraction and separation of Co(II) and Ni(II) from acidic sulfate solutions using Aliquat 336. *J. Hazard. Mater.* **2010**, *173*, 223–230. [CrossRef]
40. Sakr, A.K.; Snelling, H.V.; Young, N.A. Experimental evidence for the molecular molybdenum fluorides MoF to MoF6: A matrix isolation and DFT investigation. *New J. Chem.* **2022**, *46*, 9666–9684. [CrossRef]
41. Chang, S.H.; Teng, T.T.; Ismail, N.; Alkarkhi, A.F.M. Selection of design parameters and optimization of operating parameters of soybean oil-based bulk liquid membrane for Cu(II) removal and recovery from aqueous solutions. *J. Hazard. Mater.* **2011**, *190*, 197–204. [CrossRef]
42. Sayed, A.S.; Abdelmottaleb, M.; Cheira, M.F.; Abdel-Aziz, G.; Gomaa, H.; Hassanein, T.F. Date seed as an efficient, eco-friendly, and cost-effective bio-adsorbent for removal of thorium ions from acidic solutions. *Aswan Univ. J. Environ. Stud.* **2020**, *1*, 106–124. [CrossRef]
43. Gado, M.A.; Atia, B.M.; Cheira, M.F.; Abdou, A.A. Thorium ions adsorption from aqueous solution by amino naphthol sulphonate coupled chitosan. *Int. J. Environ. Anal. Chem.* **2021**, *101*, 1419–1436. [CrossRef]
44. Gado, M.; Rashad, M.; Kassab, W.; Badran, M. Highly Developed Surface Area Thiosemicarbazide Biochar Derived from Aloe Vera for Efficient Adsorption of Uranium. *Radiochemistry* **2021**, *63*, 353–363. [CrossRef]
45. Atia, B.M.; Gado, M.A.; Abd El-Magied, M.O.; Elshehy, E.A. Highly efficient extraction of uranyl ions from aqueous solutions using multi-chelators functionalized graphene oxide. *Sep. Sci. Technol.* **2020**, *55*, 2746–2757. [CrossRef]
46. Acosta-Rodríguez, I.; Rodríguez-Pérez, A.; Pacheco-Castillo, N.C.; Enríquez-Domínguez, E.; Cárdenas-González, J.F.; Martínez-Juárez, V.-M. Removal of Cobalt (II) from Waters Contaminated by the Biomass of Eichhornia crassipes. *Water* **2021**, *13*, 1725. [CrossRef]
47. Cheira, M.F.; Atia, B.M.; Kouraim, M.N. Uranium(VI) recovery from acidic leach liquor by Ambersep 920U SO4 resin: Kinetic, equilibrium and thermodynamic studies. *J. Radiat. Res. Appl. Sci.* **2017**, *10*, 307–319. [CrossRef]
48. Mercier, L.; Pinnavaia, T.J. Direct Synthesis of Hybrid Organic−Inorganic Nanoporous Silica by a Neutral Amine Assembly Route: Structure−Function Control by Stoichiometric Incorporation of Organosiloxane Molecules. *Chem. Mater.* **2000**, *12*, 188–196. [CrossRef]
49. Asefa, T.; MacLachlan, M.J.; Coombs, N.; Ozin, G.A. Periodic mesoporous organosilicas with organic groups inside the channel walls. *Nature* **1999**, *402*, 867–871. [CrossRef]
50. Lim, M.H.; Blanford, C.F.; Stein, A. Synthesis and Characterization of a Reactive Vinyl-Functionalized MCM-41: Probing the Internal Pore Structure by a Bromination Reaction. *J. Am. Chem. Soc.* **1997**, *119*, 4090–4091. [CrossRef]
51. Lim, M.H.; Blanford, C.F.; Stein, A. Synthesis of Ordered Microporous Silicates with Organosulfur Surface Groups and Their Applications as Solid Acid Catalysts. *Chem. Mater.* **1998**, *10*, 467–470. [CrossRef]
52. Asefa, T.; Kruk, M.; MacLachlan, M.J.; Coombs, N.; Grondey, H.; Jaroniec, M.; Ozin, G. Sequential Hydroboration–Alcoholysis and Epoxidation–Ring Opening Reactions of Vinyl Groups in Mesoporous Vinylsilica. *Adv. Funct. Mater.* **2001**, *11*, 447–456. [CrossRef]
53. Wahab, M.A.; Imae, I.; Kawakami, Y.; Kim, I.; Ha, C.-S. Functionalized periodic mesoporous organosilica fibers with longitudinal pore architectures under basic conditions. *Microporous Mesoporous Mater.* **2006**, *92*, 201–211. [CrossRef]

54. Moreau, J.J.E.; Vellutini, L.; Wong Chi Man, M.; Bied, C. Shape-Controlled Bridged Silsesquioxanes: Hollow Tubes and Spheres. *Chem. A Eur. J.* **2003**, *9*, 1594–1599. [CrossRef]
55. Wahab, M.A.; Guo, W.; Cho, W.-J.; Ha, C.-S. Synthesis and Characterization of Novel Amorphous Hybrid Silica Materials. *J. Sol-Gel Sci. Technol.* **2003**, *27*, 333–341. [CrossRef]
56. Park, S.S.; Ha, C.-S. Organic–inorganic hybrid mesoporous silicas: Functionalization, pore size, and morphology control. *Chem. Rec.* **2006**, *6*, 32–42. [CrossRef]
57. Awual, M.R. A novel facial composite adsorbent for enhanced copper(II) detection and removal from wastewater. *Chem. Eng. J.* **2015**, *266*, 368–375. [CrossRef]
58. Awual, M.R.; Suzuki, S.; Taguchi, T.; Shiwaku, H.; Okamoto, Y.; Yaita, T. Radioactive cesium removal from nuclear wastewater by novel inorganic and conjugate adsorbents. *Chem. Eng. J.* **2014**, *242*, 127–135. [CrossRef]
59. Awual, M.R.; Yaita, T.; Shiwaku, H. Design a novel optical adsorbent for simultaneous ultra-trace cerium(III) detection, sorption and recovery. *Chem. Eng. J.* **2013**, *228*, 327–335. [CrossRef]
60. Yuan, G.; Tu, H.; Li, M.; Liu, J.; Zhao, C.; Liao, J.; Yang, Y.; Yang, J.; Liu, N. Glycine derivative-functionalized metal-organic framework (MOF) materials for Co(II) removal from aqueous solution. *Appl. Surf. Sci.* **2019**, *466*, 903–910. [CrossRef]
61. Fan, G.; Lin, R.; Su, Z.; Lin, X.; Xu, R.; Chen, W. Removal of Cr (VI) from aqueous solutions by titanate nanomaterials synthesized via hydrothermal method. *Can. J. Chem. Eng.* **2017**, *95*, 717–723. [CrossRef]
62. Yuan, G.; Tu, H.; Liu, J.; Zhao, C.; Liao, J.; Yang, Y.; Yang, J.; Liu, N. A novel ion-imprinted polymer induced by the glycylglycine modified metal-organic framework for the selective removal of Co(II) from aqueous solutions. *Chem. Eng. J.* **2018**, *333*, 280–288. [CrossRef]
63. Negm, S.H.; Abd El-Hamid, A.A.M.; Gado, M.A.; El-Gendy, H.S. Selective uranium adsorption using modified acrylamide resins. *J. Radioanal. Nucl. Chem.* **2019**, *319*, 327–337. [CrossRef]
64. Gado, M.A. Sorption of thorium using magnetic graphene oxide polypyrrole composite synthesized from natural source. *Sep. Sci. Technol.* **2018**, *53*, 2016–2033. [CrossRef]
65. Cho, G.; Fung, B.M.; Glatzhofer, D.T.; Lee, J.-S.; Shul, Y.-G. Preparation and Characterization of Polypyrrole-Coated Nanosized Novel Ceramics. *Langmuir* **2001**, *17*, 456–461. [CrossRef]
66. Tian, B.; Zerbi, G. Lattice dynamics and vibrational spectra of polypyrrole. *J. Chem. Phys.* **1990**, *92*, 3886–3891. [CrossRef]
67. Sakr, A.K.; Al-Hamarneh, I.F.; Gomaa, H.; Abdel Aal, M.M.; Hanfi, M.Y.; Sayyed, M.I.; Khandaler, M.U.; Cheira, M.F. Removal of uranium from nuclear effluent using regenerated bleaching earth steeped in β-naphthol. *Radiat. Phys. Chem.* **2022**, 110204. [CrossRef]
68. Mahmud, H.N.M.E.; Kassim, A.; Zainal, Z.; Yunus, W.M.M. Fourier transform infrared study of polypyrrole–poly(vinyl alcohol) conducting polymer composite films: Evidence of film formation and characterization. *J. Appl. Polym. Sci.* **2006**, *100*, 4107–4113. [CrossRef]
69. Gado, M.; Atia, B.; Morcy, A. The role of graphene oxide anchored 1-amino-2-naphthol-4-sulphonic acid on the adsorption of uranyl ions from aqueous solution: Kinetic and thermodynamic features. *Int. J. Environ. Anal. Chem.* **2019**, *99*, 996–1015. [CrossRef]
70. Hassanin, M.A.; Negm, S.H.; Youssef, M.A.; Sakr, A.K.; Mira, H.I.; Mohammaden, T.F.; Al-Otaibi, J.S.; Hanfi, M.Y.; Sayyed, M.I.; Cheira, M.F. Sustainable Remedy Waste to Generate $SiO_2$ Functionalized on Graphene Oxide for Removal of U(VI) Ions. *Sustainability* **2022**, *14*, 2699. [CrossRef]
71. Gado, M.A.; Atia, B.M.; Cheira, M.F.; Elawady, M.E.; Demerdash, M. Highly efficient adsorption of uranyl ions using hydroxamic acid-functionalized graphene oxide. *Radiochim. Acta* **2021**, *109*, 743–757. [CrossRef]
72. Cheira, M.F.; Mira, H.I.; Sakr, A.K.; Mohamed, S.A. Adsorption of U(VI) from acid solution on a low-cost sorbent: Equilibrium, kinetic, and thermodynamic assessments. *Nucl. Sci. Tech.* **2019**, *30*, 156. [CrossRef]
73. Sakr, A.K.; Cheira, M.F.; Hassanin, M.A.; Mira, H.I.; Mohamed, S.A.; Khandaker, M.U.; Osman, H.; Eed, E.M.; Sayyed, M.I.; Hanfi, M.Y. Adsorption of Yttrium Ions on 3-Amino-5-Hydroxypyrazole Impregnated Bleaching Clay, a Novel Sorbent Material. *Appl. Sci.* **2021**, *11*, 10320. [CrossRef]
74. Allam, E.M.; Lashen, T.A.; Abou El-Enein, S.A.; Hassanin, M.A.; Sakr, A.K.; Cheira, M.F.; Almuqrin, A.; Hanfi, M.Y.; Sayyed, M.I. Rare Earth Group Separation after Extraction Using Sodium Diethyldithiocarbamate/Polyvinyl Chloride from Lamprophyre Dykes Leachate. *Materials* **2022**, *15*, 1211. [CrossRef]
75. Radwan, H.A.; Gado, M.A.; El-Wahab, Z.H.A.; El-Sheikh, E.M.; Faheim, A.A.; Taha, R.H. Recovery of uranium from ferruginous Shale mineralization from Um Bogma formation, Egypt, via Duolite ES-467 chelating resin. *Z. Anorg. Allg. Chem.* **2021**, *647*, 396–412. [CrossRef]
76. Garoub, M.; Gado, M. Separation of Cadmium Using a new Adsorbent of Modified Chitosan with Pyridine Dicarboxyamide derivative and application in different samples. *Z. Anorg. Allg. Chem.* **2022**, *648*, e202100222. [CrossRef]
77. Allam, E.M.; Lashen, T.A.; Abou El-Enein, S.A.; Hassanin, M.A.; Sakr, A.K.; Hanfi, M.Y.; Sayyed, M.I.; Al-Otaibi, J.S.; Cheira, M.F. Cetylpyridinium Bromide/Polyvinyl Chloride for Substantially Efficient Capture of Rare Earth Elements from Chloride Solution. *Polymers* **2022**, *14*, 954. [CrossRef]
78. Radwan, H.A.; Faheim, A.A.; El-Sheikh, E.M.; Abd El-Wahab, Z.H.; Gado, M.A. Optimization of the leaching process for uranium recovery and some associated valuable elements from low-grade uranium ore. *Int. J. Environ. Anal. Chem.* **2021**, 1–23. [CrossRef]

79. Mahmoud, N.S.; Atwa, S.T.; Sakr, A.K.; Abdel Geleel, M. Kinetic and thermodynamic study of the adsorption of Ni (II) using Spent Activated clay Mineral. *N. Y. Sci. J.* **2012**, *5*, 62–68. [CrossRef]
80. Korkmaz, K.; Alemrajabi, M.; Rasmuson, Å.; Forsberg, K. Recoveries of Valuable Metals from Spent Nickel Metal Hydride Vehicle Batteries via Sulfation, Selective Roasting, and Water Leaching. *J. Sustain. Metall.* **2018**, *4*, 313–325. [CrossRef]
81. Kumbasar, R.A. Selective extraction and concentration of cobalt from acidic leach solution containing cobalt and nickel through emulsion liquid membrane using PC-88A as extractant. *Sep. Purif. Technol.* **2009**, *64*, 273–279. [CrossRef]
82. Pietrelli, L.; Bellomo, B.; Fontana, D.; Montereali, M. Characterization and leaching of NiCd and NiMH spent batteries for the recovery of metals. *Waste Manag.* **2005**, *25*, 221–226. [CrossRef]
83. Vassura, I.; Morselli, L.; Bernardi, E.; Passarini, F. Chemical characterisation of spent rechargeable batteries. *Waste Manag.* **2009**, *29*, 2332–2335. [CrossRef]
84. Atia, B.M.; Sakr, A.K.; Gado, M.A.; El-Gendy, H.S.; Abdelazeem, N.M.; El-Sheikh, E.M.; Hanfi, M.Y.; Sayyed, M.I.; Al-Otaibi, J.S.; Cheira, M.F. Synthesis of a New Chelating Iminophosphorane Derivative (Phosphazene) for U(VI) Recovery. *Polymers* **2022**, *14*, 1687. [CrossRef]
85. Li, C.; Liu, J.; Shi, X.; Yang, J.; Yang, Q. Periodic Mesoporous Organosilicas with 1,4-Diethylenebenzene in the Mesoporous Wall: Synthesis, Characterization, and Bioadsorption Properties. *J. Phys. Chem. C* **2007**, *111*, 10948–10954. [CrossRef]

*Article*

# Adjusting Some Properties of Poly(methacrylic acid) (Nano)Composite Hydrogels by Means of Silicon-Containing Inorganic Fillers

Claudia Mihaela Ninciuleanu [1,2], Raluca Ianchiș [1], Elvira Alexandrescu [1], Cătălin Ionuț Mihăescu [1], Sabina Burlacu [1], Bogdan Trică [1], Cristina Lavinia Nistor [1], Silviu Preda [3], Cristina Scomoroscenco [1], Cătălina Gîfu [1], Cristian Petcu [1] and Mircea Teodorescu [2,*]

[1] National Institute for Research and Development in Chemistry and Petrochemistry-ICECHIM, Spl. Independentei 202, 060021 Bucharest, Romania
[2] Department of Bioresources and Polymer Science, Faculty of Applied Chemistry and Materials Science, Politehnica University of Bucharest, 1-7 Gh. Polizu Street, 011061 Bucharest, Romania
[3] Institute of Physical Chemistry "Ilie Murgulescu", Romanian Academy, Spl. Independentei 202, 6th District, P.O. Box 194, 060021 Bucharest, Romania
* Correspondence: mircea.teodorescu@upb.ro; Tel.: +40-745907871

**Citation:** Ninciuleanu, C.M.; Ianchiș, R.; Alexandrescu, E.; Mihăescu, C.I.; Burlacu, S.; Trică, B.; Nistor, C.L.; Preda, S.; Scomoroscenco, C.; Gîfu, C.; et al. Adjusting Some Properties of Poly(methacrylic acid) (Nano)Composite Hydrogels by Means of Silicon-Containing Inorganic Fillers. *Int. J. Mol. Sci.* **2022**, 23, 10320. https://doi.org/10.3390/ijms231810320

Academic Editor: Dippong Thomas

Received: 8 August 2022
Accepted: 4 September 2022
Published: 7 September 2022

**Publisher's Note:** MDPI stays neutral with regard to jurisdictional claims in published maps and institutional affiliations.

**Copyright:** © 2022 by the authors. Licensee MDPI, Basel, Switzerland. This article is an open access article distributed under the terms and conditions of the Creative Commons Attribution (CC BY) license (https://creativecommons.org/licenses/by/4.0/).

**Abstract:** The present work aims to show how the main properties of poly(methacrylic acid) (PMAA) hydrogels can be engineered by means of several silicon-based fillers (Laponite XLS/XLG, montmorillonite (Mt), pyrogenic silica (PS)) employed at 10 wt% concentration based on MAA. Various techniques (FT-IR, XRD, TGA, SEM, TEM, DLS, rheological measurements, UV-VIS) were used to comparatively study the effect of these fillers, in correlation with their characteristics, upon the structure and swelling, viscoelastic, and water decontamination properties of (nano)composite hydrogels. The experiments demonstrated that the nanocomposite hydrogel morphology was dictated by the way the filler particles dispersed in water. The equilibrium swelling degree (SDe) depended on both the pH of the environment and the filler nature. At pH 1.2, a slight crosslinking effect of the fillers was evidenced, increasing in the order Mt < Laponite < PS. At pH > pKa$_{MAA}$ (pH 5.4; 7.4; 9.5), the Laponite/Mt-containing hydrogels displayed a higher SDe as compared to the neat one, while at pH 7.4/9.5 the PS-filled hydrogels surprisingly displayed the highest SDe. Rheological measurements on as-prepared hydrogels showed that the filler addition improved the mechanical properties. After equilibrium swelling at pH 5.4, G' and G" depended on the filler, the Laponite-reinforced hydrogels proving to be the strongest. The (nano)composite hydrogels synthesized displayed filler-dependent absorption properties of two cationic dyes used as model water pollutants, Laponite XLS-reinforced hydrogel demonstrating both the highest absorption rate and absorption capacity. Besides wastewater purification, the (nano)composite hydrogels described here may also find applications in the pharmaceutical field as devices for the controlled release of drugs.

**Keywords:** nanocomposites; poly(methacrylic acid); hydrogel; montmorillonite; Laponite; pyrogenic silica; water decontamination

## 1. Introduction

Composite materials have increasingly attracted the attention of researchers within the last few decades due to their properties being far superior to simple materials [1–5]. A special class of such materials is represented by composite hydrogels, which may generally be defined as hydrophilic polymer networks capable of retaining certain quantities of aqueous solutions in the presence of small particles acting as reinforcing agents [6–8]. Composite hydrogels have been and are being studied a great deal due to their improved mechanical, electrical, thermal, and optical properties, absorption capacity, and sensitivity to different stimuli as compared to their unreinforced counterparts [9,10]. The improved properties of the composite hydrogels are a consequence of the synergistic effect of the individual phases, i.e., the polymer matrix and the inorganic filler [11]. Various nano- and microparticles have been used as reinforcing agents for hydrogels over the years, such as graphene and its derivatives (graphene oxide, carbon nanotubes) [12,13], bioactive glass [14], metal nanoparticles (Ag, Au), and metal oxides ($Fe_3O_4$, $Fe_2O_3$, alumina, etc.) [15], the distinct properties of the reinforcing agents allowing the design of the final hydrogel in agreement with the required characteristics.

A special class of fillers for hydrogels is represented by materials that contain Si-O groups within their structure. This category includes various types of silica-based nanoparticles [16], such as pyrogenic silica [17–20] and layered clays such as montmorillonite (Mt) [21] and Laponite (Lap) [22], which have the advantage of being cheaper than the agents listed above. Among them, clays have attracted more and more attention as fillers due to their remarkable properties, such as high specific surface area and adsorption capacity, optimal rheological properties, chemical inertia, and low toxicity, Lap and Mt being the most used for hydrogel reinforcement [6,21–24]. From a structural point of view, they have similarities, as both belong to the smectite class. Clays are usually micro- or nanometric particles made up of layered sheets of 2D silicate (Figure S1). The empirical formula of Mt is $(Na,Ca)_{0.33}(Al,Mg)_2(Si_4O_{10})(OH)_2 \cdot nH_2O$ [25]. Its layers, approximately 1 nm thick and 100 nm × 100 nm width × length, are composed of two tetrahedral sheets formed by O-Si-O bonds and an octahedral sheet formed by O-Al(Mg)-O bonds [26]. The Lap layers are instead in the form of disks of approximately 25 nm diameter and 0.92 nm thickness [26]. The empirical formula of Lap reported in the literature is $Na_{0.7}^+[(Si_8Mg_{5.5}Li_{0.3})O_2(OH)_4]^{-0.7}$ [27]. Commercial Lap is available in different varieties: XLG, XLS, RD, RDS, XLG, and XLS possessing a higher purity and a lower content of heavy metals than Mt. In aqueous dispersions with concentrations higher than 2%, Laponite XLG can form the so-called "house of cards" structure, leading to gelation due to the ionic interactions among the positively-charged edges of some sheets and the negatively-charged faces of others [27]. Laponite XLS does not form the "house of cards" structure because it is modified with pyrophosphate ions [28,29].

Pyrogenic (fumed) silica nanoparticles (Figure S1) are obtained by $SiCl_4$ pyrolysis at temperatures above 1000 °C and are in the form of silica spheres with 5–30 nm diameter, forming particle chains of 100–1000 nm length. These particle chains lead eventually to porous networks that extend up to 250 μm. The primary particles are composed of $SiO_4$ tetrahedra, while on the surface they display both oxygen atoms belonging to siloxane groups (Si-O-Si) and silicon atoms from silanol groups (Si-OH) [30]. Pure pyrogenic silica is hydrophilic and has a high surface energy due to the presence of these groups [31]. Pyrogenic silica nanoparticles are stable in aqueous solutions [32], but unlike layered silicates, they do not dissociate with the formation of free ions [17].

Both clays [24,33] and pyrogenic silica nanoparticles [17–19] have proven their effect on hydrogels, especially by increasing their mechanical and/or thermal properties or changing the swelling degree, but silica nanoparticles are less studied as reinforcing agents for hydrogels as compared to clays. Incorporating clays into the hydrogel matrix is also a good way to increase the final absorption properties of the hydrogel because they can absorb positively-charged pollutants, such as heavy metals and cationic dyes [34]. A study in this regard was conducted by Peng et al., who synthesized a hydrogel from cellulose and Mt that was tested for the adsorption of methylene blue, obtaining an absorption capacity of over 90% [35]. For methylene blue absorption as well, Yi et al. synthesized hybrid hydrogels based on polyacrylamide, sodium humate, and Lap [36].

Acrylic hydrogels are used in many biomedical fields such as drug delivery, intraocular and contact lenses, bone cement for orthopedics, dressings, and implants for regenerative medicine [37]. Another possible application is for removing pollutants from wastewater through possible interactions between the polymer functional groups and the pollutant. It is well known that one of the big challenges of the environment is the pollution of water by dyes, and various studies have shown that hydrogels represent a promising way to solve this problem [38]. Among the acrylic hydrogels, those based on methacrylic acid (MAA) have an important share due to their remarkable properties, such as pH-sensitive character, mucoadhesive characteristics [33], good absorption properties [39], lack of toxicity, etc. [40]. Composite hydrogels based on poly(methacrylic acid) (PMAA) were obtained by reinforcing hydrogels with different types of clays: bentonite [41], Mt [42–44], modified Mt [11], Lap [45,46], kaolin [47], gold nanoparticles [48], pyrogenic silica [17], and carbon nanotubes [49]. By introducing different nanoparticles into the PMAA matrix, properties such as the mechanical and thermal ones, and water absorption have been improved. Potential applications for PMAA-based composite hydrogels reported in the literature include controlled drug release [48,50,51] or treatment of contaminated water [39,44,49]. Regarding the employment of PMAA hydrogels as dye absorbents, there are only two studies on this subject, as far as we know, describing the use of both unreinforced and zeolite-reinforced hydrogels in connection with Yellow 28 dye [39,52].

The present work aims to show how the main properties of the PMAA hydrogels can be engineered by means of several silicon-based fillers, namely Laponite XLS (XLS), Laponite XLG (XLG), montmorillonite (Mt), and two commercial brands of pyrogenic silica differing mainly by the particle size and zeta potential of their 1.5% aqueous dispersion. For this purpose, the effect of these fillers on the structure and swelling, viscoelastic properties, and water decontamination ability of the (nano)composite hydrogels was comparatively studied in correlation with the filler characteristics. The hydrogels obtained were structurally characterized by FT-IR spectroscopy, X-ray diffractometry, thermogravimetric analysis, and electron microscopy (SEM, TEM). The viscoelastic properties were studied for both as-prepared and after equilibrium swelling hydrogels, while their swelling capacity was analyzed in aqueous media with various pHs, as a function of the nature of the reinforcing agent. Finally, the absorption properties of two cationic dyes, namely methylene blue and crystal violet, which are often found in the wastewater from several industries, were studied. To the best of our knowledge, this is the first report of such a comparative study of the influence of various reinforcing nano-agents, in particular clays and pyrogenic silica, in correlation with their characteristics, on the properties of some pH-sensitive PMAA-crosslinked hydrogels. This study also comparatively shows, for the first time, how the cationic dye absorption ability of a composite hydrogel depends on the reinforcing agent used. It should also be mentioned that the nanocomposite PMAA hydrogels reinforced with pyrogenic silica have already been reported in a paper that analyzed only the physical interactions occurring in the nanocomposite hydrogel [17]. Unlike that paper, the present work provides, for the first time, the extended characterization of these hydrogels, both structurally and in terms of rheological and swelling properties. The results presented in this paper may be useful in selecting an appropriate filler in order to adjust the properties of a PMAA hydrogel in agreement with the intended application. Previously, Zhang and Wang [Zhang, J; Wang, A. Study on superabsorbent composites. IX: Synthesis, characterization and swelling behaviours of polyacrylamide/clay composites based on various clays. React. Func. Polym. 2007, 67, 737–745] compared the effect of five clays (attapulgite, kaolinite, mica, vermiculite and Na$^+$-montmorillonite) upon the thermal stability and swelling properties of some superabsorbent polyacrylamide hydrogels.

## 2. Results and Discussion

The studied hydrogels were obtained by the radical copolymerization of MAA with N,N'-methylenebisacrylamide (BIS) in aqueous solution, in the presence of ammonium persulfate (APS) as the initiator, by using different reinforcing agents (XLG, XLS, and Mt clays and two types of pyrogenic silica nanoparticles—HDK and FS—differing mainly by the particle size and zeta potential of their 1.5% aqueous dispersion). The characteristics of both reinforcing agents in water and their aqueous dispersions, determined by us, are displayed in Table 1. According to the measured zeta potential (Table 1), the aqueous dispersions of these fillers are stable over time and allow the acquisition of hydrogels with a uniform distribution of the reinforcing agent. Measurement of their particle size in 1.5 wt% aqueous dispersion showed that only XLG and XLS were exfoliated and dispersed as individual sheets. Mt was in the form of particles composed of non-exfoliated stacked clay layers, possibly mixed with individual sheets, while FS and HDK appeared as agglomerations of silica nanospheres (Table 1). In addition, the measurement of the pH of the filler aqueous dispersion indicated the presence of ionized basic functional groups in the case of clays, while the Si-OH functional groups on the surface of the pyrogenic silica particles were shown to be practically non-ionized in DI (pH ≈ 5.4).

Table 1. Characteristics of the investigated fillers and their aqueous dispersions [1].

| Filler Code | Average Particle Size (nm) | Polydispersity Index | Zeta Potential (mV) | pH of Aqueous Dispersion |
|---|---|---|---|---|
| FS | 367 | 0.417 | −22.8 | 5.8 |
| HDK | 570 | 0.481 | −17.5 | 5.9 |
| Mt | 724.5 | 0.535 | −37.7 | 9.6 |
| XLG | 45.8 | 0.591 | −41.9 | 9.6 |
| XLS | 38.2 | 0.521 | −49.2 | 9.5 |

[1] Concentration of the aqueous dispersion = 1.5 wt%; room temperature; dispersing time in deionized water = 24 h.

The control hydrogel was prepared from an aqueous solution containing 15 wt% MAA and 2 mol% BIS based on MAA, plus the initiator (APS, 1 mol% to MAA), whereas in the case of (nano)composite hydrogels, reinforcing agents in a proportion of 1.5 wt% to the whole reaction mass were added. Hydrogels are indicated by an "H" followed by the abbreviation for the reinforcing agent. For example, "H" indicates the control hydrogel without reinforcing agent, while "HXLG" and "HHDK" stand for the hydrogels having Laponite XLG and HDK N20 pyrogenic silica as reinforcing agents, respectively.

The hydrogels were structurally characterized, and their viscoelastic and swelling properties were also investigated. The viscoelastic properties were investigated both in as-prepared state, when the composition of the hydrogels was the same in all cases, as well as after swelling at equilibrium. In addition, for a correct interpretation of the results, the water absorption/swelling degree was determined after the purification of the hydrogels and calculated only in relation to the amount of polymer in the hydrogel, excluding the mass of reinforcing agent incorporated in the case of composite hydrogels (Equation (6)). Because both the viscoelastic properties of hydrogels and their swelling degree depend on monomer conversion, this was determined in each case (Equation (1)). The results showed very high conversions (92–95%), which allowed comparison of the rheological and swelling measurements in all cases.

*2.1. Hydrogel Structure*

The structure of the synthesized hydrogels was studied by FT-IR spectroscopy, X-ray diffraction (XRD), electron microscopy (SEM, TEM), and thermogravimetric analysis (TGA) measurements. FT-IR spectroscopy was used to identify the interactions between the polymer matrix and each reinforcing agent, and also to observe the influence of the pH of the swelling aqueous medium upon the hydrogel structure. By comparing the FT-IR spectra of the (nano)composite hydrogels swelled in deionized water (pH 5.4) with both the control hydrogel and corresponding reinforcing agent (Figure S2), a strong solid-state interaction between the polymeric matrix and the inorganic agent was revealed by the shift of the Si-O band, characteristic to both clays and pyrogenic silica, to higher wavenumber values. The Si-O band was present in the spectrum of xerogels as a shoulder more (HFS, HDK, Figure S2d,e) or less (HXLG, HXLS, Figure S2a,b) pronounced or as a well-defined peak (HMt, Figure S2c).

Due to the pH-sensitive nature of PMAA hydrogels, their contact with aqueous solutions with different pHs causes changes of the contained functional groups, leading to different swelling behavior. To observe these changes, the hydrogels were analyzed by FT-IR after swelling at pH 1.2, 5.4, 7.4, and 9.5; drying; and grinding (Figure 1). The

FT-IR spectra of the hydrogels swelled at pH 7.4 and 9.5 displayed a band characteristic of the carboxylate group at 1537 cm$^{-1}$, simultaneously with the decrease of the COOH band from 1650 cm$^{-1}$. This may be explained by the basic character of the swelling medium [53], whose pH was appreciably higher than the pKa value of the MAA units in the hydrogel. No notable differences were seen between the spectra at pH 7.4 and 9.5, although the hydrogels behaved differently at these pH values from a swelling point of view (see below). The decrease of pH to 1.2 led to the conversion of all hydrogel groups to COOH, and therefore, to the disappearance of the COO$^-$ band in all spectra. At pH 5.4, which is the pH of the deionized water (DI) in which the hydrogels were synthesized and purified, different situations were encountered in the case of various nanocomposite hydrogels.

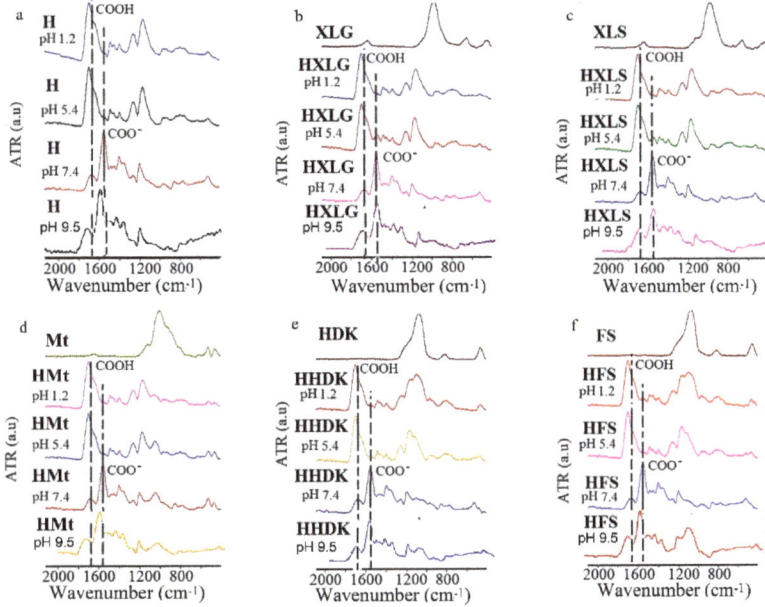

**Figure 1.** FT-IR spectra of (nano)composite hydrogels swelled at different pH values. (**a**) H; (**b**) HXLG; (**c**) HXLS; (**d**) HMt; (**e**) HHDK; (**f**) HFS.

Thus, the FT-IR spectra of HXLG and HXLS (Figure 1b,c) displayed the COO$^-$ group characteristic band, but of a much lower intensity than at pH 7.4/9.5, which can be explained by the basic groups contained by Lap clays, as proven by the basic pH of their aqueous dispersions (Table 1). They reacted with MAA in the hydrogel synthesis step to form COO$^-$ groups, which were then preserved in the hydrogel structure after the purification step [45]. Very interesting is the fact that, although the aqueous dispersion of Mt also had a basic pH, practically identical to that of Lap, the FT-IR spectrum of HMt swelled to pH 5.4 did not show the COO$^-$ characteristic band (Figure 1d). This may be explained by the smaller number of COO$^-$ groups in the hydrogel formed in the synthesis stage as compared to the Lap-reinforced hydrogels. The smaller number of carboxylate groups was probably due to the much weaker exfoliated structure of Mt in water, leading to a lower contact with the monomer. The COO$^-$ group band was not observed at pH 5.4 in the case of HHDK and HFS, due to the lack of basic character of the pyrogenic silica aqueous dispersions (Table 1). As it will be shown below, some of the COOH groups were ionized at pH 5.4 because of the lower pKa of the MAA units, but their concentration was probably too small to become visible in the FT-IR spectrum of H hydrogel (Figure 1a). The presence of COO$^-$ groups in the hydrogels, as indicated by the FT-IR spectra, correlated very well with the swelling experiments that will be presented later in Section 2.2.

The XRD analysis revealed the complete disappearance of the clay characteristic peaks in the case of Lap-reinforced hydrogels (Figure 2a,b), proving that exfoliated nanocomposite hydrogels were obtained [54]. In the case of Mt-reinforced xerogels (Figure 2c), the Mt characteristic reflection at $2\theta = 6.6°$ was present, indicating the formation of intercalated composite hydrogels [42]. Pyrogenic silica-reinforced hydrogels were also characterized by XRD analysis (Figure 2d,e). The X-ray diffraction spectra of the silica nanoparticles showed only a wide band, indicative of the amorphous nature of commercial silica [55]. This band was no longer present in the corresponding hydrogel spectra, suggesting a strong polymer matrix-silica interaction.

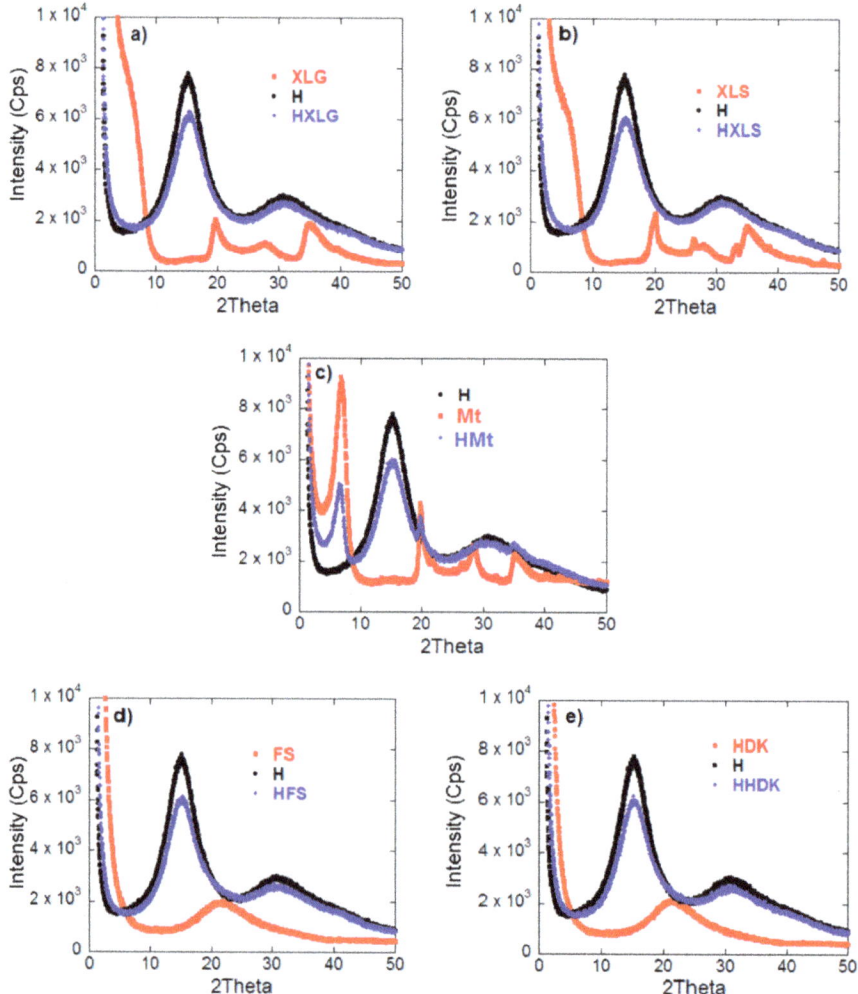

**Figure 2.** XRD spectra of both hydrogels investigated and reinforcing agents used. (**a**) HXLG; (**b**) HXLS; (**c**) HMt; (**d**) HFS; (**e**) HHDK.

The TEM analysis of HHDK and HFS xerogels showed that the fillers were dispersed within the xerogel in the form of approximately spherical particles of several tens of nanometers diameter, associated in large groups (Figure 3). In the case of the Mt-reinforced xerogel, the multi-layer clay particles were visible in the polymer matrix, suggesting the formation of intercalated nanocomposites, while for the XLS- and XLG-filled xerogels the TEM micrographs proved the exfoliation of the clay, as indicated by the single Lap layers with random orientation (Figure 3). Therefore, the TEM analysis supported the conclusions obtained by the XRD measurements. It should also be noted that the filler particles preserved in the hydrogel the structure they had in aqueous dispersion, i.e., isolated sheets in the case of XLG and XLS, stacked clay layers for Mt, and agglomerated spherical nanoparticles in the case of FS and HDK, as revealed by the DLS measurements.

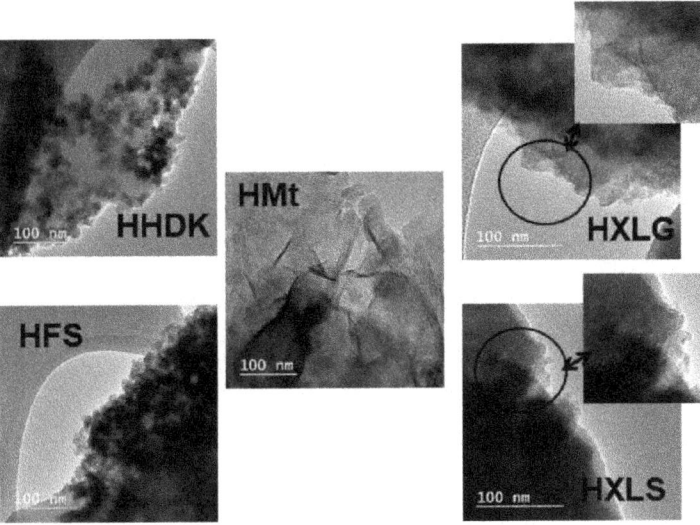

**Figure 3.** TEM images for hydrogels reinforced with various reinforcing agents.

The SEM analysis of the hydrogels swollen in DI and lyophilized revealed a structure with large open and interconnected pores in the case of both unreinforced and reinforced hydrogels (Figure S3).

The TGA investigation of the obtained (nano)composite xerogels showed an increased residue amount at 700 °C as compared with the unreinforced one (Figures 4, S4, and S5, and Table 2), thus proving once again the presence of the inorganic filler within the polymer matrix. The thermogravimetric curves (Figures 4, S3, and S4) showed three decomposition steps. The first step (0–120 °C) was characteristic for the evaporation of the moisture existing in the xerogel and the second step (120–300 °C) was ascribed to both polymer dehydration by inter- and intramolecular water removal and decarboxylation of COOH groups, while the total decomposition of the PMAA chains took place in the third stage (300–700 °C) [50]. The inclusion of the reinforcing agents into the PMAA matrix did not lead to an appreciable change of the decomposition temperature of the polymer. However, some small differences could still be observed regarding the third stage, which seemed to start earlier for Lap samples, followed by the Mt sample and then the pyrogenic silica samples. This may be explained by the basic character of both Laponite and Mt (Table 1), which promoted the decomposition of PMAA chains, unlike the pyrogenic silica samples. As Lap was better dispersed within the xerogels than Mt, its effect on the polymer decomposition was more pronounced than for Mt, and therefore the PMAA chains started to decompose at a lower temperature in the case of Lap-filled samples. One should mention that, according to literature data, the presence of the inorganic component into organic–inorganic nanocomposites does not necessarily increase the thermal resistance of the material, being reported both increase [56] and decrease [57] of the thermal stability of the material, depending on its composition.

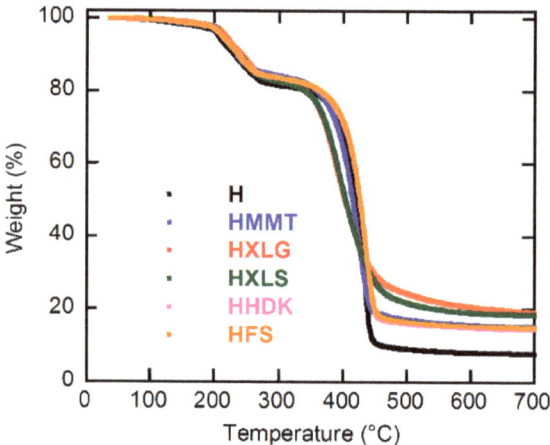

**Figure 4.** TGA curves for the composite xerogels obtained.

**Table 2.** TGA/DTG results obtained for the investigated composite xerogels.

| Hydrogel | Weight Loss (%) | | | Residue at 700 °C (%) |
|---|---|---|---|---|
| | 0–120 °C | 120–300 °C | 300–700 °C | |
| H | 0.90 | 17.38 | 73.78 | 7.92 |
| HMt | 0.59 | 15.07 | 68.87 | 15.42 |
| HXLG | 0.40 | 16.32 | 63.90 | 19.36 |
| HXLS | 0.60 | 16.73 | 64.13 | 18.50 |
| HHDK | 0.62 | 15.51 | 69.18 | 14.65 |
| HFS | 0.66 | 15.65 | 68.59 | 15.08 |

## 2.2. Hydrogel Swelling

The investigation of the swelling properties of the PMAA composite hydrogels at four different pHs (1.2, 5.4, 7.4, 9.5) at 25 °C showed a different behavior of the hydrogels, depending on both pH and reinforcing agent. Due to the pH-sensitive nature of PMAA-based hydrogels, the equilibrium swelling degree increased with increasing pH from 1.2 to 5.4 and further to 7.4/9.5, regardless of the nature of the filler, as expected, but at constant pH, hydrogels with different reinforcing agents behaved differently (Figure 5).

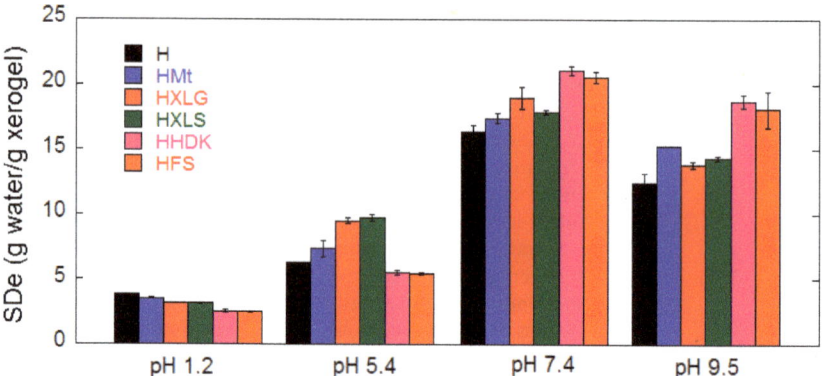

**Figure 5.** Equilibrium swelling degree in four different pH media.

In a strong acidic environment (pH 1.2), the reinforced hydrogels swelled less than H, most likely because of the slight crosslinking effect of the fillers [44]. The additional crosslinking effect induced by the filler could be best identified at this pH because no $COO^-$ groups formed due to the basicity of the medium or the reinforcing agent being present (Figure 1) [43,45,58]. An estimation of the structural parameters of the hydrogels at this pH (Equations (2)–(5)) showed an increase in the crosslinking density ($\rho_c$) as a function of the filler (H < HMt < HXLG ≈ HXLS < HHDK ≈ HFS, Table 3), probably determined by the number of additional PMAA network-reinforcing agent interactions.

**Table 3.** Dependence of the structural parameters $\overline{M}_c$ and $\rho_c$ of the composite hydrogels on the reinforcing agent at pH = 1.2.

| Hydrogel | $\overline{M}_c$ (Da) | $\rho_c \times 10^{-4}$ (mol/cm$^3$) |
|---|---|---|
| H | 643.0 | 20.0 |
| HMt | 539.7 | 23.8 |
| HXLG | 440.3 | 29.2 |
| HXLS | 423.3 | 30.4 |
| HHDK | 271.2 | 47.4 |
| HFS | 266.2 | 48.3 |

The pH increase to 5.4 (deionized water) led to a general SDe increase and also to a different swelling behavior of hydrogels as a function of the reinforcing agent in comparison with their behavior at pH 1.2 (Figure 5). The general SDe increase was due to the ionization of a part of the COOH groups (pKa$_{MAA}$ ≈ 4.5 [59] < 5.4 = pH), which led to both electrostatic repulsion among the $COO^-$ groups and increase of osmotic pressure due to the formation of electrical charges inside the hydrogel [60]. This was the only phenomenon occurring in the case of H, HFS, and HHDK, and therefore, the SDe of silica-reinforced hydrogels was lower than that of the control hydrogel due to the crosslinking effect of inorganic particles (Figure 5).

Unlike FS and HDK, the other three reinforcing agents had a basic reaction in water (Table 1) due to the functional groups contained. As a result, HXLS, HXLG, and HMt contained $COO^-$ groups from the synthesis step, which added to those formed at pH 5.4, leading to higher SDe as compared to H, HFS, and HHDK (Figure 5). The lower amount of $COO^-$ groups of HMt (Figure 1) led to a smaller SDe than for HXLS and HXLG.

Further increase of the swelling medium pH to 7.4 determined a strong SDe increase in all cases, due to the formation of an increased number of $COO^-$ groups within the hydrogels (Figure 5). It can be seen that, while the clay-reinforced hydrogels have maintained their advance in terms of SDe as compared to the control hydrogel for the reason discussed above, the SDe of silica-reinforced hydrogels displayed a dramatic increase at pH 7.4, exceeding the SDe of all the other hydrogels. This may be explained by their weakly acidic character (pKa ≈ 6–6.5 of the Si-OH groups on the surface of pyrogenic silica) [30], which meant they were not ionized at pH 5.4 and therefore they did not alter the water absorption of HFS and HHDK, but at the weak basic pH 7.4 they became ionized, which resulted in an increased water absorption.

The pH rise to 9.5 surprisingly led to a decrease in the SDe values for all hydrogels, while the hydrogel SDe order noticed at pH 7.4 was generally preserved (Figure 5). This phenomenon has been previously reported [61,62] in the case of PMAA hydrogels, but no explanation was provided. We do not have a clear explanation of this effect at this moment. As the FTIR spectra at pH 7.4 and 9.5 did not differ too much (Figure 1), we may only assume that an ionization degree reduction of the hydrogel $COO^-$ groups may have happened, although the ionic strength of the swelling media was the same at these pH values.

## 2.3. Viscoelastic Properties

The viscoelastic properties of the (nano)composite hydrogels were investigated, both in the as-prepared state, when the composition of the hydrogel was the same as in the precursor solution, and after equilibrium swelling in deionized water (pH 5.4). The frequency sweep measurements showed, in all cases, G' higher than G" over the whole investigated range, thus confirming the crosslinked character of the hydrogels (Figure 6). In addition, G' and G" increased with frequency, which is characteristic for networks with wide meshes [63].

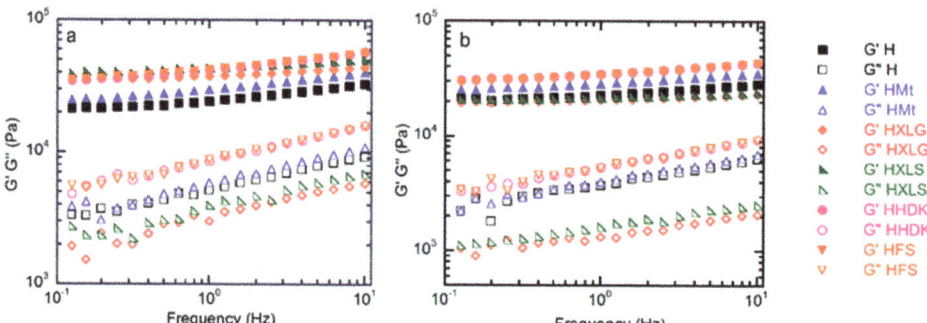

**Figure 6.** Frequency sweep rheological measurements on the (nano)composite hydrogels (**a**) in the as-prepared state and (**b**) after swelling at equilibrium.

The inclusion of reinforcing agents within the PMAA hydrogel matrix improved the mechanical properties, as confirmed by the rheological measurements on as-prepared hydrogels. The G' values of these hydrogels (Figures 6a and 7a) followed roughly the same order as the crosslinking density calculated for the swelled hydrogels at pH 1.2, thus confirming the previous results.

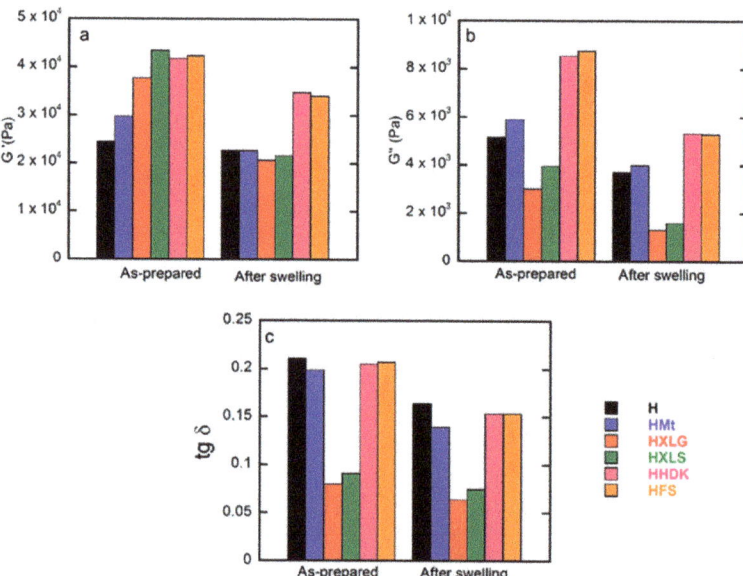

**Figure 7.** Frequency sweep rheological measurements on both as-prepared and equilibrium-swelled (pH = 5.4) hydrogels: (**a**) storage modulus; (**b**) loss modulus; (**c**) loss factor. Values obtained at 1 Hz.

The same measurements on hydrogels swelled at equilibrium in DI (pH 5.4) showed, as expected, a decrease in G' due to increased swelling (Figures 6b and 7b) [64]. The G' values obtained for the hydrogels reinforced with various fillers displayed an inverse tendency in relation to their swelling degree, i.e., the hydrogels with the lowest swelling degree (HFS, HHDK, Figure 5) showed the highest G' (Figure 6a), while the hydrogels with the largest SDe (HXLG, HXLS, Figure 5) showed the lowest G' values (Figure 7a).

The G' and G" values at 1 Hz were used to calculate the loss factor (tan δ = G"/G', Figure 7c). The tan δ values less than 1 obtained for all our hydrogels represented an additional confirmation of their crosslinked character. Additionally, tan δ greater than 0.15 in the case of H, HMt, HHDK, and HFS (Figure 7c), combined with the frequency dependency of G' and G" (Figure 6), indicated their weak gel character [65], in both as-prepared and equilibrium-swollen (pH 5.4) states, although the G' values were in the range of tens of kPa. Stronger hydrogels, with a more elastic character than the others, were obtained with XLG and XLS as the fillers, as indicated by the tan δ values less than 0.1 in both as-prepared and equilibrium-swollen states.

*2.4. Absorption Properties of Cationic Dyes*

Composite hydrogels have begun to gain more and more ground over time as potential candidates for use in the treatment of contaminated water [23,38]. Contamination with dyes from various fields, such as paper printing and textiles, plastics, and food and cosmetics industries is one of the biggest problems of environmental pollution [66]. Methylene blue (MB) and crystal violet (CV) are two of the most-used dyes in these industries. They belong to the category of cationic dyes containing the azo chromophore group and represent a concern for society due to their complex structure and non-biodegradable nature. They also hinder the penetration of sunlight into water, which affects living organisms [67].

MB and CV were used within the present work as cationic dye models in aqueous solutions of 20 mg/L concentration in order to investigate the absorption properties of the synthesized composite hydrogels. The study of dye absorption by the (nano)composite hydrogels showed similar results in the case of both dyes, but the behavior of the hydrogels was different. Thus, HXLS showed the highest rate of dye absorption among all hydrogels, practically the entire amount of dye being absorbed from the solution in less than 30 min (Figure 8), a performance that was not achieved by the other hydrogels, even after 30 h (Figure 9).

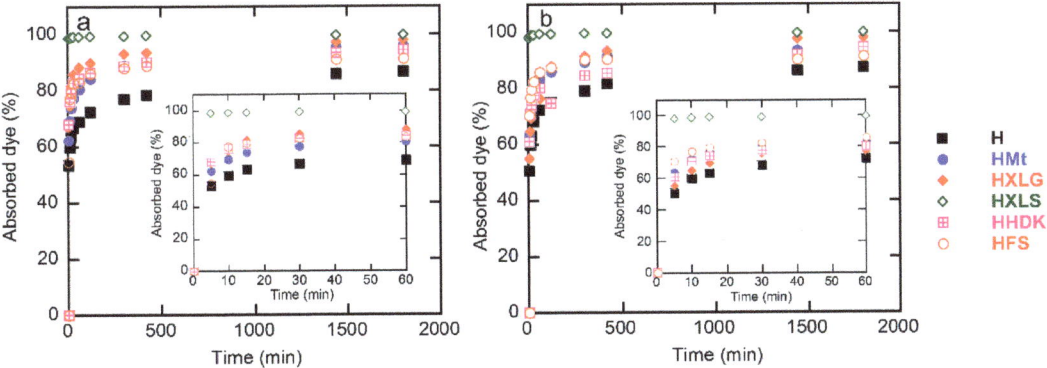

**Figure 8.** Time dependence of the percentage of dye absorbed by the (nano)composite hydrogels: (**a**) methylene blue; (**b**) crystal violet. Initial dye concentration = 20 mg/L; xerogel amount = 0.003 g; dye solution volume = 30 mL; temperature = 25 ± 0.5 °C.

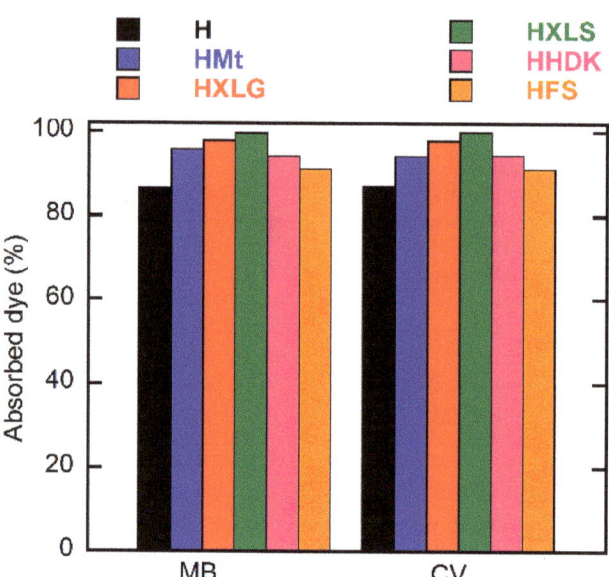

**Figure 9.** Dependence of the percentage of dye absorbed on reinforcing agent. Initial dye concentration = 20 mg/L; xerogel amount = 0.003 g; dye solution volume = 30 mL; temperature = 25 ± 0.5 °C; time = 30 h.

For both dyes, a lowest absorption rate was displayed by the control hydrogel. This demonstrated the overall contribution of the reinforcing agent to the absorption process, through both its chemical structure and effect upon the hydrogel swelling degree.

In the case of PMAA hydrogels, the absorption of the cationic dyes was determined mainly by the carboxylate groups formed in the hydrogel in deionized water (pH 5.4), as explained earlier, which attract the positively-charged dye [68]. The control hydrogel, HFS, and HHDK contained only carboxylate groups from the ionization of MAA units in water, while hydrogels reinforced with clays (HMt, HXLG, HXLS) comprised additional carboxylate groups due to the clay basic reaction in water. In addition, the clay network also contained anionic groups that rapidly ionize in water, leading to the fast absorption of the cationic dye by an ion exchange process involving the positively-charged inorganic counterions of the clay [36]. Due to its preparation procedure, Laponite XLS also contains tetrasodium pyrophosphate, which represents additional anionic charges within the hydrogel, leading to an increased rate of dye absorption (Figure 8).

Although FS and HDK do not influence the pH of deionized water (Table 1), the weakly acidic groups on the surface may contribute to the absorption of the dye. As a result, HFS and HHDK displayed a higher absorption rate as compared to the control hydrogel. The absorption curves showed in all cases a high absorption rate of the dye in the first 60 min, after which it considerably decreased (Figure 8). At the end of the investigated period (30 h), the absorption of the dye was over 85% in all cases, the lowest value being recorded for the control sample and the highest in the case of HXLS (Figure 9). The amount of dye $\rho$ (mg) absorbed by 1 g of hydrogel after 30 h also depended on the nature of the reinforcing agent, decreasing for both dyes in the order HXLS > HXLG > HMt ≈ HHDK > HFS > H, from approximately 199.5 mg/g (MB)/196 mg/g (CV) in the case of HXLS to approximately 173 mg/g (MB)/174 mg/g (CV) for H (Figure 10). This represents a further confirmation of the beneficial contribution of the reinforcing agents described here to the absorption of cationic dyes.

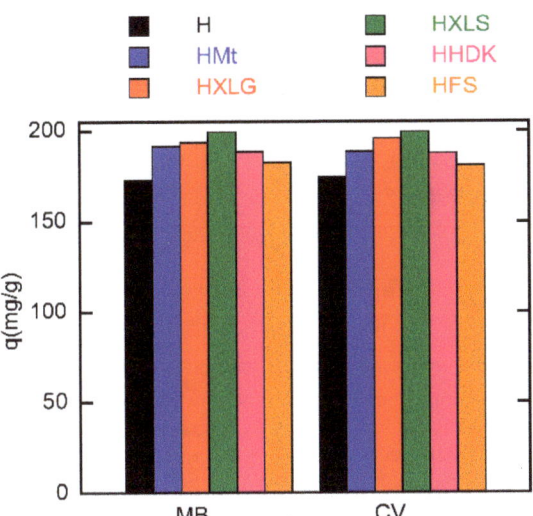

**Figure 10.** Dependence of the absorption capacity on the filler of the (nano)composite hydrogels. Initial dye concentration = 20 mg/L; xerogel amount = 0.003 g; dye solution volume = 30 mL; temperature = 25 ± 0.5 °C; time = 30 h.

## 3. Materials and Methods

### 3.1. Materials

Laponite XLG and Laponite XLS (BYK Additives & Instruments, kindly donated by Cosichem & Analytical, Romania), the two commercial brands of hydrophilic pyrogenic silica (fumed silica, FS, Sigma and HDK N20 pyrogenic silica, HDK, Wacker Silicones), methacrylic acid (MAA, Jansen Chemistry, 99%), N,N′-methylenebisacrylamide (BIS, 99%, Sigma Aldrich), ammonium persulfate (APS, Sigma Aldrich, 98%), methylene blue (MB, Loba Chemie, 95%), and crystal violet (CV, Loba Chemie, 96%) were used as received. Unmodified montmorillonite (Mt, Cloisite Na, a gift from Southern Clay Products Inc. Gonzales, TX, USA) was purified by the method previously described [43]. The characteristics of both reinforcing agents in water and their aqueous dispersions, determined by us, are displayed in Table 1. Deionized water (DI, 18.2 MΩ resistivity) was employed as the solvent for all syntheses. The swelling investigations were performed in DI (pH 5.4), HCl solution (pH 1.2), PBS solution (0.01 M, pH 7.4) obtained from tablets (Phosphate Buffered Saline, Sigma) and $NaHCO_3$-$Na_2CO_3$ buffer (0.01 M, pH 9.5), whose ionic strength was adjusted by adding NaCl to make it equal to that of PBS (I = 0.17).

### 3.2. Synthesis of Hydrogels

The procedure employed was similar to the one previously described in reference [43]. The hydrogels were obtained by the radical copolymerization of MAA (15 wt% to the whole reaction mass) and BIS (2 mol% based on MAA) in the presence of APS (1 mol% based on MAA) as an initiator in aqueous solution. For the preparation of composite hydrogels, the calculated amount of reinforcing agent (1.5 wt% based on the whole hydrogel composition) was dispersed in DI for 24 h, and then MAA and BIS were added, followed by the initiator. The final reaction mixture was injected into a mold made up of two glass plates separated by a 1-mm thick Teflon gasket and glued with silicone. The mold was then placed in a thermostatted bath at 60 °C for 5 h. At the end of the polymerization time, the mold was removed from the thermostatted bath and allowed to cool to room temperature. Discs with diameters of 20 mm and 8 mm were cut from the material resulted after opening the mold. Part of the 20-mm discs were immediately subjected to rheological analysis, while the rest were analyzed after undergoing a purification-swelling process in DI for 7 days, together with the 8-mm discs, with a daily change of water.

To determine the monomer conversion, part of the hydrogel was first dried in atmosphere and then over anhydrous $CaCl_2$ to a constant mass and weighed ($W_o$). The dry xerogel was then immersed for purification in excess DI, which was changed daily for 7 days, dried again, then the final mass ($W_{ext}$) was recorded. The monomer conversion (C%) was calculated according to Equation (1):

$$C\% = (W_{ext} - W_{AR})/(W_o - W_{AR}) \cdot 100 \tag{1}$$

where $W_{AR}$ is the amount of reinforcing agent contained in the material employed to determine the conversion.

### 3.3. Crosslinking Density and Average Molecular Weight between Crosslinks

The average molecular weight between crosslinks ($\overline{M}_c$) and the crosslinking density ($\rho_c$) were estimated in the case of hydrogels swelled at pH 1.2 in order to avoid the influence of the basic character of the filler aqueous dispersion on the swelling degree of the composite hydrogels. At pH 1.2, all the MAA units within the hydrogel were in acid form, and therefore the hydrogels can be considered as non-ionic. In the case of non-ionic hydrogels and neglecting the contribution of the chain-end defects, $\overline{M}_c$ can be calculated according to Equation (2) [64,69]:

$$\overline{M}_c = -\frac{(1 - \frac{2}{f})V_1 v_{2r}^{2/3} v_{2m}^{1/3}}{\overline{v}[\ln(1 - v_{2m}) + v_{2m} + \chi v_{2m}^2]} \tag{2}$$

where f is the functionality of the crosslinking site (f = 4), $V_1$ represents the molar volume of the solvent (for water $V_1$ = 18 cm$^3$/mol), $v_{2r}$ and $v_{2m}$ represent the polymer volume fraction of the as-prepared (relaxed state) hydrogel and the hydrogel swelled at equilibrium, respectively, $\overline{v}$ is the specific volume of the polymer, calculated as $1/\rho_p$ (cm$^3$/g), while $\chi$ represents the Flory polymer–solvent interaction factor (0.48 for water—PMAA at 0% degree of ionization [70]). Both $v_{2r}$ and $v_{2m}$ were calculated by excluding the amount of filler contained by hydrogel and xerogel.

The polymer volume fractions ($v_{2r}, v_{2m}$) were calculated using Equation (3) [71]:

$$v_{2r/m} = \left[1 + \frac{\rho_p}{\rho_s} \times \left(\frac{W_h}{W_x} - 1\right)\right]^{-1} \tag{3}$$

where $\rho_p$ and $\rho_s$ are the polymer and solvent densities (1.00 g/cm$^3$ in the case of water), respectively, while $W_h$ and $W_x$ represent the mass of the hydrogel at equilibrium swelling ($v_{2m}$) or under as-prepared conditions ($v_{2r}$), and the mass of the corresponding xerogel, respectively, after excluding the filler in both cases. The polymer density was determined by the picnometer method, using Equation (4) and toluene as the non-solvent [72]. The polymer density represents the density of the H xerogel (corresponding to the hydrogel without filler).

$$\rho_X = \frac{m_X \rho_T}{m_a + m_X - m_b} \tag{4}$$

where $m_X$ is the xerogel weight, $\rho_T$ is the density of toluene, determined by us with the picnometer, while $m_a$ and $m_b$ represent the weights of the picnometer filled with toluene and of the picnometer filled with toluene and containing the xerogel sample, respectively.

The crosslinking density of the hydrogels was calculated according to Equation (5) [71]:

$$\rho_c = \frac{1}{\overline{v} \cdot \overline{M}_c} \tag{5}$$

*3.4. Swelling Degree*

To determine the swelling degree, pre-weighed xerogels ($W_X$) were immersed in 40 mL of swelling solution and kept at 25 ± 0.5 °C for 72 h. At the end of the time interval, the hydrogels were removed from the solution, wiped superficially with filter paper, and weighed ($W_H$). The equilibrium swelling degree (SDe) was calculated as the ratio of the amount of water absorbed to the mass of polymer in xerogel according to Equation (6):

$$\text{SDe (g/g)} = (W_H - W_X)/(W_X \cdot (100 - \%A_R)/100) \quad (6)$$

where $\%A_R$ is the percentage of the reinforcing agent in xerogel, calculated based on the amount of monomers and filler employed in the polymerization and the overall monomer conversion. Each experiment was carried out in duplicate, and the average value ± the error was calculated for each point and reported.

*3.5. Absorption Properties*

To test the absorption properties of the synthesized materials, two cationic dyes, methylene blue (MB, maximum UV absorption at $\lambda$ = 664 nm [73]) and crystal violet (CV, $\lambda$ = 591 nm [74]), were used as models. The absorption capacity of hydrogels was determined using xerogel powders obtained by grinding in a ball mill, followed by drying over anhydrous $CaCl_2$. A precisely weighed amount of approximately 0.003 g of xerogel was immersed in a centrifuge tube containing approximately 30 mL of a 20 mg/L aqueous dye solution and kept at 25 ± 0.5 °C with continuous orbital stirring on a heating–cooling dry plate (Torrey Pines Scientific Inc., USA) for 30 h. The pH of the dye solutions was 5.0 in the case of MB and 5.1 for crystal violet. The xerogel amount/dye solution volume ratio was kept constant in order to be able to compare the results obtained for the different xerogels used. Solution samples of 2 mL volume were withdrawn at predetermined time intervals, and the dye concentration was determined by UV-VIS spectrometry (UV-VIS Thermo instrument) by comparison with a calibration curve. To keep the solution volume constant, the volume withdrawn was reintroduced into the centrifuge tube after each analysis. The percentage of dye absorbed at various time intervals ($PC_t$) and the amount of dye absorbed by 1 g of hydrogel ($\rho$) were determined using Equations (7) and (8):

$$PC_t\ (\%) = \frac{C_o - C_t}{C_o} \times 100 \quad (7)$$

$$\rho\ (\text{mg/g}) = \frac{C_o - C_t}{m} \times V \quad (8)$$

where $C_o$ and $C_t$ (mg/L) represent the initial dye concentration and the concentration at time t, respectively, m = amount of xerogel used (g), and V = volume of solution used (L).

*3.6. Characterizations*

The viscoelastic properties of the hydrogels were investigated in both as-prepared and equilibrium-swollen state by means of a Kinexus Pro rheometer (Malvern Instruments, UK, software 1.60), equipped with a Peltier element for temperature control. The rheological measurements were performed at 25 °C using a geometry with 20-mm parallel plates with rough faces to avoid slippage. The applied normal force was 0.5 N. The amplitude sweep measurements were performed at 1 Hz constant frequency, and a strain within the linear viscoelasticity region was selected to be used in the frequency sweep measurements.

FT-IR analyses were run on a Tensor 37 Bruker (Woodstock, NY, USA) equipment with a resolution of 4 $cm^{-1}$ and accumulation of 16 scans. The X-ray diffraction measurements were carried out in continuous mode, at room temperature and atmospheric pressure, on a Rigaku Ultima IV (Tokyo, Japan) instrument with CuK$\alpha$ radiation ($\lambda$ = 1.5406 Å), operated at 40 kV and 30 mA. The data were collected over the 2$\theta$ range 1–50° and a scanning speed of 1°/min. Xerogel powders were used in both cases.

The thermogravimetric analyses (TGA) were performed on a TA Q5000 IR (TA Instruments) equipment under nitrogen atmosphere (40 mL/min), the samples being heated from room temperature to 700 °C at a rate of 10 °C/min.

SEM micrographs were obtained using an environmental scanning electron microscope (ESEM-FEI Quanta 200, Eindhoven, The Netherlands). The analyses were carried out under low vacuum at 30 kV acceleration voltage. The swollen samples were previously freeze-dried in an ALPHA 1-2 LDplus lyophilizer, Martin Christ, Germany. The morphology of the nanocomposite xerogels were obtained by transmission electron microscopy (TEM) by means of a TECNAI F20 G$^2$ TWIN Cryo-TEM (FEI, USA) equipment. The dried hydrogels were milled to a fine powder. Lacey formvar/carbon 200 mesh copper grids from Ted Pella, Inc., Redding, CA, USA, were gently contacted with each powder and thoroughly shaken afterwards to remove large powder particles. The samples were analyzed in bright field mode at an acceleration voltage of 120 kV.

The particle size of the reinforcing agents was measured by dynamic light scattering (DLS). The zeta potential of each aqueous dispersion of nanoparticles with a concentration of 1.5% was also measured. The measurements were performed by using a Nano ZS ZEN3600 Zetasizer instrument, Malvern Instruments (Malvern, UK). The pH of the aqueous dispersions of the reinforcing agents was measured with a pH-meter (Hanna Instruments), which was previously calibrated.

## 4. Conclusions

The present study showed that both the structure of hydrogels and some of their properties, such as viscoelastic, swelling, or absorbent/depolluting properties, can be adjusted to meet the requirements of a certain application by the correct choice of the reinforcing agent. To demonstrate this, the effect of five silicon-based fillers (Laponite XLS and XLG, Mt, and two commercial brands of pyrogenic silica), which differed mainly in particle size and pH of the aqueous dispersion, upon the PMAA crosslinked hydrogels was compared. Thus, XRD and TEM analyses showed that the two Lap clays and Mt led to exfoliated and respectively intercalated nanocomposites, while pyrogenic silica formed agglomerations of spherical nanoparticles within the hydrogel. The distribution of the nanoparticles within hydrogels was in agreement with the way the filler particles dispersed in water, i.e., isolated sheets in the case of XLG and XLS, stacked clay layers for Mt, and agglomerated spherical nanoparticles in the case of FS and HDK, as revealed by the DLS measurements. The structure of PMAA hydrogels was also affected by the reinforcing agent through the pH of its aqueous dispersion. Thus, Lap and Mt, displaying a basic reaction in water, determined the formation of carboxylate groups on the PMAA chains during the hydrogel synthesis stage. This led to an increase of the SDe of the nanocomposite hydrogels as compared to the one without filler at pH values larger than the pKa of the MAA monomer units.

The effect of the filler on SDe depended on both its nature and the pH of the swelling medium. At strong acidic pH (pH 1.2), a slight crosslinking effect of the filler was noticed, the most pronounced effect being displayed by pyrogenic silica. At pH 5.4, SDe significantly increased for all samples, the most pronounced increase occurring for the hydrogels synthesized in the presence of Laponite XLS and XLG. The crosslinking effect of pyrogenic silica particles was still evident at this pH. At basic pH, i.e., 7.4 and 9.5, the SDe of the pyrogenic silica-reinforced hydrogels surprisingly exceeded the SDe of all the other hydrogels. Also unexpectedly, the SDe at pH 9.5 was lower than at 7.4 for all hydrogels.

At approximately constant swelling degree, the filler addition improved th mechanical properties of the (nano)composite hydrogels, while after equilibrium swelling (pH 5.4), the viscoelastic moduli values depended on the filler. The rheological measurements also showed that the strongest hydrogels were obtained in the case of Lap as the reinforcing agent.

The (nano)composite hydrogels synthesized displayed a different ability to decontaminate cationic dye-containing waters as a function of the filler included, with both the highest absorption rate and absorption capacity being displayed by the Laponite XLS-reinforced hydrogel. Additionally, the presence of the filler within the hydrogel increased, in all cases, the amount of dye absorbed.

**Supplementary Materials:** The following supporting information can be downloaded at: https://www.mdpi.com/article/10.3390/ijms231810320/s1.

**Author Contributions:** Conceptualization, C.M.N. and M.T.; investigation, C.M.N., R.I., E.A., C.I.M., S.B., B.T., C.L.N., S.P., C.S. and C.G.; methodology, C.M.N., R.I. and M.T.; supervision, C.P. and M.T.; writing—original draft, C.L.N. and M.T.; writing—review and editing, C.M.N., C.S., C.G., C.P. and M.T. All authors have read and agreed to the published version of the manuscript.

**Funding:** This work was funded by POC 2016 SECVENT nr. 81/2016 project, subsidiary contract 1817/2020, MySMIS: 105684.

**Institutional Review Board Statement:** Not applicable.

**Informed Consent Statement:** Not applicable.

**Data Availability Statement:** The data presented in this study are available on request from the corresponding author.

**Conflicts of Interest:** The authors declare no conflict of interest.

# References

1. Egbo, M.K. A Fundamental Review on Composite Materials and Some of Their Applications in Biomedical Engineering. *J. King Saud Univ. Eng. Sci.* **2020**, *33*, 557–568. [CrossRef]
2. Tishkevich, D.I.; Vorobjova, A.I.; Trukhanov, A.V. Thermal Stability of Nano-Crystalline Nickel Electrodeposited into Porous Alumina. *Solid State Phenom.* **2020**, *299*, 281–286. [CrossRef]
3. Tishkevich, D.I.; Grabchikov, S.S.; Grabchikova, E.A.; Vasin, D.S.; Yakushevich, A.S.; Vinnik, D.A.; Zubar, T.I.; Kalagin, I.V.; Yakimchuk, D.V.; Trukhanov, A.V. Modeling of Paths and Energy Losses of High-Energy Ions in Single-Layered and Multilayered Materials. *IOP Conf. Ser. Mater. Sci. Eng.* **2020**, *848*, 012089. [CrossRef]
4. Tishkevich, D.I.; Vorobjova, A.I.; Vinnik, D.A. Formation and Corrosion Behavior of Nickel/Alumina Nanocomposites. *Solid State Phenom.* **2020**, *299*, 100–106. [CrossRef]
5. Tishkevich, D.I.; Zubar, T.I.; Zhaludkevich, A.L.; Razanau, I.U.; Vershinina, T.N.; Bondaruk, A.A.; Zheleznova, E.K.; Dong, M.; Hanfi, M.Y.; Sayyed, M.I.; et al. Isostatic Hot Pressed W–Cu Composites with Nanosized Grain Boundaries: Microstructure, Structure and Radiation Shielding Efficiency against Gamma Rays. *Nanomaterials* **2022**, *12*, 1642. [CrossRef]
6. Schexnailder, P.; Schmidt, G. Nanocomposite Polymer Hydrogels. *Colloid. Polym. Sci.* **2009**, *287*, 1–11. [CrossRef]
7. Gaharwar, A.K.; Peppas, N.A.; Khademhosseini, A. Nanocomposite Hydrogels for Biomedical Applications: Nanocomposite Hydrogels. *Biotechnol. Bioeng.* **2014**, *111*, 441–453. [CrossRef]
8. Pavlyuchenko, V.N.; Ivanchev, S.S. Composite Polymer Hydrogels. *Polym. Sci. Ser. A* **2009**, *51*, 743–760. [CrossRef]
9. Khan, M.; Shah, L.A.; Khan, M.A.; Khattak, N.S.; Zhao, H. Synthesis of an Un-Modified Gum Arabic and Acrylic Acid Based Physically Cross-Linked Hydrogels with High Mechanical, Self-Sustainable and Self-Healable Performance. *Mater. Sci. Eng. C* **2020**, *116*, 111278. [CrossRef]
10. Subhan, H.; Alam, S.; Shah, L.A.; Khattak, N.S.; Zekker, I. Sodium Alginate Grafted Hydrogel for Adsorption of Methylene Green and Use of the Waste as an Adsorbent for the Separation of Emulsified Oil. *J. Water Proc. Eng.* **2022**, *46*, 102546. [CrossRef]
11. Ianchis, R.; Ninciuleanu, C.; Gifu, I.; Alexandrescu, E.; Somoghi, R.; Gabor, A.; Preda, S.; Nistor, C.; Nitu, S.; Petcu, C.; et al. Novel Hydrogel-Advanced Modified Clay Nanocomposites as Possible Vehicles for Drug Delivery and Controlled Release. *Nanomaterials* **2017**, *7*, 443. [CrossRef] [PubMed]
12. Yi, J.; Choe, G.; Park, J.; Lee, J.Y. Graphene Oxide-Incorporated Hydrogels for Biomedical Applications. *Polym J.* **2020**, *52*, 823–837. [CrossRef]
13. Zhao, Y.; Terai, W.; Hoshijima, Y.; Gotoh, K.; Matsuura, K.; Matsumura, K. Development and Characterization of a Poly(Vinyl Alcohol)/Graphene Oxide Composite Hydrogel as An Artificial Cartilage Material. *Appl. Sci.* **2018**, *8*, 2272. [CrossRef]
14. Killion, J.A.; Kehoe, S.; Geever, L.M.; Devine, D.M.; Sheehan, E.; Boyd, D.; Higginbotham, C.L. Hydrogel/Bioactive Glass Composites for Bone Regeneration Applications: Synthesis and Characterisation. *Mater. Sci. Eng. C* **2013**, *33*, 4203–4212. [CrossRef]
15. Tan, H.-L.; Teow, S.-Y.; Pushpamalar, J. Application of Metal Nanoparticle–Hydrogel Composites in Tissue Regeneration. *Bioengineering* **2019**, *6*, 17. [CrossRef]
16. Dannert, C.; Stokke, B.T.; Dias, R.S. Nanoparticle-Hydrogel Composites: From Molecular Interactions to Macroscopic Behavior. *Polymers* **2019**, *11*, 275. [CrossRef]

17. Terziyan, T.V.; Safronov, A.P.; Belous, Y.G. Interaction of Aerosil Nanoparticles with Networks of Polyacrylamide, Poly (Acrylic Acid), and Poly(Methacrylic Acid) Hydrogels. *Polym. Sci. Ser. A* **2015**, *57*, 200–208. [CrossRef]
18. Schiraldi, C.; D'Agostino, A.; Oliva, A.; Flamma, F.; De Rosa, A.; Apicella, A.; Aversa, R.; De Rosa, M. Development of Hybrid Materials Based on Hydroxyethylmethacrylate as Supports for Improving Cell Adhesion and Proliferation. *Biomaterials* **2004**, *25*, 3645–3653. [CrossRef]
19. Slisenko, O.V. Synthesis and Swelling Behaviour of Polyacrylamide/Modified Silica Hybrid Gels. In Proceedings of the 2017 IEEE 7th International Conference Nanomaterials: Application & Properties (NAP), Odessa, UKraine, 10–15 September 2017.
20. Kertkal, N.; Jinawong, P.; Rithiyong, A.; Kusuktham, B. Hybrid Hydrogels for PH Indicator. *Silicon* **2021**, *14*, 2609–2624. [CrossRef]
21. Bao, Y.; Ma, J.; Yang, Z. Preparation and Application of Poly (Methacrylic Acid)/Montmorillonite Nanocomposites. *Mater. Manuf. Processes.* **2011**, *26*, 604–608. [CrossRef]
22. Zhao, L.Z.; Zhou, C.H.; Wang, J.; Tong, D.S.; Yu, W.H.; Wang, H. Recent Advances in Clay Mineral-Containing Nanocomposite Hydrogels. *Soft. Matter.* **2015**, *11*, 9229–9246. [CrossRef] [PubMed]
23. Khan, S.A.; Khan, T.A. Clay-Hydrogel Nanocomposites for Adsorptive Amputation of Environmental Contaminants from Aqueous Phase: A Review. *J. Environ. Chem. Eng.* **2021**, *9*, 105575. [CrossRef]
24. Li, P.; Kim, N.H.; Hui, D.; Rhee, K.Y.; Lee, J.H. Improved Mechanical and Swelling Behavior of the Composite Hydrogels Prepared by Ionic Monomer and Acid-Activated Laponite. *Appl. Clay Sci.* **2009**, *46*, 414–417. [CrossRef]
25. Uddin, F. Montmorillonite: An introduction to properties and utilization. In *Current Topics in the Utilization of Clay in Industrial and Medical Applications*; Zoveidavianpoor, M., Ed.; IntechOpen: London, UK, 2018.
26. Zhou, C.; Tong, D.; Yu, W. 7—Smectite Nanomaterials: Preparation, Properties, and Functional Applications. In *Nanomaterials from Clay Minerals*; Wang, A., Wang, W., Eds.; Micro and Nano Technologies; Elsevier: Amsterdam, The Netherlands, 2019; pp. 335–364. ISBN 978-0-12-814533-3.
27. Tomás, H.; Alves, C.S.; Rodrigues, J. Laponite®: A Key Nanoplatform for Biomedical Applications? *Nanomed Nanotechnol. Biol. Med.* **2018**, *14*, 2407–2420. [CrossRef] [PubMed]
28. Liu, Y.; Zhu, M.; Liu, X.; Zhang, W.; Sun, B.; Chen, Y.; Adler, H.-J.P. High Clay Content Nanocomposite Hydrogels with Surprising Mechanical Strength and Interesting Deswelling Kinetics. *Polymer* **2006**, *47*, 1–5. [CrossRef]
29. Afghah, F.; Altunbek, M.; Dikyol, C.; Koc, B. Preparation and Characterization of Nanoclay-Hydrogel Composite Support-Bath for Bioprinting of Complex Structures. *Sci. Rep.* **2020**, *10*, 5257. [CrossRef]
30. Vansant, E.F.; Voort, P.V.D.; Vrancken, K.C. *Characterization and Chemical Modification of the Silica Surface*; Elsevier: Amsterdam, The Netherlands, 1995; Volume 93, ISBN 978-0-08-052895-3.
31. Whitby, C.P. Structuring Edible Oils With Fumed Silica Particles. *Front. Sustain. Food Syst.* **2020**, *4*, 201–209. [CrossRef]
32. Yang, J.; Han, C.-R.; Duan, J.-F.; Xu, F.; Sun, R.-C. In Situ Grafting Silica Nanoparticles Reinforced Nanocomposite Hydrogels. *Nanoscale* **2013**, *5*, 10858–10863. [CrossRef]
33. Güler, M.A.; Gök, M.K.; Figen, A.K.; Özgümüş, S. Swelling, Mechanical and Mucoadhesion Properties of Mt/Starch-g-PMAA Nanocomposite Hydrogels. *Appl. Clay Sci.* **2015**, *112–113*, 44–52. [CrossRef]
34. Shabtai, I.A.; Lynch, L.M.; Mishael, Y.G. Designing Clay-Polymer Nanocomposite Sorbents for Water Treatment: A Review and Meta-Analysis of the Past Decade. *Water Res.* **2021**, *188*, 116571. [CrossRef]
35. Peng, N.; Hu, D.; Zeng, J.; Li, Y.; Liang, L.; Chang, C. Superabsorbent Cellulose–Clay Nanocomposite Hydrogels for Highly Efficient Removal of Dye in Water. *ACS Sustain. Chem. Eng.* **2016**, *4*, 7217–7224. [CrossRef]
36. Yi, J.-Z.; Zhan, L.-M. Removal of Methylene Blue Dye from Aqueous Solution by Adsorption onto Sodium Humate/Polyacrylamide/Clay Hybrid Hydrogels. *Bioresour. Technol.* **2008**, *99*, 2182–2186. [CrossRef]
37. Serrano-Aroca, Á.; Deb, S. Acrylic-Based Hydrogels as Advanced Biomaterials. In *Acrylate Polymers for Advanced Applications*; Serrano-Aroca, Á., Deb, S., Eds.; IntechOpen: London, UK, 2020; ISBN 978-1-78985-183-0.
38. Pereira, A.G.B.; Rodrigues, F.H.A.; Paulino, A.T.; Martins, A.F.; Fajardo, A.R. Recent Advances on Composite Hydrogels Designed for the Remediation of Dye-Contaminated Water and Wastewater: A Review. *J. Cleaner. Prod.* **2021**, *284*, 124703. [CrossRef]
39. Panic, V.V.; Velickovic, S.J. Removal of Model Cationic Dye by Adsorption onto Poly(Methacrylic Acid)/Zeolite Hydrogel Composites: Kinetics, Equilibrium Study and Image Analysis. *Sep. Purif. Technol.* **2014**, *122*, 384–394. [CrossRef]
40. Munteanu, T.; Ninciuleanu, C.M.; Gifu, I.C.; Trica, B.; Alexandrescu, E.; Gabor, A.R.; Preda, S.; Petcu, C.; Nistor, C.L.; Nitu, S.G.; et al. The Effect of Clay Type on the Physicochemical Properties of New Hydrogel Clay Nanocomposites. In *Current Topics in the Utilization of Clay in Industrial and Medical Applications*; Zoveidavianpoor, M., Ed.; InTech: London, UK, 2018; ISBN 978-1-78923-728-3.
41. Zhumagaliyeva, S.N.; Iminova, R.S.; Kairalapova, G.Z.; Beysebekov, M.M.; Beysebekov, M.K.; Abilov, Z.A. Composite Polymer-Clay Hydrogels Based on Bentonite Clay and Acrylates: Synthesis, Characterization and Swelling Capacity. *Eurasian Chem. Technol. J.* **2017**, *19*, 279–288. [CrossRef]
42. Junior, C.R.F.; de Moura, M.R.; Aouada, F.A. Synthesis and Characterization of Intercalated Nanocomposites Based on Poly(Methacrylic Acid) Hydrogel and Nanoclay Cloisite-Na$^+$ for Possible Application in Agriculture. *J. Nanosci. Nanotechnol.* **2017**, *17*, 5878–5883. [CrossRef]
43. Ninciuleanu, C.M.; Ianchiş, R.; Alexandrescu, E.; Mihăescu, C.I.; Scomoroşcenco, C.; Nistor, C.L.; Preda, S.; Petcu, C.; Teodorescu, M. The Effects of Monomer, Crosslinking Agent, and Filler Concentrations on the Viscoelastic and Swelling Properties of Poly(Methacrylic Acid) Hydrogels: A Comparison. *Materials* **2021**, *14*, 2305. [CrossRef] [PubMed]

44. Khan, S.A.; Siddiqui, M.F.; Khan, T.A. Synthesis of Poly(Methacrylic Acid)/Montmorillonite Hydrogel Nanocomposite for Efficient Adsorption of Amoxicillin and Diclofenac from Aqueous Environment: Kinetic, Isotherm, Reusability, and Thermodynamic Investigations. *ACS Omega* **2020**, *5*, 2843–2855. [CrossRef] [PubMed]
45. Ninciuleanu, C.; Ianchis, R.; Alexandrescu, E.; Mihaescu, C.; Trica, B.; Scomoroscenco, C.; Petcu, C.; Preda, S.; Teodorescu, M. Nanocomposite Hydrogels Based on Poly(Methacrylic Acid) and Laponite XLG. *UPB Sci. Bull. Ser. B-Chem. Mater. Sci.* **2021**, *83*, 43–58.
46. Junior, C.R.F.; Fernandes, R.d.S.; de Moura, M.R.; Aouada, F.A. On the Preparation and Physicochemical Properties of PH-Responsive Hydrogel Nanocomposite Based on Poly(Acid Methacrylic)/Laponite RDS. *Mater. Today Commun.* **2020**, *23*, 100936. [CrossRef]
47. Sadeghi, M. Synthesis and Swelling Behavior of Protein-g-poly Methacrylic Acid/Kaolin Superabsorbent Hydrogel Composites. *AIP Conf. Proc.* **2008**, *1042*, 318–320. [CrossRef]
48. Mandal, B.; Rameshbabu, A.P.; Dhara, S.; Pal, S. Nanocomposite Hydrogel Derived from Poly (Methacrylic Acid)/Carboxymethyl Cellulose/AuNPs: A Potential Transdermal Drugs Carrier. *Polymer* **2017**, *120*, 9–19. [CrossRef]
49. Sun, X.-F.; Ye, Q.; Jing, Z.; Li, Y. Preparation of Hemicellulose-g-Poly(Methacrylic Acid)/Carbon Nanotube Composite Hydrogel and Adsorption Properties. *Polym. Compos.* **2014**, *35*, 45–52. [CrossRef]
50. Qi, X.; Wei, W.; Li, J.; Liu, Y.; Hu, X.; Zhang, J.; Bi, L.; Dong, W. Fabrication and Characterization of a Novel Anticancer Drug Delivery System: Salecan/Poly(Methacrylic Acid) Semi-Interpenetrating Polymer Network Hydrogel. *ACS Biomater. Sci. Eng.* **2015**, *1*, 1287–1299. [CrossRef] [PubMed]
51. Liu, T.; Liu, H.; Wu, Z.; Chen, T.; Zhou, L.; Liang, Y.; Ke, B.; Huang, H.; Jiang, Z.; Xie, M.; et al. The Use of Poly(Methacrylic Acid) Nanogel to Control the Release of Amoxycillin with Lower Cytotoxicity. *Mater. Sci. Eng. C.* **2014**, *43*, 622–629. [CrossRef] [PubMed]
52. Panic, V.V.; Madzarevic, Z.P.; Volkov-Husovic, T.; Velickovic, S.J. Poly(Methacrylic Acid) Based Hydrogels as Sorbents for Removal of Cationic Dye Basic Yellow 28: Kinetics, Equilibrium Study and Image Analysis. *Chem. Eng. J.* **2013**, *217*, 192–204. [CrossRef]
53. Infrared Spectroscopy. Available online: http://www.umsl.edu/~orglab/documents/IR/IR2.html (accessed on 17 August 2021).
54. Ma, J.; Xu, Y.; Fan, B.; Liang, B. Preparation and Characterization of Sodium Carboxymethylcellulose/Poly(N-Isopropylacrylamide)/Clay Semi-IPN Nanocomposite Hydrogels. *Eur. Polym. J.* **2007**, *43*, 2221–2228. [CrossRef]
55. Santos, J.; Barreto, Â.; Nogueira, J.; Daniel-da-Silva, A.L.; Trindade, T.; Amorim, M.J.B.; Maria, V.L. Effects of Amorphous Silica Nanopowders on the Avoidance Behavior of Five Soil Species—A Screening Study. *Nanomaterials* **2020**, *10*, 402. [CrossRef]
56. Chang, C.-J.; Tzeng, H.-Y. Preparation and Properties of Waterborne Dual Curable Monomers and Cured Hybrid Polymers for Ink-Jet Applications. *Polymer* **2006**, *47*, 8536–8547. [CrossRef]
57. Tsai, M.-H.; Chang, C.-J.; Chen, P.-J.; Ko, C.-J. Preparation and Characteristics of Poly(Amide–Imide)/Titania Nanocomposite Thin Films. *Thin Solid Film.* **2008**, *516*, 5654–5658. [CrossRef]
58. Pacelli, S.; Paolicelli, P.; Moretti, G.; Petralito, S.; Di Giacomo, S.; Vitalone, A.; Casadei, M.A. Gellan Gum Methacrylate and Laponite as an Innovative Nanocomposite Hydrogel for Biomedical Applications. *Eur. Polym. J.* **2016**, *77*, 114–123. [CrossRef]
59. PubChem Methacrylic Acid. Available online: https://pubchem.ncbi.nlm.nih.gov/compound/4093 (accessed on 16 July 2021).
60. Teodorescu, M.; Cursaru, B.; Stanescu, P.; Draghici, C.; Stanciu, N.D.; Vuluga, D.M. Novel Hydrogels from Diepoxy-Terminated Poly(Ethylene Glycol)s and Aliphatic Primary Diamines: Synthesis and Equilibrium Swelling Studies. *Polym. Adv. Technol.* **2009**, *20*, 907–915. [CrossRef]
61. Liu, Y.-Y.; Liu, W.-Q.; Chen, W.-X.; Sun, L.; Zhang, G.-B. Investigation of Swelling and Controlled-Release Behaviors of Hydrophobically Modified Poly(Methacrylic Acid) Hydrogels. *Polymer* **2007**, *48*, 2665–2671. [CrossRef]
62. Zhang, K.; Luo, Y.; Li, Z. Synthesis and Characterization of a PH- and Ionic Strength-Responsive Hydrogel. *Soft Mater.* **2007**, *5*, 183–195. [CrossRef]
63. Mezger, T.G. *The Rheology Handbook: For Users of Rotational and Oscillatory Rheometers*; Vincentz Network GmbH & Co KG: Hanover, Germany, 2006; ISBN 978-3-87870-174-3.
64. Richbourg, N.R.; Peppas, N.A. The Swollen Polymer Network Hypothesis: Quantitative Models of Hydrogel Swelling, Stiffness, and Solute Transport. *Prog. Polym. Sci.* **2020**, *105*, 101243. [CrossRef]
65. Okay, O.; Oppermann, W. Polyacrylamide−Clay Nanocomposite Hydrogels: Rheological and Light Scattering Characterization. *Macromolecules* **2007**, *40*, 3378–3387. [CrossRef]
66. Natarajan, S.; Bajaj, H.C.; Tayade, R.J. Recent Advances Based on the Synergetic Effect of Adsorption for Removal of Dyes from Waste Water Using Photocatalytic Process. *J. Environ. Sci.* **2018**, *65*, 201–222. [CrossRef]
67. Tara, N.; Siddiqui, S.I.; Rathi, G.; Chaudhry, S.A.; Inamuddin; Asiri, A.M. Nano-Engineered Adsorbent for the Removal of Dyes from Water: A Review. *CAC* **2020**, *16*, 14–40. [CrossRef]
68. Van Tran, V.; Park, D.; Lee, Y.-C. Hydrogel Applications for Adsorption of Contaminants in Water and Wastewater Treatment. *Environ. Sci. Pollut. Res.* **2018**, *25*, 24569–24599. [CrossRef]
69. Sen, M.; Yakar, A.; Güven, O. Determination of Average Molecular Weight between Cross-Links (Mc) from Swelling Behaviours of Diprotic Acid-Containing Hydrogels. *Polymer* **1999**, *40*, 2969–2974. [CrossRef]
70. Safronov, A.P.; Adamova, L.V.; Blokhina, A.S.; Kamalov, I.A.; Shabadrov, P.A. Flory-Huggins Parameters for Weakly Crosslinked Hydrogels of Poly(Acrylic Acid) and Poly(Methacrylic Acid) with Various Degrees of Ionization. *Polym. Sci. Ser. A* **2015**, *57*, 33–42. [CrossRef]

71. Jerca, F.A.; Anghelache, A.M.; Ghibu, E.; Cecoltan, S.; Stancu, I.-C.; Trusca, R.; Vasile, E.; Teodorescu, M.; Vuluga, D.M.; Hoogenboom, R.; et al. Poly(2-Isopropenyl-2-Oxazoline) Hydrogels for Biomedical Applications. *Chem. Mater.* **2018**, *30*, 7938–7949. [CrossRef]
72. Panic, V.; Adnadjevic, B.; Velickovic, S.; Jovanovic, J. The Effects of the Synthesis Parameters on the Xerogels Structures and on the Swelling Parameters of the Poly(Methacrylic Acid) Hydrogels. *Chem. Eng. Process.* **2010**, *156*, 206–214. [CrossRef]
73. Huang, H.; Liu, Z.; Yun, J.; Yang, H.; Xu, Z. Preparation of Laponite Hydrogel in Different Shapes for Selective Dye Adsorption and Filtration Separation. *Appl. Clay Sci.* **2021**, *201*, 105936. [CrossRef]
74. Parisi, F. Adsorption and Separation of Crystal Violet, Cerium(III) and Lead(II) by Means of a Multi-Step Strategy Based on K10-Montmorillonite. *Minerals* **2020**, *10*, 466. [CrossRef]

Article

# Sulfonated Pentablock Copolymer Coating of Polypropylene Filters for Dye and Metal Ions Effective Removal by Integrated Adsorption and Filtration Process

Simona Filice [1], Viviana Scuderi [1,*], Sebania Libertino [1], Massimo Zimbone [2], Clelia Galati [3], Natalia Spinella [3], Leon Gradon [4], Luciano Falqui [5] and Silvia Scalese [1,*]

[1] Consiglio Nazionale delle Ricerche, Istituto per la Microelettronica e Microsistemi (CNR-IMM), Ottava Strada n.5, 95121 Catania, Italy
[2] Consiglio Nazionale delle Ricerche, Istituto per la Microelettronica e Microsistemi (CNR-IMM), Via S. Sofia 64, 95123 Catania, Italy
[3] STMicroelectronics Stradale Primosole 50, 95121 Catania, Italy
[4] Faculty of Chemical and Process Engineering, Warsaw University of Technology, ul. Waryńskiego 1, 00-645 Warsaw, Poland
[5] Plastica Alfa SpA, Zona Industriale, C. da S.M. Poggiarelli, 95041 Caltagirone (CT), Italy
* Correspondence: viviana.scuderi@imm.cnr.it (V.S.); silvia.scalese@imm.cnr.it (S.S.)

**Abstract:** In this work, we coated polypropylene (PP) fibrous filters with sulfonated pentablock copolymer (s-PBC) layers and tested them for the removal of cationic organic dyes, such as methylene blue (MB), and heavy metal ions ($Fe^{3+}$ and $Co^{2+}$) from water by adsorption and filtration experiments. Some of the coated filters were irradiated by UV light before being exposed to contaminated water and then were tested with unirradiated filters in the same adsorption and filtration experiments. Polymer-coated filters showed high efficiency in removing MB from an aqueous solution in both absorption and filtration processes, with 90% and 80% removal, respectively. On the other hand, for heavy metal ions ($Fe^{3+}$ and $Co^{2+}$), the coated filters showed a better removal performance in the filtration process than for the adsorption one. In fact, in the adsorption process, controlled interaction times allow the ionic species to interact with the surface of the filters leading to the formation and release of new species in solution. During filtration, the ionic species are easily trapped in the filters, in particular by UV modified filters, and we observed for $Fe^{3+}$ ions a total removal (>99%) in a single filtration process and for $Co^{2+}$ ions a larger removal with respect to the untreated filter. The mechanisms involved in the removal of the contaminants processes were investigated by characterizing the filters before and after use by means of scanning electron microscopy (SEM) combined with energy-dispersive X-ray (EDX) analysis and Fourier transform infrared spectroscopy (FT-IR).

**Keywords:** sulfonated pentablock copolymer; ions removal; adsorption; filtration

## 1. Introduction

Nowadays, a large part of the world's inhabitants has limited access to clean water, and this is one of the most dramatic effects deriving from anthropogenic activities. This problem is expected to increase due to the high rate of world population growth and, in the coming years, it will also affect regions currently considered to be rich in water. Harmful and toxic waste produced by industrial and agricultural activities ends up in the waterbodies and, therefore, new effective strategies are required to prevent pollution or to remove it from water.

For this reason, much effort is dedicated to find new remediation methods to purify water, efficiently, with less energy, minimizing the consumption of chemical compounds. Much attention is also given to the impact that the various techniques/materials used have on the environment.

Research is mainly focused on four fronts: disinfection, decontamination, water reuse and desalination [1].

Several conventional techniques were used to reduce toxic dye compounds in wastewater, including separation [2], reverse osmosis, precipitation, ion exchange method, and ultra-filtration adsorption on activated carbon [3]. The greater disadvantage of these techniques is the formation of secondary waste that cannot be reprocessed.

Instead, photocatalytic methods consist of highly advanced oxidation processes that are used for the photodegradation of toxic compounds, showing high efficiency, simple operation, low cost, and low energy consumption [4]. Many photocatalytic materials are known, including ZnO [5], $TiO_2$ [6], BiOBr [7], and $WO_3$ [8].

In this context, membrane technology plays an important role in water treatment due to its interesting features, easy operation, no addition of chemical additives (or less), cost-efficiency, high productivity, easy scaling, and suitable removal capacity. A membrane is a selective system that allows wanted materials to pass through and unwanted to be retained on the membrane surface [9].

The current membranes are based on some conventional polymeric materials, and they frequently suffer from limited chemical, mechanical and thermal stabilities [10]. Inorganic membranes that can achieve high selectivity and/or high permeability are also weak, expensive and difficult to expand [11]. In this context, the development of innovative new generation membrane material, such as two-dimensional (2D) porous organic polymers (POPs), is receiving great attention due to highly tunable pores/channels, robust frameworks/networks, intrinsic flexibility, and light weight for multiple separation purposes. It has recently been shown that the filtering properties of a membrane, as well as the resistance to acidic environments, can be increased either by modifying the surface of the polymer or by incorporating polymers/molecules with different structures and properties [12–14].

In our previous works [15–18], we showed that sulfonilic groups of sulfonated pentablock copolymer (s-PBC) are highly hydrophilic, negatively charged, and acidic, making this polymer a suitable high capacitive adsorbent for the removal of cationic species. In particular, membranes and s-PBC/graphene oxide (GO) nanocomposite membranes show suitable adsorption abilities for the removal of different heavy metal ions ($Ni^{2+}$, $Co^{2+}$, $Cr^{3+}$, and $Pb^{2+}$) from aqueous solutions containing the corresponding metal salts at different concentrations, due to the presence of sulfonic groups that play a fundamental role in the adsorption process of metal ions. Starting from these considerations, we proposed for the first time the application of s-PBC (Nexar$^{TM}$) as a multifunctional coating for water filters [19,20]: due to hydrophilicity, acidity, and smoothness of the coating layer, the adhesion, and proliferation of planktonic *P. aeruginosa* was contrasted by a combined repulsion and contact killing mechanism avoiding the biofilm formation. This polyvalent antimicrobial nature of Nexar$^{TM}$ could provide a substantial improvement to the surface coatings for multi-purpose filters with antifouling properties and simultaneous removal of bacteria and cationic contaminants from water. Furthermore, this strategy allows reducing the use of chemical and toxic agents to avoid the biofouling effect, increasing the filter's lifespan.

To take advantage of the described properties, in this work polypropylene (PP) fibrous filters were coated by a s-PBC layer (s-PBC@PP) and tested in adsorption and filtration processes for a fast and efficient removal of water contaminants. The PP filter has the double role of support and porous structure. The s-PBC layer provides useful additional functionalities to the commercial filter (not only size separation, but also molecules and ions removal by adsorption). This is very important since it allows to keep the filter cost low by using only a thin active coating layer on a cheap commercial PP filter. The filtration in this integrated process is assumed (understandable) as a transport and deposition complex phenomenon. To explore the possibility of further increasing the filtering efficiency of s-PBC coating, the coated filters were irradiated by UV light. The modifications induced by irradiation were investigated by FT-IR and EDX analysis.

Aiming to study the adsorption and filtration properties of the s-PBC-coated PP filters, with and without UV-induced modification, two kinds of contaminants were chosen: a cationic dye, i.e., methylene blue (MB), and two heavy metal ions, i.e., $Fe^{3+}$ and $Co^{2+}$, using the raw PP filter as a reference. The contaminants removal rate was measured by UV-Vis absorbance spectroscopy during both adsorption and filtration processes.

Cationic dye and metal ions were efficiently removed by s-PBC-coated PP filters due to their hydrophilic, acid and negatively charged character. In particular, the polymer-coated filters showed a high efficiency in removing MB from an aqueous solution in both absorption and filtration processes.

In the presence of heavy metal ions ($Fe^{3+}$ and $Co^{2+}$), on the other hand, the coated filters show a better efficiency for the filtration process, in particular in the filters whose surface has been modified by UV treatment.

The filters were characterized before and after removal processes by scanning electron microscope (SEM) energy-dispersive X-ray (EDX) analysis and FT-IR.

## 2. Results

### 2.1. Samples Characterization

Polypropylene filters were coated by s-PBC solution (s-PBC@PP), as described in the Section 3, and some of them underwent a UV irradiation process (s-PBC@PP_UV).

Their morphology was characterized by SEM analysis, and images are reported in Figure 1. For each sample, the corresponding picture is also shown in the insets.

**Figure 1.** SEM images of PP (**left**), s-PBC@PP (**middle**) and s-PBC@PP after UV irradiation (**right**) samples.

As we showed in previous works [19,20], the casting procedure used to cover the PP surface produces a homogeneous and complete coverage of PP fibers. Indeed, the PP filter (Figure 1, on the left) is white and is formed by a three-dimensional network of randomly distributed fibers with diameters ranging between 0.2 and 5 µm. The porosity $\varepsilon_F$ is 0.98 and was determined, as described in Section 3.1 and in ref [20]. After s-PBC deposition, the sample surface turns to light yellow (Figure 1, in the middle) and the s-PBC completely covers the PP fibers as shown in the image on the right resulting in a homogeneous and smooth surface. After UV irradiation, the s-PBC layer covering PP became dark yellow (Figure 1, on the right) and the fibers become more visible under the surface, suggesting the occurrence of shrinkage or a rearrangement of the coating layer, probably due to drying, even though the sample weight does not change significantly.

The change from hydrophobic to hydrophilic behavior of the PP surface after the deposition of s-PBC and the effect of UV irradiation on s-PBC hydrophilicity was confirmed by water uptake and contact angle measurements reported in Table 1 for PP, s-PBC@PP, and s-PBC@PP_UV samples, respectively.

**Table 1.** Water uptake (%) and contact angle measured for PP, s-PBC@PP, and s-PBC@PP_UV samples.

| Sample | Water Uptake (%) | Average Contact Angle (Deg) |
| --- | --- | --- |
| PP | 3 | 129.6 |
| s-PBC@PP | 262 | 103.1 |
| s-PBC@PP_UV | 308 | 85.6 |

The starting PP surface is totally hydrophobic. The s-PBC coating results in a highly hydrophilic surface as shown by the water uptake value (262%). This value increases up to 308% for the sample after UV irradiation. The change from hydrophobic to hydrophilic behavior of the PP surface after the deposition of s-PBC was confirmed by contact angle measurements (see Figure S3 of Supplementary Materials). The average angle measured on three drops for each sample, is 129.6° for PP samples and decreases to 103.1° for s-PBC-coated PP and down to 85.6° after UV irradiation of the coating layer. In other words, the s-PBC coverage turns the PP surface character from totally hydrophobic to less hydrophobic and to hydrophilic after the UV irradiation.

The more hydrophilic character of s-PBC@PP and s-PBC@PP_UV was confirmed by FT-IR analysis in comparison with spectra acquired on PP. Previous studies [15,16] performed by FT-IR analysis on s-PBC structure demonstrate that its molecular structure in a polar solvent does not undergo chemical modification with respect to the commercial material. In fact, the IR absorption peaks characteristic of $SO_3H$ groups were unmodified after dispersion in polar solvent. These are responsible for the negative charge and acid character of s-PBC coating, as also confirmed by its ability to adsorb positively charged molecules [16–18].

Figure 2 reports the FT-IR spectra of a s-PBC film deposited on a PP filter before (red line) and after (blue line) the UV treatment in the 4000–500 $cm^{-1}$ wavelength range (Figure 2a) and in a reduced range between 1900 and 900 $cm^{-1}$ (Figure 2b) to highlight some features more in detail. The IR spectrum of an unmodified PP filter (black line) is reported for comparison.

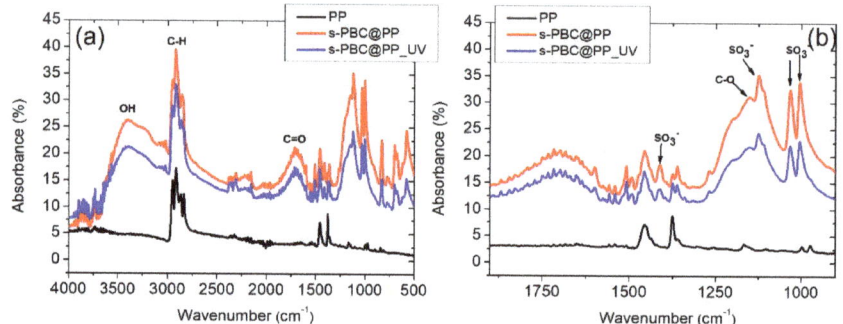

**Figure 2.** (**a**) FT-IR spectra of a PP filter (black line), s-PBC-coated PP filter before (red line) and after (blue line) the UV treatment. In (**b**), the same spectra in the wavelength range 1900–900 $cm^{-1}$ are reported to show the IR features more in detail.

There are two "types" of signals in the IR spectrum: those related to the presence of functional groups (4000–1900 $cm^{-1}$) and those considered to be fingerprint (1000–400 $cm^{-1}$) typical and characteristic of the single molecules. These fingerprint signals mean that it is not possible for different molecules to have the same IR spectrum. The characteristic peaks of functional groups instead (always) fall at the same frequencies, regardless of the structure of the molecule in which the group is present.

A wide band corresponding to stretching vibrations of the hydroxyl group (OH group, from adsorbed water) occurs in the wavenumber range 3700–3000 $cm^{-1}$ only for the filters coated with s-PBC, confirming the hydrophilicity of the polymeric coating and the totally hydrophobic behavior of the PP filter. Negligible differences in the shape of this band before and after the UV treatment were evidenced. The three peaks between 2800 and 3000 $cm^{-1}$ are associated with asymmetric and symmetric CH- stretching vibrations from the methyl groups. The region between 2400 and 2000 $cm^{-1}$ is strongly influenced by the presence of carbon dioxide, and the intense and broad band centered at about 1707 $cm^{-1}$ is attributable to the C=O (carboxylic groups) stretching vibrations. Peaks evident at 1411, 1126, 1035, and 1006 $cm^{-1}$ constitute the characteristic peaks of the $SO_3^-$ (Figure 2, right panel), whereas the peak at 1151 $cm^{-1}$ is the C-O stretching [21,22]. The wavelength region below 900 $cm^{-1}$

is characterized by the presence of double bonds C=C bending and C-H bending. The fingerprint area, from 1300 to 400 cm$^{-1}$, is characterized by the presence of specific bands of each molecule, originating from vibrations of the entire molecular skeleton. The absence of new peaks after UV treatment suggests that no change of the polymer main structure occurs [23].

*2.2. Adsorption and Filtration Tests*

PP, s-PBC@PP, and s-PBC@PP_UV samples were tested for the removal of cationic contaminants, i.e., Methylene Blue, $Fe^{3+}$, and $Co^{2+}$ ions, both by adsorption and filtration.

Before testing the filters for contaminants removal, water flux through each filter was measured by filtering 40 mL of water in the same apparatus used for contaminants removal. The water flux for each filter, calculated as the mean value of five consecutive tests, is 18.8 mL/s for the PP filter and 4.1 and 4.9 mL/s for the ones coated by s-PBC before and after UV irradiation, respectively. The strong decrease in the water flow observed after depositing s-PBC on the PP surface is due to a reduction in the porosity of the PP filter. By comparing Figure 1a–c, we observe that the s-PBC layer completely covers PP fibers forming a compact layer, thus reducing PP porosity (also reported in [20]).

2.2.1. MB Adsorption and Filtration

Figure 3 reports the UV-Visible absorbance spectra of MB solutions (a) after adsorption or (b) after filtration for PP (black curve), s-PBC@PP (red curve), and s-PBC@PP_UV (green curve) samples. The reference spectrum of MB solution with the initial concentration is also reported (blue curve). In the inset, a photo of each sample after three hours of adsorption is reported.

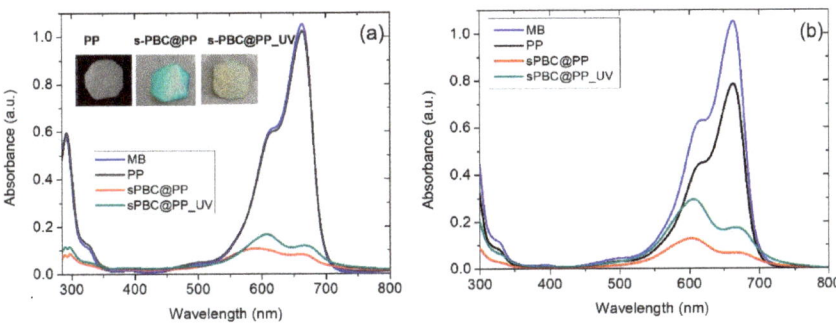

**Figure 3.** (a) UV-Visible absorbance spectra of MB solutions after dipping PP (black curve), s-PBC@PP (red curve), and s-PBC@PP_UV (green curve) for 3 h. In the inset a photo of each sample after three hours of adsorption; (b) UV-Visible absorbance spectra of MB solutions after filtration for PP (black curve), s-PBC@PP (red curve) and s-PBC@PP_UV (green curve) samples. The blue curve in both graphs represents the UV-Vis spectrum of MB at the initial concentration.

Methylene blue is a cationic, thiazine dye, which absorbs light at 664 nm (n-δ*) (monomer) with a shoulder at 610 nm corresponding to the dimer. In concentrated aqueous solutions, aggregation occurs, resulting in a shift of absorbance peak to lower wavelengths with respect to the monomer [17]. The covering of the PP surface with a layer of s-PBC makes the adsorption of MB possible due to the presence of sulfonic groups giving the surface a hydrophilic and acid character with negatively charged adsorption sites. The MB absorbance peak in solution decreased by increasing the contact time with coated samples: the main adsorption occurred suddenly, and low variations occurred in the following hours up to being constant after three hours. Thus, we report the spectra of MB solutions after three hours of contact with samples. At this time, for a solution in which s-PBC-coated PP samples were immersed, the absorbance peak of the monomer slightly shifts to higher wavelengths while the one related to dimer shifts to lower wavelengths. Their relative ratio is reduced, indicating that MB higher aggregates do not reduce linearly with monomers, or even they may form while an MB adsorption–desorption equilibrium is reached. This is not observed for the PP sample for which the adsorption is neglectable. MB removal (%) for all the samples and the amount (mg) of adsorbed MB molecules per g of deposited s-PBC layer for the coated samples are reported in Table 2 after 3 h of dipping.

**Table 2.** MB removal (%) and amount of adsorbed dye per gram of adsorbent ($Q_t$, mg/g), considering the peak at 664 nm after adsorption and filtration tests.

| Sample | MB Removal by Adsorption (%) | $Q_{t\_ads}$ (mg/g) | MB Removal by Filtration (%) | $Q_{t\_filt}$ (mg/g) |
|---|---|---|---|---|
| PP | 5 | - | 20 | - |
| s-PBC@PP | 90 | 0.58 | 80 | 0.45 |
| s-PBC@PP-UV | 80 | 0.56 | 70 | 0.38 |

The PP filter adsorbs only ~5% of MB. By covering its surface with s-PBC, a strong increase in MB removal efficiency is observed after three hours, i.e., up to 90% and 80% for s-PBC@PP and s-PBC@PP_UV, respectively. This result is also clearly visible by observing the color of samples after being immersed in MB solutions for three hours (see inset of Figure 2). The reference sample remained unaltered after being in contact with MB solution, while the s-PBC@PP sample changed its color to blue, indicating a strong MB adsorption. The UV-irradiated s-PBC@PP sample shows a removal efficiency and Qt values such as s-PBC@PP; however, its color does not turn blue after MB adsorption. This suggests that MB is converted into solution or adsorbed on the sample surface in its colorless form (i.e., leuco) as a result of a reduction process of MB molecules; this reduction could be ascribed to a modification induced on the sample surface by UV irradiation that enriches its surface of electrons. When MB is reduced to its leuco form, the main absorbance peak of MB at 664 nm disappears, and a new absorbance peak in the UV range appears, i.e., 254 nm [24]. This peak is not observed in the absorbance spectra of the residual solution, thus suggesting that MB is reduced to its leuco form only after adsorption on the s-PBC@PP_UV sample.

The leuco MB species is not observed when MB is removed by filtration using PP, s-PBC@PP, and s-PBC@PP_UV filters. Indeed, the s-PBC@PP filters treated or not by UV irradiation became blue after filtration while the eluate was uncolored. This could be ascribed to the processing time: filtration occurred in less than one minute while adsorption occurred in three hours. In the last case, the contaminant molecules have more time to interact with radical species generated within the polymeric matrix during its irradiation [25].

PP alone (black curve) removes about 20% of MB after the first filtration process, and then this percentage decreases with subsequent filtrations, possibly due to a release of MB from the filter (spectra not shown here). For the s-PBC@PP (red curve) sample, the filtration removal efficiency reaches 80% for the first process, and for the s-PBC@PP_UV (green curve) sample, the removal is lower (about 70%). In both cases, the peak associated with dimers and oligomers at about 600 nm is more intense than the one due to the monomer at 664 nm.

Table 2 reports the $Q_t$ values also for filtration processes: s-PBC@PP shows a $Q_t$ value of 0.45 mg/g that reduces to 0.38 mg/g after UV irradiation of the s-PBC layer. These values are lower but quite similar to the one reported for adsorption tests suggesting that a longer contact time between filter and contaminants (3 h for adsorption process vs. 1 min for filtration) results in better removal efficiency. However, in terms of time required for removal, filtration is clearly more advantageous and can be repeated consecutively to improve the final result.

### 2.2.2. $Fe^{3+}$ Adsorption and Filtration

Figure 4 reports the UV-Visible absorbance spectra of $FeCl_3$ solutions where filters were immersed in the dark for three hours (Figure 4a) or after filtration (Figure 4b). The red curve represents the PP filter whose adsorption is neglectable, and only a small decrease in $Fe^{3+}$ concentration is observed after filtration (i.e., 20% $Fe^{3+}$ reduction). After covering the filter with s-PBC, the adsorption ability increased up to 75% with respect to PP itself (see the blue curve of Figure 4a), and this is due to the hydrophilicity and negative charge of the s-PBC layer surface. A similar result was obtained for filtration using the s-PBC@PP filter: the $Fe^{3+}$ concentration was reduced by 50% at the first filtration (see the blue curve of Figure 4b) and up to 80% by filtering the same solution for the second time through the same filter (see the magenta curve of Figure 4b). The UV irradiation of s-PBC layers worsened the adsorption ability (i.e., 28% $Fe^{3+}$ reduction, see the green curve of Figure 4a) but highly increased the filtration efficiency since the total removal of $Fe^{3+}$ ions was observed immediately after the first filtration process (see the green curve of Figure 4b), due to the higher hydrophilicity of the sample.

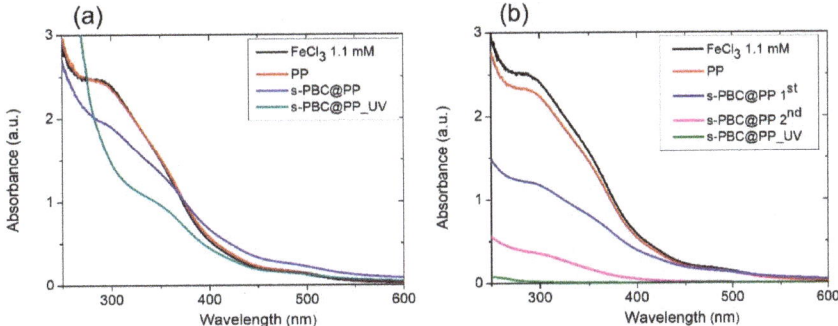

**Figure 4.** UV-Visible absorbance spectra of $FeCl_3$ solutions after (**a**) adsorption on the different filters or (**b**) filtration through them. The reference spectrum of the initial $FeCl_3$ solution is reported in both graphs.

However, observing the shape of the spectra after absorption and filtration of the solution, they show a different shape for the sample treated with UV. In fact, for the s-PBC@PP filter, the spectra of the absorbed and filtered solutions are similar in shape, both with each other and with the initial solution. The only difference is represented by the intensity, which is lower for the filtered solution, indicating a release of $Fe^{3+}$ ions from the sample during the three hours of absorption. For the s-PBC@PP_UV filter, the spectra of the absorbed and filtered solution show different shapes, both with each other and with the initial solution. In particular, while the spectrum after filtration shows only a strong variation in intensity (>99% of the ions were removed), the spectrum after adsorption shows the appearance of a very intense absorbance signal for wavelengths lower than 300 nm. From the literature [26], it is known that Fe ions show absorption peaks whose position varies according to the oxidation number ($Fe^{2+}$ and $Fe^{3+}$) and the group to which the Fe ion is bound ($Fe(OH)^{2+}$, $Fe(SO_4)_2^-$, $FeSO_4^+$, $FeCl_2^+$, $FeCl^{2+}$, ...). So, considering that the shape of the UV-Vis spectrum changed for the s-PBC@PP_UV filter after three hours of contact with the iron solution in the dark, we believe that the change is due to the formation of different iron species in the solution. Accordingly, the UV-irradiated layer reacts with iron ions in solution instead of adsorbing them.

The s-PBC coating of the PP filter increased its $Fe^{3+}$ removal efficiency with respect to PP by both adsorption and filtration. Higher adsorption efficiency is shown by the s-PBC@PP filter (i.e., 7.75 mg/g), while UV irradiation reduces this performance (i.e., 2.8 mg/g). On the contrary, the UV-irradiated filter showed a higher removal efficiency by the first filtration test (up to 8.48 mg/g). A similar result was shown by the s-PBC@PP filter after filtering twice the same solution through the same filter (i.e., 7.96 mg/g).

### 2.2.3. $Co^{2+}$ Adsorption and Filtration

Figure 5 reports the UV-Visible spectra of (i) as-prepared $CoCl_2$ solution and the same after contact with different filters in the dark for three hours (Figure 5a); (ii) as-prepared $CoCl_2$ solution and the same after filtration (Figure 5b). The main peak associated with $Co^{2+}$ is positioned at 520 nm. The red curve represents the initial PP filter whose adsorption is neglectable. After coating the filter with s-PBC, the adsorption ability increases, and 15% of the initial $Co^{2+}$ concentration is removed (see the blue curve of Figure 5a) due to the more negatively charged surface of the s-PBC layer with respect to the totally hydrophobic PP surface. For the UV-irradiated s-PBC layers, 14% $Co^{2+}$ reduction is observed (see the green curve of Figure 5a), indicating that the UV treatment did not significantly affect the adsorption ability with respect to the s-PBC@PP filter. However, similarly to the case of $Fe^{3+}$ ions described in the previous section, a significant variation in the shape of the absorbance spectra occurs at wavelengths below 400 nm after the prolonged interaction of the filters with the $Co^{2+}$ ions solution.

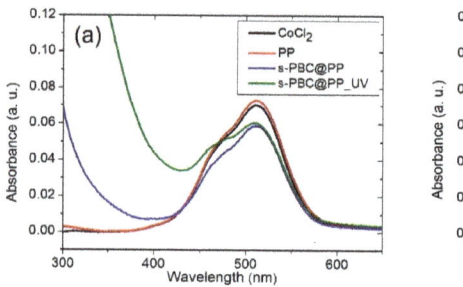 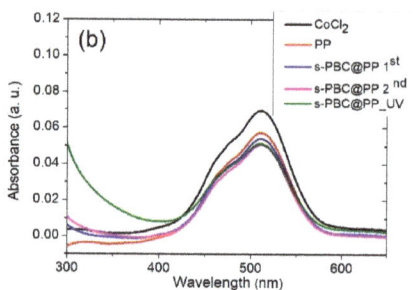

**Figure 5.** UV-Visible absorbance spectra of $CoCl_2$ solutions after (**a**) adsorption on the different filters for two hours or (**b**) filtration through them. The reference spectrum of the initial $CoCl_2$ solution is reported in both graphs.

After filtration through the PP filter, $Co^{2+}$ concentration decreases by 15% (red curve in Figure 5b). A slightly better result was obtained using the s-PBC@PP filter: 20% of the initial $Co^{2+}$ concentration was removed after the first process (see the blue curve of Figure 5b) and 28% by re-filtering the same solution twice through the same filter (see the magenta curve of Figure 5b). The same decrease (28%) of the 520 nm peak was observed for the s-PBC@PP_UV filter, along with a moderate absorbance increase at low wavelengths (below 400 nm).

The higher removal efficiency observed for filtration may be mainly ascribed to the higher hydrophilicity of polymer-coated filters after UV irradiation.

We believe that the large increase in the absorbance at low wavelength values observed for s-PBC@PP_UV, more evident in the case of adsorption but also present after filtration, might be due to the formation of different cobalt species in solution as byproducts of a reaction occurring between the UV-irradiated polymer layer and the cobalt ions in solution. This could explain the absorbance increase observed in the UV range of the spectrum and the very small decrease in the 520 nm peak.

The $Q_t$ values for cobalt removal by adsorption and filtration were calculated according to the definition given in the Section 3. Coating the PP filter by an s-PBC layer allows a moderate $Co^{2+}$ removal by adsorption, i.e., $Q_t$ = 21.9 mg/g (no adsorption is observed for raw PP filter) that slightly increases ($Q_t$ = 24.0 mg/g) by filtration. UV irradiation lowers the removal efficiency of the filter to 20.0 mg/g and 17.3 mg/g for adsorption and filtration, respectively.

To understand in more detail these results, we have performed FT-IR and EDX analysis on the three kinds of filters after adsorption and filtration of $Fe^{2+}$ and $Co^{2+}$ ions.

## 2.3. Post-Process Filters Characterization

### 2.3.1. FT-IR Analysis

Figure 6 shows the FT-IR spectra of an s-PBC@PP filter after UV treatment (black line) and the s-PBC@PP filters without and with UV treatment after being dipped in a 1.1 mM $FeCl_3$ solution (red and blue line, respectively, Figure 6a,b) or after being dipped in a 17.5 mM $CoCl_2$ solution (red and blue line, respectively, Figure 6c,d). As a reference, we report the s-PBC@PP after UV treatment because it shows the same features as the untreated s-PBC@PP, as shown in Figure 3.

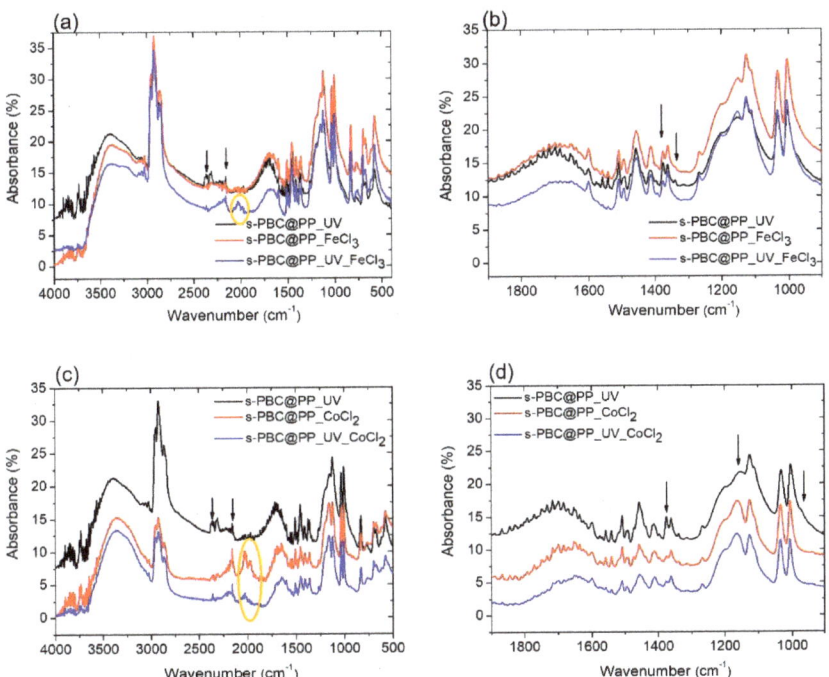

**Figure 6.** FT-IR spectra of (**a**,**b**) s-PBC@PP filter (red line) and UV-treated s-PBC@PP filter (blue line) after being dipped in a $FeCl_3$ solution; (**c**,**d**) s-PBC@PP filter (red line) and UV-treated s-PBC@PP filter (blue line) after being dipped in a $CoCl_2$ solution. The (**b**,**d**) figures show more in detail the 1900–900 $cm^{-1}$ range. FT-IR spectra of the s-PBC@PP filter after UV treatment is reported in all the graphs as a reference (black line).

After the adsorption of $Fe^{3+}$, the main differences observed among the spectra are in the ranges 1900–2350 $cm^{-1}$ and 1400–1100 $cm^{-1}$: we observe the presence of two large features in the range 2250–1900 $cm^{-1}$ only in the case of the UV-treated s-PBC@PP sample. The peaks centered at approximately 2350 and 2157 $cm^{-1}$ (indicated by the arrows in Figure 6a) are present in all the filters and are related to the O=C=O and C=C=O bonds, respectively. The peaks centered at 2027 e 1973 $cm^{-1}$ and circled in yellow have not been identified. Further differences concern the band centered at 1707 $cm^{-1}$ (related to the C=O stretching vibrations) that shifts to 1679 $cm^{-1}$, and the decrease in relative intensities of the peaks at 1375 $cm^{-1}$ (S=O), 1338 (S=O) $cm^{-1}$ indicated by the arrows in Figure 6b.

In the case of $Co^{2+}$ adsorption, the two features in the range 2250–1900 $cm^{-1}$ are present for both s-PBC@PP with and without UV treatment, and, in particular, the peaks centered at 2027 e 1973 $cm^{-1}$ and circled in yellow are present (Figure 6c), more evident for the untreated sample. The band centered at 1707 $cm^{-1}$ (related to the C=O stretching vibrations) shifts to 1650 $cm^{-1}$, and the intensity of the peaks at 1375 $cm^{-1}$ and 1338 $cm^{-1}$ (both related to S=O stretching) decrease, such as in the case of $Fe^{3+}$ adsorption. Furthermore, the 1153 cm$^{-1}$ peak (related to C-O stretching) shifts to 1161 cm$^{-1}$, and its relative intensity increases after the interaction with $CoCl_2$, as indicated by the arrows in Figure 6d.

The results of our adsorption/filtration tests indicate that after the contact between s-PBC@PP and s-PBC@PP_UV samples with $FeCl_3$ and $CoCl_2$ solutions: (i) the amount of S=O bonds decreases; (ii) the peaks related to C=O (1707 cm$^{-1}$) and C-O (1153 cm$^{-1}$) stretching shift, respectively, to a lower and to a higher value, suggesting that there is a change in the electron distribution of the molecular bonds, probably due to strong interaction with cationic ions. All the observed changes (change of relative intensity and shifts) are much more evident in the case of contact with the $CoCl_2$ solution, highlighting a stronger interaction of sulfonilic groups with $Co^{2+}$ with respect to $Fe^{3+}$ ions.

### 2.3.2. EDX Microanalysis

Figure 7 reports the C/O, S/O, and S/C weight ratios obtained by EDX analysis carried out on s-PBC@PP and s-PBC@PP_UV samples before and after being dipped for 180 min in a 1.1 mM $FeCl_3$ solution or in a 17.5 mM $CoCl_2$ solution. The comparison between the s-PBC layer before and after UV irradiation (black and red curves, respectively) indicates that the UV treatment slightly changes the relative amounts of O and S with respect to C. In particular, we observe a decrease in the S/C ratio and an increase in the C/O ratio, while S/O remains almost unvaried.

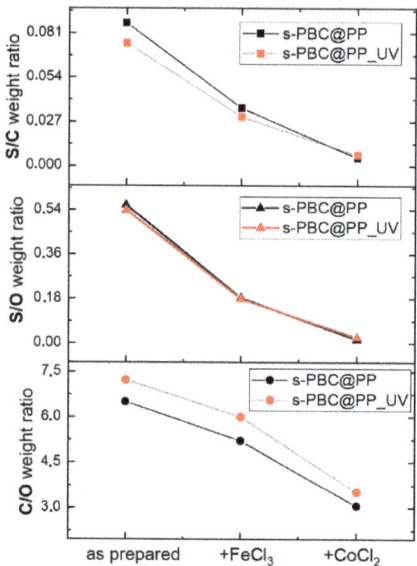

**Figure 7.** C/O, S/O, and S/C weight ratios measured by EDX analysis for s-PBC@PP and s-PBC@PP_UV samples before and after dipping in $FeCl_3$ solution or in $CoCl_2$ solution for 180 min.

The S/C and S/O weight ratios measured for the samples exposed to $FeCl_3$ and $CoCl_2$ strongly decrease with respect to the values measured for the corresponding as-prepared samples, and this effect is enhanced in the case of immersion in the $CoCl_2$ solution. The C/O weight ratio decreases moderately for the samples in contact with the $FeCl_3$ solution and much stronger for the samples in contact with the $CoCl_2$ solution. No significant difference in the general trend is observed between s-PBC@PP and s-PBC@PP treated by UV irradiation.

A decrease in the S content in the s-PBC coating is in agreement with the FT-IR results, where a reduction in the S=O stretching features (1376 cm$^{-1}$, 1338 cm$^{-1}$) is observed. EDX provides the additional information that this decrease is strongly enhanced by the presence of $CoCl_2$ with respect to $FeCl_3$.

An increase in O content seems to be supported by smaller C/O and S/O ratios after contact with both solutions and by the increase in the IR peak at 1153 cm$^{-1}$ assigned to C-O stretching. Changes in the electron distribution of the molecular bonds between C and O are also suggested by the downshifts observed in the FT-IR features from 1707 cm$^{-1}$ to 1679–1650 cm$^{-1}$ (C=O stretching) and by the upshift from 1153 to 1161 cm$^{-1}$ (C-O).

These results suggest that adsorption/filtration of ions, in particular, cobalt, occurred by coordination with sulfonilic groups that dissolve in solution as sulfate salts, as demonstrated by the presence of byproducts shown in Figures 4 and 5.

## 3. Materials and Methods

### 3.1. Chemicals

Multilayer polypropylene (PP) filters were produced using the melt-blown technology process, as reported in [19,27].

The produced PP fibers have diameters smaller than 5 mm and form a three-dimensional network with a range distribution of the pore size. The filter porosity $\varepsilon_F$ is 0.98 and was determined using the formula: $\varepsilon_F = 1 - \rho_{SF}/(\rho_F \cdot L)$, where $r_{SF}$ is the surface density calculated as the mass of the filter divided by the surface area ($m_F/A_F$); $\rho_F$ is the fibers material density (910 Kg/m$^3$ for polypropylene was used); L is the PP filter thickness.

A sulfonated pentablock copolymer poly(tBS–HI–sS:S–HI–tBS) solution, or s-PBC, with 10–12 wt% polymer in a cyclohexane/heptane mixed solvent was provided courtesy of Kraton Polymers LLC. A scheme of this copolymer, commercially available as Nexar$^{TM}$, is reported in Figure S1 of Supplementary Materials [15–18]. The sulfonation degree is 52 mol%, and these groups confer to the polymer an ion exchange capacity (IEC) value of 2.0 meq/g. The molecular weight is 112,500 g mol$^{-1}$ and the volume fraction (tBS–[sS:S]–HI) is 0.300–[0.226:0.208]–0.266 [28]. The structure of this polymer is composed of an alternation of hydrophobic and hydrophilic domains within a micellar morphology in solution, depending on the solvent polarity [28]. This morphology affects the morphology of casted film: discrete, spherical ion-rich microdomains are formed in films cast from nonpolar solvent, whereas an apparently mixed morphology with a continuous ion-rich pathway and channels is generated when the casting solvent is more highly polar [28]. The presence of these pathways and channels could facilitate the diffusion of ions and/or other polar species through the nanostructured medium. Taking into consideration these observations, we prepared an s-PBC solution (3 wt%) by dispersing the commercial s-PBC after evaporation of commercial solvents in a polar solvent, i.e., isopropyl alcohol (IPA), since the obtained structure could be more suitable for filtration purpose.

### 3.2. Samples Preparation

For adsorption tests, polypropylene filters were cut into circular coupons of ~0.3 cm thickness and ~2.25 cm diameter. A 1 mL volume of s-PBC solution was spotted on the PP coupon to obtain a homogeneous coverage of its surface. Each coupon was air-dried for 24 h before weighing. The amount of deposited s-PBC was ≈0.030 ± 0.01 g for each coupon. Similarly, we prepared larger coated filters for filtration processes: polypropylene filters were cut into circular coupons of ~0.3 cm thickness and ~5 cm in diameter. These were coated with 5 mL of s-PBC solution, and each filter was air-dried for 24 h before weighing. The amount of deposited s-PBC was ≈0.1468 ± 0.0030 g for each filter as measured by a precision weight scale. PP samples covered by s-PBC are here named s-PBC@PP samples.

Some s-PBC@PP samples underwent a UV irradiation process for 7 h, respectively (here named s-PBC@PP_UV). The irradiation was performed by an 18 W UVA/blue DULUX n.78 OSRAM lamp (producing mainly UV emission at 365 nm and a few narrow lines in the visible). The emission spectrum of the lamp is reported in Figure S2 of Supplementary Materials.

The irradiated and unirradiated s-PBC@PP samples were characterized and tested for water contaminants removal as described below. These were compared with unmodified PP samples.

*3.3. Samples Characterization*

Morphological characterization and chemical mapping of the samples were performed using a field emission scanning electron microscope (Supra 35 FE-SEM by Zeiss, Oberkochen, Germany) equipped with an energy-dispersive X-ray (EDX) microanalysis system (X-MAX, 80 mm$^2$ by Oxford Instruments, Abingdon, UK).

The hydrophilicity of the sample's surface was investigated by contact angle measurements using an Optical compact angle meter (CAM 200 model by KSV Instruments LTD, Helsinki, Finland). Moreover, their water uptake values were calculated according to the following simple mass balance equation [15–18]:

$$\text{Uptake}\% = [(m_{\text{wet}} - m_{\text{dry}})/m_{\text{dry}}] \times 100 \qquad (1)$$

where $m_{\text{dry}}$ is the mass of the sample air-dried at least for 24 h and then put into a desiccator; $m_{\text{wet}}$ is the weight of the sample soaked in distilled water at room temperature for 48 h and quickly wiped with a paper tissue in order to remove most of the free surface water. Both masses were measured by a microbalance.

IR measurements on irradiated and not covered PP filters were performed by Jasco FT-IR-4700 spectrophotometer. Equipped with an ATR (ATR-PRO ONE) with a diamond prism. Clamps ensure sample contact with the crystal.

Water flux through the initial PP filter and s-PBC-coated filters before and after UV irradiation, respectively, was measured by the ratio of water volume on filtration time for five consecutive filtration processes of a 40 mL water sample using a vacuum filtration unit with a diaphragm vacuum pump. The same system was used for filtration experiments, as reported below.

*3.4. Adsorption and Filtration Tests*

s-PBC@PP and s-PBC_UV samples were tested for the adsorption and filtration of a cationic molecule (i.e., methylene blue MB) and cationic metals such as $Fe^{3+}$ and $Co^{2+}$ from water. For adsorption experiments, samples were immersed in the dark in 5 mL of contaminant aqueous solutions with an initial concentration of $10^{-5}$ M for MB, 1.1 mM for FeCl$_3$, and 17.5 mM for CoCl$_2$, respectively, for three hours. For each filtration process, three different 20 mL aliquots of initial contaminant solution were filtered, and each eluted aliquot was filtered for consecutive cycles. The filtration tests were conducted through commercial vacuum filtration systems (Nalgene®). In the systems, the membrane in PES Supor® machV was removed and replaced with the filters to be tested. The mean flow (mL/s) was calculated for all filters by filtering 40 mL of H$_2$O 4 times. The mean values obtained were 4.18, 4.11, and 4.87 mL/s for PP, s-PBC@PP, and s-PB@/PP_UV, respectively. Filtration time was recorded for each process. All the solutions before and after adsorption and filtration were analyzed by recording the absorbance spectra variations using a UV/Vis AGILENT Cary 50 spectrophotometer in a wavelength range between 200 and 800 nm. The contaminant removal was evaluated by the Lambert–Beer law via the absorbance peak at 664 nm for MB, 286 nm for $Fe^{3+}$, and 520 nm for $Co^{2+}$. Adsorption and filtration processes on PP filters used as references were also evaluated to point out the role acted by s-PBC covering.

The efficiency of adsorption and filtration tests is measured as an amount (%) of removed ions or as a weight ratio ($Q_t$, mg/g) between removed ions and the s-PBC amount of the coating layer.

## 4. Conclusions

New technical solutions for functionalizing materials by assigning them defined structural features create the possibility of integrating processes to increase the separation efficiency and reduce the operational costs of the process. This paper describes an example of such an operation for the case of water purification using coupled adsorption and filtration processes.

In this work, we coated PP filters with s-PBC layers and tested them for the removal of cationic organic dyes, such as methylene blue (MB), and heavy metal ions ($Fe^{3+}$ and $Co^{2+}$) from water by adsorption and filtration experiments. Some of the coated filters were irradiated by UV light before being exposed to contaminated water and then were tested with the unirradiated filters in the same adsorption and filtration experiments.

S-PBC coating made PP filters partially or totally able to remove the cationic dye and metal ions from water solutions due to their hydrophilic, acid, and negatively charged character.

In particular, the polymer-coated filters showed high efficiency in removing MB from an aqueous solution in both adsorption and filtration processes, with efficiencies of 90% and 80%, respectively.

In the presence of heavy metal ions ($Fe^{3+}$ and $Co^{2+}$), on the other hand, the coated filters showed a better performance for the filtration process rather than in the adsorption processes. In particular, the UV treatment improves the removal ability of $Co^{2+}$ ions (about 28% rather than 20% obtained for the untreated s-PBC@PP filters) and allows a total removal of the $Fe^{3+}$ ions (>99% instead of 50% achieved by the untreated s-PBC@PP filters) after just one single filtration step.

The characterization of the filters before and after use by FT-IR and EDX analysis allowed us to investigate the mechanisms involved in the metal ions removal processes: in the adsorption processes, the interaction times are long enough to allow the ionic species to interact with the surface of the filters, leading to the formation and release of new species in solution. The filtration processes are faster, and, therefore, the interaction time is not enough to release reaction byproducts to the solution; the ionic species are easily trapped in the filters, in particular for the filters whose surface was modified by UV treatment. It has also been shown that the treatment increases the hydrophilicity of the filters, enhancing their filtration capacity. Although further work is needed to extensively investigate the lifetime and regeneration processes of such kinds of filters, we have shown that functional polymeric coating of commercial and low-cost filters is a promising and low-cost strategy for the effective removal of pollutants from water.

**Supplementary Materials:** The following supporting information can be downloaded at: https://www.mdpi.com/article/10.3390/ijms231911777/s1.

**Author Contributions:** Conceptualization, S.F., V.S. and S.S.; data curation, S.F. and V.S.; formal analysis, S.F., V.S. and S.S.; funding acquisition, S.S.; investigation, S.F., V.S., S.S., S.L., M.Z., C.G. and N.S.; resources, S.S., L.G. and L.F.; supervision, S.S.; visualization, S.F., V.S. and N.S.; writing—original draft, S.F., V.S., and S.S.; writing—review and editing, S.F., V.S., S.S., S.L., L.G., L.F., C.G. and M.Z. All authors have read and agreed to the published version of the manuscript.

**Funding:** This research received no external funding.

**Institutional Review Board Statement:** Not applicable.

**Informed Consent Statement:** Not applicable.

**Data Availability Statement:** The data presented in this study are available on request from the corresponding authors.

**Acknowledgments:** The authors wish to thank Kraton Polymers LLC for providing s-PBC (Nexar™). The authors thank Markus Italia for technical support.

**Conflicts of Interest:** The authors declare no conflict of interest.

# References

1. Shannon, M.A.; Bohn, P.W.; Elimelech, M.; Georgiadis, J.G.; Marinas, B.J.; Mayes, A.M. Science and technology for water purification in the coming decades. *Nature* **2008**, *452*, 301–310. [CrossRef] [PubMed]
2. Wang, T.; Wu, H.; Zhao, S.; Zhang, W.; Tahir, M.; Wang, Z.; Wang, J. Interfacial polymerized and pore-variable covalent organic framework composite membrane for dye separation. *Chem. Eng. J.* **2020**, *384*, 123347. [CrossRef]

3. Wang, Z.; Wang, G.; Li, W.; Cui, Z.; Wu, J.; Akpinar, I.; Yu, L.; He, G.; Hu, J. Loofah activated carbon with hierarchical structures for high-efficiency adsorption of multi-level antibiotic pollutants. *Appl. Surf. Sci.* **2021**, *550*, 149313. [CrossRef]
4. Bedia, J.; Muelas-Ramos, V.; Peñas-Garzón, M.; Gómez-Avilés, A.; Rodríguez, J.J.; Belver, C. A Review on the Synthesis and Characterization of Metal Organic Frameworks for Photocatalytic Water Purification. *Catalysts* **2019**, *9*, 52. [CrossRef]
5. Chankhanittha, T.; Komchoo, N.; Senasu, T.; Piriyanon, J.; Youngme, S.; Hemavibool, K.; Nanan, S. Silver decorated ZnO photocatalyst for effective removal of reactive red azo dye and ofloxacin antibiotic under solar light irradiation. *Colloids Surf. A Physicochem. Eng. Asp.* **2021**, *626*, 127034. [CrossRef]
6. Scuderi, V.; Amiard, G.; Sanz, R.; Boninelli, S.; Impellizzeri, G.; Privitera, V. $TiO_2$ coated CuO nanowire array: Ultrathin p–n heterojunction to modulate cationic/anionic dye photo-degradation in water. *Appl. Surf. Sci.* **2017**, *416*, 885–890. [CrossRef]
7. Senasu, T.; Nijpanich, S.; Juabrum, S.; Chanlek, N.; Nanan, S. CdS/BiOBr heterojunction photocatalyst with high performance for solar-light-driven degradation of ciprofloxacin and norfloxacin antibiotics. *Appl. Surf. Sci.* **2021**, *567*, 150850. [CrossRef]
8. Piriyanon, J.; Takhai, P.; Patta, S.; Chankhanittha, T.; Senasu, T.; Nijpanich, S.; Juabrum, S.; Chanlek, N.; Nanan, S. Performance of sunlight responsive WO3/AgBr heterojunction photocatalyst toward degradation of Rhodamine B dye and ofloxacin antibiotic. *Opt. Mater.* **2021**, *121*, 111573. [CrossRef]
9. Mezher, T.; Fath, H.; Abbas, Z.; Khaled, A. Techno-economic assessment and environmental impacts of desalination technologies. *J. Des.* **2011**, *266*, 263–273. [CrossRef]
10. Yuan, S.; Li, X.; Zhu, J.; Zhang, G.; Van Puyvelde, P.; Van der Bruggen, B. Covalent organic frameworks for membrane separation. *Chem. Soc. Rev.* **2019**, *48*, 2665–26815. [CrossRef]
11. Caro, J. Hierarchy in inorganic membranes. *Chem. Soc. Rev.* **2016**, *45*, 3468–3478. [CrossRef] [PubMed]
12. Bai, Y.; Gao, P.; Fang, R.; Cai, J.; Zhang, L.D.; He, Q.Y.; Zhou, Z.H.; Sun, S.P.; Cao, X.L. Constructing positively charged acid-resistant nanofiltration membranes via surface postgrafting for efficient removal of metal ions from electroplating rinse wastewater. *Sep. Purif. Technol.* **2022**, *297*, 121500. [CrossRef]
13. Lu, Y.; Liu, W.; Liu, J.; Li, X.; Zhang, S. A review on 2D porous organic polymers for membrane-based separations: Processing and engineering of transport channels. *Adv. Membr.* **2021**, *1*, 100014. [CrossRef]
14. Wang, Y.; Yang, Z.; Liu, L.; Chen, Y. Construction of high performance thin-film nanocomposite nanofiltration membrane by incorporation of hydrophobic MOF-derived nanocages. *Appl. Surf. Sci.* **2021**, *570*, 151093. [CrossRef]
15. Filice, S.; Urzì, G.; Milazzo, R.G.; Privitera, S.M.S.; Lombardo, S.A.; Compagnini, G.; Scalese, S. Applicability of a New Sulfonated Pentablock Copolymer Membrane and Modified Gas Diffusion Layers for Low-Cost Water Splitting Processes. *Energies* **2019**, *12*, 2064. [CrossRef]
16. Filice, S.; Mazurkiewicz-Pawlicka, M.; Malolepszy, A.; Stobinski, L.; Kwiatkowski, R.; Boczkowska, A.; Gradon, L.; Scalese, S. Sulfonated Pentablock Copolymer Membranes and Graphene Oxide Addition for Efficient Removal of Metal Ions from Water. *Nanomaterials* **2020**, *10*, 1157. [CrossRef]
17. Filice, S.; D'Angelo, D.; Scarangella, A.; Iannazzo, D.; Compagnini, G.; Scalese, S. Highly effective and reusable sulfonated pentablock copolymer nanocomposites for water purification applications. *RSC Adv.* **2017**, *7*, 45521–45534. [CrossRef]
18. D'Angelo, D.; Filice, S.; Scarangella, A.; Iannazzo, D.; Compagnini, G.; Scalese, S. $Bi_2O_3$/Nexar® polymer nanocomposite membranes for azo dyes removal by UV–vis or visible light irradiation. *Catal. Today* **2019**, *321*, 158–163. [CrossRef]
19. Sciuto, E.L.; Filice, S.; Coniglio, M.A.; Faro, G.; Gradon, L.; Galati, C.; Spinella, N.; Libertino, S.; Scalese, S. Antimicrobial s-PBC Coatings for Innovative Multifunctional Water Filters. *Molecules* **2020**, *25*, 5196. [CrossRef]
20. Filice, S.; Sciuto, E.L.; Scalese, S.; Faro, G.; Libertino, S.; Corso, D.; Timpanaro, R.M.; Laganà, P.; Coniglio, M.A. Innovative Antibiofilm Smart Surface against Legionella for Water Systems. *Microorganisms* **2022**, *10*, 870. [CrossRef]
21. Dai, Z.; Ansaloni, L.; Ryan, J.J.; Spontak, R.J.; Deng, L. Incorporation of an ionic liquid into a midblock-sulfonated multiblock polymer for $CO_2$ capture. *J. Membr. Sci.* **2019**, *588*, 117193. [CrossRef]
22. Available online: https://www.sigmaaldrich.com/IT/it/technical-documents/technical-article/analytical-chemistry/photometry-and-reflectometry/ir-spectrum-table (accessed on 29 August 2022).
23. Chen, H.; Chang, K.; Men, X.; Sun, K.; Fang, X.; Ma, C.; Zhao, Y.; Yin, S.; Qin, W.; Wu, C. Covalent Patterning and Rapid Visualization of Latent Fingerprints with Photo-Cross-Linkable Semiconductor Polymer Dots. *ACS Appl. Mater. Interfaces* **2015**, *7*, 14477–14484. [CrossRef] [PubMed]
24. Leea, S.-K.; Mills, A. Novel photochemistry of leuco-Methylene Blue. *Chem. Commun.* **2003**, *18*, 2366–2367. [CrossRef] [PubMed]
25. Harito, C.; Bavykin, D.V.; Yuliarto, B.; Dipojono, H.K.; Walsh, F.C. Inhibition of Polyimide Photodegradation by Incorporation of Titanate Nanotubes into a Composite. *J. Polym. Environ.* **2019**, *27*, 1505–1515. [CrossRef]
26. Le, T.G.; Nguyen, N.T.; Nguyen, Q.T.; De Laat, J.; Dao, H.Y. Effect of Chloride and Sulfate Ions on the Photoreduction Rate of Ferric Ion in UV Reactor Equipped with a Low Pressure Mercury Lamp. *J. Adv. Oxid. Technol.* **2014**, *17*, 305–330. [CrossRef]
27. Sikorska, E.; Gradoń, L. Biofouling reduction for improvement of depth Water filtration. Filter production and testing. *Chem. Process Eng.* **2016**, *37*, 319–330. [CrossRef]
28. Mineart, K.P.; Jiang, X.; Jinnai, H.; Takahara, A.; Spontak, R.J. Morphological Investigation of Midblock-Sulfonated Ionomers Prepared from Solvents Differing in Polarity. *Macromol. Rapid Commun.* **2015**, *36*, 432–438. [CrossRef]

Article

# Granulation of Bismuth Oxide by Alginate for Efficient Removal of Iodide in Water

Tae-Hyun Kim [1], Chihyun Seo [1], Jaeyoung Seon [1,2], Anujin Battulga [1] and Yuhoon Hwang [1,*]

1. Department of Environmental Engineering, Seoul National University of Science and Technology, Seoul 01811, Korea
2. Water Quality Center, Chemicals & Environment Research Institute, Korea Testing & Research Institute, Gyeonggi-do 13810, Korea
* Correspondence: yhhwang@seoultech.ac.kr; Tel.: +82-2-970-6626; Fax: +82-2-971-5776

**Abstract:** The granulation of bismuth oxide (BO) by alginate (Alg) and the iodide adsorption efficacy of Alg–BO for different initial iodide concentrations and contact time values were examined. The optimal conditions for Alg–BO granulation were identified by controlling the weight ratio between Alg and BO. According to the batch iodide adsorption experiment, the Alg:BO weight ratio of 1:20 was appropriate, as it yielded a uniform spherical shape. According to iodide adsorption isotherm experiments and isotherm model fitting, the maximum sorption capacity ($q_m$) was calculated to be 111.8 mg/g based on the Langmuir isotherm, and this value did not plateau even at an initial iodide concentration of 1000 mg/L. Furthermore, iodide adsorption by Alg–BO occurred as monolayer adsorption by the chemical interaction and precipitation between bismuth and iodide, followed by physical multilayer adsorption at a very high concentration of iodide in solution. The iodide adsorption over time was fitted using the intraparticle diffusion model. The results indicated that iodide adsorption was proceeded by boundary layer diffusion during 480 min and reached the plateau from 1440 min to 5760 min by intraparticle diffusion. According to the images obtained using cross-section scanning electron microscopy assisted by energy-dispersive spectroscopy, the adsorbed iodide interacted with the BO in Alg–BO through Bi–O–I complexation. This research shows that Alg–BO is a promising iodide adsorbent owing to its high adsorption capacity, stability, convenience, and ability to prevent secondary pollution.

**Keywords:** bismuth oxide; alginate; granulation; iodide adsorption

## 1. Introduction

At present, nuclear energy is being widely used as a reliable and clean energy source for electricity. Despite those advantages, nuclear accidents, such as those of the Chernobyl and Fukushima nuclear power plants, may release dangerous radioisotopes such as Cs-134, Cs-137, I-129, and I-131 into the environment [1,2]. The recovery of radioactive iodine (I-129 and I-131) from the environments is challenging owing to its high radioactive toxicity, solubility, and mobility [3,4]. Moreover, the considerably longer half-life of I-129 ($1.57 \pm 0.006 \times 10^7$ years) than that of I-131 (8 days) means that it may expose humans and other living organisms to chronic toxicity [5,6]. In aqueous systems, iodine ($I_2$) mainly occurs as iodide ($I^-$) and iodate ($IO_3^-$), depending on the pH, with iodide being the dominant species at neutral pH [7–9]. The release of radioactive iodide is attributable to the dissolution of CsI, which is typically used as nuclear fuel in light-water reactors [10]. The iodine is mainly released as a gaseous contaminant due to its high volatility, but it can be easily dissolved in the resulting water form of radioactive iodide ions. Therefore, effective strategies to remove radioactive iodide from the aqueous medium must be established to ensure the safety of the nuclear industry and environment.

Many researchers have examined iodide adsorption with mineral-based, metal-based, polymer-based, and carbon-based adsorbents in aqueous systems [11–16]. Bismuth-based adsorbents, e.g., bismuth oxide (BO), basic bismuth nitrate, and bismuth subcarbonate, have attracted considerable research interest because of their high selectivity for iodide and low toxicity and cost [17–19]. The high iodide selectivity of bismuth-based adsorbents is attributable to the formation of effective Bi–O–I compounds as a novel waste form [20]. It was reported that ~76% of the removal capacity was still maintained in the presence of chloride when iodide adsorption on mesoporous bismuth oxide was performed with 6 mM chloride ions [17]. The stable adsorption behavior under a wide pH range of 4–11 was also verified using Microrosette-like δ-$Bi_2O_3$ [21]. However, powdered adsorbents are challenging to separate and recycle. Granulation of the powdered form of particulate matter was considered an effective process for enhancing its practical applicability [22–25]. Many researchers have attempted to prepare granules of adsorbents through pellet molding methods by applying strong compressive force as a facile methodology. Notably, the equipment for pellet molding is expensive, and the process requires the application of high pressure, which may block the pores on the surface or change the internal structure of the adsorbents [26]. To address these problems, bismuth-based adsorbents were fixed on a substance or granulated using polymers [27]. Among various polymers, alginate (Alg) extracted from brown algae is a promising candidate owing to its environmental friendliness, low cost, and easy preparation method by simply dropping it on a crosslinking agent such as $Ca^{2+}$ ions [28]. Owing to the ionotropic gelation of spherical drops, the polyguluronate units in the alginate molecules form a chelated structure with metal ions. Then, the chelate structure is transformed to become kinetically stable toward dissociation while the polymannuronate units show normal cations binding [29]. According to the two interactions, granulation by Alg leads to the formation of spherical-shaped beads [30]. Spherical Alg granules are convenient to use as adsorbents in practical applications. They can help avoid secondary pollution resulting from dissolution, as reported in several articles with various powder adsorbents, such as iron oxide, clay, and activated carbon [31–33].

Considering these aspects, in this study, BO was prepared using the solvothermal method and then granulated with Alg to realize iodide adsorption in an aqueous system. Specifically, Alg–BO was prepared in a facile manner by dropping the Alg and BO suspension into a calcium chloride ($CaCl_2$) solution to achieve a uniform spherical shape. The granulation of BO by Alg was characterized by powder X-ray diffraction (PXRD), Fourier-transform infrared (FT-IR), Brunauer–Emmett–Teller (BET) surface area analysis, and microscopic analysis. The optimal conditions for granulation were identified by controlling the weight ratio of Alg to BO through batch iodide adsorption experiments. Moreover, iodide adsorption isotherm experiments were conducted, and the results were fitted using the Langmuir and Freundlich models for the adsorption isotherm. According to the kinetic experiment results fitted with the intraparticle diffusion model, iodide adsorption occurred through boundary layer diffusion in the initial stage and then through intraparticle diffusion. The effect of pH on iodide adsorption, as well as the integrity of granules, was also investigated. After iodide adsorption, the Alg–BO sample was characterized via PXRD, FT-IR, and scanning electron microscopy (SEM) assisted by energy-dispersive spectroscopy (EDS) to evaluate the structural changes and adsorbed iodide distribution.

## 2. Results and Discussion

### 2.1. Optimization of Alg–BO Preparation Conditions for Iodide Adsorption

The granulation conditions were optimized by preparing Alg–BO considering five weight ratios of Alg to BO (1:5, 1:10, 1:20, 1:30, and 1:40). As shown in Figure S1, the beads prepared with weight ratios ranging from 1:5 and 1:20 were spherical with a diameter of approximately 0.3 mm. When the ratio was increased to 1:30 and 1:40, the shape of Alg–BO was slightly elongated and irregular. At higher BO ratios, the Alg was inadequate to establish a stable structure once it reacted with the calcium ions in the bath. Similar results have been observed for halloysite–alginate and organoclay–alginate nanocomposites [33,34]. The sphericity of grains of the filtration bed applied for water treatment was considered an important parameter for column design as it affects the bed porosity. Siwiec (2007) reported that high sphericity could bring lower porosity in the filter bed [35]; therefore, a denser adsorbent bed could be expected. Furthermore, as adsorption performance is affected by the adsorbent mass in the unit bed volume, the bed filled with spherical granules could expect a higher performance and lifetime. These results indicated that Alg:BO ratios of 1:5 to 1:20 were suitable for stable bead formation.

The iodide adsorption capacity of the prepared Alg beads with/without BO was evaluated through simple batch adsorption experiments (Figure 1). The Alg beads without BO exhibited 0% iodide adsorption capacity even after 24 h. When BO was introduced, the iodide adsorption capacity gradually increased with the weight ratio (from 3.7 mg/g (1:5) to 6.9 mg/g (1:30)) after 24 h. However, when the weight ratio was increased to 1:40, the adsorption capacity decreased by approximately 25%. These results were attributable to the aggregation of BO particles when the amount of BO was increased in the Alg matrix for granulation [36,37]. Considering these preliminary iodide adsorption results for different ratios of Alg and prepared BO, Alg–BO with a weight ratio of 1:20 was selected for further study.

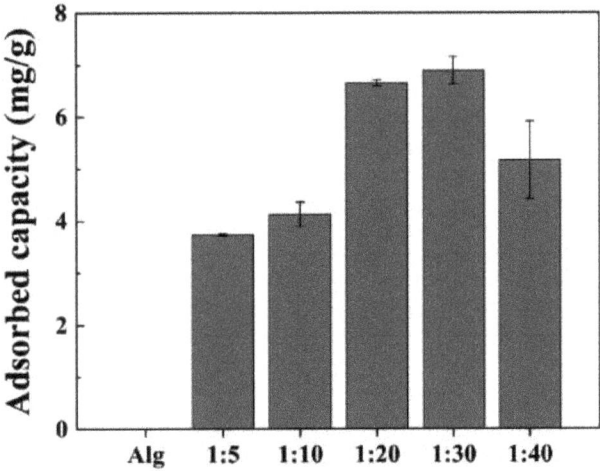

**Figure 1.** Iodide adsorption capacity of parent alginate (Alg) and beads prepared with different Alg and bismuth oxide (BO) weight ratios (initial iodide concentration: 20 mg/L, initial adsorbent concentration: 1 g/L, and contact time: 24 h).

*2.2. Characterization of Prepared Alg–BO*

The Alg–BO (1:20) selected in the previous analysis was characterized by PXRD, FT-IR, and SEM. The PXRD patterns of parent Alg exhibited small diffraction at 31.8°, corresponding to (111) diffraction (asterisk in Figure 2), and broad, amorphous diffraction in the range of 20°–45°, consistent with the previously reported result [38]. The prepared BO consisted of two types of bismuth oxide forms: $\gamma$-$Bi_2O_3$ (PDF No. 00-027-0052) and $Bi_2O_{2.33}$ (PDF No. 00-027-0051). The major diffractions of the two types of BO—(110), (200) for $\gamma$-$Bi_2O_3$ and (101), (111) for $Bi_2O_{2.33}$—were well developed through the solvothermal preparation method. After the granulation of BO with Alg, the characteristic diffractions associated with the two BO forms were preserved, and the small diffractions from Alg disappeared owing to the small amount of Alg used (weight ratio of 1:20). The PXRD patterns indicated that the granulation of the prepared BO with Alg did not significantly influence the BO crystal structure.

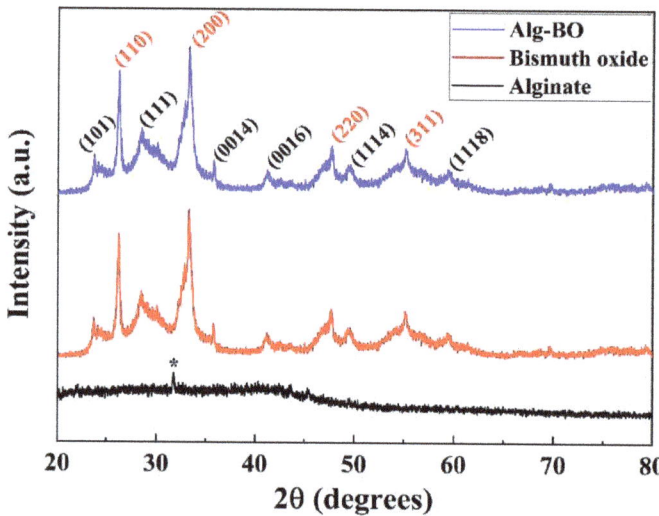

**Figure 2.** Powder X-ray diffraction patterns of Alg, BO, and Alg–BO (asterisk indicates (111) diffraction from Alg; black and red Miller indices indicate two bismuth oxide forms: $Bi_2O_{2.33}$ and $\gamma$-$Bi_2O_3$, respectively).

Moreover, FT-IR spectroscopy was performed to identify the changes in the chemical properties of BO, Alg, and Alg–BO after granulation (Figure 3). In the FT-IR spectra of BO, characteristic stretching vibrations of O−H, C−H, and −$CH_2$ were observed at approximately 3500–3200 $cm^{-1}$ and 2800–3200 $cm^{-1}$. The vibrations of $(CH_2)_n$, C−O, and C−O−C groups appeared at 700–1000 and 1100–1200 $cm^{-1}$. The peaks at 1285 and 1641 $cm^{-1}$ were attributable to the –COOH and C=O ester functional groups, respectively. The vibrations at 700–650 $cm^{-1}$ originated from the metal-oxygen (Bi–O) vibrations. All vibrations attributable to BO were consistent with those observed in previous studies in which BO was prepared using the solvothermal method with an organic solvent such as ethanol or ethylene glycol [39,40]. Moreover, the intense vibration at 1543 $cm^{-1}$ was attributable to the $NO_3$ group, which indicated the existence of $NO_3$ functional groups on the BO surface [39,41]. The spectra of the sodium alginate powder exhibited characteristic asymmetric and symmetric stretching vibrations of the carboxylate group ($COO^-$) in Alg at 1407 and 1596 $cm^{-1}$, respectively. The Alg–BO prepared by polymerization of Alg with $CaCl_2$ exhibited characteristic vibrations attributable to both BO and sodium alginate [33]. Moreover, the C=O vibrations resulting from the ionic bonding between calcium ions and Alg was observed at approximately 1641 $cm^{-1}$ as a shoulder, owing to the intense symmetric stretching vibrations of the carboxylate groups [42]. According to the PXRD and FT-IR results, BO was successfully incorporated with Alg by the polymerization and appeared in a bead form.

To investigate the morphology of Alg–BO, SEM images of the surface and cross-section were obtained (Figure 4). As shown in Figure 4A,B, Alg–BO appeared as a spherical bead with a diameter of approximately 0.2–0.3 mm. The Alg–BO surface was smooth with BO particles (approximately 2–4 μm) aggregated with polymeric alginate. According to the cross-sectional images (Figure 4C,D), spherical BO particles (2–4 μm) were packed homogeneously in the Alg–BO, with irregular nanoparticles sized tens to hundreds of nanometers. The aggregates of rod or plate-like nanoparticles in the intraparticle pores likely enhanced the iodide adsorption.

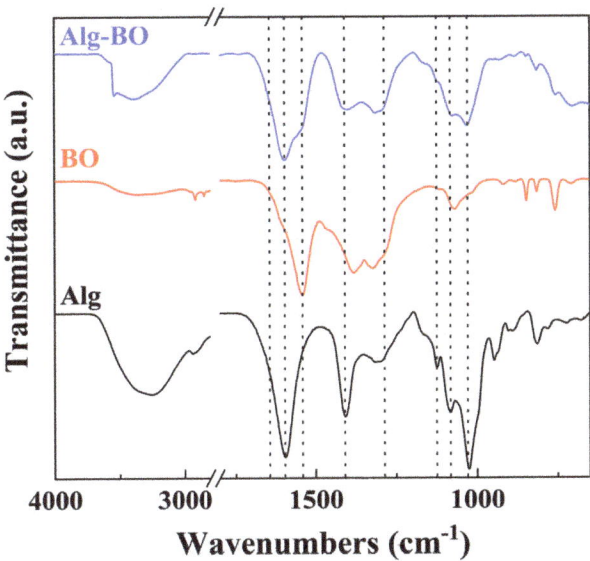

**Figure 3.** FT-IR spectra of Alg, BO, and Alg–BO (dotted lines: 1641, 1596, 1543, 1407, 1285, 1124, 1082, and 1025 cm$^{-1}$).

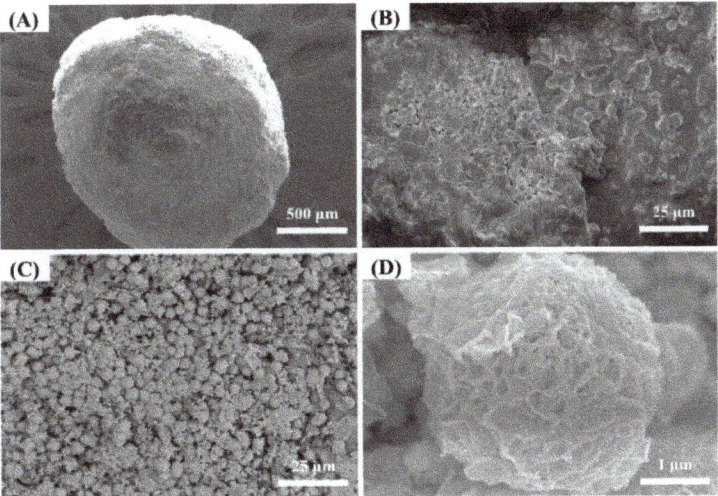

**Figure 4.** SEM images of the (**A**,**B**) surface and (**C**,**D**) cross-section of prepared Alg–BO.

According to the N$_2$ adsorption–desorption hysteresis loop of Alg–BO (Figure 5), it can be classified as an H3 hysteresis loop based on the IUPAC classification [43]. This H3 hysteresis loop is known to be attributed to aggregates or agglomerates of particles with a nonuniform size and/or shape. This result corresponds quite well to the SEM images. The obtained specific surface area (SSA) was determined as 61.15 m$^2$/g. This value shows an 18-44-times-higher SSA value than the previously reported SSA of bismuth oxide powder [44]. The increment in SSA might be attributed to Alg polymer, which makes it possible to form intraparticle pores between BO particles. In addition, the detailed pore width and pore volume calculated by the BJH method showed that the pores of Alg–BO were distributed in the range of 50 Å to 500 Å and mainly formed around 250 Å.

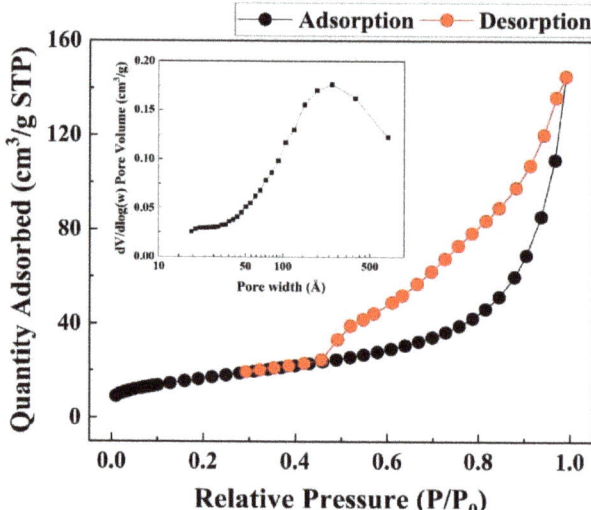

**Figure 5.** Nitrogen adsorption–desorption hysteresis loop of Alg–BO (inset graph shows the pore volume distribution).

*2.3. Iodide Adsorption Performance of Alg–BO*

2.3.1. Adsorption Kinetics

The iodide adsorption efficacy as a function of the contact time (5–5760 min) was evaluated with the initial iodide concentration being 20 mg/L. As shown in Figure 6, the prepared Alg–BO gradually adsorbed iodide for 480 min, and the capacity plateaued after 1440 min. This sustained adsorption, unlike that of powdered BO, was attributable to the granulation with Alg, which covered the BO surface. The kinetic results were fitted using pseudo-first-order (Equation (5)) and second-order (Equation (6)) kinetic models. According to the $R^2$ values obtained using the two kinetic models, Alg–BO followed the pseudo-second-order instead of the pseudo-first-order kinetic model. Moreover, the kinetic parameters indicated that the adsorption rate and capacity of Alg–BO were 0.0007 g/mg·min and 6.792 mg/g, respectively.

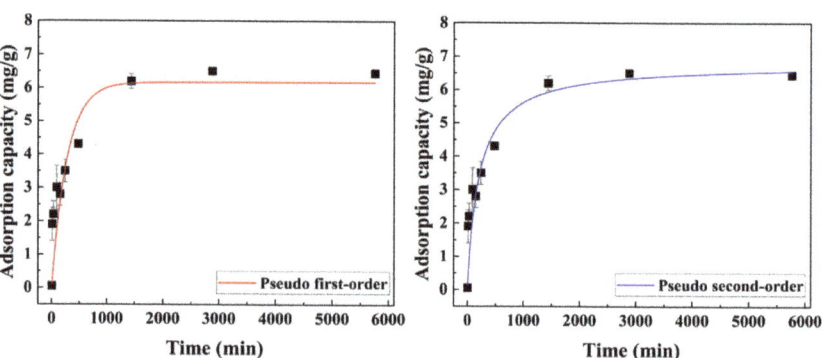

**Figure 6.** Iodide adsorption kinetics of Alg–BO and results of nonlinear fitting using pseudo-first- (red line) and pseudo-second-order (blue line) models. Initial iodide concentration = 20 mg/L; adsorbent concentration ≈ 1.4 g/L; contact time = 5–5760 min.

The intraparticle diffusion model was applied to clarify the adsorption mechanism between iodide and Alg–BO, and the results are summarized in Figure S2 and Table S1.

The intraparticle diffusion model involves three steps: (i) film diffusion of the liquid adsorbate onto the adsorbent surface, (ii) diffusion of the surface adsorbate into pores, and (iii) adsorption onto the inner surfaces of the pores [45]. As shown in Figure S2, the adsorption kinetics corresponded to intraparticle diffusion during 5760 min. The Alg–BO exhibited gradual adsorption through boundary layer diffusion during 480 min and reached the plateau from 1440 min to 5760 min, indicating intraparticle diffusion. This result implies that iodide ions steadily diffused into the intraparticle space of the Alg–BO beads for 5760 min. These results were consistent with those derived from the cross-section SEM images (Figure 4C,D), in which intraparticle pores were observed between spherical BO particle aggregates.

2.3.2. Adsorption Isotherms

To evaluate the iodide adsorption capacity of the prepared Alg–BO, iodide adsorption isotherm experiments were carried out with iodide solutions having various initial concentrations (10, 20, 50, 100, 200, and 1000 mg/L) (Figure 7). The obtained results were analyzed using Langmuir (Equation (3)) and Freundlich (Equation (4)) isotherm models (Figure 7 and Table 1). As summarized in Table 1, the $q_m$ for the Langmuir isotherm model was 111.8 mg/g, with high correlation coefficients ($R^2$ values; 0.9561). It is worth noting that this $q_m$ was approximately 9.0 times higher than the reported powdered bismuth oxide and 510 times larger than that for the polyacrylonitrile encapsulated bismuth oxyhydroxide nanocomposite, as summarized in Table 2. Moreover, the separation factor ($R_L$; dimensionless constant) was calculated as the following equation [46]:

$$R_L = \frac{1}{(1 + a_L C_0)} \quad (1)$$

where $C_0$ is the highest initial adsorbate concentration (mg/L) and $a_L$ is the Langmuir constant (L/mg). The calculated $R_L$ indicated the adsorption isotherm and could be interpreted as unfavorable ($R_L > 1$), linear ($R_L = 1$), favorable ($0 < R_L < 1$), or irreversible ($R_L = 0$). The $R_L$ of Alg–BO was 0.385, implying that the iodide adsorption on Alg–BO was favorable. Additionally, the Freundlich isotherm model was used to interpret the adsorption data considering a heterogeneous adsorption system. According to the fitting parameters and $n$ value (1.750), Alg–BO represented a favorable iodide adsorbent. The $R^2$ value for the Freundlich model was higher than that of the Langmuir model (0.9921). Therefore, the iodide adsorption by Alg–BO has been noted to correspond to monolayer iodide adsorption through the chemical interaction and precipitation between bismuth and iodide [47,48], followed by physical multilayer adsorption at a very high concentration of iodide in solution.

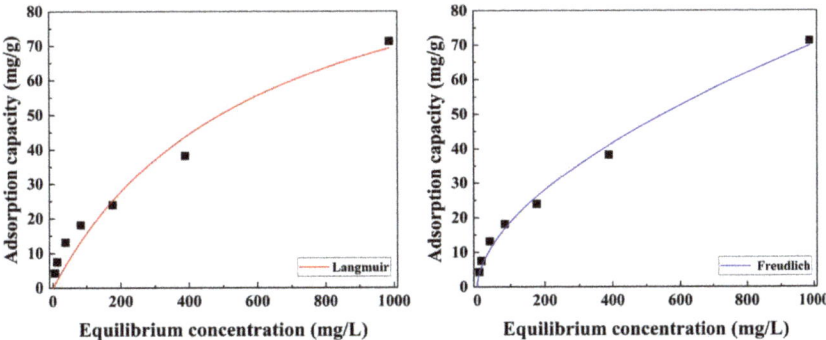

**Figure 7.** Iodide adsorption isotherm of Alg–BO and results of nonlinear fitting using Langmuir (red line) and Freundlich (blue line) models. Initial iodide concentration = 10–1000 mg/L; adsorbent concentration = 1 g/L; contact time = 24 h.

Table 1. Isotherm fitting and kinetic fitting parameters of iodide adsorption by Alg–BO.

| | | | Iodide isotherm models | | | | |
|---|---|---|---|---|---|---|---|
| | | Langmuir | | | Freundlich | | |
| | $q_m$ (mg/g) | $a_L$ (L/mg) | $R^2$ | | $K_F$ | $n$ | $R^2$ |
| | 111.8 | 0.0016 | 0.9561 | | 1.358 | 1.750 | 0.9921 |
| Alg–BO | | | Iodide kinetic models | | | | |
| | | Pseudo-first-order equation | | | Pseudo-second-order equation | | |
| | $q_e$ (mg/g) | $k$ (min$^{-1}$) | $R^2$ | | $q_e$ (mg/g) | $k_2$ (g/mg·min) | $R^2$ |
| | 6.166 | 0.00358 | 0.8743 | | 6.792 | 0.0007 | 0.9048 |

Table 2. Summarized iodine species adsorption capacity of various adsorbents.

| Material | Sample Type | Iodine Species | Adsorption Capacity (mg/g) | Initial Iodide Concentration (mg/L) | Sample Dosing(g/L) | Contact Time | Ref |
|---|---|---|---|---|---|---|---|
| Activated bismuth oxide | Powder | I$^-$ | 100 | 200 | 1.0 | 4 h | [49] |
| Bismuth oxide | Powder | I$^-$ | 12.3 | 200 | 1.0 | 4 h | [49] |
| Cu$_2$O | Powder | I$^-$ | 0.3 | 13 | 50 | 5 d | [50] |
| Cu$_2$S | Powder | I$^-$ | 2.54 | 13 | 20 | 8 d | [51] |
| Mg-Al (NO$_3$) LDH | Powder | I$^-$ | 10.1 | 342.33 | 20 | 4 h | [52] |
| Modified zeolite | Powder | I$^-$ | 3.6 | 10–500 | 10 | 24 h | [53] |
| Ag/Cu$_2$O | Powder | I$^-$ | 25.4 | 2.6–26 | 1.0 | 12 h | [54] |
| Cu/Cu$_2$O | Powder | I$^-$ | 22.9 | 2.6–39 | 1.0 | 12 h | [55] |
| silver nanoparticles-impregnated zeolite | Powder | I$^-$ | 19.54–20.44 | 75–450 | 5.0 | 900 min | [56] |
| Polyacrylonitrile-bismuth oxyhydroxide | Bead | IO$_3^-$ | 0.216 | 1.0 | 5.0 | 24 h | [57] |
| Polyacrylonitrile-bismuth subnitrate | Bead | IO$_3^-$ | 0.199 | 1.0 | 5.0 | 24 h | [58] |
| Cu/Cu$_2$O-immobilized cellulosic filter | Filter | I$^-$ | 10.32 | 1–25 | 2.0 | 15 h | [59] |
| 3D Graphene-Formicary-like δ-Bi$_2$O$_3$ Aerogels | Filter | I$^-$ | 259.08 | 13–130 | 1.0 | 12.5 min | [60] |
| Nano-cellulose hydrogel coated flexible titanate-bismuth oxide membrane | Filter | I$^-$ | 225.9 | 500 | - | 360 min | [61] |
| Alg–BO | Bead | I$^-$ | 111.8 | 10–1000 | 1.0 | 24 h | This study |

The obtained q$_m$ value was compared to the reported value in the literature in Table 2. For the powder-type adsorbents, the copper-based adsorbents showed ≤2 mg/g of adsorption capacity, while zeolite and LDH showed ≤10 mg/g of adsorption capacity with a low selectivity toward iodide. The adsorption capacity could be enhanced by combining elemental silver or elemental Cu (22.9–25.4 mg/g). On the other hand, bismuth oxide (BO), which can be prepared in a simple procedure, showed 12.3–100 mg/g of iodide adsorption capacity.

For the structured BO, polyacrylonitrile-based BO beads showed a very low iodide adsorption capacity due to their low BO content and low tested initial concentration. Even though the reaction condition was different, our result using Alg–BO showed that 111.8 mg/g of iodide adsorption capacity was one of the best results among bead-type adsorbents. This enhanced iodide adsorption capacity might be due to the 18-44 times larger SSA of Alg–BO than the SSA of powdered BO, which makes it possible to form intraparticle pores as a result of granulation with Alg [44]. Of course, several articles present higher adsorption performances using cellulose nanofiber and graphene, but our work still presents a meaningful result with a comparably high adsorption performance using a simple preparation method based on natural polymer, alginate.

2.3.3. Effect of pH

The iodide adsorption capacity as a function of initial pH was evaluated by simple batch adsorption experiments (Figure 8). At pH 4, the iodide adsorption capacity of Alg–BO exhibited a statically similar iodide adsorption capacity based on a t-test with a 95% confidence level. However, when the initial pH was increased to 10, the iodide adsorption capacity decreased by around 15%, from 6.42 mg/g to 5.47 mg/g. This decrement in iodide adsorption capacity at higher pH (pH 10) was attributed to the surface charge of bismuth oxide in Alg–BO. The point of zero charge ($pH_{pzc}$) for BO was around pH 9.4, meaning that BO's surface charge was shifted from positive to negative [62] at pH 10. Due to the negatively charged BO in Alg–BO, the adsorption of iodide could be interrupted by charge–charge repulsion, and it might lead to a decrement in iodide adsorption capacity.

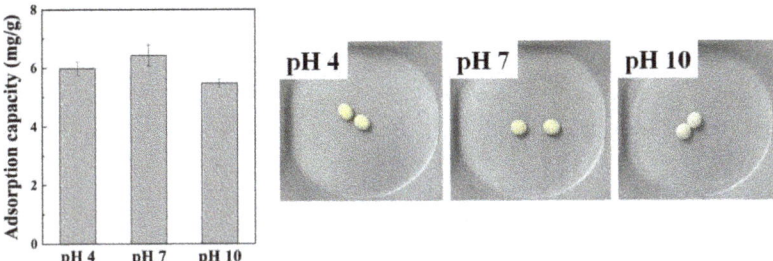

**Figure 8.** Iodide adsorption capacity as a function of initial pH of iodide solution (**left**) and photographs of Alg–BO after iodide adsorption (**right**). Initial iodide concentration = 20 mg/L; initial pH 4, 7, and 10; adsorbent concentration = 1 g/L; contact time = 24 h.

*2.4. Characterization of Alg–BO after Iodide Adsorption*

The crystal structure and morphological changes of Alg–BO after iodide adsorption were investigated with PXRD and EDS-mapping-assisted SEM. In the PXRD patterns (Figure S3), characteristic diffractions attributable to two different types of BO ($Bi_2O_{2.33}$ and $\gamma$-$Bi_2O_3$) were observed. The relative intensity of diffractions slightly decreased (approximately 20%), but the crystal structure of Alg–BO was preserved even after 24 h.

To visualize the distribution of the adsorbed iodide ion in Alg–BO, EDS-mapping-assisted SEM was performed (Figure 9). Before iodide adsorption, bismuth (yellow dot) and oxygen (blue dot) were homogeneously distributed on the BO particles, as observed in the SEM images. After exposure to the iodide solution, iodide ions (magenta dots) appeared not only on the surface but also in the cross-section. On the surface of Alg–BO, iodide ions appeared homogenously with bismuth and oxygen. Interestingly, the cross-sectional EDS mapping images indicated that the iodide was distributed only with bismuth. According to the EDS images, the adsorbed iodide ions interacted with BO in Alg–BO. In addition, from the FT-IR spectra of Alg–BO after iodide adsorption, the characteristic vibrations at 1641, 1596, 1543, 1407, 1285, 1124, 1082, and 1025 $cm^{-1}$ attributed by alginate and BO were well-maintained after iodide adsorption (Figure S4). From the FT-IR spectra, iodide adsorption did not affect the chemical properties of Alg–BO.

Based on the previous study, bismuth oxide can form Bi–O–I bonding directly through chemisorption via Bi–O–I complexation (Figure 10) [48,59,63]. The chemisorption mechanism could lead to less desorption of bound iodide from the adsorbent; therefore, it is promising for handling radioactive contaminants.

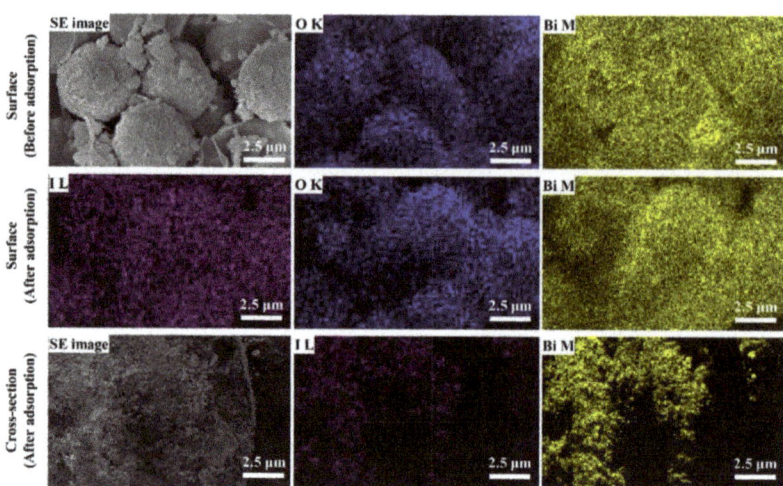

**Figure 9.** Element mapping images obtained by scanning electron microscopy of Alg–BO before and after iodide adsorption (blue: oxygen, yellow: bismuth, and magenta: iodide).

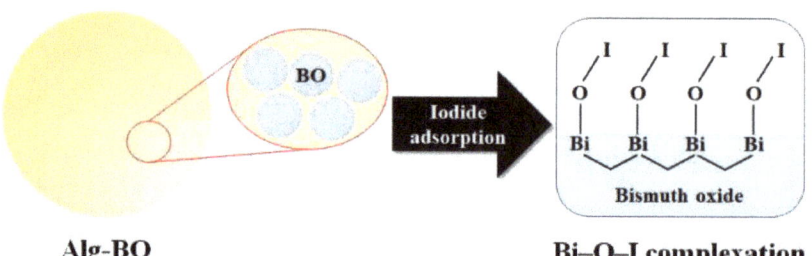

**Figure 10.** Schematic adsorption mechanism between Alg–BO and iodide.

Furthermore, the bismuth ion concentration in the supernatant was quantified to evaluate the stability of Alg–BO. The amount of dissolved bismuth ions was 2.6 µg/L, which was significantly low, indicating that the BO in Alg–BO was highly stable in the iodide solution for 24 h. Therefore, according to the iodide adsorption and characterization results, Alg–BO represents a promising iodide adsorbent, which has a controllable size and contents, can be easily managed, and can, thus, be applied in various fields.

## 3. Materials and Methods

### 3.1. Materials

Bismuth (III) nitrate pentahydrate (Bi(NO$_3$)$_3$·5H$_2$O; 98%), potassium iodide (KI; 99.5%), sodium hydroxide (NaOH; 97%), hydrochloric acid (HCl; 35%), and ethanol (94.5%) were purchased from Samchun Chemicals Co., Ltd. (Seoul, Korea). Sodium alginate was obtained from Junsei Chemical Co., Ltd. (Tokyo, Japan). CaCl$_2$ was obtained from Dongyang Chemical Co., Ltd. (Yeongam-gun, Korea). Ethylene glycol (C$_2$H$_4$(OH)$_2$; 99.8%) was purchased from Sigma-Aldrich Co., LLC (St. Louis, MO, USA). All chemicals were used without purification. Ultrapure water (deionized water; DI) was produced using a water purification system (Synergy®, Merck, Kenilworth, NJ, USA).

### 3.2. Synthesis of Bismuth Oxide (BO)

To prepare BO, 0.97 g of bismuth (III) nitrate pentahydrate, ethanol (34 mL), and ethylene glycol (17 mL) (ethanol: ethylene glycol = 2:1 $v/v$%) were added to a 100 mL glass beaker and stirred with a magnetic stirrer for approximately 30 min until the bismuth nitrate completely dissolved. The prepared solution was transferred to a stainless-steel autoclave with a Teflon liner and hydrothermally treated in an oven at 160 °C for 10 h. The product (50 mL) was collected by centrifugation (6000 rpm for 5 min) in a conical tube. The obtained slurry was washed four times with a mixed solution of DI water and ethanol (1:1 $v/v\%$) and dried in an oven at 60 °C.

*3.3. Granulation of Bismuth Oxide by Alginate (Alg–BO)*

BO was granulated by dropping the Alg/BO mixed slurry into a $CaCl_2$ solution, as described in our previous work [33]. First, the sodium alginate was dissolved in DI (10 mg/mL) by stirring for over 30 min with a mechanical stirrer. Subsequently, powdered BO was added to the Alg solution, with five weight ratios of Alg to BO considered (1:5, 1:10, 1:20, 1:30, and 1:40). This suspension was stirred for 2 h, transferred (5 mL) to a syringe, and added to a 2 $w/v\%$ $CaCl_2$ solution dropwise through a syringe pump (NE4000, NEW ERA; 1.5 mL/min) with vigorous stirring. The beads generated in the $CaCl_2$ solution were stabilized by stirring for 30 min, washed with DI water, and stored in a conical tube with DI.

*3.4. Characterization*

The PXRD patterns were obtained in the range of 20° to 80°, using the Bruker DE/D8 Advance (Bruker AXS GmbH, Berlin, Germany) with a 5 mm air-scattering slit and 2.6 mm equatorial slit, in timestep increments of 3.9 °/min. The FT-IR attenuated total-reflectance (ATR) spectroscopy (Spectrum two, Perkin Elmer, UK) results for a dried bead were obtained in the range of 450–4000 $cm^{-1}$ with eight scans and a resolution of 4 $cm^{-1}$. The size and morphology of the Alg–BO were determined through high-resolution field emission SEM (HR-SEM) using a Hitachi SU8010 (Hitachi High-Technologies Corporation, Tokyo, Japan) assisted by EDS (X-Max, Horiba, Kyoto, Japan) along with a 10 kV accelerated electron beam and a working distance of 8 mm. To perform the SEM/EDS analysis, the prepared Alg–BO was lyophilized and attached to a piece of carbon tape. To obtain cross-section images, the lyophilized Alg–BO was sliced using a surgical knife. Subsequently, the sample surface was coated with a Pt/Pd layer (approximately 10 nm thickness) by using a high-resolution sputter coater. Inductively coupled plasma-mass spectrometry (ICP-MS; Agilent 7900, Agilent Technologies, Inc., CA, USA) was performed to quantify the BO released in the supernatant after iodide adsorption. The nitrogen adsorption–desorption isotherm hysteresis loop and Brunauer–Emmet–Teller (BET) surface area were obtained by a 3Flex physisorption analyzer (Micromeritics, Norcross, GA, USA). The average pore volume and width were determined using the Barrett–Joyner–Halenda (BJH) method.

*3.5. Iodide Adsorption Experiments*

3.5.1. Optimization of Granulation Conditions for Alg–BO

The variation in the iodide adsorption efficacy with the Alg:BO ratio was determined through a simple batch test to optimize the granulation conditions for Alg–BO. Approximately 40 mg of Alg–BO (1:5, 1:10, 1:20, 1:30, and 1:40 of wt%) was added to a 40 mL potassium iodide solution (20 mg I/L) and continuously agitated using a vertical shaker for 24 h. The initial pH of the iodide solution was adjusted to 7.0 using HCl and NaOH. The supernatant was collected using a syringe filter (polyethersulfone (PES), 0.45 μm). The iodide concentration in the supernatant was quantified by ultraviolet (UV) absorbance at a wavelength of 225 nm using a UV–visible spectrometer (UV-vis spectrometer, Genesys 50, Thermo Fisher Scientific, USA). After the adsorption experiments, the amount of iodide ions adsorbed per weight of adsorbent $q_e$ (mg/g) was determined using Equation (2).

$$q_e \left(\frac{mg}{g}\right) = \frac{(C_0 - C_e)}{\left(\frac{m}{V}\right)} \quad (2)$$

where $C_0$ is the initial iodide concentration (mg/L), $C_e$ is the equilibrium concentration after adsorption (mg/L), $m$ is the adsorbent weight (g), and $V$ is the volume of the solution (L).

### 3.5.2. Iodide Adsorption Isotherm and Kinetic Experiments

Iodide adsorption isotherm experiments were conducted with initial iodide concentrations of 10, 20, 50, 100, 200, 400, and 1000 mg/L (pH 7). Approximately 30 mg of Alg–BO was dispersed in 30 mL of each iodide solution (1 g/L) and continuously shaken using a vertical shaker for 24 h. The sample was collected and quantified, as described in Section 3.5.1. The obtained isotherm result was fitted using the Langmuir (Equation (3)) [64] and Freundlich (Equation (4)) [65] isotherm models.

$$q_e = \frac{(q_m a_L C_e)}{(1 + a_L C_e)} \quad (3)$$

$$q_e = K_F \cdot C_e^{(\frac{1}{n})} \quad (4)$$

where $q_e$ is the quantity of adsorbate adsorbed per unit weight of solid adsorbent, $q_m$ is the maximum sorption capacity of the adsorbent (mg/g), $C_e$ is the equilibrium concentration of the adsorbate in solution (mg/L), and $a_L$ (L/mg) is the Langmuir affinity constant. $K_F$ is the Freundlich constant indicating adsorption capacity, and $n$ is the Freundlich constant related to the favorability of the adsorption process.

Moreover, iodide adsorption kinetic experiments were conducted using 20 mg/L iodide solutions (pH 7) and 1.4 g/L of adsorbent dose. The reaction vessel was closed and gently stirred using a magnetic stirrer at 25 °C. The supernatant was collected and filtrated through a 0.45 μm PES syringe filter at designed time points (5, 10, 30, 90, 150, 240, 480, 1440, 2880, and 5760 min). The obtained supernatant was analyzed through a UV–vis spectrometer. The kinetic results were analyzed using pseudo-first-order (Equation (5)) [66] and pseudo-second-order [67] kinetic models (Equation (6)). Moreover, the intraparticle diffusion kinetics model (Equation (7)) [68] was used to investigate the adsorption mechanisms.

$$q_t = q_e \left(1 - e^{-kt}\right) \quad (5)$$

$$q_t = \frac{k_2 q_e^2 t}{1 + k_2 q_e t} \quad (6)$$

$$q_t = K_{id} t^{\frac{1}{2}} + c \quad (7)$$

where $q_t$ is the adsorbed amount at time $t$ (mg/g), $q_e$ is the equilibrium concentration (mg/g), $k$ is the first-order rate constant (1/min), and $k_2$ is the second-order rate constant (g/mg·min). Moreover, $K_{id}$ (mg/g·min$^{1/2}$) is the intraparticle rate constant, and c (mg/g) is the thickness of the boundary layer formed in the first interval.

### 3.5.3. Effect of pH

To evaluate the effect of initial iodide solution pH, the pH of 20 mg/L of iodide solution was adjusted to pH 4 and 10 by HCl and NaOH, respectively. Around 30 mg of Alg–BO was placed into 30 mL of pH-adjusted iodide solution (1 g/L of adsorbent dose) and continuously agitated using a vertical shaker for 24 h. The sample was collected and quantified, as described in Section 3.5.1.

## 4. Conclusions

The BO was successfully granulated with Alg. The optimal condition for the granulation was determined considering different weight ratios of BO to Alg (1:5–1:40) in batch iodide adsorption experiments. The weight ratio of 1:20 wt% was selected as the optimal condition. According to the characterization results obtained through PXRD, FT-IR, and SEM analyses, BO appeared in two forms: $Bi_2O_{2.33}$ and $\gamma$-$Bi_2O_3$, and was successfully granulated with Alg, yielding spherical beams with a diameter of 3 mm. According to the cross-sectional SEM images, irregular nanoparticles sized tens to hundreds of nanometers were packed into a few millimeters of the granulated adsorbent. The intraparticle pores in the granule could enhance the iodide adsorption. The iodide adsorption capacity of Alg–BO gradually increased and did not reach a plateau even at an initial iodide concentration of 1000 mg/L. Moreover, the calculated $q_m$ was 111.8 mg/g. According to the isotherm model analysis, iodide adsorption occurred as monolayer adsorption through the chemical interaction and precipitation between bismuth and iodide, followed by physical multilayer adsorption at a very high concentration of iodide in solution. Furthermore, the iodide adsorption as a function of contact time was analyzed by fitting with the intraparticle diffusion model through boundary layer diffusion during 480 min, reaching the plateau from 1440 min to 5760 min by intraparticle diffusion. EDS mapping images of the surface and cross-section after iodide adsorption indicated that the adsorbed iodide interacted with BO in Alg–BO through Bi–O–I complexation. This research shows that Alg–BO is a promising iodide adsorbent with a high absorption capacity, stability, and convenience, and it can help prevent secondary pollution.

**Supplementary Materials:** The following supporting information can be downloaded at: https://www.mdpi.com/article/10.3390/ijms232012225/s1.

**Author Contributions:** Conceptualization, T.-H.K. and Y.H.; methodology T.-H.K., C.S., J.S. and A.B.; validation C.S., J.S., A.B. and T.-H.K.; formal analysis, T.-H.K. and J.S.; investigation, C.S., J.S. and A.B.; resources, Y.H.; data curation, T.-H.K. and J.S.; writing—original draft preparation, C.S., A.B. and T.-H.K.; writing—review and editing, T.-H.K. and Y.H.; visualization, C.S. and A.B.; supervision, Y.H.; project administration, Y.H.; funding acquisition, Y.H. All authors have read and agreed to the published version of the manuscript.

**Funding:** This work was supported by the National Research Foundation of Korea (NRF) grant funded by the Korea government (MSIT) (NRF-2021M3E8A2100648). This study was also supported by the Human Resource Development Programs for Green Convergence Technology funded by the Korea Ministry of Environment (MOE), the Republic of Korea.

**Data Availability Statement:** The data presented in this study are available on request from the corresponding author.

**Conflicts of Interest:** The authors declare no conflict of interest.

# References

1. Tanaka, K.; Takahashi, Y.; Sakaguchi, A.Y.A.; Umeo, M.; Hayakawa, S.; Tanida, H.; Saito, T.; Kanai, Y. Vertical profiles of Iodine-131 and Cesium-137 in soils in Fukushima Prefecture related to the Fukushima Daiichi Nuclear Power Station Accident. *Geochem. J.* **2012**, *46*, 73–76. [CrossRef]
2. Hou, X.L.; Fogh, C.L.; Kucera, J.; Andersson, K.G.; Dahlgaard, H.; Nielsen, S.P. Iodine-129 and Caesium-137 in Chernobyl contaminated soil and their chemical fractionation. *Sci. Total Environ.* **2003**, *308*, 97–109. [CrossRef]
3. Strickert, R.; Friedman, A.M.; Fried, S. The sorption of technetium and iodine radioisotopes by various minerals. *Nucl. Technol.* **1980**, *49*, 253–266. [CrossRef]
4. Sazarashi, M.; Ikeda, Y.; Seki, R.; Yoshikawa, H. Adsorption of I$^-$ ions on minerals for $^{129}$I waste management. *J. Nucl. Sci. Technol.* **1994**, *31*, 620–622. [CrossRef]
5. Pravdivtseva, O.; Meshik, A.; Hohenberg, C.M.; Krot, A.N. I-Xe systematics of the impact plume produced chondrules from the CB carbonaceous chondrites: Implications for the half-life value of $^{129}$I and absolute age normalization of $^{129}$I–$^{129}$Xe chronometer. *Geochim. Cosmochim. Acta* **2017**, *201*, 320–330. [CrossRef] [PubMed]
6. Takahashi, T.; Schoemaker, M.; Trott, K.; Simon, S.; Fujimori, K.; Nakashima, N.; Fukao, A.; Saito, H. The relationship of thyroid cancer with radiation exposure from nuclear weapon testing in the Marshall islands. *J. Epidemiol.* **2003**, *13*, 99–107. [CrossRef] [PubMed]
7. Fuge, R.; Johnson, C.C. The geochemistry of iodine—A review. *Environ. Geochem. Health* **1986**, *8*, 31–54. [CrossRef] [PubMed]
8. Kodama, S.; Takahashi, Y.; Okumura, K.; Uruga, T. Speciation of iodine in solid environmental samples by iodine K-edge XANES: Application to soils and ferromanganese oxides. *Sci. Total Environ.* **2006**, *363*, 275–284. [CrossRef] [PubMed]
9. Shimamoto, Y.S.; Takahashi, Y.; Terada, Y. Formation of organic iodine supplied as iodide in a soil–water system in Chiba, Japan. *Environ. Sci. Technol.* **2011**, *45*, 2086–2092. [CrossRef]
10. Tietze, S.; Foreman, M.R.S.J.; Ekberg, C. Synthesis of I-131 labelled iodine species relevant during severe nuclear accidents in light water reactors. *Radiochim. Acta* **2013**, *101*, 675–680. [CrossRef]
11. Zhang, H.; Gao, X.; Guo, T.; Li, Q.; Liu, H.; Ye, X.; Guo, M.; Wu, Z. Adsorption of iodide ions on a calcium alginate–silver chloride composite adsorbent. *Colloids Surf. A Physicochem. Eng. Asp.* **2011**, *386*, 166–171. [CrossRef]

12. Choung, S.; Kim, M.; Yang, J.-S.; Kim, M.-G.; Um, W. Effects of radiation and temperature on iodide sorption by surfactant-modified bentonite. *Environ. Sci. Technol.* **2014**, *48*, 9684–9691. [CrossRef]
13. Choung, S.; Um, W.; Kim, M.; Kim, M.-G. Uptake mechanism for iodine species to black carbon. *Environ. Sci. Technol.* **2013**, *47*, 10349–10355. [CrossRef]
14. Theiss, F.L.; Ayoko, G.A.; Frost, R.L. Iodide removal using LDH technology. *Chem. Eng. J.* **2016**, *296*, 300–309. [CrossRef]
15. Zhang, W.; Li, Q.; Mao, Q.; He, G. Cross-linked chitosan microspheres: An efficient and eco-friendly adsorbent for iodide removal from waste water. *Carbohydr. Polym.* **2019**, *209*, 215–222. [CrossRef] [PubMed]
16. Phanthuwongpakdee, J.; Babel, S.; Kaneko, T. Screening of new bio-based materials for radioactive iodide adsorption from water environment. *J. Water Process Eng.* **2021**, *40*, 101955. [CrossRef]
17. Zhang, L.; Jaroniec, M. SBA-15 templating synthesis of mesoporous bismuth oxide for selective removal of iodide. *J. Colloid Interface Sci.* **2017**, *501*, 248–255. [CrossRef]
18. Ng, C.H.B.; Fan, W.Y. Shape-controlled preparation of basic bismuth nitrate crystals with high iodide-removal capacities. *ChemNanoMat* **2016**, *2*, 133–139. [CrossRef]
19. Cheng, G.; Yang, H.; Rong, K.; Lu, Z.; Yu, X.; Chen, R. Shape-controlled solvothermal synthesis of bismuth subcarbonate nanomaterials. *J. Solid State Chem.* **2010**, *183*, 1878–1883. [CrossRef]
20. Krumhansl, J.L.; Nenoff, T.M. Hydrotalcite-like layered bismuth–iodine–oxides as waste forms. *Appl. Geochem.* **2011**, *26*, 57–64. [CrossRef]
21. Liu, L.; Liu, W.; Zhao, X.; Chen, D.; Cai, R.; Yang, W.; Komarneni, S.; Yang, D. Selective capture of iodide from solutions by Microrosette-like δ-$Bi_2O_3$. *ACS Appl. Mater. Interfaces* **2014**, *6*, 16082–16090. [CrossRef]
22. Milferstedt, K.; Hamelin, J.; Park, C.; Jung, J.; Hwang, Y.; Cho, S.-K.; Jung, K.-W.; Kim, D.-H. Biogranules applied in environmental engineering. *Int. J. Hydrogen Energy* **2017**, *42*, 27801–27811. [CrossRef]
23. Mines, P.D.; Thirion, D.; Uthuppu, B.; Hwang, Y.; Jakobsen, M.H.; Andersen, H.R.; Yavuz, C.T. Covalent organic polymer functionalization of activated carbon surfaces through acyl chloride for environmental clean-up. *Chem. Eng. J.* **2017**, *309*, 766–771. [CrossRef]
24. Wu, H.-X.; Wang, T.-J.; Chen, L.; Jin, Y.; Zhang, Y.; Dou, X.-M. Granulation of Fe–Al–Ce hydroxide nano-adsorbent by immobilization in porous polyvinyl alcohol for fluoride removal in drinking water. *Powder Technol.* **2011**, *209*, 92–97. [CrossRef]
25. Zhang, X.; Liu, W.; Zhou, S.; Li, Z.; Sun, J.; Hu, Y.; Yang, Y. A review on granulation of CaO-based sorbent for carbon dioxide capture. *Chem. Eng. J.* **2022**, *446*, 136880. [CrossRef]
26. Shim, W.G.; Lee, J.W.; Moon, H. Adsorption equilibrium and column dynamics of VOCs on MCM-48 depending on pelletizing pressure. *Microporous Mesoporous Mater.* **2006**, *88*, 112–125. [CrossRef]
27. Zhao, Q.; Chen, G.; Wang, Z.; Jiang, M.; Lin, J.; Zhang, L.; Zhu, L.; Duan, T. Efficient removal and immobilization of radioactive iodide and iodate from aqueous solutions by bismuth-based composite beads. *Chem. Eng. J.* **2021**, *426*, 131629. [CrossRef]
28. Kwon, O.-H.; Kim, J.-O.; Cho, D.-W.; Kumar, R.; Baek, S.H.; Kurade, M.B.; Jeon, B.-H. Adsorption of As(III), As(V) and Cu(II) on zirconium oxide immobilized alginate beads in aqueous phase. *Chemosphere* **2016**, *160*, 126–133. [CrossRef]
29. Smidsrød, O. Molecular basis for some physical properties of alginates in the gel state. *Faraday Discuss. Chem. Soc.* **1974**, *57*, 263–274. [CrossRef]
30. Bajpai, S.K.; Sharma, S. Investigation of swelling/degradation behaviour of alginate beads crosslinked with $Ca^{2+}$ and $Ba^{2+}$ ions. *React. Funct. Polym.* **2004**, *59*, 129–140. [CrossRef]
31. Alamin, N.U.; Khan, A.S.; Nasrullah, A.; Iqbal, J.; Ullah, Z.; Din, I.U.; Muhammad, N.; Khan, S.Z. Activated carbon-alginate beads impregnated with surfactant as sustainable adsorbent for efficient removal of methylene blue. *Int. J. Biol. Macromol.* **2021**, *176*, 233–243. [CrossRef]
32. Zouboulis, A.I.; Katsoyiannis, I.A. Arsenic removal using iron oxide loaded alginate beads. *Ind. Eng. Chem. Res.* **2002**, *41*, 6149–6155. [CrossRef]
33. Son, Y.; Kim, T.-H.; Kim, D.; Hwang, Y. Porous Clay Heterostructure with Alginate Encapsulation for Toluene Removal. *Nanomaterials* **2021**, *11*, 388. [CrossRef]
34. Huang, B.; Liu, M.; Long, Z.; Shen, Y.; Zhou, C. Effects of halloysite nanotubes on physical properties and cytocompatibility of alginate composite hydrogels. *Mater. Sci. Eng. C-Mater. Biol. Appl.* **2017**, *70*, 303–310. [CrossRef]
35. Siwiec, T. The Sphericity of grains filtration beds applied for water treatment on examples of selected minerals. *Electron. J. Pol. Agric. Univ.* **2007**, *10*, 30.
36. Jin, H.-H.; Lee, C.-H.; Lee, W.-K.; Lee, J.-K.; Park, H.-C.; Yoon, S.-Y. In-situ formation of the hydroxyapatite/chitosan-alginate composite scaffolds. *Mater. Lett.* **2008**, *62*, 1630–1633. [CrossRef]
37. Zvulunov, Y.; Radian, A. Alginate composites reinforced with polyelectrolytes and clay for improved adsorption and bioremediation of formaldehyde from water. *ACS ES&T Water* **2021**, *1*, 1837–1848.
38. Sundarrajan, P.; Eswaran, P.; Marimuthu, A.; Subhadra Lakshmi, B.; Kannaiyan, P. One pot synthesis and characterization of alginate stabilized semiconductor nanoparticles. *Bull. Korean Chem. Soc.* **2012**, *33*, 3218–3224. [CrossRef]
39. Chen, M.-S.; Chen, S.-H.; Lai, F.-C.; Chen, C.-Y.; Hsieh, M.-Y.; Chang, W.-J.; Yang, J.-C.; Lin, C.-K. Sintering temperature-dependence on radiopacity of $Bi_{(2-x)}ZrO_{(3+x/2)}$ powders prepared by sol-gel process. *Materials* **2018**, *11*, 1685. [CrossRef]

40. Mohammadi, M.; Tavajjohi, A.; Ziashahabi, A.; Pournoori, N.; Muhammadnejad, S.; Delavari, H.; Poursalehi, R. Toxicity, morphological and structural properties of chitosan-coated $Bi_2O_3$–$Bi(OH)_3$ nanoparticles prepared via DC arc discharge in liquid: A potential nanoparticle-based CT contrast agent. *Micro Nano Lett.* **2019**, *14*, 239–244. [CrossRef]
41. Li, W. Facile synthesis of monodisperse $Bi_2O_3$ nanoparticles. *Mater. Chem. Phys.* **2006**, *99*, 174–180. [CrossRef]
42. Kusuktham, B.; Prasertgul, J.; Srinun, P. Morphology and property of calcium silicate encapsulated with alginate beads. *Silicon* **2014**, *6*, 191–197. [CrossRef]
43. Meshkani, F.; Rezaei, M. Effect of process parameters on the synthesis of nanocrystalline magnesium oxide with high surface area and plate-like shape by surfactant assisted precipitation method. *Powder Technol.* **2010**, *199*, 144–148. [CrossRef]
44. Hu, Y.; Li, D.; Sun, F.; Weng, Y.; You, S.; Shao, Y. Temperature-induced phase changes in bismuth oxides and efficient photodegradation of phenol and p-chlorophenol. *J. Hazard. Mater.* **2016**, *301*, 362–370. [CrossRef]
45. Weber, W.J.; Morris, J.C. Kinetics of Adsorption on Carbon from Solution. *J. Sanit. Eng. Div.* **1963**, *89*, 31–59. [CrossRef]
46. Hameed, B.H.; Salman, J.M.; Ahmad, A.L. Adsorption isotherm and kinetic modeling of 2,4-D pesticide on activated carbon derived from date stones. *J. Hazard. Mater.* **2009**, *163*, 121–126. [CrossRef]
47. Liu, S.; Kang, S.; Wang, H.; Wang, G.; Zhao, H.; Cai, W. Nanosheets-built flowerlike micro/nanostructured $Bi_2O_{2.33}$ and its highly efficient iodine removal performances. *Chem. Eng. J.* **2016**, *289*, 219–230. [CrossRef]
48. Lee, S.-H.; Takahashi, Y. Selective immobilization of iodide onto a novel bismuth-impregnated layered mixed metal oxide: Batch and EXAFS studies. *J. Hazard. Mater.* **2020**, *384*, 121223. [CrossRef]
49. Wang, C.; Hu, H.; Yan, S.; Zhang, Q. Activating $Bi_2O_3$ by ball milling to induce efficiently oxygen vacancy for incorporating iodide anions to form BiOI. *Chem. Phys.* **2020**, *533*, 110739. [CrossRef]
50. Lefèvre, G.; Walcarius, A.; Ehrhardt, J.J.; Bessière, J. Sorption of Iodide on Cuprite ($Cu_2O$). *Langmuir* **2000**, *16*, 4519–4527. [CrossRef]
51. Lefèvre, G.; Bessière, J.; Ehrhardt, J.-J.; Walcarius, A. Immobilization of iodide on copper(I) sulfide minerals. *J. Environ. Radioact.* **2003**, *70*, 73–83. [CrossRef]
52. Kentjono, L.; Liu, J.C.; Chang, W.C.; Irawan, C. Removal of boron and iodine from optoelectronic wastewater using Mg–Al ($NO_3$) layered double hydroxide. *Desalination* **2010**, *262*, 280–283. [CrossRef]
53. Warchoł, J.; Misaelides, P.; Petrus, R.; Zamboulis, D. Preparation and application of organo-modified zeolitic material in the removal of chromates and iodides. *J. Hazard. Mater.* **2006**, *137*, 1410–1416. [CrossRef]
54. Mao, P.; Liu, Y.; Jiao, Y.; Chen, S.; Yang, Y. Enhanced uptake of iodide on Ag@$Cu_2O$ nanoparticles. *Chemosphere* **2016**, *164*, 396–403. [CrossRef]
55. Mao, P.; Qi, L.; Liu, X.; Liu, Y.; Jiao, Y.; Chen, S.; Yang, Y. Synthesis of Cu/$Cu_2O$ hydrides for enhanced removal of iodide from water. *J. Hazard. Mater.* **2017**, *328*, 21–28. [CrossRef]
56. Tauanov, Z.; Inglezakis, V.J. Removal of iodide from water using silver nanoparticles-impregnated synthetic zeolites. *Sci. Total Environ.* **2019**, *682*, 259–270. [CrossRef]
57. Cordova, E.A.; Garayburu-Caruso, V.; Pearce, C.I.; Cantrell, K.J.; Morad, J.W.; Gillispie, E.C.; Riley, B.J.; Colon, F.C.; Levitskaia, T.G.; Saslow, S.A.; et al. Hybrid Sorbents for [129]I Capture from Contaminated Groundwater. *ACS Appl. Mater. Interfaces* **2020**, *12*, 26113–26126. [CrossRef]
58. Seon, J.; Hwang, Y. Cu/$Cu_2O$-immobilized cellulosic filter for enhanced iodide removal from water. *J. Hazard. Mater.* **2021**, *409*, 124415. [CrossRef]
59. Tesfay Reda, A.; Pan, M.; Zhang, D.; Xu, X. Bismuth-based materials for iodine capture and storage: A review. *J. Environ. Chem. Eng.* **2021**, *9*, 105279. [CrossRef]
60. Xiong, Y.; Dang, B.; Wang, C.; Wang, H.; Zhang, S.; Sun, Q.; Xu, X. Cellulose fibers constructed convenient recyclable 3D graphene-formicary-like δ-$Bi_2O_3$ aerogels for the selective capture of iodide. *ACS Appl. Mater. Interfaces* **2017**, *9*, 20554–20560. [CrossRef]
61. Xiong, Y.; Wang, C.; Wang, H.; Jin, C.; Sun, Q.; Xu, X. Nano-cellulose hydrogel coated flexible titanate-bismuth oxide membrane for trinity synergistic treatment of super-intricate anion/cation/oily-water. *Chem. Eng. J.* **2018**, *337*, 143–151. [CrossRef]
62. Kosmulski, M. Isoelectric points and points of zero charge of metal (hydr)oxides: 50years after Parks' review. *Adv. Colloid Interface Sci.* **2016**, *238*, 1–61. [CrossRef]
63. Zhang, L.; Gonçalves, A.A.S.; Jiang, B.; Jaroniec, M. Capture of iodide by bismuth vanadate and bismuth oxide: An insight into the process and its aftermath. *ChemSusChem* **2018**, *11*, 1486–1493. [CrossRef]
64. Langmuir, I. The adsorption of gases on plane surface of glass, mica and platinum. *J. Am. Chem. Soc.* **1918**, *40*, 1361–1403. [CrossRef]
65. Haring, M.M. Colloid and capillary chemistry (Freundlich, Herbert). *J. Chem. Educ.* **1926**, *3*, 1454. [CrossRef]
66. Lagergren, S. Zur theorie der sogenannten adsorption gelöster stoffe. *Kungliga Svenska Vetenskapsakademiens Handlingar* **1898**, *24*, 1–39.
67. Ho, Y.S.; McKay, G. Pseudo-second order model for sorption processes. *Process Biochem.* **1999**, *34*, 451–465. [CrossRef]
68. Meng, L.; Zhang, X.; Tang, Y.; Su, K.; Kong, J. Hierarchically porous silicon-carbon-nitrogen hybrid materials towards highly efficient and selective adsorption of organic dyes. *Sci. Rep.* **2015**, *5*, srep07910.

*Article*

# One-Step Carbonization Synthesis of Magnetic Biochar with 3D Network Structure and Its Application in Organic Pollutant Control

Xiaoxin Chen [1,2], Jiacheng Lin [1,2], Yingjie Su [1,2,*] and Shanshan Tang [1,2,*]

1. College of Life Sciences, Jilin Agricultural University, Changchun 130118, China
2. Key Laboratory of Straw Comprehensive Utilization and Black Soil Conservation, Ministry of Education, Jilin Agricultural University, Changchun 130118, China
* Correspondence: yjsu@jlau.edu.cn (Y.S.); tangshanshan81@163.com (S.T.)

**Abstract:** In this study, a magnetic biochar with a unique 3D network structure was synthesized by using a simple and controllable method. In brief, the microbial filamentous fungus *Trichoderma reesei* was used as a template, and $Fe^{3+}$ was added to the culture process, which resulted in uniform recombination through the bio-assembly property of fungal hyphae. Finally, magnetic biochar ($BMFH/Fe_3O_4$) was synthesized by controlling different heating conditions in a high temperature process. The adsorption and Fenton-like catalytic performance of $BMFH/Fe_3O_4$ were investigated by using the synthetic dye malachite green (MG) and the antibiotic tetracycline hydrochloride (TH) as organic pollutant models. The results showed that the adsorption capacity of $BMFH/Fe_3O_4$ for MG and TH was 158.2 and 171.26 mg/g, respectively, which was higher than that of most biochar adsorbents, and the Fenton-like catalytic degradation effect of organic pollutants was also better than that of most catalysts. This study provides a magnetic biochar with excellent performance, but more importantly, the method used can be effective in further improving the performance of biochar for better control of organic pollutants.

**Keywords:** water; biochar; organic pollutants; adsorption; Fenton-like catalysis

## 1. Introduction

As one of Earth's natural resources, water is essential for all aspects of life. Except in various forms in the oceans, glaciers, and the atmosphere, less than 0.3 per cent of the water on Earth is fresh enough to meet the survival and development needs of all living things [1]. Furthermore, nearly one-fifth of the world's population still does not have access to safe drinking water [2]. Therefore, water safety has always been a major, widespread issue. With the rapid development of society, the production and discharge of a large number of industrial processes containing various dyes, drugs, heavy metals, and other harmful wastewater has gradually become one of the main risks threatening the safety of water bodies [3]. Among them, organic pollutants significantly reduce water quality, and in addition to affecting ecosystems by disrupting photosynthesis of aquatic plants, they also have a significant impact on human health [4,5]. Studies have pointed out that most organic pollutants are persistent forms of pollution (synthetic dyes, antibiotics, etc.), which are difficult to degrade naturally, are toxic and carcinogenic, have high tolerance, and have begun to threaten the global water environment [6,7]. As highlighted by the United Nations Sustainable Development Goals, providing clean water remains a major challenge, especially in developing regions of the world [8], so the need for more efficient and advanced sewage purification technologies is becoming more urgent.

Bio-remediation [9], membrane treatment [10], precipitation [11], microbial degradation [12], and other technologies have been used to remove pollutants from water, but most of the treatment technologies often make it difficult to effectively and thoroughly treat

these organic pollutants. Adsorption and catalysis technologies have attracted extensive attention because of their low cost, easy operation, safety, and control [13]. Adsorption is a surface phenomenon where organic contaminants adhere to the surface of an adsorbent by physical or chemical attraction, whereas catalysis refers to the acceleration of chemical conversion of pollutants by the addition of catalysts [14,15]. The traditional catalytic treatment of pollutants can be divided into photo-catalysis and electro-catalysis; the former requires expensive precious metals (such as Ti or Ag), while the latter requires a lot of electricity. These disadvantages have made these methods less attractive for the treatment of pollutants [16,17]. Conversely, the Fenton/Fenton-like catalytic process is one of the most effective and promising advanced catalytic processes due to its low cost, mild operating conditions, and easy recovery [18–20]. In addition, iron-based catalysts have gradually become a research hotspot due to their abundant resources, environmental friendliness, and high efficiency. In recent years, it has become an effective method for loading or embedding iron-based nanoparticles with biochar to form magnetic biochar materials [21,22]. On the one hand, the loading of iron-based nanoparticles means that biochar can easily be recycled by magnets. On the other hand, biochar materials can maintain a high concentration of local pollutants through adsorption, thus improving the catalytic efficiency of iron-based catalysis.

In order to pursue the extreme performance of materials so as to better cope with water security crises, many scientists have developed a large number of materials with unique structures and excellent performance [23,24]. Among many advanced materials, biochar has received much attention because of its attractive structure, huge specific surface area, rich functional groups, and so on [25–27]. In general, the characteristics and properties of biochar often determine its application. Among them, the main parameters affecting biochar should include physicochemical properties, performance, production process, and cost factors. Its physical and chemical properties and properties depend on the synthesis technology (that is, the production process), and the source of carbon and more advanced processes are the main basis for reducing the production cost. Therefore, how to develop new and cheaper carbon sources to synthesize biochar materials with higher performance through simpler or more efficient processes is the focus of many researchers. Research into the production of biochar from cheaper and more readily available biomass is, of course, important, but the development of new carbon sources is also essential. Recently, micro-sized biochar materials based on microorganisms such as bacteria, fungi, and viruses have become one of the most popular hotspots [28]. Due to their strong life activity, high reproduction rate, unique structure, and cheap source, microorganisms, especially fungal hyphae, are widely present in nature as one of the important carbon resources used to prepare biochar materials [29]. Fungal hypha is the structural unit of most fungi, which is formed by the continuous division of fungal spores after germination, finally forming a three-dimensional (3D) network structure [30,31]. This fascinating 3D network is often used as a template for biomass composites, which provides a great motivation for us to continue exploring this interesting resource.

In this work, *Trichoderma reesei* (*T. reesei*), a filamentous fungus which has short growth cycles and large fungal hyphae (FH) volumes, was used as a template to support the micro-sized 3D network structure. In the process of fungal culture, $Fe^{3+}$ can complete uniform biological assembly with the FH through the characteristics of step-wise growth. Then, a magnetic biochar with a 3D network structure (BMFH/$Fe_3O_4$) was prepared by fractional heating. Here, the presence of magnetic nanoparticles not only makes it easier to recover biochar from water but also endows biochar with the effect of Fenton-like catalytic reagents, thus further improving the treatment effect of pollutants. Synthesized BMFH/$Fe_3O_4$ can be used as the ideal biochar materials for removing organic pollutants including malachite green (MG) and tetracycline hydrochloride (TH). The adsorption performance of BMFH/$Fe_3O_4$ on organic pollutants was investigated by batch adsorption experiments, including concentration, pH, and temperature. Subsequently, $H_2O_2$ was used to construct a heterogeneous Fenton-like system to effectively catalyze the degradation of organic pollutants. Moreover, the stability and reusability of the BMFH/Fe$_3$O$_4$ were also investigated. The focus of this study is to develop a novel strategy for the synthesis of magnetic biochar materials through a one-step process of biological assembly and fractional heating for the effective removal of organic pollutants from water.

## 2. Results and Discussion

### 2.1. Preparation of BMFH/Fe$_3$O$_4$

*T. reesei* is a multicellular eukaryotic microorganism, the asexual form of *Hypocrea jecorina*, belonging to the genus *Penicillium* (*Moniliales*), which is a typical filamentous fungus [32,33]. *T. reesei* was cultured on PDA plate medium to observe the colony biomorphology during its growth, and the results are shown in Figure 1A. *T. reesei* filaments presented white dense flat hyphae at the initial stage of growth, and then became white flocculent, and grew diffusely to the periphery. With the progress of culture, a pale-green spore-producing cluster area appeared at the edge of the colony. The growth curve of *T. reesei* was measured by the dry weight method through liquid PDL culture medium, and the results are shown in Figure 1B. It can be seen that *T. reesei* began to enter the logarithmic growth phase at 24 h, the fastest growth was achieved at 72 h, and it entered the stable phase after 96 h. The complete culture of *T. reesei* was placed under a light microscope to observe its mycelial morphology, and the results are shown in Figure 1C. We can see that *T. reesei* is composed of hyphae with diameters of 1–5 um, which interweave and intertwine with each other, presenting a classical 3D network structure, indicating that *T. reesei* has great potential as a carbon source for high-performance biochar materials.

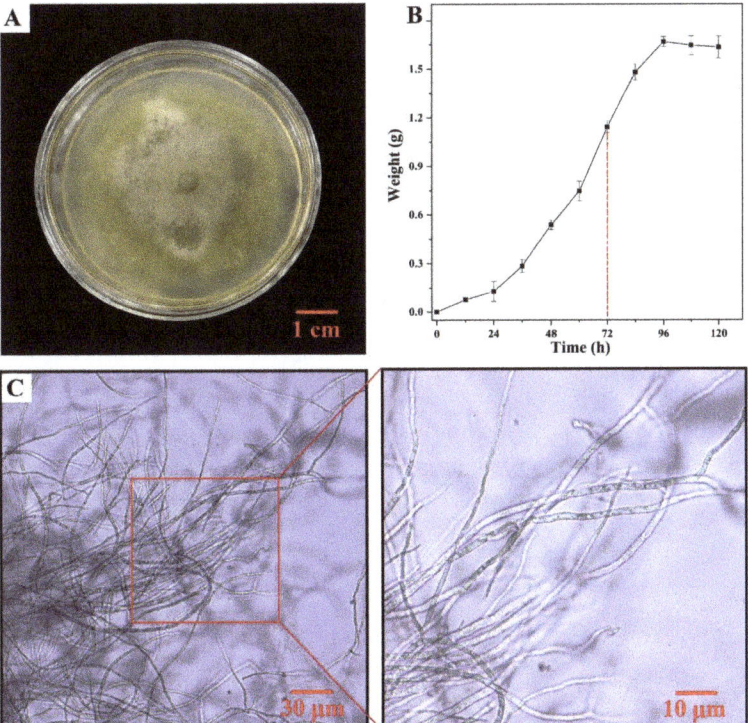

**Figure 1.** Biological morphology (**A**), growth curve test (**B**), and optical microscope structure observation results (**C**) of *T. reesei*.

The preparation process of BMFH/Fe$_3$O$_4$ is shown in Scheme 1 and can be divided into two stages. First, the filamentous fungus *T. reesei* was inoculated in a triangle flask containing PDL medium and incubated in a constant temperature shaker (29 °C, 150 RPM). The shear force of the liquid maintains the continuous growth of mycelia, thus forming a stable 3D network structure. Fe$^{3+}$ was added when the mycelium grew the fastest (72 h), and the Fe$^{3+}$ combined evenly with the mycelium by shaking the flask to form a load. After culture (96 h), the mycelium was collected and washed with ultrapure water to remove excess Fe$^{3+}$ on the surface. Finally, Fe$^{3+}$ loaded and uniformly dispersed mycelial composite FH/Fe$^{3+}$ was obtained. Then, in order to obtain magnetic BMFH/Fe$_3$O$_4$, the obtained FH/Fe$^{3+}$ was heated at a high temperature to complete the carbonization and compound magnetic process. In the process of this one-step synthesis method, we controlled the final product by dividing the heating into three stages (100 °C for 30 min, 200 °C for 30 min, and 600 °C for 60 min). In the first stage, when the temperature reached 100 °C, the moisture adsorbed by the mycelium evaporated due to heat, and at this time, Fe$^{3+}$ reacted with water vapor to form ferric hydroxide colloid (Fe(OH)$_3$). As the temperature rose above 200 °C, the colloidal ferric hydroxide was thermally decomposed to form ferric oxide (Fe$_2$O$_3$) and water (H$_2$O). At this time, the water vapor and the volatile substances formed by the decomposition of oxygenated components during the thermal lysis of mycelium were simultaneously discharged from the reaction system. When the temperature continued to rise to 600 °C, the carbonation process of mycelium continued, and the iron oxide decomposed into ferric tetroxide (Fe$_3$O$_4$) magnetic nanoparticles due to a high-temperature reaction. Finally, the magnetic mycelial biochar BMFH/Fe$_3$O$_4$ with 3D network structure after carbonization and composite magnetism was successfully obtained.

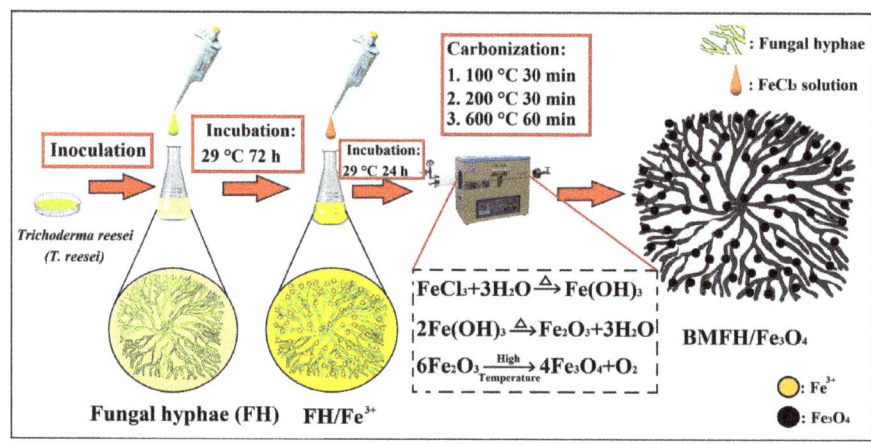

**Scheme 1.** Schematic plot of the preparation of BMFH/Fe$_3$O$_4$.

*2.2. Characterization Results*

The morphologies of the samples were characterized by SEM, TEM, and EDS, as shown in Figure 2. It is clearly seen from Figure 2A that many slender hyphae are interwoven to form this fascinating 3D network. BMFH, which is made from carbonized mycelia, still retains its structure, but its constituent mycelia are much finer than those without carbonization (Figure 1C), indicating that the carbonization process is quite complete and successful. However, after adding Fe$^{3+}$, BMFH/Fe$_3$O$_4$ also has a 3D network structure, but it does appear to be somewhat fragmented compared with BMFH. This may be because the preparation process involves the synthesis of Fe$_3$O$_4$ at a high temperature. During this process, a large amount of water vapor and oxygen are released during the generation of magnetic nanoparticles, which causes the carbonation process to become more violent, resulting in the fracture of a large number of mycelial structures. In addition, as seen in

Figure 2C,D, a certain amount of obvious $Fe_3O_4$ nanoparticles was successfully loaded onto FH, which demonstrated the feasibility and effectiveness of the one-step method. Energy dispersive spectrometer (EDS) was used to analyze the elemental composition of samples, and the results are shown in Figure 2E,F and Table S1 (Supplementary Materials). Clearly, BMFH mainly comprises C (83.16%), O (14.63%), and N (2.21%) elements (Figure 2E and Table S1). There was also a large amount of Fe on the surface of BMFH/$Fe_3O_4$ compared with BMFH, which once again proved that the one-step method can be used to synthesize $Fe_3O_4$ and load magnetic nanoparticles onto the surface of mycelium through the carbonization process.

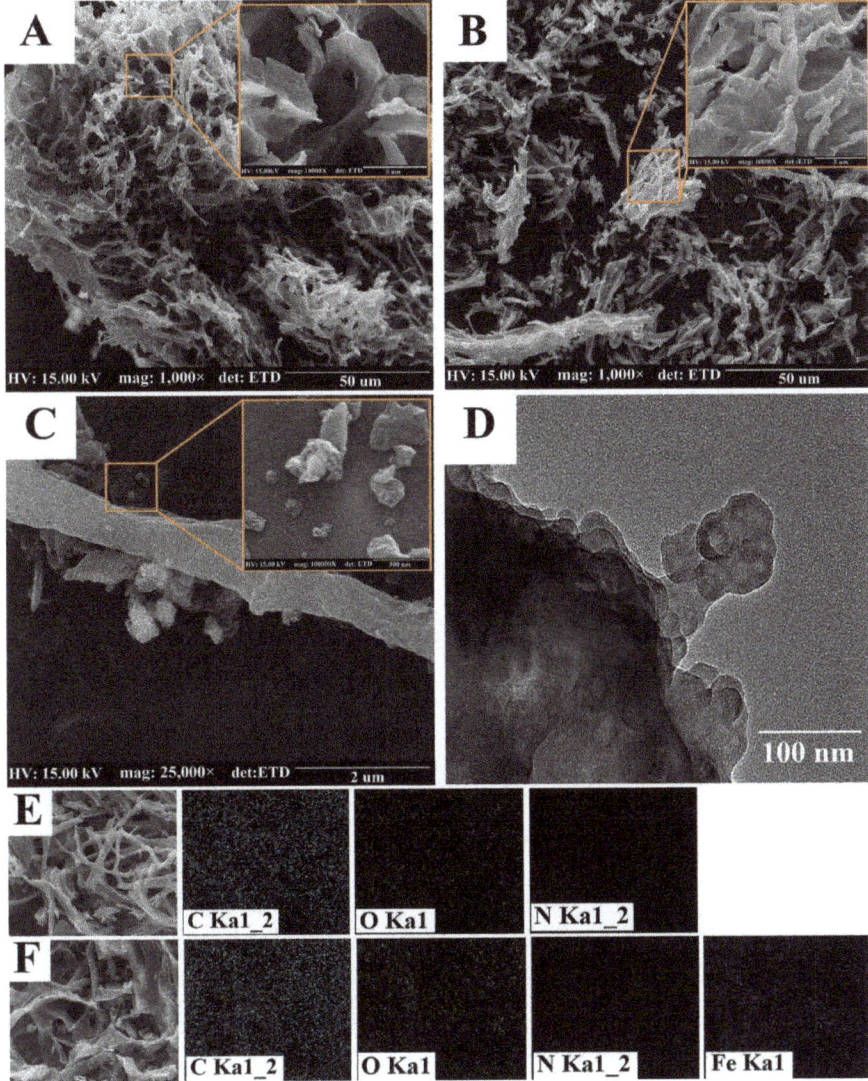

**Figure 2.** SEM images of (**A**) BMFH, (**B**) BMFH/$Fe_3O_4$-0.01, and (**C**) $Fe_3O_4$ nanoparticles loaded on the BMFH. (**D**) TEM image of BMFH/$Fe_3O_4$-0.01. EDS mapping of (**E**) BMFH and (**F**) BMFH/$Fe_3O_4$-0.01.

The functional groups of the samples were analyzed by FT-IR, as shown in Figure 3A. The wide band corresponding to the stretching vibration of the hydroxyl functional group at a range of 3400–3500 cm$^{-1}$ indicates that a large number of oxygen-containing functional groups [23–27] exist on all samples. In the range of 2800–2900 cm$^{-1}$, symmetric and asymmetric tensile vibration peaks of -CH groups were found [30,31], which may be caused by the thermal cracking of macromolecules from FH at high temperature and the removal of a large amount of hydrogen [31,34]. Bands in the range of 1600–1700 cm$^{-1}$ represent C=C stretching of the aromatic ring, which can be interpreted as the decomposition of the C-H bond into a more stable aromatic C=C bond after high-temperature treatment [31]. The overlapping band in the 900–1300 cm$^{-1}$ region corresponds to the stretching vibration of C-O and C-N heterocycles [31,34]. The band at 600–750 cm$^{-1}$ in BMFH represents $SiO_2$ [25]. Interestingly, after $Fe^{3+}$ compound is formed, the band there becomes weak and even disappears gradually compared with the BMFH/$Fe_3O_4$ samples. The characteristic band at 470–570 cm$^{-1}$ representing the Fe-O group from $Fe_3O_4$ [25] can be observed, which means the $Fe_3O_4$ particles were successfully loaded onto the BMFH.

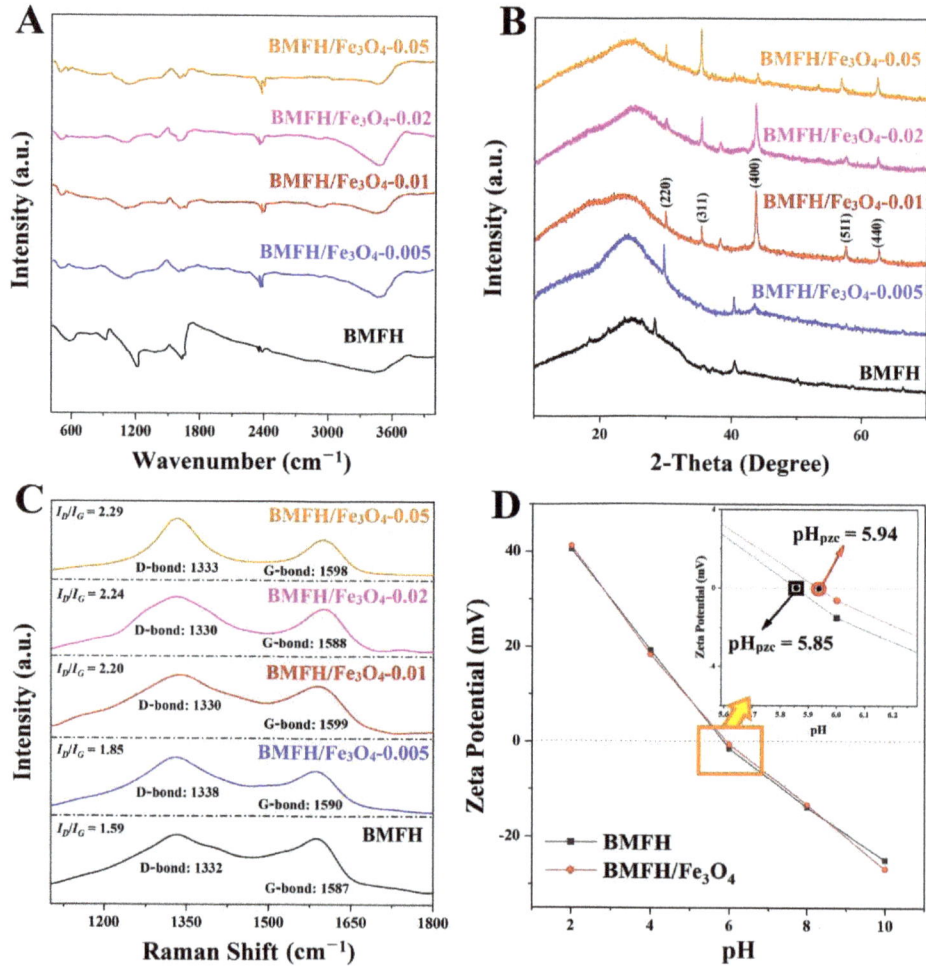

**Figure 3.** (**A**) FT-IR spectra, (**B**) XRD pattern, (**C**) Raman spectra of samples, and (**D**) Zeta potential tests of BMFH/$Fe_3O_4$-0.01.

The crystal structures of the samples were analyzed by XRD, as shown in Figure 3B. BMFH and BMFH/Fe$_3$O$_4$ prepared with different concentrations of Fe$^{3+}$ were both amorphous structures. The peaks at 19, 27, and 40° of BMFH were related to inorganic salts, which were observed probably due to the salting-out effect by water loss or cracking of FH during the high-temperature pyrolysis [24]. From all the BMFH/Fe$_3$O$_4$ samples, we can clearly see that there are many characteristic peaks, such as (220), (311), (400), (511), and (440), which belong to Fe$_3$O$_4$ with a classical structure [25]. It is worth noting that the Fe$^{3+}$ concentration used in the preparation of BMFH/Fe$_3$O$_4$-0.005 is the lowest, so the characteristic peak signal of Fe$_3$O$_4$ is the weakest in the prepared samples, with many peak positions not even capable of being easily captured compared with BMFH/Fe$_3$O$_4$-0.01.

Defects in carbon-based materials are usually caused by atomic losses or lattice distortions; here, Raman spectra were used to test the BMFH and BMFH/Fe$_3$O$_4$ samples, as shown in Figure 3C. Two typical characteristic peaks were the D-band associated with amorphous carbon, approximately $1334 \pm 4$ cm$^{-1}$, and the G-band associated with graphitic carbon, approximately $1593 \pm 6$ cm$^{-1}$ [24–26]. The intensity ratio of D-band ($I_D$) and G-band ($I_G$) is an important parameter to measure the degree of defect and disorder in carbon, shown as $I_D/I_G$. The increase in $I_D/I_G$ values of all BMFH/Fe$_3$O$_4$ samples (1.85 for BMFH/Fe$_3$O$_4$-0.005, 2.20 for BMFH/Fe$_3$O$_4$-0.01, 2.24 for BMFH/Fe$_3$O$_4$-0.02, and 2.29 for BMFH/Fe$_3$O$_4$-0.05) compared with BMFH (1.59) indicates the presence of more amorphous structures in the biochar material, either as amorphous carbon or due to Fe$_3$O$_4$.

A zeta potential test was used to evaluate the surface charges of BMFH and BMFH/Fe$_3$O$_4$-0.01, as shown in Figure 3D. Like many biochar materials [29–31], when pH changes from acid to basic (2 to 10), the surface charges of samples change from positive (40.5 and 41.2 for BMFH and BMFH/Fe$_3$O$_4$-0.01, respectively) to negative (−25.0 and −26.8 for BMFH and BMFH/Fe$_3$O$_4$-0.01, respectively). When the pH value is lower than pH$_{pzc}$, the sample surface is positively charged, and vice versa. The pH$_{pzc}$ of BMFH is 5.85, which is lower than BMFH/Fe$_3$O$_4$-0.01 (5.94), indicating that Fe$_3$O$_4$ loaded onto the samples may have a slight effect on the charge of magnetic biochars. The pH$_{pzc}$ of all samples was less than 7.00, indicating that they are more suitable for the water treatment in an alkaline environment.

N$_2$ adsorption–desorption isotherms were used to evaluate the specific surface area ($S_{BET}$) and pore structure of biochar materials, as shown in Figure 4 and Table 1. Both BMFH and BMFH/Fe$_3$O$_4$ prepared under different conditions exhibited typical type IV isotherms with H3 hysteresis loop (Figure 4A), which indicated that they were both typical mesoporous materials [29,30]. The specific surface area and the total pore volume of BMFH were 27.91 m$^2$/g and 0.0369 cm$^3$/g, respectively. With the increase in Fe$^{3+}$ concentration in the preparation of BMFH/Fe$_3$O$_4$, the specific surface area decreased significantly. When the concentration of Fe$^{3+}$ increased to 0.05 M (BMFH/Fe$_3$O$_4$-0.05), the specific surface area sharply decreased to 0.89 m$^2$/g. At this time, the pore volume of BMFH/Fe$_3$O$_4$-0.05 was 0.0071 cm$^3$/g, which can be said to have lost most of the pore structure compared with BMFH. Figure 4B shows that the mean pore sizes of BMFH and magnetic biochar range from 2 to 30 nm, which again indicate that they are mesoporous materials. In addition, the adsorption capacities of BMFH and magnetic biochars prepared under different conditions for organic pollutants (MG and TH) were compared (Figure S3, Supplementary Materials). The results showed that the adsorption capacity was positively correlated with the specific surface area and total pore volume to some extent [31]. Interestingly, there was no significant difference in adsorption capacity and specific surface area between BMFH/Fe$_3$O$_4$-0.01 and BMFH/Fe$_3$O$_4$-0.005. Therefore, on the basis of comprehensive considerations, BMFH/Fe$_3$O$_4$-0.01 was selected as the model representative of magnetic biochar in order to facilitate subsequent studies (adsorption, Fenton-like catalysis, and cyclic stability studies).

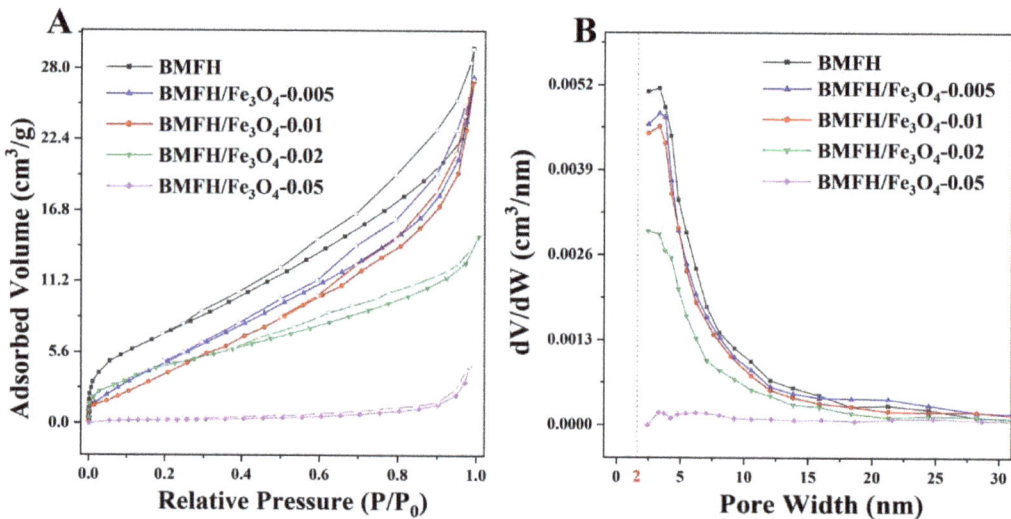

**Figure 4.** (**A**) $N_2$ adsorption–desorption isotherms and (**B**) pore size distribution curves of BMFH, BMFH/Fe$_3$O$_4$-0.005, BMFH/Fe$_3$O$_4$-0.01, BMFH/Fe$_3$O$_4$-0.02, and BMFH/Fe$_3$O$_4$-0.05.

**Table 1.** The data of $N_2$ adsorption–desorption for BMFH/Fe$_3$O$_4$ prepared under different conditions and BMFH.

| Samples | $S_{BET}$ (m$^2$/g) | $P_m$ (nm) | $V_{total}$ (cm$^3$/g) |
|---|---|---|---|
| BMFH | 27.91 | 5.69 | 0.0369 |
| BMFH/Fe$_3$O$_4$-0.005 | 22.77 | 5.34 | 0.0236 |
| BMFH/Fe$_3$O$_4$-0.01 | 21.59 | 5.16 | 0.0214 |
| BMFH/Fe$_3$O$_4$-0.02 | 13.40 | 6.05 | 0.0202 |
| BMFH/Fe$_3$O$_4$-0.05 | 0.89 | 31.93 | 0.0071 |

$P_m$ (nm) is the mean pore size and $V_{total}$ (cm$^3$/g) is the total pore volume.

The surface chemical properties and electronic states of the BMFH and BMFH/Fe$_3$O$_4$-0.01 were analyzed by XPS, as shown in Figure 5. It can be seen that BMFH contains mainly C (81.22%), O (10.34%), and N (2.13%) elements, and BMFH/Fe$_3$O$_4$-0.01 contains C (71.86%), O (13.07%), N (6.64%), and Fe (2.13) elements. The C/O ratio of BMFH/Fe$_3$O$_4$-0.01 (5.98) is significantly lower than that of BMFH (7.85), indicating that loaded Fe$_3$O$_4$ introduced more O element, which also proved the reliability of this one-step method. The high-resolution C1s spectrum of the BMFH (Figure 5B) and BMFH/Fe$_3$O$_4$-0.01 (Figure 5E) showed two peaks, including C-C at approximately 283.70 ± 0.25 eV and C-O at approximately 285.62 ± 0.06 eV [23,24]. The high-resolution O1s spectrum of the BMFH (Figure 5C) and BMFH/Fe$_3$O$_4$-0.01 (Figure 5F) showed three peaks, including the quinones at approximately 530.10 ± 0.22 eV, C=O at approximately 531.21 ± 0.18 eV, and C-O at approximately 532.55 ± 0.10 eV [24,26]. The high-resolution N1s spectrum of the BMFH (Figure 5D) and BMFH/Fe$_3$O$_4$-0.01 (Figure 5G) showed two peaks: pyridinic-N at approximately 398.64 ± 0.30 eV and pyrrolic-N at approximately 400.32 ± 0.02 eV [27]. The high-resolution Fe 2p spectrum of BMFH/Fe$_3$O$_4$-0.01 (Figure 5H) showed three peaks: Fe 2p$_{3/2}$ corresponding to 711.02 eV, Fe 2p$_{1/2}$ corresponding to 724.28 eV, and shakeup satellite peak at approximately 716.04 eV, which revealed the characteristic peaks of Fe in Fe$_3$O$_4$ [25]. On the basis of all the above data, BMFH/Fe$_3$O$_4$-0.01 was selected as the model BMFH/Fe$_3$O$_4$ for subsequent study.

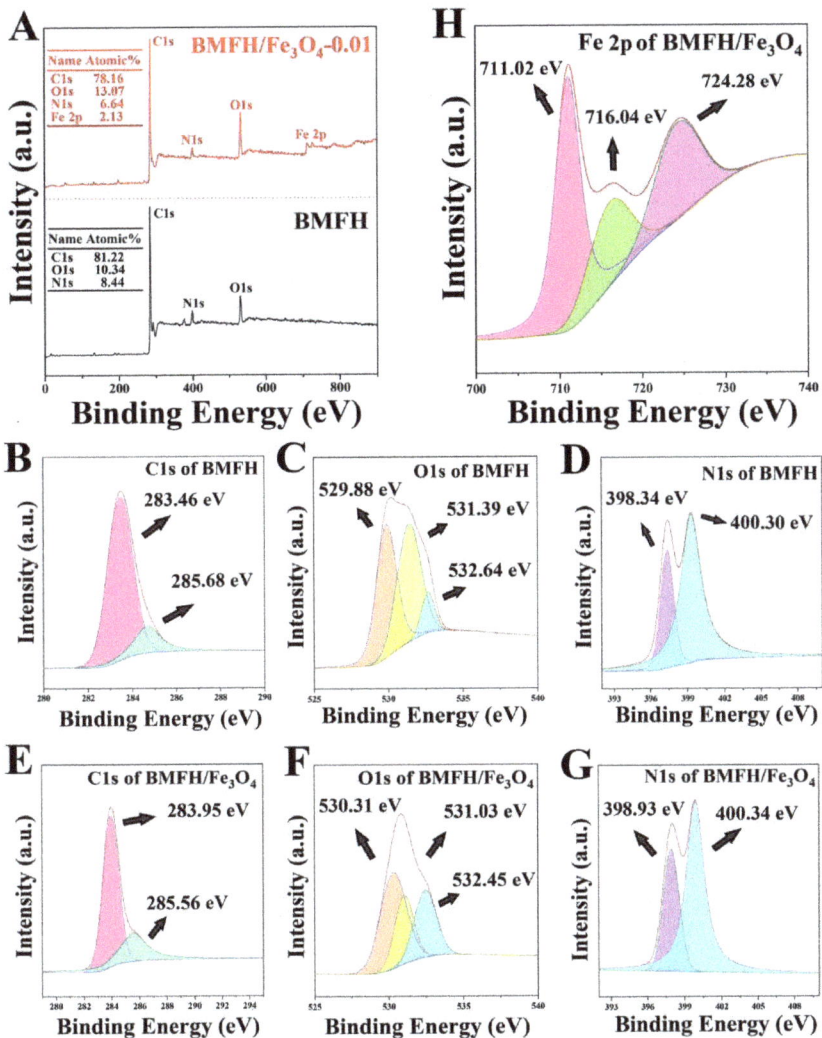

**Figure 5.** (**A**) XPS spectra of BMFH and BMFH/Fe$_3$O$_4$. The (**B**) C1s, (**C**) O1s, and (**D**) N1s of BMFH. The (**E**) C1s, (**F**) O1s, (**G**) N1s, and (**H**) Fe 2p of BMFH/Fe$_3$O$_4$.

### 2.3. Adsorption Performances

#### 2.3.1. Adsorption Kinetic

Adsorption kinetics is often used to study the influence of the adsorbate concentration on the adsorption process and the control mechanism of chemical reactions in this process. Generally, the study of adsorption kinetics can help us to better understand the behavior of adsorbent in the adsorption process and help to study the mechanism of this process [14]. The effect of contact time on the adsorption of MG and TH (50, 100, and 200 mg/L) by BMFH and BMFH/Fe$_3$O$_4$ is shown in Figure 6. It can be seen that all the adsorption processes have the same trend. The first is a rapid adsorption process that dramatically increases the adsorption capacity in the first 10 min and gradually reaches adsorption saturation in 60 min. In addition, the adsorption capacity increases with the increase in the initial concentration of adsorbed substance. These results indicate that both BMFH and BMFH/Fe$_3$O$_4$ have great potential to remove organic pollutants (MG and TH).

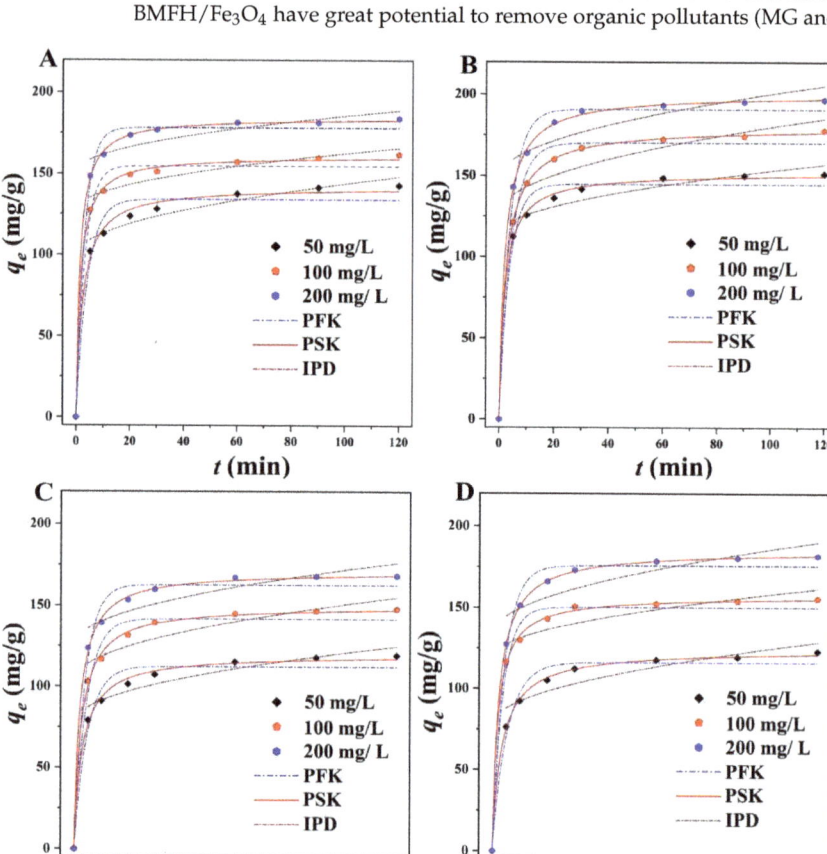

**Figure 6.** PFK, PSK, and IPD plots of BMFH (**A**,**B**) and BMFH/Fe$_3$O$_4$ (**C**,**D**) for MG and TH at 303 K.

Models that control surface adsorption are defined as adsorption reactions and empirical models including the pseudo-first-order kinetic (PFK, Equation (1)), pseudo-second-order kinetic (PSK, Equation (2)), and the intra-particle diffusion model (IPD, Equation (3)). The models are expressed below:

$$q_t = q_e - \frac{q_e}{e^{k_1 t}} \qquad (1)$$

$$q_t = \frac{k_2 q_e^2 t}{1 + k_2 q_e t} \qquad (2)$$

$$q_t = k_3 t^{1/2} + B \qquad (3)$$

where $k_1$, $k_2$, and $k_3$ represent the rate constants of the kinetic models, $q_t$ is the adsorption capacity of a sample at different time points, and $B$ represents the constant of the boundary layer thickness.

These three adsorption kinetics models were fitted with the experimental data, and the fitting parameters are shown in Table 2. For both the MG and the TH adsorption of BMFH and BMFH/Fe$_3$O$_4$, the curves of the PSK matched the relationship between the equilibrium adsorption capacities and the equilibrium concentrations better than the curves of the PFK and IPD. The correlation coefficients ($R^2$) of the PFK were 0.9732–0.9926 and 0.9847–0.9909 for BMFH adsorption of MG and TH, respectively. When the adsorbent was BMFH/Fe$_3$O$_4$, the correlation coefficients ($R^2$) of the PFK were 0.9737–0.9862 and 0.9821–0.9912 for MG and TH, respectively. All the $q_{e.cat}$ values calculated theoretically according to PFK were lower than the experimental $q_e$, indicating that the adsorption process may not belong to Lagergren's model. On the contrary, the correlation coefficients ($R^2$) of the PSK were 0.9949–0.9996 and 0.9966–0.9991 for BMFH and BMFH/Fe$_3$O$_4$, respectively, when the adsorbate was MG, and 0.9986–0.9998 and 0.9990–0.9999, respectively, when the adsorbate was TH. At the same time, the $q_{e.cat}$ values calculated theoretically according to PSK were more suitable for the experimental quantification $q_e$, indicating the applicability and potential advantages of Ho–McKay's model in the adsorption process; that is, the adsorption rate may be controlled by chemical reactions, and similar reactions that form chemisorption bonds between adsorbent and adsorbent through transfer, exchange, or sharing can promote the adsorption process [23–27]. For the IPD model, the correlation coefficients ($R^2$) were 0.7582–0.9050 (MG) and 0.7178–0.8271 (TH) for BMFH and 0.7824–0.8741 (MG) and 0.7341–0.8163 (Th) for BMFH/Fe$_3$O$_4$, demonstrating that the IPD model was not able to explain the whole adsorption process.

2.3.2. Adsorption Isotherm

Adsorption isotherms are often used to study the effect of concentration on adsorption capacity and to help explore how adsorbents interact with adsorbates during the adsorption process [29]. The Langmuir model (Equation (4)) applies to materials with uniform energy adsorption sites and monolayer adsorption layer coverage, and isotherms assume that all adsorption sites are equivalent to uniform surface coverage. In contrast, the Freundlich model (Equation (5)) describes the adsorption process on a non-uniform surface with different energy adsorption sites. They are expressed below:

$$q_e = \frac{q_m K_L C_e}{1 + K_L C_e} \quad (4)$$

$$q_e = K_F C_e^{1/n_F} \quad (5)$$

where $q_m$ (mg/g) represents the maximum adsorption capacity of a sample, $C_e$ (mg/L) is the solution concentration at equilibrium, $K_L$ and $K_F$ represent the constants of the Langmuir and Freundlich models, respectively, and $n_F$ represents the constants of the Freundlich isotherm models.

In this study, the effect of initial concentrations on the adsorption of MG and TH (50, 100, 150, 200, and 250 mg/L) by BMFH and BMFH/Fe$_3$O$_4$ are shown in Figure 7, and the data fitted using Langmuir and Freundlich isotherm models are shown in Table 3. With the increase in initial MG and TH concentrations, the adsorption capacities of BMFH and BMFH/Fe$_3$O$_4$ increased gradually. For BMFH, the correlation coefficients ($R^2$) of the Langmuir model were 0.8905 for MG and 0.9378 for TH. For BMFH/Fe$_3$O$_4$, the correlation coefficients ($R^2$) of the Langmuir model were 0.9380 for MG and 0.9659 for TH. These results indicated that the processes were not homogeneous monolayer adsorption. The correlation coefficients ($R^2$) of the Freundlich were all higher than 0.99 (0.9928 and 0.9945 of BMFH for MG and TH, 0.9965 and 0.9915 of BMFH/Fe$_3$O$_4$ for MG and TH, respectively). Furthermore, the intensity factors $n_F$ of BMFH and BMFH/Fe$_3$O$_4$ related to the adsorption intensity or surface uniformity were 8.3675 and 6.0368 for MG and 8.4253 and 5.1808 for TH, respectively, suggesting that the adsorption process may belong to multilayer adsorption with non-uniform surface [30,31].

2.3.3. Adsorption Thermodynamic Results

Temperature is an important parameter affecting the adsorption process. On the one hand, it can promote the thermal movement of molecules in the reaction system, and on the other hand, it can directly affect the endothermic or exothermic reaction through the thermal energy [27]. The thermodynamic parameters standard entropy ($\Delta S$), standard Gibbs free energy ($\Delta G$), and standard enthalpy ($\Delta H$) were analyzed to study the adsorption process. $\Delta H$, $\Delta S$, and $\Delta G$ could be calculated by Equations (6)–(8):

$$\ln(K_T) = \frac{\Delta S}{R} - \frac{\Delta H}{RT} \tag{6}$$

$$K_T = \frac{q_e}{C_e} \tag{7}$$

$$\Delta G = \Delta H - T\Delta S \tag{8}$$

where $T$ is the temperature (K), $K_T$ is the thermodynamic equilibrium constant, and $R$ represents the gas constant (8.314 J/K mol).

**Table 2.** Fitting parameters of adsorption kinetic models for MG and TH at 303 K.

| Adsorbates | Adsorbents | Models | Parameters | $C_0$ (mg L$^{-1}$) | | |
|---|---|---|---|---|---|---|
| | | | | 50 | 100 | 200 |
| MG | BMFH | PFK | $q_e$ (mg/g) | 142.99 | 161.89 | 184.05 |
| | | | $k_1$ (min$^{-1}$) | 0.2438 | 0.3179 | 0.3319 |
| | | | $q_{e.cat}$ (mg/g) | 134.21 | 154.69 | 178.15 |
| | | | $R^2$ | 0.9732 | 0.9888 | 0.9926 |
| | | PSK | $k_2$ (g mg$^{-1}$ min$^{-1}$) | 0.0031 | 0.0042 | 0.0042 |
| | | | $q_{e.cat}$ (mg/g) | 144.36 | 162.15 | 185.38 |
| | | | $R^2$ | 0.9949 | 0.9988 | 0.9996 |
| | | IPD | $k_3$ (mg g$^{-1}$ min$^{-0.5}$) | 4.4963 | 3.4926 | 3.4629 |
| | | | $B$ | 98.96 | 127.57 | 150.89 |
| | | | $R^2$ | 0.9050 | 0.8452 | 0.7582 |
| | BMFH/Fe$_3$O$_4$ | PFK | $q_e$ (mg/g) | 119.33 | 147.90 | 168.44 |
| | | | $k_1$ (min$^{-1}$) | 0.2039 | 0.2208 | 0.2526 |
| | | | $q_{e.cat}$ (mg/g) | 112.33 | 141.29 | 162.65 |
| | | | $R^2$ | 0.9737 | 0.9819 | 0.9862 |
| | | PSK | $k_2$ (g mg$^{-1}$ min$^{-1}$) | 0.0029 | 0.0027 | 0.0027 |
| | | | $q_{e.cat}$ (mg/g) | 119.99 | 150.15 | 171.13 |
| | | | $R^2$ | 0.9966 | 0.9984 | 0.9991 |
| | | IPD | $k_3$ (mg g$^{-1}$ min$^{-0.5}$) | 4.2913 | 4.6966 | 4.6213 |
| | | | $B$ | 77.76 | 103.57 | 125.51 |
| | | | $R^2$ | 0.8741 | 0.8037 | 0.7824 |
| TH | BMFH | PFK | $q_e$ (mg/g) | 151.20 | 177.81 | 196.65 |
| | | | $k_1$ (min$^{-1}$) | 0.2674 | 0.2241 | 0.2476 |
| | | | $q_{e.cat}$ (mg/g) | 144.76 | 170.19 | 190.80 |
| | | | $R^2$ | 0.9847 | 0.9899 | 0.9909 |
| | | PSK | $k_2$ (g mg$^{-1}$ min$^{-1}$) | 0.0034 | 0.0023 | 0.0024 |
| | | | $q_{e.cat}$ (mg/g) | 152.12 | 179.98 | 200.38 |
| | | | $R^2$ | 0.9986 | 0.9998 | 0.9996 |
| | | IPD | $k_3$ (mg g$^{-1}$ min$^{-0.5}$) | 4.0254 | 5.2583 | 5.3479 |
| | | | $B$ | 112.90 | 126.09 | 148.21 |
| | | | $R^2$ | 0.8271 | 0.7566 | 0.7178 |
| | BMFH/Fe$_3$O$_4$ | PFK | $q_e$ (mg/g) | 123.03 | 155.47 | 181.75 |
| | | | $k_1$ (min$^{-1}$) | 0.1807 | 0.2642 | 0.2336 |
| | | | $q_{e.cat}$ (mg/g) | 116.14 | 150.24 | 175.74 |
| | | | $R^2$ | 0.9821 | 0.9883 | 0.9912 |
| | | PSK | $k_2$ (g mg$^{-1}$ min$^{-1}$) | 0.0024 | 0.0033 | 0.0023 |
| | | | $q_{e.cat}$ (mg/g) | 124.42 | 157.48 | 185.32 |
| | | | $R^2$ | 0.9991 | 0.9990 | 0.9999 |
| | | IPD | $k_3$ (mg g$^{-1}$ min$^{-0.5}$) | 4.6731 | 3.9191 | 5.2245 |
| | | | $B$ | 77.47 | 118.65 | 132.95 |
| | | | $R^2$ | 0.8163 | 0.7471 | 0.7341 |

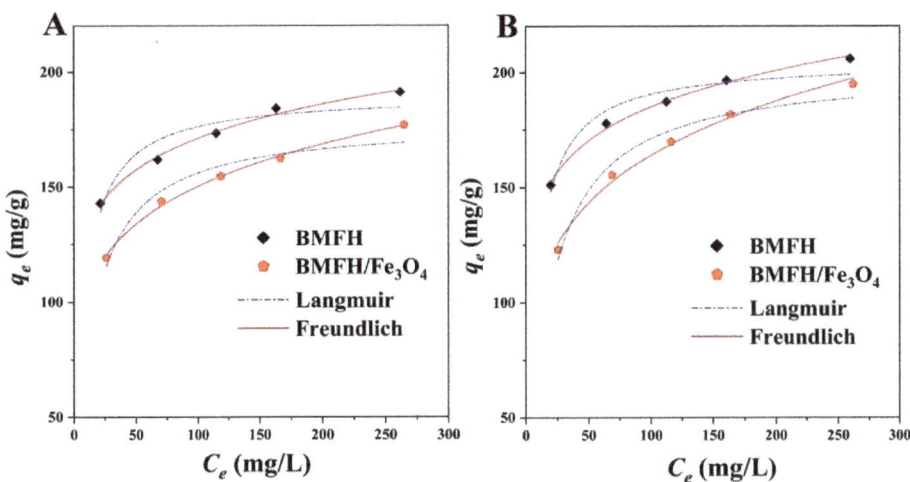

**Figure 7.** Langmuir and Freundlich isotherm plots of BMFH and BMFH/Fe$_3$O$_4$ for (**A**) MG and (**B**) TH at 303 K.

**Table 3.** Fitting parameters of adsorption isotherm models for MG and TH at 303 K.

| Adsorbates | Adsorbents | Types | Parameters | |
|---|---|---|---|---|
| MG | BMFH | Langmuir | $q_m$ (mg/g) | 190.01 |
| | | | $K_L$ (L/mg) | 0.1279 |
| | | | $R^2$ | 0.8905 |
| | | Freundlich | $K_F$ (mg g$^{-1}$(L mg$^{-1}$)$^{1/n}$) | 98.73 |
| | | | $n_F$ | 8.3675 |
| | | | $R^2$ | 0.9928 |
| | BMFH/Fe$_3$O$_4$ | Langmuir | $q_m$ (mg/g) | 178.28 |
| | | | $K_L$ (L/mg) | 0.0707 |
| | | | $R^2$ | 0.9380 |
| | | Freundlich | $K_F$ (mg g$^{-1}$(L mg$^{-1}$)$^{1/n}$) | 70.12 |
| | | | $n_F$ | 6.0368 |
| | | | $R^2$ | 0.9965 |
| TH | BMFH | Langmuir | $q_m$ (mg/g) | 204.94 |
| | | | $K_L$ (L/mg) | 0.1324 |
| | | | $R^2$ | 0.9378 |
| | | Freundlich | $K_F$ (mg g$^{-1}$(L mg$^{-1}$)$^{1/n}$) | 107.11 |
| | | | $n_F$ | 8.4253 |
| | | | $R^2$ | 0.9945 |
| | BMFH/Fe$_3$O$_4$ | Langmuir | $q_m$ (mg/g) | 201.47 |
| | | | $K_L$ (L/mg) | 0.0567 |
| | | | $R^2$ | 0.9659 |
| | | Freundlich | $K_F$ (mg g$^{-1}$(L mg$^{-1}$)$^{1/n}$) | 67.44 |
| | | | $n_F$ | 5.1808 |
| | | | $R^2$ | 0.9915 |

In this study, the temperature conditions were chosen according to the reported references and our previous work [23–27]. The results of effect of distinct temperatures (293, 298, 303, 308, and 313 K) on the adsorption capacities of the BMFH and BMFH/Fe$_3$O$_4$ for MG and TH are shown in Figure 8 and the parameters are shown in Table 4. As the temperature increased from 293 to 303 K, the adsorption capacities of BMFH and BMFH/Fe$_3$O$_4$ for RhB and TH increased. While the temperature further increased to 313 K, the increasing trend of adsorption capacities followed the same pattern. Overtly, the increase in temperature promoted the process of adsorption, and high temperature was conducive to the adsorption of MG and TH by BMFH and BMFH/ $Fe_3O_4$. In addition, it is worth noting that the adsorption capacities of BMFH increased by 11.14 mg/g (156.12 to 167.26 mg/g) for MG and 20.36 mg/g (165.79 to 186.15 mg/g) for TH as the temperature increased from 293 to 313 K; in the meantime, the adsorption capacities of BMFH/$Fe_3O_4$ increased by 22.90 mg/g (135.35 to 158.25 mg/g) for MG and 33.71 mg/g (137.55 to 171.26 mg/g) for TH. These results indicate that the temperature effect was higher on the BMFH/$Fe_3O_4$ than on the BMFH. The ΔG and ΔH were negative and positive, indicating that adsorption is a spontaneous endothermic process [31]. The ΔS (20.06 and 30.80 J $mol^{-1}$ $K^{-1}$ of BMFH for MG and TH, 33.92 and 46.53 J $mol^{-1}$ $K^{-1}$ of BMFH/$Fe_3O_4$ for MG and TH, respectively) were all positive, indicating that the confusion and randomness of the interface between adsorbents and adsorbates increased with the increase in temperature [29,30].

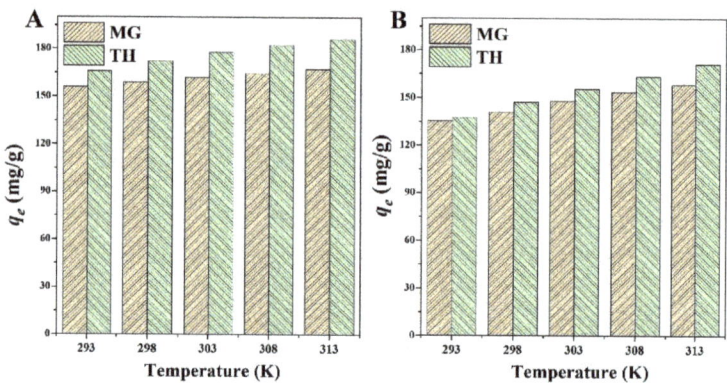

**Figure 8.** Effect of different temperatures on the adsorption capacities of (**A**) BMFH and (**B**) BMFH/$Fe_3O_4$ for MG and TH.

**Table 4.** Fitting adsorption thermodynamic parameters for MG and TH.

| Adsorbents | Adsorbates | T (K) | ΔG (kJ/mol) | ΔH (kJ/mol) | ΔS (J $mol^{-1}$ $K^{-1}$) |
|---|---|---|---|---|---|
| BMFH | MG | 293 | −2.00 | 3.88 | 20.06 |
|  |  | 303 | −2.19 |  |  |
|  |  | 313 | −2.39 |  |  |
|  | TH | 293 | −2.21 | 6.81 | 30.80 |
|  |  | 303 | −2.56 |  |  |
|  |  | 313 | −2.83 |  |  |
| BMFH/$Fe_3O_4$ | MG | 293 | −1.51 | 8.43 | 33.92 |
|  |  | 303 | −1.87 |  |  |
|  |  | 313 | −2.18 |  |  |
|  | TH | 293 | −1.56 | 12.07 | 46.53 |
|  |  | 303 | −2.05 |  |  |
|  |  | 313 | −2.49 |  |  |

*2.4. Effect of pH*

The pH value is one of the most important factors affecting the adsorption process, mainly by changing the surface properties of adsorbent and adsorbate chemical properties to promote or inhibit the adsorption process [30]. The results of the effect of pH change on the adsorption of MG and TH for BMFH and BMFH/$Fe_3O_4$ in the pH value range from 2 to 10 are shown in Figure 9.

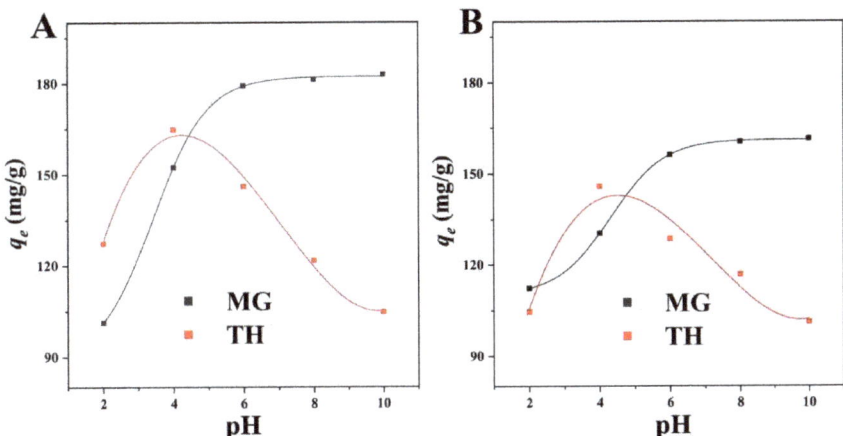

**Figure 9.** Effect of pH on the adsorption capacities of (**A**) BMFH and (**B**) BMFH/Fe$_3$O$_4$ for MG and TH.

For both BMFH and BMFH/Fe$_3$O$_4$, the adsorption capacities of MG increased with the increase in pH (6 ≥ pH ≥ 2) and tended to be stable (pH ≥ 6). For TH, the adsorption properties of BMFH and BMFH/Fe$_3$O$_4$ generally increased first (4 ≥ pH ≥ 2) and then decreased (pH ≥ 4). This can be explained by the existence of TH in different forms at different pH levels (TH$^+$ at pH ≤ 3.30, TH$^0$ at pH = 3.30–7.68, TH$^-$ and TH$^{2-}$ at pH ≥ 7.68) [29]. When the pH was 6, the maximum adsorption capacities of BMFH and BMFH/Fe$_3$O$_4$ for MG were 179.24 and 156.20 mg/g, respectively; when the pH was 4, the maximum adsorption capacities for TH were 168.79 and 145.81 mg/g, respectively. Electrostatic attraction may play a role in this process; for example, higher pH results in more negative charges on BMFH and BMFH/Fe$_3$O$_4$ surfaces, which increases the adsorption capacity of positively charged cationic dye MG and decreases the adsorption capacity of the negatively charged form of TH.

*2.5. Fenton-like Catalysis Performance Results*

2.5.1. Effect of BMFH/Fe$_3$O$_4$ and H$_2$O$_2$ on Catalytic Performance

The Fenton-like catalytic performance of BMFH/Fe$_3$O$_4$ on organic pollutants (MG and TH) was evaluated by control experiments, and the effect of dosage of BMFH/Fe$_3$O$_4$ and concentration of H$_2$O$_2$ on the degradation of MG and TH is shown in Figure 10 (A and B for MG, C and D for TH). While the $C_0$ was 50 mg/L, the concentration of H$_2$O$_2$ was 50 mmol/L, the pH was 6, and the T was 303 K; the effect of the dosage of BMFH/Fe$_3$O$_4$ on the removal efficiency of MG was studied. With the increase in BMFH/Fe$_3$O$_4$ dosage from 0.1 g/L to 0.2 g/L, the removal rate of MG increased from 55% to 99% within 60 min. This is mainly because the higher the amount of BMFH/Fe$_3$O$_4$ added, the more catalyst active sites provided, and the higher the probability of collision between H$_2$O$_2$ and the catalyst active sites [19,20]. Therefore, producing more hydroxyl radicals will increase the degradation rate. However, when the amount of BMFH/Fe$_3$O$_4$ was further increased (0.3 g/L), the removal efficiency of MG was not significantly improved, which was due to the saturation of the active site [19]. Using a dosage of 0.2 g/L of BMFH/Fe$_3$O$_4$, the effect of H$_2$O$_2$ concentration on the removal efficiency of MG was studied. When the concentration of H$_2$O$_2$ increased from 10 mmol/L to 50 mmol/L, the removal efficiency of MG increased from 50% to 99% within 60 min. In a Fenton-like oxidation catalytic reaction system, H$_2$O$_2$ is the main source of hydroxyl radicals [21]. Increasing the concentration of H$_2$O$_2$ can enhance the accessibility of H$_2$O$_2$ to the active site, which is conducive to the generation of active species, which improves the catalytic degradation rate [22]. However, further increase in the concentration of H$_2$O$_2$ (100 mmol/L) did not provide significant improvement because the BMFH/Fe$_3$O$_4$ catalytic active site was saturated. Therefore, when the $C_0$ of MG was 50 mg/L, the optimal ratio of BMFH/Fe$_3$O$_4$ and H$_2$O$_2$ was 50 mmol/L and 0.2 g/L, respectively. When the $C_0$ of TH was 50 mg/L, the pH was 4, and the T was 303 K; when the dosage of BMFH/Fe$_3$O$_4$ was 0.2 g/L, the concentration of H$_2$O$_2$ was 50 mmol/L, and the removal efficiency was 99% within 60 min.

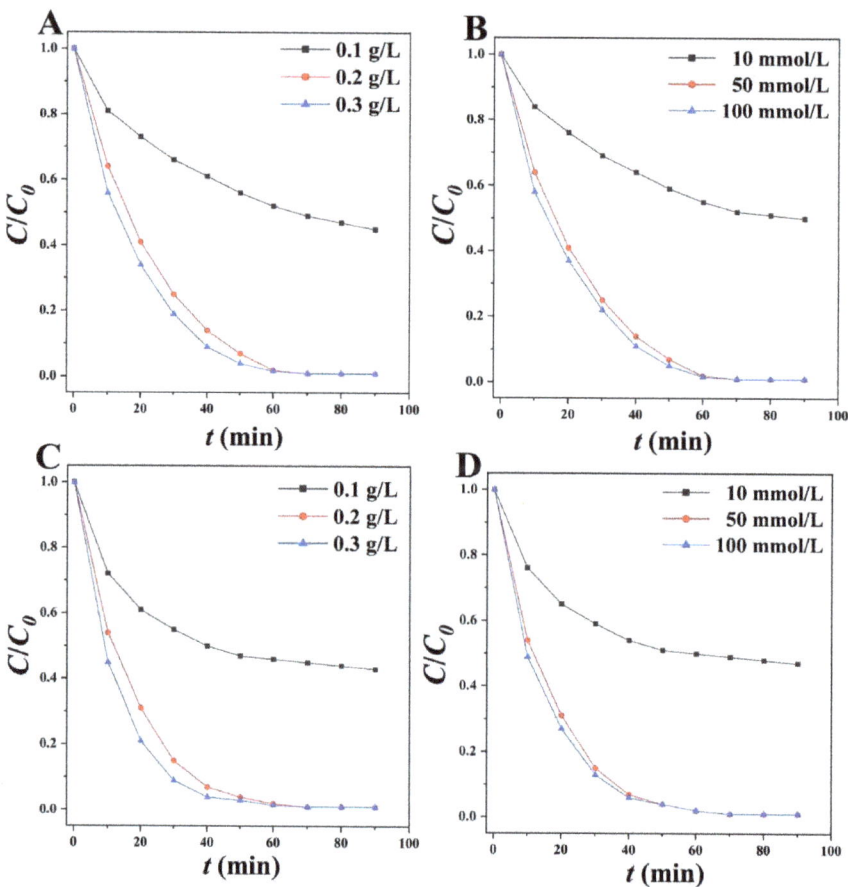

**Figure 10.** Effect of dosage of BMFH/Fe$_3$O$_4$ and concentration of H$_2$O$_2$ on the removal efficiency of (**A**,**B**) MG and (**C**,**D**) TH.

2.5.2. Oxidative Radicals Quenching Experiment Results

Oxidative radicals produced by H$_2$O$_2$ play an important role in the catalytic degradation of organic pollutants (MG and TH). The catalyst used to activate H$_2$O$_2$ has a great influence on the types of radical. *t*-Butanol and *p*-Benzoquinone were used as scavengers of OH· and HO$_2$· to quench the active substances to determine the major oxidative radicals formed in the Fenton-like catalytic reactions [19,20]. The effect of radical scavengers on the removal efficiency of MG and TH are shown in Figure 11. The removal efficiency of both MG and TH was 99% within 60 min without scavenger, while it decreased to only 30% for MG and 33% for TH within 60 min in the presence of 50 mmol/L *t*-Butanol. After adding 50 mmol/L of *p*-Benzoquinone, the removal efficiency of MG and TH was 72% and 83%, respectively. It can be inferred that the ·OH and HO$_2$· radicals were generated in the Fenton-like catalytic system of BMFH/Fe$_3$O$_4$ and H$_2$O$_2$. In the oxidative catalysis degradation of MG and TH, the ·OH radical may play a more important role compared with HO$_2$· [19,20].

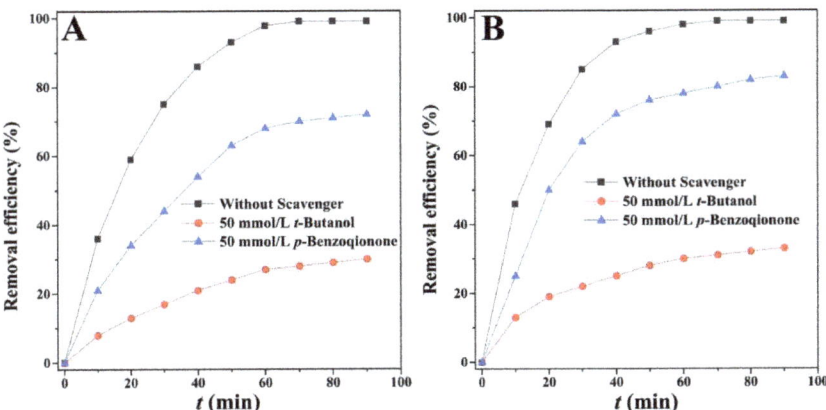

**Figure 11.** Effect of radical scavenger on the removal efficiency of (**A**) MG and (**B**) TH.

### 2.6. Cycling Stability and Performance Comparison of BMFH and BMFH/Fe$_3$O$_4$

Cycling stability is very important for evaluating the effectiveness of biochar so that the environmental material can be used repeatedly [29,30]. Compared with BMFH, BMFH/Fe$_3$O$_4$ could be easily separated from the pollutant water by a magnet due to the magnetic property of the nano-particles of Fe$_3$O$_4$. In this investigation, 10 cycle experiments were used to assess regenerative capacities of BMFH and BMFH/Fe$_3$O$_4$ (Figure 12). With the cycles increased the removal rate of BMFH decreased clearly and could only be maintained at 59.1% for MG and 55.4% for TH after 10 cycles. Meanwhile, the removal rate of BMFH/Fe$_3$O$_4$ could still be maintained above 80% after 10 cycles in the experiment (80.9% for MG and 80.2% for TH). This can be explained by the fact that every time high-temperature carbonization occurs in the regeneration process, it makes the biochar brittle and more fragile, thus affecting its subsequent cycle testing performance. Compared with BMFH, the presence of Fe$_3$O$_4$ particles may provide a more stable support for BMFH/Fe$_3$O$_4$ in the process of recycling and regeneration. On the basis of the above data, it can be concluded that BMFH and BMFH/Fe$_3$O$_4$ both have high stability and great potential for the control of organic pollutants in water environments.

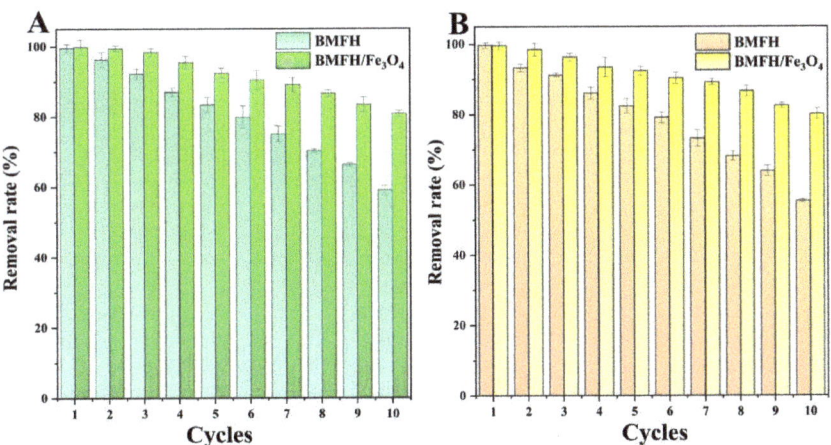

**Figure 12.** Cycling stability of the BMFH and BMFH/Fe$_3$O$_4$ to remove (**A**) MG and (**B**) TH.

## 2.7. Comparison with Other Biochars

It can be seen from Table S2 (Supplementary Material) that different biochars have different adsorption capacities for organic pollutants (MG and TH). According to the current data, the adsorption capacity of magnetic biochar BMFH/$Fe_3O_4$ (171.26 mg/g) was slightly lower than that of BMFH (186.15 mg/g), but compared with most biochar adsorbents [35–54], BMFH and BMFH/$Fe_3O_4$ still showed good adsorption performance for the removal of MG and TH. We speculate that, despite their low surface area, their unique 3D network structure and rich surface chemistry may play a major role. This result also indicates that not only the specific surface area but also the surface chemistry and spatial structure of biochar may affect the adsorption process. We will explore this further in a follow-up study. Due to the presence of $Fe_3O_4$, BMFH/$Fe_3O_4$ also has the ability of Fenton-like catalytic degradation of organic pollutants (MG and TH). Compared with some catalysts (Table S3) [55–64], although there is still a certain gap in catalytic performance, it is enough to demonstrate the potential of magnetic biochars prepared by one-step carbonization synthesis in the control of organic pollutants.

## 2.8. Possible Mechanisms

In this study, BMFH/$Fe_3O_4$ removes organic pollutants (MG and TH) via adsorption and the Fenton-like catalytic method; therefore, the possible mechanisms can be divided into two types, adsorption mechanism and catalytic reaction mechanism (Figure 13).

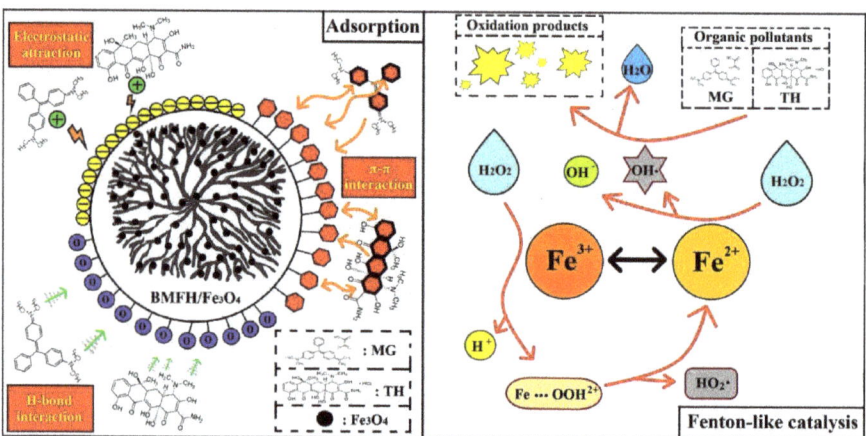

**Figure 13.** Possible mechanisms of BMFH/$Fe_3O_4$ for removal of MG and TH via adsorption and Fenton-like catalysis.

The possible mechanism in the adsorption process may include electrostatic attraction, H-bond interaction, and π-π interaction. There is a strong electrostatic interaction between organic pollutants and BMFH/$Fe_3O_4$ with negative zeta potential, which may be beneficial to adsorption in a water environment with higher pH. It can be found from FT-IR and XPS spectroscopy test results that there are various functional groups on the surface of BMFH/$Fe_3O_4$ that are conducive to adsorption; for example, the C-O group may form an H-bond with the C-H group in organic pollutants, and the O-H group may form an H-bond with $N^+$ or -O- in organic pollutants. FT-IR and Raman spectra indicated that the aromatic rings in organic pollutants may interact π-π with aromatic rings in BMFH/$Fe_3O_4$, thus enhancing the adsorption capacity.

In BMFH/$Fe_3O_4$ synthesized by the one-step method, ferric ions exist in two oxidation states, $Fe^{3+}$ and $Fe^{2+}$, which was also confirmed by XPS results. During the Fenton-like catalytic reaction process, $H_2O_2$ will react with $Fe^{3+}$ to form peroxo-intermediate. Then, peroxo-intermediate will generate $HO_2·$ and $Fe^{2+}$, which will happen again on reacting with the $H_2O_2$ molecule to form $OH·$. Finally, $OH·$ is involved in catalytic degradation of organic pollutants (MG and TH) to oxidation products. Thus, the best-proposed mechanism is as follows (Equations (9)–(12)):

$$Fe^{3+} + H_2O_2 \leftrightarrow Fe \cdots OOH^{2+} + H^+ \quad (9)$$

$$Fe \cdots OOH^{2+} \rightarrow HO_2· + Fe^{2+} \quad (10)$$

$$H_2O_2 + Fe^{2+} \rightarrow OH· + OH^- + Fe^{3+} \quad (11)$$

$$OH· + Pollutants \rightarrow Products + H_2O \quad (12)$$

## 3. Materials and Methods

### 3.1. Materials and Reagents

*T. reesei* was extracted and preserved from decayed straw by the Key Laboratory of Straw Comprehensive Utilization and Black Soil Conservation, Ministry of Education, China.

The potato was bought from the local vegetable market. D-glucose monohydrate, $FeCl_3$, HCl, $H_2O_2$, and NaOH were supplied by Beijing Chemical Works (Beijing, China). Malachite green (MG), tetracycline hydrochloride (TH), *t*-Butanol, and *p*-Benzoquinone were supplied by Aladdin Chemical (Shanghai) Co., Ltd. (Shanghai, China). All the above reagents were of analytical purity grade and did not require further purification. The water involved in the rinsing and preparation process was deionized water.

### 3.2. Preparation of BMFH/Fe$_3$O$_4$

*T. reesei* was seeded onto potato-glucose-agar (PDA, 200.0 g/L of potato extract, 20.0 g/L of D-glucose, and 20.0 g/L of agar) solid medium and cultured at 29 °C for 96 h. They were then isolated from PDA plates and inoculated into 200.0 mL potato-glucose-liquid (PDL, PDA medium without agar) medium for growth, resulting in the formation of numerous FH. After incubation for 72 h in a constant temperature incubator at 29 °C, different concentrations of $FeCl_3$ (0.001–0.5 mol/L) were added and incubated for another 24 h. After separation from the medium, it was washed with deionized water to remove water-soluble impurities and excess ferric chloride, denoted as FH/$Fe^{3+}$, for subsequent high-temperature experiments.

FH/$Fe^{3+}$ was calcined under nitrogen atmosphere protection in a horizontal tube furnace. To synthesize magnetic FH biochar materials (BMFH/$Fe_3O_4$) in one step, we divided the calcination process into three temperatures: 100 °C for 30 min, 200 °C for 30 min, and 600 °C for 60 min (Figure S1, Supplementary Materials). The abbreviations BMFH/$Fe_3O_4$-0.001 and BMFH represent the magnetic biochar material from FH/$Fe^{3+}$ containing 0.001 mol/L $FeCl_3$ by a one-step calcination method and FH-based biochar, respectively. Biochar yield (BY, %) is a valuable parameter used to evaluate the practicability of production methods, which is calculated by Equation (13):

$$BY = \frac{M_{biomass}}{M_{biochar}} \times 100\% \quad (13)$$

where $M_{biomass}$ (g) and $M_{biochar}$ (g) represent the mass of biomass used for preparation and the mass of obtained biochar, respectively. The results are shown in Table S4 (Supplementary Materials). The BYs of BMFH and BMFH/$Fe_3O_4$-0.001 were 43.85 ± 0.55% and 45.54 ± 1.13%, respectively, indicating the feasibility of this method for biochar production.

### 3.3. Adsorption Performances

In a batch adsorption experiment, 0.20 g/L of BMFH or BMFH/$Fe_3O_4$ was added to a flask containing organic pollutant solutions (MG or TH); the structural formulas of MG and TH are shown in Figure S2 (Supplementary Materials). When the adsorption process reached equilibrium, 1.0 mL of the suspension was taken out, and BMFH or BMFH/$Fe_3O_4$ was filtered through a 0.22 μm filter film. Next, the suspension was centrifuged at 12,000 RPM for 10 min and the supernatant was diluted with deionized water. The concentration of the solution was determined by Agilent Cary-300 UV-vis spectrophotometer. The adsorption capacities of BMFH/$Fe_3O_4$ were calculated by Equation (14):

$$q_e = \frac{(C_0 - C_e) \times V}{M} \tag{14}$$

where $q_e$ (mg/g) is the adsorption capacity of samples; $C_e$ and $C_0$ (mg/L) denote the equilibrium and initial concentrations of the dye solution, respectively; and $M$ (g) and $V$ (L) denote the mass of the samples and the volume of the solutions, respectively.

3.3.1. Adsorption Kinetic Performances

The organic pollutant solutions were prepared at different concentrations (50, 100, and 200 mg/L). Then, 0.2 g/L of BMFH or BMFH/$Fe_3O_4$ was dispersed into flasks containing MG or TH solutions and shaken at 150 RPM in the dark at 303 K. Finally, the concentrations of the solutions were determined at preset time intervals.

3.3.2. Adsorption Isotherm Experiments

The organic pollutant solutions at different initial concentrations (50, 100, 150, 200, and 250 mg/L) were prepared and used to test the adsorption isotherm at 303 K. After adsorption saturation, the absorbance of the solutions was measured by UV-vis spectrophotometer

3.3.3. Adsorption Thermodynamic Experiments

The effect of temperatures (293, 298, 303, 308, and 313 K) on the adsorption capacity of the adsorbates was investigated at an initial concentration of 100 mg/L with 0.2 g/L of BMFH or BMFH/$Fe_3O_4$.

3.3.4. The Effect of pH on Adsorption Capacities

The variation of the adsorption capacity of the samples with pH (2, 4, 6, 8, and 10) was also investigated. The solutions were adjusted to different pH values by HCl and NaOH.

3.4. Fenton-like Catalysis Performances

3.4.1. Catalytic Activity Test of BMFH/$Fe_3O_4$

In a batch adsorption experiment, BMFH/$Fe_3O_4$ (0.1, 0.2, and 0.3 g/L) was added to a flask containing organic pollutant solutions (MG or TH). The flask was placed in a constant temperature shaker at 150 RPM. After the adsorption process reached equilibrium, $H_2O_2$ (10, 50, and 100 mmol/L) was added to the flask to initiate the catalysis degradation reaction at 303 K. At regular intervals, 1.0 mL of the suspension was taken out, and BMFH/$Fe_3O_4$ was separated by using a magnet and filtered through a 0.22 μm filter film. After that, the suspension of organic pollutants was centrifuged at 12,000 RPM for 10 min, and the supernatant was diluted with deionized water. Then, the concentration of the solutions was determined by using an Agilent Cary-300 UV-vis spectrophotometer. The removal efficiency ($R\%$) of organic pollutants can be determined by using Equation (15):

$$R\% = \frac{C_0 - C}{C_0} \times 100\% \tag{15}$$

where $C_0$ and $C$ are the initial and non-degraded adsorbates concentrations (mg/L), respectively, in the supernatant after centrifuging.

3.4.2. Oxidative Radicals Quenching Experiments

At the beginning of the catalysis experiment, 50 mmol/L of $t$-Butanol or $p$-Benzoquinone was added to study the effect of different oxidation radicals. The concentration of the solutions was determined by using an Agilent Cary-300 UV-vis spectrophotometer.

*3.5. Cycling Stability Studies*

A total of 1.0 g/L of BMFH/Fe$_3$O$_4$ was added to 100 mL of contaminants in each cycle (100 mg/L). When the experiments finished, the recycled samples were collected by a magnet. After that, the BMFH/Fe$_3$O$_4$ was washed with water and carbonized for 60 min at 600 °C under nitrogen atmosphere protection. In the next cycle, reused samples were employed as fresh adsorbents. The details of the characterization methods are shown in Supporting Information.

## 4. Conclusions

In this study, magnetic biochar (BMFH/Fe$_3$O$_4$) with a unique three-dimensional network structure was synthesized by a simple and controllable method, which has both adsorption and Fenton-like catalytic properties. The adsorption capacities of BMFH/Fe$_3$O$_4$ for organic pollutants (158.2 mg/g for MG and 171.26 mg/g for TH) were higher than those of most biochars. The catalytic degradation of organic pollutants reached 99% in 60 min, which was better than most catalysts. After 10 cycles, the removal ability of BMFH/Fe$_3$O$_4$ in relation to MG and TH remained above 80%. This work not only prepared magnetic biochar materials with excellent performance, which can better control organic pollutants in water, but more importantly, it also provided a new method for the development of other biomass. In the future, we will continue to explore and optimize this synthesis method and further investigate its role in other biomass (such as agricultural waste).

**Supplementary Materials:** The following supporting information can be downloaded at: https://www.mdpi.com/article/10.3390/ijms232012579/s1.

**Author Contributions:** Conceptualization, X.C. and Y.S.; Data curation, Y.S.; Formal analysis, X.C.; Funding acquisition, J.L. and Y.S.; Investigation, S.T.; Methodology, X.C., J.L. and S.T.; Project administration, S.T.; Resources, J.L.; Writing—original draft, X.C.; Writing—review & editing, Y.S. and S.T. All authors have read and agreed to the published version of the manuscript.

**Funding:** This research was funded by [Jilin Scientific and Technological Development Program] grant number [20220203108SF] and the National College Students Innovation and Entrepreneurship Project in China. The APC was funded by [20220203108SF].

**Institutional Review Board Statement:** Not applicable.

**Informed Consent Statement:** Not applicable.

**Data Availability Statement:** The study did not report any data.

**Conflicts of Interest:** The authors declare no conflict of interest.

## References

1. Agasti, N.; Gautam, V.; Priyanka; Manju; Pandey, N.; Genwa, M.; Meena, P.L.; Tandon, S.; Samantaray, R. Carbon nanotube based magnetic composites for decontamination of organic chemical pollutants in water: A review. *Appl. Surf. Sci. Adv.* **2022**, *10*, 100270. [CrossRef]
2. Schwarzenbach, R.P.; Escher, B.I.; Fenner, K.; Hofstetter, T.B.; Johnson, C.A.; Gunten, U.V.; Wehrli, B. The challenge of micropollutants in aquatic systems. *Science* **2006**, *313*, 1072–1077. [CrossRef] [PubMed]
3. Tufail, M.A.; Iltaf, J.; Zaheer, T.; Tariq, L.; Amir, M.B.; Fatima, R.; Asbat, A.; Kabeer, T.; Fahad, M.; Naeem, H.; et al. Recent advances in bioremediation of heavy metals and persistent organic pollutants: A review. *Sci. Total Environ.* **2022**, *850*, 157961. [CrossRef]
4. Ang, W.L.; McHugh, P.J.; Symes, M.D. Sonoelectrochemical processes for the degradation of persistent organic pollutants. *Chem. Eng. J.* **2022**, *444*, 136573. [CrossRef]
5. Qiu, B.; Shao, Q.; Shi, J.; Yang, C.; Chu, H. Application of biochar for the adsorption of organic pollutants from wastewater: Modification strategies, mechanisms and challenges. *Sep. Purif. Technol.* **2022**, *300*, 121925. [CrossRef]
6. Ruan, Y.; Kong, L.; Zhong, Y.; Diao, Z.; Shih, K.; Hou, L.; Wang, S.; Chen, D. Review on the synthesis and activity of iron-based catalyst in catalytic oxidation of refractory organic pollutants in wastewater. *J. Clean. Prod.* **2021**, *321*, 128924. [CrossRef]
7. Qu, J.; Shi, J.; Wang, Y.; Tong, H.; Zhu, Y.; Xu, L.; Wang, Y.; Zhang, B.; Tao, Y.; Dai, X.; et al. Applications of functionalized magnetic biochar in environmental remediation: A review. *J. Hazard. Mater.* **2022**, *434*, 128841. [CrossRef]

8. Adeola, A.O.; Abiodun, B.A.; Adenuga, D.O.; Nomngongo, P.N. Adsorptive and photocatalytic remediation of hazardous organic chemical pollutants in aqueous medium: A review. *J. Contam. Hydrol.* **2022**, *248*, 104019. [CrossRef]
9. Khan, A.H.; Khan, N.A.; Zubair, M.; Shaida, M.A.; Manzar, M.S.; Abutaleb, A.; Naushad, M.; Iqbal, J. Sustainable green nanoadsorbents for remediation of pharmaceuticals from water and wastewater: A critical review. *Environ. Res.* **2022**, *204*, 112243. [CrossRef]
10. Laqbaqbi, M.; García-Payo, M.C.; Khayet, M.; Kharraz, J.E.; Chaouch, M. Application of direct contact membrane distillation for textile wastewater treatment and fouling study. *Sep. Purif. Technol.* **2019**, *209*, 815–825. [CrossRef]
11. Dotto, J.; Fagundes-Klen, M.R.; Veit, M.T.; Palacio, S.M.; Bergamasco, R. Performance of different coagulants in the coagulation/flocculation process of textile wastewater. *J. Clean. Prod.* **2019**, *208*, 656–665. [CrossRef]
12. Gou, Z.; Hopla, G.A.; Yao, M.; Cui, B.; Su, Y.; Rinklebe, J.; Sun, C.; Chen, G.; Ma, N.L.; Sun, Y. Removal of dye pollution by an oxidase derived from mutagenesis of the *Deuteromycete Myrothecium* with high potential in industrial applications. *Environ. Pollut.* **2022**, *310*, 119726. [CrossRef] [PubMed]
13. Khan, S.; Naushad, M.; Al-Gheethi, A.; Iqbal, J. Engineered nanoparticles for removal of pollutants from wastewater: Current status and future prospects of nanotechnology for remediation strategies. *J. Environ. Chem. Eng.* **2021**, *9*, 106160. [CrossRef]
14. Chen, S.; Xia, Y.; Zhang, B.; Chen, H.; Chen, G.; Tang, S. Disassembly of lignocellulose into cellulose, hemicellulose, and lignin for preparation of porous carbon materials with enhanced performances. *J. Hazard. Mater.* **2021**, *408*, 124956. [CrossRef] [PubMed]
15. Wang, L.; Zhang, Q.; Chen, B.; Bu, Y.; Chen, Y.; Ma, J.; Rosario-Ortiz, F.L.; Zhu, R. Some issues limiting photo(cata)lysis application in water pollutant control: A critical review from chemistry perspectives. *Water Res.* **2020**, *174*, 115605. [CrossRef]
16. Hou, C.; Jiang, X.; Chen, D.; Zhang, X.; Liu, X.; Mu, Y.; Shen, J. Ag-$TiO_2$/biofilm/nitrate interface enhanced visible light-assisted biodegradation of tetracycline: The key role of nitrate as the electron acceptor. *Water Res.* **2022**, *215*, 118212. [CrossRef]
17. Hong, X.; Zhang, R.; Tong, S.; Ma, C. Preparation of Ti/PTFE-F-$PbO_2$ electrode with a long life from the sulfamic acid bath and its application in organic degradation. *Chin. J. Chem. Eng.* **2011**, *19*, 1033–1038. [CrossRef]
18. Dihingia, H.; Tiwari, D. Impact and implications of nanocatalyst in the Fenton-like processes for remediation of aquatic environment contaminated with micro-pollutants: A critical review. *J. Water Process Eng.* **2022**, *45*, 102500. [CrossRef]
19. Yang, R.; Peng, Q.; Yu, B.; Shen, Y.; Cong, H. Yolk-shell $Fe_3O_4$@MOF-5 nanocomposites as a heterogeneous Fenton-like catalyst for organic dye removal. *Sep. Purif. Technol.* **2021**, *267*, 118620. [CrossRef]
20. Tang, J.; Wang, J. Fenton-like degradation of sulfamethoxazole using Fe-based magnetic nanoparticles embedded into mesoporous carbon hybrid as an efficient catalyst. *Chem. Eng. J.* **2018**, *351*, 1085–1094. [CrossRef]
21. Ke, P.; Zeng, D.; Wang, R.; Cui, J.; Li, X.; Fu, Y. Magnetic carbon microspheres as a reusable catalyst in heterogeneous Fenton system for the efficient degradation of phenol in wastewater. *Colloids Surf. A* **2022**, *638*, 128265. [CrossRef]
22. Shin, J.; Bae, S.; Chon, K. Fenton oxidation of synthetic food dyes by Fe-embedded coffee biochar catalysts prepared at different pyrolysis temperatures: A mechanism study. *Chem. Eng. J.* **2021**, *421*, 129943. [CrossRef]
23. Zhang, B.; Jin, Y.; Qi, J.; Chen, H.; Chen, G.; Tang, S. Porous carbon materials based on *Physalis alkekengi L.* husk and its application for removal of malachite green. *Environ. Technol. Innov.* **2021**, *21*, 101343. [CrossRef]
24. Jin, Y.; Zhang, B.; Chen, G.; Chen, H.; Tang, S. Combining biological and chemical methods to disassemble of cellulose from corn straw for the preparation of porous carbons with enhanced adsorption performance. *Int. J. Biol. Macromol.* **2022**, *209*, 315–329. [CrossRef] [PubMed]
25. Chen, S.; Chen, G.; Chen, H.; Sun, Y.; Yu, X.; Su, Y.; Tang, S. Preparation of porous carbon-based material from corn straw via mixed alkali and its application for removal of dye. *Colloids Surf. A* **2019**, *568*, 173–183. [CrossRef]
26. Chen, S.; Zhang, B.; Xia, Y.; Chen, H.; Chen, G.; Tang, S. Influence of mixed alkali on the preparation of edible fungus substrate porous carbon material and its application for the removal of dye. *Colloids Surf. A* **2021**, *609*, 125675. [CrossRef]
27. Chen, X.; Yu, G.; Chen, Y.; Tang, S.; Su, Y. Cow dung-based biochar materials prepared via mixed base and its application in the removal of organic pollutants. *Int. J. Mol. Sci.* **2022**, *23*, 10094. [CrossRef]
28. Zhong, Y.; Xia, X.; Deng, S.; Xie, D.; Shen, S.; Zhang, K.; Guo, W.; Wang, X.; Tu, J. Spore carbon from *Aspergillus Oryzae* for advanced electrochemical energy storage. *Adv. Mater.* **2018**, *30*, 1805165. [CrossRef]
29. Zhang, B.; Jin, Y.; Huang, X.; Tang, S.; Chen, H.; Su, Y.; Yu, X.; Chen, S.; Chen, G. Biological self-assembled hyphae/starch porous carbon composites for removal of organic pollutants from water. *Chem. Eng. J.* **2022**, *450*, 138264. [CrossRef]
30. Chen, S.; Wang, Z.; Xia, Y.; Zhang, B.; Chen, H.; Chen, G.; Tang, S. Porous carbon material derived from fungal hyphae and its application for the removal of dye. *RSC Adv.* **2019**, *9*, 25480. [CrossRef]
31. Xia, Y.; Jin, Y.; Qi, J.; Chen, H.; Chen, G.; Tang, S. Preparation of biomass carbon material based on *Fomes fomentarius* via alkali activation and its application for the removal of brilliant green in wastewater. *Environ. Technol. Innov.* **2021**, *23*, 101659. [CrossRef]
32. Pant, S.; Ritika; Nag, P.; Ghati, A.; Chakraborty, D.; Maximiano, M.R.; Franco, O.L.; Mandal, A.K.; Kuila, A. Employment of the CRISPR/Cas9 system to improve cellulase production in *Trichoderma reesei*. *Biotechnol. Adv.* **2022**, *60*, 108022. [CrossRef] [PubMed]
33. Liu, X.; Yu, X.; He, A.; Xia, J.; He, J.; Deng, Y.; Xu, N.; Qiu, Z.; Wang, X.; Zhao, P. One-pot fermentation for erythritol production from distillers grains by the co-cultivation of *Yarrowia lipolytica* and *Trichoderma reesei*. *Bioresour. Technol.* **2022**, *351*, 127053. [CrossRef] [PubMed]
34. Shen, Y.; Zhang, N. A facile synthesis of nitrogen-doped porous carbons from lignocellulose and protein wastes for VOCs sorption. *Environ. Res.* **2020**, *189*, 109956. [CrossRef]

35. Giri, B.S.; Sonwani, R.K.; Varjani, S.; Chaurasia, D.; Varadavenkatesan, T.; Chaturvedi, P.; Yadav, S.; Katiyar, V.; Singh, R.S.; Pandey, A. Highly efficient bio-adsorption of Malachite green using Chinese Fan-Palm Biochar (*Livistona chinensis*). *Chemosphere* **2022**, *287*, 132282. [CrossRef] [PubMed]
36. Ali, F.; Bibi, S.; Ali, N.; Ali, Z.; Said, A.; Wahab, Z.U.; Bilal, M.; Iqbal, H.M.N. Sorptive removal of malachite green dye by activated charcoal: Process optimization, kinetic, and thermodynamic evaluation. *Case Stud. Chem. Environ. Eng.* **2020**, *2*, 100025. [CrossRef]
37. Vigneshwaran, S.; Sirajudheen, P.; Karthikeyan, P.; Meenakshi, S. Fabrication of sulfur-doped biochar derived from tapioca peel waste with superior adsorption performance for the removal of Malachite green and Rhodamine B dyes. *Surf. Interfaces* **2021**, *23*, 100920. [CrossRef]
38. Leng, L.; Yuan, X.; Zeng, G.; Shao, J.; Chen, X.; Wu, Z.; Wang, H.; Peng, X. Surface characterization of rice husk bio-char produced by liquefaction and application for cationic dye (Malachite green) adsorption. *Fuel* **2015**, *155*, 77–85. [CrossRef]
39. Motaghi, H.; Arabkhani, P.; Parvinnia, M.; Asfaram, A. Simultaneous adsorption of cobalt ions, azo dye, and imidacloprid pesticide on the magnetic chitosan/activated carbon@UiO-66 bio-nanocomposite: Optimization, mechanisms, regeneration, and application. *Sep. Purif. Technol.* **2022**, *284*, 120258. [CrossRef]
40. Ahmad, M.A.; Alrozi, R. Removal of malachite green dye from aqueous solution using rambutan peel-based activated carbon: Equilibrium, kinetic and thermodynamic studies. *Chem. Eng. J.* **2011**, *171*, 510–516. [CrossRef]
41. Altintig, E.; Onaran, M.; Sarı, A.; Altundag, H.; Tuzen, M. Preparation, characterization and evaluation of bio-based magnetic activated carbon for effective adsorption of malachite green from aqueous solution. *Mater. Chem. Phys.* **2018**, *220*, 313–321. [CrossRef]
42. Sharma, G.; Sharmac, S.; Kumar, A.; Naushad, M.; Du, B.; Ahamad, T.; Ghfar, A.A.; Alqadami, A.A.; Stadler, F.J. Honeycomb structured activated carbon synthesized from Pinus roxburghii cone as effective bioadsorbent for toxic malachite green dye. *J. Water Process Eng.* **2019**, *32*, 100931. [CrossRef]
43. Tsai, C.; Lin, P.; Hsieh, S.; Kirankumar, R.; Patel, A.K.; Singhania, R.; Dong, C.; Chen, C.; Hsieh, S. Engineered mesoporous biochar derived from rice husk for efficient removal of malachite green from wastewaters. *Bioresour. Technol.* **2022**, *347*, 126749. [CrossRef] [PubMed]
44. Eltaweil, A.S.; Mohamed, H.A.; El-Monaem, E.M.A.; El-Subruiti, G.M. Mesoporous magnetic biochar composite for enhanced adsorption of malachite green dye: Characterization, adsorption kinetics, thermodynamics and isotherms. *Adv. Powder Technol.* **2020**, *31*, 1253–1263. [CrossRef]
45. Hoslett, J.; Ghazal, H.; Katsou, E.; Jouhara, H. The removal of tetracycline from water using biochar produced from agricultural discarded material. *Sci. Total Environ.* **2021**, *751*, 141755. [CrossRef]
46. Zhang, X.; Li, Y.; Wu, M.; Pang, Y.; Hao, Z.; Hu, M.; Qiu, R.; Chen, Z. Enhanced adsorption of tetracycline by an iron and manganese oxides loaded biochar: Kinetics, mechanism and column adsorption. *Bioresour. Technol.* **2021**, *320*, 124264. [CrossRef]
47. Liu, P.; Liu, W.; Jiang, H.; Chen, J.; Li, W.; Yu, H. Modification of bio-char derived from fast pyrolysis of biomass and its application in removal of tetracycline from aqueous solution. *Bioresour. Technol.* **2012**, *121*, 235–240. [CrossRef]
48. Mu, Y.; He, W.; Ma, H. Enhanced adsorption of tetracycline by the modified tea-based biochar with the developed mesoporous and surface alkalinity. *Bioresour. Technol.* **2021**, *342*, 126001. [CrossRef]
49. Liu, H.; Xu, G.; Li, G. The characteristics of pharmaceutical sludge-derived biochar and its application for the adsorption of tetracycline. *Sci. Total Environ.* **2020**, *747*, 141492. [CrossRef]
50. Dai, J.; Meng, X.; Zhang, Y.; Huang, Y. Effects of modification and magnetization of rice straw derived biochar on adsorption of tetracycline from water. *Bioresour. Technol.* **2020**, *311*, 123455. [CrossRef]
51. Mei, Y.; Xu, J.; Zhang, Y.; Li, B.; Fan, S.; Xu, H. Effect of Fe–N modification on the properties of biochars and their adsorption behavior on tetracycline removal from aqueous solution. *Bioresour. Technol.* **2021**, *325*, 124732. [CrossRef] [PubMed]
52. Khanday, W.A.; Hameed, B.H. Zeolite-hydroxyapatite-activated oil palm ash composite for antibiotic tetracycline adsorption. *Fuel* **2018**, *215*, 499–505. [CrossRef]
53. Wang, Y.; Lei, S.; Liang, L. Preparation of porous activated carbon from semi-coke by high temperature activation with KOH for the high-efficiency adsorption of aqueous tetracycline. *Appl. Surf. Sci.* **2020**, *530*, 147187. [CrossRef]
54. Liu, H.; Xu, G.; Li, G. Preparation of porous biochar based on pharmaceutical sludge activated by NaOH and its application in the adsorption of tetracycline. *J. Colloid Interface Sci.* **2021**, *587*, 271–278. [CrossRef]
55. Hu, Y.; Li, Y.; He, J.; Liu, T.; Zhang, K.; Huang, X.; Kong, L.; Liu, J. EDTA-Fe(III) Fenton-like oxidation for the degradation of malachite green. *J. Environ. Manag.* **2018**, *226*, 256–263. [CrossRef]
56. Wu, Y.; Zeng, S.; Wang, F.; Megharaj, M.; Naidu, R.; Chen, Z. Heterogeneous Fenton-like oxidation of malachite green by iron-based nanoparticles synthesized by tea extract as a catalyst. *Sep. Purif. Technol.* **2015**, *154*, 161–167. [CrossRef]
57. Elhalil, A.; Tounsadi, H.; Elmoubarki, R.; Mahjoubi, F.Z.; Farnane, M.; Sadiq, M.; Abdennouri, M.; Qourzal, S.; Barka, N. Factorial experimental design for the optimization of catalytic degradation of malachite green dye in aqueous solution by Fenton process. *Water Resour. Ind.* **2016**, *15*, 41–48. [CrossRef]
58. Yuan, M.; Fu, X.; Yu, J.; Xu, J.; Huang, J.; Li, Q.; Sun, D. Green synthesized iron nanoparticles as highly efficient fenton-like catalyst for degradation of dyes. *Chemosphere* **2020**, *261*, 127618. [CrossRef]
59. Lu, J.; Zhou, Y.; Lei, J.; Ao, Z.; Zhou, Y. $Fe_3O_4$/graphene aerogels: A stable and efficient persulfate activator for the rapid degradation of malachite green. *Chemosphere* **2020**, *251*, 126402. [CrossRef]

60. Huang, X.; Xiao, J.; Yi, Q.; Li, D.; Liu, C.; Liu, Y. Construction of core-shell $Fe_3O_4$@GO-CoPc photo-Fenton catalyst for superior removal of tetracycline: The role of GO in promotion of $H_2O_2$ to OH conversion. *J. Environ. Manag.* **2022**, *308*, 114613. [CrossRef]
61. Jafari, A.J.; Kakavandi, B.; Jaafarzadeh, N.; Kalantary, R.R.; Ahmadi, M.; Babaei, A.A. Fenton-like catalytic oxidation of tetracycline by AC@$Fe_3O_4$ as a heterogeneous persulfate activator: Adsorption and degradation studies. *J. Ind. Eng. Chem.* **2017**, *45*, 323–333. [CrossRef]
62. Zhang, X.; Ren, B.; Li, X.; Liu, B.; Wang, S.; Yu, P.; Xu, Y.; Jiang, G. High-efficiency removal of tetracycline by carbon-bridge-doped g-$C_3N_4$/$Fe_3O_4$ magnetic heterogeneous catalyst through photo-Fenton process. *J. Hazard. Mater.* **2021**, *418*, 126333. [CrossRef]
63. Zhang, X.; Wang, F.; Wang, C.; Wang, P.; Fu, H.; Zhao, C. Photocatalysis activation of peroxodisulfate over the supported $Fe_3O_4$ catalyst derived from MIL-88A(Fe) for efficient tetracycline hydrochloride degradation. *Chem. Eng. J.* **2021**, *426*, 131927. [CrossRef]
64. Khodadadi, M.; Panahi, A.H.; Al-Musawi, T.J.; Ehrampoush, M.H.; Mahvi, A.H. The catalytic activity of $FeNi_3$@$SiO_2$ magnetic nanoparticles for the degradation of tetracycline in the heterogeneous Fenton-like treatment method. *J. Water Process Eng.* **2019**, *32*, 100943. [CrossRef]

Article

# Spark Plasma Sintering-Assisted Synthesis of $Bi_2Fe_4O_9/Bi_{25}FeO_{40}$ Heterostructures with Enhanced Photocatalytic Activity for Removal of Antibiotics

Zhifei Liu, Yaqi Tan, Xuefeng Ruan, Jing Guo, Wei Li, Jiajun Li, Hongyu Ma, Rui Xiong * and Jianhong Wei *

Key Laboratory of Artificial Micro- and Nano-Structures of Ministry of Education,
School of Physics and Technology, Wuhan University, Luojiashan Road, Wuhan 430072, China
* Correspondence: xiongrui@whu.edu.cn (R.X.); jhwei@whu.edu.cn (J.W.)

**Abstract:** Bismuth ferrite-based heterojunction composites have been considered as promising visible-light responsive photocatalysts because of their narrow band gap structure; however, the synthetic methods reported in the literature were usually time-consuming. In this study, we report a facile and quick preparation of bismuth ferrite-based composites by the hydrothermal method, combined with spark plasma sintering (SPS), a technique that is usually used for the high-speed consolidation of powders. The result demonstrated that the SPS-assisted synthesized samples possess significant enhanced photoelectric and photocatalytic performance. Specifically, the SPS650 (sintered at the 650 °C for 5 min by SPS) exhibits a 1.5 times enhancement in the photocurrent density and a 3.8 times enhancement in the tetracycline hydrochloride photodegradation activity than the unmodified bismuth ferrite samples. The possible influence factors of SPS on photoelectric and photocatalytic performance of bismuth ferrite-based composites were discussed carefully. This study provides a feasible method for the facile and quick synthesis of a highly active bismuth ferrite-based visible-light-driven photocatalyst for practical applications.

**Keywords:** bismuth ferrite; spark plasma sintering; heterostructure; visible-light irradaition; antibiotics

Citation: Liu, Z.; Tan, Y.; Ruan, X.; Guo, J.; Li, W.; Li, J.; Ma, H.; Xiong, R.; Wei, J. Spark Plasma Sintering-Assisted Synthesis of $Bi_2Fe_4O_9/Bi_{25}FeO_{40}$ Heterostructures with Enhanced Photocatalytic Activity for Removal of Antibiotics. *Int. J. Mol. Sci.* **2022**, *23*, 12652. https://doi.org/10.3390/ijms232012652

Academic Editor: Dippong Thomas

Received: 9 September 2022
Accepted: 10 October 2022
Published: 21 October 2022

**Publisher's Note:** MDPI stays neutral with regard to jurisdictional claims in published maps and institutional affiliations.

**Copyright:** © 2022 by the authors. Licensee MDPI, Basel, Switzerland. This article is an open access article distributed under the terms and conditions of the Creative Commons Attribution (CC BY) license (https://creativecommons.org/licenses/by/4.0/).

## 1. Introduction

Over the past few decades, the development of sustainable energy resources, such as solar energy, wind energy, hydrogen energy, and geothermal energy, etc., has become one of the most important issues around the world because of the shortage of conventional energy resources and environmental pollution problem. Among various sustainable energies, solar energy has attracted much attention, due to its inexhaustibility, environmental friendliness, and wide availability, etc. [1,2]. Semiconductor photocatalytic technology, which can directly convert solar energy into chemical energy through semiconductor photo-catalysts, has been recognized as one of the most attractive solutions for achieving sustainable energy supply and environmental restoration [3–6].

The common semiconductor photocatalytic materials include $TiO_2$, ZnO, $SrTiO_3$, $Bi_2S_3$, GaN, etc. Compared with the other photocatalytic materials, $TiO_2$ possesses the advantages of stable, non-toxic, high activity, photo-corrosion resistance, etc., and these characteristics make it one of the most representative semiconductor photocatalysts. However, the low utilization of visible light and high electron-hole recombination rate limit its wide application. In order to expand the spectral absorption range of $TiO_2$ photocatalyst and improve its photocatalytic efficiency, scientists have tried a variety of methods, such as metal/non-metal doping, hetero-junction construction, dye sensitization, etc. [7–10]. On the other hand, much effort had been paid to the development of novel, visible-light-responsive photocatalytic materials, such as metal oxides, metal sulfide, metal nitride, and organic compounds, etc. Recently, bismuth ferrite (BFO), such as $BiFeO_3$, $Bi_2Fe_4O_9$, $Bi_{25}FeO_{40}$, etc., have attracted much attention as a kind of potential candidate for visible-light responsive photocatalysts because of their narrower band gap structure (1.5~2.2 eV) and ability to degrade organic pollutants under visible light illumination [11–15]. However, there are still two issues, such as fast electron-hole recombination and low quantum yield, that limit its practical application. To further improve their photocatalytic performance, much effort had been made, such as synthesis with different methods and the control of process parameters, etc. However, most synthetic methods reported in the literature are time-consuming. Thus, developing a more economical and facile method to obtain highly efficient BFO-based photocatalyst is desirable.

Spark plasma sintering (SPS) is a typical field-assisted sintering technique, which is usually used for a high-speed consolidation of powders [16–18]. Recently, M.A. Lange reported that the heat treatment of $WO_{3-x}F_x$ by SPS on a minute (<10 min) scale can obtain higher photocatalytic performance [19]. Yang reported that self-doping $TiO_2$ with the use of directly treated commercial P25 at a desired temperature for only 5 min through SPS technology exhibited significantly high photoelectric and photocatalytic performance [20].

Compared with the other high-temperature heat treatment technologies, SPS possesses the following advantages. Firstly, its temperature can be raised quickly in a short time, while conventional methods usually take much longer to reach the same temperature. Secondly, the direct pulse current applied in the SPS process tends to activate the surface of the particles, thereby generating a large amount of hydroxyl radicals and atomic oxygen or oxygen-free radicals, which is profitable for the promotion of photocatalytic activity. Although the effects of SPS heat treatment on the ferroelectric, piezoelectric, and dielectric properties of bismuth ferrite (BFO) materials have been reported [21–23], the effect of SPS heat treatment on the photoelectric and photocatalytic activity of BFO materials is still lacking.

To explore the influence of SPS on the BFO photocatalysts and expand its application range, in this study, we reported a facile and rapid preparation of BFO-based composites by the direct use of hydrothermal-prepared BFO at a desired SPS temperature X (X = 300, 400, 500, 600, 650, 700) °C for only 5 min, and the corresponding products were named SPSX. The overall procedure for the synthesis of heterostructured BFO-based composites is shown in Figure 1a.

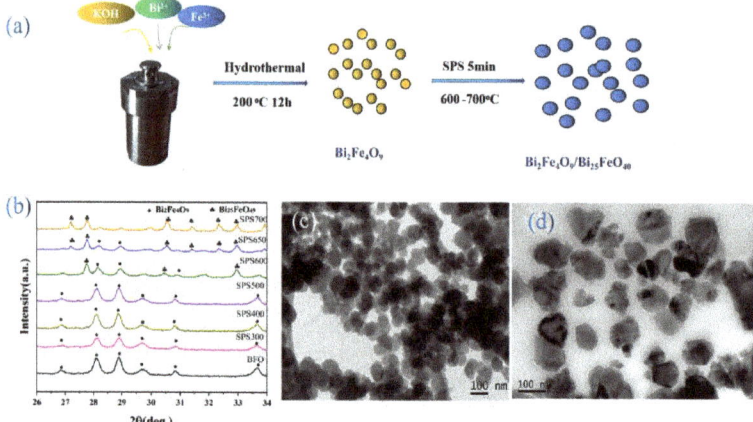

Figure 1. (a) Preparation scheme for bismuth ferrites, (b) X-ray diffraction (XRD) spectra of as-prepared samples, (c) SEM image of untreated BFO, (d) SEM image of SPS650.

## 2. Results and Discussion

The phase structure of the as-prepared samples was determined by X-ray diffraction (XRD) (Figures 1b and S1). As shown, the unmodified sample (BFO) and low-temperature SPS treatment samples (SPS300, SPS400, and SPS500) exhibited similar diffraction patterns, all diffraction peak patterns can be indexed to orthorhombic $Bi_2Fe_4O_9$ (JCPDS#25-0090) [24–26], and no other impurity phase appeared, indicating that low-temperature SPS treatment has no obvious effect on its phase structure. When the SPS heating temperature increased to 600 °C, two strong diffraction peaks at 28.2° and 28.9°, corresponding to the (121) and (211) crystal planes of orthorhombic $Bi_2Fe_4O_9$, weakened, and two new characteristic peaks appeared at 27.8° and 32.9°, which can be attributed to the (310) and (321) crystal planes of cubic $Bi_{25}FeO_{40}$, according to the PDF standard cards (JCPDS#46-0416) [27,28], which means that the $Bi_{25}FeO_{40}$ and $Bi_2Fe_4O_9$ phase coexistence was at 600 °C. With the increasing of the SPS processing temperature (SPS 650), the content of $Bi_{25}FeO_{40}$ phase gradually increased. At 700 °C, $Bi_2Fe_4O_9$ was almost entirely transformed into $Bi_{25}FeO_{40}$. Anyway, the sample treated at 300, 400, and 500 °C under SPS for 5 min exhibited similar photocatalytic activity as the unmodified BFO (omitted here). For the sake of simplicity, in this work, we mainly reported the unmodified BFO sample and the samples treated at 600, 650, and 700 °C under SPS for 5 min and the percentage of $Bi_2Fe_4O_9$ and $Bi_{25}FeO_{40}$ in the BFO, SPS600, SPS650, SPS700 were shown in Table S1. The morphology of the BFO and SPS650 samples was observed by scanning electron microscope (SEM) (Figure 1c,d). As indicated in Figure 1c, the unmodified BFO particles are near spherical with average diameter of ca. 70 nm. After being treated by SPS 5 min at 650 °C, the shape of BFO became irregularly spherical, with an average diameter reaching 80–90 nm (Figures 1d, S2 and S3). In Figure S4, the peaks of Bi 4f, Fe 2p, and O 1s can be observed in the survey XPS spectra of the BFO samples before and after SPS modification, indicating that the elements contained in the sample were consistent with the target product.

For clarity, the HRTEM of as-prepared BFO and SPS650 samples were performed to further elucidate the microstructure of the samples. For unmodified BFO (Figure 2a), the lattice fringes of the (001) planes with interplanar spacing of approximately 0.600 nm corresponded to $Bi_2Fe_4O_9$ [29,30]. For SPS650 (Figure 2b–d), the fringe of the (212) plane with interplanar spacing of approximately 0.220 nm corresponded to $Bi_2Fe_4O_9$, while the fringes of the (321) planes with interplanar spacing of approximately 0.273 nm corresponded to $Bi_{25}FeO_{40}$ [31–33], indicating some of $Bi_2Fe_4O_9$ transformed into $Bi_{25}FeO_{40}$ when treated with SPS for 5 min, and the $Bi_2Fe_4O_9/Bi_{25}FeO_{40}$ heterojunction was synchronously formed during this process.

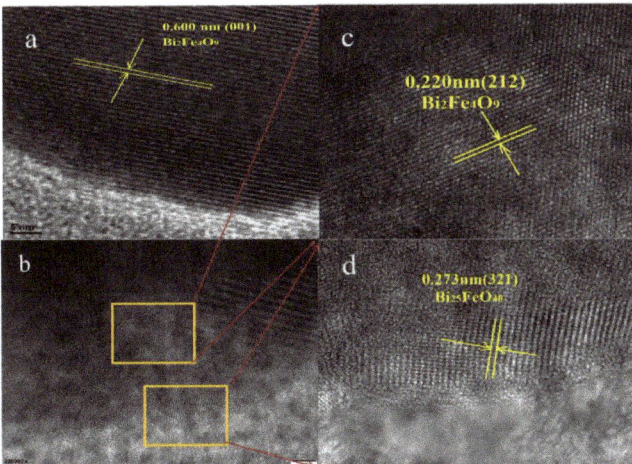

**Figure 2.** HRTEM image of (**a**) untreated BFO, (**b**) SPS650, (**c**,**d**) corresponding magnified image of (**d**).

The optical absorption properties of the as-prepared samples were investigated by UV–Vis DRS technology, as shown in Figure 3a,b. As indicated, all the samples can response to visible light; however, in comparison with the unmodified BFO, the samples with SPS treatment exhibited a significant enhancement in absorption intensities, along with the absorption redshift. The absorption edges of BFO located at ~570 nm corresponds to an optical bandgap ($Eg$) of 2.18 eV, which was calculated from the tangent line in the plot of the K-M function $(\alpha h v)^2$ vs. photo energy ($hv$) by extrapolating the tangent line to another small absorption edge at ~810 nm, which corresponds to an optical band gap of 1.60 eV. The absorption edge at ~570 nm can be ascribed to the electronic transfer from the valence band to the conduction band, and the absorption at ~810 nm is ascribed to the d-d transitions of Fe [34–36]. For the $Bi_2Fe_4O_9/Bi_{25}FeO_{40}$ samples, the absorption edge of SPS700 was ~775 nm, and the absorption edges of SPS600 and SPS650 lied somewhere in between. According to Kubelka–Munk function, the band gaps for SPS600, SPS650, and SPS700 were 2.10 eV, 2.02 eV, and 1.60 eV, respectively (Figure 3b). In view of the preparation process, the close contact and strong interaction between the $Bi_2Fe_4O_9$ and $Bi_{25}FeO_{40}$ existence in the SPS-treated samples is suggested. The valence band position (VB) of semiconductors was further determined by X-ray photoelectron spectroscopy, as shown in Figure 3c, which indicates that the VB (Ev) of $Bi_2Fe_4O_9$ (BFO) and $Bi_{25}FeO_{40}$ (SPS700) were about 1.55 eV and 0.27 eV, respectively. Correspondingly, the conductor band positions (CB, Ec) of $Bi_2Fe_4O_9$ (BFO) and $Bi_{25}FeO_{40}$ (SPS700) were estimated to be −0.63 eV and −1.33 eV, respectively, according to the equation: Ec = Eg − Ev.

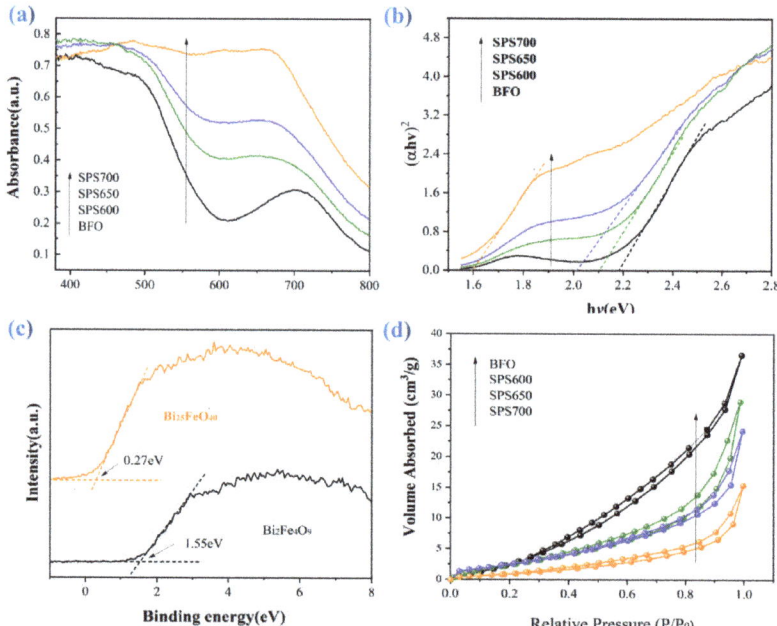

**Figure 3.** (a) The UV–Vis absorption spectrum of the as-prepared sample; (b) the calculation bandgap diagram of the as-prepared sample, according to Kubelka–Munk function. (c) The XPS VB spectra of as-synthesized $Bi_2Fe_4O_9$ and $Bi_{25}FeO_{40}$ samples; (d) $N_2$ adsorption and desorption isotherms of the as-prepared samples.

The nitrogen adsorption isotherms ($S_{BET}$) of the as-prepared catalysts were shown in Figure 3d, and the specific surface area ($S_{BET}$) was calculated according to the Brunauer–Emmet–Teller method. The order of $S_{BET}$ was as follows: SPS650 (23.1 $m^2 \cdot g^{-1}$) > SPS600 (19.2 $m^2 \cdot g^{-1}$) > SPS700 (15.8 $m^2 \cdot g^{-1}$) > BFO (11.7 $m^2 \cdot g^{-1}$). In our experiment, with the increasing of heat-treated temperature, the $Bi_2Fe_4O_9$ phase gradually changed into the $Bi_{25}FeO_{40}$ phase. The specific surface area is connected to the crystal structure and particle stacking way, so the SPS650 sample exhibited the optimum specific surface area in our case. A larger surface, in general, means more surface-active sites and faster interfacial charge transfer for the reaction; thus, the SPS650 sample was expected to exhibit the higher photocatalytic activity.

The separation and transfer rate of photogenerated electrons and holes are also the main factors affecting the photocatalytic performance. Herein, to explore the photoelectric separation and transferring performance of the as-prepared samples, the photocurrent response (I-t), photoluminescence emission spectra (PL), electrochemical impedance spectroscopy (EIS), and transient photoluminescence spectra of the as-obtained samples were carefully explored (Figure 4). Generally, the transient photocurrent reflects the charge carrier density and charge mobility. The stronger the photocurrent, the greater the density of the photo-generated carrier, and the more efficiencies of the charge separation [37–40]. As indicated in Figure 4a, the SPS650 sample exhibited the highest photocurrent density of 2.78 µA/cm$^2$, which is about 5.56 times that of unmodified BFO (0.5 µA/cm$^2$), and far larger than that of SPS600 (2.20 µA/cm$^2$) and SPS700 (1.75 µA/cm$^2$).

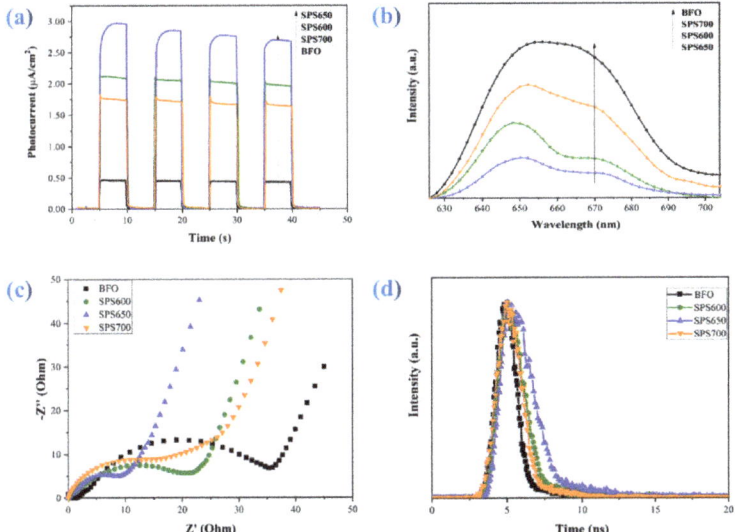

**Figure 4.** (a) The curve of photocurrent density versus time; (b) photoluminescence spectrum (PL), (c) electrochemical impedance spectroscopy (EIS), (d) transient photoluminescence spectrum (TRPL).

Figure 4b indicates the PL spectra of the as-prepared samples at an excitation wavelength of 325 nm. In general, PL emission intensity is related to the recombination of excited electrons and holes—the lower the PL peak intensity, the smaller the probability of recombination [41,42]. As indicated, the PL intensity of the SPS650 sample is the weakest among all the samples, inferring that the recombination rate of charge carries is the smallest among all the samples. The possible reason is maybe due to the fact that the moderate heterogeneous junctions produced in the SPS650 sample effectively accelerate the charge transfer and correspondingly reduces the charge's recombination. As a result, the problem of fast charge recombination, as a historical intrinsic drawback of BFO photocatalysts, was effectively restrained by simply modifying the BFO by SPS heat-treatment for 5 min at a set temperature. Additionally, the diameters of the arc radius on the EIS Nynquist plot of the SPS-treated BFO samples were smaller to that of the unmodified BFO, while the SPS650 sample shows the least arc radius (Figure 4c). The smaller the internal resistance of the charge transfer means the higher the migration rate of the photo-generated electron-hole pairs and the higher the carrier separation rate [43–45]. Anyway, the time-resolved PL

(TRPL) measurement results of the as-prepared samples indicate that the SPS650 composites have longer lifetimes than all the other samples (Figure 4d). In other words, compared with the other samples, the photoinduced electron-hole pairs in SPS650 are easier separated and transferred to the sample surface through an interfacial interaction between two different bismuth oxides, correspondingly resulting in a higher photocatalytic performance.

Tetracycline hydrochloride (TCH) is a commonly used, but difficult to self-degrade, antibiotic [46]. Therefore, the photocatalytic performance of the as-prepared samples was evaluated by TCH as the target degradation product under visible light irradiation. As shown in Figure 5a, the characteristic absorption peaks at 275 nm and 356 nm decreased with the increasing of irradiation time, indicating the photodegradation of TCH under visible-light irradiation in the presence of catalysts. Figure 5b indicates that about 58% of TCH are degraded by BFO for 2 h, while the SPS-treated samples exhibited significantly enhanced photocatalytic performance, and the degradation ratios of TCH reached 82%, 96%, and 74% for the SPS600, SPS650, and SPS700 samples, respectively. It has been found that the SPS650 sample demonstrates the optimum photocatalytic activity, revealing the synergistic effect between $Bi_{25}FeO_{40}$ and $Bi_2Fe_4O_9$, which is beneficial to the separation of photo-generated carriers, correspondingly resulting in a higher photocatalytic performance. This value is also higher than some previously reported catalysts (Table S3) [47–52]. The linear plots of ($C_0/C$) versus irradiation time (t) suggest a pseudo-first order kinetic (Figure 5c). The rate constants k are estimated to be 0.006 $min^{-1}$, 0.014 $min^{-1}$, 0.025 $min^{-1}$, and 0.012 $min^{-1}$ for the samples BFO, SPS600, SPS650, and SPS700, respectively. The apparent quantum yield ($AQY\%$) was calculated to be 4.66%, according to equation [47,48], and the detailed calculation can be found in Table S2:

$$AQY\% = \frac{\text{the number of degraded moleculars}}{N} \times 100\%$$

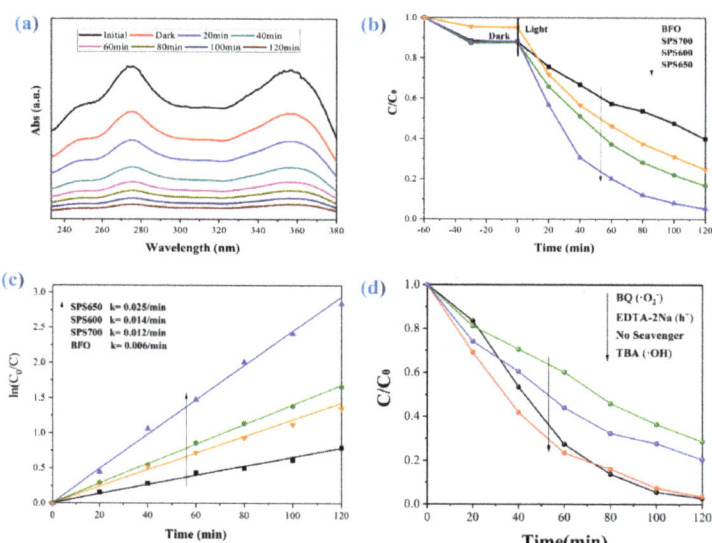

**Figure 5.** (**a**) Absorption spectra of the TCH solution over SPS650 under visible light irradiation; (**b**) photocatalytic degradation of TCH under visible light irradiation for as-prepared samples; (**c**) the corresponding photodegradation kinetics curves for as-prepared samples; (**d**) photocatalytic activities over the SPS650 sample with different scavengers for the degradation of TCH under visible-light irradiation.

Obviously, with the increasing of SPS heat-treat temperature, the photocatalytic efficiencies of the composites firstly increase to a maximal value and then decrease. The photo-reactive rate constant of the sample SPS650 was at 0.025 min$^{-1}$, which is about 4.2 times that of unmodified BFO, 1.8 times that of the sample SPS600, and 2.1 times that of the sample SPS700. The reason for the relatively weaker photocatalytic performance for SPS600 and SPS700 can be ascribed to the fact that the amount of $Bi_2Fe_4O_9/Bi_{25}FeO_{40}$ heterojunction was not formed enough. In addition, the photodegradation efficiency of SPS650 had slightly decreased after 5 cycle times (Figure S5), inferring its excellent stability, which is beneficial for the practice application.

The total organic carbon (TOC) removal ratio was used to evaluate the mineralization efficiency of TCH by SPS650 under visible light irradiation. As shown in Figure S6, the TOC decreased gradually, and the efficiency of TOC removal was 67% under 2 h irradiation. It is worth noting that the efficiency of TOC removal was lower than that of photodegradation, which can be attributed to the incomplete mineralization of the TCH or generation of a low molecular weight organic.

To confirm the speculation about the main active species, the agents of 1,4-benzoquinone (BQ, 0.1 mmol), disodium ethylenediamine tetraacetate (EDTA-2Na, 0.1 mmol), and tert-butyl alcohol (TBA, 0.1 mmol) were introduced as the superoxide radical ($\bullet O^{2-}$) scavenger, hole ($h^+$) scavenger, and hydroxyl radical ($\bullet OH$) scavenger, respectively. As indicated in Figure 5d, the degradation ratio of TCH over SPS650, without adding scavengers, was 96%. With the adding of BQ and EDTA-2Na, the degradation efficiencies of TCH were reduced to 63%, 72%, and 96%, respectively, indicating that superoxide radical was main active species, and the hole plays a relatively important role in the photocatalytic oxidation process. On the contrary, the photocatalytic activity of SPS650 had no obvious change by the addition of TBA, which infers that the hydroxyl radical was not the main active species to the degradation process.

On the basis of the above analysis, the photodegradation mechanism for the $Bi_2Fe_4O_9/Bi_{25}FeO_{40}$ composites was proposed (Scheme 1). Generally, the performance of photocatalyst depends on a series of parameters, such as the surface area, adsorption capacity, light absorption, carrier recombination rate, and energy band structure [53–58]. In our case, the significant improvement of the photodegradation performance for SPS650 may be related to the following factors: firstly, due to the high-temperature sintering of the spark plasma, the surface of the modified BFO sample has more separated electrons, holes, and OH radicals. These active materials react with pollutants in the solution to show better photocatalytic activity. Secondly, the $Bi_{25}FeO_{40}$ produced by SPS treatment results in the formation of $Bi_2Fe_4O_9/Bi_{25}FeO_{40}$ heterojunction. Under visible light irradiation, the photoinduced electrons tend to migrate from $Bi_{25}FeO_{40}$ to $Bi_2Fe_4O_9$, which leads to the faster transfer and separation of photogenerated carriers at the heterojunction interface, thereby improving the photodegradation performance.

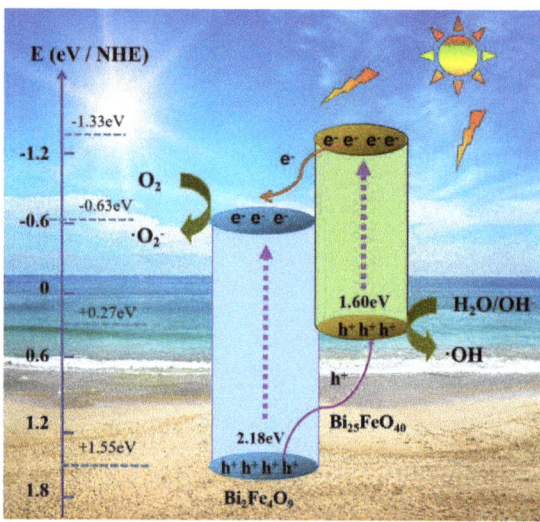

**Scheme 1.** The schematic illustration of photo-reactive mechanism for $Bi_2Fe_4O_9/Bi_{25}FeO_{40}$ heterostructure composites.

## 3. Materials and Methods

### 3.1. Synthesis

BFO powders were firstly synthesized by the hydrothermal method, as reported previously, with a little change. Typically, $Bi(NO_3)_3 \cdot 5H_2O$ and $Fe(NO_3)_3 \cdot 9H_2O$ with the molar mass ratio $Bi^{3+}$: $Fe^{3+}$ = 1:1 were first dissolved and stirred at 1 M $HNO_3$ solution to form an aqueous solution, and pH is ~2. Then, 4 M KOH solution was slowly added into the aforementioned solution with vigorously stirring for 1 h to tune the pH value to 9 (here, $HNO_3$ was used as dissolvent, and KOH as precipitator and mineralizer). After that, it was quickly transferred into a teflon-lined stainless-steel autoclave and heated at 200 °C for 12 h. Finally, the product was collected and washed with deionized water and alcohol several times to about pH 7, followed by drying at 80 °C overnight. According to the XRD analysis (Figure 1), the obtained sample belonged to orthorhombic $Bi_2Fe_4O_9$ (JCPDS 25-0090) and was named BFO.

Next, the obtained BFO sample was further treated using the SPS technology. In a typical SPS heat-treatment process, 1.0 g of the as-prepared BFO powder was first put into a cylindrical carbon die with an inner diameter of 15 mm and then transferred into the SPS device (SPS-3.20 MKII, Sumitomo Coal Mining Co. Ltd., Tokyo, Japan) for sintering. The temperature was raised to the set temperature with a heating rate of 50 °C/min and kept at the temperature for 5 min under 60 MPa, then naturally cooled to room temperature. During the sintering process, the sample temperature was measured using an infrared camera through a hole in the middle of the cylindrical carbon die. The modified BFO sample after SPS heat treatment were denoted as SPSX, and X (X = 300–700) represents the set temperature from 300 °C to 700 °C.

### 3.2. Characterization

The phase composition of the samples was analyzed by Bruker D8 X-ray diffractometer (XRD, Cu Kα, λ = 1.5406 Å), the sampling interval was 0.02°, the sampling rate was 5 °/min, and the scanning range of the sample was 10°–70°. TEM (TEM, JEOL JEM-2010 FEF, Japan) was used to further confirm the morphology of the obtained samples. Ultraviolet-visible diffuse reflectance spectroscopy (UV–Vis DRS) analysis adopted U-4100 solid spectrophotometer, and the test wavelength range was 200–800 nm. The valence bands of the sample surfaces were characterized using X-ray photoelectron spectroscopy

(XPS, Thermo Scientific Escalab 250, Waltham, MA, USA) with a monochromatic Al K X-ray source. All binding energies were referenced to the C 1s peak (284.6 eV) arising from adventitious carbon. The specific surface areas of the catalysts were determined by applying the Brunauer–Emmett–Teller (BET) method to the adsorption of nitrogen at 77 K. All the samples were degassed at 180 °C prior to nitrogen adsorption measurements. The photocurrents, electrochemical impedance spectroscopy (EIS), and Mott–Schottky were measured by an electrochemical analyzer (CHI660A, CH Instruments Co. Shanghai, China) at room temperature. The photoluminescence (PL) spectroscopy was carried out with the FLs980 full-function steady-state/transient fluorescence spectrometer in Edinburgh, UK, with an excitation wavelength of 325 nm. Total organic carbon (TOC) was measured by a Multi N/C 3100 TOC analysis instrument.

Photocatalytic Performance

The photocatalytic performance was measured by degrading the 10 mg/L tetracycline hydrochloride (TCH) aqueous solution under visible light. Typically, 50 mg of catalyst was added to 50 mL of TCH solution and kept in the dark for 1 h to reach adsorption-desorption equilibrium. After that, a 350 W xenon lamp and 420 nm cut-off filter were used as the light source to simulate sunlight, and samples were taken every 20 min.

## 4. Conclusions

In summary, the self-doped $Bi_2Fe_4O_9/Bi_{25}FeO_{40}$ heterostructure composites were successfully obtained by spark plasma sintering technology, combined with a facile wet chemical process. The ratio of $Bi_2Fe_4O_9$ to $Bi_{25}FeO_{40}$ in the composites was controlled by the spark plasma sintering temperature. The as-prepared sample (SPS650) exhibited significantly higher photoelectric performance and photocatalytic activity than that of $Bi_2Fe_4O_9$ and $Bi_{25}FeO_{40}$ on the degradation of TCH. It was found that formation of well-defined heterojunction between $Bi_2Fe_4O_9$ and $Bi_{25}FeO_{40}$, which effectively sped the transformation and separation of photoinduced carriers, correspondingly resulted in the enhancement of the photoelectric property and photocatalytic activity. Anyway, it could be easily recycled without an obvious decrease of photocatalytic activity. This study provides a simple and economic method for the facile and quick synthesis of a highly active bismuth ferrite-based visible-light-driven photocatalyst for practical applications.

**Supplementary Materials:** The following supporting information can be downloaded at: https://www.mdpi.com/article/10.3390/ijms232012652/s1.

**Author Contributions:** Z.L., Y.T. and J.G. methodology and investigation; X.R. investigation and resources; W.L., J.L., H.M. and X.R. data curation, and formal analysis; R.X. and J.W. writing, review, and editing, and supervision. All authors have read and agreed to the published version of the manuscript.

**Funding:** This research was funded by the National Science Foundation of China, grant numbers 91963207 and 12075174.

**Institutional Review Board Statement:** Not applicable.

**Informed Consent Statement:** Not applicable.

**Data Availability Statement:** Not applicable.

**Conflicts of Interest:** The authors declare no conflict of interest.

## References

1. Zhang, P.; Wang, T.X.; Chang, X.; Gong, J.L. Effective charge carrier utilization in photocatalytic conversions. *Acc. Chem. Res.* **2016**, *49*, 911–921. [CrossRef] [PubMed]
2. Armaroli, N.; Balzani, V. The future of energy supply: Challenges and Opportunities. *Angew. Chem. Int. Ed.* **2007**, *46*, 52–66. [CrossRef] [PubMed]
3. Chong, M.N.; Jin, B.; Chow, C.W.K.; Saint, C. Recent developments in photocatalytic water treatment technology: A review. *Water Res.* **2010**, *44*, 2997–3027. [CrossRef] [PubMed]

4. Bahnemann, D. Photocatalytic Water Treatment: Solar energy applications. *Sol. Energy* **2004**, *77*, 445–459. [CrossRef]
5. Wang, H.L.; Zhang, L.S.; Chen, Z.G.; Hu, J.Q.; Li, S.J.; Wang, Z.H.; Liu, J.S.; Wang, X.C. Semiconductor heterojunction photocatalysts: Design, construction, and photocatalytic performances. *Chem. Soc. Rev.* **2014**, *43*, 5234–5244. [CrossRef]
6. Chen, F.; Ma, T.; Zhang, T.; Zhang, Y.; Huang, H. Atomic-level charge separation strategies in semiconductor-based photocatalysts. *Adv. Mater.* **2021**, *33*, 2005256. [CrossRef]
7. Schneider, J.; Matsuoka, M.; Takeuchi, M.; Zhang, J.L.; Horiuchi, Y.; Anpo, M.; Bahnemann, D.W. Understanding $TiO_2$ photocatalysis: Mechanisms and materials. *Chem. Rev.* **2014**, *114*, 9919–9986. [CrossRef]
8. Lee, S.Y.; Park, S.J. $TiO_2$ Photocatalyst for water treatment application. *J. Ind. Eng. Chem.* **2013**, *19*, 1761–1768. [CrossRef]
9. Choi, S.K.; Kim, S.; Lim, S.K.; Park, H. Photocatalytic Comparison of $TiO_2$ nanoparticles and electrospun $TiO_2$ nanofibers: Effects of mesoporosity and interparticle charge transfer. *J. Phys. Chem. C* **2010**, *114*, 16474–16480. [CrossRef]
10. Asahi, R.; Morikawa, T.; Ohwaki, T.; Aoki, K.; Taga, Y. Visible-light photocatalysis in nitrogen-dDoped titanium oxides. *Science* **2001**, *293*, 269–271. [CrossRef]
11. Gadhi, T.A.; Mahar, R.B.; Bonelli, B. Chapter 12—Actual mineralization versus partial egradation of wastewater contaminants in nanomaterials for the detection and removal of wastewater pollutants. *Micro Nano Tech.* **2020**, 331–350.
12. Hua, Z.T.; Liu, J.C.; Yan, X.L.; Oh, W.D.; Lim, T.T. Low-temperature synthesis of graphene/$Bi_2Fe_4O_9$ composite for synergistic adsorption-photocatalytic degradation of hydrophobic pollutant under solar irradiation. *Chem. Eng. J.* **2015**, *262*, 1022–1032. [CrossRef]
13. Li, F.H.; Zhou, J.K.; Gao, C.J.; Qiu, H.X.; Gong, Y.L.; Gao, J.H.; Liu, Y. A green method to prepare magnetically recyclable Bi/$Bi_{25}FeO_{40}$-C nanocomposites for photocatalytic hydrogen generation. *Appl. Surf. Sci.* **2020**, *521*, 146342. [CrossRef]
14. Wu, Y.; Chen, Y.C.; Huang, S.Q.; Li, G.B.; Sun, S.P.; Jiang, Y.; Liang, G.B.; Zhao, S.J.; Liu, W.Q. Comparison of bismuth ferrites for chloride removal: Removal efficiency, stability, and structure. *Appl. Surf. Sci.* **2022**, *576*, 151804. [CrossRef]
15. Zhang, T.; Shen, Y.; Qiu, Y.H.; Liu, Y.; Xiong, R.; Shi, J.; Wei, J.H. Facial Synthesis and Photoreaction Mechanism of $BiFeO_3$/$Bi_2Fe_4O_9$ Heterojunction Nanofibers. *ACS Sustain. Chem. Eng.* **2017**, *5*, 4630–4636. [CrossRef]
16. Guillon, O.; Julian, J.G.; Dargatz, B.; Kessel, T.; Schierning, G.; Rathel, J.; Herrmann, M. Field-assisted sintering technology/spark plasma sintering: Mechanisms, materials, and technology developments. *Adv. Eng. Mater.* **2014**, *16*, 830–849. [CrossRef]
17. Dai, Z.H.; Akishige, Y. $BiFeO_3$ ceramics synthesized by spark plasma sintering. *Ceram. Inter.* **2012**, *38*, S403–S406. [CrossRef]
18. Delbari, S.A.; Nayebi, B.; Ghasali, E.; Shokouhimehr, M.; Asl, M.S. Spark plasma sintering of TiN ceramics codoped with SiC and CNT. *Ceram. Int.* **2019**, *45*, 3325–3332. [CrossRef]
19. Lange, M.A.; Krysiak, Y.; Hartmann, J.; Dewald, G.; Cerretti, G.; Tahir, M.N.; Panthöfer, M.; Barton, B.; Reich, T.; Zeier, W.G.; et al. Solid state fluorination on the minute scale: Synthesis of $WO_{3-x}F_x$ with photocatalytic activity. *Adv. Funct. Mater.* **2020**, *30*, 1909051. [CrossRef]
20. Yang, Y.C.; Zhang, T.; Le, L.; Ruan, X.F.; Fang, P.F.; Pan, C.X.; Xiong, R.; Shi, J.; Wei, J.H. Quick and facile preparation of visible Light-Driven $TiO_2$ photocatalyst with high absorption and photocatalytic activity. *Sci. Rep.* **2014**, *4*, 7045. [CrossRef]
21. Wang, T.; Song, S.-H.; Ma, Q.; Tan, M.-L.; Chen, J.-J. Highly improved multiferroic properties of Sm and Nb co-doped $BiFeO_3$ ceramics prepared by spark plasma sintering combined with sol-gel powders. *J. Alloys Compd.* **2019**, *795*, 60–68. [CrossRef]
22. Nandy, S.; Mocherla, P.S.V.; Sudakar, C. Photoconductivity induced by nanoparticle segregated grain-boundary in spark plasma sintered $BiFeO_3$. *J. Appl. Phys.* **2017**, *121*, 203102. [CrossRef]
23. Zhang, L.W.; Ke, H.; Zhang, H.J.; Luo, H.; Li, F.Z.; Cao, L.; Wang, W.; Jia, D.; Zhou, C.Y. Ferroelectric domain structures in strained $BiFeO_3$ ceramics synthesizd by spark plasma sintering. *Mater. Charact.* **2020**, *159*, 110044. [CrossRef]
24. Yang, Y.C.; Liu, Y.; Wei, J.H.; Pan, C.X.; Xiong, R.; Shi, J. Electrospun Nanofibers of p-type $BiFeO_3$/n-type $TiO_2$ heterojunctions with enhanced visible-light photocatalytic activity. *RSC Adv.* **2014**, *4*, 31941–31947. [CrossRef]
25. Ruan, Q.J.; Zhang, W.D. Tunable morphology of $Bi_2Fe_4O_9$ crystals for photocatalytic oxidation. *J. Phys. Chem. C* **2009**, *113*, 4168–4173. [CrossRef]
26. Hu, Z.T.; Liu, J.W.; Zhao, J.; Ding, Y.; Jin, Z.Y.; Chen, J.H.; Dai, Q.Z.; Pan, B.J.; Chen, Z.; Chen, J.M. Enhanced $BiFeO_3$/$Bi_2Fe_4O_9$/$H_2O_2$ heterogeneous system for sulfamethoxazole decontamination: System optimization and degradation pathways. *J. Colloid. Interface Sci.* **2020**, *577*, 54–65. [CrossRef]
27. Jiang, T.J.; Wang, Y.; Guo, Z.C.; Luo, H.P.; Zhan, C.X.; Wang, Y.J.; Wang, Z.; Jiang, F.; Chen, H. $Bi_{25}FeO_{40}$/$Bi_2O_2CO_3$ piezoelectric catalyst with built-in electric fields that was prepared via photochemical self-etching of $Bi_{25}FeO_{40}$ for 4-chlorophenol degradation. *J. Clean. Prod.* **2022**, *341*, 130908. [CrossRef]
28. Ren, L.; Lu, S.Y.; Fang, J.Z.; Wu, Y.; Chen, D.Z.; Huang, L.Y.; Chen, Y.F.; Cheng, C.; Liang, Y.; Fang, Z.Q. Enhanced degradation of organic pollutants using $Bi_{25}FeO_{40}$ microcrystals as an efficient reusable heterogeneous photo-fenton like catalyst. *Catal. Today* **2017**, *281*, 656–661. [CrossRef]
29. Wang, Y.Q.; Daboczi, M.; Mesa, C.A.; Ratnasingham, S.R.; Kim, J.S.; Durrant, J.R.; Dunn, S.; Yan, H.X.; Briscoe, J. $Bi_2Fe_4O_9$ thin films as novel visible-light-active photoanodes for solar water splitting. *J. Mater. Sci. Chem. A* **2019**, *7*, 9537–9541. [CrossRef]
30. Liu, H.H.; Li, L.; Guo, C.F.; Ning, J.Q.; Zhong, Y.J.; Hu, Y. Thickness-dependent carrier separation in $Bi_2Fe_4O_9$ nanoplates with enhanced photocatalytic water oxidation. *Chem. Eng. J.* **2020**, *385*, 123929. [CrossRef]
31. Wang, G.M.; Cheng, D.; He, T.C.; Hu, Y.Y.; Deng, Q.R.; Mao, Y.W.; Wang, S.G. Enhanced visible-light responsive photocatalytic activity of $Bi_{25}FeO_{40}$/$Bi_2Fe_4O_9$ composites and mechanism investigation. *J. Mater. Sci. Mater.* **2019**, *30*, 10923–10933. [CrossRef]

32. Köferstein, R.; Buttlar, T.; Ebbinghaus, S.G. Investigations on $Bi_{25}FeO_{40}$ powders synthesized by hydrothermal and combustion-like processes. *J. Solid. State Chem.* **2014**, *217*, 50–56. [CrossRef]
33. Liu, Y.; Guo, H.G.; Zhang, Y.L.; Tang, W.H.; Cheng, X.; Li, W. Heterogeneous activation of peroxymonosulfate by sillenite $Bi_{25}FeO_{40}$: Singlet oxygen generation and degradation for aquatic levofloxacin. *Chem. Eng. J.* **2018**, *343*, 128–137. [CrossRef]
34. Ameer, S.; Jindal, K.; Tomar, M.; Jha, P.K.; Gupta, V. Insight into electronic, magnetic and optical properties of magnetically ordered $Bi_2Fe_4O_9$. *J. Magn. Magn. Mater.* **2019**, *475*, 695–702. [CrossRef]
35. Ji, W.; Yao, K.; Lim, Y.F.; Liang, Y.C.; Suwardi, A. Epitaxial ferroelectric $BiFeO_3$ thin films for unassisted photocatalytic water splitting. *Appl. Phys Lett.* **2013**, *103*, 062901. [CrossRef]
36. Yin, S.M.; Li, W.Q.; Cheng, R.S.; Yuan, Y.F.; Guo, S.Y.; Ren, Z.H. Hydrothermal synthesis, photocatalytic and magnetic properties of pure-phase $Bi_2Fe_4O_9$ Microstructures. *J. Electron. Mater.* **2021**, *50*, 954–959. [CrossRef]
37. Zhong, X.; Sun, Y.Y.; Chen, X.L.; Zhuang, G.L.; Li, X.N.; Wang, J.G. Mo doping induced more active sites in urchin-like $W_{18}O_{49}$ nanostructure with remarkably enhanced performance for hydrogen evolution reaction. *Adv. Funct. Mater.* **2016**, *26*, 5778–5786. [CrossRef]
38. Pan, L.; Zhang, J.W.; Jia, X.; Ma, Y.H.; Zhang, X.W.; Wang, L.; Zou, J.J. Highly efficient Z-scheme $WO_{3-x}$ quantum dots/$TiO_2$ for photocatalytic hydrogen generation. *Chin. J. Catal.* **2017**, *38*, 253–259. [CrossRef]
39. Khoomortezaei, S.; Abdizadeh, H.; Golobostanfard, M.R. Triple Layer Heterojunction $WO_3$/$BiVO_4$/$BiFeO_3$ porous photoanode for efficient photoelectrochemical water splitting. *ACS Appl. Energy Mater.* **2019**, *2*, 6428–6439. [CrossRef]
40. Huang, S.Q.; Li, L.; Zhu, N.W.; Lou, Z.Y.; Liu, W.Q.; Cheng, J.H.; Wang, H.M.; Luo, P.X.; Wang, H. Removal and recovery of chloride ions in concentrated leachate by Bi(III) containing oxides quantum dots/two-dimensional flakes. *J. Hazard. Mater.* **2020**, *382*, 121041. [CrossRef]
41. Rafiq, U.; Majid, K. Mitigating the charge recombination by the targeted synthesis of $Ag_2WO_4$/$Bi_2Fe_4O_9$ composite: The facile union of orthorhombic semiconductors towards efficient photocatalysis. *J. Alloy. Compd.* **2020**, *842*, 155876. [CrossRef]
42. Zhang, J.; Xu, Q.; Feng, Z.C.; Li, M.J.; Li, C. Importance of the relationship between surface phases and photocatalytic activity of $TiO_2$. *Angew. Chem. Inter. Ed.* **2008**, *47*, 1766–1769. [CrossRef] [PubMed]
43. Godin, R.; Wang, Y.; Zwijnenburg, M.A.; Tang, J.W.; Durrant, J.R. Time-resolved spectroscopic investigation of charge trapping in carbon nitrides photocatalysts for hydrogen generation. *J. Am. Chem. Soc.* **2017**, *139*, 5216–5224. [CrossRef]
44. Zhao, G.; Hao, S.H.; Guo, J.H.; Xing, Y.P.; Zhang, L.; Xu, X.J. Design of p-n homojunctions in metal-free carbon nitride photocatalyst for overall water splitting. *Chin. J. Catal.* **2021**, *42*, 501–509. [CrossRef]
45. Indra, A.; Acharjya, A.; Menezes, P.W.; Merschjann, C.; Hollmann, D.; Schwarze, M.; Aktas, M.; Friedrich, A.; Lochbrunner, S.; Thomas, A.; et al. Boosting Visible-Light-Driven Photocatalytic Hydrogen Evolution with an Integrated Nickel Phosphide-Carbon Nitride System. *Angew. Chem. Int. Ed.* **2017**, *56*, 1653–1657. [CrossRef]
46. Han, W.M.; Wu, T.; Wu, Q.S. Fabrication of $WO_3$/$Bi_2MoO_6$ heterostructures with efficient and highly selective photocatalytic degradation of tetracycline hydrochloride. *J. Colloid Inter. Sci.* **2021**, *602*, 544–552. [CrossRef]
47. Zhang, Y.; Shi, J.; Xu, Z.; Chen, Y.; Song, D. Degradation of tetracycline in a schorl/$H_2O_2$ system: Proposed mechanism and intermediates. *Chemosphere* **2018**, *202*, 661–668. [CrossRef]
48. Xie, Z.; Feng, Y.; Wang, F.; Chen, D.; Zhang, Q.; Zeng, Y.; Lv, W.; Liu, G. Construction of carbon dots modified $MoO_3$/g-$C_3N_4$ Z-scheme photocatalyst with enhanced visible-light photocatalytic activity for the degradation of tetracycline. *Appl. Catal. B Environ.* **2018**, *229*, 96–104. [CrossRef]
49. Liu, Y.; Kong, J.; Yuan, J.; Zhao, W.; Zhu, X.; Sun, C.; Xie, J. Enhanced photocatalytic activity over flower-like sphere Ag/$Ag_2CO_3$/$BiVO_4$ plasmonic heterojunction photocatalyst for tetracycline degradation. *Chem. Eng. J.* **2018**, *331*, 242–254. [CrossRef]
50. Deng, F.; Zhao, L.; Luo, X.; Luo, S.; Dionysiou, D.D. Highly efficient visible-light photocatalytic performance of Ag/$AgIn_5S_8$ for degradation of tetracycline hydrochloride and treatment of real pharmaceutical industry wastewater. *Chem. Eng. J.* **2018**, *333*, 423–433. [CrossRef]
51. Wang, W.; Xiao, K.; Zhu, L.; Yin, Y.; Wang, Z. Graphene oxide supported titanium dioxide & ferroferric oxide hybrid, a magnetically separable photocatalyst with enhanced photocatalytic activity for tetracycline hydrochloride degradation. *RSC Adv.* **2017**, *7*, 21287–21297.
52. Barhoumi, H.; Olvera-Vargas, N.; Oturan, D.; Huguenot, A.; Gadri, S.; Ammar, E.; Brillas, M.A. Kinetics of oxidative degradation/mineralization pathways of the antibiotic tetracycline by the novel heterogeneous electro-Fenton process with solid catalyst chalcopyrite. *Appl. Catal. B Environ.* **2017**, *209*, 637–647. [CrossRef]
53. Shraddha, J.; Ricardo, N.M.; Paulina, L.S.; Ignacio, R.G.E.; Oracio, S.; Juan, M.P.H. Enhanced photocatalytic activity of $TiO_2$ modified with GaI toward environmental application. *Inorg. Chem.* **2020**, *59*, 1315–1322.
54. Jadhav, S.; Hasegaw, S.; Hisatom, T.; Wang, Z.; Seo, J.; Higashi, T.; Katayama, M.; Minegishi, T.; Takata, T.; Peralta-Hernández, J.M.; et al. Efficient photocatalytic oxygen evolution using $BaTaO_2N$ obtained from nitridation of perovskitetype oxide. *J. Mater. Chem. A* **2020**, *8*, 1127–1130. [CrossRef]
55. Zhang, L.Y.; Zhang, J.J.; Yu, H.G.; Yu, J.G. Emerging S-Scheme Photocatalyst. *Adv. Mater.* **2022**, *34*, 2107668. [CrossRef]
56. Kumar, R.; Raizada, P.; Verma, N.; Bandegharaei, A.H.; Thakur, V.K.; Le, Q.V.; Nguyen, V.H.; Selvasembian, R.; Singh, P. Recent advances on water disinfection using bismuth based modified photocatalysts: Strategies and challenges. *J. Clean. Prod.* **2021**, *297*, 126617. [CrossRef]

57. Xu, C.P.; Anusuyadevi, P.R.; Aymonier, C.; Luque, R.; Marre, S. Nanostructured materials for photocatalysis. *Chem. Soc. Rev.* **2019**, *48*, 3868–3902. [CrossRef] [PubMed]
58. Gadhi, T.A.; Hernández, S.; Castellino, M.; Chiodoni, A.; Husak, T.; Barrerad, G.; Allia, P.; Russo, N.; Tagliaferro, A. Single $BiFeO_3$ and mixed $BiFeO_3/Fe_2O_3/Bi_2Fe_4O_9$ ferromagnetic photocatalysts for solar light driven water oxidation and dye pollutants degradation. *J. Ind. Eng. Chem.* **2018**, *63*, 437–448. [CrossRef]

Article

# Preparation of a Z-Type g-C₃N₄/(A-R)TiO₂ Composite Catalyst and Its Mechanism for Degradation of Gaseous and Liquid Ammonia

Jiaming Zhu [1,2], Zuohua Liu [1], Hao Wang [1,2], Yue Jian [1,2], Dingbiao Long [1,2,*] and Shihua Pu [1,2,*]

1. Chongqing Academy of Animal Sciences, Chongqing 402460, China
2. Scientific Observation and Experiment Station of Livestock Equipment Engineering in Southwest, Ministry of Agriculture, Chongqing 402460, China
* Correspondence: longdb@cqaa.cn (D.L.); push@cqaa.cn (S.P.)

**Abstract:** In this study, an (A-R)TiO₂ catalyst (ART) was prepared via the sol–gel method, and g-C₃N₄ (CN) was used as an amendment to prepare the g-C₃N₄/(A-R)TiO₂ composite catalyst (ARTCN). X-ray diffraction (XRD), scanning electron microscopy (SEM), transmission electron microscopy (TEM), Raman spectroscopy, $N_2$ adsorption–desorption curves (BET), UV–Vis diffuse absorption spectroscopy (UV–Vis DRS), and fluorescence spectroscopy (PL) were used to evaluate the structure, morphology, specific surface area, optical properties, and photocarrier separation ability of the catalysts. The results showed that when the modifier CN content was 0.5 g, the dispersion of the ARTCN composite catalyst was better, with stronger light absorption performance, and the forbidden band width was smaller. Moreover, the photogenerated electrons in the conduction band of ART transferred to the valence band of CN and combined with the holes in the valence band of CN, forming Z-type heterostructures that significantly improved the efficiency of the photogenerated electron-hole migration and separation, thus increasing the reaction rate. Gaseous and liquid ammonia were used as the target pollutants to investigate the activity of the prepared catalysts, and the results showed that the air wetness and initial concentration of ammonia had a great influence on the degradation of gaseous ammonia. When the initial concentration of ammonia was 50 mg/m³ and the flow rate of the moist air was 0.9 mL/min, the degradation rate of gaseous ammonia by ARTCN-0.5 reached 88.86%, and it had good repeatability. When the catalytic dose was 50 mg and the initial concentration of $NH_4^+$ was 100 mg/L, the degradation rate of liquid ammonia by ARTCN-0.5 was 71.60% after 3 h of reaction, and small amounts of $NO_3^-$ and $NO_2^-$ were generated. The superoxide anion radical ($\cdot O_2^-$) and hydroxyl radical ($\cdot OH$) were the main active components in the photocatalytic reaction process.

**Keywords:** photocatalysis; ammonia; degradation; mechanism

## 1. Introduction

Ammonia is a colorless alkaline gas that causes strong irritation, mainly from agricultural fertilization, animal husbandry, and the use of antifreeze [1]. Estimates have shown that China emits 10 to 15 million tons of ammonia into the air every year, almost double the total of the United States and the European Union [2]. Large-scale ammonia emissions will not only cause diseases in animals and humans, such as central muscle paralysis and bronchitis, but also cause global climate change [3]. Ammonia reacts with oxides in the air, producing particulate matter such as ammonium sulfate and ammonium nitrate, which are the main factors that form haze [4]; thus, controlling ammonia emissions is beneficial to controlling haze [5]. In addition, ammonia gas will return to the surface through atmospheric dry and wet deposition, leading to the eutrophication of water and affecting the stability of the ecosystem [6]. Traditional ammonia treatment processes have high costs and poor stability, which may aggravate the secondary pollution of the environment [7].

In 1972, Fujishima discovered that $TiO_2$ could photolyze water under ultraviolet light, and this photocatalytic reaction has been widely used in environmental governance, energy development, biological applications, self-cleaning materials, antibacterial applications, sensors, and other fields due to its thorough reaction and lack of secondary pollution [8]. For example, A. Enesca et al. [9] prepared doped tin oxide films with different dopant concentrations through spray pyrolysis deposition and found that the photodegradation efficiency of the $SnO_2$ film could reach about 30% under the condition of zinc doping. X. Hu et al. [10] introduced and discussed the current challenges and future development prospects of $CO_2$ photoreduction for hydrocarbon fuels. X. Liu et al. [11] prepared CdS@ZIS-SV, and its hydrogen production rate reached 18.06 mmol/g/h, which was 16.9 and 19.6 times that of original CdS (1.16 mmol/g/h) and ZIS (0.92 mmol/g/h) materials, respectively. In ammonia gas degradation, P.A. Kolinko et al. [12] and H.M. Wu et al. [13] pointed out that the N element had various valence states. Herein, its main product was $N_2$, and its by-products were $N_2O$, $NO_2^-$, and $NO_3^-$. Among the many photocatalytic materials, $TiO_2$ has become photocatalyst with the most potential due to its advantages such as good chemical stability, safety, non-toxicity, low cost, and strong REDOX ability [14]. Most studies have indicated that the degradation performance of anatase (A-$TiO_2$) is better than rutile (R-$TiO_2$), and reports on $TiO_2$ have mainly used the anatase phase [15]. Shen et al. [16] stated that when the A-$TiO_2$ and R-$TiO_2$ phases formed a heterogeneous structure, an internal electric field could form, thus promoting the transfer of charges on the interface and improving photocatalytic activity. Das et al. [17] prepared a mixed anatase/rutile crystal structure that had a higher photocatalytic performance than commercial P25 under visible light. Xiong et al. [18] showed that mixed crystalline anatase and rutile $TiO_2$ nanoparticles exhibited a high photocatalytic carbon dioxide reduction capacity. However, its high bandgap width (3.2 eV for anatase and 3.0 eV for rutile) resulted in a response to only high-energy UV light, and its photogenerated charge carriers were easy to recombine, thus limiting its catalytic activity [19]. For example, Guarino et al. [20] sprayed $TiO_2$ on a wall with a total area of 150 $m^2$ at a spray quantity of 70 $g/m^2$ and using a 36 W UV light as the light source, and the degradation rate of ammonia was found to be about 30%.

Photocatalytic materials can be used to form heterogeneous structures. For example, Shihua Pu et al. [21] achieved the degradation of ammonia under sunlight for the first time through the $Cu_2O$ improvement of {001}$TiO_2$, and the degradation rate of ammonia was found to be more than 80% within 2 h. However, due to the photocorrosion of $Cu_2O$ itself, the degradation rate of ammonia gas was only maintained at 40% after four repeated uses, making the choice of amendment very important. $g-C_3N_4$ has shown a narrow band gap, with a wide range of light responses and good thermal stability, chemical stability, and strong corrosion resistance [22]. Most studies have found that the formation of heterostructures through the preparation of $g-C_3N_4/TiO_2$ composite catalysts could broaden the solar spectral response of the catalyst, and photogenerated charge carriers are not easy to recombine [23]. For example, Li et al. [24] constructed a $g-C_3N_4/TiO_2$ heterostructure that achieved effective photoinduced electron-hole separation in the photocatalytic process and showed a good photocatalytic effect and cyclic stability. Zhao S et al. [25] used synthesized $g-C_3N_4/TiO_2$ to degrade phenol with 2.41 and 3.12 times $g-C_3N_4$ and $TiO_2$ contents, respectively. Sun et al. [26] used synthetic $g-C_3N_4/TiO_2$ to degrade methylene blue with 1.85 and 4 times pure $g-C_3N_4$ and $TiO_2$ content, respectively; this showed that it was feasible to improve $TiO_2$ degradation capacity by using $g-C_3N_4$ as an amendment, but a study on ammonia degradation has not been previously reported.

In this study, an (A-R)$TiO_2$ catalyst (ART) was prepared via the sol–gel method, where $g-C_3N_4$ (CN) was used as an amendment to prepare $g-C_3N_4/(A-R)TiO_2$ (ARTCN) with a Z-type heterostructure, which improved the efficiency of photogenerated electron-hole migration and separation. In this study, using gaseous ammonia and ammonia as the target pollutants, we explored the ARTCN degradation of ammonia, the mechanism of performance improvement, the intermediate, and the reaction process. We also studied the repeated use of ARTCN, as evidence on the properties and mechanism noted in this research regarding the effective governance of ammonia pollution has been relatively scarce, and this research could provide a certain theoretical basis for the management of ammonia.

## 2. Results
*2.1. Analysis of the Characterization Results*
### 2.1.1. XRD Analysis

When 2θ = 25.28°, 36.95°, 37.80°, 38.55°, 48.04°, 53.89°, 55.06°, 62.68°, 70.31°, 75.03°, and 76.02°, the characteristic peaks corresponding to the (101), (103), (004), (112), (200), (105), (211), (204), (220), (215), and (301) crystal planes, respectively, were almost consistent with the anatase $TiO_2$ (JCPDS No. 21-1272) standard cards [27]. Furthermore, 2θ = 27.45°, 36.09°, 39.19°, 41.26°, 44.05°, 54.32°, 56.64°, 64.04°, and 69.01° corresponded to the (110), (101), (200), (111), (210), (211), (220), (310), and (301) crystal planes, respectively, which was almost consistent with the rutile $TiO_2$ (JCPDS No. 21-1276) standard card [28]. The characteristic peaks at 2θ = 13.1° and 27.5° belonged to the (100) and (002) planes, respectively which was largely in line with g-$C_3N_4$ (JCPDS87-1526) [29].

Figure 1 shows the XRD patterns of the prepared samples. We found that ART contained not only the characteristic peaks of the anatase phase but also the characteristic peaks of the rutile phase, indicating that it was a mixed crystal type, which was the prepared (A-R)$TiO_2$ catalyst. CN had a weak diffraction peak near 13.1°, corresponding to the (100) crystal plane of CN, which was formed by the 3-S monotriazine structural unit of the plane [30]. There was a strong diffraction peak at 27.50°, which was caused by the layered accumulation of graphite in the conjugate plane. The diffraction peak of the ARTCN-X composite corresponded to pure ART, which indicated that CN did not enter the ART lattice and was only attached to its surface. There was no diffraction peak in the composite samples of ART and CN at 13.1°, because the amount of CN was small and had a weak peak. Although the peaks of CN at 27.50° and R-$TiO_2$ at 27.45° coincided, we observed that as the composite ratio increased from 0.1 to 1, the composite gradually widened around 27°, indicating that the composite material was successfully prepared.

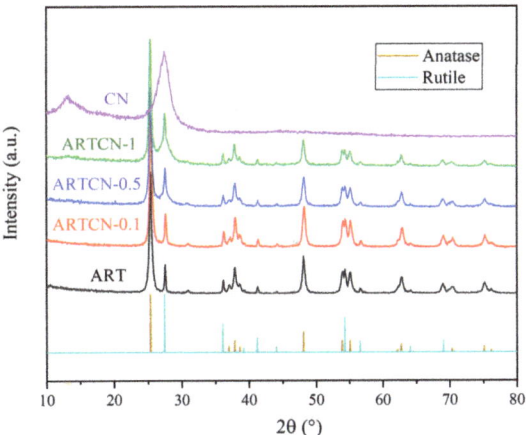

**Figure 1.** XRD patterns of the samples.

The accuracy was further verified by calculating the rutile content of several materials [31], and the calculation results are shown in Table 1. We found that the rutile contents of ART, ARTCN-0.1, ARTCN-0.5, and ARTCN-1 were 23.49%, 25.06%, 29.43%, and 34.61%, respectively. High temperatures were conducive to the conversion of A-$TiO_2$ into R-$TiO_2$, while ART, ARTCN-0.1, ARTCN-0.5, and ARTCN-1 were prepared at the same temperature; thus, the increased rutile content was not rutile in the real sense but rather the added amendment of CN.

**Table 1.** Rutile content of the catalysts.

| Samples | $I_R$ (2θ = 27.24°) | $I_A$ (2θ = 25.28°) | $X_R$ |
|---|---|---|---|
| ART | 523.48 | 2131.93 | 23.49% |
| ARTCN-0.1 | 605.19 | 2260.33 | 25.06% |
| ARTCN-0.5 | 704.02 | 2110.14 | 29.43% |
| ARTCN-1 | 953.81 | 2251.77 | 34.61% |

Table 1 shows the rutile content of each catalyst. The formula (Equation (1)) is given as follows:

$$X_R = \frac{1}{1 + 0.8 \times \frac{I_A}{I_R}} \quad (1)$$

where $X_R$ is the rutile content, $I_A$ is the peak intensity of anatase at 2θ = 25.28°, and $I_R$ is the peak intensity of rutile at 2θ = 27.24°.

2.1.2. Raman Analysis

The crystal form and structure of the catalyst were further determined by Raman spectroscopy, and Figure 2 shows the Raman spectra of the prepared catalyst. The anatase phase of $TiO_2$ corresponded to ν = 144 cm$^{-1}$, ν = 197 cm$^{-1}$, ν = 392 cm$^{-1}$, ν = 514 cm$^{-1}$, and ν = 635 cm$^{-1}$ [32], while ν = 437 cm$^{-1}$ corresponded to rutile $TiO_2$. The reason why other characteristic peaks did not appear was that the content of rutile in ART was low (as indicated by the peak intensity of XRD). ART and composite ARTCN-X both had anatase and rutile peaks, especially the CN peak of ARTCN-1, which further validated the XRD results.

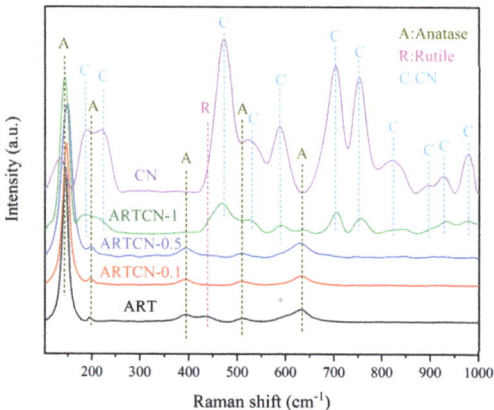

**Figure 2.** Raman spectra of the samples.

2.1.3. Morphology and Lattice Spacing Analysis

According to Figure 3a,h, CN was a curved and folded film with certain holes. As shown in Figure 3b,i, ART consisted of a particle with a uniform size but serious agglomeration, and Figure 3c shows that CN in ARTCN-0.5 was no longer a whole film but fragmented into many small pieces; thus, its position relative to ART could not be clearly observed. Combined with Figure 3d–g, we clearly observed that elements O, Ti, and N were evenly distributed in ARTCN-0.5. This showed that a composite material with a heterogeneous structure was synthesized. In addition, as clearly shown in Figure 3j, the ART agglomeration phenomenon was significantly improved after the improvement of

CN, which indicated that the utilization of light by ART in the composite ARTCN-0.5 was further enhanced.

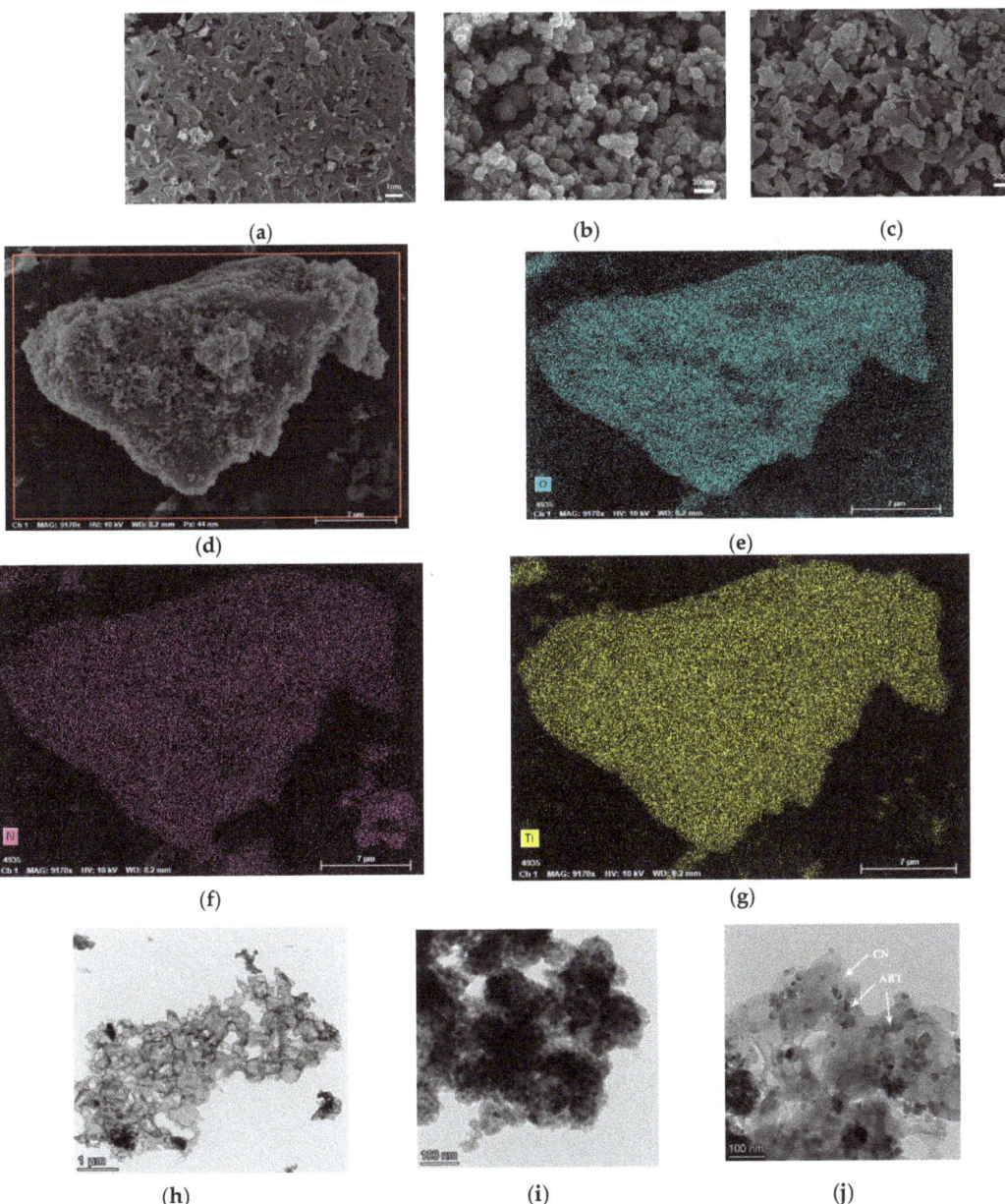

**Figure 3.** SEM images of CN (**a**), ART (**b**), and ARTCN-0.5 (**c**); images and corresponding EDS elemental mapping images (**d–g**) of the ARTCN-0.5, TEM of CN (**h**), ART (**i**), and ARTCN-0.5 (**j**).

2.1.4. Analysis of Adsorption–Desorption of $N_2$

The Brunauer–Emmett–Teller (BET) method was used to calculate the surface area of the catalyst, and the Barrett–Joyner–Halenda (BJH) method was used to analyze the pore size and pore volume. The catalysts shown in Figure 4a all exhibited typical nitrogen adsorption–desorption isotherms of type IV, and the hysteresis loops of the catalysts all showed obvious openings, indicating the formation of mesoporous catalysts [33]. Figure 4b shows the pore size distribution of the catalyst, which showed that the pore size of the prepared catalyst was mainly concentrated at 0–20 nm, indicating that the particle size distribution of the catalyst was narrow. Combined with the data in Table 2, we found that the pores of the catalysts showed little difference, while the specific surface area and pore diameter of the catalysts increased with increasing CN content.

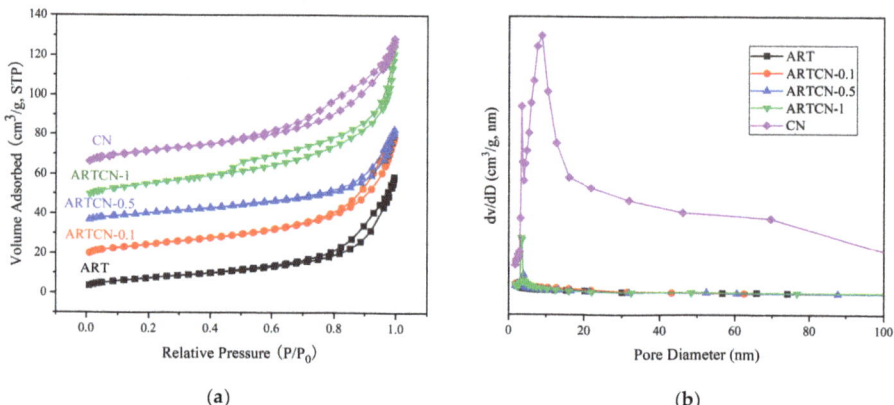

**Figure 4.** $N_2$ adsorption–desorption curve (**a**) and pore size distribution (**b**) of the samples.

**Table 2.** Structural parameters of the samples.

| Samples | Specific Surface Area (m$^2$/g) | Pore Volume (cm$^3$/g) | Pore Diameter (nm) |
|---|---|---|---|
| ART | 27.28 | 0.08 | 11.28 |
| ARTCN-0.1 | 33.92 | 0.09 | 10.13 |
| ARTCN-0.5 | 36.16 | 0.08 | 9.65 |
| ARTCN-1 | 38.08 | 0.12 | 9.29 |
| CN | 42.09 | 0.11 | 8.67 |

2.1.5. Optical Performance Analysis

The optical absorption properties of the prepared samples were investigated via the UV–Vis absorption spectra, and the results are shown in Figure 5a. The optical absorption intensity of the ARTCN-X composite material improved by CN widened in the visible light range because the specific ART surface area improved as CN increased, and the agglomeration phenomenon also significantly improved. These results indicated that CN could effectively expand the optical absorption range of ART, thus improving the response and utilization efficiency of visible light. In addition, the band gap width of the prepared material was calculated with the Kubelka-Munk method [34], as shown in Figure 5b. The bandgap widths of ART, ARTCN-0.1, ARTCN-0.5, ARTCN-1, and CN were 3.01, 2.92, 2.86, 2.85, and 2.84 eV, respectively. First, we clearly observed that the band gap width of ART was smaller than the 3.12 eV value reported in the literature, because heterostructures would form between the anatase phase $TiO_2$ and rutile phase $TiO_2$, enhancing the ability of photogenerated carrier separation [35]. Secondly, we found that with the increase in the CN content of the amendment, the band gap width of the composite ARTCN-X became significantly smaller than that of ART and approached that of CN, which was consistent with other studies [36].

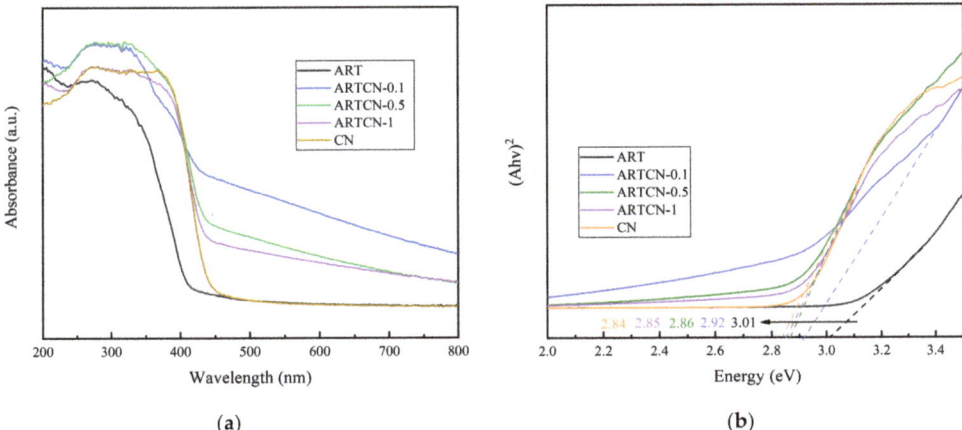

**Figure 5.** Optical properties (**a**) and the band gap width (**b**) of the samples.

2.1.6. PL and EPR Analysis

A PL spectrum can be used to study the separation of the photogenerated carriers in semiconductors, where the lower the peak intensity, the lower the recombination rate of the photogenerated carriers and the higher the photocatalytic activity [37]. As shown in Figure 6a, ARTCN-0.5 had the lowest peak intensity, indicating that its photocatalytic activity was the highest, which further indicated that as an amendment, the amount of CN addition did not follow the logic of "the more content the better" but had an appropriate ratio with ART. In addition, we observed that the wide wavelength range of ARTCN-X after CN modification in the interval of 451.8–468.8 nm was attributed to the oxygen vacancy that contained two captured electrons, which both promoted the formation of superoxide radicals ($\cdot O_2^-$) and hydroxyl radicals ($\cdot OH$) and was favorable for photocatalytic degradation.

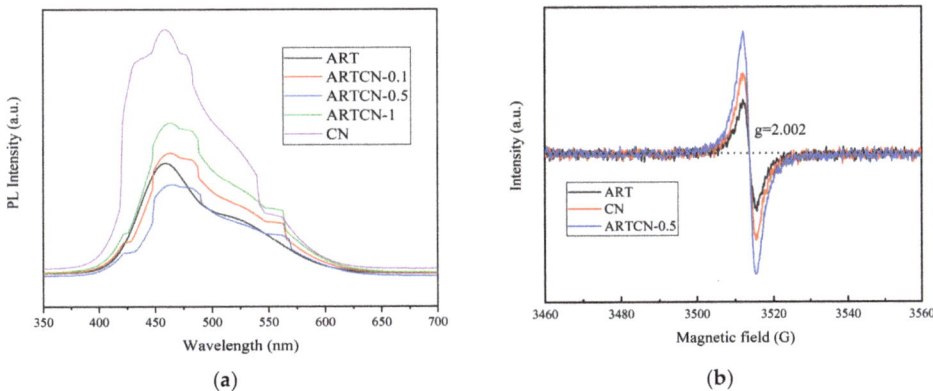

**Figure 6.** PL (**a**) and (**b**) EPR of the samples.

Additional EPR spectra that were collected at room temperature provided information regarding the oxygen vacancies (Ov). As shown in Figure 6b, the signal at g = 2.002 corresponded to Ov [38]. We found that all samples had oxygen vacancy signals, and the intensity of ARTCN-0.5 was stronger than that of single ART and CN, which indicated that there was a large amount of Ov that could effectively inhibit the recombination of electrons and holes and improve the photocatalytic activity, thus confirming the results shown in Figure 6a.

### 2.2. Photocatalytic Performance Test Results

#### 2.2.1. Study on the Photocatalytic Degradation of Gaseous Ammonia

The initial concentration of ammonia was 50 mg/m$^3$, and the flow rate of the moist air was 0.9 mL/min. Different catalysts were used to study their influence on the degradation of gaseous ammonia, and the results are shown in Figure 7a. This indicated that PET itself was not good at removing gaseous ammonia, and the average degradation rates of gaseous ammonia by ART, ARTCN-0.1, ARTCN-0.5, ARTCN-1, and CN were 52.35%, 61.43%, 88.86%, 63.90%, and 33.51%, respectively. These results indicated that the degradation performance of gaseous ammonia by the ARTCN-X composite catalyst modified by CN was improved, and the effect was most obvious when the amount of CN was 0.5 g because ARTCN-0.5 had the strongest photogenerated carrier separation ability.

The initial concentration of ammonia was 50 mg/m$^3$, and the flow rate of moist air was adjusted. ARTCN-0.5 was selected to study its influence on the degradation of gaseous ammonia, and the results are shown in Figure 7b. At 1.2 mL/min, the degradation rates of gaseous ammonia by ARTCN-0.5 were 57.93%, 63.98%, 78.26%, 83.38%, 88.86%, 87.10%, and 67.04%. The degradation rates of gaseous ammonia by ARTCN-0.5 first increased and then decreased with the flow rate of moist air. This was because moist air was conducive to the deposition of gaseous ammonia. However, ARTCN-0.5 could oxidize water molecules into hydroxyl radicals (·OH) with strong oxidation in the photocatalytic reaction process, thus improving the efficiency of the photocatalytic reaction. However, ammonia molecules could not tightly bind to the catalyst and flow out of the system before being reacted, resulting in a decrease in its degradation rate.

When the flow rate of moist air was 0.9 mL/min, ARTCN-0.5 was selected to study its influence on the degradation of gaseous ammonia, and the results are shown in Figure 7c. When the concentrations of gaseous ammonia were 30 mg/m$^3$, 50 mg/m$^3$, and 70 mg/m$^3$, the average degradation rates of gaseous ammonia by ARTCN-0.5 were 80.38%, 88.86%, and 73.15%, respectively. The degradation rate decreased if the concentration of gaseous ammonia was too low or too high. This was because ARTCN-0.5 released a limited number of active free radicals after it degraded ammonia when the concentration of gaseous ammonia was too large. Many gaseous ammonia molecules went along with the airflow, and when the concentration of gaseous ammonia was too small, the active radicals could not completely combine with the gaseous ammonia.

The initial concentration of ammonia was 50 mg/m$^3$, and the flow rate of moist air was 0.9 mL/min. After the degradation performance of ARTCN-0.5 was tested, PET loaded with ARTCN-0.5 was removed and kept in an oven at 70 °C for 2 h. Then, the test was repeated to explore the reuse performance of ARTCN-0.5; the results are shown in Figure 7d. The average degradation rates for the five reuse cycles were 88.86%, 85.00%, 80.40%, 76.48%, and 70.05%. Given the inevitable catalyst loss in the process of repeated testing, ARTCN-0.5 demonstrated a good overall repeatable degradation performance.

Figure 7e shows the XRD patterns of ARTCN-0.5 before and after repeated use (five times). We found that the crystal shape and structure of ARTCN-0.5 did not significantly change after five repeated uses, indicating that its properties were stable.

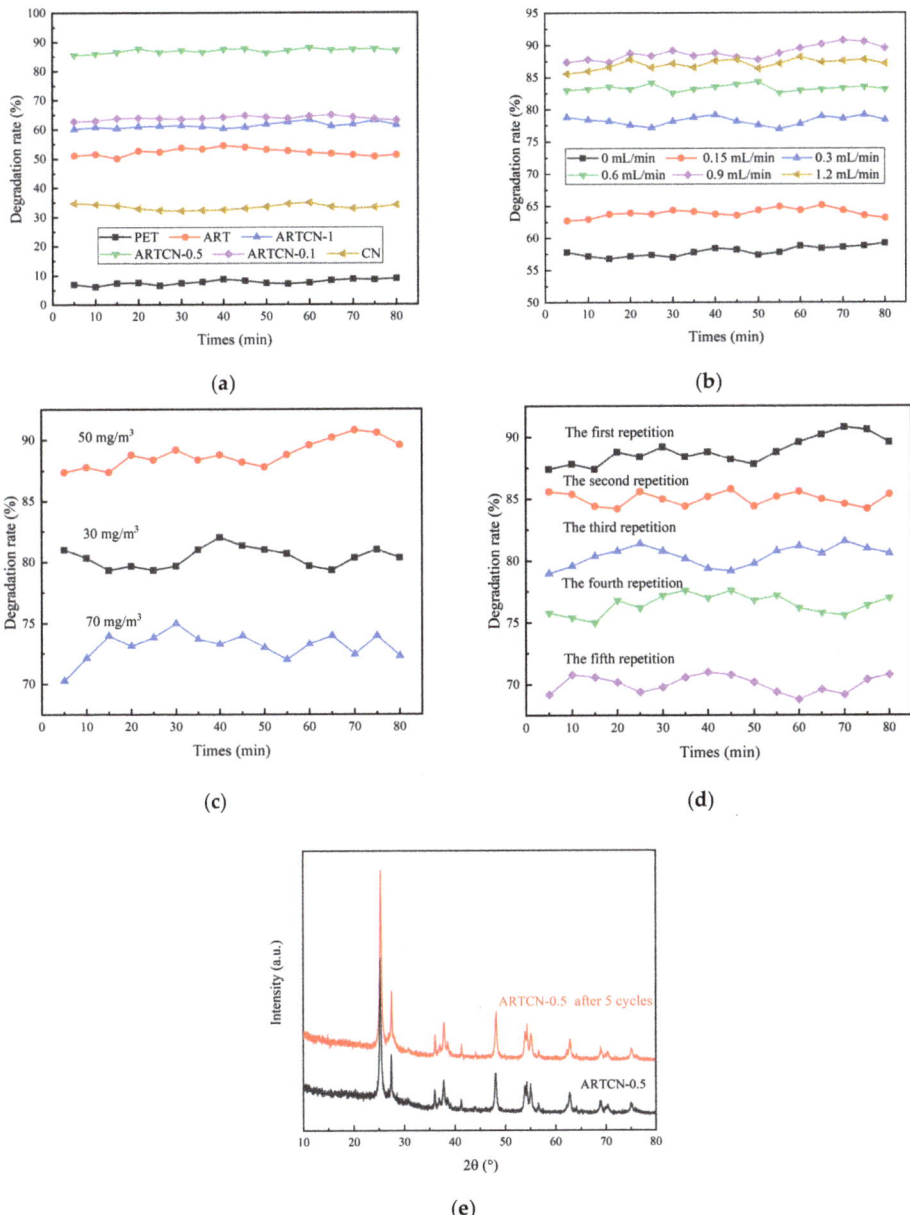

**Figure 7.** Influence of different catalysts on the degradation of gaseous ammonia (**a**), the effect of different humid air flow rates on the degradation of gaseous ammonia (**b**), the effects of different ammonia concentrations on the degradation of gaseous ammonia (**c**), the repeated degradation performance of ARTCN-0.5 (**d**), and XRD patterns before and after the repeated use (5 times) of ARTCN-0.5 (**e**).

2.2.2. Study on the Photocatalytic Degradation of Liquid Ammonia

When pH > 10, $NH_4^+$ hydrolyzes in an alkaline solution and then exists in the form of $NH_3·H_2O$ [39]. Therefore, we chose to adjust the pH of ammonia nitrogen to 10.1, and then we tested the concentration of ammonia nitrogen in the solution. We found that it decreased by 3.82% because a small amount of $NH_3 \cdot H_2O$ escaped in the gaseous form. The degradation results of ammonia nitrogen by different catalysts are shown in Figure 8a, where the catalytic dose was 50 mg and the initial concentration of $NH_4^+$ was 100 mg/L. The degradation rates of liquid ammonia by ART, ARTCN-0.1, ARTCN-0.5, ARTCN-1, and CN after 3 h of reaction were 50.54%, 63.18%, 71.60%, 53.55%, and 37.91%, respectively, indicating that the degradation performance of the gaseous ammonia by the ARTCN-X composite catalyst after CN improvement was improved.

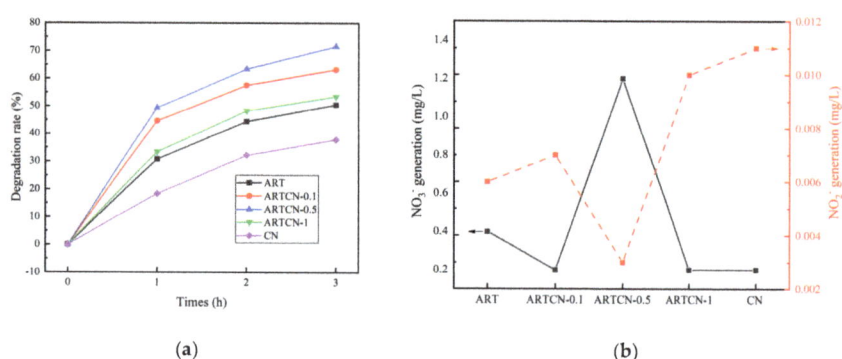

**Figure 8.** Degradation of the liquid ammonia by samples (**a**); $NO_3^-$ and $NO_2^-$ production after the reaction (**b**).

When the initial concentration of ammonia nitrogen was 100 mg/L, 50 mg of ARTCN-0.5 catalyst was added, the pH was 10.1, and the reaction was performed after 3 h. The results are shown in Figure 8b, which shows that the $NO_3^-$ and $NO_2^-$ concentrations were not greater than 1.4 and 0.012 mg/L, respectively. According to the literature reports, there are three main reaction products of ammonia nitrogen, nitrate, nitrite, and nitrogen, and ammonia nitrogen is mainly oxidized to $N_2$ [40].

## 3. Discussion

### 3.1. Charge Transfer Mechanism Discussion

To understand the reason for the observed improved photocatalytic performance, we proposed a charge separation and transfer mechanism. The conduction band position of a semiconductor could be calculated by the empirical formulas shown in Equations (2) and (3) [41]:

$$Ec = \chi - Ee - Eg/2 \quad (2)$$

$$Ev = Ec + Eg \quad (3)$$

where $\chi$ is the geometric mean of the absolute electronegativity of each atom in the semiconductor, with $TiO_2$ [42] and g-$C_3N_4$ [43] $\chi$ values of 5.81 eV and 4.82 eV, respectively; Ee is a constant relative to the standard hydrogen electrode of about 4.5 eV [44]; Eg is the semiconductor band gap width; Ev is the semiconductor valence band energy; and Ec is the semiconductor conduction band energy.

According to the analysis presented in Figure 5b, the Eg values of $TiO_2$ and g-$C_3N_4$ were 3.01 and 2.84 eV, respectively. We calculated that the Ev and Ec values of ART were 2.815 and −0.195 eV, respectively, the Ev and Ec of g-$C_3N_4$ had a value of 1.59, and Ec was −1.25 V(vs NHE). CN was more negative than ART, and the Ev of ART was higher than that of CN. The high photocatalytic activity of the heterojunctions between ARTCN could be explained by the following mechanism.

Two possibilities existed for the photocatalytic mechanism of the composite catalyst: (1) conventional type II heterojunctions and (2) Z-type heterojunctions [45]. Under simulated solar irradiation, ART and CN were excited and generated electron-hole pairs. If the type II heterojunction mechanism was followed, the photogenerated electrons in the conduction band of CN would be transferred to the conduction band of ART and the photogenerated holes in the valence band of ART would be transferred to the valence band of CN [46], as shown in Figure 9a. However, the hole in the valence band of CN could not generate ·OH by reacting with $H_2O$. This was because the valence band potential of CN (1.59 eV) was lower than the standard oxidation potential E ($H_2O$/·OH) (2.38 eV) [47], resulting in a decrease in ·OH content, which was inconsistent with the results shown in Figure 9b. If the charge transfer mechanism of the Z-type heterostructure was followed, the photogenerated electrons in the ART conduction band would transfer to the valence band of CN and combine with the photogenerated holes in the CN valence band. This would result in a reduction in electrons in ARTCN and the accumulation of electrons in the CN conduction band and holes in the ART valence band, which was why the content of ·$O_2^-$ produced by CN shown in Figure 10a was higher than that of ARTCN. The holes that accumulated in the valence band of ART had strong oxidability and could directly degrade ammonia molecules. However, the Ec (−1.10 V) of CN was more negative than $E_O$ ($O_2$/·$O_2^-$), and the electrons that accumulated in the conduction band of CN could react with $O_2$ to generate ·$O_2^-$. This z-type heterostructure was more consistent with the characterization results of ESR (Figure 10).

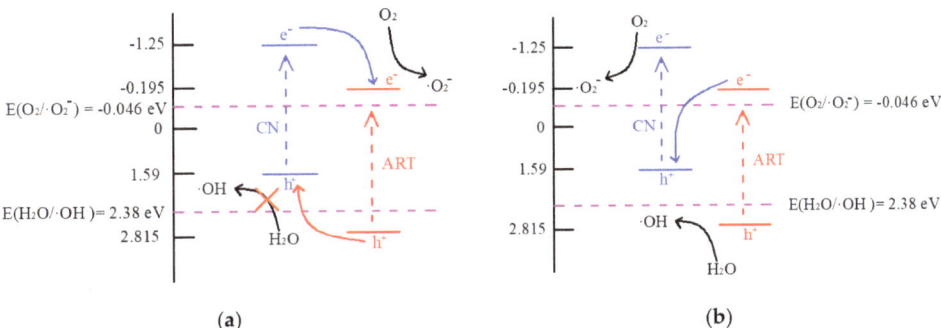

**Figure 9.** Charge transfer mechanism of the photocatalytic degradation conventional type II heterojunction (**a**) and Z-type heterostructure (**b**).

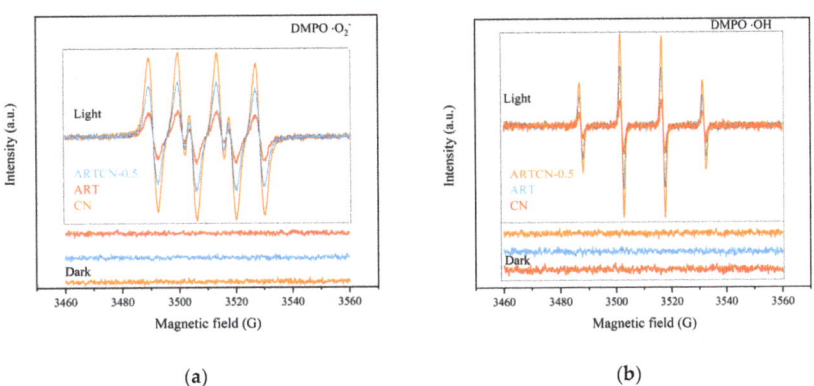

**Figure 10.** ESR profiles of DMPO ·$O_2^-$ (**a**) and DMPO ·OH (**b**) of the samples.

To further verify the accuracy of the above results, the test results of EIS are presented in Figure 11, which shows the impedance diagrams of ART, CN, and ARTCN. The electron transfer resistance in the sample was equivalent to the semicircle diameter on the EIS diagram, where the smaller the arc radius, the lower the charge transfer resistance of the composite sample [48]. Hence, due to the formation of the Z-type heterostructure, the charge-transfer resistance of ARTCN-0.5 was lower than ART and CN, which significantly improved the efficiency of photogenerated electron-hole migration and separation, which was consistent with the work of Y.Y. Wang [49].

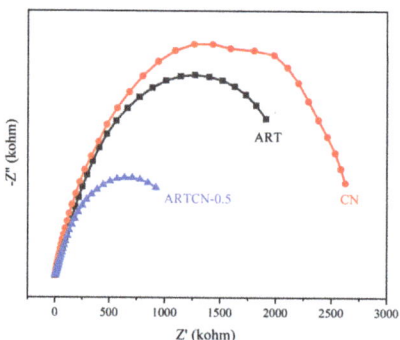

**Figure 11.** EIS spectra of the samples.

*3.2. Photocatalytic Degradation Mechanism Discussion*

ESR was used to detect the types of free radicals in the catalysts under light, and then we explored their degradation mechanism of ammonia. As shown in Figure 10a,b, ART, CN, and ARTCN-0.5 did not produce free radicals in the absence of light, and superoxide free radicals ($\cdot O_2^-$) and hydroxyl free radicals ($\cdot OH$) were detected after light exposure. $\cdot O_2^-$ and $\cdot OH$ played decisive roles in the entire reaction [50]. The degradation mechanism was as follows. After gaseous ammonia combined with moist air, $NH_4^+$ was present as $NH_3 \cdot H_2O$ under alkaline conditions. First, when the catalyst was illuminated, electrons and holes were generated [51]. The holes oxidized the water molecules on the surface of the catalyst, forming $\cdot OH$, while the electrons and dissolved oxygen underwent a series of reactions to form $\cdot O_2^-$, following Equations (4)–(6) [52]. $\cdot OH$ and $\cdot O_2^-$ were the main active components in the photocatalytic reaction process, and they could rapidly oxidize the $NH_4^+/NH_3/NH_3 \cdot H_2O$ adsorbed on the surface of the catalyst [53]. Equations (9)–(13) show that ammonia was directly and completely oxidized to $N_2$ [54], and the $NO_3^-$ and $NO_2^-$ contents were very low after the reaction. Equations (12)–(16) show that ammonia was not completely oxidized to $NO_3^-$ and $NO_2^-$, which was consistent with the research of Sun et al. [55].

$$g\text{-}C_3N_4/(A\text{-}R)TiO_2 + h\nu \rightarrow g\text{-}C_3N_4/(A\text{-}R)TiO_2\ (h^+ + e^-) \tag{4}$$

$$h^+ + H_2O \rightarrow H^+ + \cdot OH \tag{5}$$

$$e^- + O_2 \rightarrow \cdot O_2^- \tag{6}$$

$$NH_3 + \cdot OH \rightarrow NH_2 + 2H_2O \tag{7}$$

$$NH_2 + \cdot OH \rightarrow NH + H_2O \tag{8}$$

$$NH + \cdot OH \rightarrow N + H_2O \tag{9}$$

$$NH_x + NH_y \rightarrow N_2H_{x+y}\ (x, y = 0, 1, 2) \tag{10}$$

$$N_2H_{x+y} + (x+y)OH \rightarrow N_2 + (x+y) H_2O \tag{11}$$

$$NH_3 + \cdot OH\ (h^+) \rightarrow NH_2OH + H^+ \tag{12}$$

$$NH_2OH + \cdot O_2^- \rightarrow \cdot O_2NHOH \tag{13}$$

$$O_2NHOH + OH \rightarrow NO_2^- + H_2O + \cdot OH \quad (14)$$

$$NO_2^- + OH \rightarrow HONO_2 \quad (15)$$

$$HONO_2 \rightarrow NO_3^- + H^+ \quad (16)$$

## 4. Materials and Methods

### 4.1. Materials

Ammonium chloride (AR, Chengdu Colon Chemicals Co., Ltd. Chengdu, Sichuan province, China), butyl titanate (AR, Chengdu Colon Chemicals Co., Ltd. Chengdu, Sichuan province, China), absolute ethanol (AR, Chongqing Chuandong Chemical Co., Ltd. Nanan district, Chongqing, China), urea (AR, Sinopharm Chemical Reagent Co., Ltd. Huangpu district, Shanghai, China), and NaOH (AR, Chongqing Chuandong Chemical Co., Ltd. Nanan district, Chongqing, China) were used in this study.

### 4.2. Preparation of Catalysts

We added 30 mL of absolute ethanol to 35 mL of butyl titanate, which was denoted as solution A. Then, we added 30 mL of absolute ethanol to 100 mL of distilled water, which was denoted as solution B. Solution A was dropwise added to solution B at 4 drops per second, and then it was mixed and stirred at a low speed for 2 h to obtain the $TiO_2$ gel, which was aged at room temperature. The aged $TiO_2$ gel was transferred to a stainless steel reaction kettle with a polytetrafluoron liner and maintained at 100 °C for 2 h. After cooling, it was centrifuged and settled, washed with deionized water and ethanol 3 times, and dried in a 100 °C air-drying oven. Then, the $TiO_2$ catalyst powder was obtained after grinding. The $TiO_2$ powder was maintained at 600 °C for 2 h at a heating rate of 10 °C/min, and then it was cooled to room temperature and ground to obtain ART.

Subsequently, 10 g of urea and 50 mL of water were added to the crucible and evenly stirred. The crucible was placed in a muffle furnace and maintained at 500 °C for 2 h at a heating rate of 10 °C/min. After the heating program was finished and the muffle furnace naturally cooled, the obtained bulk particles were ground to obtain CN.

We mixed 0.1 g of CN, 0.5 g of CN, 1 g of CN, and 1 g of ART; added 5 mL of absolute ethanol; stirred evenly; and then separated and dispersed the mixture using a cell fragmentation apparatus (Ningbo, Zhejiang Province, China. Xinzhi Biotechnology Co., Ltd., SCIENTZ-IID, 65 Hibiscus Road, Ningbo National High-tech Zone). Maintaining the temperature at 300 °C for 2 h at a heating rate of 10 °C/min, ARTCN-X was obtained and is denoted as ARTCN-0.1, ARTCN-0.5, and ARTCN-1.

### 4.3. Fixation of Photocatalytic Materials

We washed the polyester fiber cotton (PET) with 1 mol/L of NaOH to remove the surface impurities, and then we dried and set it aside. Subsequently, 100 mg of catalyst was dissolved in water, the treated PET was added and shaken in a shaker for 30 min, and the water on the surface and the excess catalyst were drained before the catalyst was dried at 70 °C for later use. The PET scanning electron microscope results before and after catalyst loading are shown in Figure 12. We clearly observed that the photocatalytic materials werare evenly loaded on PET.

 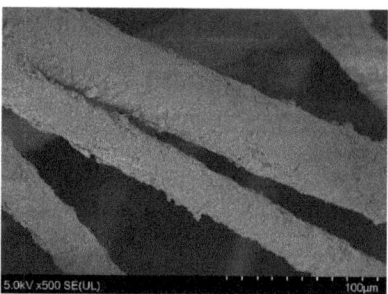

**Figure 12.** SEM of the PET scan before and after catalyst support.

### 4.4. Catalyst Characterization

A D8 Advance model X-ray diffractometer (XRD, Bruker, Germany) was used to analyze the crystal characteristics of the catalyst. The operating parameters were a Cu X-ray tube target and a scanning range of 10–80°. An HR800 laser confocal Raman spectrometer (Raman, Horiba Jobin Yvon, France) was used to detect the sample structures, where the excitation wavelength was 633 nm. A TriStar II 3020 series automatic specific surface analyzer (BET, GA, USA) was used to determine the specific surface area and porosity of the catalyst. The catalyst was pretreated under vacuum degassing at 200 °C for 5 h, and high-purity nitrogen was used as the adsorbent at 77 K. Then, a UV–Vis diffuse reflectance (UV–Vis DRS, Hitachi, Japan) instrument (model U-3010) was used to test the optical properties of the catalyst in the range of 200–800 nm. An F-2700 fluorescence spectrophotometer (PL, Japan, Hitachi) was used to measure the electron hole recombination, with an operating voltage of 250 V, a wavelength of 5 nm, and an excitation wavelength of 300 nm. The surface morphology and elemental distribution of the samples were analyzed by scanning electron microscopy (SEM, Sigma500, Germany) and an energy dispersive spectrometer (EDS, Bruker, Germany). Transmission electron microscopy (TEM, FEI TalOS F200S, USA) was used to analyze the lattice spacings of the samples. Fluorescence spectrophotometry (PL, HORIBA, Osaka, Japan) was used to test the photogenerated carrier separation of the catalyst with a hydrogen light source with a pulse width of 1.0 to 1.6 ns, which was used to test the fluorescence lifetime of the sample. Electron paramagnetic resonance (EPR, Bruke, Germany) was performed at room temperature using an A300 spectrometer, and 5,5-dimethyl-1-pyrrolidine N-oxide (DMPO) was used as the spin capture agent for ESR analysis. A CHI-660C electrochemical workstation was used for transient photocurrent measurements. A $Na_2SO_4$ aqueous solution (1 M) was used as the electrolyte solution, and electrochemical impedance spectroscopy (EIS, Chenhua, China) measurements were performed under visible light irradiation with frequencies ranging from $4 \times 10^6$ to $1 \times 10^{-2}$ Hz.

### 4.5. Photocatalytic Activity Tests

#### 4.5.1. Degradation of Gaseous Ammonia

Figure 13 shows the device diagram for the photocatalytic degradation of gaseous ammonia. The light source consisted of a 300 W xenon lamp (30.2 mW/cm$^2$), which was installed in a quartz water pipe with a condensation cycle to absorb the heat generated by illumination. Just below the xenon lamp was a photocatalytic quartz reaction tube, and PET loaded with the photocatalytic materials was placed in the tube. After exiting the cylinder, ammonia entered the reaction tube through a flowmeter. The different concentrations of standard ammonia (mixed with nitrogen) were 30 mg/m$^3$, 50 mg/m$^3$, and 70 mg/m$^3$, with a flow rate of 100 mL/min, and the stability test results of the different concentrations of ammonia are shown in Figure 14. Air entered the photocatalytic reaction tube through the water, needle valve, and flowmeter in turn (moist air was the only source of oxygen). Ammonia was mixed with air at a certain humidity to simulate gaseous ammonia. After connecting each pipeline, the standard gas was ventilated to determine whether there was air leakage. The system was run for 10 min until it was stable, and the degradation efficiency of the gaseous ammonia in the entire process was $\eta_1$ (Equation (17)):

$$\eta_1 = (C_{01} - C_1) \times 100\%/C_{01} \qquad (17)$$

where $C_{01}$ is the standard concentration of gaseous ammonia (mg/m$^3$) and $C_1$ is the concentration of gaseous ammonia in the reaction process (mg/m$^3$).

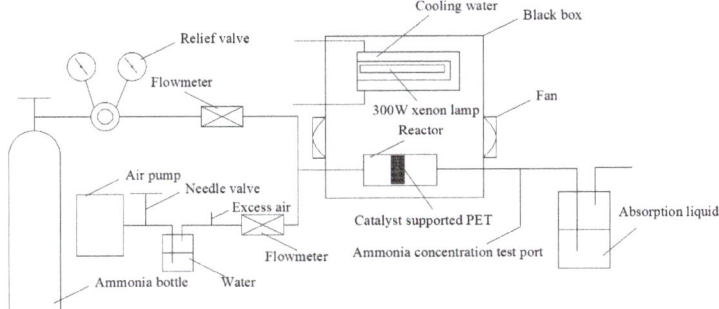

**Figure 13.** Device diagram of the photocatalytic degradation of gaseous ammonia.

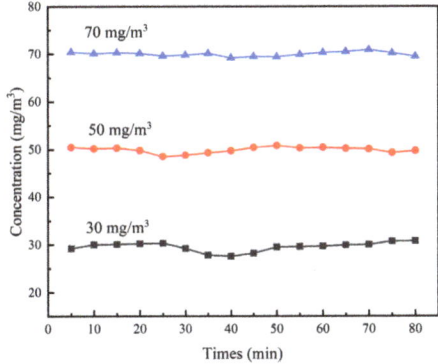

**Figure 14.** System stability test.

4.5.2. Degradation of Liquid Ammonia

The photoreactor consisted of a 200 mL double-layer quartz beaker, for which the outer layer was permeated with cooling water to ensure a constant reaction temperature and the inner layer consisted of 100 mg/L of an ammonia nitrogen solution. The pH was adjusted to 10.1 by sodium hydroxide. The 300 W xenon lamp (30.2 mW/m$^2$) was used as the light source to simulate sunlight while ensuring that the distance between the xenon lamp and the liquid level was 15 cm. Then, 50 mg (0.5 g/L) of catalyst was added to the solution and stirred at medium speed at room temperature; 2 mL of the sample from the reaction solution was extracted every 60 min and centrifuged for 10 min with a 10,000 r/min high-speed centrifuge, and the supernatant was obtained to determine the concentration of ammonia nitrogen and calculate its degradation rate $\eta_2$ (Equation (18)). The nitrate and nitrite concentrations were determined after the reaction:

$$\eta_2 = (C_{02} - C_2) \times 100\%/C_{02} \qquad (18)$$

where $C_{02}$ is the initial concentration of ammonia nitrogen and $C_2$ is the concentration of ammonia nitrogen in the reaction process (mg/L).

## 5. Conclusions

A g-$C_3N_4$/(A-R)$TiO_2$ composite catalyst (ARTCN) was prepared by using g-$C_3N_4$ (CN) as the amendment, and the amount of CN had a great influence on the performance of ARTCN. When the amount of CN was 0.5 g, ARTCN-0.5 had a better dispersion, a smaller band gap width, a larger specific surface area, a stronger light absorption capacity, and a stronger photogenerated carrier separation ability than ART.

The air wetness and initial concentration of the ammonia had a great influence on the degradation of the gaseous ammonia. When the initial concentration of ammonia was 50 mg/m$^3$ and the flow rate of the moist air was 0.9 mL/min, the degradation rate of gaseous ammonia by ARTCN-0.5 reached 88.86%, and it had good repeatability. When the catalytic dose was 50 mg and the initial concentration of $NH_4^+$ was 100 mg/L, the degradation rate of the liquid ammonia by ARTCN-0.5 was 71.60% at 3 h, and small amounts of $NO_3^-$ and $NO_2^-$ were generated. Subsequently, ·OH and ·$O_2^-$ were the main active components in the photocatalytic reaction process.

The photogenerated electrons in the conduction band of ART transferred to the valence band of CN and combined with the photogenerated holes in the valence band of CN, forming a Z-type heterostructure that significantly improved the efficiency of the photogenerated electron-hole migration and separation, thus increasing the reaction rate.

**Author Contributions:** Conceptualization: J.Z. and S.P.; data curation: Y.J.; investigation and methodology: H.W.; project administration: Z.L.; resources, J.Z. and Z.L.; supervision, J.Z. and D.L.; writing—original draft: J.Z. and S.P.; writing—review and editing: D.L. and J.Z. All authors have read and agreed to the published version of the manuscript.

**Funding:** Financial support was provided by the Chongqing scientific research institution performance Incentive and guidance special project (cstc2020jxjl20006), National Center of Technology Innovation for Pigs, Chongqing 402460, China. The Modern agroindustry technology research system (CARS-35, 21203).

**Institutional Review Board Statement:** Not applicable.

**Informed Consent Statement:** Not applicable.

**Data Availability Statement:** Not applicable.

**Acknowledgments:** We thank the editors of MDPI for their patience during the processing of this manuscript.

**Conflicts of Interest:** There are no conflict of interest in the manuscript submission, and it has been approved by all authors for publication. All authors listed have approved the enclosed manuscript.

## References

1. Clarisse, L.; Clerbaux, C.; Dentener, F.J.; Hurtmans, D.; Coheur, P.F. Global ammonia distribution derived from infrared satellite observations. *Nat. Geosci.* **2009**, *2*, 479. [CrossRef]
2. Van Damme, M.; Clarisse, L.; Whitburn, S.; Hadji-Lazaro, J.; Hurtmans, D.; Clerbaux, C.; Coheur, P.F. Industrial and agricultural ammonia point sources exposed. *Nature* **2018**, *564*, 99. [CrossRef] [PubMed]
3. Schauberger, G.; Piringer, M.; Mikovits, C.; Zollitsch, W.; Hörtenhuber, S.; Baumgartner, J.; Niebuhr, K.; Anders, I.; Andre, K.; Hennig, P.I.; et al. Impact of global warming on the odour and ammonia emissions of livestock buildings used for fattening pigs. *Biosyst. Eng.* **2018**, *175*, 106. [CrossRef]
4. Derwent, R.; Witham, C.; Redington, A.; Jenkin, M.; Stedman, J.; Yardley, R.; Hayman, G. Particulate matter at a rural location in southern England during 2006: Model sensitivities to precursor emissions. *Atmos. Environ.* **2013**, *69*, 211. [CrossRef]
5. Bessagnet, B.; Beauchamp, M.; Guerreiro, C.D.B.B.; de Leeuw, F.; Tsyro, S.; Colette, A.; Meleux, F.; Rouïl, L.; Ruyssenaars, P.; Sauter, F.; et al. Can further mitigation of ammonia emissions reduce exceedances of particulate matter air quality standards? *Environ. Sci. Policy* **2014**, *44*, 149. [CrossRef]
6. Pearson, J.; Stewart, G.R. The deposition of atmospheric ammonia and its effects on plants. *New Phytol.* **1993**, *125*, 283. [CrossRef]
7. Pu, S.; Long, D.; Liu, Z.; Yang, F.; Zhu, J. Preparation of RGO-P25 Nanocomposites for the Photocatalytic Degradation of Ammonia in Livestock Farms. *Catalysts* **2018**, *8*, 189. [CrossRef]

8. Zhu, J.; Jian, Y.; Long, D.; Wang, H.; Zeng, Y.; Li, J.; Xiao, R.; Pu, S. Degradation of ammonia gas by $Cu_2O/\{001\}TiO_2$ and its mechanistic analysis. *RSC Adv.* **2021**, *11*, 3695. [CrossRef]
9. Enesca, A.; Andronic, L.; Duta, A. Optimization of Opto-Electrical and Photocatalytic Properties of $SnO_2$ Thin Films Using $Zn^{2+}$ and $W^{6+}$ Dopant Ions. *Catal. Lett.* **2012**, *142*, 224. [CrossRef]
10. Hu, X.; Guo, R.T.; Chen, X.; Bi, Z.X.; Wang, J.; Pan, W.-G. Bismuth-based Z-scheme structure for photocatalytic $CO_2$ reduction: A review. *J. Environ. Chem. Eng.* **2022**, *10*, 108582. [CrossRef]
11. Liu, X.; Xu, J.; Jiang, Y.; Du, Y.; Zhang, J.; Lin, K. In-situ construction of CdS@ZIS Z-scheme heterojunction with core-shell structure: Defect engineering, enhance photocatalytic hydrogen evolution and inhibit photo-corrosion. *Int. J. Hydrog. Energy* **2022**, *47*, 35241. [CrossRef]
12. Kolinko, P.A.; Kozlov, D.V. Products distribution during the gas phase photocatalytic oxidation of ammonia over the various titania based photocatalysts. *Appl. Catal. B Environ.* **2009**, *90*, 126. [CrossRef]
13. Wu, H.; Ma, J.; Li, Y.; Zhang, C.; He, H. Photocatalytic oxidation of gaseous ammonia over fluorinated $TiO_2$ with exposed (001) facets. *Appl. Catal. B Environ.* **2014**, *152*, 82. [CrossRef]
14. Wang, J.Y.; Liu, B.S.; Nakata, K. Effects of crystallinity, {001}/{101} ratio, and Au decoration on the photocatalytic activity of anatase $TiO_2$ crystals. *Chin. J. Catal.* **2019**, *40*, 403. [CrossRef]
15. Hosono, E.; Fujihara, S.; Kakiuchi, A.K.; Imai, H. Growth of submicrometer-scale rectangular parallelepiped rutile $TiO_2$ films in aqueous $TiCl_3$ solutions under hydrothermal conditions. *J. Am. Chem. Soc.* **2004**, *126*, 7790. [CrossRef] [PubMed]
16. Shen, S.; Wang, X.; Chen, T.; Feng, Z.; Li, C. Transfer of photoinduced electrons in anatase-rutile $TiO_2$ determined by time-resolved midinfrared spectroscopy. *J. Phys. Chem. C* **2014**, *118*, 12661. [CrossRef]
17. Das, B.; Nair, R.G.; Rajbongshi, B.; Samdarshi, S. Investigation of the Photoactivity of Pristine and Mixed Phase N-doped Titania under Visible and Solar Irradiation. *Mater. Charact.* **2013**, *83*, 145–151. [CrossRef]
18. Xiong, J.Y.; Zhang, M.M.; Cheng, G. Facile polyol triggered anatase-rutile heterophase $TiO_2$-x nanoparticles for enhancing photocatalytic $CO_2$ reduction. *J. Colloid Interface Sci.* **2020**, *579*, 872. [CrossRef]
19. Rani, S.; Garg, A.; Singh, N. Highly efficient photo-degradation of cetirizine antihistamine with $TiO_2$-$SiO_2$ photocatalyst under ultraviolet irradiation. *Int. J. Chem. React. Eng.* **2022**, *20*, 183. [CrossRef]
20. Guarino, M.; Costa, A.; Porro, M. Photocatalytic $TiO_2$ coating to reduce ammonia and greenhouse gases concentration and emission from animal husbandries. *Bioresour. Technol.* **2008**, *99*, 2650. [CrossRef]
21. Pu, S.; Wang, H.; Zhu, J.; Li, L.; Long, D.; Jian, Y.; Zeng, Y. Heterostructure $Cu_2O/(001)TiO_2$ Effected on Photocatalytic Degradation of Ammonia of Livestock Houses. *Catalysts* **2019**, *9*, 267. [CrossRef]
22. Liu, C.H.; Dai, H.L.; Tan, C.Q.; Pan, Q.Y.; Hu, F.P.; Peng, X.M. Photo-Fenton degradation of tetracycline over Z-scheme Fe-g-$C_3N_4$/$Bi_2WO_6$ heterojunctions: Mechanism insight, degradation pathways and DFT calculation. *Appl. Catal. B Environ.* **2022**, *310*, 121326. [CrossRef]
23. Luan, S.; Qu, D.; An, L.; Jiang, W.; Gao, X.; Hua, S.; Miao, X.; Wen, Y.; Sun, Z. Enhancing photocatalytic performance by constructing ultrafine $TiO_2$ nanorods/g-$C_3N_4$ nanosheets heterojunction for water treatment. *Sci. Bull.* **2018**, *63*, 683. [CrossRef]
24. Al-Hajji, L.; Ismail, A.A.; Atitar, M.F.; Abdelfattah, I.; El-Toni, A.M. Construction of mesoporous g-$C_3N_4$/$TiO_2$ nanocrystals with enhanced photonic efficiency. *Ceram. Int.* **2019**, *45*, 1265. [CrossRef]
25. Zhao, S.; Chen, S.; Yu, H.; Quan, X. g-$C_3N_4$/$TiO_2$ hybrid photocatalyst with wide absorption wavelength range and effective photogenerated charge separation. *Sep. Purif. Technol.* **2012**, *99*, 50. [CrossRef]
26. Sun, M.; Shen, S.; Wu, Z.; Tang, Z.; Shen, J.; Yang, J. Rice spike-like g-$C_3N_4$/$TiO_2$ heterojunctions with tight-binding interface by using sodium titanate ultralong nanotube as precursor and template. *Ceram. Int.* **2018**, *44*, 8125. [CrossRef]
27. Li, Y.; Zhang, M.Q.; Liu, Y.F.; Sun, Y.X.; Zhao, Q.H.; Chen, T.L.; Chen, Y.F.; Wang, S.F. In Situ Construction of Bronze/Anatase $TiO_2$ Homogeneous Heterojunctions and Their Photocatalytic Performances. *Nanomaterials* **2022**, *12*, 1122. [CrossRef]
28. Chouhan, L.; Srivastava, S.K. Observation of room temperature d0 ferromagnetism, band-gap widening, zero dielectric loss and conductivity enhancement in Mg doped $TiO_2$ (rutile + anatase) compounds for spintronics applications. *J. Solid State Chem.* **2022**, *307*, 122828. [CrossRef]
29. Mirhosseini, H.; Mostafavi, A.; Shamspur, T.; Sargazi, G. Fabrication of an efficient ternary $TiO_2$/$Bi_2WO_6$ nanocomposite supported on g-$C_3N_4$ with enhanced visible light photocatalytic activity: Modeling and systematic optimization procedure. *Arab. J. Chem.* **2022**, *15*, 103729. [CrossRef]
30. Guan, Y.; Wu, J.; Wang, L.; Shi, C.; Lv, K.; Lv, Y. Application of g-$C_3N_4$/CNTs nanocomposites in energy and environment. *Energy Rep.* **2022**, *8*, 1190. [CrossRef]
31. Hou, C.T.; Hu, B.; Zhu, J.M. Photocatalytic Degradation of Methylene Blue over $TiO_2$ Pretreated with Varying Concentrations of NaOH. *Catalysts* **2018**, *8*, 575. [CrossRef]
32. Chen, Y.; Mao, G.; Tang, Y.; Wu, H.; Wang, G.; Zhang, L.; Liu, Q. Synthesis of core-shell nanostructured $Cr_2O_3$/C@$TiO_2$ for photocatalytic hydrogen production. *Chin. J. Catal.* **2021**, *42*, 225. [CrossRef]
33. Wang, J.; Wang, G.; Jiang, J.; Wan, Z.; Su, Y.; Tang, H. Insight into charge carrier separation and solar light utilization: rGO decorated 3D ZnO hollow microspheres for enhanced photocatalytic hydrogen evolution. *J. Colloid Interface Sci.* **2020**, *564*, 322. [CrossRef] [PubMed]

34. Li, L.; Ouyang, W.; Zheng, Z.; Ye, K.; Guo, Y.; Qin, Y.; Wu, Z.; Lin, Z.; Wang, T.; Zhang, S. Synergetic photocatalytic and thermocatalytic reforming of methanol for hydrogen production based on Pt@TiO$_2$ catalyst. *Chin. J. Catal.* **2022**, *43*, 1258. [CrossRef]
35. Chen, M.; Chen, J.; Chen, C.; Zhang, C.; He, H. Distinct photocatalytic charges separation pathway on CuOx modified rutile and anatase TiO$_2$ under visible light. *Appl. Catal. B Environ.* **2022**, *300*, 120735. [CrossRef]
36. Wei, T.; Xu, J.; Kan, C.; Zhang, L.; Zhu, X. Au tailored on g-C$_3$N$_4$/TiO$_2$ heterostructure for enhanced photocatalytic performance. *J. Alloy. Compd.* **2022**, *894*, 162338. [CrossRef]
37. Kubiak, A.; Grzegórska, A.; Zembrzuska, J.; Zielińska, J.A.; Siwińska, C.K.; Janczarek, M.; Krawczyk, P.; Jesionowski, T. Design and Microwave-Assisted Synthesis of TiO$_2$ Lanthanides Systems and Evaluation of Photocatalytic Activity under UV-LED Light Irradiation. *Catalysts* **2021**, *12*, 8. [CrossRef]
38. Jian, Y.; Liu, H.; Zhu, J.; Zeng, Y.; Liu, Z.; Hou, C.; Pu, S. Transformation of novel TiOF$_2$ nanoparticles to cluster TiO$_2$-{001/101} and its degradation of tetracycline hydrochloride under simulated sunlight. *RSC Adv.* **2020**, *10*, 42860. [CrossRef]
39. Zhu, J.; Liu, Z.; Yang, F.; Long, D.; Jian, Y.; Pu, S. The Preparation of {001}TiO$_2$/TiOF$_2$ via a One-Step Hydrothermal Method and its Degradation Mechanism of Ammonia Nitrogen. *Materials* **2022**, *15*, 6465. [CrossRef]
40. Yao, F.; Fu, W.; Ge, X.; Wang, L.; Wang, J.; Zhong, W. Preparation and characterization of a copper phosphotungstate/titanium dioxide (Cu-H$_3$PW$_{12}$O$_{40}$/TiO$_2$) composite and the photocatalytic oxidation of high concentration ammonia nitrogen. *Sci. Total Environ.* **2020**, *727*, 138425. [CrossRef]
41. Zhang, H.; Zhang, G.; Zhang, H.; Tang, Q.; Xiao, Y.; Wang, Y.; Cao, J. Facile synthesis hierarchical porous structure anatase–rutile TiO$_2$/g-C$_3$N$_4$ composite for efficient photodegradation tetracycline hydrochloride. *Appl. Surf. Sci.* **2021**, *567*, 150833. [CrossRef]
42. Hao, R.; Wang, G.; Tang, H.; Sun, L.; Xu, C.; Han, D. Template-free preparation of macro/mesoporous g-C$_3$N$_4$/TiO$_2$ heterojunction photocatalysts with enhanced visible light photocatalytic activity. *Appl. Catal. B Environ.* **2016**, *187*, 47. [CrossRef]
43. Zhang, B.; Hu, X.; Liu, E.; Fan, J. Novel S-scheme 2D/2D BiOBr/g-C$_3$N$_4$ heterojunctions with enhanced photocatalytic activity. *Chin. J. Catal.* **2021**, *42*, 1519. [CrossRef]
44. Hong, Y.; Jiang, Y.; Li, C.; Fan, W.; Yan, X.; Yan, M.; Shi, W. In-situ synthesis of direct solid-state Z-scheme V$_2$O$_5$/g-C$_3$N$_4$ heterojunctions with enhanced visible light efficiency in photocatalytic degradation of pollutants. *Appl. Catal. B Environ.* **2016**, *180*, 663. [CrossRef]
45. Wang, S.; Zhao, T.; Tian, Y.; Yan, L.; Su, Z. Mechanistic insight into photocatalytic CO$_2$ reduction by a Z-scheme g-C$_3$N$_4$/TiO$_2$ heterostructure. *New J. Chem.* **2021**, *45*, 11474. [CrossRef]
46. Yang, D.; Zhao, X.; Chen, Y.; Wang, W.; Zhou, Z.; Zhao, Z.; Jiang, Z. Synthesis of g-C$_3$N$_4$ Nanosheet/TiO$_2$ Heterojunctions Inspired by Bioadhesion and Biomineralization Mechanism. *Ind. Eng. Chem. Res.* **2019**, *58*, 5516–5525. [CrossRef]
47. Xia, J.; Di, J.; Li, H.; Xu, H.; Li, H.; Guo, S. Ionic liquid-induced strategy for carbon quantum dotsbiox (x = Br, Cl) hybrid nanosheets with superior visible light driven photocatalysis. *Appl. Catal. B Environ.* **2016**, *181*, 260. [CrossRef]
48. Abdel, K.A.M.; Fadlallah, S.A. Fabrication of titanium nanotubes array: Phase structure, hydrophilic properties, and electrochemical behavior approach. *J. Appl. Electrochem.* **2022**, *52*, 17. [CrossRef]
49. Wang, Y.; Yang, W.; Chen, X.; Wang, J.; Zhu, Y. Photocatalytic activity enhancement of core-shell structure g-C$_3$N$_4$@TiO$_2$ via controlled ultrathin g-C$_3$N$_4$ layer. *Appl. Catal. B Environ.* **2018**, *220*, 337. [CrossRef]
50. Sun, X.Y.; Zhang, X.; Qian, N.X.; Wang, M.; Ma, Y.-Q. Improved adsorption and degradation performance by S-doping of (001)-TiO$_2$. *Beilstein J. Nanotechnol.* **2019**, *10*, 2116. [CrossRef]
51. Rej, S.; Bisetto, M.; Naldoni, A.; Fornasiero, P. Well-defined Cu$_2$O photocatalysts for solar fuels and chemicals. *J. Mater. Chem. A* **2021**, *9*, 5915. [CrossRef]
52. Rafiee, E.; Noori, E.; Zinatizadeh, A.A.; Zanganeh, H. A new visible driven nanocomposite including Ti-substituted polyoxometalate/TiO$_2$: Synthesis, characterization, photodegradation of azo dye process optimization by RSM and specific removalrate calculations. *J. Mater. Sci.-Mater. Electron.* **2018**, *29*, 20668. [CrossRef]
53. Peng, X.; Wang, M.; Hu, F.; Qiu, F.; Zhang, T.; Dai, H.; Cao, Z. Multipath fabrication of hierarchical CuAl layered double hydroxide/carbon fiber composites for the degradation of ammonia nitrogen. *J. Environ. Manag.* **2018**, *220*, 173. [CrossRef] [PubMed]
54. Li, H.; Cao, Y.; Liu, P.; Li, Y.; Zhou, A.; Ye, F.; Xue, S.; Yue, X. Ammonia-nitrogen removal from water with g-C$_3$N$_4$-rGO-TiO$_2$ Z-scheme system via photocatalytic nitrification denitrification process. *Environ. Res.* **2022**, *205*, 112434. [CrossRef]
55. Sun, D.; Sun, W.; Yang, W.; Li, Q.; Shang, J.K. Efficient photocatalytic removal of aqueous NH$_4^+$-NH$_3$ by palladium-modified nitrogen-doped titanium oxide nanoparticles under visible light illumination, even in weak alkaline solutions. *Chem. Eng. J.* **2015**, *264*, 728. [CrossRef]

Article

# Size Effect in Hybrid TiO$_2$:Au Nanostars for Photocatalytic Water Remediation Applications

Fangyuan Zheng [1], Pedro M. Martins [2,3], Joana M. Queirós [2,3,4], Carlos J. Tavares [4,5], José Luis Vilas-Vilela [1,6], Senentxu Lanceros-Méndez [1,7] and Javier Reguera [1,*]

1. BCMaterials, Basque Center for Materials, Applications and Nanostructures, UPV/EHU Science Park, 48940 Leioa, Spain
2. Centre of Molecular and Environmental Biology (CBMA), University of Minho, 4710-057 Braga, Portugal
3. Institute for Research and Innovation on Bio-Sustainability (IB-S), University of Minho, 4710-057 Braga, Portugal
4. Physics Centre of Minho and Porto Universities (CF-UM-UP), University of Minho, 4710-057 Braga, Portugal
5. LaPMET—Laboratory of Physics for Materials and Emergent Technologies, University of Minho, 4710-057 Braga, Portugal
6. Macromolecular Chemistry Research Group (LABQUIMAC), Department of Physical Chemistry, Faculty of Science and Technology, University of the Basque Country (UPV/EHU), 48940 Leioa, Spain
7. Ikerbasque, Basque Foundation for Science, 48009 Bilbao, Spain
* Correspondence: javier.reguera@bcmaterials.net

**Abstract:** TiO$_2$:Au-based photocatalysis represents a promising alternative to remove contaminants of emerging concern (CECs) from wastewater under sunlight irradiation. However, spherical Au nanoparticles, generally used to sensitize TiO$_2$, still limit the photocatalytic spectral band to the 520 nm region, neglecting a high part of sun radiation. Here, a ligand-free synthesis of TiO$_2$:Au nanostars is reported, substantially expanding the light absorption spectral region. TiO$_2$:Au nanostars with different Au component sizes and branching were generated and tested in the degradation of the antibiotic ciprofloxacin. Interestingly, nanoparticles with the smallest branching showed the highest photocatalytic degradation, 83% and 89% under UV and visible radiation, together with a threshold in photocatalytic activity in the red region. The applicability of these multicomponent nanoparticles was further explored with their incorporation into a porous matrix based on PVDF-HFP to open the way for a reusable energy cost-effective system in the photodegradation of polluted waters containing CECs.

**Keywords:** antibiotic degradation; hybrid TiO$_2$:Au nanoparticles; visible photocatalysis; water remediation

## 1. Introduction

Photocatalysis has received considerable attention in water remediation applications to degrade contaminants of emerging concern (CECs) [1–3] such as pesticides, personal care products, or pharmaceuticals [4–6]. The prospect of using sunlight as a light source [1,2] makes them highly relevant in the current world energetic crisis [7,8]. One requirement of photocatalysis for this possibility to happen is that the absorbed photons by the photocatalyst should own higher energy than its bandgap under light radiation so that electron-hole (e$^-$-h$^+$) pairs are generated. Thus, the photogenerated e$^-$ and h$^+$ migrate to the surface of photocatalysts to react with H$_2$O and O$_2$ producing highly reactive species such as hydroxyl radical (•OH) and superoxide radical (•O$_2^-$) which react with the pollutants to eventually degrade them into harmless compounds (e.g., CO$_2$ and H$_2$O) [1,8].

Among several photocatalysts, titanium dioxide (TiO$_2$) is one of the most studied due to its remarkable properties: low cost, high stability, large abundance, biocompatibility, and high photocatalytic efficiency [1,2,5,8,9]. In spite of the aforementioned advantages, its wide bandgap (3.0–3.2 eV), which is only excited under UV radiation or near UV region [8], limits its applicability [1,2,10]. Thus, sunlight radiation cannot be used efficiently since just less than 4% of this radiation corresponds to UV [2]. At the same time, the visible and infrared (IR) radiation of the sunlight spectrum remains unused for this purpose [2]. Another setback of using $TiO_2$ is the fast recombination of the $e^-$ $h^+$ pair, reducing its photocatalytic efficiency [1,8]. Different strategies have been used for extending the photocatalytic efficiency of $TiO_2$ under sunlight, such as metal and non-metal doping, metal loading, semiconductor combination, co-catalyst loading, and nanocomposite materials [2,8,10]. It has been shown that the functionalization of the $TiO_2$ surface with plasmonic nanoparticles allows efficient photocatalytic activities under visible light because of the Schottky junction development and localized surface plasmon resonance (LSPR) [10,11]. The former reduces the recombination rate of the electron-hole pair, whereas the latter contributes to the strong absorption of visible light and the excitation of active charge carriers (hot electrons) [10,11]. Au has been extensively studied due to its excellent optical properties, low toxicity, and physical and chemical stability [8,10,12]. Moreover, the plasmonic resonance of Au is highly tunable depending on the size and shape of the nanoparticles [10–12]. Nevertheless, most work uses spherical Au, limiting its sensitizing use to a relatively narrow plasmonic band around 520 nm [13], with just a few studies [13–23] expanding this response to other wavelengths of the visible spectrum by using gold with different morphologies (mainly combined with $TiO_2$ macroscopic substrates) such as nanorods [13–18,20,22], nanostar [19,21,23] and trigonal nanoprisms [17], or hexagonal nanoprisms [17]. Nanostars (Au with branched morphology), which are particles with the morphology of multiple highly sharp branches protruding from a central core [23], are ideally suited platforms for the synthesis of nanostructured photocatalysts due to their multiple plasmonic electromagnetic hot-spots and high light absorption cross-section [21,23]. Furthermore, they allow LSPR tunability by changing the size of the Au nanostar, concomitant with a change of the nanostar spikes aspect ratio, which can be used to enhance its light absorption from the visible to NIR region [21,24].

Among several wet chemistry-based synthesis methods for Au nanostar preparation, the seed-mediated-growth process is a common method for the synthesis of monodisperse nanostars [25]. Many of these synthesis methods use surfactants or polymers as shape-directing agents [21,24]. However, their presence can reduce the photocatalytic efficiency by blocking the active sites of the photocatalysts [26]. In this sense, surfactant-free methods, mainly based on the use of Ag as a shape-directing agent [27] to synthesize the Au nanostar are highly suited for catalytic applications with variable plasmon resonance from visible to NIR.

In this work, we developed a multistep approach to produce new hybrid Au-sensitised $TiO_2$ nanoparticles that are surfactant-free and where the gold component has a size-tunable nanostar morphology. In this multistep approach, Au spherical nanoparticles were initially generated onto $TiO_2$ nanoparticles through a deposition–precipitation method, and then further modified (growth) to induce a change in shape by a surfactant-free nanostars synthesis, generating a branched morphology. By changing the synthesis conditions (seeds to growth Au ratio) different sizes of Au NSs were produced which support broadening the absorption band to the whole visible region and a part of the NIR region.

The different versions of these nanoparticles, with different nanostar sizes, were evaluated and compared for their photocatalytic activity under UV and visible light radiation to degrade the antibiotic, ciprofloxacin (CIP). CIP was chosen as the target contaminant as it is one of the most detected antibiotics in different water matrixes [1,28,29] due to its inefficient removal by conventional wastewater treatment plants and, like all antibiotics, due to their risk to generate antimicrobial resistance bacteria in water reservoirs [1]. We prove that these novel nanoparticles are efficient in the removal of CIP under UV and visible light. Moreover, the photocatalytic assay under different wavelengths from the visible to NIR region was also carried out to understand the effect of the increase of size of Au branched nanoparticles. The results show a threshold in the maximum usable illumination wavelength.

Finally, it is essential to incorporate the nanoparticles into a support material [30,31] for a cost-effective way to degrade the pollutants in water treatment and to avoid the possible secondary pollution coming from nanoparticles [32,33]. Poly (vinylidene fluoride) (PVDF) and its copolymers have been widely used as a polymeric substrate to produce membranes, mainly due to their high chemical, mechanical, thermal, and UV stability, related to the stable C-F bonds of the polymer chain [32–34]. Here, the best nanoparticles in terms of performance were selected to be incorporated into a porous polymer matrix of poly(vinylidene fluoride-co-hexafluoropropylene) (PVDF-HFP). The excellent performance of the photocatalytic membranes is proven, opening the way to advanced water remediation strategies on broadband pollutant removal.

## 2. Results and Discussion

### 2.1. Nanoparticle Synthesis and Characterisation

Hybrid nanoparticles of Au and $TiO_2$ were synthesised following a multistep approach (Section 3.2). Starting with commercial $TiO_2$ nanoparticles, Au nanoparticles (NPs) were grown from those nanoparticles through a deposition–precipitation method. As previously described [8], this step generates the formation of small randomly distributed Au spherical nanoparticles attached to the $TiO_2$ nanoparticles ($TiO_2$:Au-NSph) (Figure S2a,b). EDX mapping was used to confirm the homogeneous distribution of small Au NPs (red colour) on the surface of $TiO_2$ (blue colour) in $TiO_2$:Au-NSph (Figure S3). In a second step, the Au component of the hybrid nanoparticles was further grown through a seed-mediated-growth process and using Ag, a branch-inducing agent, to generate Au nanostars attached to the $TiO_2$ nanoparticles ($TiO_2$:Au-NSs). The added Au grows from the Au component of the $TiO_2$:Au-NSph because it is a more energetically favorable process as has been shown in other hybrid nanoparticles [35–37]. This branched-induction mechanism has the advantage that it is produced through a surfactant-free method offering a non-coated nanoparticle surface, which is advantageous for the catalytic function of the nanoparticles. Based on this method, three different versions of nanoparticles, Sample A, B, and C, were generated with an increasing quantity of Au and, therefore, with the expanding size of the Au nanostar parts.

STEM-HAADF was used to assess the morphology of the synthesised nanoparticles. As presented in Figure 1a,b,d,e,g,h, non-spherical Au nanoparticles on the $TiO_2$ surface can be detected as high contrast areas in the STEM-HAADF. The shape of the Au particle depends on the synthesis conditions. Increasing the volume ratio of the gold solution to the seed solution generated bigger Au nanostars with more developed tips exhibiting higher aspect ratios (Figure 1b,e,h). Moreover, EDX mapping was used to confirm the presence of Au (red colour) on the surface of $TiO_2$ (blue colour) in $TiO_2$:Au-NSs-A, B, and –C (Figure 1c,f,i).

Regarding the crystal structure, XRD (Figure 2a) showed the presence of anatase (peaks at 25.3, 37.8, and 48.0°) and rutile (peaks at 27.49°) in all the samples, in good agreement with the literature [8,38]. Moreover, there was no significant difference between the intensities or positions of the peaks from these samples, independent of the Au presence and size. On the other hand, no diffraction peaks of Au were detected in $TiO_2$:Au-NSph and $TiO_2$:Au-NSs-A, -B and -C, which can be explained by the low amount of Au present in these samples. In addition, XRF was used to detect the amount of Au in the synthesised hybrid nanoparticles (Table S1, Figure S4). The amount of Au with respect to $TiO_2$ was 2.38 and 5.83 wt.% in $TiO_2$:Au-NSs-B, and –C, respectively, very similar to the theoretical values, considering the quantities of the reagents, indicating a high yield of the Au reduction. Due to the quantification limit of the technique, <1 wt.%, the amount of Au could not be determined for $TiO_2$:Au-NSph and $TiO_2$:Au-NSs-A although similar results are expected.

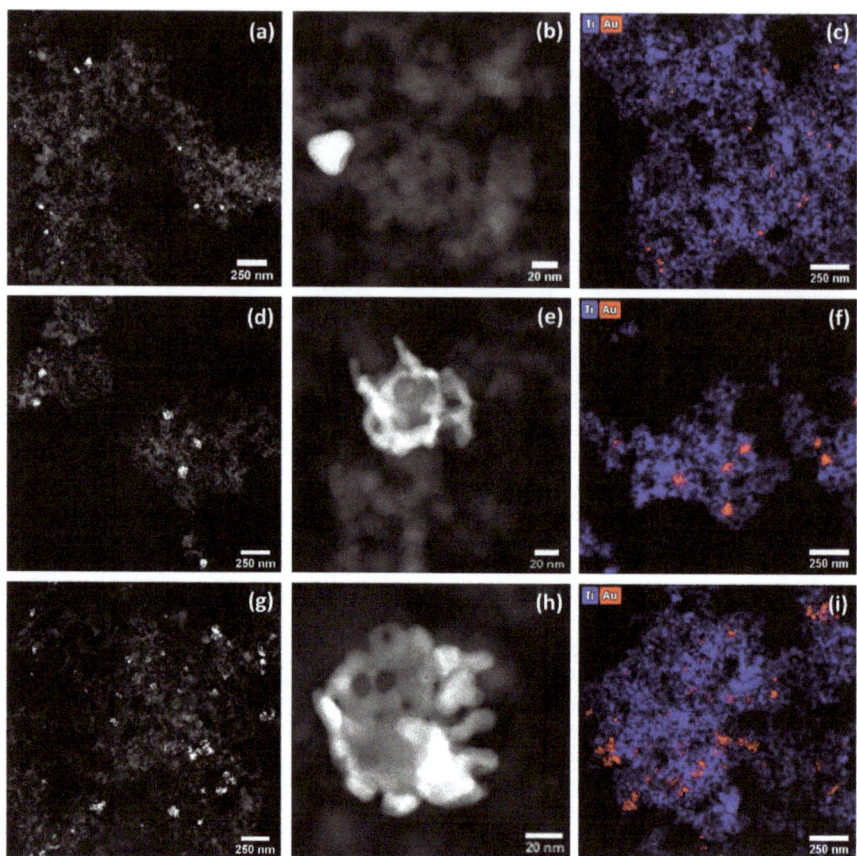

**Figure 1.** STEM-HAADF micrographs with the different magnifications (**a,b,d,e,g,h**) and EDX mapping (**c,f,i**) of $TiO_2$:Au-NSs, Sample A (**a–c**), Sample B (**d–f**), and Sample C (**g–i**).

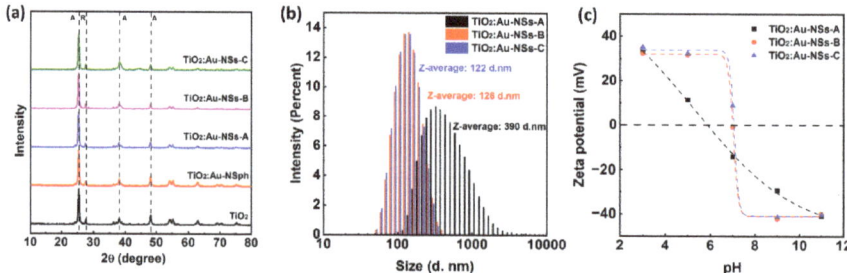

**Figure 2.** X-ray diffraction spectra of pristine $TiO_2$, $TiO_2$:Au-NSph, $TiO_2$:Au-NSs-A, B, and C nanocomposites and identification of the representative diffraction peaks for anatase (A) and rutile (R) phases (**a**). The intensity size distribution of the $TiO_2$:Au-NSs-A, B, and C nanocomposite and respective Z-average hydrodynamic size (**b**). Zeta potential measurements performed at different pHs (3, 5, 7, 9, and 11) for $TiO_2$:Au-NSs-A, B, and C nanocomposite (**c**).

The hydrodynamic size for Samples A, B, and C was studied by DLS, (Figure 2b). The results show diameters of 390 ± 13.4, 126 ± 2.9, and 122 ± 0.8 nm for A, B, and C, respectively. These diameters were much smaller than the one of the pristine $TiO_2$ nanoparticles previously reported [8] and decreased when increasing the Au-NS size. This is probably due to the different processing that takes place during the growth of Au on the hybrid nanoparticles, with respect to other methods, and the presence of Au nanoparticles over the $TiO_2$ surface, preventing the formation of big nanoparticles' aggregates [8]. Concerning Z-potential, the measurements were performed as a function of the pH and it is presented in Figure 2c. In general, they presented a zero potential around a pH range of 6–7 and a Z-potential modulus that rapidly increased when separating from that point to reach values higher than 30 mV for pH below 3 or above 9 where the nanoparticles show superior electrostatic stability [39,40]. When comparing the different nanoparticles, a slight increase of the pH at zero potential was observed when increasing the Au-NS size from Sample A to Samples B and C together with a more pronounced slope, probably due to the smaller aggregate size as indicated in the DLS measurements.

This change in Au morphology affected the optical properties of the nanoparticles. DRS was used to evaluate the optical properties of pure $TiO_2$ nanoparticles, $TiO_2$:Au-NSph and $TiO_2$:Au-NSs (Figure 3a,c,e), and $TiO_2$:Au-NSs at different Au-NSs sizes (Figure 3b,d,f).

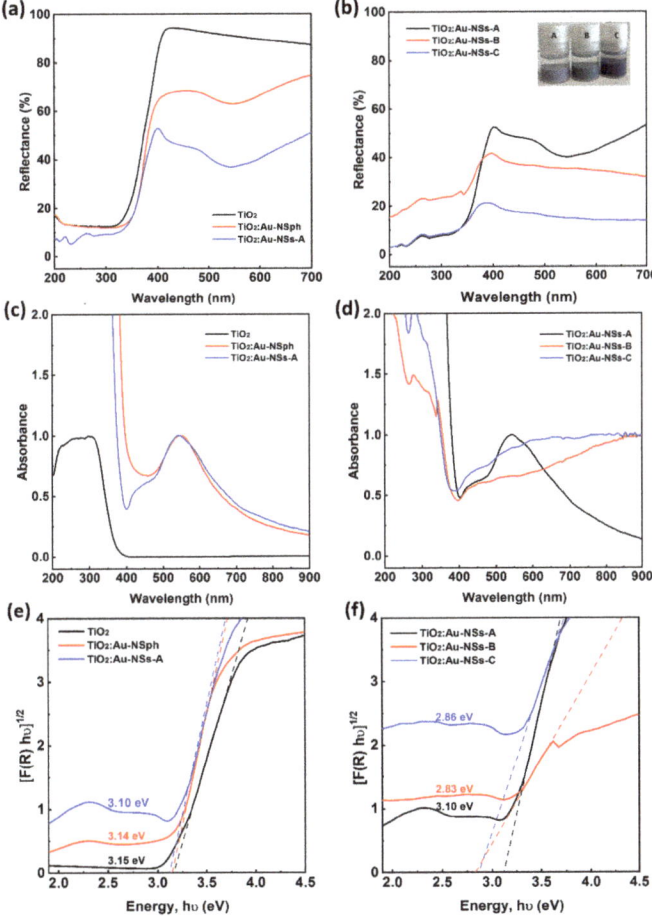

**Figure 3.** UV-Vis reflectance spectra (**a**,**b**) and UV-Vis absorption spectra (**c**,**d**) of the nanoparticles in the different synthesis steps (**a**,**c**) and nanoparticles with different Au NS sizes (**b**,**d**). Estimation of the bandgap for nanoparticles at different steps of the synthesis (**e**) and nanoparticles with different Au-NS sizes (**f**). (The bandgap is taken as the extrapolation of the linear part at $[F(R)h\upsilon]^{1/2} = 0$).

When observing the reflectance spectra of the three types of nanoparticles (Figure 3a), all the samples showed similar low reflectance in the UV range (200–400 nm) mainly due to the $TiO_2$ high cross-section at the UV. On the other hand, in the visible range (400–700 nm), the pure $TiO_2$ nanoparticles had a reflection of ≈87% of the radiation, while the nanohybrids $TiO_2$:Au-NSph presented a reflectance below 75% for the same range with a minimum reflectance (≈63%) at 544 nm due to the surface plasmon of Au spherical nanoparticles, which are in line with the literature [8,41]. In the same range, $TiO_2$:Au-NSs-A showed a reflectance below 51% due to the higher content of Au and its branched morphology. When comparing the $TiO_2$:Au-NSs at different Au-NSs sizes (Figure 3b), reflectance decreased further with the increasing amount and size of gold nanostars. $TiO_2$:Au-NSs-B and C, in the same range, presented a reflectance below 32% and 14%, respectively.

The complementary graph of absorbance shows that this one changed significantly when Au was included, with the appearance of a plasmonic peak in the visible region that extended to near IR (Figure 3c). This absorption increased with the concentration of Au and the size of Au branched NPs, generating a broader peak that extended to much higher wavelengths (Samples B and C in Figure 3d). This redshift is in agreement with the literature for homocomponent nanostars [24,27], while the broadening can be due to the interaction with the excess of $TiO_2$ nanoparticles affecting its uniformity and morphology.

The bandgap of the samples was estimated from the DRS spectrum by applying Equations (1) and (2) and after line fitting in the linear region 3.3–3.6 eV, as shown in Figure 3e,f. The pure $TiO_2$ nanoparticle showed a bandgap of 3.15 eV, typical for $TiO_2$ (3.0 to 3.2 eV depending on the ratio of crystalline phases) [42]. The $TiO_2$:Au-NSph and $TiO_2$:Au-NSs-A showed a lower bandgap than pure $TiO_2$: 3.14 and 3.10 eV, respectively. This decrease of the bandgap is related to the absorption of longer wavelengths and has been previously reported for other hybrid systems [8,38,43]. The bandgap reduction was more evident when increasing the Au nanostar size due to their higher absorption in the visible region with values of 2.83 and 2.86 for $TiO_2$:Au-NSs-B and C, respectively (Figure 3f).

## 2.2. Photocatalytic Degradation under UV and Visible Radiation

The photocatalytic activity of the synthesized $TiO_2$:Au-NSs-A, B, and C nanoparticles was evaluated and compared with $TiO_2$:Au-NSph under both UV and visible light radiation in the degradation of CIP under colloidal suspension conditions. Figure 4a,b show the results of photocatalytic experiments under UV and visible radiation, respectively. Table 1 shows the apparent reaction rate constant ($k$) calculated by Equation (3) for the different synthesized nanoparticles. As a control procedure, it should be noted that under the same irradiation conditions of UV or visible light and in the absence of nanoparticles, there was very low photolysis of CIP (Figures S5 and S6).

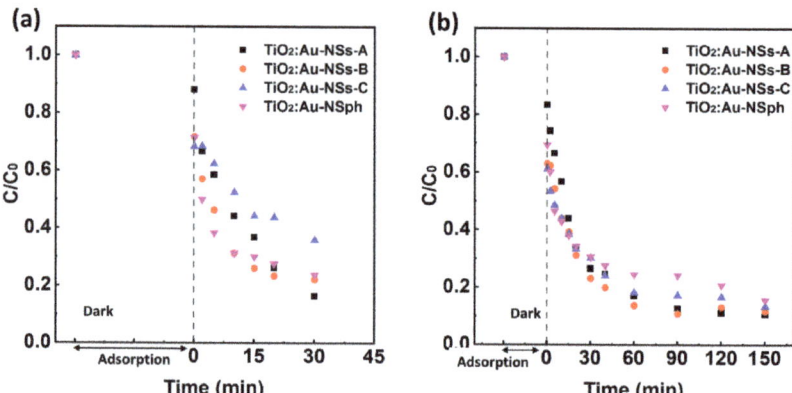

**Figure 4.** Photocatalytic degradation of CIP (5 mg/L) with $TiO_2$:Au-NSph, $TiO_2$:Au-NSs-A, B, and C nanoparticles under 30 min of UV radiation (**a**) and 150 min of visible radiation (**b**).

**Table 1.** CIP degradation efficiencies (DE, %) and corresponding apparent reaction rate constants (*k*) under 30 min of UV radiation and 150 min of visible radiation for TiO$_2$:Au-NSs-A, B, and C nanoparticles.

| Sample | UV | | Xenon | |
|---|---|---|---|---|
| | $k$ (min$^{-1}$) | DE (%) | $k$ (min$^{-1}$) | DE (%) |
| TiO$_2$:Au-NSs-A | 0.053 | 83 | 0.014 | 89 |
| TiO$_2$:Au-NSs-B | 0.040 | 78 | 0.012 | 88 |
| TiO$_2$:Au-NSs-C | 0.023 | 64 | 0.009 | 86 |
| TiO$_2$:Au-NSph | 0.031 | 76 | 0.008 | 84 |

The photocatalytic assays under UV light had a degradation efficiency of 83, 78, and 64% for A, B, and C nanoparticles, respectively, under the same experimental conditions. The reaction rate constant showed a similar tendency, $k$ = 0.053, 0.040, and 0.023 min$^{-1}$ for A, B, and C nanoparticles, respectively. Nanoparticles A presented the best degradation efficiency under UV of the three samples. This result can be rationalized by the lower quantity of TiO$_2$ (total nanoparticle mass is kept constant) and the lower active area presented by the photocatalyst when adding the Au to its surface for an increasing quantity of Au. Among these samples, A showed better photocatalytic activity than TiO$_2$:Au-NSph, which presented a degradation efficiency of 76%, with $k$ = 0.031 min$^{-1}$. The change of the Au spherical morphology to branched morphology reduced the bandgap of TiO$_2$ and improved the photocatalytic efficiency.

On the other hand, in the adsorption process in the dark (before irradiation), Samples A, B, and C adsorbed 12%, 28%, and 32% of CIP, respectively. The higher amount of Au on TiO$_2$ surface led to a higher CIP adsorption, which agrees with previously reported work [8].

Interestingly, the addition of Au onto the TiO$_2$ surface made it possible to produce the photocatalytic degradation of CIP by the nanoparticles under visible light (Figure 4b) due to the improvement of the absorption of longer solar wavelengths as well as the lower bandgap (Figure 3e,f). Under this illumination, all the TiO$_2$:Au-NSs nanoparticles presented very similar degradation efficiency, 89%, 88%, and 86%, with $k$ = 0.014, 0.013, and 0.009 min$^{-1}$, for A, B, and C, respectively. The nanoparticle TiO$_2$:Au-NSph showed slightly lower photocatalytic performance with a degradation efficiency of 84% with $k$ = 0.008 min$^{-1}$ compared with A, B, and C. Note here that although TiO$_2$:Au-NSs-A, B, and C showed a similar effect under visible radiation, Samples B and C showed a broader plasmonic band that extended deeper into the near IR and could be beneficial for the light-harvesting of higher wavelengths of the sunlight radiation. Finally, the TiO$_2$:Au-NSs-A, B, and C after the photocatalytic tests were recovered and assessed by XRD to confirm their stability after photocatalytic application. There was no difference observed in the crystal structure of nanoparticles before (Figure 2a) and after photocatalysis (Figure S7).

Based on the photocatalytic activity results presented, Sample A—with the best photocatalytic performance—was selected as the best candidate to immobilise into a PVDF-HFP-based nanocomposite membrane. Additionally, further experiments were carried out for highly reactive oxygen species (ROS) detection, •OH and $^1O_2$, in Sample A. Figure S8 shows that the generation of hydroxyl radical (•OH) increased during the 30 min of UV radiation. The result also indicates that the concentration of $^1O_2$ achieved a maximum at an irradiation time of 15 min, after which the generation of $^1O_2$ became constant. Therefore, •OH played the dominant role during the photocatalytic degradation of CIP.

On the other hand, a comparison between TiO$_2$:Au-NSs-A and previous work using TiO$_2$-based plasmonic photocatalysts was performed (Table 2). Due to the different experimental conditions applied in each work, this comparison is not straightforward but allows contextualizing of our results. Although the previous works showed a slightly higher degradation of CIP than our results, they used a much higher amount of plasmonic nanoparticles and intensity of visible radiation (less cost-effective process) or longer degradation time than the one we used, which makes the comparison less straightforward.

**Table 2.** Comparison of results between the present work and previous work that used plasmonic nanoparticles to functionalize $TiO_2$ for ciprofloxacin (CIP) degradation under visible light.

| CIP Concentration (ppm) | Photocatalyst | Cu, Ag or Au Amount (wt.%) | Photocatalyst Concentration (mg/mL) | Irradiation | Efficiency (%) | Time (min) | Ref. |
|---|---|---|---|---|---|---|---|
| 30 | $Cu/TiO_2$ | 1.0 | 0.5 | 500 W/m$^2$ | 99 | 180 | [44] |
| 80 | $Cu/TiO_2$ | 1.0 | 0.25 | 500 W/m$^2$ | 85 | 240 | [45] |
| 3.3 | $Ag/TiO_2$ | 5.0 | 0.5 | 60 W | 87 | 60 | [46] |
| 30 | $Ag/TiO_2$ | 1.5 | 0.5 | 500 W/m$^2$ | 99 | 240 | [44] |
| 30 | $Au/TiO_2$ | 1.5 | 0.5 | 500 W/m$^2$ | 99 | 180 | [44] |
| 5 | $Au/TiO_2$ | 0.5 | 1.0 | 98 W/m$^2$ | 45 | 180 | [8] |
| 5 | $TiO_2$:Au-NSs-A | 0.68 | 1.0 | 300 W/m$^2$ | 89 | 150 | Present work |

### 2.3. Photocatalytic Degradation under Different Wavelengths

To understand the photocatalytic behavior of the nanoparticles after increasing the Au branched morphology, the synthesized $TiO_2$:Au-NSs-A, B, and C nanoparticles were evaluated under different wavelengths of light radiation in the visible and NIR region for the degradation of CIP.

Figure 5a–d show the results of photocatalytic experiments under blue (460 nm), green (530 nm), red (630 nm), and NIR (730 nm) light radiation, respectively. It should be noted that under the same irradiation conditions of these types of light and in the absence of nanoparticles, there was no photolysis of CIP (Figure S9). Table 3 shows the apparent reaction rate calculated by Equation (3).

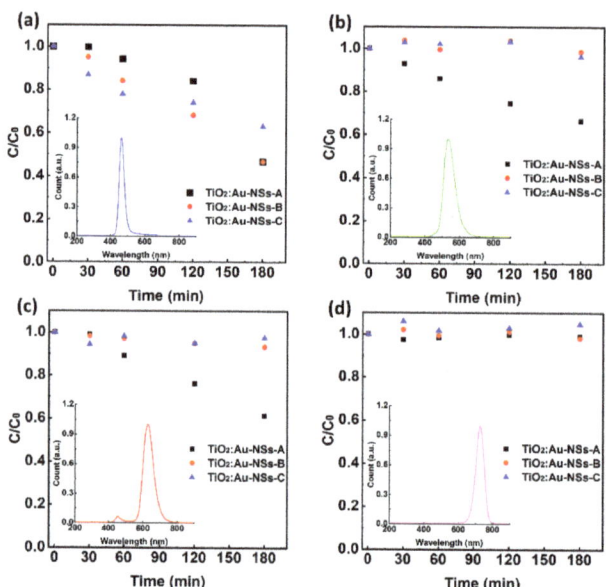

**Figure 5.** Photocatalytic degradation of CIP (5 mg/L) with $TiO_2$:Au-NSs-A, B, and C nanoparticles under 180 min of different wavelengths of light (inset images): blue (**a**), green (**b**), red (**c**), and NIR light (**d**) radiation.

Table 3. CIP degradation efficiencies (DE, %) and corresponding apparent reaction rate constants ($k$) under 150 min blue, green, red and NIR radiation for TiO$_2$:Au-NSs-A, B, and C nanoparticles.

| Sample | Blue | | Green | | Red | | NIR | |
|---|---|---|---|---|---|---|---|---|
| | $k$ (min$^{-1}$) | DE (%) | $k$ (min$^{-1}$) | DE (%) | $k$ (min$^{-1}$) | DE (%) | $k$ (min$^{-1}$) | DE (%) |
| TiO$_2$:Au-NSs-A | 0.0031 | 53 | 0.0024 | 34 | 0.0025 | 39 | - | - |
| TiO$_2$:Au-NSs-B | 0.0038 | 53 | - | - | - | - | - | - |
| TiO$_2$:Au-NSs-C | 0.0027 | 37 | - | - | - | - | - | - |

Regarding the photocatalytic assays under blue light, all nanoparticles showed an excellent degradation activity. Degradation efficiency of 53, 53, and 37% was observed for A, B, and C nanoparticles, respectively, under the same experimental conditions. The rate constant presented a similar tendency, $k$ = 0.031, 0.038, and 0.027 min$^{-1}$ for A, B, and C nanoparticles, respectively. This wavelength is lower than the plasmonic band of the Au NSs, however as observed in Figure 3d, the nanoparticles still showed a high absorbance. Interestingly, the nanoparticles showing the highest absorbance are the ones with the lowest degradation activity. This counterintuitive trend can be rationalized by the high absorbance of light by the nanoparticles that block the pass of light deeper in the cuvette and produce the catalytic effect only in the first part of the optical path. In fact, sample C, the one with the highest absorbance, completely blocked the light in a few millimeters. In addition, the highest content of gold, as shown above, produced higher absorption in the dark, and reduced the TiO$_2$ surface area (total mass of nanoparticles constant for all experiments) contributing to the lower performance. As the wavelength was increased to green and red light, only TiO$_2$:Au-NSs- Sample A was activated in the CIP degradation. A degradation efficiency of 34% and 39% of TiO$_2$:Au-NSs-A, with $k$ = 0.0024 and 0.0025 min$^{-1}$ for the green and red light radiation was found, respectively. According to the results of photocatalytic assays under NIR light, none of these three nanoparticles could be activated.

For the catalytic process to become activated, electrons from the Au part should gain enough energy from the incident absorbed photon to overpass the Shottky barrier, i.e., the difference between the work function of the Au and the electron affinity in the TiO$_2$ as determined by the Schottky–Mott equation ($\Phi_{SB} = \phi_M - \chi_{SM}$) [47]. Then a lower limit is expected in the absorbed photon energies, i.e., a higher limit in the wavelength. On the other hand, several studies have shown that $\Phi_{SB}$ is not only determined by $\phi_M$ and $\chi_{SM}$ but also is significantly influenced by interfacial chemistry, giving rise to differences between different types of nanoparticles, in this case with a limit in the 635–735 nm region [48,49].

## 2.4. Membrane Processing and Characterisation

Nonsolvent induces phase separation, NIPS, combined with salt leaching was used to incorporate the synthesized TiO$_2$:Au-NSs nanoparticles into the PVDF-HFP polymer matrix and to obtain a porous microstructure. The successful incorporation of the nanoparticles in the polymer matrix extends their reuse, assuring the recovery of the catalyst and, therefore, allowing for a more sustainable and cost-effective application.

The porous morphology and the presence of nanoparticles were analyzed by SEM (Figure 6a,b). The thickness of the membranes was 271 μm and 217 μm for 0 and 10 wt.% TiO$_2$:Au-NSs/PVDF-HFP, respectively. After the incorporation of nanoparticles, the thickness of the membranes was slightly reduced.

High porosity and well-distributed and interconnected pores were observed in both membranes. Notably, the incorporated nanoparticles with 10 wt.% amount did not produce significant changes in the morphology of the pristine membranes. Both membranes presented two ranges of porous distribution due to the polydisperse NaCl grains [50] located mainly in the lower part of the membrane and the additional porous formation by the simultaneous NIPS mechanism [50]. The prepared membrane with 10 wt.% TiO$_2$:Au-NSs nanoparticles was tested in the degradation of CIP under visible light (Figure 6c). The membrane did not show a release of nanoparticles to the solution and a CIP degradation efficiency of 69% at 600 min, with $k = 0.002$ min$^{-1}$. Table S2 presents the detected intermediate products of CIP after the photocatalytic assay, confirming the photocatalytic degradation of CIP in the presence of the prepared membrane.

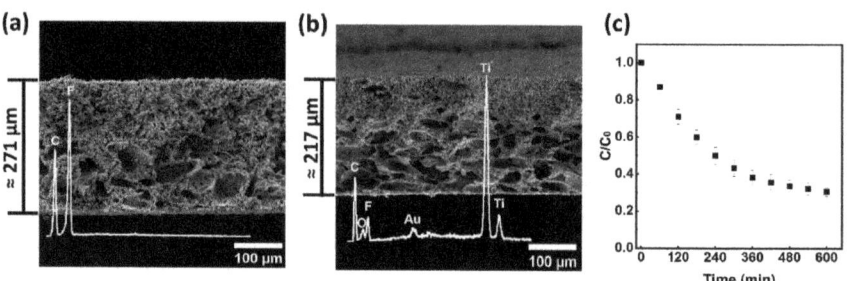

**Figure 6.** Cross-section SEM images of 0 wt.% (**a**) and 10 wt.% TiO$_2$:Au-NSs/PVDF-HFP (**b**). Photocatalytic degradation of CIP (5 mg/L) with 10 wt.% TiO$_2$:Au-NSs/PVDF-HFP membrane under 600 min of visible radiation (**c**).

The photocatalytic degradation process was lower when compared to the corresponding colloidal dispersion mainly due to the blocking of part of the nanoparticle surface by the polymer; however, it was enough to obtain a good degradation of the antibiotic and its application facilitates the catalyst recovery and the reusability, and decreases the secondary pollution by nanoparticles to the medium, which makes it an excellent system for environmental application.

## 3. Materials and Methods

### 3.1. Materials

Poly(vinylidene fluoride-co-hexafluoropropylene) (PVDF-HFP, SOLEF® 21216/1001) was purchased from Solvay (Brussels, Belgium). N,N-dimethylformamide (DMF, ≥99%) and sodium chloride (NaCl, analytical reagent grade) were supplied by Fisher-Scientific (Illkirch, France). Titanium dioxide (TiO$_2$) nanoparticles were provided by Evonik Industries AG (Essen, Germany). Hydrogen tetrachloroaurate (III) trihydrate (HAuCl$_4$·3H$_2$O, 99.99%) was supplied by Alfa Aesar (Stoughton, MA, USA). Sodium hydroxide (NaOH, 98.0–100.5%) was obtained from Panreac (Barcelona, Spain). Hydrochloric Acid (HCl, 37%) was supplied by LABKEM. L-ascorbic acid (AA, ≥99%) and silver nitrate (AgNO$_3$, ≥99%) were purchased from Sigma-Aldrich (St. Louis, MO, USA). Milli-Q ultrapure water (resistivity 18.2 MΩ·cm) was used in all experiments. Ciprofloxacin (CIP, ≥98% (HPLG), C$_{17}$H$_{18}$FN$_3$O$_3$) with maximum light absorption at a wavelength of 277 nm was supplied by Sigma-Aldrich. Absolute ethanol (C$_2$H$_5$OH, ≥99.5%) were purchased from Sigma-Aldrich.

### 3.2. TiO$_2$:Au-NSs Hybrid Nanoparticles Synthesis

The synthesis of TiO$_2$:Au-NSs hybrid nanoparticles was divided into two steps. In the first step, Au NPs were grown into TiO$_2$ nanoparticles by a deposition–precipitation method (DP) as described by Martins et al. [8] to obtain TiO$_2$:Au nanoparticles, where the Au part formed small spherical particles (NSph) of around 5 nm. Briefly, 200 mg of TiO$_2$ nanoparticles were dispersed in 40 mL of ultrapure water in a sonication bath for 30 min. Afterwards, a specific volume of HAuCl$_4$ (1 mM) was added to achieve an Au loading of 0.05 wt.% and stirred at room temperature for 10 min to disperse the gold precursor homogeneously. Later, NaOH (0.1 M) was added dropwise to obtain a pH = 9 and then stirred for 10 min. Finally, the solution was centrifuged and washed twice with ultrapure water. In the last step, the nanoparticles were dried overnight in an oven at 80 °C and then grounded with a mortar to obtain a fine powder (TiO$_2$:Au-NSph).

In a second step, the Au morphology in the TiO$_2$:Au-NSph was modified by the seed-mediated-growth process, from spherical Au to star-shaped (NS) Au, using the modified surfactant-free method with the assistance of Ag as a shape-directing agent [51]. Then, 150 mg of TiO$_2$:Au-NSph nanoparticles were dispersed in 3.8 mL of ultrapure water in a sonication bath for 30 min as the seed solution. Afterwards, a growth solution was prepared by mixing 18.9 mL of ultrapure water, 19 µL of HCl (1 M), and 95 µL of HAuCl$_4$ (50 mM), considering a volume ratio between the gold solution (HAuCl$_4$, 50 mM) and seed solution of 0.025. Then, the prepared seed solution was added to this growth solution at room temperature and under moderate stirring. According to the used growth solution volume, 57 µL of AgNO$_3$ (10 mM) and 95 µL of AA (100 mM) solution were simultaneously and quickly added to the above mixture under vigorous stirring. The solution rapidly turned from light pink to purplish-grey, indicating the modification of Au morphology from Au sphere to star. This color tended to be more bluish when the ratio of the gold solution to seed solution increased, indicative of the formation of bigger nanostars with higher aspect ratio branches. The obtained samples (TiO$_2$:Au-NSs) were centrifuged and washed twice with ultrapure water, resuspended in ultrapure water, and named Sample A for the final application.

In terms of NP size tuning, the final Au star size was controlled by modifying the volume ratio between the gold solution and seed solution in the seed-mediated-growth synthesis step. Samples B and C were obtained by maintaining the volume of the seed solution but modifying the volume of the gold solution (HAuCl$_4$, 50 mM), with a volume ratio between gold solution and seed solution of 0.1 and 0.25, for Samples B and C, respectively. The volume of the other reagents in the growth solution, AgNO$_3$ and AA solution, were prepared proportionally to the volume of gold solution (HAuCl$_4$, 50 mM).

*3.3. TiO$_2$:Au-NSs/PVDF-HFP Membranes Preparation*

The most photocatalytic efficient TiO$_2$:Au-NSs of the previously synthesized nanoparticles were used to prepare nanoparticle-loaded membranes through a salt leaching technique combined with a non-solvent-induced phase separation (NIPS) following the main guidelines previously described [50] but using different type of coagulation bath. 111 mg of TiO$_2$:Au-NSs nanoparticles were dispersed in 9 mL of DMF to obtain a TiO$_2$:Au-NSs/PVDF-HFP final mass ratio of 0 and 10 wt.% in an ultrasonication bath for 2 h with control of temperature to achieve a good nanoparticles dispersion. Later, 1 g PVDF-HFP polymer was added to the solution to obtain a PVDF-HFP/DMF concentration of 1:9 $v/v$. After dissolving the polymer completely, 5 g of NaCl particles with a diameter of 90 µm were added and stirred for 1 h to achieve a homogeneous distribution of the NaCl particles. Then, the mixed solution was spread onto a glass substrate by a doctor blade with a defined gap of 950 µm. Afterwards, the glass substrate was immersed in an absolute ethanol coagulation bath at room temperature and detached the films. Then, the films were immersed in a distilled water bath at 45 °C to remove possible traces of solvent and dried at room temperature for 24 h. Finally, the film was washed in deionized water for 1 week to remove NaCl particles and then dried at room temperature.

*3.4. Characterisation Techniques*

Transmission electron microscopy (TEM) images were acquired with a JEOL JEM 1400 Plus set up operating at 100 kV in bright field and a Talos (Thermo Scientific, Waltham, MA, USA) system working at 200 kV for the HAADF-STEM and EDX-STEM measurements. To prepare the samples, the nanoparticle powder was dispersed in ultrapure water and sonicated for 1 min, and then 2 µL of the suspension was placed on a 400-mesh carbon-coated copper grid and left to dry at room temperature. The analysis of the images was performed using the ImageJ software package.

To perform diffuse reflectance spectroscopy (DRS), a UV-Visible-NIR Jasco V-770 spectrometer equipped with a 150 mm diameter integrating sphere coated with Spectralon with 1 nm spectral resolution was used. DRS was carried out in the 250–2200 nm wavelength range. A Spectralon reference was used to measure the 100% reflectance, and internal attenuators were used to determine zero reflectance to remove background and noise. The samples saved in ultrapure water were placed in the support and dried at room temperature and the powder samples were placed in a quartz cuvette, sealed, and mounted on a Teflon sample holder before the DRS measurement. The measured reflectance spectra were subsequently converted to Kubelka–Munk (K-M) absorption factors to evaluate the absorption spectra of the samples. This conversion was performed using the K-M Equation (1) [52]:

$$F(R) = (1 - R_\infty)^2/(2R_\infty) \qquad (1)$$

where $R_\infty$ ($R_{Sample}/R_{Spectralon}$) corresponds to the reflectance of the sample and $F(R)$ is the absorbance.

The sample bandgap was estimated using the Tauc plot Equation (2):

$$[F(R)hV]^{1/n} \text{ versus } hV \qquad (2)$$

where $h$ is Planck's constant, $\upsilon$ the frequency, and $n$ the nature of the electronic bandgap transition type, taken as $n$ = 2 for indirect transition [53].

Dynamic light scattering (DLS) and Z-potential were measured with a Zetasizer NANO ZS-ZEN3600 (Malvern Instruments Limited, Malvern, UK), equipped with a He–Ne laser (wavelength 633 nm) and backscatter configuration (173°). The nanoparticles were dispersed (1 mg/mL) in ultrapure water in an ultrasonication bath at room temperature for 1 h to avoid aggregation, and each sample was measured five times at pH = 3 to obtain the hydrodynamic diameter. The Z- potential was evaluated at different pH (3, 5, 7, 9, and 11), and each sample was measured five times. HCl (0.1 M) and NaOH (0.1 M) solutions were used to adjust the pH. The resulting particle size was determined using the Smoluchowski model [54]. The manufacturer software (Zetasizer 7.13) was used to estimate the hydrodynamic diameter of the nanoparticles (cumulant diameter), the polydispersity index (PDI), and zeta potential values.

The crystal structure of the nanoparticles was assessed by X-ray diffraction (XRD) using a Philips X'pert PRO automatic diffractometer operating at 40 kV and 40 mA, in theta-theta configuration, secondary monochromator with Cu-Kα radiation (λ = 1.5418 Å) and a PIXcel solid-state detector (active length in 2θ 3.347°). Data were collected from 5 to 80° 2θ, step size 0.026° and time per step of 60 s at room temperature, total measurement time 10 min. Then, 1° fixed soller slit and divergence slit which provided a constant volume of sample illumination, was used.

X-ray fluorescence (XRF) was used to quantify the ratio of Au:$TiO_2$ (wt.%). The measurements were obtained by using a MIDEX SD (Spectro, Kleve, Germany) X-ray microfluorescence spectrometer, energy dispersion ED-XRF for elemental analysis. Automatic XYZ tray and collimator changer, X-ray with Mo tube with maximum power 40 W/voltage 48 kV and silicon drift detector (SDD) with 30 $mm^2$ area. The calibration and calculations were done by fundamental parameters FP Plus.

A field emission gun scanning electron microscope (FEG-SEM) Hitachi S-4800N operating at 10 kV voltage was used to image the membranes. The samples were coated with a thin layer of gold (≈15 nm) in an Emitech K550X ion-sputter before measurement.

Elemental analysis was carried out by Energy Dispersive X-ray spectroscopy (EDX) using a Carl Zeiss EVO 40 (Oberkochen, Germany) SEM equipped with an EDX Oxford Instrument X-Max detector (Abingdon, UK). The measurement were performed in a high vacuum condition, at a voltage of 20 kV, a current of 100–400 pA and a working distance of 9–10 mm.

Ultrahigh-performance liquid chromatography (UHPLC), coupled with a time-of-flight high-resolution mass spectrometry (TOF-HRMS, Synapt G2 from Waters Cromatografia S.A, Barcelona, Spain) by an electrospray ionisation source in positive mode (ESI+), was used for detecting the products in liquid solution. The chromatographic separation was achieved using an Acquity UPLC BEH C18 column (1.7 μm, 2.1 × 50 mm i.d.) with an Acquity UPLC BEH C18 1.7 μm VanGuard pre-column (2.1 × 5 mm) (Waters Cromatografia S.A.) and a binary solvent A/B gradient (A: water with 0.1% formic acid and B: methanol).

The gradient program was as follows: initial conditions were 5% B, raised to 99% B over 2.5 min, held at 99% B until 4 min, decreased to 5% B over the next 0.1 min, and held at 5% B until 5 min for re-equilibration of the system prior to the next injection. A flow rate of 0.25 mL/min was used, with the column temperature at 30 °C, the autosampler temperature at 4 °C, and the injection volume of 5 µL.

*3.5. Photocatalytic Degradation of Ciprofloxacin under UV and Visible Radiation*

The photocatalytic activity of the produced $TiO_2$:Au-NSs nanoparticles, A, B, and C, were tested under both UV and visible light radiation. Firstly, the CIP solution of 5 mg/L was prepared and adjusted to pH = 3. Before the degradation assays, 50 mg of nanoparticles as photocatalysts were stirred in 50 mL of CIP solution in the dark for 30 min to achieve an adsorption–desorption equilibrium.

The photocatalytic activity of the produced membranes was tested under visible light radiation using the same CIP solution. Before the degradation assays, the membrane, with a sample area of 18 cm$^2$, was immersed and stirred in 50 mL of CIP solution in the dark for 30 min to achieve an adsorption–desorption equilibrium.

The UV degradation of CIP was performed in a photoreactor with eight UV lamps of 8 W, with an emission peak at 365 nm, over 30 min. The suspensions of photocatalysts and CIP were kept stirred in a 100 mL beaker under illumination from the top. The distance between the solution and the lamp was 13.5 cm, and the irradiance at the solution was 3.3 W/m$^2$.

For the visible light degradation, a filtered Xenon lamp (sun emulator) with an excitation peak at 550 nm and irradiance of 300 W/m$^2$ (spectra in Figure S1) was used, over 150 min for degradation in suspension and 600 min for degradation using membranes. The suspensions of photocatalysts and CIP were stirred in a 100 mL beaker under lateral illumination. The distance between the CIP solution and the lamp was 21 cm.

Aliquots as samples were taken out at different periods during the degradation assays and centrifuged to remove the photocatalysts. Afterwards, the 200 µL of the supernatant in each sample after centrifugation was analyzed by UV-Vis spectroscopy. The absorbance variation of the 277 nm peak of the CIP spectrum was monitored using a microplate reader Infinite 200 Pro in the range of 230 to 450 nm.

The photocatalytic degradation rate was fit to a pseudo-first-order reaction, which is based on the Langmuir–Hinshelwood model described by Equation (3):

$$\ln\left(\frac{C}{C_0}\right) = -kt \qquad (3)$$

where $C$ and $C_0$ represent the pollutant concentration at time $t$ and at the beginning of the photocatalytic assessment respectively, and $k$ is the first-order rate constant of the reaction [8].

*3.6. Photocatalytic Degradation under the Different Wavelengths*

The photocatalytic activity of the produced $TiO_2$:Au-NSs nanoparticles, A, B, and C was also assessed under different wavelengths of light radiation: blue light (emission peak at 460 nm), green light (emission peak at 530 nm), red light (emission peak at 630 nm), NIR light (emission peak at 730 nm), for ciprofloxacin degradation.

Firstly, the CIP solution of 5 mg/L was prepared and adjusted to pH = 3. After achieving the adsorption–desorption equilibrium of the photocatalysts and CIP solution as described previously (Section 3.5), the degradation of CIP was carried out in a cuvette under different wavelengths of light radiation with an intensity of 0.5 W, over 180 min. The suspensions of photocatalysts and CIP were kept stirred under lateral illumination. The distance between the CIP solution and the lamp was 1 cm.

Aliquots were withdrawn at different times during the degradation assessment, and centrifuged and analyzed using an Agilent (Santa Clara, CA, USA) Cary 60 UV-Vis Spectrophotometer.

## 4. Conclusions

Novel hybrid nanoparticles, $TiO_2$:Au-NSs, with an Au branched morphology were synthesized successfully through the surfactant-free method and characterized and tested in photocatalytic assays for ciprofloxacin (CIP) degradation. The characterization results of TEM and DRS show that different sizes of Au NPs with branched morphology were produced by modifying the synthesis conditions, which allowed the tuning of the optical properties of hybrid nanoparticles. When increasing the size of Au NPs with branched morphology, the reflectance of the hybrid nanoparticles decreased from 57% to 13% in the visible region. Additionally, the increase of the size of the Au branched nanoparticles extended the light absorption to the whole visible and part of the NIR region and reduced the bandgap from 3.10 eV to 2.86 eV, respectively.

The photocatalytic assays confirmed that all the synthesized nanoparticles degraded target compound CIP under both UV and white radiation. It was also possible to understand the impact of the size of Au branched nanoparticles in the photocatalytic response. $TiO_2$:Au-NSs nanoparticles with smaller Au branched morphology and lowest amount of added Au among these hybrid nanoparticles showed a better photocatalytic performance degrading 83% and 89% ciprofloxacin under UV and visible radiation, respectively.

According to the results, under the different wavelengths of light in the visible and NIR region, the nanoparticles could be activated under blue, green, and red light radiation showing a CIP degradation efficiency of 57%, 34%, and 39%, respectively. However, there was no photocatalytic degradation of CIP under NIR radiation. The bigger size of Au branched nanoparticles limited the light-harvesting of $TiO_2$, and reduced the photocatalytic activity, although they showed a broader light absorption in the whole visible and part of the NIR region of sunlight radiation.

The nanoparticles $TiO_2$:Au-NSs with lower branching and best performance were selected as the best candidates to incorporate into a PVDF-HFP polymer matrix through the NIPS technique. The membranes were produced successfully and presented high porosity and a well-distributed porous structure.

In short, these results indicate the outstanding performance of the synthesized nanoparticles for water remediation applications in the degradation of persistent contaminants such as ciprofloxacin. Moreover, their successful incorporation into a polymer matrix opens the door to future application in a cost-effective way to degrade a high number of contaminants of emerging concern and other possible applications.

**Supplementary Materials:** The following supporting information can be downloaded at: https://www.mdpi.com/article/10.3390/ijms232213741/s1.

**Author Contributions:** Investigation, formal analysis, writing—original draft preparation, F.Z.; supervision, methodology, writing—review and editing, P.M.M.; investigation, J.M.Q.; resources, C.J.T.; resources, J.L.V.-V.; conceptualization, supervision, funding acquisition, writing—review and editing, S.L.-M.; conceptualization, methodology, formal analysis, writing—review and editing, J.R. All authors have read and agreed to the published version of the manuscript.

**Funding:** This research was funded by Spanish State Research Agency (AEI) through the project PID2019-106099RB-C43/AEI/10.13039/501100011033 and the Basque Government under the ELKARTEK program. P.M.M. thanks the FCT for contract 2020.02802.CEECIND. C.J.T. acknowledges the funding from FCT/PIDDAC through the Strategic Funds project reference UIDB/04650/2020-2023. F.Z. thanks the University of the Basque Country (UPV/EHU) for the PhD fellowship.

**Institutional Review Board Statement:** Not applicable.

**Informed Consent Statement:** Not applicable.

**Data Availability Statement:** Not applicable.

**Acknowledgments:** The authors thank the technical support of SGIker (UPV/EHU) group.

**Conflicts of Interest:** The authors declare no conflict of interest.

## References

1. Kutuzova, A.; Dontsova, T.; Kwapinski, W. Application of $TiO_2$-Based Photocatalysts to Antibiotics Degradation: Cases of Sulfamethoxazole, Trimethoprim and Ciprofloxacin. *Catalysts* **2021**, *11*, 728. [CrossRef]
2. Byrne, C.; Subramanian, G.; Pillai, S.C. Recent Advances in Photocatalysis for Environmental Applications. *J. Environ. Chem. Eng.* **2018**, *6*, 3531–3555. [CrossRef]
3. Hunge, Y.M.; Yadav, A.A.; Kang, S.; Jun, S.; Kim, H. Visible Light Activated $MoS_2$/ZnO Composites for Photocatalytic Degradation of Ciprofloxacin Antibiotic and Hydrogen Production. *J. Photochem. Photobiol. A Chem.* **2023**, *434*, 114250. [CrossRef]
4. Dey, S.; Bano, F.; Malik, A. Pharmaceuticals and Personal Care Product (PPCP) Contamination—A Global Discharge Inventory. In *Pharmaceuticals and Personal Care Products: Waste Management and Treatment Technology*; Elsevier Inc.: Amsterdam, The Netherlands, 2019; pp. 1–26. ISBN 9780128161890.
5. Vasilachi, I.C.; Asiminicesei, D.M.; Fertu, D.I.; Gavrilescu, M. Occurrence and Fate of Emerging Pollutants in Water Environment and Options for Their Removal. *Water* **2021**, *13*, 181. [CrossRef]
6. Tang, Y.; Peng, X.; Yang, W.; Zhang, Y.; Yin, M.; Liang, Y. Emerging Pollutants—Part I: Occurrence, Fate and Transport. *Water Environ. Res.* **2017**, *89*, 1810–1828. [CrossRef]
7. Cuerda-correa, E.M.; Alexandre-franco, M.F.; Fern, C. Antibiotics from Water. An Overview. *Water* **2020**, *12*, 102. [CrossRef]
8. Martins, P.; Kappert, S.; Le, H.N.; Sebastian, V.; Kühn, K.; Alves, M.; Pereira, L.; Cuniberti, G.; Melle-Franco, M.; Lanceros-Méndez, S. Enhanced Photocatalytic Activity of $Au/TiO_2$ Nanoparticles against Ciprofloxacin. *Catalysts* **2020**, *10*, 234. [CrossRef]
9. Hunge, Y.M.; Yadav, A.A.; Khan, S.; Takagi, K.; Suzuki, N.; Teshima, K.; Terashima, C.; Fujishima, A. Photocatalytic Degradation of Bisphenol A Using Titanium Dioxide@nanodiamond Composites under UV Light Illumination. *J. Colloid Interface Sci.* **2021**, *582*, 1058–1066. [CrossRef]
10. Zhang, X.; Chen, Y.L.; Liu, R.-S.; Tsai, D.P. Plasmonic Photocatalysis. *Rep. Prog. Phys.* **2013**, *76*, 046401. [CrossRef]
11. Wang, M.; Ye, M.; Iocozzia, J.; Lin, C.; Lin, Z. Plasmon-Mediated Solar Energy Conversion via Photocatalysis in Noble Metal/Semiconductor Composites. *Adv. Sci.* **2016**, *3*, 1600024. [CrossRef]
12. Wang, C.; Astruc, D. Nanogold Plasmonic Photocatalysis for Organic Synthesis and Clean Energy Conversion. *Chem. Soc. Rev.* **2014**, *43*, 7188–7216. [CrossRef] [PubMed]
13. Liu, L.; Ouyang, S.; Ye, J. Gold-Nanorod-Photosensitized Titanium Dioxide with Wide-Range Visible-Light Harvesting Based on Localized Surface Plasmon Resonance. *Angew. Chemie Int. Ed.* **2013**, *52*, 6689–6693. [CrossRef] [PubMed]
14. Ying, L.; Shuo, C.; Xie, Q.; Hongtao, Y. Fabrication of a $TiO_2$/Au Nanorod Array for Enhanced Photocatalysis. *Chin. J. Catal.* **2011**, *32*, 1838–1843. [CrossRef]
15. Xiao, Y.; Huang, H.; Xue, S.; Zhao, J. Light Switching of Amine Oxidation Products from Oximes to Imines: Superior Activity of Plasmonic Gold Nanorods-Loaded $TiO_2$(B) Nanofibers under Visible-near IR Light. *Appl. Catal. B Environ.* **2020**, *265*, 118596. [CrossRef]
16. Sharma, V.; Kumar, S.; Krishnan, V. Clustered Au on $TiO_2$ Snowman-Like Nanoassemblies for Photocatalytic Applications. *ChemistrySelect* **2016**, *1*, 2963–2970. [CrossRef]
17. Sharma, V.; Kumar, S.; Krishnan, V. Shape Selective $Au-TiO_2$ Nanocomposites for Photocatalytic Applications. *Mater. Today Proc.* **2016**, *3*, 1939–1948. [CrossRef]
18. Sun, H.; Zeng, H.; He, Q.; She, P.; Xu, K.; Liu, Z. Spiky $TiO_2$/Au Nanorod Plasmonic Photocatalysts with Enhanced Visible-Light Photocatalytic Activity. *Dalt. Trans.* **2017**, *46*, 3887–3894. [CrossRef]
19. Wang, L.; Wang, Y.; Schmuki, P.; Kment, S.; Zboril, R. Nanostar Morphology of Plasmonic Particles Strongly Enhances Photoelectrochemical Water Splitting of $TiO_2$ Nanorods with Superior Incident Photon-to-Current Conversion Efficiency in Visible/near-Infrared Region. *Electrochim. Acta* **2018**, *260*, 212–220. [CrossRef]
20. Si, Y.; Cao, S.; Wu, Z.; Ji, Y.; Mi, Y.; Wu, X.; Liu, X.; Piao, L. What Is the Predominant Electron Transfer Process for Au $NRs/TiO_2$ Nanodumbbell Heterostructure under Sunlight Irradiation? *Appl. Catal. B Environ.* **2018**, *220*, 471–476. [CrossRef]
21. Atta, S.; Pennington, A.M.; Celik, F.E.; Fabris, L. $TiO_2$ on Gold Nanostars Enhances Photocatalytic Water Reduction in the Near-Infrared Regime $TiO_2$ on Gold Nanostars Enhances Photocatalytic Water Reduction in the Near-Infrared Regime. *CHEM* **2018**, *4*, 2140–2153. [CrossRef]
22. Liu, Y.; Xiao, Z.; Cao, S.; Li, J.; Piao, L. Controllable Synthesis of $Au-TiO_2$ Nanodumbbell Photocatalysts with Spatial Redox Region. *Chin. J. Catal.* **2020**, *41*, 219–226. [CrossRef]
23. Zhang, H.; Li, X.; Chooi, K.S.; Jaenicke, S.; Chuah, G. $TiO_2$ Encapsulated Au Nanostars as Catalysts for Aerobic Photo-Oxidation of Benzyl Alcohol under Visible Light. *Catal. Today* **2020**, *375*, 558–564. [CrossRef]
24. Khoury, C.G.; Vo-dinh, T. Gold Nanostars For Surface-Enhanced Raman Scattering: Synthesis, Characterization and Optimization. *J. Phys. Chem. C* **2008**, *112*, 18849–18859. [CrossRef] [PubMed]
25. Guerrero-Martínez, A.; Barbosa, S.; Pastoriza-Santos, I.; Liz-Marzán, L.M. Current Opinion in Colloid & Interface Science Nanostars Shine Bright for You Colloidal Synthesis, Properties and Applications of Branched Metallic Nanoparticles. *Curr. Opin. Colloid Interface Sci.* **2011**, *16*, 118–127. [CrossRef]
26. Koczkur, K.M.; Mourdikoudis, S.; Polavarapu, L.; Skrabalak, S.E. Polyvinylpyrrolidone (PVP) in Nanoparticle Synthesis. *Dalt. Trans.* **2015**, *44*, 17883–17905. [CrossRef]

27. Ramsey, J.D.; Zhou, L.; Almlie, C.K.; Lange, J.D.; Burrows, S.M. Achieving Plasmon Reproducibility from Surfactant Free Gold Nanostar Synthesis. *New J. Chem.* **2015**, *39*, 9098–9108. [CrossRef]
28. Reyes, N.J.D.G.; Geronimo, F.K.F.; Yano, K.A.V.; Guerra, H.B.; Kim, L.-H. Pharmaceutical and Personal Care Products in Different Matrices: Occurrence, Pathways, and Treatment Processes. *Water* **2021**, *13*, 1159. [CrossRef]
29. Shurbaji, S.; Huong, P.T.; Altahtamouni, T.M. Review on the Visible Light Photocatalysis for the Decomposition of Ciprofloxacin, Norfloxacin, Tetracyclines, and Sulfonamides Antibiotics in Wastewater. *Catalysts* **2021**, *11*, 437. [CrossRef]
30. Singh, S.; Mahalingam, H.; Singh, P.K. Polymer-Supported Titanium Dioxide Photocatalysts for Environmental Remediation: A Review. *Appl. Catal. A Gen.* **2013**, *462–463*, 178–195. [CrossRef]
31. Akerdi, A.G.; Bahrami, S.H. Application of Heterogeneous Nano-Semiconductors for Photocatalytic Advanced Oxidation of Organic Compounds: A Review. *J. Environ. Chem. Eng.* **2019**, *7*, 103283. [CrossRef]
32. Martins, P.M.; Ribeiro, J.M.; Teixeira, S.; Petrovykh, D.Y.; Cuniberti, G.; Pereira, L.; Lanceros-Méndez, S. Photocatalytic Microporous Membrane against the Increasing Problem of Water Emerging Pollutants. *Materials* **2019**, *12*, 1649. [CrossRef] [PubMed]
33. Salazar, H.; Martins, P.M.; Santos, B.; Fernandes, M.M.; Reizabal, A.; Sebastian, V.; Botelho, G.; Tavares, C.J.; Vilas-Viela, J.L.; Lanceros-Mendez, S. Photocatalytic and Antimicrobial Multifunctional Nanocomposite Membranes for Emerging Pollutants Water Treatment Applications. *Chemosphere* **2020**, *250*, 126299. [CrossRef] [PubMed]
34. Liu, F.; Hashim, N.A.; Liu, Y.; Abed, M.R.M.; Li, K. Progress in the Production and Modification of PVDF Membranes. *J. Memb. Sci.* **2011**, *375*, 1–27. [CrossRef]
35. Reguera, J.; Jiménez De Aberasturi, D.; Henriksen-Lacey, M.; Langer, J.; Espinosa, A.; Szczupak, B.; Wilhelm, C.; Liz-Marzán, L.M. Janus Plasmonic-Magnetic Gold-Iron Oxide Nanoparticles as Contrast Agents for Multimodal Imaging. *Nanoscale* **2017**, *9*, 9467–9480. [CrossRef] [PubMed]
36. Reguera, J.; de Aberasturi, D.J.; Winckelmans, N.; Langer, J.; Bals, S.; Liz-Marzán, L.M. Synthesis of Janus Plasmonic-Magnetic, Star-Sphere Nanoparticles, and Their Application in SERS Detection. *Faraday Discuss.* **2016**, *191*, 47–59. [CrossRef]
37. Espinosa, A.; Reguera, J.; Curcio, A.; Muñoz-Noval, Á.; Kuttner, C.; Van De Walle, A.; Liz-Marzán, L.M.; Wilhelm, C. Janus Magnetic-Plasmonic Nanoparticles for Magnetically Guided and Thermally Activated Cancer Therapy. *Small* **2020**, *16*, 1904960. [CrossRef]
38. Fonseca-Cervantes, O.R.; Alejandro, P.; Romero, H.; Sulbaran-Rangel, B. Effects in Band Gap for Photocatalysis in $TiO_2$ Support by Adding Gold and Ruthenium. *Processes* **2020**, *8*, 1032. [CrossRef]
39. Sentein, C.; Guizard, B.; Giraud, S.; Yé, C.; Ténégal, F. Dispersion and Stability of $TiO_2$ Nanoparticles Synthesized by Laser Pyrolysis in Aqueous Suspensions. *J. Phys.* **2009**, *170*, 012013. [CrossRef]
40. Israelachvili, J.N. *Intermolecular and Surface Forces*, 3rd ed.; Elsevier: Amsterdam, The Netherlands, 2011; ISBN 9780123751829.
41. Chen, W.; Zhang, J.; Cai, W. Sonochemical Preparation of Au, Ag, Pd/$SiO_2$ Mesoporous Nanocomposites. *Scr. Mater.* **2003**, *48*, 1061–1066. [CrossRef]
42. Fujishima, A.; Zhang, X.; Tryk, D.A. $TiO_2$ Photocatalysis and Related Surface Phenomena. *Surf. Sci. Rep.* **2008**, *63*, 515–582. [CrossRef]
43. Khore, S.K.; Kadam, S.R.; Naik, S.D.; Kale, B.B.; Sonawane, R.S. Solar Light Active Plasmonic Au@$TiO_2$ Nanocomposite with Superior Photocatalytic Performance for $H_2$ Production and Pollutant Degradation. *New J. Chem.* **2018**, *42*, 10958–10968. [CrossRef]
44. Durán-Álvarez, J.C.; Avella, E.; Ramírez-Zamora, R.M.; Zanella, R. Photocatalytic Degradation of Ciprofloxacin Using Mono- (Au, Ag and Cu) and Bi- (Au–Ag and Au–Cu) Metallic Nanoparticles Supported on $TiO_2$ under UV-C and Simulated Sunlight. *Catal. Today* **2016**, *266*, 175–187. [CrossRef]
45. Gan, Y.; Zhang, M.; Xiong, J.; Zhu, J.; Li, W.; Zhang, C.; Cheng, G. Impact of Cu Particles on Adsorption and Photocatalytic Capability of Mesoporous Cu@ $TiO_2$ Hybrid towards Ciprofloxacin Antibiotic Removal. *J. Taiwan Inst. Chem. Eng.* **2019**, *96*, 229–242. [CrossRef]
46. Photocatalysts, A.T.; Mach, A.; Font, K.; Garc, D.; Sampayo, P.; Col, C.; Claudio-serrano, G.J.; Sotov, L.; Resto, E.; Petrescu, F.I.; et al. Hydrogen Production and Degradation of Ciprofloxacin by Ag@$TiO_2$-$MoS_2$ Photocatalysts. *Catalysts* **2022**, *12*, 267. [CrossRef]
47. Zhang, Y.; He, S.; Guo, W.; Hu, Y.; Huang, J.; Mulcahy, J.R.; Wei, W.D. Surface-Plasmon-Driven Hot Electron Photochemistry. *Chem. Rev.* **2018**, *118*, 2927–2954. [CrossRef]
48. Jiang, W.; Bai, S.; Wang, L.; Wang, X.; Yang, L.; Li, Y.; Liu, D.; Wang, X.; Li, Z.; Jiang, J.; et al. Integration of Multiple Plasmonic and Co-Catalyst Nanostructures on $TiO_2$ Nanosheets for Visible-Near-Infrared Photocatalytic Hydrogen Evolution. *Small* **2016**, *12*, 1640–1648. [CrossRef]
49. Nishijima, Y.; Ueno, K.; Yokota, Y.; Murakoshi, K.; Misawa, H. Plasmon-Assisted Photocurrent Generation from Visible to near-Infrared Wavelength Using a Au-Nanorods/$TiO_2$ Electrode. *J. Phys. Chem. Lett.* **2010**, *1*, 2031–2036. [CrossRef]
50. Ribeiro, C.; Costa, C.M.; Correia, D.M.; Nunes-Pereira, J.; Oliveira, J.; Martins, P.; Gonçalves, R.; Cardoso, V.F.; Lanceros-méndez, S. Electroactive Poly(Vinylidene Fluoride)-Based Structures for Advanced Applications. *Nat. Protoc.* **2018**, *13*, 681–704. [CrossRef]
51. Serrano-montes, A.B.; Langer, J.; Henriksen-Lacey, M.; de Aberasturi, D.J.; Solís, D.M.; Taboada, J.M.; Obelleiro, F.; Bals, S.; Bekdemir, A.; Stellacci, F.; et al. Gold Nanostar-Coated Polystyrene Beads as Multifunctional Nanoprobes for SERS Bioimaging. *J. Phys. Chem. C* **2016**, *120*, 20860–20868. [CrossRef]
52. Abdullahi, S.S.; Güner, S.; Koseoglu, Y.; Musa, I.M.; Adamu, B.I.; Abdulhamid, M.I. Simple Method for the Determination of Band Gap of a Nanopowdered Sample Using Kubelka Munk Theory. *J. Niger. Assoc. Math. Phys.* **2016**, *35*, 241–246.

53. Sakthivel, S.; Hidalgo, M.C.; Bahnemann, D.W.; Geissen, S.U.; Murugesan, V.; Vogelpohl, A. A Fine Route to Tune the Photocatalytic Activity of TiO$_2$. *Appl. Catal. B Environ.* **2006**, *63*, 31–40. [CrossRef]
54. Karmakar, S. Particle Size Distribution and Zeta Potential Based on Dynamic Light Scattering: Techniques to Characterize Stability and Surface Charge Distribution of Charged Colloids. In *Recent Trends in Materials: Physics and Chemistry*; Studium Press: New Delhi, India, 2019; pp. 117–159.

### Article

# The Synergistic Effect of Adsorption-Photocatalysis for Removal of Organic Pollutants on Mesoporous $Cu_2V_2O_7/Cu_3V_2O_8/g$-$C_3N_4$ Heterojunction

Jian Feng, Xia Ran, Li Wang, Bo Xiao, Li Lei, Jinming Zhu, Zuoji Liu, Xiaolan Xi, Guangwei Feng, Zeqin Dai * and Rong Li *

Engineering Research Center for Molecular Medicine, School of Basic Medical Sciences, Guizhou Medical University, Guiyang 550025, China
* Correspondence: daizeqin300@163.com (Z.D.); lirong1@gmc.edu.cn (R.L.); Tel.: +86-851-88174017 (Z.D. & R.L.)

**Abstract:** $Cu_2V_2O_7/Cu_3V_2O_8/g$-$C_3N_4$ heterojunctions (CVCs) were prepared successfully by the reheating synthesis method. The thermal etching process increased the specific surface area. The formation of heterojunctions enhanced the visible light absorption and improved the separation efficiency of photoinduced charge carriers. Therefore, CVCs exhibited superior adsorption capacity and photocatalytic performance in comparison with pristine g-$C_3N_4$ (CN). CVC-2 (containing 2 wt% of $Cu_2V_2O_7/Cu_3V_2O_8$) possessed the best synergistic removal efficiency for removal of dyes and antibiotics, in which 96.2% of methylene blue (MB), 97.3% of rhodamine B (RhB), 83.0% of ciprofloxacin (CIP), 86.0% of tetracycline (TC) and 80.5% of oxytetracycline (OTC) were eliminated by the adsorption and photocatalysis synergistic effect under visible light irradiation. The pseudo first order rate constants of MB and RhB photocatalytic degradation on CVC-2 were 3 times and 10 times that of pristine CN. For photocatalytic degradation of CIP, TC and OTC, it was 3.6, 1.8 and 6.1 times that of CN. DRS, XPS VB and ESR results suggested that CVCs had the characteristics of a Z-scheme photocatalytic system. This study provides a reliable reference for the treatment of real wastewater by the adsorption and photocatalysis synergistic process.

**Keywords:** $Cu_2V_2O_7/Cu_3V_2O_8/g$-$C_3N_4$ heterojunctions; adsorption; photocatalysis; degradation; Z-scheme

## 1. Introduction

In recent decades, with the rapid development of industrialization, environmental pollution has become increasingly serious. More and more organic chemicals have been released into the environment. Water pollution has become one of the major obstacles to the sustainable development of human society [1]. Various organic pollutants, such as dyes and antibiotics, are seriously harmful to the ecological environment and human health [2]. Many traditional techniques have been developed to remove the organic pollutants in wastewater, including the bioelectrochemical method [3], electrochemical advanced oxidation processes [4], flocculent precipitation [5], physisorption [6], biological degradation, incineration, membrane filtration, etc. [7]. However, these traditional techniques suffer from the drawbacks, such as not being suitable for low concentrations of pollutant, high operation costs, low removal efficiency and secondary pollution. To date, the integration of adsorption and photocatalysis has been regarded as the most promising technology for the elimination of low-concentration contaminants [8–10]. This technology can combine the advantages of adsorption and photocatalysis, such as high efficiency, low cost, wide availability to adsorbates, superior recoverability and less secondary pollution. The adsorption process can ameliorate the accumulation of contaminants on the catalyst surface from the wastewater and conduce to improving the photocatalytic degradation efficiency.

The photocatalysis can ultimately mineralize the contaminants to $H_2O$ and $CO_2$ under light irradiation at room temperature, which reproduces the surface of the catalyst for the next adsorption process [11–13]. Therefore, the development of semiconductor materials with superior adsorption and photocatalysis performance in wastewater treatment has extensively aroused research interests [14–16].

At present, MXenes [17], hydrogels [18,19], graphene-like nanomaterials [20] and their composites have been extensively studied in the photocatalysis and adsorption fields. Among them, graphitic carbon nitride (CN) is considered as one of the most promising 2D materials for environmental remediation based on its impressive merits of low cost, high stability, proper band gap for visible light harvesting and low toxicity [21,22]. Many researchers have reported the adsorption capability of CN for wastewater treatment [16,23,24]. However, the adsorption and photocatalysis performance of pristine CN still sustains small specific surface area, fast recombination of photoinduced charge carriers and narrow visible light absorption [25]. Accordingly, lots of technologies such as elemental doping [25], heterostructure construction [26–28], morphology control [29,30] and defect engineering [31] have been devoted to ameliorating its adsorption and photocatalysis performance. Wherein, controlling the morphology of CN and constructing CN-based heterostructures are especially regarded as effective strategies to extend the visible light absorption range, increase the specific surface area and adsorption sites and accelerate the separation of photoinduced charge carriers [32]. Ultrathin CN nanosheets acquired by the exfoliation process have a much larger specific surface area than bulk CN, which is instrumental in promoting the adsorption capability [30]. Moreover, the diffusion distance of photoinduced charge carriers is shortened in ultrathin CN nanosheets. It can reduce the recombination of photoinduced charge carriers and boost the photocatalytic performance [33]. Additionally, the construction of CN-based heterojunctions via loaded metal oxides is another method to improve the photocatalytic performance [32]. The spatial separation of photoinduced charge carriers at the interface of two semiconductors will prohibit their recombination. Moreover, the formation of CN-based heterojunctions can promote band matching and light absorption, and enhance photocatalytic reaction activity [21]. Therefore, manufacturing CN-based heterojunctions with ultrathin structures should be a feasible strategy to raise the adsorption and photocatalysis synergy for removal of contaminants.

Copper vanadates are a class of catalysts with band gap of ~2 eV, making them suitable for visible light absorption [34]. They thus exhibit dye degradation activity as photocatalysts [35,36] and water splitting property as photoanode candidates [37–39]. High-throughput research results show that copper vanadates should be a novel class of materials for photocatalytic application [38,40]. Copper vanadates with different stoichiometric ratios, such as $CuV_2O_6$ [37], $Cu_2V_2O_7$ [34,36] and $Cu_3V_2O_8$ [34,41], can be prepared by changing the molar ratio of Cu:V and the synthesis method. Khan et al. synthesized $CuV_2O_6$ and $Cu_2V_2O_7$ via a sonication assisted sol–gel method with a band gap of 1.84 eV and 2.2 eV, respectively [35]. $CuV_2O_6$, $Cu_2V_2O_7$ and $Cu_5V_2O_{10}$ were prepared using a sol–gel method by adjusting the Cu:V [39]. Then, these copper vanadates were applied as photocatalysts on the selective oxidation of cyclohexane. Keerthana et al. synthesized β-$Cu_2V_2O_7$, CTAB-β-$Cu_2V_2O_7$ and PVP-$Cu_3V_2O_8$ by a hydrothermal method [36]. The bandgaps were 3.09 eV, 2.97 eV and 2.28 eV, respectively. PVP-$Cu_3V_2O_8$ presented 96%, 77% and 96% photocatalytic removal efficiency for MB, RhB and malachite green dyes. β-$Cu_2V_2O_7$ nanorods were also synthesized and the photocatalytic degradation property was assessed, with 81.85% of MB degraded within 60 min of visible light irradiation [42]. $Cu_3V_2O_8$ nanoparticles were produced via a precipitation approach using Schiff base as the ligand. They exhibited 79% MB degradation efficiency under UV irradiation [43]. Actually, despite the favorable band gaps for visible light harvesting, copper vanadates still manifested low photocatalytic efficiencies [37].

To date, element doping and heterojunction constructing are the major approaches that have been investigated to enhance photocatalytic efficiency of copper vanadates. Indium-doped $CuV_2O_6$ was prepared by the hydrothermal method and it revealed efficient optical absorption from the UV to visible region with gap energy of 1.96 eV [44]. About 95% of RhB was removed during visible light irradiation for 120 min over $CuV_2O_6$:$In^{3+}$. Contrastively, only 57% of RhB was eliminated over $CuV_2O_6$ under the same conditions. Furthermore, copper vanadate-based heterojunctions, involving in $Cu_2V_2O_7/g-C_3N_4$ [45], $Cu_2V_2O_7/CoFe_2O_4/g-C_3N_4$ [46], $Cu_2O/Cu_2V_2O_7$ [47], r-GO/β-$Cu_2V_2O_7$/$TiO_2$ [48], β-$Cu_2V_2O_7/Zn_2V_2O_6$ [49] and $Cu_2V_2O_7/Cu_3V_2O_8$ [41] were manufactured to enhance photocatalytic activities. Although the relevant studies have shown important progress, more extensive efforts should be made to enhance photocatalytic efficiency of copper vanadates. Herein, $Cu_2V_2O_7/Cu_3V_2O_8$ (CV) was loaded on ultrathin CN nanosheets to form CVC heterojunctions. The exfoliation of bulk CN increased the specific surface area and shortened the diffusion distance of photoinduced charge carriers. The construction of CVCs further accelerated the separation of charge carriers. Thus, the adsorption capability and photocatalytic degradation activity could be remarkably enhanced. The removal of CIP, TC, OTC, MB and RhB was investigated to discuss the synergistic effect of adsorption and photocatalysis.

## 2. Results and Discussion

The XRD patterns of CN, CVC-2, CVC-5, CVC-10 and CVC-20 heterojunctions are depicted in Figure 1a. All samples presented two peaks at 13.0° and 27.6°, corresponding to (110) and (022) crystal planes of CN. They demonstrated the existence of CN in these samples [16]. The diffraction peaks of CV were undetectable in CVC-2, CVC-5 and CVC-10. This may have resulted from the low dosage of CV in these heterojunctions, which could not cause the change of the chemical skeleton and structure of CN. This result was previously clarified by the literature [50,51]. Moreover, the diffraction peak intensity of CV in CVC-2, CVC-5 and CVC-10 were weaker while CV nanoparticles were covered by CN nanosheets [51]. Comparatively, the diffraction peaks of both CN and CV were all found in CVC-20. In fact, the existence of CV in CVC-2, CVC-5 and CVC-10 could be corroborated by EDS and elemental mapping results. SEM images and corresponding EDS of CVC-2 are displayed in Figure S1. It was revealed that the mass percentage of CV in CVC-2 was about 2.11%, which was well in accordance with the value calculated from the content of $Cu(NO_3)_3$ and $NH_4VO_3$ in the precursor. The chemical composition and uniformity of CVC-2 were confirmed by the elemental mappings of C, N, O, Cu and V. All elements can be observed in Figure S2b–f. The uniformity of C, N, O, Cu and V demonstrated the homogeneous distribution of CV on CN nanosheets. In addition, the XRD patterns of CV conformed well to the monoclinic phase $Cu_2V_2O_7$ (PDF#73-1032) and monoclinic $Cu_3V_2O_8$ (PDF#74-1503), indicating that CV was the composite of $Cu_2V_2O_7$ and $Cu_3V_2O_8$ (Figure 1b). These XRD results definitely proved the successful formation of CVCs.

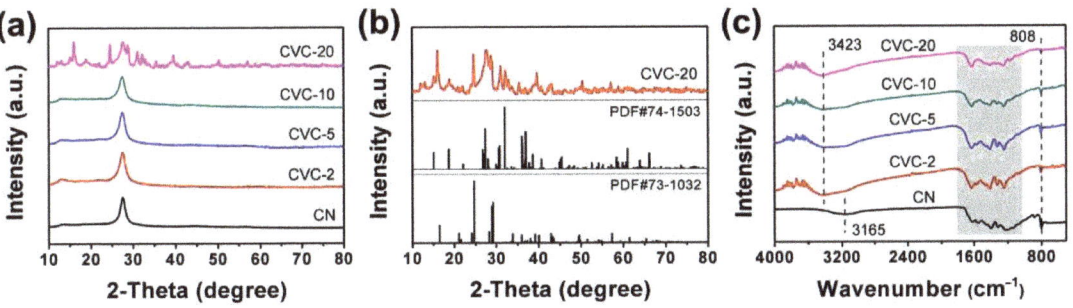

**Figure 1.** (**a**) XRD patterns of CN, CVC-2, CVC-5, CVC-10 and CVC-20, (**b**) XRD patterns of CVC-20, monoclinic $Cu_3V_2O_8$ (PDF#74-1503) and monoclinic $Cu_2V_2O_7$ (PDF#73-1032), (**c**) FTIR spectra of CN, CVC-2, CVC-5, CVC-10 and CVC-20.

The FTIR spectra of CN, CVC-2, CVC-5, CVC-10 and CVC-20 are shown in Figure 1c. The sharp adsorption peaks at 806 and 886 cm$^{-1}$ were assigned to breathing vibration and N-H deformation vibration mode of CN s-triazine, respectively. The characteristic adsorption band in the range of 1200–1700 cm$^{-1}$ was ascribed to the stretching vibrations of C–N and C=N of aromatic CN heterocycles [52]. The broad adsorption band at 2900–3500 cm$^{-1}$ was attributed to N-H stretching vibration of residual unpolymerized amino groups. This adsorption band of CN centered at 3165 cm$^{-1}$ shifted to 3423 cm$^{-1}$ while CV was loaded on the CN nanosheets. This change of the peak position implied the strong interaction between CV nanoparticles and CN nanosheets [32].

TEM and HRTEM images of CVC-2 are displayed in Figure 2. As shown in Figure 2a,b, there was an ultrathin layered structure of CN nanosheets. It profited from the thermal etching effect in the reheating synthesis process. Especially, the porous structure of CVC-2 could also be observed (Figure 2a,b). The pore diameter was between several and tens of nanometers [50]. This porous structure might originate from the generation of NH$_3$ and HCl in the polymerization process of dicyandiamine and NH$_4$Cl [53]. The ultrathin and porous structure increased the specific surface area of CVC-2, which was proved by BET results. The increased specific surface area could improve the adsorption capability of CVC-2. CV nanoparticles with evident aggregation were also observed in Figure 2a,b. Therefore, the diameters of CV nanoparticles could not be measured accurately from the TEM image. They were estimated to be 50–100 nm in diameter. From Figure 2a,b, it can also be seen that CV nanoparticles were located on CN nanosheets. The spacing of lattice fringes of 0.360 and 0.325 nm are distinctly observed in HRTEM images illustrated in Figure 2c,d. This was consistent with the (200) and (111) crystal planes of monoclinic phase $Cu_2V_2O_7$ (PDF#73-1032) and monoclinic $Cu_3V_2O_8$ (PDF#74-1503), respectively. Moreover, the intimate contact of $Cu_2V_2O_7$ and $Cu_3V_2O_8$ noticeably appeared in Figure 2d, demonstrating the formation of $Cu_2V_2O_7/Cu_3V_2O_8$ composite. Based on the above XRD, FTIR, TEM and HRTEM results, it was concluded that CVC heterojunctions were successfully constructed [54].

The adsorption and photocatalysis performance of dyes (MB and RhB) over CN and CVCs were investigated and the results are depicted in Figure 3. As shown in Figure 3a,d, the adsorption–desorption equilibrium was achieved within 30 min for all samples. The adsorption capacities of MB and RhB on CVC-2 were 3 times that of CN. The enhanced adsorption capacity of CVC-2 indicated the stronger interaction between CVCs and dyes. All CVCs had much higher adsorption capacity of MB than CN. There was no necessary relation between the adsorption capacity and CV content, suggesting that the changed surface charge that resulted from the addition of CV nanoparticles was not the main factor to affect the interaction between CVCs and dyes. The larger specific surface area profited from the thermal etching effect in the reheating synthesis process which improved the adsorption capacity. The adsorption and photocatalytic degradation results revealed in Figure 3a,d manifest that CVC-2 exhibited the best overall performance for removal of dyes, although it removed MB mainly by adsorption and eliminated RhB primarily via photocatalysis (Figure 3c,f). The kinetic data of MB and RhB photocatalytic degradation on CN and CVCs were well fitted by a pseudo first order rate equation (Figure 3b,e). The degradation rate constants of MB and RhB on CVC-2 were, respectively, 0.036 min$^{-1}$ and 0.061 min$^{-1}$, which were about 3 times and 10 times that of pristine CN. The total removal efficiency of MB and RhB on CVC-2 was 96.2% and 97.3%, respectively. It was much higher than that of CN (55.3% for MB and 34.3% for RhB). This could be considered as the result of the synergistic effect of adsorption and photocatalysis.

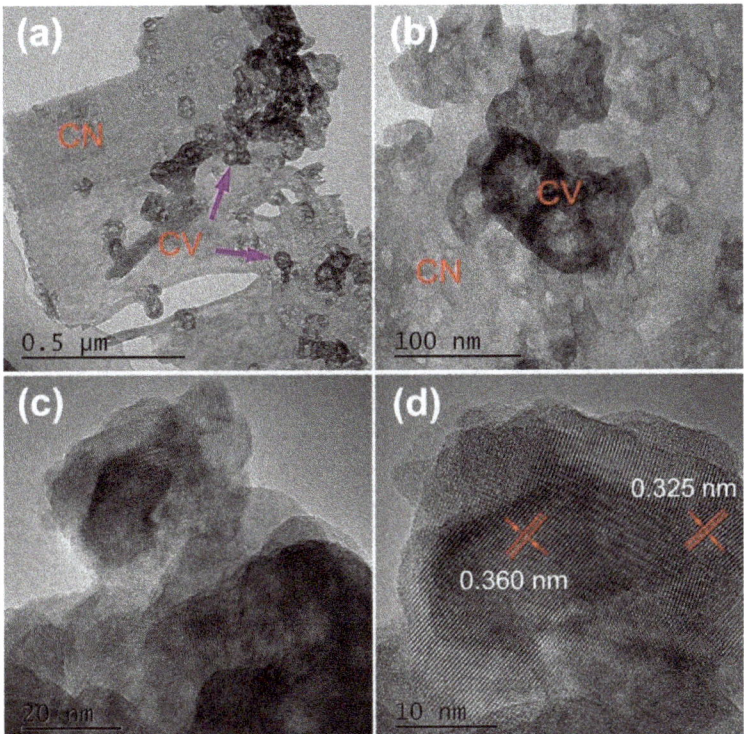

**Figure 2.** (**a**,**b**) TEM and (**c**,**d**) high-resolution TEM images of CVC-2, the scale bar represented 0.5 μm, 100 nm, 20 nm and 10 nm, respectively.

Antibiotics (CIP, TC and OTC) were also selected as the target contaminants to further investigate the synergistic removal effect of adsorption and photocatalysis on CVCs. The results are displayed in Figure 4. As presented in Figure 4a,d,g, the adsorption–desorption equilibrium was achieved within 30 min for all samples. The adsorption capacity of CVCs was much higher than that of pristine CN. The larger specific surface area improved the adsorption capacity. CVC-2 exhibited the best overall performance for removal of antibiotics, although it removed TC mainly by adsorption and eliminated CIP and OTC primarily via photocatalysis (Figure 4c,f,i). The adsorption capacity of CVC-2 was 16.9 (CIP), 58.6 (TC) and 4.2 times (OTC) that of CN, respectively. The total removal efficiency of CIP, TC and OTC on pristine CN was 30.2%, 17.5% and 28.0% during the adsorption and photocatalytic degradation process. In contrast, it was 83.0%, 86.0% and 80.5% for CVC-2, respectively. This was considered as the result of the synergistic effect of adsorption and photocatalysis. All CVCs exhibited considerably higher total removal efficiency than CN, indicating the excellent adsorption and photocatalysis performance of CVCs. From Figure 4c,f,i, we found that CVC-2 possessed highest photocatalytic activity. By comparison, the photocatalytic activity of CVC-5, CVC-10 and CVC-20 was decreased with the increase in CV content. This might be derived from the reaction active sites on the surface of CN nanosheets that were excessively occupied by CV nanoparticles in CVC-5, CVC-10 and CVC-20 [16,32]. Moreover, the removal efficiency of dyes and antibiotics on CVC-2 was compared with various recent reported results. As depicted in Table S1, the diversity degradation activities were higher than literature values [25,28,52,55–59], signifying that CVC-2 is probably a valuable catalyst of practical application in environmental sewage treatment. The kinetic data of CIP, TC and OTC photocatalytic degradation on CN and CVCs were well fitted by a pseudo first order rate equation (Figure 4b,e,h). The degradation rate constants of CVC-2 were 0.017 min$^{-1}$ (CIP), 0.0049 min$^{-1}$ (TC) and 0.014 min$^{-1}$ (OTC). These were 3.6, 1.8 and 6.1 times that of pristine CN, respectively. This result indicated that the loaded CV nanoparticles on the surface of CN nanosheets could improve the photocatalytic activity.

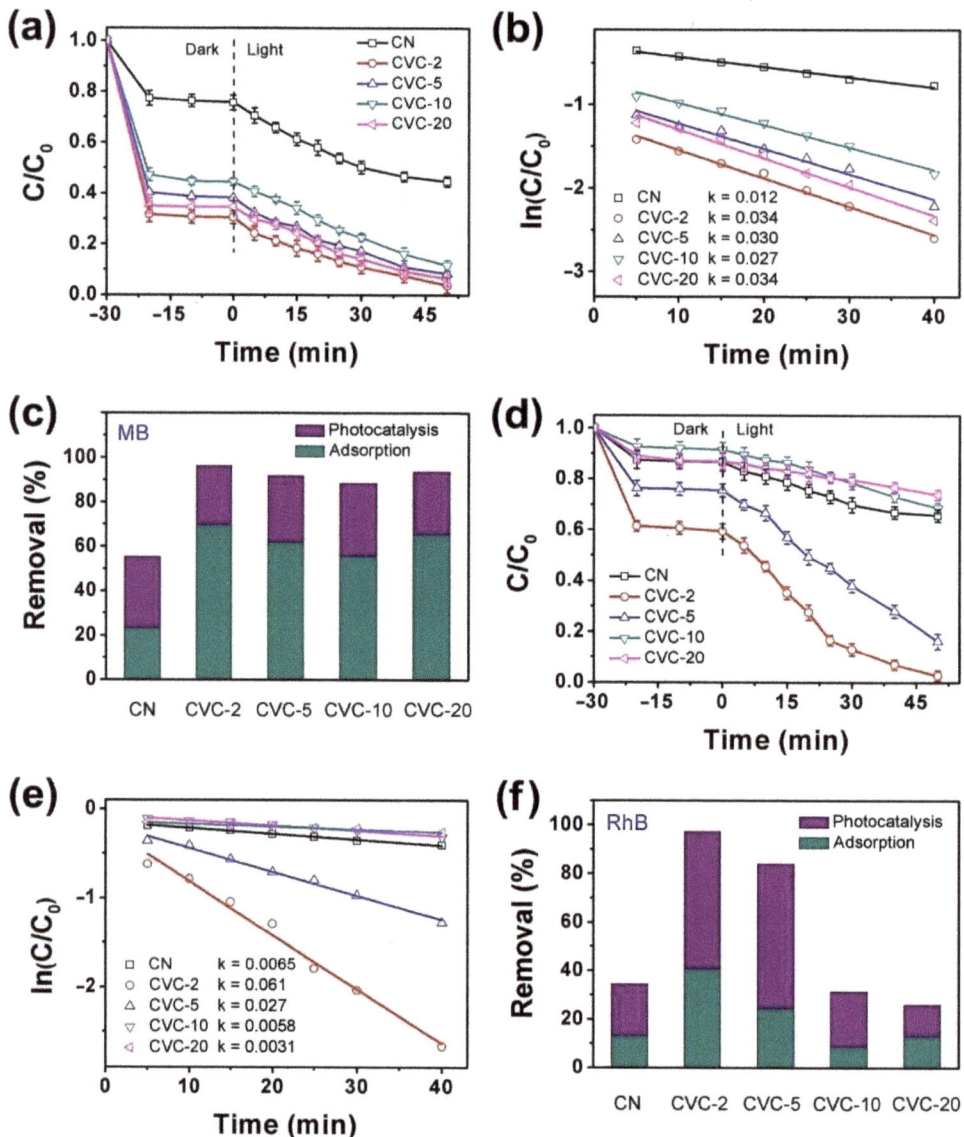

**Figure 3.** Adsorption and photocatalysis synergistic removal of (**a**) MB (concentration: 5 mg/L) and (**d**) RhB (concentration: 5 mg/L) over CN and CVCs (dosage: 100 mg) under 40 W white LED irradiation, fitted by pseudo first order rate equation and the rate constants (k) for photocatalytic degradation of (**b**) MB and (**e**) RhB, the total adsorption and photocatalysis degradation efficiency of (**c**) MB and (**f**) RhB.

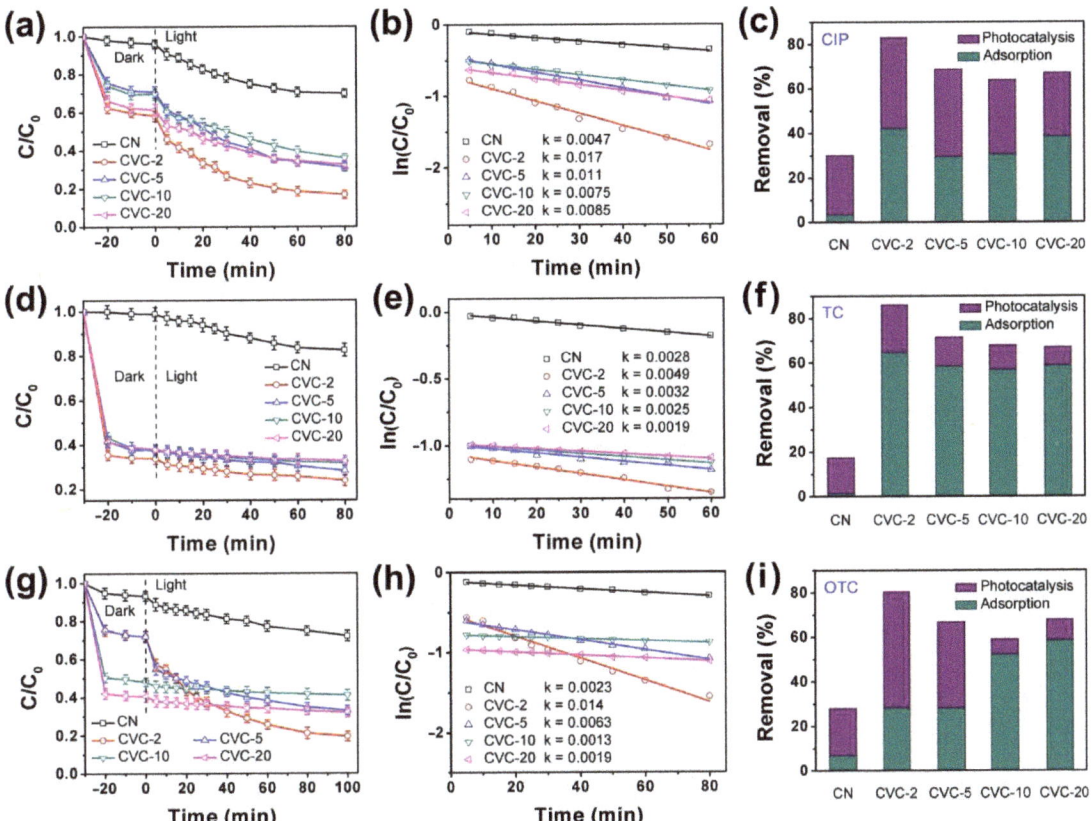

**Figure 4.** Adsorption and photocatalysis synergistic removal of (**a**) CIP (concentration: 4 mg/L), (**d**) TC (concentration: 40 mg/L) and (**g**) OTC (concentration: 20 mg/L) over CN and CVCs (dosage: 100 mg) under 40 W white LED irradiation, fitted by pseudo first order rate equation and the rate constants (k) for photocatalytic degradation of (**b**) CIP, (**e**) TC and (**h**) OTC, the total adsorption and photocatalysis degradation efficiency of (**c**) CIP, (**f**) TC and (**i**) OTC.

The BET surface area, pore volume and average pore diameter of CN and CVC-2 were investigated by $N_2$ adsorption–desorption isotherms. As displayed in Figure 5a, the adsorption–desorption isotherms possessed the features of type IV curves, suggesting the samples had a mesoporous structure [60]. The H3 hysteresis loop at high $P/P_0$ manifested that the mesopores of CN and CVC-2 were irregular. The BET surface area, pore volume and average pore diameter of CVC-2 were evidently greater than those of CN (Figure 5a, inset). They were, respectively, around 4.4, 15.0 and 1.1 times those of CN, which might result from the thermal etching in the reheating synthesis process [61]. The BJH pore size distribution of CN and CVC-2 revealed the wide pore size distribution from 20 to 60 nm, which probably resulted from the aggregation of CN in the reheating synthesis process as displayed previously in the SEM results (Figure S1a). The pore size between 10 and 20 nm originated from the porous structure of CN, which was observed previously in TEM images (Figure 2a,b). Therefore, the increased specific surface area, pore volume and average pore diameter could be instrumental in providing more adsorption and photocatalytic reaction active sites, and thus finally improve the synergistic effect of adsorption and photocatalysis.

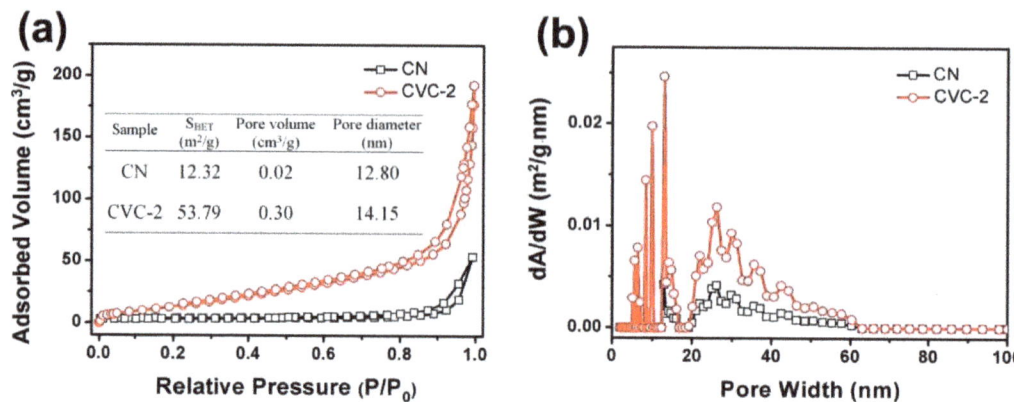

**Figure 5.** (a) the $N_2$ adsorption–desorption isotherms and (b) the corresponding BJH pore size distribution curves of CN and CVC-2.

UV–Vis diffuse reflectance spectra of CN and CVC-2 are exhibited in Figure 6a. The absorption edge of pristine CN was at 457 nm, corresponding to a band gap of 2.71 eV. It was in agreement with the result reported previously [32,52], confirming that pristine CN had proper band gap for visible light harvesting. The absorption edge of CVC-2 was at 488 nm. The red shift of the absorption edge of CVC-2 suggested that the incorporation of CV nanoparticles onto CN nanosheets was instrumental in extending the visible light absorption range. From the DRS spectra in Figure 6a, it could be observed distinctly that CVC-2 possessed stronger light absorption in the wavelength range of 400–800 nm than pristine CN nanosheets. This could result in the improvement of the visible light-driven photocatalytic activity, which was clarified by the adsorption and photocatalysis degradation of CIP, TC and OTC depicted in Figure 4. The photoluminescence spectra (PL) were used to investigate the separation and recombination process of photogenerated charge carriers in CN and CVC-2 (Figure 6b). The higher PL intensity commonly indicates that photogenerated charge carriers have lower separation efficiency and faster recombination rate [32]. The PL emission peak intensity of CVC-2 was remarkably weaker than that of CN. It corroborated that the separation and recombination of photogenerated charge carriers was effectively ameliorated while CVC heterostructure was formed by incorporating CV nanoparticles into CN nanosheets. It could boost the photocatalytic activity of CVCs. The property of separation and recombination of the charge carriers in CN and CVC-2 could be further acquired from the photochemical measurements. TPC spectra and EIS Nyquist plots of CN and CVC-2 are exhibited in Figure 6c,d. The TPC spectra revealed that CVC-2 had better photostability and higher photocurrent density than CN (Figure 6c). The photocurrent density of CVC-2 was about 2.6 times that of pristine CN. It suggested that the photogenerated charge carriers possessed higher transfer rate in CVC-2, which was beneficial to the migration of the charge carriers to the surface of CVC-2 and consequently to improving its photocatalytic activity. EIS Nyquist plots of CN and CVC-2 are shown in Figure 6d. The arc radius of CN and CVC-2 under visible light irradiation was smaller than that detected in the dark. It demonstrated that the photoelectrode conductivity would be increased under the light irradiation condition. In addition, the arc radius of CVC-2 was less than that of pristine CN in either case, indicating that CVC-2 had higher photogenerated charge transfer efficiency. It was consistent with the previous PL and TPC results, demonstrating that the formation of CVC heterostructure could effectively facilitate the separation and transfer of the photogenerated charge carriers, suppress their recombination and finally improve the photocatalytic activity. The deduction obtained from the DRS, PL, TPC and EIS measurements conformed to the photocatalytic degradation experimental results of dyes and antibiotics.

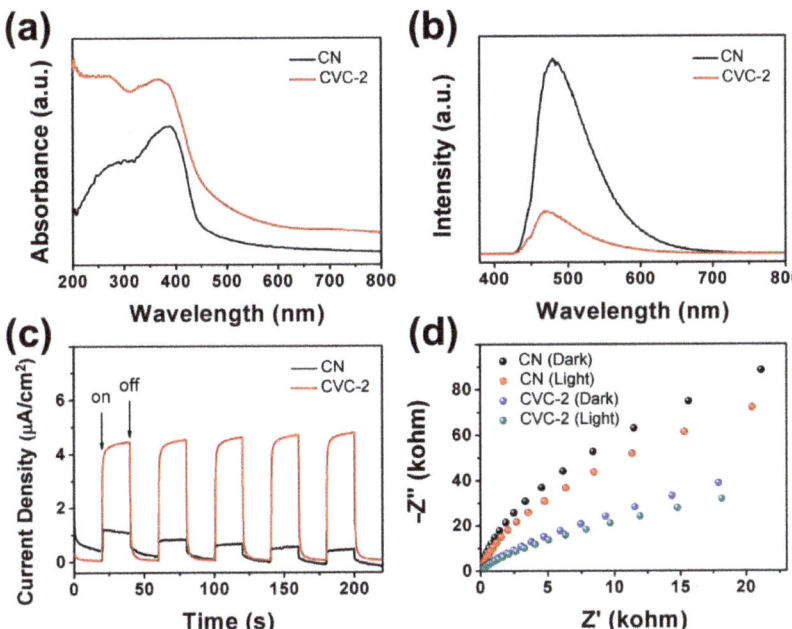

**Figure 6.** (a) UV–Vis diffuse reflectance spectra, (b) PL spectra, (c) TPC spectra and (d) EIS Nyquist plots of CN and CVC-2.

The reactive species involved in the photocatalytic degradation reaction were evaluated by radical scavenger experiments. NaNO$_3$, ammonium oxalate(AO), isopropyl alcohol(IPA) and p-benzoquinone (pBQ) were adopted as the scavengers to trap e$^-$, h$^+$, •OH and •O$_2^-$ in the photocatalytic degradation of CIP over CVC-2, respectively. As revealed in Figure 7a, the removal efficiency of CIP was reduced dramatically after adding AO and pBQ, suggesting that h+ and •O$_2^-$ were the major reactive species in the photocatalytic process. Comparatively, just a slight reduction of degradation efficiency was observed with the addition of NaNO$_3$ and IPA. It indicated that e$^-$ and •OH hardly participated in the photocatalytic reaction. To further confirm the production of •O$_2^-$, ESR spectra of DMPO-•O$_2^-$ over CVC-2 in the CIP photocatalytic system were detected under different light irradiation times and the result is presented in Figure 7b. No obvious ESR signals of •O$_2^-$ were found in the dark, but they appeared after 15 min of visible light irradiation. These results demonstrated that •O$_2^-$ was generated under the visible light irradiation and participated in the photocatalytic degradation process of CIP.

The stability was assessed by four cycles of the adsorption and photocatalysis synergistic removal experiments of MB and CIP on CVC-2. CVC-2 was centrifuged and desorbed in deionized water several times to eliminate absorbed MB and CIP after each cycle experiment. As shown in Figure 8a,b, the removal efficiency of MB and CIP was well maintained after four degradation runs, confirming that CVC-2 had superior stability and excellent potential application prospects in wastewater treatment. TEM, XRD and FTIR were used to characterize the differences of structure and morphology of CVC-2 before and after the synergistic removal experiment. As shown by the TEM, XRD and FTIR results depicted in Figures S3–S5, the phase structure and morphology characteristics were not changed evidently after four degradation runs, which further illustrated the excellent stability of CVC-2.

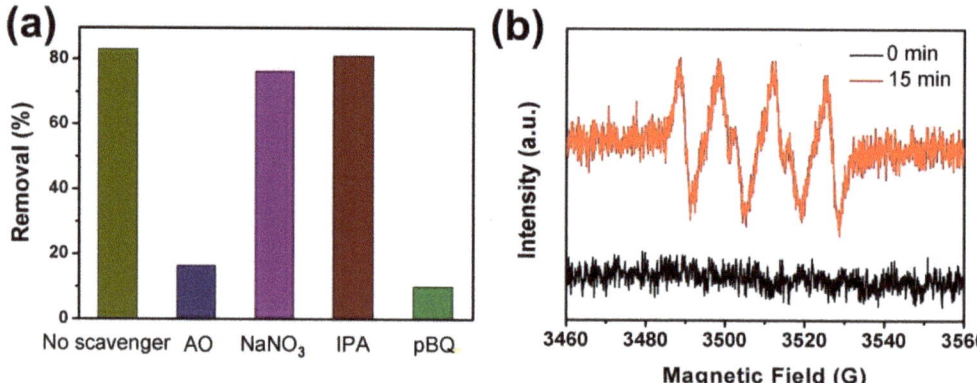

**Figure 7.** (a) The effect of radical scavengers on CIP removal efficiency over CVC-2 under 40 W LED irradiation and (b) ESR spectra of DMPO-•$O_2^-$ in methanol with CVC-2 in CIP removal system under different light irradiation times.

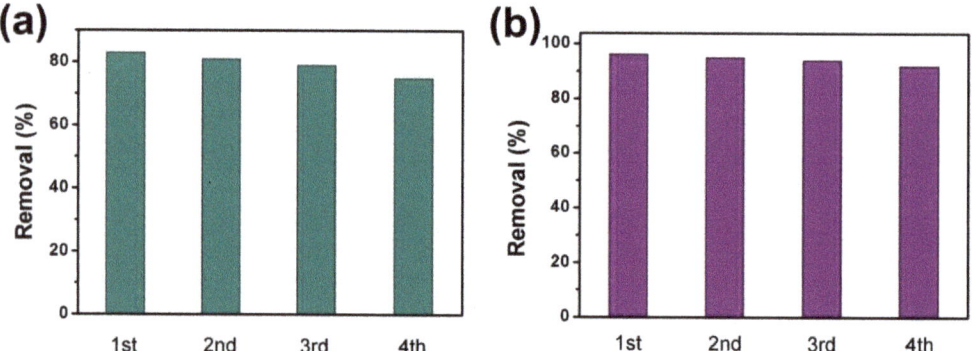

**Figure 8.** The stability experiments for the synergistic removal of (a) CIP and (b) MB over CVC-2.

The energy band structure of CN and CV was obtained by DRS and XPS VB spectra. The DRS spectrum of CV is displayed in Figure S6. It exhibited the absorption edge of 628 nm, corresponding to the band gap of 1.97 eV. It was in agreement with that β-$Cu_2V_2O_7$ in the literature [62]. XPS VB spectra of CN and CV are shown in Figure S7. The $E_{VB}$ of CN and CV was 1.58 eV and 2.20 eV, respectively. The $E_{CB}$ of CN and CV could be estimated based on the following equation: $E_g = E_{VB} - E_{CB}$. It was -1.13 eV and 0.23 eV, respectively. The $E_{CB}$ of CV was more positive than the redox potential of $O_2$/•$O_2^-$ (−0.33 V) [52], meaning that $O_2$ adsorbed on CV could not be reduced to •$O_2^-$. Therefore, the reduction reaction of $O_2$ could only occur on the CB of CN. ESR spectra of DMPO-•OH were also measured under visible light irradiation with the presence of CVC-2 (Figure S8). They confirmed the production of •OH although the radical scavenger experiments indicated that •OH radicals did not participate in the photocatalytic reaction. Furthermore, the $E_{VB}$ of CN was more negative than the redox potential of •OH/OH⁻ (+1.99 V) and •OH/$H_2O$ (+2.37 V) [63]. The •OH radicals could not be generated on CN, and only formed on the VB of CV. These results clearly suggested that CVC heterojunctions possessed the characteristics of Z-scheme photocatalytic systems.

The possible photocatalytic mechanism of dyes and antibiotics on CVCs is consequently proposed in Figure 9. Firstly, the photoinduced electrons (e⁻) were excited and transferred to the CB of CN and CV under visible light irradiation. The photoinduced holes (h⁺) were still retained in the VB. Secondly, the electrons in the CB of CV transferred to CN and recombined with the holes located in the VB of CN. This recombination could significantly accelerate the separation of the photoinduced electrons and holes of CN and CV (Equation(1)), which was the advantage of Z-scheme heterojunctions [64,65]. Thirdly, the electrons on CN reduced oxygen to generate $\bullet O_2^-$ and then $\bullet O_2^-$ further oxidized dyes and antibiotics (Equations(2) and (3)). The holes on CV directly oxidized dyes and antibiotics to small molecules (Equation(4)). The photocatalytic degradation process of dyes and antibiotics could be elaborated as follows:

$$CVC\text{-}2 + h\nu \rightarrow CN\ (e^-) + CV\ (h^+), \tag{1}$$

$$CN\ (e^-) + O_2 \rightarrow \bullet O_2^-, \tag{2}$$

$$\text{Dyes and antibiotics} + \bullet O_2^- \rightarrow \text{degraded products}, \tag{3}$$

$$\text{Dyes and antibiotics} + CV\ (h^+) \rightarrow \text{degraded products}. \tag{4}$$

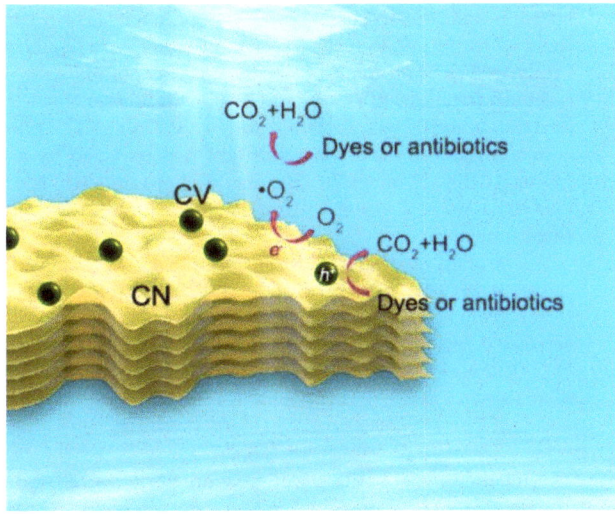

**Figure 9.** The possible mechanism for photocatalytic degradation of dyes and antibiotics on CVCs.

## 3. Materials and Methods

### 3.1. Synthesis of CVCs

Preparation of CN nanosheets: 1:1 mole ratio of dicyandiamine and NH$_4$Cl was mixed and calcinated for 4 h at 550 °C [61]. One hundred micrograms of obtained powder was ground and dispersed in 100 mL deionized water. The mixture was treated ultrasonically for 8 h and then centrifuged for 10 min at 5000 rpm. The as-prepared CN nanosheets in the supernatant were separated by vacuum freeze-drying.

Preparation of CVCs: 100 mg of CN nanosheets, 1:1 mole ratio of Cu(NO$_3$)$_3$ and NH$_4$VO$_3$ was dispersed in 100 mL of deionized water. Then, the water was evaporated and annealed at 500 °C for 2 h. A series of CVCs were synthesized by adjusting the mass ratio of CV and CN. The as-prepared CVC-2, CVC-5, CVC-10 and CVC-20 contained 2, 5, 10 and 20 wt% of CV, respectively. CV nanoparticles were synthesized under the same conditions except for the absence of CN nanosheets.

### 3.2. Catalyst Characterization

X-ray diffraction (XRD) patterns of CN, CV and CVCs were tested on a Rigaku Smartlab diffractometer equipped with a Cu-Kα radiation source. Transmission electron microscopy (TEM) and high-resolution TEM (HRTEM) images of CVCs were taken on a

JEOL-2100F system. Fourier-transform infrared spectra (FTIR) of CN and CVCs were detected on a Nicolet NEXUS 470 spectrometer in the range of 4000–500 cm$^{-1}$. Scanning electron microscopy (SEM) images, energy dispersion spectrum (EDS) and elemental mapping images of CVC-2 were characterized on a JSM-4800F scanning electron microscope. X-ray photoelectron spectroscopy (XPS) spectra of CN, CV and CVCs were obtained on a Thermo ESCALAB 250XI spectrometer equipped with an AlKα X-ray source. The BET specific surface area and $N_2$ adsorption desorption isotherms of CN and CVCs were recorded on a Micromeritics ASAP 2460 analyzer at 77 K. The ultraviolet–visible diffuse reflectance spectra (DRS) of CN and CVCs were taken on a Shimadzu UV-2401 spectrophotometer equipped with an integrating sphere accessory.

*3.3. Adsorption and Photocatalytic Experiments*

The adsorption of contaminants on CN, CV and CVCs was implemented in the dark before the photocatalysis process. Typically, 50 mg CN, CV or CVCs was dispersed in 100 mL antibiotic (CIP, TC and OTC) or dye (MB and RhB) solution. The pH value of the solution was adjusted using 0.1 M HCl or NaOH. The mixture was stirred continuously in the dark for 30 min to reach adsorption–desorption equilibrium of the contaminants on the catalysts. The photocatalytic degradation of contaminants was evaluated under the visible light irradiation of 40 W white LED. Five milliliters of solution was fetched out at intervals and the catalyst was removed through a 0.22 μm PTFE filter membrane. The contaminant concentration in adsorption solution was detected by a UV–Vis spectrophotometer at 277 nm (CIP), 357 nm (TC), 352 nm (OTC), 664 nm (MB) and 552 nm (RhB). The temperature was fixed at 25 °C in the adsorption and photocatalysis process. The tests were repeated three times.

The stability of CVC-2 was appraised by 4 cycles of adsorption photocatalytic degradation experiments. CVC-2 was centrifuged and desorbed in deionized water several times after each cycle experiment to eliminate absorbed contaminant. Then, CVC-2 was separated and freeze-dried for the usage in the next cycle experiment. The reproducibility of CVC-2 was evaluated by repeating the above process twice. $NaNO_3$, AO, IPA and pBQ were selected as the scavengers to trap $e^-$, $h^+$, •OH and •$O_2^-$, respectively.

*3.4. Photoelectrochemical Measurement*

The photoelectrochemical measurements were conducted on the CHI 660E electrochemical workstation. Ag/AgCl was the reference electrode and Pt foil was the counter electrode. A 0.2 M $Na_2SO_4$ solution was the electrolyte. For the preparation of working electrode, 5 mg of CN or CVCs was mixed with 0.2 mL Nafion and 1.8 mL ethanol ultrasonically. Then, the mixture was dropped on 1 cm$^2$ of FTO glass and dried naturally. The transient photocurrent response (TPC) was tested under the irradiation of 40 W white LED. The irradiation intervals were realized using a mechanical light chopper. The electrochemical impedance spectroscopy (EIS) was measured at a frequency from 100 to 0.01 Hz.

## 4. Conclusions

In summary, CVC heterojunctions with ultrathin structure were successfully prepared by the reheating synthesis process for the adsorption and photocatalysis synergistic removal of various dyes and antibiotics. The thermal etching process increased the specific surface area of CVCs. The formation of heterojunctions enhanced the visible light absorption and improved the separation efficiency of photoinduced charge carriers. These factors simultaneously ameliorated the adsorption capacity and photocatalytic degradation performance of CVCs. CVC-2 exhibited the best synergistic removal efficiency of MB (96.2%), RhB (97.3%), CIP (83.0%), TC (86.0%) and OTC (80.5%). These photocatalytic degradation processes followed the pseudo first order equation. The pseudo first order rate constants of MB, RhB, CIP, TC and OTC photocatalytic degradation on CVC-2 were 3, 10, 3.6, 1.8 and 6.1 times those of pristine CN, respectively. DRS, XPS VB and ESR results suggested that CVCs had the characteristics of Z-scheme photocatalytic systems. Moreover, superoxide radicals and photoinduced holes were proved to be the major active species in the photocatalytic degradation process. This work provides a reliable reference for environmental sewage treatment by the adsorption and photocatalysis synergistic process.

**Supplementary Materials:** The following supporting information can be downloaded at: https://www.mdpi.com/article/10.3390/ijms232214264/s1.

**Author Contributions:** Methodology, Conceptualization, Writing—review and editing, Resources, Funding acquisition, J.F.; Investigation, Visualization, X.R.; Investigation, Data curation, Visualization, L.W.; Project administration, B.X.; Funding acquisition, L.L.; Investigation, Data curation, J.Z.; Investigation, Z.L.; Supervision, X.X.; Resources, Funding acquisition, G.F.; Investigation, Resources, Z.D.; Resources, Funding acquisition, R.L. All authors have read and agreed to the published version of the manuscript.

**Funding:** This research was funded by the National Natural Science Foundations of China, grant number 21865006, the Key Project of Basic Research Program of Guizhou Province, China (No. ZK[2021]022), the Science and Technology Project of Guizhou Province, China (No. [2018]5779-18), and the National Training Programs of Innovation and Entrepreneurship for Undergraduates (S202010660010).

**Institutional Review Board Statement:** Not applicable.

**Informed Consent Statement:** Not applicable.

**Data Availability Statement:** The original data are available from the corresponding author upon reasonable request.

**Conflicts of Interest:** The authors declare no conflict of interest.

# References

1. He, Y.Q.; Zhang, F.; Ma, B.; Xu, N.; Junior, L.B.; Yao, B.; Yang, Q.; Liu, D.; Ma, Z. Remarkably enhanced visible-light photocatalytic hydrogen evolution and antibiotic degradation over g-$C_3N_4$ nanosheets decorated by using nickel phosphide and gold nanoparticles as cocatalysts. *Appl. Surf. Sci.* **2020**, *517*, 146187. [CrossRef]
2. Guo, S.; Yang, W.; You, L.; Li, J.; Chen, J.; Zhou, K. Simultaneous reduction of Cr(VI) and degradation of tetracycline hydrochloride by a novel iron-modified rectorite composite through heterogeneous photo-Fenton processes. *Chem. Eng. J.* **2020**, *393*, 124758. [CrossRef]
3. Ahmad, A.; Priyadarshani, M.; Das, S.; Ghangrekar, M.M. Role of bioelectrochemical systems for the remediation of emerging contaminants from wastewater: A review. *J. Basic Microbiol.* **2022**, *62*, 201–222. [CrossRef] [PubMed]
4. Raj, R.; Tripathi, A.; Das, S.; Ghangrekar, M.M. Removal of caffeine from wastewater using electrochemical advanced oxidation process: A mini review. *Case Stud. Chem. Environ. Eng.* **2021**, *4*, 100129. [CrossRef]
5. Wang, G.; Chang, Q.; Han, X.; Zhang, M. Removal of Cr(VI) from aqueous solution by flocculant with the capacity of reduction and chelation. *J. Hazard. Mater.* **2013**, *248–249*, 115–121. [CrossRef]
6. Uddin, M.K. A review on the adsorption of heavy metals by clay minerals, with special focus on the past decade. *Chem. Eng. J.* **2017**, *308*, 438–462. [CrossRef]
7. Li, M.; Liu, Y.; Zeng, G.; Liu, N.; Liu, S. Graphene and graphene-based nanocomposites used for antibiotics removal in water treatment: A review. *Chemosphere* **2019**, *226*, 360–380. [CrossRef]
8. Zou, W.; Gao, B.; Ok, Y.S.; Dong, L. Integrated adsorption and photocatalytic degradation of volatile organic compounds (VOCs) using carbon-based nanocomposites: A critical review. *Chemosphere* **2019**, *218*, 845–859. [CrossRef]
9. Huang, Y.; Xu, H.; Yang, H.; Lin, Y.; Liu, H.; Tong, Y. Efficient Charges Separation Using Advanced BiOI-Based Hollow Spheres Decorated with Palladium and Manganese Dioxide. *ACS Sustain. Chem. Eng.* **2018**, *6*, 2751–2757. [CrossRef]
10. Anantha, M.S.; Olivera, S.; Hu, C.; Jayanna, B.K.; Reddy, N.; Venkatesh, K.; Muralidhara, H.B.; Naidu, R. Comparison of the photocatalytic, adsorption and electrochemical methods for the removal of cationic dyes from aqueous solutions. *Environ. Technol. Inno.* **2020**, *17*, 100612. [CrossRef]
11. Li, Y.; Lai, Z.; Huang, Z.; Wang, H.; Zhao, C.; Ruan, G.; Du, F. Fabrication of BiOBr/MoS2/graphene oxide composites for efficient adsorption and photocatalytic removal of tetracycline antibiotics. *Appl. Surf. Sci.* **2021**, *550*, 149342. [CrossRef]
12. Kassahun, S.K.; Kiflie, Z.; Kim, H.; Baye, A.F. Process optimization and kinetics analysis for photocatalytic degradation of emerging contaminant using N-doped $TiO_2$-$SiO_2$ nanoparticle: Artificial Neural Network and Surface Response Methodology approach. *Environ. Technol. Inno.* **2021**, *23*, 101761. [CrossRef]
13. Liu, L.; Su, G.; Liu, X.; Dong, W.; Niu, M.; Kuang, Q.; Tang, A.; Xue, J. Fabrication of magnetic core–shell $Fe_3O_4$@$SiO_2$@$Bi_2O_2CO_3$–sepiolite microspheres for the high-efficiency visible light catalytic degradation of antibiotic wastewater. *Environ. Technol. Inno.* **2021**, *22*, 101436. [CrossRef]

14. Wang, Y.; Guan, Y.; Li, Y.; Li, Z.; Wan, J.; Zhang, Y.; Fu, J. High adsorption behavior and photo regeneration of modified graphiteoxide-titanium dioxide nanocomposites for tetracycline removal in water. *Process Saf. Environ.* **2021**, *149*, 123–134. [CrossRef]
15. Wu, H.; Liu, X.; Wen, J.; Liu, Y.; Zheng, X. Rare-earth oxides modified Mg-Al layered double oxides for the enhanced adsorption-photocatalytic activity. *Colloid. Surface. A* **2021**, *610*, 125933. [CrossRef]
16. Xu, K.; Yang, X.; Ruan, L.; Qi, S.; Liu, J.; Liu, K.; Pan, S.; Feng, G.; Dai, Z.; Yang, X.; et al. Superior Adsorption and Photocatalytic Degradation Capability of Mesoporous $LaFeO_3/g-C_3N_4$ for Removal of Oxytetracycline. *Catalysts* **2020**, *10*, 301. [CrossRef]
17. Sun, Y.; Li, Y. Potential environmental applications of MXenes: A critical review. *Chemosphere* **2021**, *271*, 129578. [CrossRef]
18. Ponce, J.; Peña, J.; Román, J.; Pastor, J.M. Recyclable photocatalytic composites based on natural hydrogels for dye degradation in wastewaters. *Sep. Purif. Technol.* **2022**, *299*, 121759. [CrossRef]
19. Salati, M.A.; Khazai, J.; Tahmuri, A.M.; Samadi, A.; Taghizadeh, A.; Taghizadeh, M.; Zarrintaj, P.; Ramsey, J.D.; Habibzadeh, S.; Seidi, F.; et al. Agarose-Based Biomaterials: Opportunities and Challenges in Cartilage Tissue Engineering. *Polymers* **2020**, *12*, 1150. [CrossRef]
20. Yu, M.; Yuan, X.; Guo, J.; Tang, N.; Ye, S.; Liang, J.; Jiang, L. Selective graphene-like metal-free 2D nanomaterials and their composites for photocatalysis. *Chemosphere* **2021**, *284*, 131254. [CrossRef]
21. Fronczak, M. Adsorption performance of graphitic carbon nitride-based materials: Current state of the art. *J. Environ. Chem. Eng.* **2020**, *8*, 104411. [CrossRef]
22. Yousefi, M.; Villar-Rodil, S.; Paredes, J.I.; Moshfegh, A.Z. Oxidized graphitic carbon nitride nanosheets as an effective adsorbent for organic dyes and tetracycline for water remediation. *J. Alloys Compd.* **2019**, *809*, 151783. [CrossRef]
23. Jiang, L.; Yuan, X.; Pan, Y.; Liang, J.; Zeng, G.; Wu, Z.; Wang, H. Doping of graphitic carbon nitride for photocatalysis: A review. *Appl. Catal. B-Environ.* **2017**, *217*, 388–406. [CrossRef]
24. Yi, F.; Ma, J.; Lin, C.; Wang, L.; Zhang, H.; Qian, Y.; Zhang, K. Insights into the enhanced adsorption/photocatalysis mechanism of a $Bi_4O_5Br_2/g-C_3N_4$ nanosheet. *J. Alloy. Compd.* **2020**, *821*, 153557. [CrossRef]
25. Xu, C.; Wang, J.; Gao, B.; Dou, M.; Chen, R. Synergistic adsorption and visible-light catalytic degradation of RhB from recyclable 3D mesoporous graphitic carbon nitride/reduced graphene oxide aerogels. *J. Mater. Sci.* **2019**, *54*, 8892–8906. [CrossRef]
26. Sharma, S.; Dutta, V.; Raizada, P.; Khan, A.A.P.; Le, Q.V.; Thakur, V.K.; Biswas, J.K.; Selvasembian, R.; Singh, P. Controllable functionalization of $g-C_3N_4$ mediated all-solid-state (ASS) Z-scheme photocatalysts towards sustainable energy and environmental applications. *Environ. Technol. Inno.* **2021**, *24*, 101972. [CrossRef]
27. Tian, C.; Zhao, H.; Sun, H.; Xiao, K.; Wong, P.K. Enhanced adsorption and photocatalytic activities of ultrathin graphitic carbon nitride nanosheets: Kinetics and mechanism. *Chem. Eng. J.* **2020**, *381*, 122760. [CrossRef]
28. Shi, J.; Chen, T.; Guo, C.; Liu, Z.; Feng, S.; Li, Y.; Hu, J. The bifunctional composites of AC restrain the stack of $g-C_3N_4$ with the excellent adsorption-photocatalytic performance for the removal of RhB. *Colloid. Surface. A* **2019**, *580*, 123701. [CrossRef]
29. Liu, W.; Zhang, D.; Han, J.; Liu, C.; Ding, Y.; Wang, Z.; Wang, A. Adsorption enhanced photocatalytic degradation sulfadiazine antibiotic using porous carbon nitride nanosheets with carbon vacancies. *Chem. Eng. J.* **2020**, *382*, 123017. [CrossRef]
30. Shi, W.; Ren, H.; Huang, X.; Li, M.; Tang, Y.; Guo, F. Low cost red mud modified graphitic carbon nitride for the removal of organic pollutants in wastewater by the synergistic effect of adsorption and photocatalysis. *Sep. Purif. Technol.* **2020**, *237*, 116477. [CrossRef]
31. Xia, P.; Zhu, B.; Yu, J.; Cao, S.; Jaroniec, M. Ultra-thin nanosheet assemblies of graphitic carbon nitride for enhanced photocatalytic CO2 reduction. *J. Mater. Chem. A* **2017**, *5*, 3230–3238. [CrossRef]
32. Kim, M.; Joshi, B.; Yoon, H.; Ohm, T.Y.; Kim, K.; Al-Deyab, S.S.; Yoon, S.S. Electrosprayed copper hexaoxodivanadate ($CuV_2O_6$) and pyrovanadate ($Cu_2V_2O_7$) photoanodes for efficient solar water splitting. *J. Alloy. Compd.* **2017**, *708*, 444–450. [CrossRef]
33. Lin, N.; Pei, L.Z.; Wei, T.; Yu, H.Y. Synthesis of Cu vanadate nanorods for visible-light photocatalytic degradation of gentian violet. *Cryst. Res. Technol.* **2015**, *50*, 255–262. [CrossRef]
34. Keerthana, S.P.; Yuvakkumar, R.; Kumar, P.S.; Ravi, G.; Velauthapillai, D. Surfactant induced copper vanadate ($β-Cu_2V_2O_7$, $Cu_3V_2O_8$) for different textile dyes degradation. *Environ. Res.* **2022**, *211*, 112964. [CrossRef] [PubMed]
35. Khan, I.; Qurashi, A. Shape Controlled Synthesis of Copper Vanadate Platelet Nanostructures, Their Optical Band Edges, and Solar-Driven Water Splitting Properties. *Sci. Rep.* **2017**, *7*, 14370. [CrossRef] [PubMed]
36. Wiktor, J.; Reshetnyak, I.; Strach, M.; Scarongella, M.; Buonsanti, R.; Pasquarello, A. Sizable Excitonic Effects Undermining the Photocatalytic Efficiency of $β-Cu_2V_2O_7$. *J. Phys. Chem. Lett.* **2018**, *9*, 5698–5703. [CrossRef] [PubMed]
37. Yan, Q.; Yu, J.; Suram, S.K.; Zhou, L.; Shinde, A.; Newhouse, P.F.; Chen, W.; Li, G.; Persson, K.A.; Gregoire, J.M. Solar Fuels Photoanode Materials Discovery by Integrating High-Throughput Theory and Experiment. *Proc. Natl. Acad. Sci. USA* **2017**, *114*, 3040–3043. [CrossRef] [PubMed]
38. Ghiyasiyan-Arani, M.; Masjedi-Arani, M.; Salavati-Niasari, M. Facile synthesis, characterization and optical properties of copper vanadate nanostructures for enhanced photocatalytic activity. *J. Mater. Sci. Mater. Electron.* **2016**, *27*, 4871–4878. [CrossRef]
39. Zhou, L.; Shinde, A.; Newhouse, P.F.; Guevarra, D.; Wang, Y.; Lai, Y.; Kan, K.; Suram, S.K.; Haber, J.A.; Gregoire, J.M. Quaternary Oxide Photoanode Discovery Improves the Spectral Response and Photovoltage of Copper Vanadates. *Matter* **2020**, *3*, 1614–1630. [CrossRef]
40. Feng, Y.; Jia, C.; Zhao, H.; Wang, K.; Wang, X. Phase-dependent photocatalytic selective oxidation of cyclohexane over copper vanadates. *New J. Chem.* **2022**, *46*, 4082–4089. [CrossRef]

41. Seabold, J.A.; Neale, N.R. All First Row Transition Metal Oxide Photoanode for Water Splitting Based on $Cu_3V_2O_8$. *Chem. Mater.* **2015**, *27*, 1005–1013. [CrossRef]
42. Muthamizh, S.; Yesuraj, J.; Jayavel, R.; Contreras, D.; Varman, K.A.; Mangalaraja, R.V. Microwave synthesis of β-$Cu_2V_2O_7$ nanorods: Structural, electrochemical supercapacitance, and photocatalytic properties. *J. Mater. Sci. Mater. Electron.* **2021**, *32*, 2744–2756. [CrossRef]
43. Ghiyasiyan-Arani, M.; Masjedi-Arani, M.; Salavati-Niasari, M. Novel Schiff base ligand-assisted in-situ synthesis of $Cu_3V_2O_8$ nanoparticles via a simple precipitation approach. *J. Mol. Liq.* **2016**, *216*, 59–66. [CrossRef]
44. Wang, F.; Zhang, H.; Liu, L.; Shin, B.; Shan, F. Synthesis, surface properties and optical characteristics of $CuV_2O_6$ nanofibers. *J. Alloys Compd.* **2016**, *672*, 229–237. [CrossRef]
45. Truc, N.T.T.; Hanh, N.T.; Nguyen, M.V.; Chi, N.T.P.L.; Noi, N.V.; Tran, D.T.; Ha, M.N.; Trung, D.Q.; Pham, T.-D. Novel direct Z-scheme $Cu_2V_2O_7$/g-$C_3N_4$ for visible light photocatalytic conversion of $CO_2$ into valuable fuels. *Appl. Surf. Sci.* **2018**, *457*, 968–974. [CrossRef]
46. Paul, A.; Dhar, S.S. Designing $Cu_2V_2O_7$/$CoFe_2O_4$/g-$C_3N_4$ ternary nanocomposite: A high performance magnetically recyclable photocatalyst in the reduction of 4-nitrophenol to 4-aminophenol. *J. Solid State Chem.* **2020**, *290*, 121563. [CrossRef]
47. Kumar, A.; Sharma, S.K.; Sharma, G.; Guo, C.; Vo, D.-V.N.; Iqbal, J.; Naushad, M.; Stadler, F.J. Silicate glass matrix@$Cu_2O$/$Cu_2V_2O_7$ p-n heterojunction for enhanced visible light photo-degradation of sulfamethoxazole: High charge separation and interfacial transfer. *J. Hazard. Mater.* **2021**, *402*, 123790. [CrossRef]
48. Shuang, S.; Girardi, L.; Rizzi, G.A.; Sartorel, A.; Marega, C.; Zhang, Z.; Granozzi, G. Visible Light Driven Photoanodes for Water Oxidation Based on Novel r-GO/β-$Cu_2V_2O_7$/$TiO_2$ Nanorods Composites. *Nanomaterials* **2018**, *8*, 544. [CrossRef]
49. Ashraf, M.; Khan, I.; Baig, N.; Hendi, A.H.; Ehsan, M.F.; Sarfraz, N. A Bifunctional 2D Interlayered β-$Cu_2V_2O_7$/$Zn_2V_2O_6$ (CZVO) Heterojunction for Solar-Driven Nonsacrificial Dye Degradation and Water Oxidation. *Energy Technol.* **2021**, *9*, 2100034. [CrossRef]
50. Wang, L.; Ran, X.; Xiao, B.; Lei, L.; Zhu, J.; Xi, X.; Feng, G.; Li, R.; Feng, J. Visible light assisted Fenton degradation of oxytetracycline over perovskite $ErFeO_3$/porous g-$C_3N_4$ nanosheets p-n heterojunction. *J. Environ. Chem. Eng.* **2022**, *10*, 108330. [CrossRef]
51. Xiao, J.; Xie, Y.; Nawaz, F.; Wang, Y.; Du, P.; Cao, H. Dramatic coupling of visible light with ozone on honeycomb-like porous g-$C_3N_4$ towards superior oxidation of water pollutants. *Appl. Catal. B-Environ.* **2016**, *183*, 417–425. [CrossRef]
52. Zhang, J.; Mei, J.; Yi, S.; Guan, X. Constructing of Z-scheme 3D g-$C_3N_4$-ZnO@graphene aerogel heterojunctions for high-efficient adsorption and photodegradation of organic pollutants. *Appl. Surf. Sci.* **2019**, *492*, 808–817. [CrossRef]
53. Cai, C.; Zhang, Z.; Liu, J.; Shan, N.; Zhang, H.; Dionysiou, D.D. Visible light-assisted heterogeneous Fenton with $ZnFe_2O_4$ for the degradation of Orange II in water. *Appl. Catal. B-Environ.* **2016**, *182*, 456–468. [CrossRef]
54. Wang, Q.; Liu, Z.; Liu, D.; Wang, W.; Zhao, Z.; Cui, F.; Li, G. Oxygen vacancy-rich ultrathin sulfur-doped bismuth oxybromide nanosheet as a highly efficient visible-light responsive photocatalyst for environmental remediation. *Chem. Eng. J.* **2019**, *360*, 838–847. [CrossRef]
55. Guo, F.; Chen, Z.; Huang, X.; Cao, L.; Cheng, X.; Shi, W.; Chen, L. Ternary Ni2P/Bi2MoO6/g-C3N4 composite with Z-scheme electron transfer path for enhanced removal broad-spectrum antibiotics by the synergistic effect of adsorption and photocatalysis. *Chin. J. Chem. Eng.* **2022**, *44*, 157–168. [CrossRef]
56. Jiang, F.; Yan, T.; Chen, H.; Sun, A.; Xu, C.; Wang, X. A g-$C_3N_4$–CdS composite catalyst with high visible-light-driven catalytic activity and photostability for methylene blue degradation. *Appl. Surf. Sci.* **2014**, *295*, 164–172. [CrossRef]
57. Chen, F.; Huang, S.; Xu, Y.; Huang, L.; Wei, W.; Xu, H.; Li, H. Novel ionic liquid modified carbon nitride fabricated by in situ pyrolysis of 1-butyl-3-methylimidazolium cyanamide to improve electronic structurefor efficiently degradation of bisphenol A. *Colloids Surf. A Physicochem. Eng. Asp.* **2021**, *610*, 125648. [CrossRef]
58. Shia, W.; Li, M.; Huang, X.; Ren, H.; Guo, F.; Tang, Y.; Lu, C. Construction of CuBi2O4/Bi2MoO6 p-n heterojunction with nanosheets-onmicrorods structure for improved photocatalytic activity towards broadspectrum antibiotics degradation. *Chem. Eng. J.* **2020**, *394*, 125009. [CrossRef]
59. Ruan, X.; Hu, H.; Che, H.; Che, G.; Li, C.; Liu, C.; Dong, H. Facile fabrication of $Ag_2O$/$Bi_{12}GeO_{20}$ heterostructure with enhanced visible-light photocatalytic activity for the degradation of various antibiotics. *J. Alloy. Compd.* **2019**, *773*, 1089–1098. [CrossRef]
60. Wu, Y.; Wang, H.; Tu, W.; Liu, Y.; Tan, Y.Z.; Yuan, X.; Chew, J.W. Quasi-polymeric construction of stable perovskite-type $LaFeO_3$/g-$C_3N_4$ heterostructured photocatalyst for improved Z-scheme photocatalytic activity via solid p-n heterojunction interfacial effect. *J. Hazard. Mater.* **2018**, *347*, 412–422. [CrossRef]
61. Jing, J.; Cao, C.; Ma, S.; Li, Z.; Qu, G.; Xie, B.; Jin, W.; Zhao, Y. Enhanced defect oxygen of $LaFeO_3$/GO hybrids in promoting persulfate activation for selective and efficient elimination of bisphenol A in food wastewater. *Chem. Eng. J.* **2021**, *407*, 126890. [CrossRef]
62. Turnbull, R.; González-Platas, J.; Rodríguez, F.; Liang, A.; Popescu, C.; He, Z.; Santamaría-Pérez, D.; Rodríguez-Hernández, P.; Muñoz, A.; Errandonea, D. Pressure-Induced Phase Transition and Band Gap Decrease in Semiconducting β-$Cu_2V2O7$. *Inorg. Chem.* **2022**, *61*, 3697–3707. [CrossRef] [PubMed]
63. Dong, F.; Wang, Z.; Li, Y.; Ho, W.-K.; Lee, S.C. Immobilization of Polymeric g-$C_3N_4$ on Structured Ceramic Foam for Efficient Visible Light Photocatalytic Air Purification with Real Indoor Illumination. *Environ. Sci. Technol.* **2014**, *48*, 10345–10353. [CrossRef] [PubMed]

64. Xu, Q.; Zhang, L.; Yu, J.; Wageh, S.; Al-Ghamdi, A.A.; Jaroniec, M. Direct Z-scheme photocatalysts: Principles, synthesis, and applications. *Mater. Today* **2018**, *21*, 1042–1063. [CrossRef]
65. Rej, S.; Bisetto, M.; Naldoni, A.; Fornasiero, P. Well-defined Cu2O photocatalysts for solar fuels and chemicals. *J. Mater. Chem. A* **2021**, *9*, 5915–5951. [CrossRef]

International Journal of
*Molecular Sciences*

Article

# One-Step Synthesis of Nitrogen-Doped Porous Biochar Based on N-Doping Co-Activation Method and Its Application in Water Pollutants Control

Yingjie Su [1,2], Yuqing Shi [1,2], Meiyi Jiang [1,2] and Siji Chen [1,2,*]

1. College of Life Sciences, Jilin Agricultural University, Changchun 130118, China
2. Key Laboratory of Straw Comprehensive Utilization and Black Soil Conservation, Ministry of Education, Jilin Agricultural University, Changchun 130118, China
* Correspondence: 18638342679@163.com

**Abstract:** In this work, birch bark (BB) was used for the first time to prepare porous biochars via different one-step methods including direct activation (BBB) and N-doping co-activation (N-BBB). The specific surface area and total pore volume of BBB and N-BBB were 2502.3 and 2292.7 $m^2/g$, and 1.1389 and 1.0356 $cm^3/g$, respectively. When removing synthetic methyl orange (MO) dye and heavy metal $Cr^{6+}$, both BBB and N-BBB showed excellent treatment ability. The maximum adsorption capacities of BBB and N-BBB were 836.9 and 858.3 mg/g for MO, and 141.1 and 169.1 mg/g for $Cr^{6+}$, respectively, which were higher than most previously reported biochar adsorbents. The probable adsorption mechanisms, including pore filling, π–π interaction, H-bond interaction, and electrostatic attraction, supported the biochars' demonstrated high performance. In addition, after five recycles, the removal rates remained above 80%, which showed the high stability of the biochars. This work verified the feasibility of the one-step N-doping co-activation method to prepare high-performance biochars, and two kinds of biochars with excellent performance (BBB and N-BBB) were prepared. More importantly, this method provides new directions and ideas for the development and utilization of other biomasses.

**Keywords:** one-step synthesis; N-doping co-activation; biochar; water pollutants; adsorption performance

**Citation:** Su, Y.; Shi, Y.; Jiang, M.; Chen, S. One-Step Synthesis of Nitrogen-Doped Porous Biochar Based on N-Doping Co-Activation Method and Its Application in Water Pollutants Control. *Int. J. Mol. Sci.* **2022**, *23*, 14618. https://doi.org/10.3390/ijms232314618

Academic Editor: Dippong Thomas

Received: 18 October 2022
Accepted: 17 November 2022
Published: 23 November 2022

**Publisher's Note:** MDPI stays neutral with regard to jurisdictional claims in published maps and institutional affiliations.

**Copyright:** © 2022 by the authors. Licensee MDPI, Basel, Switzerland. This article is an open access article distributed under the terms and conditions of the Creative Commons Attribution (CC BY) license (https://creativecommons.org/licenses/by/4.0/).

## 1. Introduction

Water pollution is a serious threat to the ecosystem and human safety [1]. Organic pollutants (aromatic species) and heavy metals in water have become the focus of global attention due to their strong teratogenicity, carcinogenicity, and non-degradability [2,3]. $Cr^{6+}$ is one of the main forms of Cr. Because of its high solubility, easy accumulation, and difficulty degrading in the ecosystem, $Cr^{6+}$ is one of the main toxic heavy metals that pollute water bodies [4]. $Cr^{6+}$ ions are discharged into the water in large quantities during industrial processes such as metallurgy and electroplating [5]. Similarly, synthetic dyes are widely used in the printing, dyeing, and textile industries [6]. Many synthetic dyes have caused irreversible threats to human safety, such as skin allergy and cancer, and have become a research hotspot in the field of water pollution remediation [7,8]. Therefore, it is important to remove $Cr^{6+}$ and synthetic dyes to purify water.

At present, the removal methods for $Cr^{6+}$ and synthetic dyes from polluted water mainly include biological treatment, chemical treatment, and physical treatment [9–11]. As one of the physical treatment methods, the adsorption method is widely used in water pollution treatment because of its low cost, lack of by-products, and mild conditions [12,13]. It is also considered an ideal method to effectively remove $Cr^{6+}$ and synthetic dyes. Many materials, such as graphene oxide and carbon nanotubes, are used to treat water pollution due to their excellent adsorption properties on pollutants, but the high cost often limits their application [14,15]. In contrast, biochar is attracting increasing attention as an excellent adsorbent with a large specific surface area, abundant functional groups, and a well-developed pore structure [16,17]. Unfortunately, due to the limited adsorption sites, the adsorption performance is often limited. Appropriately modifying and improving the adsorption efficiency of biochar is the key to determining whether it can become a widely used adsorbent.

As an effective strategy to improve biochar performance, nitrogen doping (N-doping) technology has been applied in environmental remediation, such as adsorption, catalysis, and electrochemistry [18,19]. N hetero-atoms provide more binding sites on the biochar surface, which enhances the hydrophilicity and adsorption capacity of the compound by forming a more stable complex with pollutants [20]. At the same time, the N-doping method can adjust part of the electronic structure in the carbon skeleton, and further improve the reactivity of the biochar [21]. The pore structure, specific surface area, and other physical and chemical properties of biochars are also crucial factors that not only affect the adsorption capacity but also determine the degree of adsorbed pollutant diffusion [22,23]. The pore-forming effect of common activators (such as NaOH and KOH) in biochar preparation is beyond doubt [24,25]. However, there are few studies on biochar prepared with one-step pyrolysis and N-doping.

To remedy this, we explored a novel biochar preparation strategy to improve the performance of biochars, that is, N-doping co-activation. N-doping co-activation refers to the incorporation of nitrogen sources in the activation process in order to complete nitrogen doping and activation at the same time and successfully prepare nitrogen-doped porous biochars in one step. In this study, Birch bark was used as the carbon source for biochar preparation for the first time. N-doped porous biochar was synthesized with one-step co-activation pyrolysis by adding an activator (NaOH) and a nitrogen source (urea) in the preparation process. The focus of this study is to determine the feasibility of one-step N-doping co-activation in biochar preparation and its superiority in water pollution treatment, in order to provide a more effective way to deal with water pollution.

## 2. Results and Discussion
### 2.1. Preparation of CBB, BBB, and N-BBB

The sample preparation process is shown in Figure 1 and can be divided into three processes: direct carbonization, activation, and N-doping co-activation. The probable chemical reactions during the processes are summarized as follows:

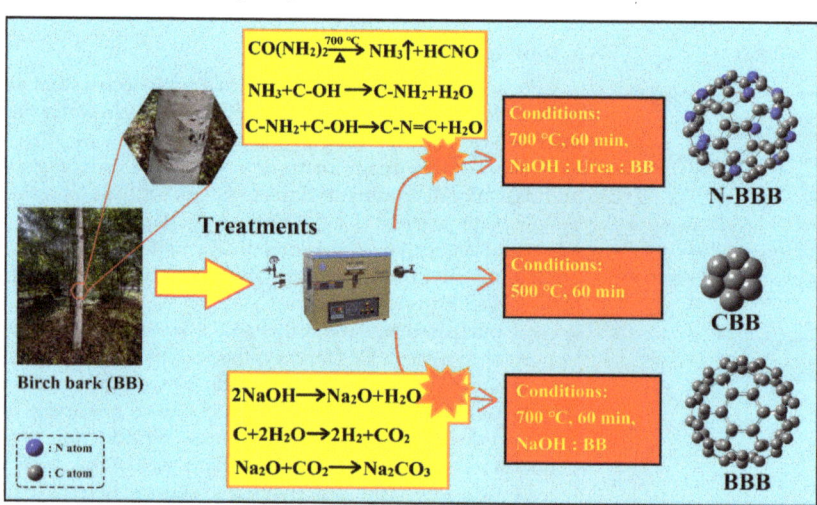

**Figure 1.** Schematic diagram of the preparation.

*Activation process* [26–28]: In the high-temperature pyrolysis process, NaOH will generate $Na_2O$ and water, and then be further ionized to form $Na^+$ and $OH^-$. These ions will migrate or insert carbon precursors and react with them to generate carbon dioxide and water. The carbon dioxide further reacts with the oxide produced by NaOH ($Na_2O$) to form carbonate ($Na_2CO_3$), which etches the carbonized sample. Finally, microporous, or mesoporous, structures are formed in the carbon.

*N-doping process* [29–31]: The decomposition of $CO(NH_2)_2$ at a high temperature produces solid and gaseous substances. Through the diffusion of the gas, the surface of the biochar develops various wrinkles and pores. The $NH_3$ released during the pyrolysis process combines with the -OH group in lignocellulose (the main component of BB) to form C-NH2, and further generate C=N-C [20]. Eventually, these nitrogenous groups form various N-bond configurations including pyrrolic-N, pyridinic-N, and graphite-N.

*2.2. Results of Characterizations*

The microscopic morphology and elemental composition of the samples were observed by SEM and EDS, as shown in Figure 2. BB had a rough, irregular surface and uneven lamellar structure. After carbonization (Figure 2B), CBB began to show a dense and smooth surface morphology. This can be explained by the fact that the main composition of BB is lignocellulose, consisting mainly of cellulose composed of glucose and hemicellulose composed of xylose [32]. During the high-temperature pyrolysis process, denatured glucose and xylose changed the morphology of CBB [33]. With the activation process treatment, especially in the presence of NaOH, BBB, and N-BBB (Figure 2C,E), many cracks and fragmentation occurred on the surface due to a severe reaction phenomenon. BBB and N-BBB were mainly composed of C, O, and N elements. However, because of different treatments, their nitrogen content was different. The C, O, and N contents of BBB and N-BBB were 84.62%, 13.52%, and 1.87%, and 84.16%, 12.14%, and 3.70%, respectively. Compared with BBB, the content of N in N-BBB nearly doubled. The above results not only proved the success of the activation process but also proved the feasibility of the N-doping co-activation process.

The influence of temperature on BB was measured by a TGA test under $N_2$ protection, as shown in Figure 3A. The TGA curves corresponding to three main stages of weight loss ranged from room temperature to 1200 °C. The first stage occurred at room temperature to 275 °C and was caused by evaporation and loss of residual water from physical surfaces and internal pores [34,35]. The second stage, which started at 275 °C and ended at 500 °C, was the most significant stage of weight loss. It can be explained by the fact that the main oxygenated component of BB was lignocellulose, which can be cracked into gases and tar at higher temperatures. With the removal of these pyrolytic substances, weight loss in the second stage was induced. Therefore, 500 °C was selected as the carbonization temperature for CBB preparation, and the yield was 20.28% at this time. As for the third stage (500–1200 °C), the TG curve was relatively stable, indicating that there was no obvious weight loss phenomenon. However, when the temperature exceeded 1000 °C (especially when it was close to 1200 °C), CBB exhibited a slight mass change, which may have been caused by the pyrolysis of some minerals [36,37].

The FT-IR spectra of the functional groups of the samples were analyzed, as shown in Figure 2B. The broad band at around 3450 and 2930 $cm^{-1}$ represented the stretching vibrations of hydroxyl functional groups (O-H) and -CH, -$CH^2$, and -$CH^3$ groups [38]. The band around 1730 $cm^{-1}$ represented the stretching vibration of C=O. The peak at 1630 $cm^{-1}$ represented the axial deformation of the carbonyl group. The band at around 1375 and 1420 $cm^{-1}$ represented the C-H symmetric bending vibration of the methyl group and the deformation vibration of methylene [38,39]. In common with many biomass materials [34,35,38,39], the bands at around 1047–1159 $cm^{-1}$ represented the tensile vibrations of C-O from alcohols, phenols, acids, or esters. After N-doping co-activation, a new vibration peak appeared at 1690 $cm^{-1}$, which indicated that the C=N was formed on N-BBB [20].

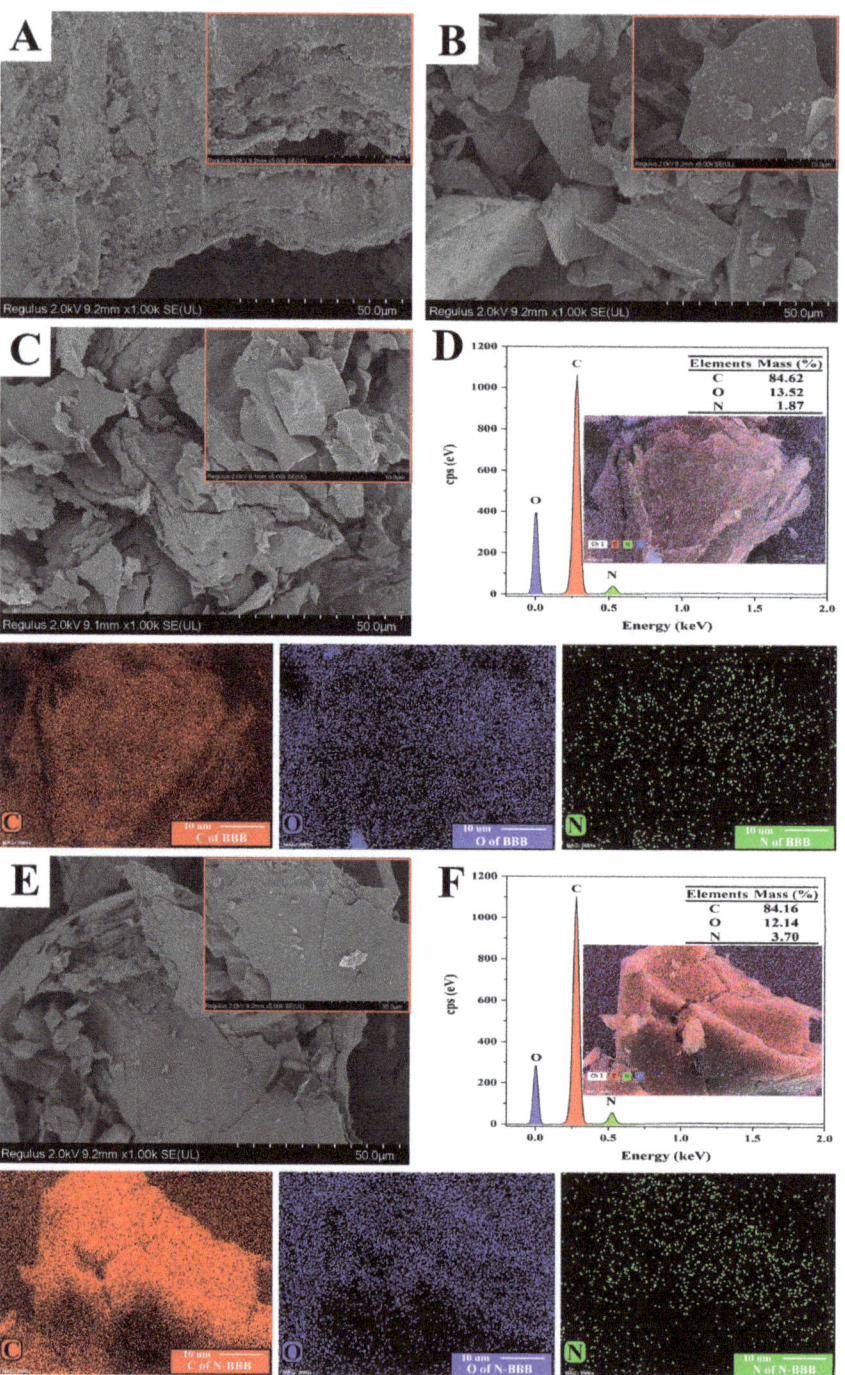

**Figure 2.** SEM images of (**A**) BB, (**B**) CBB, (**C**) BBB, and (**E**) N-BBB. EDS mapping of (**D**) BBB and (**F**) N-BBB.

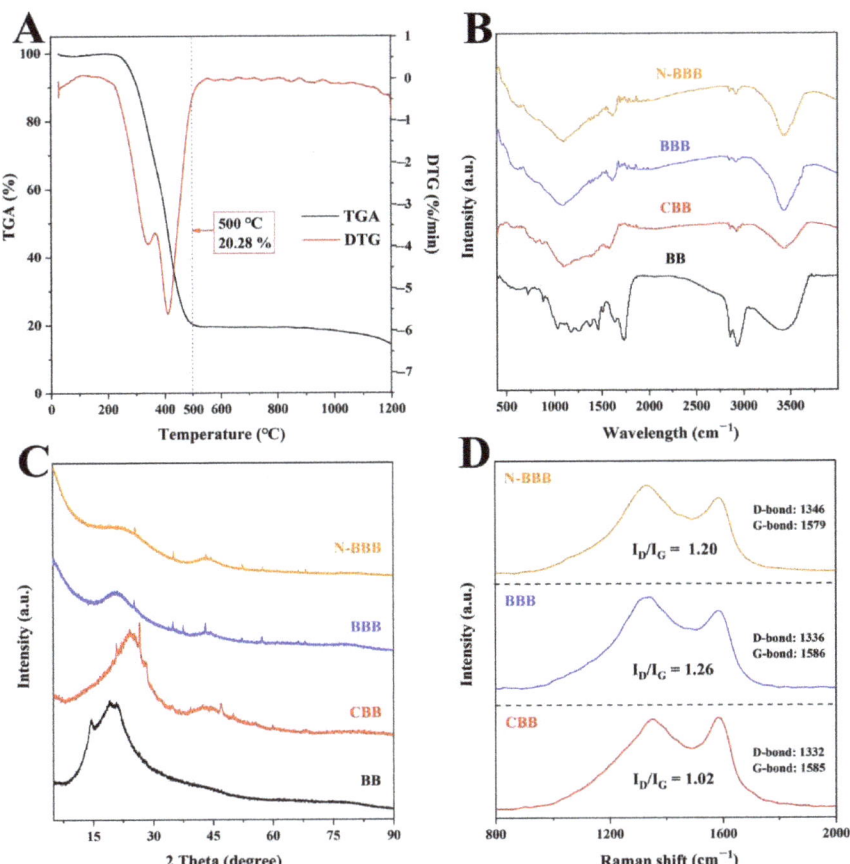

**Figure 3.** (**A**) TGA curves of BB. (**B**) FT-IR spectra and (**C**) XRD of BB, CBB, BBB, and N-BBB. (**D**) Raman spectra of CBB, BBB, and N-BBB.

The crystal structure of the samples was tested by XRD, as shown in Figure 3C. The peaks at 17° and 23° represent the cellulose from the lignocellulose of raw BB [32,33]. The irregular peaks represent minerals that can be interpreted as inorganic salts. These peaks became sharper and more distinct after carbonization. After direct activation and N-doped co-activation, these peaks were significantly weakened, which can be explained by the removal of a large quantity of soluble mineral salts during the washing process. In addition, we noted that the diffraction peaks of CBB, BBB, and N-BBB were in the range of 10–30° and 38–45°, indicating that the prepared biochars had the local structure of typical carbon materials; that is, they contained both amorphous carbon and a 2D graphite planar structure [40].

The presence of defects in the carbon was determined by Raman spectra, as shown in Figure 3D. Two typical peaks obtained from the results include the D-band with amorphous carbon at around $1339 \pm 7$ cm$^{-1}$ and the G-band with graphitic carbon at around $1583 \pm 3$ cm$^{-1}$ [38–40]. To measure the degree of defect and disorder in the carbons, the intensity ratio of the D-band and G-band ($I_D/I_G$) was used as an important index. The $I_D/I_G$ value of CBB was 1.02. After activation, the $I_D/I_G$ value of BBB and N-BBB were 1.26 and 1.20, which indicated that more amorphous carbon structures were generated in the biochars.

The specific surface area and porosity of the samples were tested by $N_2$ adsorption-desorption isotherms, as shown in Figure 4, and the data are shown in Table 1. The specific surface area and the total pore volume of CBB were 49.5 $m^2/g$ and 0.0222 $cm^3/g$, respectively, which were not enough to support CBB as a porous biochar for adsorption. Therefore, further activation treatment is needed to greatly improve its properties and enhance its application performance. After direct activation and N-doping co-activation, the specific surface areas of BBB and N-BBB were 2502.3 and 2292.7 $m^2/g$, respectively, and their total pore volumes were 1.1389 and 1.0356 $cm^3/g$, respectively. Compared with CBB, the values were greatly improved, which indicated the success and effectiveness of the further activation treatment. Moreover, both BBB and N-BBB showed typical type IV isotherms with a slight H3 hysteresis loop, indicating that some mesoporous structures existed in the prepared biochars [39,40]. The volumes of micropores in BBB and N-BBB were 1.1118 and 1.0170 $cm^3/g$, respectively, accounting for 97.6% and 98.2% of the total pore volume, respectively, indicating that they were porous biochars with mainly micropore structure and partly mesoporous structure [24,25]. Pore size distribution was used to further analyze the porosity of the samples. The results based on the NLDFT method also showed that both BBB and N-BBB had microporous and mesoporous structures. The BJH and H-K methods were used to study the pore distribution, which not only re-analyzed the multi-pore structure of BBB and N-BBB but also further proved the existence of both mesoporous and microporous structures.

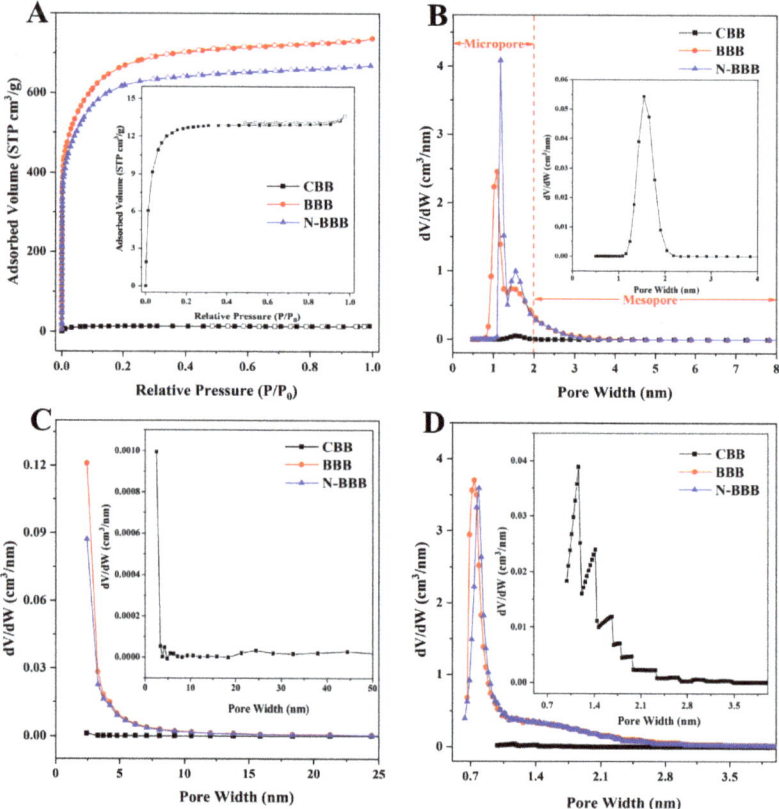

**Figure 4.** (**A**) $N_2$ adsorption-desorption isotherms of samples. Pore distribution of samples based on (**B**) NLDFT method, (**C**) BJH method, and (**D**) H-K method.

Table 1. The data of $N_2$ adsorption-desorption for CBB, BBB, and N-BBB.

| Samples | $S_{BET}$ (m$^2$/g) | $V_{micro}$ (cm$^3$/g) | $V_{total}$ (cm$^3$/g) |
|---|---|---|---|
| CBB | 49.5 | 0.0187 | 0.0222 |
| BBB | 2502.3 | 1.1118 | 1.1389 |
| N-BBB | 2292.7 | 1.0170 | 1.0356 |

$S_{BET}$, $V_{micro}$, and $V_{total}$ represent the BET specific surface area, the volume of micropores, and the total pore volume.

In addition, it is noteworthy that the specific surface area and total pore volume of N-BBB were smaller than those of BBB. There are three possible reasons for this. First, the addition of the nitrogen source urea in the activation process will compete with the activator NaOH for the contact area with the carbon precursor, thus affecting the specific surface area and pore structure of N-BBB. Second, it may be caused by the chemical reaction between the ammonia gas generated by the nitrogen source urea at high temperature and carbon dioxide produced in the activation process and water vapor, thus consuming part of the activator and reducing the activation efficiency. Third, the nitrogen source urea and activator simultaneously etched the surface of the carbon material and generated more gases [29–31], which diffused into the carbon precursor, resulting in partial fragmentation, affecting the pore structure and resulting in the reduction in the specific surface area and total pore volume.

The surface chemical and electronic states of the samples were determined by XPS spectra, as shown in Figure 5. Both BBB and N-BBB contained mainly C, O, and N elements, which was consistent with the EDS. The high-resolution C1s spectra of BBB and N-BBB showed three classical peaks at 283.87–283.88, 284.82–284.98, and 287.74–288.22 eV corresponding to C-C, C-O, and C=O, respectively [38–40]. The high-resolution O1s spectra of BBB and N-BBB both had three peaks at 530.74–530.93, 532.36–532.43, and 533.76–534.28 eV corresponding to C=O, C-O, and -OH, respectively [38–40]. The high-resolution N1s spectra of ITGB and MITGB both had peaks at 397.41–37.42 and 399.33–399.38 eV, corresponding to pyridinic-N and pyrrolic-N, respectively [38–40]. In addition, N-BBB had a unique peak corresponding to graphite-N at 403.53 eV [20], indicating that the N-doped co-activation method had indeed successfully doped N into N-BBB.

*2.3. Results of Adsorption Performances*

2.3.1. Adsorption Kinetics

Adsorption kinetics describes the adsorption capacity of the adsorbent at different initial solution concentrations as a function of contact time [32–35]. Therefore, the effect of time on the adsorption of MO and $Cr^{6+}$ by BBB and N-BBB at a temperature of 303 K was explored, as shown in Figure 6. Whether the adsorbent was BBB or N-BBB, or the adsorbate was MO or $Cr^{6+}$, all the adsorption process trends were similar. The adsorption capacities increased sharply in the first 30 min, and then gradually reached equilibrium at 60 min. The extension of contact time did not further significantly improve the adsorption capacities. It can be speculated that the adsorption capacities would increase with the increase in the initial concentration of the solution, and the high concentration of solution promoted the adsorption process to some certain extent [40]. In order to study the control mechanism of reactions in the process of adsorption, three common adsorption kinetic models, Lagergren's PFK model based on surface physical adsorption [40], Ho–McKay's PSK model based on chemical adsorption [39], and Weber–Morris's IPD model based on molecular diffusion [38] were used to analyze the experimental data, as shown in Table 2.

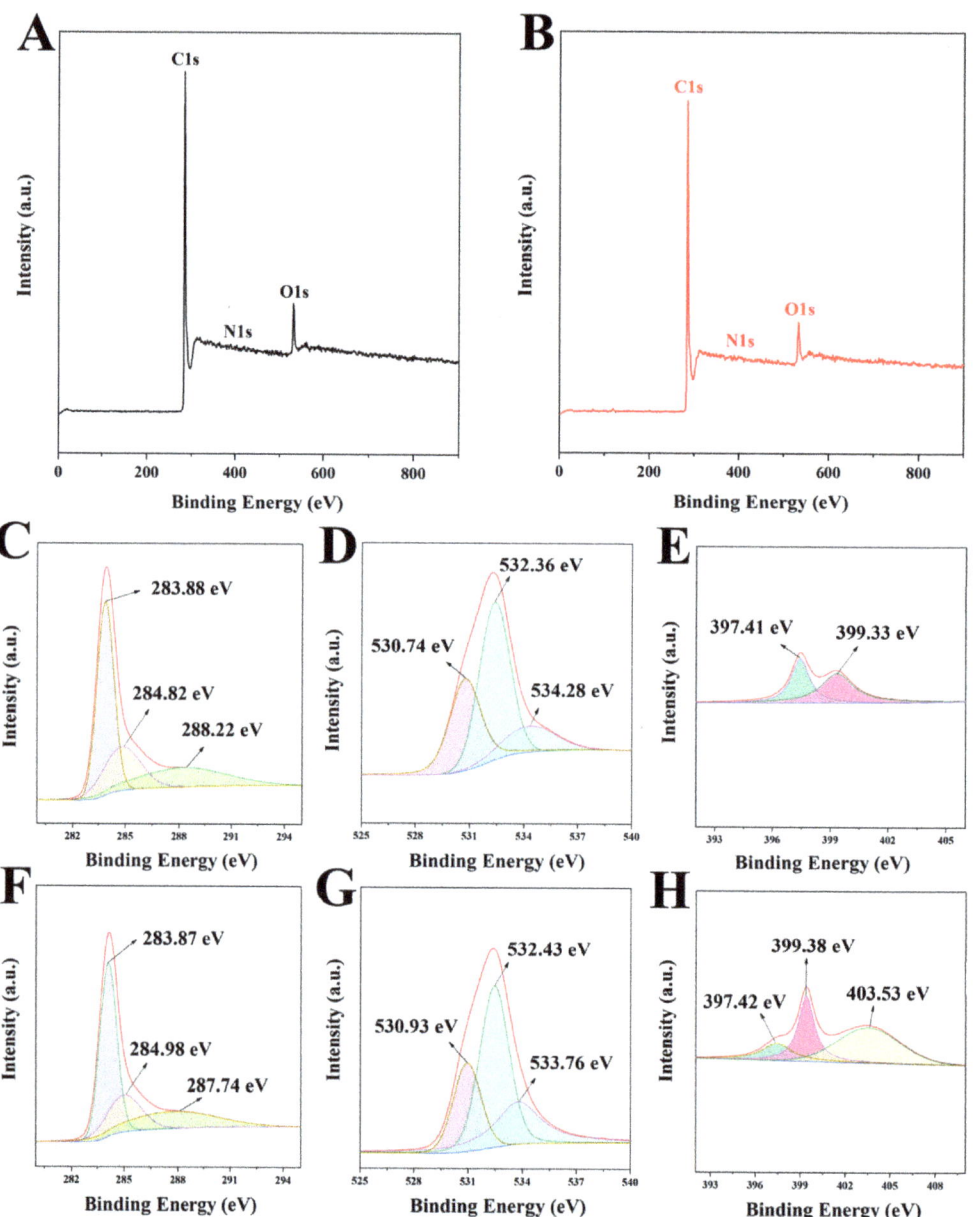

**Figure 5.** XPS spectra of (**A**) BBB and (**B**) N-BBB. The C1s, O1s, and N1s high-resolution spectra of BBB (**C–E**) and N-BBB (**F–H**).

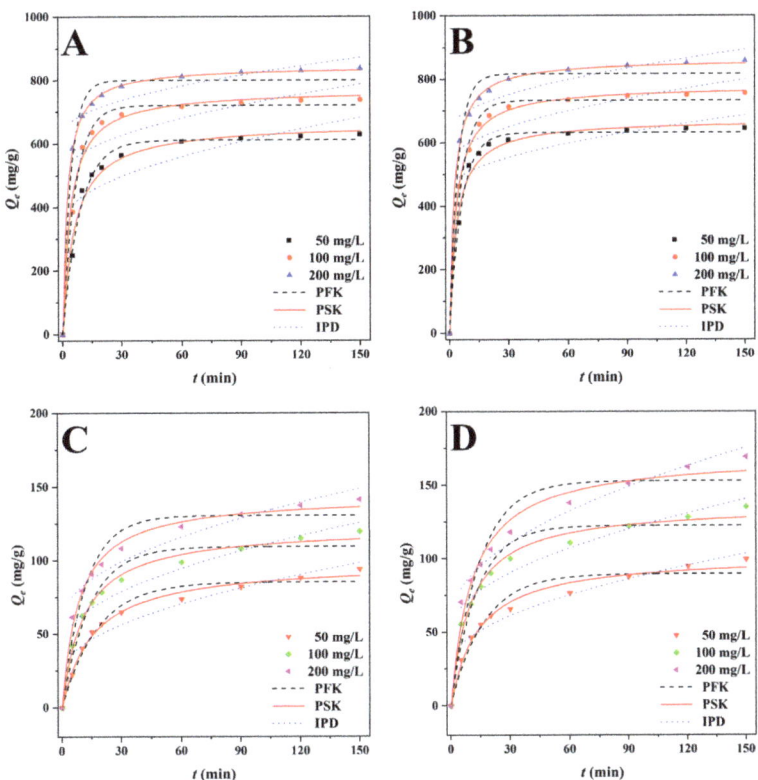

**Figure 6.** PFK, PSK, and IPD plots of MO for (**A**) BBB and (**B**) N-BBB, and plots of $Cr^{6+}$ for (**C**) BBB and (**D**) N-BBB at 303 K.

**Table 2.** Fitting parameters of adsorption kinetic models for MO and $Cr^{6+}$ at 303 K.

| Adsorbates | Adsorbents | Models | Parameters | $C_0$ (mg L$^{-1}$) | | |
|---|---|---|---|---|---|---|
| | | | | 50 | 100 | 200 |
| MO | BBB | PFK | $Q_e$ (mg/g) | 627.9 | 737.2 | 836.9 |
| | | | $k_1$ (min$^{-1}$) | 0.0028 | 0.0074 | 0.0058 |
| | | | $Q_{e.cat}$ (mg/g) | 611.8 | 720.4 | 800.3 |
| | | | $R^2$ | 0.9880 | 0.9944 | 0.9780 |
| | | PSK | $k_2$ (g mg$^{-1}$ min$^{-1}$) | 0.0003 | 0.0003 | 0.0005 |
| | | | $Q_{e.cat}$ (mg/g) | 664.5 | 769.0 | 844.2 |
| | | | $R^2$ | 0.9848 | 0.9896 | 0.9997 |
| | | IPD | $k_3$ (mg g$^{-1}$ min$^{-0.5}$) | 27.12 | 23.62 | 19.90 |
| | | | $C$ | 350.1 | 497.9 | 628.2 |
| | | | $R^2$ | 0.6464 | 0.5822 | 0.7517 |
| | N-BBB | PFK | $Q_e$ (mg/g) | 644.4 | 755.1 | 858.3 |
| | | | $k_1$ (min$^{-1}$) | 0.0044 | 0.0054 | 0.0051 |
| | | | $Q_{e.cat}$ (mg/g) | 631.3 | 732.9 | 817.1 |
| | | | $R^2$ | 0.9952 | 0.9885 | 0.9721 |
| | | PSK | $k_2$ (g mg$^{-1}$ min$^{-1}$) | 0.0004 | 0.0004 | 0.0005 |
| | | | $Q_{e.cat}$ (mg/g) | 672.1 | 777.9 | 862.8 |
| | | | $R^2$ | 0.9873 | 0.9980 | 0.9989 |
| | | IPD | $k_3$ (mg g$^{-1}$ min$^{-0.5}$) | 19.4 | 22.0 | 20.9 |
| | | | $C$ | 448.7 | 530.3 | 637.3 |
| | | | $R^2$ | 0.5471 | 0.6549 | 0.7830 |

Table 2. Cont.

| Adsorbates | Adsorbents | Models | Parameters | $C_0$ (mg L$^{-1}$) 50 | 100 | 200 |
|---|---|---|---|---|---|---|
| $Cr^{6+}$ | BBB | PFK | $Q_e$ (mg/g) | 93.8 | 119.6 | 141.1 |
| | | | $k_1$ (min$^{-1}$) | 0.0014 | 0.0017 | 0.0020 |
| | | | $Q_{e.cat}$ (mg/g) | 85.3 | 109.2 | 130.4 |
| | | | $R^2$ | 0.9741 | 0.9614 | 0.9550 |
| | | PSK | $k_2$ (g mg$^{-1}$ min$^{-1}$) | 0.0007 | 0.0008 | 0.0008 |
| | | | $Q_{e.cat}$ (mg/g) | 98.4 | 122.3 | 143.6 |
| | | | $R^2$ | 0.9923 | 0.9919 | 0.9913 |
| | | IPD | $k_3$ (mg g$^{-1}$ min$^{-0.5}$) | 6.2 | 6.9 | 7.3 |
| | | | $C$ | 22.7 | 41.4 | 59.1 |
| | | | $R^2$ | 0.9138 | 0.9280 | 0.9279 |
| | N-BBB | PFK | $Q_e$ (mg/g) | 99.5 | 135.0 | 169.1 |
| | | | $k_1$ (min$^{-1}$) | 0.0015 | 0.0017 | 0.0017 |
| | | | $Q_{e.cat}$ (mg/g) | 89.7 | 122.4 | 153.0 |
| | | | $R^2$ | 0.9500 | 0.9506 | 0.9318 |
| | | PSK | $k_2$ (g mg$^{-1}$ min$^{-1}$) | 0.0007 | 0.0007 | 0.0006 |
| | | | $Q_{e.cat}$ (mg/g) | 102.2 | 135.9 | 170.6 |
| | | | $R^2$ | 0.9843 | 0.9869 | 0.9765 |
| | | IPD | $k_3$ (mg g$^{-1}$ min$^{-0.5}$) | 6.2 | 7.3 | 9.6 |
| | | | $C$ | 27.7 | 50.5 | 58.4 |
| | | | $R^2$ | 0.9537 | 0.9432 | 0.9746 |

The PFK correlation coefficients $R^2$ of BBB for MO ranged from 0.9780 to 0.9944; while the $Q_{e.cat}$ values (611.8, 720.4, and 800.3 mg/g) were lower than the $Q_e$ (627.9, 737.2, and 836.9 mg/g) obtained from the experiments. The PFK for $Cr^{6+}$ ranged from 0.9550 to 0.9741; while the $Q_{e.cat}$ values were 85.3, 109.2, and 130.4 mg/g for different initial concentrations, and were again lower than $Q_e$ (93.8, 119.6, and 141.1 mg/g). To summarize, the PFK model was not the best kinetic to describe the whole adsorption process. The IPD correlation coefficients $R^2$ of BBB for MO ranged from 0.5822 to 0.7517, indicating that the adsorption process of MO by BBB may not be affected by particle diffusion. For $Cr^{6+}$, on the contrary, the correlation coefficients $R^2$ ranged from 0.9138 to 0.9280, which indicated the adsorption process had a particle diffusion behavior. Ho–McKay's PSK model was used to fit the data; the PSK correlation coefficients $R^2$ of BBB ranged from 0.9848 to 0.9997 for MO and from 0.9913 to 0.9919 for $Cr^{6+}$. Meanwhile, the $Q_{e.cat}$ values (664.5, 769.0, and 844.2 mg/g for MO, and 98.4, 122.3, and 143.6 mg/g for $Cr^{6+}$) agreed with those obtained from experiments, which showed the applicability of PSK in the adsorption process.

When the models were used to fit with the data of N-BBB, the PFK correlation coefficients $R^2$ were 0.9721–0.9952 for MO and 0.9318–0.9506 for $Cr^{6+}$. The $Q_{e.cat}$ values were 631.3, 732.9, and 817.1 mg/g for MO, and 89.7, 122.4 and 153.0 mg/g for $Cr^{6+}$, which were both lower than the $Q_e$ values (644.4, 755.1 and 858.3 mg/g for MO, and 99.5, 135.0, and 169.1 mg/g for $Cr^{6+}$) obtained from experiments. It can be speculated that the PFK model may have played a role in the adsorption process, although the role was not dominant. For the IPD model, the correlation coefficients $R^2$ were 0.5471–0.7830 for MO and 0.9432–0.9746 for $Cr^{6+}$, also indicating that the adsorption process of N-BBB for $Cr^{6+}$ had a particle diffusion behavior and was largely unaffected by particle diffusion for MO. The PSK coefficients $R^2$ were 0.9873–0.9989 for MO and 0.9765–0.9869 for $Cr^{6+}$, indicating the PSK model was more suitable to describe the adsorption process. Moreover, it was found that, with the increase in the initial concentration of the solution, the rate constants $k_2$ of BBB were also increased for both MO and $Cr^{6+}$, indicating that the adsorption rate gradually accelerated, and the adsorption rate was faster at a higher concentration. However, the rate constants $k_2$ of N-BBB were different. For MO, N-BBB showed a similar trend to BBB. At the same time, the rate constants $k_2$ of N-BBB for $Cr^{6+}$ decreased gradually, which indicated that, with the increase in initial concentration, the adsorption rate gradually slowed, and the adsorption rate was slower at a higher concentration.

According to the results, we inferred that the adsorption processes of BBB and N-BBB for MO and $Cr^{6+}$ may be mainly chemical reactions (while physical adsorption and particle diffusion also had certain effects on the adsorption process), and the adsorption behaviors between adsorbent and adsorbate through transfer, exchange, or sharing to form chemisorption bonds, may control the adsorption rate [40,41].

2.3.2. Adsorption Isotherms

The effect of concentration on the adsorption capacity of the adsorbent is usually determined by investigating the adsorption isotherms. The effects of different initial concentrations of solution on the adsorption process of BBB and N-BBB were studied at a temperature of 303 K, and the results are shown in Figure 7. Increasing the initial concentration of MO or $Cr^{6+}$ solution was beneficial to the forward process of adsorption.

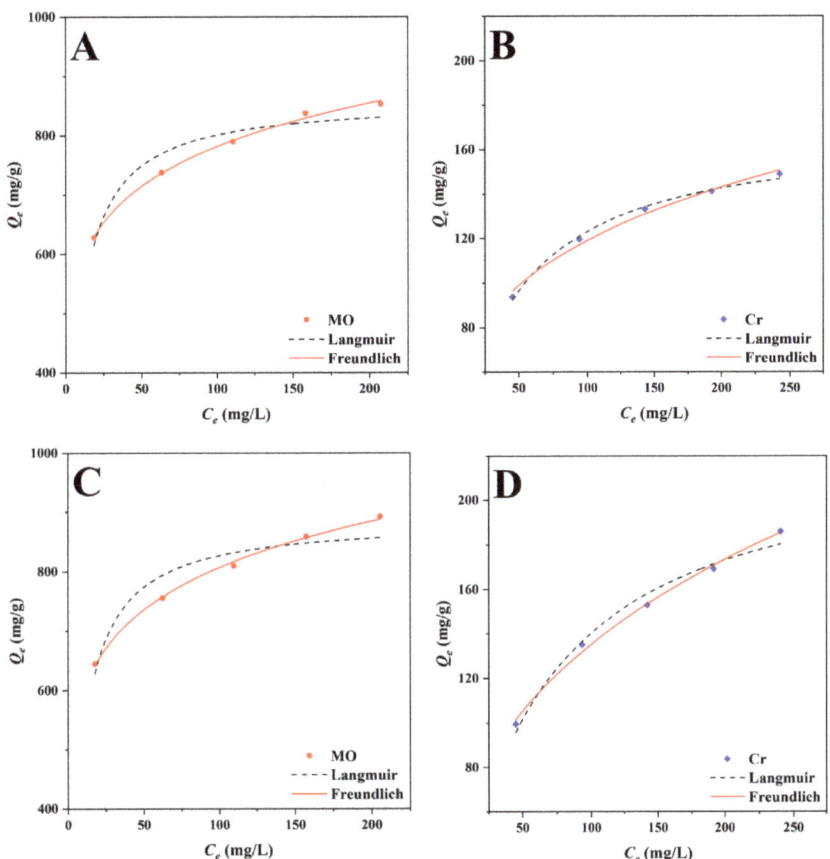

**Figure 7.** Adsorption isotherms of MO and $Cr^{6+}$ for BBB (**A**,**B**) and N-BBB (**C**,**D**) at 303 K.

Subsequently, the Langmuir and Freundlich isotherm models were used to analyze the experimental data, and the results are shown in Table 3. The Langmuir isotherm model is often used to describe the adsorption process of homogeneous molecules [42], while the Freundlich isotherm model is often used to study the heterogeneous multilayer adsorption process [43]. When the adsorbate was MO, the Langmuir isotherm correlation coefficients $R^2$ were 0.9327 for BBB and 0.9041 for N-BBB, which indicated that the adsorption processes were not uniform single-layer. The $Q_m$ of BBB and N-BBB for MO were 860.3 and 887.4 mg/g, higher than $Q_e$, indicating that the prepared biochars had higher adsorption capacities for MO. Moreover, the $K_L$ of N-BBB for MO was slightly bigger than that of BBB, which showed that N-BBB had a faster MO adsorption rate. The Freundlich isotherm correlation coefficients $R^2$ for MO were 0.9967 for BBB and 0.9883 for N-BBB. The $n_F$ values of BBB and N-BBB for MO were 7.75 and 7.51, bigger than 1.0, indicating that this adsorption model of fitting MO was appropriate. As for $Cr^{6+}$, the Langmuir isotherm correlation coefficients $R^2$ were 0.9949 for BBB and 0.9822 for N-BBB, which also showed that the uniform single-layer adsorption was not suitable to describe the adsorption process. When the Freundlich isotherm was used to fit the data, the correlation coefficients $R^2$ were 0.9976 for BBB and 0.9959 for N-BBB; while the $n_F$ values of BBB and N-BBB were 3.76 and 2.77. The results indicated that the adsorption processes were non-uniform multilayer [42–44].

**Table 3.** Fitting parameters of adsorption isotherm models for MO and $Cr^{6+}$ at 303 K.

| Adsorbates | Adsorbents | Types | Parameters | |
|---|---|---|---|---|
| MO | BBB | Langmuir | $Q_m$ (mg/g) | 860.3 |
| | | | $K_L$ (L/mg) | 0.1345 |
| | | | $R^2$ | 0.9327 |
| | | Freundlich | $K_F$ (mg g$^{-1}$(L mg$^{-1}$)$^{1/n}$) | 431.5 |
| | | | $n_F$ | 7.75 |
| | | | $R^2$ | 0.9967 |
| | N-BBB | Langmuir | $Q_m$ (mg/g) | 887.4 |
| | | | $K_L$ (L/mg) | 0.1364 |
| | | | $R^2$ | 0.9041 |
| | | Freundlich | $K_F$ (mg g$^{-1}$(L mg$^{-1}$)$^{1/n}$) | 437.1 |
| | | | $n_F$ | 7.51 |
| | | | $R^2$ | 0.9883 |
| $Cr^{6+}$ | BBB | Langmuir | $Q_m$ (mg/g) | 169.8 |
| | | | $K_L$ (L/mg) | 0.0264 |
| | | | $R^2$ | 0.9949 |
| | | Freundlich | $K_F$ (mg g$^{-1}$(L mg$^{-1}$)$^{1/n}$) | 34.9 |
| | | | $n_F$ | 3.76 |
| | | | $R^2$ | 0.9976 |
| | N-BBB | Langmuir | $Q_m$ (mg/g) | 226.2 |
| | | | $K_L$ (L/mg) | 0.0163 |
| | | | $R^2$ | 0.9822 |
| | | Freundlich | $K_F$ (mg g$^{-1}$(L mg$^{-1}$)$^{1/n}$) | 25.7 |
| | | | $n_F$ | 2.77 |
| | | | $R^2$ | 0.9959 |

### 2.3.3. Adsorption Thermodynamics

The influence of temperature (293, 303, and 313 K) on adsorption by BBB and N-BBB is shown in Figure 8. On increasing the temperature from 293 K to 313 K, the adsorption capacities of BBB for MO and $Cr^{6+}$ increased from 712.5 to 764.8 mg/g and from 105.8 to 124.1 mg/g, respectively, while the adsorption capacities of N-BBB for MO and $Cr^{6+}$ increased from 726.7 to 773.7 mg/g and from 112.6 to 148.3 mg/g, respectively. Obviously, raising the temperature increased the adsorption capacities of MO and $Cr^{6+}$—that is, a high-temperature environment promoted the adsorption processes by biochars (BBB and N-BBB).

The experimental data were analyzed by thermodynamics formulas, and the parameters are shown in Table 4. All $\Delta G$ values were negative, indicating that the adsorption occurred spontaneously, both with BBB (−5.86, −6.19, and −6.55 for MO and −0.27, −0.61, and −0.73 for $Cr^{6+}$) and N-BBB (−5.93, −6.29, and −6.60 for MO and −0.43, −0.93, and −1.23 for $Cr^{6+}$) [39,40]. The thermodynamic enthalpy values of $\Delta H$ of adsorption of MO were 4.29 and 3.82 kJ/mol for BBB and N-BBB, respectively, and the $\Delta H$ values of adsorption of $Cr^{6+}$ were 6.47 and 11.23 kJ/mol for BBB and N-BBB, respectively, which further confirmed the endothermic property of the adsorption process [38]. In addition, the positive values of thermodynamic $\Delta S$ (34.61 and 33.27 Jmol$^{-1}$ K$^{-1}$ for MO, 23.01 and 39.78 J mol$^{-1}$ K$^{-1}$ for $Cr^{6+}$) indicated that the randomness and chaos of the interface between the porous biochars and solutions increased with the increase in temperature [40].

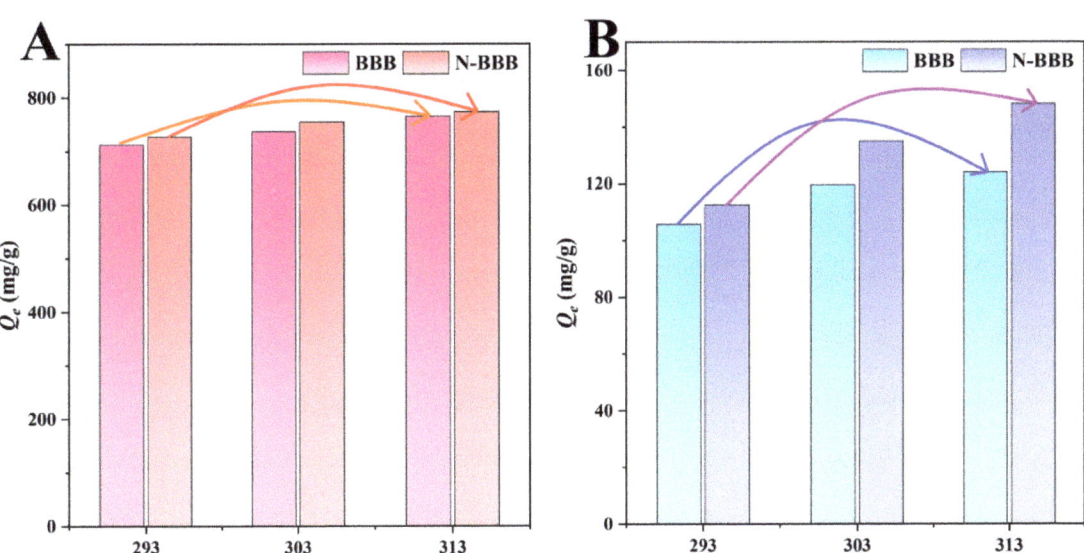

**Figure 8.** Adsorption thermodynamics of BBB and N-BBB for (**A**) MO and (**B**) $Cr^{6+}$.

**Table 4.** Fitting adsorption thermodynamic parameters for MO and $Cr^{6+}$.

| Adsorbents | Adsorbates | T (K) | ΔG (kJ/mol) | ΔH (kJ/mol) | ΔS (J mol$^{-1}$ K$^{-1}$) |
|---|---|---|---|---|---|
| BBB | MO | 293 | −5.86 | 4.29 | 34.61 |
|  |  | 303 | −6.19 |  |  |
|  |  | 313 | −6.55 |  |  |
|  | $Cr^{6+}$ | 293 | −0.27 | 6.47 | 23.01 |
|  |  | 303 | −0.61 |  |  |
|  |  | 313 | −0.73 |  |  |
| N-BBB | MO | 293 | −5.93 | 3.82 | 33.27 |
|  |  | 303 | −6.29 |  |  |
|  |  | 313 | −6.60 |  |  |
|  | $Cr^{6+}$ | 293 | −0.43 | 11.23 | 39.78 |
|  |  | 303 | −0.93 |  |  |
|  |  | 313 | −123 |  |  |

2.3.4. Effect of pH

In general, pH affects the adsorption process by changing the charge properties of the adsorbent and the adsorbate [32–35]. MO has two chemical structures, basic and acidic, and whether the chromophore was anthraquinone or azo bond depends on the pH of the solution [45]. The adsorption of MO by biochars was investigated in the pH range of 2 to 10, and the results are shown in Figure 9A. With the increase in pH, the adsorption capacity of both BBB and N-BBB decreased, which can be explained by the electrostatic attraction between the surface charge of the biochars and the ionic charge of the anionic dye MO. Compared with BBB, N-BBB had higher electronegativity due to the N-doping. When the pH value was less than 6, the affinity of the dye increased, and the negative charge on the surface of the biochar could be used as the active site to generate a strong electrostatic attraction with the dye in solution. Conversely, when the pH was higher than

6 (particularly 8), the protonation of the dye was gradually weakened, and the biochars with negative active sites on the surface were not conducive to the adsorption of anionic dyes due to electrostatic repulsion, which reduced the adsorption amount of MO. $Cr^{6+}$ also possessed various forms at different pH levels [4,5,20], where it exists as $HCrO_4^-$ and $Cr_2O_7^{2-}$ when the pH is lower than 6.5, and as $Cr_2O_7^{2-}$ and $CrO_4^{2-}$ when the pH is higher than 6.5. With the increase in pH, the adsorption capacities of both BBB and N-BBB decreased, and the maximum adsorption capacities were at a pH value of 2. This can be explained by the fact that the biochars (BBB and N-BBB) had a positive surface charge at a pH less than $pH_{pzc}$ (4.46 for BBB and 4.52 for N-BBB). When the pH was higher than 6, the adsorption capacities decreased sharply, which may be due to electrostatic repulsion. When the pH was in the range of 8 to 10, the adsorption became relatively stable, which indicated that electrostatic interaction was not the only process affecting adsorption performance [20].

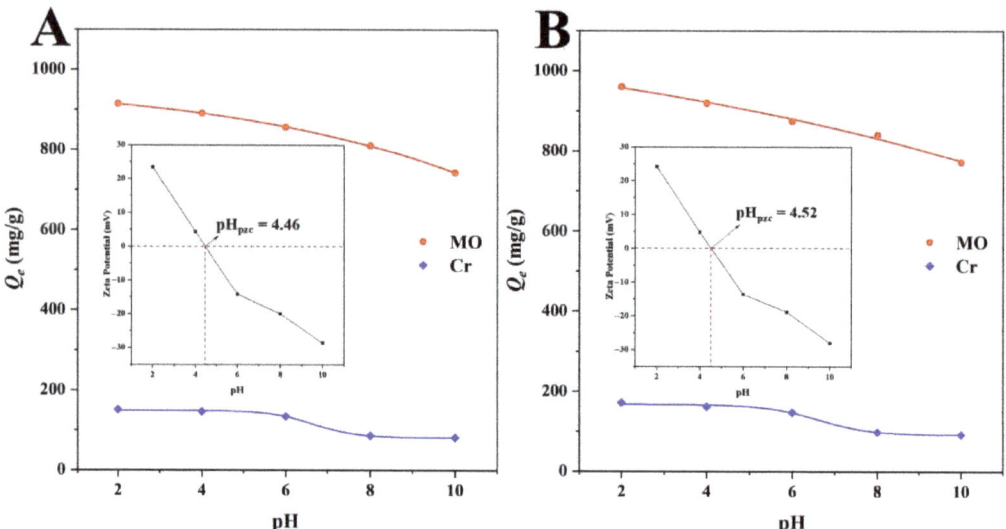

**Figure 9.** Effect of pH on the adsorption capacities of MO and $Cr^{6+}$ onto (**A**) BBB and (**B**) N-BBB (inset: Zeta potential).

### 2.4. Results of Cycle Tests

The recyclability of adsorbents is an important parameter for evaluating the practical performance of biochars [38–40]; thus, the five-cycle performance of the biochars was investigated, and the results are shown in Figure 10. The removal rates of MO and $Cr^{6+}$ by BBB and N-BBB decreased with an increase in cycle number. This can be explained as follows: on the one hand, with the treatment of the cycling experiment, adsorbed organic pollutants formed by-products on the surface of biochars [38,39]; while on the other hand, with the increase in the carbonization regeneration process, the structure of the biochars became more fragile, which further affected its regeneration. At the same time, after five cycles, the removal rate of MO and $Cr^{6+}$ by BBB and N-BBB remained above 80%, indicating that they had good stability and regeneration ability.

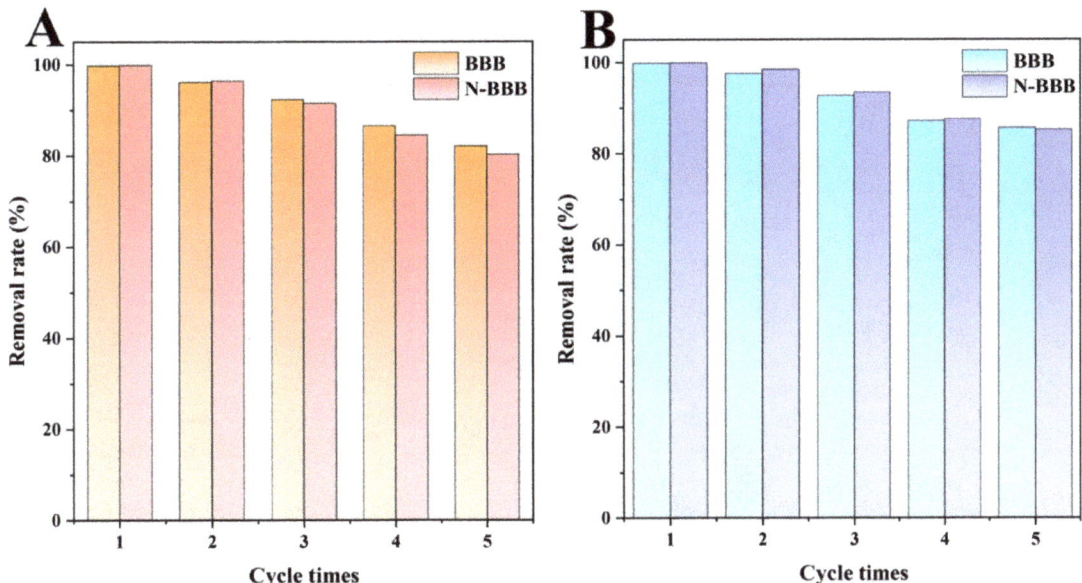

**Figure 10.** Cycling stability tests of BBB and N-BBB for (**A**) MO and (**B**) $Cr^{6+}$.

## 2.5. Probable Mechanism Analysis

In this work, the N-BBB exhibited good removal ability of MO and $Cr^{6+}$ in water solutions affected by many factors (Figure 11). Firstly, the large specific surface area and high total pore volume (2292.7 $m^2$/g and 1.0356 $cm^3$/g) of N-BBB provided many adsorption sites for the adsorption pollutant; therefore, it can be speculated that pore filling may play an important role in the adsorption process. The results based on the kinetics and isotherm showed that the adsorption process was heterogeneous multilayer adsorption with a chemical reaction, which indicated that the chemical binding force will also promote the adsorption process. In addition, from the test results with FT-IR and XPS, it can be speculated that many unsaturated functional groups containing carbon and oxygen on the biochar surface will produce hydrogen-bond interaction with the pollutant model. Moreover, FT-IR and Raman test results also showed that the prepared biochars contained aromatic rings and $sp^2$ hybridized carbon with graphite structure, and the π bond in these structures may also have π–π interaction with aromatic rings in pollutants to enhance the adsorption capacity. The charged pollutants in the appropriate pH environment formed a strong electrostatic attraction with N-BBB, which further promoted the adsorption process. Experiments under different conditions, such as the initial concentration of the solution and the temperature of the adsorption process, show that these also affect the adsorption process. To summarize, in addition to the experimental conditions, the pore filling, π–π interaction, H-bond interaction, and electrostatic attraction supported the excellent performance of N-BBB.

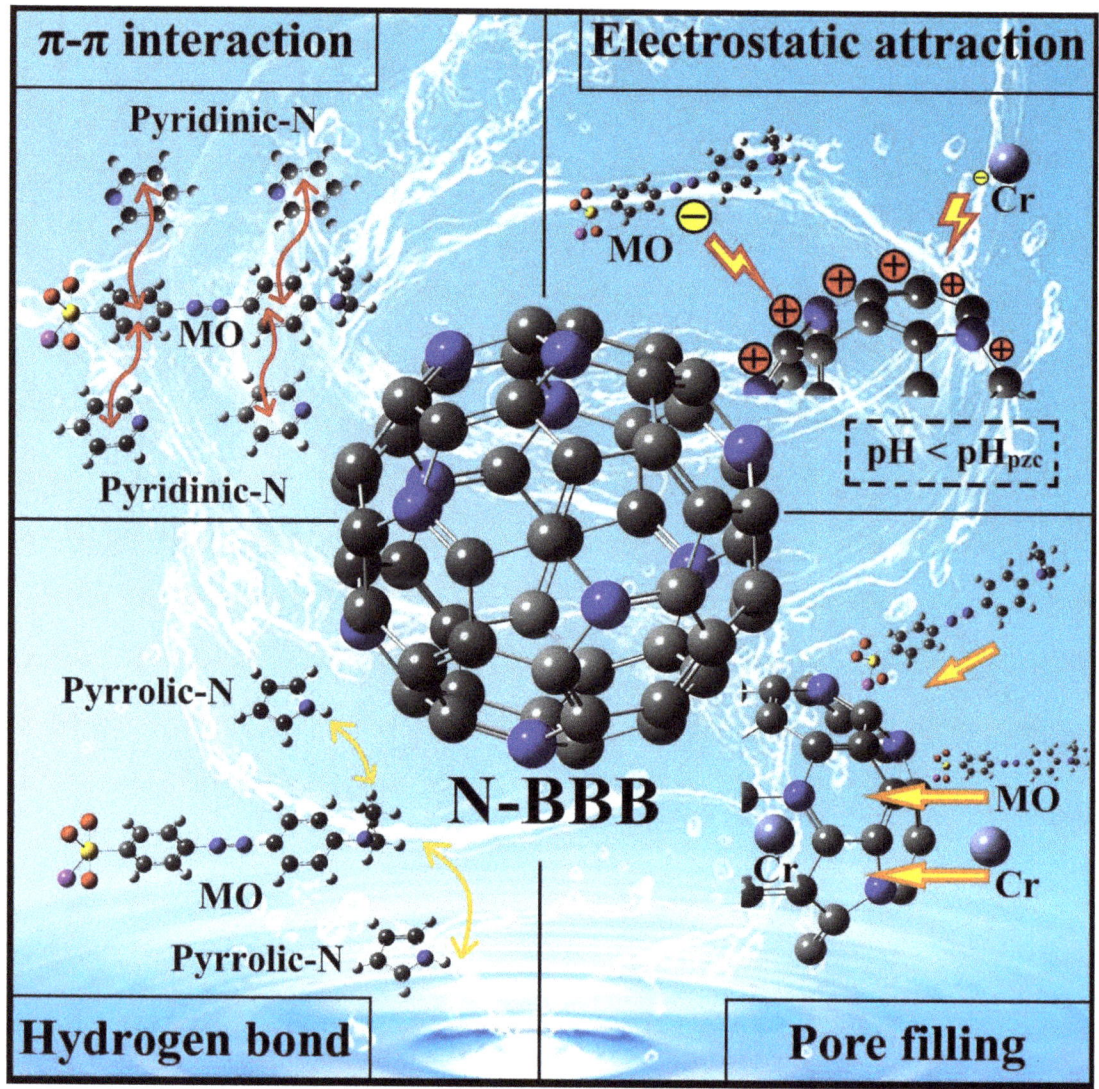

**Figure 11.** Probable mechanisms analysis for N-BBB removal of MO and $Cr^{6+}$.

*2.6. Comparison*

Adsorption capacity is an important parameter for evaluating the practical performance of an adsorbent; thus, the adsorption capacities of BBB and N-BBB are compared with other biochars, as shown in Table 5. The adsorption capacities of CBB for MO and $Cr^{6+}$ were only 25.2 and 10.4 mg/g. After activation and N-doping co-activation, the adsorption capacities of BBB and N-BBB to MO and $Cr^{6+}$ were significantly enhanced. In addition, compared with other biochars, the adsorption capacities of BBB and N-BBB were not low, which fully indicated that the prepared biochars had great potential and application prospects in the treatment of water pollutants.

**Table 5.** Comparison of BBB and N-BBB to MO and $Cr^{6+}$ with other biochars.

| Adsorbents | $Q_e$ for MO (mg/g) | $Q_e$ for $Cr^{6+}$ (mg/g) | References |
|---|---|---|---|
| BBB | 836.9 | 141.1 | This work |
| N-BBB | 858.3 | 169.1 | This work |
| Pomelo peel biochar | 147.9 | - | [46] |
| Magnetic bamboo biochar | 305.4 | - | [47] |
| Date seeds biochar | 334.0 | - | [48] |
| Lotus root biochar | 449.0 | - | [49] |
| Date palm petioles biochar | 461.0 | - | [50] |
| Landfill leachate sludge biochar | - | 17.5 | [51] |
| Zn/iron-based sludge/biochar | - | 27.0 | [52] |
| *Potamogeton crispus* biochar | - | 34.4 | [53] |
| *Egeria najas* biochar | - | 138.8 | [54] |
| Soybean protein biochar | - | 489.7 | [55] |

## 3. Materials and Methods

### 3.1. Materials and Reagents

Birch bark (BB), *Betula Mandshurica Nakai*, obtained from the campus of Jilin Agricultural University (Changchun, China) in 2022, was washed with deionized water, dried at 80 °C for 12 h, and crushed. Urea, NaOH, $H_2SO_4$, HCl, and ethanol were purchased from Beijing Chemical Works (Beijing, China) and used without further purification.

Diphenylcarbazide (CAS: 140-22-7), Methyl orange (MO, CAS: 547-58-0), and Potassium dichromate ($Cr^{6+}$, CAS: 7778-50-9) were supplied by Aladdin Chemical (Shanghai) Co., Ltd. (Shanghai, China) and the structural formulas are shown in Figure S1 (Supplementary Material).

### 3.2. Preparation of Biochars

BB was carbonized at 500 °C for 60 min with a heating rate of 10 °C/min under the protection of a nitrogen atmosphere to obtain CBB. BB was used by mixing with NaOH and Urea at a ratio of 1:4:1; meanwhile, 1.0 g BB was sufficiently ground with 4.0 g NaOH. The two mixtures were both heated at 700 °C for 60 min. After cooling to room temperature, the activated mixtures were washed with HCl and deionized water until reaching a natural pH value and dried at 180 °C for 12 h. At last, the samples including CBB, BBB, and N-BBB were kept in a desiccator prior to subsequent experiments.

### 3.3. Adsorption Performances

In a batch adsorption experiment, 0.05 g/L BBB or N-BBB was added to a flask containing pollutant solutions (MO or $Cr^{6+}$). The flask was placed in a constant temperature shaker at 150.0 RPM in the dark. After the adsorption process reached equilibrium, the suspension was centrifuged, and the supernatant was diluted with deionized water. The concentration of the solution was determined with an Agilent Cary-300 UV-vis spectrophotometer. The adsorption capacities of samples were calculated by Equation (1):

$$Q_e = \frac{(C_0 - C_e) \times V}{m} \quad (1)$$

where $Q_e$ (mg/g) represents the adsorption capacity of the sample, $C_e$ represents the equilibrium concentrations of the solution, $C_0$ (mg/L) represents the initial concentrations of the solution, $m$ (g) represents the mass of the samples, and $V$ (L) represents the volume of the solutions.

The pollutant solutions were prepared at different concentrations (50, 100, and 200 mg/L). A total of 0.05 g/L BBB or N-BBB was dispersed into flasks containing MO or $Cr^{6+}$ solutions and shaken at 150 RPM in the dark at 303 K. Then, the concentrations of the solutions were determined at preset time intervals. The pseudo-first-order kinetic (PFK, Equation (2)), the pseudo-second-order kinetic (PSK, Equation (3)), and the intra-particle diffusion model (IPD, Equation (4)) were used to analyze the adsorption kinetic data, shown as follows:

$$\ln(Q_e - Q_t) = \ln Q_e - k_1 t \tag{2}$$

$$\frac{t}{Q_t} = \frac{1}{k_2 Q_t^2} + \frac{t}{Q_e} \tag{3}$$

$$Q_t = k_3 t^{0.5} + C \tag{4}$$

where $Q_t$ represents the adsorption capacity of the sample at different time points $t$, $C$ represents the thickness of the boundary layer, $k_1$ represents the PFK adsorption kinetic rate constant, $k_2$ represents the PSK adsorption kinetic rate constant, and $k_3$ denotes the IPD adsorption kinetic rate constant.

The pollutant solutions at different initial concentrations (50, 100, 150, 200, and 250 mg/L) were prepared and used to test the adsorption isotherm at 303 K. After adsorption saturation, the absorbances of the solutions were measured using a UV-Vis spectrophotometer. The adsorption isotherm data were investigated using the Langmuir isotherm model (Equation (5)) and Freundlich isotherm model (Equation (6)), as follows:

$$\frac{C_e}{Q_e} = \frac{C_e}{Q_m} + \frac{1}{Q_m K_L} \tag{5}$$

$$\ln Q_m = \frac{1}{n} \ln C_e + \ln K_F \tag{6}$$

where $Q_m$ (mg/g) represents the maximum adsorption capacity of the sample calculated by the adsorption isotherm model, $K_L$ represents the Langmuir adsorption isotherm constant, and $K_F$ represents the Freundlich adsorption isotherm constant.

The effect of temperature (293, 303, and 313 K) on the adsorption capacity of the samples was investigated at an initial concentration of 100 mg/L BBB or 0.05 g/L N-BBB. The thermodynamic parameters were analyzed to describe the effect of temperature on the adsorption process. The calculation equations were as follows:

$$\ln(K_T) = -\frac{\Delta H}{RT} + \frac{\Delta S}{R} \tag{7}$$

$$K_T = \frac{Q_e}{C_e} \tag{8}$$

$$\Delta G = \Delta H - T\Delta S \tag{9}$$

where $\Delta S$ represents the thermodynamic parameters' standard entropy, $\Delta G$ represents the standard free Gibbs energy, $\Delta H$ represents the standard enthalpy, and $R$ represents the gas constant (8.314 J/K·mol).

The variation in the adsorption capacity of the samples with the pH (2, 4, 6, 8, and 10) was also investigated. The solutions were adjusted to different pH values by HCl and NaOH.

### 3.4. Cycle Tests

In each cycle, 1.0 g/L BBB or N-BBB was placed into a flask containing the organic pollutant solutions at a concentration of 100 mg/L at 303 K. After the adsorption of pollutants (Figure S2, Supplementary Material), the samples, BBB/MO, BBB/Cr$^{6+}$, N-BBB/MO, and N-BBB/Cr$^{6+}$, were collected and washed with deionized water. Then, the recycled samples were carbonized for 60 min at 600 °C under the protection of a nitrogen atmosphere. The re-carbonized samples were re-used as fresh adsorbent in the next cycle.

## 4. Conclusions

In this work, BB was used for the first time to prepare porous biochars via N-doping co-activation. The specific surface area and total pore volume of N-BBB were 2292.7 m$^2$/g and 1.0356 cm$^3$/g, respectively, which proved the feasibility of N-doping co-activation in pore-forming. The large specific surface area and the high total pore volume played a substantial role in the adsorption process. EDS, FT-IR, and XPS were used to characterize the samples, and the results indicated that nitrogen doping was successfully completed. In an experiment using the synthetic dye MO and heavy metal Cr$^{6+}$ as the pollutant models, N-BBB showed good removal ability. The adsorption capacity of N-BBB remained above 80% after five regenerations, which fully proved the stability of regeneration. Moreover, the excellent adsorption performance of N-BBB may have been influenced by pore filling, π–π interaction, H-bond interaction, and electrostatic attraction. This study not only provided a biochar adsorbent with excellent performance but also verified the feasibility of the one-step nitrogen-doping co-activation method to prepare high-performance biochars. In the future, we will continue to study this N-doped co-activation method and explore its application to other types of biomass to further develop more biochars with better performance.

**Supplementary Materials:** The following supporting information can be downloaded at: https://www.mdpi.com/article/10.3390/ijms232314618/s1.

**Author Contributions:** Conceptualization, S.C. and Y.S. (Yingjie Su); methodology, Y.S. (Yuqing Shi); software, M.J.; validation, Y.S. (Yingjie Su); formal analysis, Y.S. (Yuqing Shi); investigation, M.J.; resources, Y.S. (Yuqing Shi); data curation, M.J.; writing—original draft preparation, S.C.; writing—review and editing, S.C.; visualization, Y.S. (Yingjie Su); supervision, S.C.; project administration, Y.S. (Yingjie Su); funding acquisition, Y.S. (Yingjie Su). All authors have read and agreed to the published version of the manuscript.

**Funding:** This research was funded by [Jilin Scientific and Technological Development Program] grant number [20220203108SF] And The APC was funded by [20220203108SF].

**Institutional Review Board Statement:** Not applicable.

**Informed Consent Statement:** Not applicable.

**Acknowledgments:** We acknowledge the support of the Jilin Scientific and Technological Development Program (20220203108SF). This research was also supported by the China National College Students Innovation and Entrepreneurship Project.

**Conflicts of Interest:** The authors declare no conflict of interest.

## References

1. Zhou, L.; Wu, Y.; Zhang, S.; Li, Y.; Gao, Y.; Zhang, W.; Tian, L.; Li, T.; Du, Q.; Sun, S. Recent development in microbial electrochemical technologies: Biofilm formation, regulation, and application in water pollution prevention and control. *J. Water Process Eng.* **2022**, *49*, 103135. [CrossRef]
2. Yan, C.; Qu, Z.; Wang, J.; Cao, L.; Han, Q. Microalgal bioremediation of heavy metal pollution in water: Recent advances, challenges, and prospects. *Chemosphere* **2022**, *286*, 131870. [CrossRef] [PubMed]
3. Zamora-Ledezma, C.; Negrete-Bolagay, D.; Figueroa, F.; Zamora-Ledezma, E.; Ni, M.; Alexis, F.; Guerrero, V.H. Heavy metal water pollution: A fresh look about hazards, novel and conventional remediation methods. *Environ. Technol. Innov.* **2021**, *22*, 101504. [CrossRef]
4. Xu, H.; Fan, Y.; Xia, X.; Liu, Z.; Yang, S. Effect of Ginkgo biloba leaves on the removal efficiency of Cr(VI) in soil and its underlying mechanism. *Environ. Res.* **2023**, *216*, 114431. [CrossRef]
5. Njoya, O.; Zhao, S.; Shen, J.; Kong, X.; Gong, Y.; Wang, B.; Kang, J.; Chen, Z. Acetate improves catalytic performance for rapid removal of Cr(VI) by sodium borohydride in aqueous environments. *Sep. Purif. Technol.* **2022**, *301*, 122051. [CrossRef]
6. Li, X.Q.; Zhang, Q.H.; Ma, K.; Li, H.M.; Guo, Z. Identification and determination of 34 water-soluble synthetic dyes infoodstuff by high performance liquid chromatography–diode arraydetection–ion trap time–of–flight tandem mass spectrometry. *Food Chem.* **2015**, *182*, 316–326. [CrossRef]
7. Moghadas, M.R.S.; Motamedi, E.; Nasiri, J.; Naghavi, M.R.; Sabokdast, M. Proficient dye removal from water using biogenic silver nanoparticles prepared through solid-state synthetic route. *Heliyon* **2020**, *6*, e04730. [CrossRef]

8. Mandal, B.; Ray, S.K. Synthesis of interpenetrating network hydrogel from poly(acrylic acid-co-hydroxyethyl methacrylate) and sodium alginate: Modeling and kinetics study for removal of synthetic dyes from water. *Carbohyd. Polym.* **2013**, *98*, 257–269. [CrossRef]
9. Baig, U.; Kashif Uddin, M.; Gondal, M.A. Removal of hazardous azo dye from water using synthetic nano adsorbent: Facile synthesis, characterization, adsorption, regeneration and design of experiments. *Colloids Surf. A* **2020**, *584*, 124031. [CrossRef]
10. Nidheesh, P.V.; Zhou, M.; Oturan, M.A. An overview on the removal of synthetic dyes from water by electrochemical advanced oxidation processes. *Chemosphere* **2018**, *197*, 210–227. [CrossRef]
11. Ambika; Kumar, V.; Jamwal, A.; Kumar, V.; Singh, D. Green bioprocess for degradation of synthetic dyes mixture using consortium of laccase-producing bacteria from Himalayan niches. *J. Environ. Manag.* **2022**, *310*, 114764. [CrossRef] [PubMed]
12. Zaafouri, Z.; Batot, G.; Nieto-Draghi, C.; Coasne, B.; Bauer, D. Impact of adsorption kinetics on pollutant dispersion in water flowing in nanopores: A Lattice Boltzmann approach to stationary and transient conditions. *Adv. Water Resour.* **2022**, *162*, 104143. [CrossRef]
13. Tee, G.T.; Gok, X.Y.; Yong, W.F. Adsorption of pollutants in wastewater via biosorbents, nanoparticles and magnetic biosorbents: A review. *Environ. Res.* **2022**, *212*, 113248. [CrossRef]
14. Zhou, W.; Zhang, W.; Cai, Y. Enzyme-enhanced adsorption of laccase immobilized graphene oxide for micro-pollutant removal. *Sep. Purif. Technol.* **2022**, *294*, 121178. [CrossRef]
15. Ma, Y.; Li, Y.; Zhao, X.; Zhang, L.; Wang, B.; Nie, A.; Mu, C.; Xiang, J.; Zhai, K.; Xue, T.; et al. Lightweight and multifunctional superhydrophobic aramid nanofiber/multiwalled carbon nanotubes/$Fe_3O_4$ aerogel for microwave absorption, thermal insulation and pollutants adsorption. *J. Alloys Compd.* **2022**, *919*, 165792. [CrossRef]
16. Sellaoui, L.; Gomez-Aviles, A.; Dhaouadi, F.; Bedia, J.; Bonilla-Petriciolet, A.; Rtimi, S.; Belver, C. Adsorption of emerging pollutants on lignin-based activated carbon: Analysis of adsorption mechanism via characterization, kinetics and equilibrium studies. *Chem. Eng. J.* **2023**, *452*, 139399. [CrossRef]
17. Qiu, B.; Shao, Q.; Shi, J.; Yang, C.; Chu, H. Application of biochar for the adsorption of organic pollutants from wastewater: Modification strategies, mechanisms and challenges. *Sep. Purif. Technol.* **2022**, *300*, 121925. [CrossRef]
18. Zhang, B.; Wang, M.; Qu, J.; Zhang, Y.; Liu, H. Characterization and mechanism analysis of tylosin biodegradation and simultaneous ammonia nitrogen removal with strain Klebsiella pneumoniae TN-1. *Bioresour. Technol.* **2021**, *336*, 125342. [CrossRef]
19. Wu, J.; Wang, T.; Liu, Y.; Tang, W.; Geng, S.; Chen, J. Norfloxacin adsorption and subsequent degradation on ball-milling tailored N-doped biochar. *Chemosphere* **2022**, *303*, 135264. [CrossRef]
20. Qu, J.; Zhang, X.; Liu, S.; Li, X.; Wang, S.; Feng, Z.; Wu, Z.; Wang, L.; Jiang, Z.; Zhang, Y. One-step preparation of Fe/N co-doped porous biochar for chromium(VI) and bisphenol a decontamination in water: Insights to co-activation and adsorption mechanisms. *Bioresour. Technol.* **2022**, *361*, 127718. [CrossRef]
21. Li, J.; Lin, Q.; Luo, H.; Fu, H.; Wu, L.; Chen, Y.; Ma, Y. The effect of nanoscale zero-valent iron-loaded N-doped biochar on the generation of free radicals and nonradicals by peroxydisulfate activation. *J. Water Process Eng.* **2022**, *47*, 102681. [CrossRef]
22. Tan, X.; Zhu, S.; Wang, R.; Chen, Y.; Showf, P.; Zhang, F.; Ho, S. Role of biochar surface characteristics in the adsorption of aromatic compounds: Pore structure and functional groups. *Chin. Chem. Lett.* **2021**, *32*, 2939–2946. [CrossRef]
23. Zhang, J.; Huang, D.; Shao, J.; Zhang, X.; Yang, H.; Zhang, S.; Chen, H. Activation-free synthesis of nitrogen-doped biochar for enhanced adsorption of $CO_2$. *J. Clean. Prod.* **2022**, *355*, 131642. [CrossRef]
24. Liew, R.; Azwar, E.; Yek, P.; Lim, X.; Cheng, C.; Ng, J.; Jusoh, A.; Lam, W.; Ibrahim, M.; Ma, N.; et al. Microwave pyrolysis with KOH/NaOH mixture activation: A new approach to produce micro-mesoporous activated carbon for textile dye adsorption. *Bioresour. Technol.* **2018**, *266*, 1–10. [CrossRef]
25. Chen, S.; Zhang, B.; Xia, Y.; Chen, H.; Chen, G.; Tang, S. Influence of mixed alkali on the preparation of edible fungus substrate porous carbon material and its application for the removal of dye. *Colloids Surf. A* **2021**, *609*, 125675. [CrossRef]
26. Wei, M.; Marrakchi, F.; Yuan, C.; Cheng, X.; Jiang, D.; Zafar, F.F.; Fu, Y.; Wang, S. Adsorption modeling, thermodynamics, and DFT simulation of tetracycline onto mesoporous and high-surface-area NaOH-activated macroalgae carbon. *J. Hazard. Mater.* **2022**, *425*, 127887. [CrossRef] [PubMed]
27. Alfatah, T.; Mistar, E.M.; Supardan, M.D. Porous structure and adsorptive properties of activated carbon derived from Bambusa vulgaris striata by two-stage KOH/NaOH mixture activation for $Hg^{2+}$ removal. *J. Water Process Eng.* **2021**, *43*, 102294. [CrossRef]
28. Kamran, U.; Park, S. Tuning ratios of KOH and NaOH on acetic acid-mediated chitosan-based porous carbons for improving their textural features and $CO_2$ uptakes. *J. $CO_2$ Util.* **2020**, *40*, 101212. [CrossRef]
29. Yang, X.; Wang, Q.; Lai, J.; Cai, Z.; Lv, J.; Chen, X.; Chen, Y.; Zheng, X.; Huang, B.; Lin, G. Nitrogen-doped activated carbons via melamine-assisted NaOH/KOH/urea aqueous system for high performance supercapacitors. *Mater. Chem. Phys.* **2020**, *250*, 123201. [CrossRef]
30. Hou, Z.; Tao, Y.; Bai, T.; Liang, Y.; Huang, S.; Cai, J. Efficient Rhodamine B removal by N-doped hierarchical carbons obtained from KOH activation and urea-oxidation of glucose hydrochar. *J. Environ. Chem. Eng.* **2021**, *9*, 105757. [CrossRef]
31. Wang, K.; Xu, M.; Gu, Y.; Gu, Z.; Fan, Q.H. Symmetric supercapacitors using urea-modified lignin derived Ndoped porous carbon as electrode materials in liquid and solid electrolytes. *J. Power Sources* **2016**, *332*, 180–186. [CrossRef]

32. Jin, Y.; Zhang, B.; Chen, G.; Chen, H.; Tang, S. Combining biological and chemical methods to disassemble of cellulose from corn straw for the preparation of porous carbons with enhanced adsorption performance. *Int. J. Biol. Macromol.* **2022**, *209*, 315–329. [CrossRef] [PubMed]
33. Chen, S.; Xia, Y.; Zhang, B.; Chen, H.; Chen, G.; Tang, S. Disassembly of lignocellulose into cellulose, hemicellulose, and lignin for preparation of porous carbon materials with enhanced performances. *J. Hazard. Mater.* **2021**, *408*, 124956. [CrossRef] [PubMed]
34. Zhang, B.; Jin, Y.; Qi, J.; Chen, H.; Chen, G.; Tang, S. Porous carbon materials based on *Physalis alkekengi* L. husk and its application for removal of malachite green. *Environ. Technol. Innov.* **2021**, *21*, 101343. [CrossRef]
35. Xia, Y.; Jin, Y.; Qi, J.; Chen, H.; Chen, G.; Tang, S. Preparation of biomass carbon material based on Fomes fomentarius via alkali activation and its application for the removal of brilliant green in wastewater. *Environ. Technol. Innov.* **2021**, *23*, 101659. [CrossRef]
36. Fan, C.; Yan, J.; Huang, Y.; Han, X.; Jiang, X. XRD and TG-FTIR study of the effect of mineral matrix on the pyrolysis and combustion of organic matter in shale char. *Fuel* **2015**, *139*, 502–510. [CrossRef]
37. Yan, J.; Jiang, X.; Han, X.; Liu, J. A TG–FTIR investigation to the catalytic effect of mineral matrix in oil shale on the pyrolysis and combustion of kerogen. *Fuel* **2013**, *104*, 307–317. [CrossRef]
38. Zhang, B.; Jin, Y.; Huang, X.; Tang, S.; Chen, H.; Su, Y.; Yu, X.; Chen, S.; Chen, G. Biological self-assembled hyphae/starch porous carbon composites for removal of organic pollutants from water. *Chem. Eng. J.* **2022**, *450*, 138264. [CrossRef]
39. Xia, Y.; Zhang, B.; Guo, Z.; Tang, S.; Su, Y.; Yu, X.; Chen, S.; Chen, G. Fungal mycelium modified hierarchical porous carbon with enhanced performance and its application for removal of organic pollutants. *J. Environ. Chem. Eng.* **2022**, *10*, 108699. [CrossRef]
40. Chen, X.; Yu, G.; Chen, Y.; Tang, S.; Su, Y. Cow dung-based biochar materials prepared via mixed base and its application in the removal of organic pollutants. *Int. J. Mol. Sci.* **2022**, *23*, 10094. [CrossRef]
41. Chauhdary, Y.; Hanif, M.A.; Rashid, U.; Bhatti, I.A.; Anwar, H.; Jamil, Y.; Alharthi, F.A.; Kazerooni, E.A. Effective removal of reactive and direct dyes from colored wastewater using low-cost novel bentonite nanocomposites. *Water* **2022**, *14*, 3604. [CrossRef]
42. Lin, L.; Li, L.; Xiao, L.; Zhang, C.; Li, X.; Pervez, M.N.; Zhang, Y.; Nuruzzaman, M.; Mondal, M.I.H.; Cai, Y.; et al. Adsorption behaviour of reactive blue 194 on raw Ramie Yarn in palm oil and water media. *Materials* **2022**, *15*, 7818. [CrossRef] [PubMed]
43. El-Sayed, N.S.; Salama, A.; Guarino, V. Coupling of 3-Aminopropyl sulfonic acid to cellulose nanofifibers for effificient removal of cationic dyes. *Materials* **2022**, *15*, 6964. [CrossRef] [PubMed]
44. Shirendev, N.; Bat-Amgalan, M.; Kano, N.; Kim, H.-J.; Gunchin, B.; Ganbat, B.; Yunden, G. A natural zeolite developed with 3-Aminopropyltriethoxylane and adsorption of Cu(II) from aqueous media. *Appl. Sci.* **2022**, *12*, 11344. [CrossRef]
45. Jiao, Y.; Xu, L.; Sun, H.; Deng, Y.; Zhang, T.; Liu, G. Synthesis of benzxazine-based nitrogen-doped mesoporous carbon spheres for methyl orange dye adsorption. *J. Porous Mater.* **2017**, *24*, 1565–1574. [CrossRef]
46. Zhang, B.; Wu, Y.; Cha, L. Removal of methyl orange dye using activated biochar derived from pomelo peel wastes: Performance, isotherm, and kinetic studies. *J. Disper. Sci. Technol.* **2018**, *41*, 1561298. [CrossRef]
47. Zhang, H.; Li, R.; Zhang, Z. A versatile EDTA and chitosan bi-functionalized magnetic bamboo biochar for simultaneous removal of methyl orange and heavy metals from complex wastewater. *Environ. Pollut.* **2022**, *293*, 118517. [CrossRef]
48. Ouedrhiri, A.; Himi, M.A.; Youbi, B.; Lghazi, Y.; Bahar, J.; Haimer, C.E.; Aynaou, A.; Bimaghra, I. Biochar material derived from natural waste with superior dye adsorption performance. *Mater. Today. Proc.* **2022**, *66*, 259–267. [CrossRef]
49. Hou, Y.; Liang, Y.; Hu, H.; Tao, Y.; Zhou, J.; Cai, J. Facile preparation of multi-porous biochar from lotus biomass for methyl orange removal: Kinetics, isotherms, and regeneration studies. *Bioresour. Technol.* **2021**, *329*, 124877. [CrossRef]
50. Aichour, A.; Zaghouane-Boudiaf, H.; Khodja, H.D. Highly removal of anionic dye from aqueous medium using a promising biochar derived from date palm petioles: Characterization, adsorption properties and reuse studies. *Arab. J. Chem.* **2022**, *15*, 103542. [CrossRef]
51. Li, Y.; Chen, X.; Liu, L.; Liu, P.; Zhou, Z.; Huhetaoli; Wu, Y.; Lei, T. Characteristics and adsorption of Cr(VI) of biochar pyrolyzed from landfill leachate sludge. *J. Anal. Appl. Pyrolysis* **2022**, *162*, 105449. [CrossRef]
52. Zhen, Z.; Duan, X.; Tie, J. One-pot synthesis of a magnetic Zn/iron-based sludge/biochar composite for aqueous Cr(VI) adsorption. *Environ. Technol. Innov.* **2022**, *28*, 102661. [CrossRef]
53. Xu, D.; Sun, T.; Jia, H.; Sun, Y.; Zhu, X. The performance and mechanism of Cr(VI) adsorption by biochar derived from Potamogeton crispus at different pyrolysis temperatures. *J. Anal. Appl. Pyrolysis* **2022**, *167*, 105662. [CrossRef]
54. Yi, Y.; Wang, X.; Zhang, Y.; Ma, J.; Ning, P. Adsorption properties and mechanism of Cr(VI) by $Fe_2(SO_4)_3$ modified biochar derived from *Egeria najas*. *Colloids Surf. A* **2022**, *645*, 128938. [CrossRef]
55. Kuang, Q.; Liu, K.; Wang, Q.; Chang, Q. Three-dimensional hierarchical pore biochar prepared from soybean protein and its excellent Cr(VI) adsorption. *Sep. Purif. Technol.* **2023**, *304*, 122295. [CrossRef]

*Article*

# Photocatalytic Performance of Undoped and Al-Doped ZnO Nanoparticles in the Degradation of Rhodamine B under UV-Visible Light: The Role of Defects and Morphology

Alessandra Piras [1,2], Chiara Olla [3], Gunter Reekmans [4], An-Sofie Kelchtermans [2,5], Dries De Sloovere [2,5,6], Ken Elen [2,5,6], Carlo Maria Carbonaro [3], Luca Fusaro [1], Peter Adriaensens [4], An Hardy [2,5,6], Carmela Aprile [1] and Marlies K. Van Bael [2,5,6,*]

1. Laboratory of Applied Materials Chemistry, Unit of Nanomaterials Chemistry, Chemistry Department, University of Namur, NISM, Rue de Bruxelles, 61, 5000 Namur, Belgium
2. DESINe Group, Institute for Materials Research (imo-imomec), Hasselt University, Agoralaan Building D, 3590 Diepenbeek, Belgium
3. Department of Physics, University of Cagliari, Cittadella Universitaria, I-09042 Monserrato, Italy
4. Analytical and Circular Chemistry (ACC), Institute for Materials Research (imo-imomec), Hasselt University, Agoralaan Building D, 3590 Diepenbeek, Belgium
5. EnergyVille, Thor Park 8320, 3600 Genk, Belgium
6. Imec, Division Imomec, Wetenschapspark 1, 3590 Diepenbeek, Belgium
* Correspondence: marlies.vanbael@uhasselt.be

**Abstract:** Quasi-spherical undoped ZnO and Al-doped ZnO nanoparticles with different aluminum content, ranging from 0.5 to 5 at% of Al with respect to Zn, were synthesized. These nanoparticles were evaluated as photocatalysts in the photodegradation of the Rhodamine B (RhB) dye aqueous solution under UV-visible light irradiation. The undoped ZnO nanopowder annealed at 400 °C resulted in the highest degradation efficiency of ca. 81% after 4 h under green light irradiation (525 nm), in the presence of 5 mg of catalyst. The samples were characterized using ICP-OES, PXRD, TEM, FT-IR, $^{27}$Al-MAS NMR, UV-Vis and steady-state PL. The effect of Al-doping on the phase structure, shape and particle size was also investigated. Additional information arose from the annealed nanomaterials under dynamic $N_2$ at different temperatures (400 and 550 °C). The position of aluminum in the ZnO lattice was identified by means of $^{27}$Al-MAS NMR. FT-IR gave further information about the type of tetrahedral sites occupied by aluminum. Photoluminescence showed that the insertion of dopant increases the oxygen vacancies reducing the peroxide-like species responsible for photocatalysis. The annealing temperature helps increase the number of red-emitting centers up to 400 °C, while at 550 °C, the photocatalytic performance drops due to the aggregation tendency.

**Keywords:** nanomaterials; ZnO; Al-doped ZnO; Rhodamine B; photocatalysis; green light-irradiation; photoluminescence; solid-state $^{27}$Al-NMR

## 1. Introduction

Based on the world water development report 2020 [1], among the most critical contemporary global issues, the conservation of water resources associated with global climate change is of high environmental importance. Water resources are currently contaminated with various organic, inorganic and microbial pollutants. Organic contaminants, which are mainly employed in processing products such as fabrics, cosmetics, leather, plastic, ceramics, paper, ink-jet printing, pharmaceuticals, etc. [2–4], are due to the discharge of organic dyes through industrial wastewater. In aquatic life, different pollutants were identified, including textile dyes, surfactants, insecticides, pesticides and heavy metals [5]. Exposure to dyes at a small level of less than 1 mg L$^{-1}$ can seriously affect the water quality of the aquatic environment [3]. In fact, Rhodamine B (RhB), Methyl Orange (MO) and Methylene

Blue (MB) are commonly used mutagenic, toxic and non-biodegradable dyes hazardous to aquatic life. These dyes also threaten human lives due to their high potential of being carcinogenic [6]. In particular, traces of organic dyes in water can result in ailments such as abdominal disorders, irritations, anemia and many more [7,8].

In the textile industry, up to 200,000 tons of dyes are dispersed in water bodies every year during the dyeing and finishing manipulations due to the inefficiency of the dyeing process [9]. Therefore, the removal of pollutant dyes from wastewater is needed. Among the conventional treatments available nowadays, various chemical and physical processes such as chemical precipitation and separation, adsorption and coagulation methods are in use [10]. However, these methods lead to incomplete dye degradation and only transfer the contaminant from one phase to another [11]. Advanced Oxidation Processes (AOPs), based on semiconducting materials, have emerged in recent years as an alternative to conventional methods [12]. Indeed, through these processes, reactive species such as hydroxyl radicals can be generated and used as active species to oxidize the organic contaminants quickly and non-selectively. Heterogeneous photocatalysis utilizing oxide-based nanomaterials is of particular interest owing to its ability to destroy water-soluble organic pollutants in water and wastewater [13].

Ideally, the system design concept for the photocatalytic degradation reaction includes the use of particles suspended in water as photocatalysts irradiated by light. In this process, the photocatalyst plays a fundamental role. The utilized materials are semiconductors characterized by an electronic band structure, where the highest occupied energy band, called the valence band (V.B.), and the lowest empty band, the conduction band (C.B.), are separated by an energy bandgap. The absorption of photons with equivalent to or higher energies than the semiconductor bandgap leads to the generation of the electron ($e^-$)–hole ($h^+$) pairs in the semiconductor particles. A charge separation follows due to the migration of these photogenerated carriers in the semiconductor particles. Then, the surface chemical reactions between these carriers with various compounds (e.g., $H_2O$, hydroxyl radicals) occur. Electrons and holes may also recombine with each other without participating in any chemical reactions. The oxidizing agents attack the organic pollutants present on or near the catalytic surface until there is complete mineralization into harmless species [14,15].

Until now, ZnO-based nanomaterials have been studied as heterogeneous photocatalysts for dye degradation because they are abundant, environmentally friendly, non-toxic, insoluble in neutral water and economical [16]. However, the efficiency of zinc oxide as a photocatalyst is still limited due to the rapid recombination of electron–hole pairs and the lower activity in the visible region than in the UV-range [17]. Doping of ZnO is an interesting way to optimize and tune its optical, electrical, magnetic and structural performance. Zinc oxide can be observed in three polymorphs: wurtzite, zinc blend and rock salt [18]. Under ambient conditions, the thermodynamically stable phase is the wurtzite structure. Structurally, wurtzite has a hexagonal closed-packed arrangement of $O^{2-}$ anions where half of the tetrahedral sites are occupied by the $Zn^{2+}$ cations. The other half of the tetrahedral sites and all the octahedral sites are empty. These latter sites procure possible dopant sites. n-type and p-type doped ZnO materials have been extensively studied because they exhibit interesting properties for industrial applications [18,19]. Undoped ZnO displays n-type conductivity, traditionally attributed to intrinsic defects such as zinc excess at the interstitial positions and the lack of oxygen. The n-type conductivity can also be acquired by doping with post-transition metal ions such as Al [20], Ga [21] and In [22]. It is known that the trend in doping efficiency is related to the size of the trivalent dopant (In > Ga > Al) with respect to the $Zn^{2+}$ ion [21]. The greater similarity between the dopant and the host cation size allows a more favorable lattice substitution. Indeed, the ionic radii for $Zn^{2+}$, $In^{3+}$, $Ga^{3+}$ and $Al^{3+}$ in CN = 4 coordination are 0.60, 0.62, 0.47 and 0.39 Å, respectively [21]. These dopants mentioned above have fully occupied d orbitals, hence no possible internal d-d transitions. However, they stimulate the formation of native defects in the ZnO lattice (for instance, zinc interstitials and oxygen vacancies). Such defects generate mid-bandgap energy levels which are reported to increase carrier trapping leading to rapid non-radiative recombination of the electron–hole pairs within the semiconducting materials. Although doping can be used to enhance the general efficiency of the photocatalyst, many contributions to a changing performance have to be considered, such as preparation method [23], particle morphology [12,24], surface properties [25,26] and defects [27], dopant concentration [2] as well as the charge transfer dynamics of the discoloration process [28].

In this work, aluminum was selected as a dopant for the ZnO semiconductor material because of its abundancy, low price and suitability. Several articles have been published regarding the conductivity enhanced through n-type doping when Al is incorporated into the zinc oxide crystal lattice [19–21,29]. Some authors [30] also noticed that the electronic properties of the doped oxide are influenced by the crystallographic position of the aluminum dopant in the ZnO lattice. Indeed, the $Al^{3+}$ ion can occupy the empty tetrahedral sites, and also substitute a $Zn^{2+}$ ion in the tetrahedral geometry, to furnish a free electron (charge carrier) which improves the conductivity. Likewise, an interstitial octahedral coordinated site can be occupied by the $Al^{3+}$ ion. It is reported that when $Al^{3+}$ ions occupy the octahedral site, the conductivity decreases [30,31]. Thus, the $Al^{3+}$ dopant should be placed in a substitutional, tetrahedral position to optimize the ZnO host material's electronic properties.

The insertion of post-transition metals as dopants creates defects responsible for enhanced functionalities and might lead to enhanced charge separation [32]. Therefore, we synthesized, characterized and investigated undoped and Al-doped ZnO nanoparticles with increasing percentages of aluminum with respect to Zn for photocatalytic degradation under UV-visible light irradiation of Rhodamine B. Although many authors have reported on the photodegradation of Rhodamine B by ZnO [7,33–35], it is difficult, if not impossible, to directly compare their photocatalytic performances due to the many variables (e.g., reaction conditions, light source, reactor setup, etc.) that are not standardized. We also examined the effect of the thermal treatment in a dynamic nitrogen atmosphere at 400 °C and 550 °C on the catalysts, focusing on the role played by the defects.

## 2. Results and Discussion

Designing a photocatalyst requires attention to the composition (e.g., Al/Zn ratio) and the morphology. Nanosized dimensions are crucial for a material to function as an efficient photocatalyst due to a large number of atoms at the surface with their distinct optical, crystallographic and electronic properties. Thus, the strategy consisted in synthesizing undoped and Al-doped ZnO particles in the nanoscale to achieve large surface areas with consequent benefits for its performance. To this end, a method previously described by Momot et al. [19] was followed, where the authors presented an applicable solvothermal route to nanocrystalline Al-doped ZnO based on the reaction between zinc (II) acetylacetonate and aluminum (III) acetylacetonate as precursors and benzylamine used as solvent and reactant, Figure S1. The dopant concentrations range from 0 at% to 5 at% with respect to Zn. To investigate and evaluate any possible effects of the aluminum dopant on the zinc oxide structure, Inductively Coupled Plasma Optical Emission Spectroscopy (ICP-OES), Powder X-ray Diffraction (PXRD), Transmission Electron Microscopy (TEM), Fourier Transform Infrared spectroscopy (FT-IR), Solid-State $^{27}$Al-Magic Angle Spinning Nuclear Magnetic Resonance (SS $^{27}$Al-MAS NMR), UltraViolet-Visible spectroscopy (UV-vis) and steady-state Photoluminescence spectroscopy (PL) have been used. The thermal treatment under nitrogen atmosphere was applied to investigate its effect on the position of the dopant and the related performance change. In particular, the annealing in a dynamic nitrogen atmosphere was performed on the ZnO (ZO) and Al-doped ZnO (AZO-05 and AZO-5) at 400 and 550 °C. Moreover, commercial ZnO (CZO) is used as reference material. The resulting synthesized and commercial samples were coded as listed in Table 1.

**Table 1.** Undoped, Al-doped and commercial ZnO sample codes.

| Sample Type | Sample Code | Nominal Al Concentration (at%) | Annealing in $N_2$ Flow |
|---|---|---|---|
| Commercial ZnO | CZO | 0 | / |
| Undoped ZnO | ZO | 0 | / |
| Undoped ZnO | ZO-400 | 0 | 400 °C/10 min |
| Undoped ZnO | ZO-550 | 0 | 550 °C/10 min |
| Al-doped ZnO | AZO-05 | 0.5 | / |
| Al-doped ZnO | AZO-05-400 | 0.5 | 400 °C/10 min |
| Al-doped ZnO | AZO-05-550 | 0.5 | 550 °C/10 min |
| Al-doped ZnO | AZO-1 | 1 | / |
| Al-doped ZnO | AZO-2 | 2 | / |
| Al-doped ZnO | AZO-3 | 3 | / |
| Al-doped ZnO | AZO-5 | 5 | / |
| Al-doped ZnO | AZO-5-400 | 5 | 400 °C/10 min |
| Al-doped ZnO | AZO-5-550 | 5 | 550 °C/10 min |

The powder XRD patterns of commercial ZnO, undoped ZnO and Al-doped ZnO materials displayed in Figure 1, where the CZO was used as reference solid.

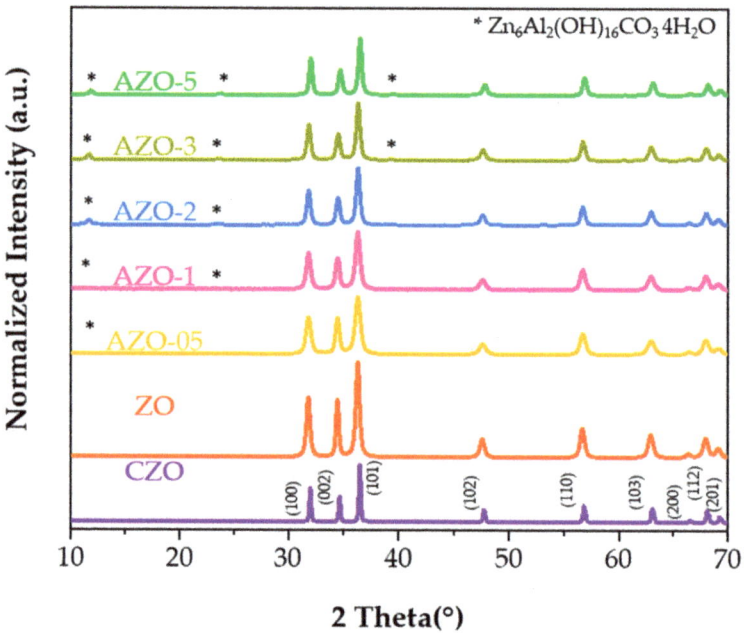

**Figure 1.** X-ray diffraction patterns of undoped ZnO and Al-doped ZnO powders as-synthesized with different aluminum content. CZO is used as reference material. Symbol: *—corresponds to the secondary phase: $Zn_6Al_2(OH)_{16}CO_3\cdot 4H_2O$.

All patterns match the hexagonal wurtzite crystal phase (PDF card 89-1397) [10]. The characteristic diffraction peaks at 2θ = 31.79°, 34.44°, 36.28°, 47.59°, 56.61°, 62.91°, 66.41°, 68.04° and 69.17° indicate the reflection from (100), (002), (101), (102), (110), (103), (200), (112) and (201) crystal planes, respectively. At an increasing amount of aluminum inserted into the zinc oxide lattice, an additional crystalline phase, identified as $Zn_6Al_2(OH)_{16}CO_3\cdot 4H_2O$ (PDF card 38-0486), is observed. Although the $Al^{3+}$ ion is smaller than the $Zn^{2+}$ ion, its incorporation is complex and incomplete [36]. Therefore, depending on slight variations of the experimental conditions [30,37], a segregated phase of aluminum external to the zinc oxide lattice might explain the presence of different diffraction peaks in the XRD patterns [20,38,39].

The doping efficiency also depends on the crystallographic position in the ZnO lattice and is majorly correlated to the post-treatments and synthesis method [30].

A quantitative determination of dopant incorporation within the ZnO structure was evaluated using ICP-OES, Table S1. Although the presence of a secondary phase (vide infra), might lead to an incorrect composition analysis, the results display a good relationship between the nominal amounts of aluminum added during the synthesis mixture and the actual aluminum found in the final solid. The missing aluminum content most probably was removed by the repeated washing procedures. This would suggest that doping ZnO quasi-spherical nanoparticles with aluminum is more efficient at lower reaction molar ratios. Similar observations for the experimental aluminum content are made in the literature [30,40].

TEM was used to evaluate the morphology and size of the nanoparticles. As displayed in Figure 2, the synthesis produced quasi-spherical nanoparticles exhibiting diameters ranging between 10 and 70 nm for ZO and AZO, highlighting the success of the synthesis method. From TEM investigation, the nominal aluminum content in the AZO nanoparticles as-synthesized does not affect the average diameter and the size distribution of the quasi-spheres when the aluminum content is low. In contrast, an increase in diameter is observed for the AZO-5 NPs. It was previously reported [41–45] that the morphology of the particles is not affected at low aluminum content while it can be strongly altered (or modified) for increased aluminum content. For instance, Kelchtermans and coworkers [46] demonstrated that at an aluminum concentration of 10 at%, the presence of both Al-doped ZnO nanoparticles and nanorods can be detected. Pinna et al. [47] observed a mixture of spheres and rods for pure ZnO, and the presence of different morphologies was particularly evident when impurities were present. Generally, when the spherical morphology is maintained, a decrease in the diameter of the AZO nanoparticles is reported.

A further structural investigation was performed after annealing the ZO and AZO nanopowders in a nitrogen atmosphere at 400 and 550 °C; see Figures S2 and S3.

The particle size distribution histograms show that thermal treatment affects the zinc oxide NPs shape and size. The particles size goes from $22 \pm 7$ nm for the bare ZO, to $42 \pm 21$ nm for the same material annealed at 400 °C and up to $51 \pm 21$ nm for the nanoparticles annealed at 550 °C. Furthermore, the aggregation tendency is more pronounced by annealing at 550 °C temperature. A broader particle size distribution together with the appearance of particles of different shape was already observed in the samples annealed at 400 °C. Indeed, the particle size distribution presents a wider standard deviation for the annealed materials due to the variety of morphologies displayed.

Solid-state $^{27}$Al MAS NMR can provide information on the local environment of the aluminum ions inserted into the zinc oxide structure [30,48]. Therefore, the technique is used to study the position of aluminum in the crystalline lattice [19,30,48]. Figure 3 presents the solid-state $^{27}$Al MAS NMR spectra of the AZO as-synthesized series, in which the signal assignments can be made based on the correlation between the observed chemical shift value and the coordination number of the aluminum atoms.

Four principal signals are often observed in the AZO $^{27}$Al MAS NMR spectra: the first signal around 10 ppm results from a 6-fold coordination, i.e., Al coordinated to 6 oxygen atoms (octahedral, $Al^{VI}$) [29]. The peak often has a narrow linewidth indicating a highly ordered, crystalline environment, implying that the aluminum is incorporated into the crystalline ZnO lattice during the synthesis. In this case, however, the general line shape of the octahedral signal is broadened by superposition with a broad contribution originating from a portion of the aluminum atoms in an amorphous, disordered environment [48]. $^{27}$Al nucleus is quadrupolar and can interact with an external magnetic field as well as with an electric field gradient generated in the surrounding environment. Therefore, in a symmetrical environment the generated signal is sharp, while in an asymmetrical environment, the signal shape is distorted and broadened [49]. The second signal around 82 ppm is referred to as the 4-fold coordination, i.e., Al coordinated to 4 oxygen atoms (tetrahedral, $Al_a^{IV}$). In agreement with results reported in the literature [30], higher aluminum contents imply a decreasing fraction of aluminum occupying tetrahedral sites. The third signal around 200 ppm is (partly) due to a spinning-side band (SS) of the octahedral signal but can also contain a Knight-shift signal ($Al_{Ks}$). In the literature, some authors [50] reported the observation of a Knight-shift signal in their NMR spectra. The mechanism of appearance of this signal in metals has been explained [51], while the exact origin of this signal in ZnO samples has been suggested [19]. These three signals are all observed in the spectra of the AZO samples in this study. Sometimes, a fourth, broad signal appears around 70 ppm indicating the presence of aluminum in a 5-fold coordination state ($Al^V$) or in a distorted 4-fold coordination ($Al_b^{IV}$) [29]. This signal is clearly observed in the spectra of the AZOs after annealing in $N_2$ at 400 and 550 °C (Figure S5).

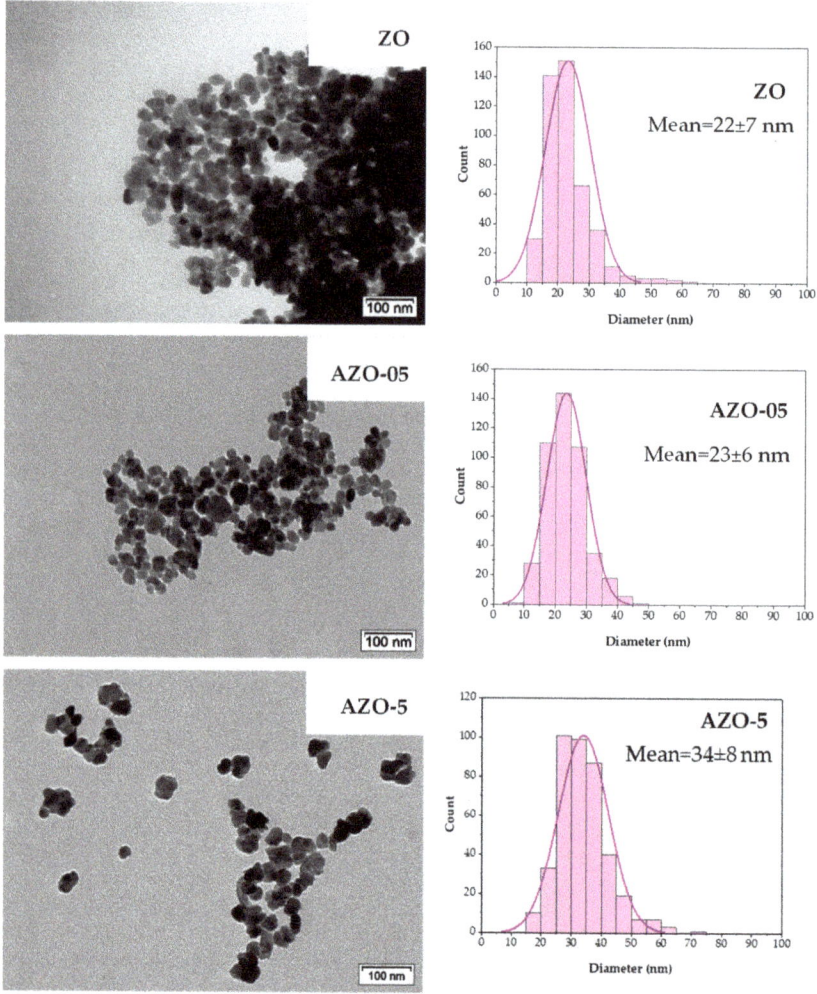

**Figure 2.** TEM micrographs of the ZO, AZO-05 and AZO-5 as-synthesized materials supported with the corresponding particle size histograms.

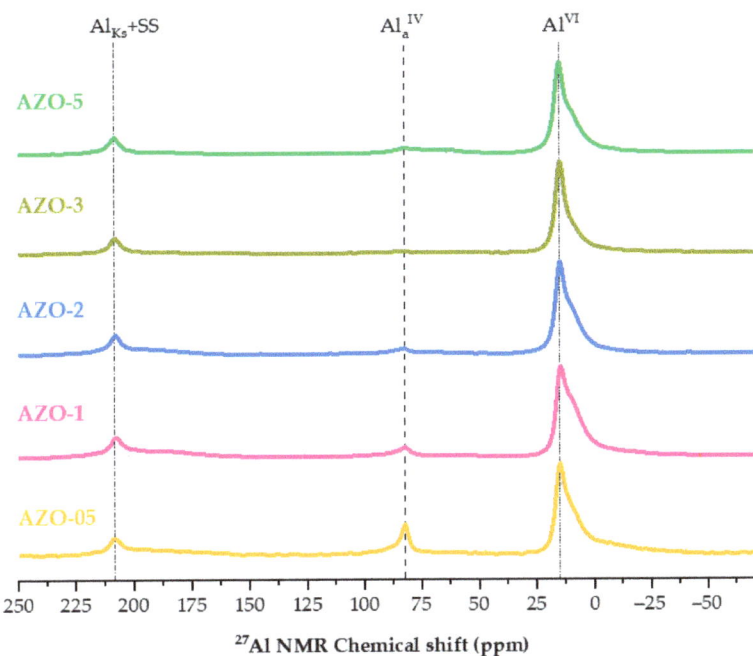

**Figure 3.** $^{27}$Al MAS NMR of AZO quasi−spherical nanoparticles as−prepared with a different nominal aluminum content. The lines highlight the peaks assigned to the 6−fold coordinated aluminum, Al$^{VI}$, 4−fold coordinated aluminum, Al$_a^{IV}$ and a spinning−side band (SS) of the octahedral signal also containing a Knight−shift signal, Al$_{Ks}$ + SS.

Based on the characterization results obtained by PXRD, ICP-OES and $^{27}$Al MAS NMR, the discussion presented below will focus mainly on three selected materials, ZO, AZO-05 and AZO-5. Indeed, the bare ZnO, AZO-05 and AZO-5 are representative for the entire AZO series.

It is reported in literature [19] that an enhanced conductivity of the AZO materials is obtained by annealing in a reductive atmosphere at increased temperatures. The authors showed that aluminum atoms occupying interstitial positions in the zinc oxide lattice can migrate to the substitutional positions, at the same time creating interstitial Zn atoms. They also provided evidence that the origin of the Knight-shift peak observed in their NMR spectra is related to the formed complex of aluminum in these substitutional positions and Zn at interstitial positions. To evaluate how the annealing procedure might affect the photocatalytic activity of the AZO semiconductors, the AZO-05 and AZO-5 samples were annealed in a tube furnace under a dynamic nitrogen atmosphere at temperatures of 400 °C and 550 °C. Figure S5 shows the $^{27}$Al MAS NMR spectra of these annealed AZO samples. Compared to the spectra of the not-annealed samples shown in Figure 3, the spectra of the annealed AZOs show a much broader signal for the octahedral Al$^{VI}$ next to the presence of a broad signal around 70 ppm from 5-fold coordinated (Al$^{V}$) or distorted 4-fold coordinated (Al$_b^{IV}$) aluminum. It emerges that a significant part of the aluminum atoms migrates from an octahedral position to a tetrahedral, as well as distorted tetrahedral (or pentahedral) position, after annealing. These results are in agreement with the literature [19]. Taking the comparison between AZO-05 and AZO-05-400 as an example, it is clear that thermal annealing favors the migration of Al atoms from octahedral environments towards tetrahedral and distorted 4-fold or 5-fold coordination sites. Moreover, by increasing the annealing temperature from 400 °C to 550 °C, aluminum atoms almost only migrate from 4-fold coordination sites towards the distorted 4-fold or 5-fold coordination sites. This behavior is also observed comparing AZO-05-400 to AZO-5-400. In addition, for the annealed AZO nanoparticles, a lower amount of aluminum doping results in an increase of aluminum in a metallic-state environment (Al$_{Ks}$).

The $^{27}$Al-NMR spectra have demonstrated that the aluminum in the zinc oxide lattice is located mainly in tetrahedral and octahedral sites. However, it is still unclear whether the tetrahedral sites arise from substituting the Zn$^{2+}$ ions. It might be possible that the signal of tetrahedral aluminum is due to Al$^{3+}$ ions occupying the vacant interstitial tetrahedral positions instead of substituting Zn$^{2+}$ ions. Therefore, to further assess the effect of aluminum as a dopant, FT-IR measurements were performed. This technique is used to obtain information about the vibrational stretching and bending of the undoped and Al-doped zinc oxide materials in the infrared region. The transmittance of the KBr pellets containing CZO, ZO and AZO materials was measured in the interval of 4000–400 cm$^{-1}$; see Figure 4.

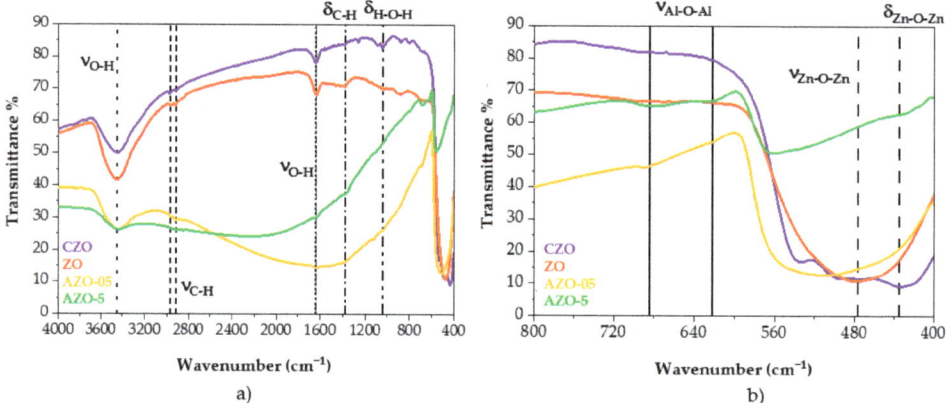

**Figure 4.** FT-IR spectra of quasi-spherical CZO, ZO, AZO-05 and AZO-5 not-annealed nanoparticles, (**a**). Magnification of the FT-IR fingerprint zone, (**b**).

The fingerprint zone (bands between 400–800 cm$^{-1}$) illustrates the stretching and bending modes of Zn-O bonds at ca. 477 and 435 cm$^{-1}$, respectively [33,52]. The bands at ca. 3400, 1637 and 1045 cm$^{-1}$ are related to the stretching and bending modes of O-H groups in physisorbed water [30]. The peaks observed at ca. 2970, 2921, and 1375 cm$^{-1}$ are due to C-H asymmetric stretching and bending vibrations of alkane groups [53,54]. These are ascribed to impurities present in the KBr powder. The emergence of the bands with low intensity at ca. 622 and 685 cm$^{-1}$ in the AZO samples are characteristic of Al-O stretching mode [53,55,56]. Together with the peaks mentioned above, a pronounced band between 800 and 3000 cm$^{-1}$ is observed for both AZO samples, while it is absent for the undoped materials. This band, called the Localized Surface Plasmon Resonance (LSPR) band [19–21,30], is a clear signature of n-doping of the AZO nanoparticles caused by the increased free electron density. The position of the LSPR band depends on the density of charge carriers. It is also affected by other parameters, such as nanoparticles size, shape, defects and segregation of dopants [21].

One of the requirements to exhibit the LSPR band is to have the Al$^{3+}$ ions dopant substituting the positions of the Zn$^{2+}$ ions instead of occupying the vacant tetrahedral positions of the wurtzite structure [30]. Therefore, the FT-IR spectra recorded of the powders of the quasi-spherical AZO nanoparticles are indicative of substitutional doping.

A closer look at the fingerprint zone in Figure 4 shows that from ZO to AZO-5, probably due to the insertion of the dopant and consequent displacement of the zinc in the oxide lattice, there is a modification of the peak's shape and a shift towards higher wavenumbers. This behavior is also observed in the literature for ZnO doped with other metals [52].

Infrared spectra of AZO and ZO nanomaterials, annealed at 400 and 550 °C under nitrogen flow, are reported in Figure S6. A comparison between the band's shape of the samples in Figure 4 and Figure S6A highlight how this thermal treatment modifies the LSPR. The band of the AZO-05 is less pronounced, while for AZO-5, it is broader and more intense when the samples are annealed at 400 °C, indicating some modification in terms of free charge carriers and their mobility. In the fingerprint zone, increasing the amount of dopant in the nanomaterials leads to a visible modification of the vibrational modes of the Zn-O peaks in terms of intensity and shape. Generally, as shown in the spectra on the right side in Figure 4, the commercial zinc oxide displays a trimodal band due to its multitude of shapes, including rod-like, quasi-spherical and agglomerated particles (see TEM, Figure S4). The ZO quasi-spherical nanoparticles present a monomodal band shape, instead, which is slightly modified for AZO-05 and more different in AZO-5. Therefore, the modification of this band shape and the shift towards a higher wavenumber could be related to the dopant insertion and a variation in morphology. In fact, in Figure S6A,B the role of dopant is principally highlighted, while Figure S6C shows the role of the annealing temperature. The TEM images show that the size of the nanoparticles starts to grow with the temperature and that they also tend to agglomerate into bigger particles, modifying the band's shape. It can be noted that ZO without annealing presents a monomodal band (fingerprint zone in Figure 4 and Figure S6C), whereas ZO-400 presents a bimodal band which we tentatively ascribe to the changes in morphology (see TEM Figure S2), and ZO-550 is again monomodal, potentially due to the substantial aggregation of the nanoparticles. These observations agree with the literature where spherical ZnO nanoparticles tend to agglomerate to bigger particles due to the calcination temperature [57].

*Photocatalytic Degradation of Rhodamine B Dye*

ZnO is one of the most widely used semiconducting materials for the photodegradation of organic pollutants in water [58–60]. Even though many scientific publications report the design of novel catalysts, the photocatalytic response of material depends mainly on the particle size distribution, morphology and crystalline phase of the selected material. Hence, the photocatalytic activity of as-synthesized ZO and AZO nanoparticles was estimated under UV-visible light irradiation by determining the photodegradation of Rhodamine B (RhB), used as a dye model compound.

The degradation of the RhB dye was verified via UV-vis analysis of the solution after removal of the solid catalyst (when needed). In the absence of photocatalyst, the degradation of Rhodamine B under UV-visible light irradiation was found negligible, whilst the dye–catalyst suspension did undergo discoloration under UV and green light irradiation.

Table 2 displays the photocatalytic activity in terms of dye degradation (%) of CZO, ZO, AZO-05 and AZO-5 nanomaterials as-synthesized. The activity was monitored after 20 min of reaction under UV-C and UV-A light irradiation. The CZO was tested as reference. Under UV-C light irradiation, RhB degradation occurred in the presence of all the catalysts. These results evidenced that the photocatalytic activity decreases while increasing the amount of aluminum in the catalysts. Among the catalysts containing aluminum, the best photocatalytic performance is shown by the AZO-05 sample. This behavior is not unexpected as the literature reports a similar behavior for Al-doped and other ZnO-based systems [61,62]. Indeed, it was reported that increasing the aluminum content of doped samples results in higher photocatalytic activity for the Al/Zn molar ratio up to 1.0%. Higher aluminum contents provoked a reduction of photocatalytic efficiency. The same considerations can be applied to the photocatalytic activities obtained under UV-A light irradiation. However, among all the catalysts synthesized in our work, the best performance has been shown by the ZO material.

**Table 2.** Degradation of RhB after 20 min of UV-C and UV-A light irradiation in the presence of commercial ZnO, undoped ZnO, 0.5 and 5 at% Al-doped ZnO nanopowders.

| Sample Codes | RhB Dye Degradation [%] UV-C 254 nm | RhB Dye Degradation [%] UV-A 368 nm |
|---|---|---|
| CZO | 94 ± 1 | 95.9 ± 0.1 |
| ZO | 44.6 ± 0.5 | 70 ± 3 |
| AZO-05 | 41 ± 2 | 53 ± 2 |
| AZO-5 | 21 ± 2 | 19 ± 3 |

Table 3 displays the RhB degradation expressed in percentage for a one-hour reaction under green light irradiation for the ZO, AZO-05, and AZO-5 nanomaterials and a four-hour reaction under green light irradiation for the ZO catalyst before and after annealing at 400 and 550 °C. Interestingly, under green light irradiation the best catalytic performance is shown by undoped ZnO solid, of which the activity overpasses that of commercial ZnO. Figure S7 shows the absorbance spectra plotted as a function of the green light exposure time for the RhB samples without the catalyst and with the catalyst that presented the best photocatalytic performance of the synthesized materials. Figure S8 displays the behavior of RhB in the absence of a catalyst, which shows photolysis effects since the dye is sensitive to visible light irradiation [63]. Therefore, the 20% degradation of CZO, shown in Table 3, could be ascribed to the thermal decomposition of the dye as a consequence of the continuous irradiation during one hour with high power lamp and the dark adsorption experiment (ca. 15%, see Figure S9). Whereas when the photocatalytic degradation of the RhB dye occurs under green light irradiation, the amount of dopant present in the zinc oxide lattice seems not to have any positive influence on the photocatalytic activity itself.

**Table 3.** Degradation of RhB under green light irradiation for one hour and four hours in the presence of CZO, ZO, AZO-05 and AZO-5 nanomaterials before and after the thermal treatment under nitrogen at 400 and 550 °C.

| Sample Codes | RhB Dye Degradation (%) under Green Light (525 nm) | | | |
|---|---|---|---|---|
| | Reaction Time | Not Annealed | 400 °C in $N_2$ | 550 °C in $N_2$ |
| CZO | 1 h | 20 ± 4 | 20 ± 4 | 19 ± 2 |
| ZO | 1 h | 39 ± 1 | 50 ± 3 | 21 ± 1 |
| AZO-05 | 1 h | 26 ± 2 | 21.7 ± 0.5 | 25 ± 1 |
| AZO-5 | 1 h | 21 ± 2 | 22 ± 1 | 23 ± 1 |
| ZO | 4 h | 64 ± 2 | 81 ± 4 | 51.3 ± 0.5 |

The data reported in Table 3 demonstrate that the annealing temperature of the (A)ZO affects the results of the photocatalytic activity. The best result is shown by the undoped ZnO material annealed at 400 °C for 10 min, presenting a photocatalytic activity value of 50% after a one-hour reaction and 81% after four hours of reaction under green light irradiation. However, when the annealing temperature reaches 550 °C, the photocatalytic activity drops, making the catalyst inactive for the same reaction.

The differences between the two annealed ZnO materials can be related to morphological and structural features.

The unexpected increase of the photocatalytic performances under the green light source of our synthesized ZnO nanoparticles (both as-synthesized and annealed at 400 °C) compared to the commercial ZnO bulk sample can be explained by the presence of different types of defects that allow the absorption of light at longer wavelengths (infra gap). Thus, this promotes the degradation process of Rhodamine B dye with less energetic radiation. This is of interest for applications where, instead of UV-light, solar light is used for the photocatalytic degradation.

To explore more in-depth and seek a hypothesis for the photocatalytic behavior of the CZO, ZO and AZO materials, the presence of these defects has been investigated using steady-state PL spectroscopy, which allows a higher sensitivity over optical absorption techniques.

UV-vis reflectance measurements of our samples cannot provide clear information on the specific characteristics of our synthesized particles. Despite the slightly different trends, all spectra show the typical ZnO pattern which is confirmed by the bandgap calculation performed by applying the Schuster–Kubelka–Munk formula [64–67] and fitting the data with the Boltzmann function according to the methodology proposed by Zanatta [68] (details in the Supplementary Figures S13 and S14 and Table S3). This analysis displays only a small variation in the absorption spectra of the samples, the band gap being estimated in the 378–388 nm range as compared to the theoretical values of 368 nm.

To bypass the high intrinsic contribution to the emission of ZnO excitons that can cover possible emission signals due to defects and excite only the latter, steady-state PL measurements were carried out in the infra-gap visible region employing 405 and 532 nm laser beam excitations (Figures 5 and S10).

Under 405 nm excitation (Figure 5), the investigated samples show different optical behavior. Figure 5a compares the normalized emission features of the commercial sample made by bulk ZnO (CZO) and the nanosized structure that we have synthesized (ZO). The main emission center in CZO is in the violet–blue region with a secondary weaker peak in the red wavelength range. At the same time, the ZO signal consists of a broad asymmetric red band extending up to the green region with a small contribution in the violet one. This difference in the emission spectra calls for a significant presence of diverse defects within the nanosized ZO responsible for infra-gap absorption and emission in the visible range. Those red-emitting defects could be involved in the observed photocatalytic process through green excitation if they absorb in the same region (vide infra). To further investigate the emission properties of nanosized ZO, we compare in Figure 5b the spectra of ZO and AZO samples. AZO-05 displays a broad emission very similar to the ZO sample, with a slight increase in the relative contribution of the green emission. The latter is further increased in the AZO-5 sample up to being the main emission band. This blue shift caused by aluminum doping could explain its deleterious effect on photocatalysis performances since the relative contribution of red-emitting defects is decreased with respect to the green ones. The opposite trend was displayed after the annealing of the sample in nitrogen (Figure 5c): at 400 °C (ZO-400), there is a slight variation of the luminescence band compared to pristine ZO, while it is red-shifted down to 675 nm after the treatment at 550 °C (ZO-550).

The reported scenario was confirmed under 532 nm excitation (Figure S10), of which the energy is comparable to the one of our photocatalytic processes: ZO nanoparticles present the same red contribution previously observed, with the AZO-5 and ZO-550 samples still slightly blue- and red-shifted, respectively, as compared to the others. These spectra demonstrate the capability of our materials to absorb infra-gap green light, suggesting that the same red-emitting defects could be responsible of our photocatalytic performances in the same energy range. Exciting the CZO sample with the green laser source recorded the absence of any red photoluminescence, supporting the observed poor photocatalytic performances. This is also an indication that the weak red emission displayed in CZO under 405 nm excitation light has a different origin as compared to our synthesized samples.

To better identify the relative contribution of the different emitting centers, a gaussian deconvolution of the spectra excited at 405 nm was performed (Figure 6).

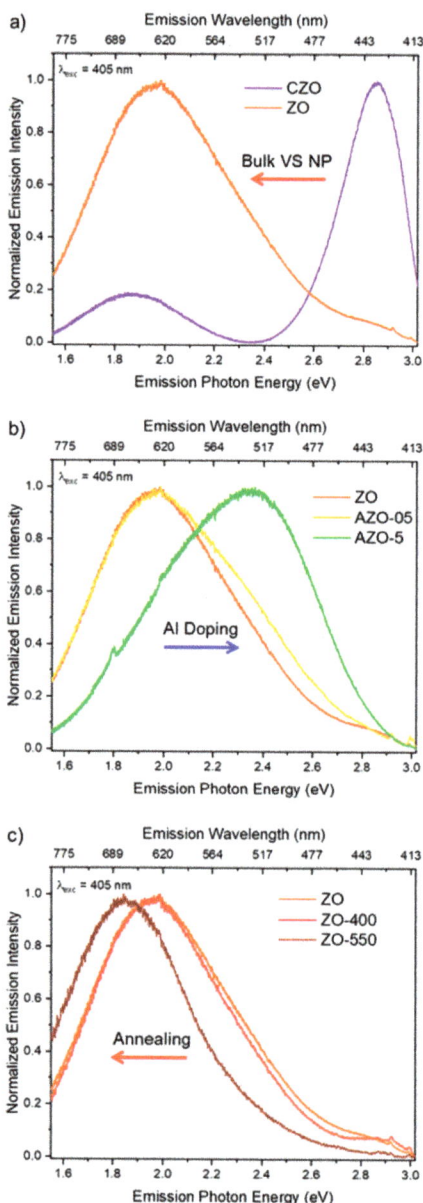

**Figure 5.** Normalized PL spectra of all samples excited at 405 nm. Comparisons among commercial (bulk) and ZnO nanoparticles (**a**), different Al-doping percentages (**b**), and annealing temperatures (**c**) are displayed.

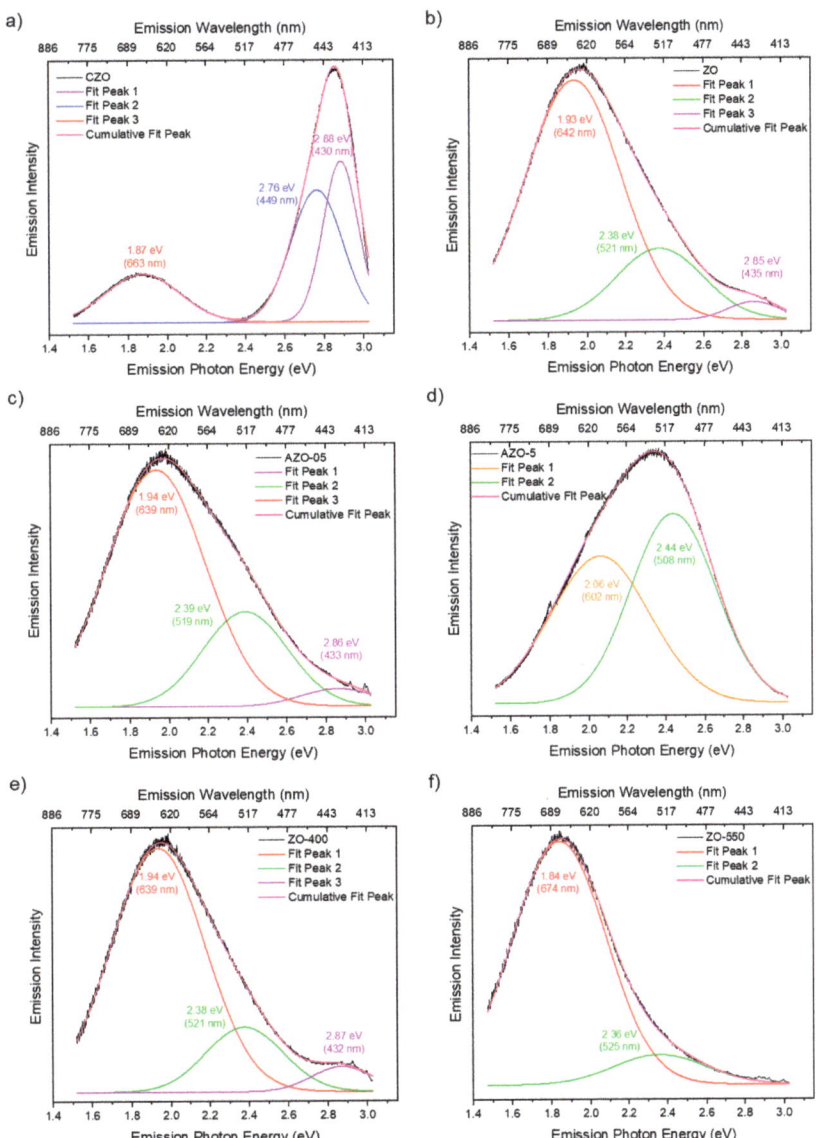

**Figure 6.** Gaussian deconvolution of emission spectra obtained under 405 nm excitation wavelength of CZO (**a**), ZO (**b**), AZO-05 (**c**), AZO-5 (**d**), ZO-400 (**e**) and ZO-550 (**f**).

All the spectra can be fitted assuming two or three contributions in the orange/red, green/blue, and violet range, as reported in Table S2. The analysis confirmed that the red emission band of CZO, centered at about 660 nm, differs from the one detected in the other samples, where the emission band peaked at about 640 nm. This red band is the primary emission band in the ZO, even after annealing at 400 °C (the relative content was about 75% of the overall emission in the undoped sample, 77% after annealing). When annealed at 550 °C, the band is still present but red-shifted by about 35 nm. It should be noted that, different morphologies of the materials may affect scattering and absorption in PL measurements [69]. Therefore, this effect could be related to a change in the environment of the emitting centers, caused by aggregation of the nanoparticles (vide infra); indeed, the TEM images in Figures S2 and S3 show the formation of larger nanoparticle agglomerates. Two other bands, in the green and violet spectral range (at about 520 and 430 nm, respectively) with decreasing relative content, were also added to complete the deconvolutions. Interestingly, the violet band in all samples (except for AZO-5 and ZO-550) peaked at the same spectral position as in the CZO sample, where a further blue band at about 450 nm is required to fit the emission spectrum, instead of the green one of our synthesized samples. Moreover, by doping the samples with a small content of aluminum and through the annealing at 400 °C, these peaks' positions remain almost unmodified, while increasing the aluminum content largely modified the spectrum, shifting the red band to about 600 nm and largely decreasing the relative contribution of this band (47%) with respect to the green one (53%). According to the detailed study by Zhang et al. [70] on ZnO nanostructures using PL and EPR (Electron Paramagnetic Resonance spectroscopy) measurements, the blue–violet peak of the CZO can be attributed to interstitial Zn in the surface and bulk position, respectively. The surface interstitial Zn band and the related defect contribution are displayed in all samples (the only exception is AZO-5) with a much weaker relative intensity that passes from 35% in CZO to 3–5% in the others. Moreover, the opposite trend of the green and red band relative intensity is worth noting. In the undoped nanoparticles, the green band represents 22% of the area, which increases to 26% in the AZO-05 sample and 53% in the AZO-5. The same authors attributed this green band to $O_2$ adsorbed on the surface or charged oxygen vacancies $V_o^+$ (not excluding $V_o$ and $V_o^{++}$) while the broad red one around 2.0 eV might be due to the presence of peroxide-like species $O_2^{2-}$ interacting with the ZnO surface. Considering these attributions, we propose to explain the photocatalytic properties of the synthesized samples as follows. As stated before, the presence of red-emitting centers promotes the photocatalytic process of our synthesized samples under green light since they can absorb in the green infra-gap region in opposition with commercial ZnO. When ZO is doped with aluminum below 0.5%, tetrahedral positions are taken. At least part of these are aluminum ions substituting zinc ions, as corroborated by the appearance of an LSPR band in FT-IR (see Figure 4a). An increase in the aluminum content up to 5 at% causes the migration of the aluminum atoms mainly towards an octahedral environment, as indicated by solid-state NMR analysis (see Figure 3). Together with the appearance of the aluminum in octahedral environment, XRD analysis (see Figure 1) shows the existence of a secondary phase, which makes us believe that this secondary phase contains an octahedral aluminum environment as well.

Doping with (small amounts of) aluminum in the substitutional $Zn^{2+}$ positions creates oxygen vacancies in the ZnO structure. Along with this, Zn ions are displaced from their lattice position. Hence, the secondary phase, observed in XRD, is probably formed by these displaced $Zn^{2+}$ ions together with the surplus (octahedral) aluminum ions and oxygen. Upon higher aluminum doping, the amount of the secondary phase increases. This may cause more oxygen vacancies to be formed in the ZnO, leading to the increase of the green emission band and, consequently, to the decrease of the photocatalytic performance.

As for the annealed samples, the emission is marginally affected by the thermal treatment at 400 °C whilst its photocatalytic efficiency is increased, from 39 to 50%. The opposite effect is displayed in the ZO-550 sample, where besides the observed red-shift, a decrease in the photocatalytic performances from 39 to 21% is also observed. The explanation of these phenomena is challenging because just a little variation in the mainly red emission spectra that we had associated with the presence of oxidative species was observed. The thermal treatment did cause a change in morphology and distribution of the particles (see Figure S2), and a variation in the aggregation of the nanoparticles, the latter being larger in ZO-550, possibly reducing the photocatalytic surface area.

## 3. Materials and Methods

### 3.1. Synthesis

All chemical reagents were used as received. Zinc acetylacetonate hydrate ((Zn(acac)$_2$ · $x$H$_2$O) 99.995% m/m% purity, Aldrich), aluminum acetylacetonate (Al(acac)$_3$ 99.999% m/m% purity, Aldrich) and benzylamine for synthesis (99% purity, ethanol absolute, for analysis, Merck).

The syntheses of undoped and Al-doped ZnO materials were carried out with 1 g of Zn(acac)$_2$ for pure ZnO and in combination with 0.0057g of Al(acac)$_3$, ranging from 0.5 to 5 mol% for the AZO nanoparticles. To obtain undoped and Al-doped ZnO nanoparticles by a solvolysis reaction in a reflux setup, Zn(acac)$_2$ hydrate was mixed with $x$ mol of Al(acac)$_3$ and 80 mL of benzylamine and heated to the boiling point (nominal temperature 185 °C). After reaching the boiling point, the mixture was refluxed for four hours under stirring. After cooling to room temperature, the mixture was centrifuged to precipitate the nanoparticles. The particles were then washed three times with ethanol and twice with water. After washing, the obtained powders were dried overnight in the oven at 60 °C. Subsequently, the catalysts were annealed under dynamic nitrogen for 10 min in a tubular oven at 400 and 550 °C.

### 3.2. Characterization Techniques

The Al/Zn ratio in the powders was determined by inductively-coupled plasma optical emission spectrometry (ICP-OES, Perkin Elmer Optima 3300 DV simultaneous spectrometer, PerkinElmer, Waltham, MA, USA). To analyze the Zn and Al content, a small portion of the powdered AZO sample was dissolved in a 5% aqueous nitric acid solution (HNO$_3$, 69.0–70.0%, J.T.Baker, for trace metal analysis). The AZO aqueous stock solutions and 1000 ppm Zn, Al standards (Merck) were diluted by 5% HNO$_3$ to 1–10 ppm and 10, 5, 2 and 1 ppm concentrations, respectively, for ICP-OES measurements. All ICP analyses were carried out twice. The measurement error was evaluated based on calibration certificates and from statistical analysis of repeated measurements. The following errors were considered: volumetric operations (volumetric flasks, measuring cylinders) and the error of concentrations/purity of commercial chemicals. For calculations, calibration certificates or information sheets from the manufacturer were used. Powder X-ray Diffraction (PXRD) patterns were collected on a Bruker AXS D8 Discover diffractometer (Cu K$\alpha$ radiation, LynxEye detector). The undoped and Al-doped ZnO powders were scanned between 2$\theta$ = 10° and 70° with a 2$\theta$ scan step size of 0.020°. Metal sample holders were used as a support for all powder samples. The profile analysis of the related diffraction patterns was carried out with the program DIFFRAC.EVA (general profile and structure analysis software for powder diffraction data, Bruker Analytical X-ray Systems). Transmission electron microscopy (TEM) images were recorded on a Philips Tecnai 10 with an acceleration voltage of 100 kV. The samples were prepared by dispersing a small quantity of nanopowders in absolute ethanol, depositing them on a carbon-film coated copper mesh and drying them. The particle size distribution was evaluated over 450 nanoparticles. Fourier transform infrared (FT-IR) spectroscopy (Bruker Vertex 70, 32 scans, scan range 4000–400 cm$^{-1}$, resolution 1 cm$^{-1}$) was performed on pellets containing trace amounts of the investigated materials diluted with KBr. The discoloration process was monitored by means of a UV-vis spectrophotometer Agilent Cary 5000. The measures were carried out in the UV-visible range of 200–800 nm, with a quartz cuvette applying baseline corrections. [27]Al solid-state MAS NMR spectra were acquired on an Agilent VNMRS DirectDrive 400 MHz spectrometer (9.4 T wide bore magnet) equipped with a T3HX 3.2 mm probe. Magic angle spinning (MAS) was performed at 20 kHz in ceramic rotors of 3.2 mm (22µL). AlCl$_3$ was used to calibrate the aluminum chemical shift scale (0 ppm). Acquisition parameters used were: a spectral width of 420 kHz, a 90° pulse length of 2.5 µs, an acquisition time of 10 ms, a recycle delay time of 20 s, a line-broadening of 200 Hz and around 10,000 accumulations. UV-vis solid-state reflectance spectra were collected (applying baseline corrections) using a Jasco V-750 spectrophotometer with a spectral bandwidth of 0.2 nm in the 300–500 nm range. Photoluminescence (PL) measurements were performed in backscattering geometry with a confocal micro-Raman system (SOL Confotec MR750) equipped with a Nikon Eclipse Ni microscope. Samples were excited with 405 and 532 nm laser diodes (IO Match-Box series), and the spectral resolution was 0.6 cm$^{-1}$ (average acquisition time 1 s, the average number of acquisitions 3, sensor temperature $-23\ °C$, objective Olympus 10×, grating with 150 grooves/mm, power excitation 3 mW).

### 3.3. Photocatalytic Assessment

The photocatalytic activity of the catalysts was estimated under UV-visible light irradiation, determining the photodegradation of Rhodamine B (RhB), used as a dye model compound. To perform the photocatalytic tests, two homemade photoreactors were used (Figures S11 and S12).

#### 3.3.1. Photodegradation in the UV Range

Quartz beakers containing 20 mL of the RhB aqueous solution (4 mg L$^{-1}$) along with the photocatalyst (5 mg) were used to carry out the photodegradation experiments. A lamp emitting in the UV (UV-C: Osram Hg lamp, 11 W, dominant wavelength 254 nm; UV-A: Phillips Hg lamp, 11 W, dominant wavelength 368 nm), was axially positioned among the four beakers and kept at 1 cm from the magnetic stirrer plate. The reactor was kept at room temperature (24 ± 3 °C) using a fan placed on the back wall of the reactor. The reaction mixtures were stirred using a magnetic stirrer bar of equal length for 30 min in the dark before irradiation to attain the adsorption–desorption equilibrium of the substrate. The reactions were stopped after 20 min. The suspensions were centrifuged to separate the catalysts from the aqueous dye solution and analysed by UV-vis spectroscopy, considering the main absorption peak of this dye in the visible range, located at 554 nm. The degradation % is calculated using Formula (1):

$$\text{Degradation \%} = \frac{(A_0 - A)}{A_0} * 100 \tag{1}$$

where $A_0$ is the absorbance of the starting solution before reaction and $A$ is the final absorbance measured after the reaction. Photolysis tests of the RhB dye were carried out under UV light irradiation. The single photocatalytic test was performed four times to ensure the reproducibility of the data.

#### 3.3.2. Photodegradation in the Visible Range

A quartz beaker containing 20 mL of the RhB aqueous solution (4 mg L$^{-1}$) and the photocatalyst (5 mg) were used to carry out the photodegradation tests. A lamp emitting in the visible range (dominant wavelength 525 nm, 18 W Evoluchem LED) was axially positioned and kept at 10 cm from the top of the beaker. To avoid the contribution of the UV-A light, a 405 nm long-pass edge filter was positioned between the beaker and the emitting lamp. The reactor was kept at room temperature (24 ± 3 °C) using a fan placed inside the reactor positioned a few centimeters aside from the beaker. The reaction mixtures were stirred using a magnetic stirrer bar of equal length for 30 min in the dark before irradiation to attain the adsorption–desorption equilibrium of the substrate. The reactions were stopped after 60 min. The suspensions were centrifuged to separate the catalysts from the dye solution and analyzed by UV-vis spectroscopy, considering the main absorption peak of this dye in the visible range, located at 554 nm. To ensure the evaluation of the photocatalytic activity measurements, tests of the RhB dye were carried out under visible light irradiation in the absence of the catalyst. The photocatalytic reactions were repeated three times to ensure the reproducibility of the data. The degradation % is calculated using Formula (1).

## 4. Conclusions

In summary, undoped ZnO and Al-doped ZnO quasi-spherical nanoparticles were successfully synthesized using a very accessible method and tested under UV and green light irradiation for the photodegradation of Rhodamine B in an aqueous solution. The ICP-OES analysis confirmed the successful incorporation of the dopant into the final solid. X-ray diffraction showed that all the samples exhibit the hexagonal wurtzite structure. After incorporating the aluminum into the zinc oxide lattice, additional peaks appeared, which correspond to the secondary phase of $Zn_6Al_2(OH)_{16}CO_3 \cdot 4H_2O$. The TEM images displayed that the quasi-spherical nanoparticles grow with the insertion of the dopant and, due to the annealing temperature, tended to agglomerate. The $^{27}Al$ NMR elucidated the position occupied by aluminum in the ZnO lattice, highlighting the migration from the octahedral coordination, which is the preferred site for the as-prepared materials, towards tetrahedral coordination after annealing in nitrogen flow. The FT-IR technique was sensitive to the variations of particle morphology and changes in the charge carriers pre- and post-annealing. PL allowed an in-depth study of the catalyst's defects.

CZO, ZO and AZO catalysts were investigated as photocatalysts for degrading Rhodamine B dye in an aqueous solution under UV-visible light irradiation. Under UV-light irradiation, all the materials underperformed with respect to commercial zinc oxide. Undoped ZnO exhibited the best photocatalytic performance when illuminated under green light (525 nm), as demonstrated by the high degree of discoloration with respect to the commercial ZnO evaluated under the same conditions. The insertion of the dopant resulted in lower photocatalytic activity under UV-light irradiation and dropped the efficiency of the AZO catalysts to zero under green light irradiation. The catalysts were also thermally treated under a nitrogen atmosphere at 400 and 550 °C and evaluated again for the discoloration of the dye. Results of the present work demonstrated that annealing undoped zinc oxide nanoparticles at 400 °C for 10 min under nitrogen flow improves the discoloration efficiency by up to 81%. These results strongly suggest the interconnection between defects, synthesis and post-synthesis route, particle size and photocatalytic activity. The photocatalytic performances under green light irradiation can be explained as follows: CZO mainly contains zinc interstitial and has emission centers in the violet–blue region that are inactive for the photodegradation of RhB dye under this excitation wavelength. Undoped ZO presents red-emitting centers, which can be associated with a high content of $O_2^{2-}$, which is the main initiator of the photodegradation and responsible for the discoloration of the dye [7,71]. Annealing ZO at 400 and 550 °C modified the morphology and the size distribution along with the aggregation. Compared to the ZO and ZO-400 samples, the degree of aggregation is larger in ZO-550. Therefore, the substance is paying the penalty of surface area reduction induced by the high temperature. The improved photocatalytic activity of ZO-400 might be ascribed to the different morphology displayed by the particles after annealing. The effect of aluminum doping was studied as well: even the insertion of a small percent of dopant in the zinc oxide lattice, such as in AZO-05, leads to a material with a significant number of oxygen vacancies (green band, PL). As shown by the PL measured at 532 nm, there is not much difference in the spectra between ZO, AZO-05 and AZO-5. Here, we suggest that a small amount (equal or less than 0.5%) of aluminum occupying substitutional tetrahedral sites of the ZnO structure, saturates the photocatalytically active sites on the catalyst surface. Higher amounts of aluminum doping surpass the solubility limit and tend to occupy octahedral positions in secondary phases as well. As a consequence, oxygen vacancies are created in the ZnO, which act as fast-charge recombination centers. Moreover, the higher amount of this type of defect, that do not absorb in the green region, and a reduction of $O_2^{2-}$ species, possibly deactivate the photocatalytic process making the AZO catalysts inactive for the photodegradation of the RhB dye in an aqueous solution under green light.

**Supplementary Materials:** The following supporting information can be downloaded at: https://www.mdpi.com/article/10.3390/ijms232415459/s1.

**Author Contributions:** Conceptualization, A.P., C.O., C.M.C., C.A., M.K.V.B., L.F., A.H. and P.A.; methodology, A.P., C.O., A.-S.K., G.R., C.A. and L.F.; validation, A.P. and C.O.; formal analysis, A.P., C.O., G.R., A.-S.K., D.D.S. and K.E.; investigation, A.P., C.O., A.-S.K., G.R., D.D.S., K.E. and P.A.; resources, A.P., C.O., C.M.C., C.A., M.K.V.B., L.F., A.H., G.R. and P.A.; data curation, A.P., C.O., A.-S.K., D.D.S., K.E., G.R. and P.A.; writing—original draft preparation, A.P., C.O.; writing—review and editing, A.P., C.O., C.M.C., A.-S.K., K.E., D.D.S., C.A., M.K.V.B., L.F., A.H. and P.A; visualization, A.P. and C.O.; supervision, C.A., M.K.V.B., L.F., A.H. and P.A.; project administration, C.A., M.K.V.B., L.F., A.H. and P.A.; funding acquisition, C.A., M.K.V.B., L.F., A.H. and P.A. All authors have read and agreed to the published version of the manuscript.

**Funding:** This research was co-funded by Hasselt University and Namur University through the BOF programme (Project R-9087) and the Fonds spécial de recherche. This work is also supported by the Research Foundation Flanders (FWO) and Hasselt University via the Hercules project AUHL/15/2-GOH3816N. This research used resources of the PC2 and the MORPH-IM platforms located at the University of Namur.

**Institutional Review Board Statement:** Not applicable.

**Informed Consent Statement:** Not applicable.

**Data Availability Statement:** Not applicable.

**Acknowledgments:** The authors acknowledge C. Charlier for his assistance with TEM and N. Billiet with N. Debusschere for the PXRD measurements.

**Conflicts of Interest:** The authors declare no conflict of interest.

## References

1. UNESCO. *The United Nations World Water Development Report 2020_ Water and Climate Change*; UNESCO: Paris, France, 2020.
2. Lim, H.; Yusuf, M.; Song, S.; Park, S.; Park, K.H. Efficient Photocatalytic Degradation of Dyes Using Photo-Deposited Ag Nanoparticles on ZnO Structures: Simple Morphological Control of ZnO. *RSC Adv.* **2021**, *11*, 8709–8717. [CrossRef] [PubMed]
3. Mahapatra, N.N. *Textile Dyes*, 1st ed.; Woodhead Publishing India Pvt Ltd.: New Delhi, India, 2016.
4. Thomas Bechtold, T.P. *Textile Chemistry*; De Gruyter: Berlin, Germany, 2019. [CrossRef]
5. Reddy, P.A.K.; Reddy, P.V.L.; Kwon, E.; Kim, K.-H.; Akter, T.; Kalagara, S. Recent Advances in Photocatalytic Treatment of Pollutants in Aqueous Media. *Environ. Int.* **2016**, *91*, 94–103. [CrossRef] [PubMed]
6. Balcha, A.; Yadav, O.P.; Dey, T. Photocatalytic Degradation of Methylene Blue Dye by Zinc Oxide Nanoparticles Obtained from Precipitation and Sol-Gel Methods. *Environ. Sci. Pollut. Res.* **2016**, *23*, 25485–25493. [CrossRef] [PubMed]
7. Dodoo-Arhin, D.; Asiedu, T.; Agyei-Tuffour, B.; Nyankson, E.; Obada, D.; Mwabora, J.M. Photocatalytic Degradation of Rhodamine Dyes Using Zinc Oxide Nanoparticles. *Mater. Today Proc.* **2021**, *38*, 809–815. [CrossRef]
8. Ahlström, L.-H.; Sparr Eskilsson, C.; Björklund, E. Determination of Banned Azo Dyes in Consumer Goods. *TrAC Trends Anal. Chem.* **2005**, *24*, 49–56. [CrossRef]
9. Khan, S.; Malik, A. Toxicity Evaluation of Textile Effluents and Role of Native Soil Bacterium in Biodegradation of a Textile Dye. *Environ. Sci. Pollut. Res.* **2018**, *25*, 4446–4458. [CrossRef]
10. Sansenya, T.; Masri, N.; Chankhanittha, T.; Senasu, T.; Piriyanon, J.; Mukdasai, S.; Nanan, S. Hydrothermal Synthesis of ZnO Photocatalyst for Detoxification of Anionic Azo Dyes and Antibiotic. *J. Phys. Chem. Solids* **2022**, *160*, 110353. [CrossRef]
11. Khataee, A.R.; Pons, M.N.; Zahraa, O. Photocatalytic Degradation of Three Azo Dyes Using Immobilized TiO2 Nanoparticles on Glass Plates Activated by UV Light Irradiation: Influence of Dye Molecular Structure. *J. Hazard. Mater.* **2009**, *168*, 451–457. [CrossRef]
12. John Peter, I.; Praveen, E.; Vignesh, G.; Nithiananthi, P. ZnO Nanostructures with Different Morphology for Enhanced Photocatalytic Activity. *Mater. Res. Express* **2017**, *4*, 124003. [CrossRef]
13. Ani, I.J.; Akpan, U.G.; Olutoye, M.A.; Hameed, B.H. Photocatalytic Degradation of Pollutants in Petroleum Refinery Wastewater by TiO2- and ZnO-Based Photocatalysts: Recent Development. *J. Clean. Prod.* **2018**, *205*, 930–954. [CrossRef]
14. Sinar Mashuri, S.I.; Ibrahim, M.L.; Kasim, M.F.; Mastuli, M.S.; Rashid, U.; Abdullah, A.H.; Islam, A.; Asikin Mijan, N.; Tan, Y.H.; Mansir, N.; et al. Photocatalysis for Organic Wastewater Treatment: From the Basis to Current Challenges for Society. *Catalysts* **2020**, *10*, 1260. [CrossRef]
15. Natarajan, T.S.; Thomas, M.; Natarajan, K.; Bajaj, H.C.; Tayade, R.J. Study on UV-LED/TiO2 Process for Degradation of Rhodamine B Dye. *Chem. Eng. J.* **2011**, *169*, 126–134. [CrossRef]
16. Ayaz, S.; Amin, R.; Samantray, K.; Dasgupta, A.; Sen, S. Tunable Ultraviolet Sensing Performance of Al-Modified ZnO Nanoparticles. *J. Alloys Compd.* **2021**, *884*, 161113. [CrossRef]
17. Saber, O.; El-Brolossy, T.A.; Al Jaafari, A.A. Improvement of Photocatalytic Degradation of Naphthol Green B Under Solar Light Using Aluminum Doping of Zinc Oxide Nanoparticles. *Water Air Soil Pollut.* **2012**, *223*, 4615–4626. [CrossRef]

18. Wu, J.; Xue, D. Progress of Science and Technology of ZnO as Advanced Material. *Sci. Adv. Mater.* **2011**, *3*, 127–149. [CrossRef]
19. Momot, A.; Amini, M.N.; Reekmans, G.; Lamoen, D.; Partoens, B.; Slocombe, D.R.; Elen, K.; Adriaensens, P.; Hardy, A.; Van Bael, M.K. A Novel Explanation for the Increased Conductivity in Annealed Al-Doped ZnO: An Insight into Migration of Aluminum and Displacement of Zinc. *Phys. Chem. Chem. Phys.* **2017**, *19*, 27866–27877. [CrossRef]
20. Buonsanti, R.; Llordes, A.; Aloni, S.; Helms, B.A.; Milliron, D.J. Tunable Infrared Absorption and Visible Transparency of Colloidal Aluminum-Doped Zinc Oxide Nanocrystals. *Nano Lett.* **2011**, *11*, 4706–4710. [CrossRef]
21. Della Gaspera, E.; Chesman, A.S.R.; van Embden, J.; Jasieniak, J.J. Non-Injection Synthesis of Doped Zinc Oxide Plasmonic Nanocrystals. *ACS Nano* **2014**, *8*, 9154–9163. [CrossRef]
22. Garcia, G.; Buonsanti, R.; Runnerstrom, E.L.; Mendelsberg, R.J.; Llordes, A.; Anders, A.; Richardson, T.J.; Milliron, D.J. Dynamically Modulating the Surface Plasmon Resonance of Doped Semiconductor Nanocrystals. *Nano Lett.* **2011**, *11*, 4415–4420. [CrossRef]
23. Jaramillo-Páez, C.; Sánchez-Cid, P.; Navío, J.A.; Hidalgo, M.C. A Comparative Assessment of the UV-Photocatalytic Activities of ZnO Synthesized by Different Routes. *J. Environ. Chem. Eng.* **2018**, *6*, 7161–7171. [CrossRef]
24. Khalid, N.R.; Hammad, A.; Tahir, M.B.; Rafique, M.; Iqbal, T.; Nabi, G.; Hussain, M.K. Enhanced Photocatalytic Activity of Al and Fe Co-Doped ZnO Nanorods for Methylene Blue Degradation. *Ceram. Int.* **2019**, *45*, 21430–21435. [CrossRef]
25. Bazzani, M.; Neroni, A.; Calzolari, A.; Catellani, A. Optoelectronic Properties of Al:ZnO: Critical Dosage for an Optimal Transparent Conductive Oxide. *Appl. Phys. Lett.* **2011**, *98*, 121907. [CrossRef]
26. Abrinaei, F.; Molahasani, N. Effects of Mn Doping on the Structural, Linear, and Nonlinear Optical Properties of ZnO Nanoparticles. *J. Opt. Soc. Am. B* **2018**, *35*, 2015. [CrossRef]
27. Calzolari, A.; Catellani, A. Doping, Co-Doping, and Defect Effects on the Plasmonic Activity of ZnO-Based Transparent Conductive Oxides. *Oxide-Based Mater. Devices VIII* **2017**, *10105*, 101050G. [CrossRef]
28. Yaqoob, A.A.; Mohd Noor, N.H.b.; Serrà, A.; Mohamad Ibrahim, M.N. Advances and Challenges in Developing Efficient Graphene Oxide-Based ZnO Photocatalysts for Dye Photo-Oxidation. *Nanomaterials* **2020**, *10*, 932. [CrossRef]
29. Damm, H.; Adriaensens, P.; De Dobbelaere, C.; Capon, B.; Elen, K.; Drijkoningen, J.; Conings, B.; Manca, J.V.; D'Haen, J.; Detavernier, C.; et al. Factors Influencing the Conductivity of Aqueous Sol(Ution)–Gel-Processed Al-Doped ZnO Films. *Chem. Mater.* **2014**, *26*, 5839–5851. [CrossRef]
30. Kelchtermans, A.; Elen, K.; Schellens, K.; Conings, B.; Damm, H.; Boyen, H.G.; D'Haen, J.; Adriaensens, P.; Hardy, A.; Van Bael, M.K. Relation between Synthesis Conditions, Dopant Position and Charge Carriers in Aluminium-Doped ZnO Nanoparticles. *RSC Adv.* **2013**, *3*, 15254–15262. [CrossRef]
31. Serier, H.; Gaudon, M.; Ménétrier, M. Al-Doped ZnO Powdered Materials: Al Solubility Limit and IR Absorption Properties. *Solid State Sci.* **2009**, *11*, 1192–1197. [CrossRef]
32. Conversion, B. *Catalysis for Clean Energy and Environmental Sustainability*; Pant, K.K., Gupta, S.K., Ahmad, E., Eds.; Springer International Publishing: Cham, Switzerland, 2021; Volume 1. [CrossRef]
33. Nandi, P.; Das, D. Photocatalytic Degradation of Rhodamine-B Dye by Stable ZnO Nanostructures with Different Calcination Temperature Induced Defects. *Appl. Surf. Sci.* **2019**, *465*, 546–556. [CrossRef]
34. Munawar, T.; Yasmeen, S.; Hussain, F.; Mahmood, K.; Hussain, A.; Asghar, M.; Iqbal, F. Synthesis of Novel Heterostructured ZnO-CdO-CuO Nanocomposite: Characterization and Enhanced Sunlight Driven Photocatalytic Activity. *Mater. Chem. Phys.* **2020**, *249*, 122983. [CrossRef]
35. Neena, D.; Kondamareddy, K.K.; Bin, H.; Lu, D.; Kumar, P.; Dwivedi, R.K.; Pelenovich, V.O.; Zhao, X.-Z.; Gao, W.; Fu, D. Enhanced Visible Light Photodegradation Activity of RhB/MB from Aqueous Solution Using Nanosized Novel Fe-Cd Co-Modified ZnO. *Sci. Rep.* **2018**, *8*, 10691. [CrossRef]
36. Burdett, J.K.; Price, G.D.; Price, S.L. Role of the Crystal-Field Theory in Determining the Structures of Spinels. *J. Am. Chem. Soc.* **1982**, *104*, 92–95. [CrossRef]
37. Kelchtermans, A.; Adriaensens, P.; Slocombe, D.; Kuznetsov, V.L.; Hadermann, J.; Riskin, A.; Elen, K.; Edwards, P.P.; Hardy, A.; Van Bael, M.K. Increasing the Solubility Limit for Tetrahedral Aluminium in ZnO:Al Nanorods by Variation in Synthesis Parameters. *J. Nanomater.* **2015**, *2015*, 5. [CrossRef]
38. Tsubota, T.; Ohtaki, M.; Eguchi, K.; Arai, H. Thermoelectric Properties of Al-Doped ZnO as a Promising Oxide Material for High-Temperature Thermoelectric Conversion. *J. Mater. Chem.* **1997**, *7*, 85–90. [CrossRef]
39. Wan Shick, H.; De Jonghe, L.C.; Xi, Y.; Rahaman, M.N. Reaction Sintering of ZnO-Al2O3. *J. Am. Ceram. Soc.* **1995**, *78*, 3217–3224.
40. Thu, T.V.; Maenosono, S. Synthesis of High-Quality Al-Doped ZnO Nanoink. *J. Appl. Phys.* **2010**, *107*, 014308. [CrossRef]
41. Serier, H.; Demourgues, A.; Majimel, J.; Gaudon, M. Infrared Absorptive Properties of Al-Doped ZnO Divided Powder. *J. Solid State Chem.* **2011**, *184*, 1523–1529. [CrossRef]
42. Suwanboon, S.; Amornpitoksuk, P.; Haidoux, A.; Tedenac, J.C. Structural and Optical Properties of Undoped and Aluminium Doped Zinc Oxide Nanoparticles via Precipitation Method at Low Temperature. *J. Alloys Compd.* **2008**, *462*, 335–339. [CrossRef]
43. Kumar, R.S.; Sathyamoorthy, R.; Sudhagar, P.; Matheswaran, P.; Hrudhya, C.P.; Kang, Y.S. Effect of Aluminum Doping on the Structural and Luminescent Properties of ZnO Nanoparticles Synthesized by Wet Chemical Method. *Phys. E Low-Dimens. Syst. Nanostruct.* **2011**, *43*, 1166–1170. [CrossRef]
44. Hartner, S.; Ali, M.; Schulz, C.; Winterer, M.; Wiggers, H. Electrical Properties of Aluminum-Doped Zinc Oxide (AZO) Nanoparticles Synthesized by Chemical Vapor Synthesis. *Nanotechnology* **2009**, *20*, 445701. [CrossRef]

45. Brehm, J.U.; Winterer, M.; Hahn, H. Synthesis and Local Structure of Doped Nanocrystalline Zinc Oxides. *J. Appl. Phys.* **2006**, *100*, 064311. [CrossRef]
46. Kelchtermans, A. Synthesis and In-Depth Characterization of Al-Doped ZnO Nanoparticles as Building Blocks for TCO Layers. Ph.D. Thesis, Hasselt University, Hasselt, Belgium, 2014.
47. Pinna, N.; Garnweitner, G.; Antonietti, M.; Niederberger, M. A General Nonaqueous Route to Binary Metal Oxide Nanocrystals Involving a C−C Bond Cleavage. *J. Am. Chem. Soc.* **2005**, *127*, 5608–5612. [CrossRef] [PubMed]
48. Avadhut, Y.S.; Weber, J.; Hammarberg, E.; Feldmann, C.; Schmedtaufder Günne, J. Structural Investigation of Aluminium Doped ZnO Nanoparticles by Solid-State NMR Spectroscopy. *Phys. Chem. Chem. Phys.* **2012**, *14*, 11610–11625. [CrossRef]
49. Titova, Y.Y.; Schmidt, F.K. What 27Al NMR Spectroscopy Can Offer to Study of Multicomponent Catalytic Hydrogenation Systems? *J. Organomet. Chem.* **2022**, *975*, 122410. [CrossRef]
50. Roberts, N.; Wang, R.; Sleight, A.W. And Impurity Nuclear Magnetic Resonance in ZnO:Al and ZnO:Ga. *Phys. Rev. B Condens. Matter Mater. Phys.* **1998**, *57*, 5734–5741. [CrossRef]
51. Knight, W.D. Nuclear Magnetic Resonance Shift in Metals. *Phys. Rev.* **1949**, *76*, 1259–1260. [CrossRef]
52. Achehboune, M.; Khenfouch, M.; Boukhoubza, I.; Leontie, L.; Doroftei, C.; Carlescu, A.; Bulai, G.; Mothudi, B.; Zorkani, I.; Jorio, A. Microstructural, FTIR and Raman Spectroscopic Study of Rare Earth Doped ZnO Nanostructures. *Mater. Today Proc.* **2022**, *53*, 319–323. [CrossRef]
53. Djelloul, A.; Aida, M.S.; Bougdira, J. Photoluminescence, FTIR and X-ray Diffraction Studies on Undoped and Al-Doped ZnO Thin Films Grown on Polycrystalline α-Alumina Substrates by Ultrasonic Spray Pyrolysis. *J. Lumin.* **2010**, *130*, 2113–2117. [CrossRef]
54. Mojumder, S.; Das, T.; Das, S.; Chakraborty, N.; Saha, D.; Pal, M. Y and Al Co-Doped ZnO-Nanopowder Based Ultrasensitive Trace Ethanol Sensor: A Potential Breath Analyzer for Fatty Liver Disease and Drunken Driving Detection. *Sens. Actuators B Chem.* **2022**, *372*, 132611. [CrossRef]
55. Zhang, X.; Chen, Y.; Zhang, S.; Qiu, C. High Photocatalytic Performance of High Concentration Al-Doped ZnO Nanoparticles. *Sep. Purif. Technol.* **2017**, *172*, 236–241. [CrossRef]
56. Munawaroh, H.; Wahyuningsih, S.; Ramelan, A.H. Synthesis and Characterization of Al Doped ZnO (AZO) by Sol-Gel Method. *IOP Conf. Ser. Mater. Sci. Eng.* **2017**, *176*, 012049. [CrossRef]
57. Sowri Babu, K.; Ramachandra Reddy, A.; Sujatha, C.; Venugopal Reddy, K.; Mallika, A.N. Synthesis and Optical Characterization of Porous ZnO. *J. Adv. Ceram.* **2013**, *2*, 260–265. [CrossRef]
58. Mihai, G.D.; Meynen, V.; Mertens, M.; Bilba, N.; Cool, P.; Vansant, E.F. ZnO Nanoparticles Supported on Mesoporous MCM-41 and SBA-15: A Comparative Physicochemical and Photocatalytic Study. *J. Mater. Sci.* **2010**, *45*, 5786–5794. [CrossRef]
59. Liao, Y.; Xie, C.; Liu, Y.; Chen, H.; Li, H.; Wu, J. Comparison on Photocatalytic Degradation of Gaseous Formaldehyde by $TiO_2$, ZnO and Their Composite. *Ceram. Int.* **2012**, *38*, 4437–4444. [CrossRef]
60. Collard, X.; El Hajj, M.; Su, B.-L.; Aprile, C. Synthesis of Novel Mesoporous $ZnO/SiO_2$ Composites for the Photodegradation of Organic Dyes. *Microporous Mesoporous Mater.* **2014**, *184*, 90–96. [CrossRef]
61. Ajala, F.; Hamrouni, A.; Houas, A.; Lachheb, H.; Megna, B.; Palmisano, L.; Parrino, F. The Influence of Al Doping on the Photocatalytic Activity of Nanostructured ZnO: The Role of Adsorbed Water. *Appl. Surf. Sci.* **2018**, *445*, 376–382. [CrossRef]
62. Hamrouni, A.; Moussa, N.; Parrino, F.; Di Paola, A.; Houas, A.; Palmisano, L. Sol-Gel Synthesis and Photocatalytic Activity of $ZnO-SnO_2$ Nanocomposites. *J. Mol. Catal. A Chem.* **2014**, *390*, 133–141. [CrossRef]
63. Amalia, F.R.; Takashima, M.; Ohtani, B. Are You Still Using Organic Dyes? Colorimetric Formaldehyde Analysis for True Photocatalytic-Activity Evaluation. *Chem. Commun.* **2022**, *58*, 11721–11724. [CrossRef]
64. Schuster, A. Radiation Through a Foggy Atmosphere. *Astrophys. J.* **1905**, *21*, 1. [CrossRef]
65. Kubelka, P.; Munk, F.; KUBELKA, P. Ein Beitrag Zur Optik Der Farbanstriche. *Z. Tech. Phys.* **1931**, *12*, 593–601.
66. Kubelka, P. New Contributions to the Optics of Intensely Light-Scattering Materials Part I. *J. Opt. Soc. Am.* **1948**, *38*, 448. [CrossRef]
67. Kubelka, P. New Contributions to the Optics of Intensely Light-Scattering Materials Part II: Nonhomogeneous Layers. *J. Opt. Soc. Am.* **1954**, *44*, 330. [CrossRef]
68. Zanatta, A.R. Revisiting the Optical Bandgap of Semiconductors and the Proposal of a Unified Methodology to Its Determination. *Sci. Rep.* **2019**, *9*, 11225. [CrossRef] [PubMed]
69. Paramo, J.A.; Strzhemechny, Y.M.; Endo, T.; Crnjak Orel, Z. Correlation of Defect-Related Optoelectronic Properties in $Zn_5(OH)_6(CO_3)_2$/ZnO Nanostructures with Their Quasi-Fractal Dimensionality. *J. Nanomater.* **2015**, *2015*, 237985. [CrossRef]
70. Zhang, M.; Averseng, F.; Haque, F.; Borghetti, P.; Krafft, J.M.; Baptiste, B.; Costentin, G.; Stankic, S. Defect-Related Multicolour Emissions in ZnO Smoke: From Violet, over Green to Yellow. *Nanoscale* **2019**, *11*, 5102–5115. [CrossRef] [PubMed]
71. Pal, M.; Bera, S.; Sarkar, S.; Jana, S. Influence of Al Doping on Microstructural, Optical and Photocatalytic Properties of Sol–Gel Based Nanostructured Zinc Oxide Films on Glass. *RSC Adv.* **2014**, *4*, 11552–11563. [CrossRef]

*Communication*

# Recovery of the *N,N*-Dibutylimidazolium Chloride Ionic Liquid from Aqueous Solutions by Electrodialysis Method

Dorota Babilas [1,*], Anna Kowalik-Klimczak [2] and Anna Mielańczyk [3]

[1] Department of Inorganic, Analytical Chemistry and Electrochemistry, Faculty of Chemistry, Silesian University of Technology, B. Krzywoustego 6, 44-100 Gliwice, Poland
[2] Bioeconomy and Eco-Innovation Centre, Łukasiewicz Research Network—The Institute for Sustainable Technologies, Pułaskiego 6/10, 26-600 Radom, Poland; anna.kowalik-klimczak@itee.lukasiewicz.gov.pl
[3] Department of Physical Chemistry and Technology of Polymers, Faculty of Chemistry, Silesian University of Technology, M. Strzody 9, 44-100 Gliwice, Poland; anna.mielanczyk@polsl.pl
* Correspondence: dorota.babilas@polsl.pl; Tel.: +48-32-237-24-90

**Abstract:** Ionic liquids (ILs), named also as liquid salts, are compounds that have unique properties and molecular architecture. ILs are used in various industries; however, due to their toxicity, the ILs' recovery from the postreaction solutions is also a very important issue. In this paper, the possibility of 1,3-dialkylimidazolium IL, especially the *N,N*-dibutylimidazolium chloride ([$C_4C_4$IM]Cl) recovery by using the electrodialysis (ED) method was investigated. The influence of [$C_4C_4$IM]Cl concentration in diluate solution on the ED efficiency was determined. Moreover, the influence of IL on the ion-exchange membranes' morphology was examined. The recovery of [$C_4C_4$IM]Cl, the [$C_4C_4$IM]Cl flux across membranes, the [$C_4C_4$IM]Cl concentration degree, the energy consumption, and the current efficiency were determined. The results showed that the ED allows for the [$C_4C_4$IM]Cl recovery and concentration from dilute solutions. It was found that the [$C_4C_4$IM]Cl content in the concentrates after ED was above three times higher than in the initial diluate solutions. It was noted that the ED of solutions containing 5–20 g/L [$C_4C_4$IM]Cl allows for ILs recovery in the range of 73.77–92.45% with current efficiency from 68.66% to 92.99%. The [$C_4C_4$IM]Cl recovery depended upon the initial [$C_4C_4$IM]Cl concentration in the working solution. The highest [$C_4C_4$IM]Cl recovery (92.45%) and ED efficiency (92.99%) were obtained when the [$C_4C_4$IM]Cl content in the diluate solution was equal 20 g/L. Presented results proved that ED can be an interesting and effective method for the [$C_4C_4$IM]Cl recovery from the dilute aqueous solutions.

**Keywords:** electrodialysis; ionic liquids recovery; 1,3-dialkylimidazolium ionic liquids; *N,N*-dibutylimidazolium chloride

**Citation:** Babilas, D.; Kowalik-Klimczak, A.; Mielańczyk, A. Recovery of the *N,N*-Dibutylimidazolium Chloride Ionic Liquid from Aqueous Solutions by Electrodialysis Method. *Int. J. Mol. Sci.* **2022**, *23*, 6472. https://doi.org/10.3390/ijms23126472

Academic Editor: Dippong Thomas

Received: 12 May 2022
Accepted: 3 June 2022
Published: 9 June 2022

**Publisher's Note:** MDPI stays neutral with regard to jurisdictional claims in published maps and institutional affiliations.

**Copyright:** © 2022 by the authors. Licensee MDPI, Basel, Switzerland. This article is an open access article distributed under the terms and conditions of the Creative Commons Attribution (CC BY) license (https://creativecommons.org/licenses/by/4.0/).

## 1. Introduction

Ionic liquids (ILs) are known as green solvents. ILs are completely composed of ions [1]. Over the past two decades, ILs have gained increasing importance in the industrial sector. Due to their unique properties, they are attractive alternatives to other organic solvents. ILs are characterized by a melting point below 100 °C, high polarity, versatile solubility, high electrical conductivity, high thermal stability, and non-volatility [2]. Due to their properties, ILs can be divided into room temperature ILs, low-temperature ILs, poly-ILs, and magnetic ILs. Room temperature ILs have a liquid or molten state below 100 °C. Low-temperature ILs are used for low-temperature applications and in electrochemical instruments to store energy. In membrane technology, because of their ability and durability to form membranes, poly-ionic liquids can be used. Magnetic ILs are characterized by paramagnetic properties, and easy dispersion in solutions [3,4]. ILs' applications have a multidisciplinary character. Generally, in chemical processes, ILs are used as reagents, catalysts, and solvents [5]. In biotechnology, ILs can be used for biocatalysis and protein purification [6,7]. In the pharmaceutical industry, the ILs can be applied in drug delivery systems and as active pharmaceutical ingredients [8,9]. Moreover, ILs have been employed in chemical engineering in extraction and separation processes [10]. ILs are often used as additives in synthesis reactions, in the physical processing of polymers, or as reaction media [10].

In last decades, the ILs are of special interest since they have proven their versatility and effectiveness in many areas of chemistry. Their application stems from their unique properties. One of the interesting ILs is $N,N$-dibutylimidazolium chloride ($[C_4C_4IM]Cl$). In the case of $[C_4C_4IM]Cl$, which is an ambient temperature IL, it was successfully used as a "green" solvent and catalyst at once in Friedlander heteroannulation reaction [11], cellulose dissolution, and the dehydration of fructose [12]. The advantages of $[C_4C_4IM]Cl$ are good thermal stability, non-volatility, solubility in water, and recyclability.

Unfortunately, due to the good miscibility of ILs with most solvents, their recovery and separation from the organic compound or polymer solutions by traditional separation processes are inefficient. Moreover, the separation of the final product requires numerous unit operations, uses environmentally hazardous reagents, leads to the unfavorable dilution of ILs to very low concentrations during individual stages of processing, and generates an increased amount of harmful waste. Due to the growing generation of wastewater by various industries, the wastewater treatment and recovery of raw materials are very important aspects [1,13,14]. Despite the unique properties of ILs, they can have a negative impact on the environment. In particular, the water-soluble, chemical, and thermal stable ILs occurring in wastewater can contaminate the soil and aquatic environment to a great extent and increase a negative effect on the environment and living organisms [15]. The imidazolium-based ILs indicate toxicity towards the aquatic system, green algae, and microorganisms [10]. In the available literature, it was also found that ILs have the ability to inhibit various enzymes. Therefore, investigations on the removal or recovery of ILs from wastewater and post-reaction mixtures should be developed [16–18].

In the available literature, the methods for recovering ILs from wastewater include adsorption, crystallization, distillation, extraction, and membrane processes [19,20]. One of the promising methods of ILs recovery is electrodialysis (ED). ED is an environmentally friendly process for solution desalination, which is easier to scale up than other wastewater treatment techniques such as adsorption or ion-exchange. In addition, ED allows for the recovery and concentration of salts from diluted solutions. In the case of ILs recovery, ED does not require the use of additional solvents. ED is known as a separation process used to separate ions using electrical potential and charged ion-exchange membranes. Thus, because of the electrolyte nature of ILs, ED could be an efficient ILs recovery method [21]. During ED, ions migrate across membranes, anions across the positively charged anion-exchange membranes, and cations across the negatively charged cation-exchange membranes. Thus, the treated solutions are desalted [22–24]. ED applications include brackish water desalination, salt pre-concentration, demineralization of food products, and wastewater treatment, especially wastewater from the electroplating industry [25–27]. ED can be also applied for ILs recovery from aqueous post-reaction solutions [28]. Effectiveness of electrodialytic ILs recovery depends on the kind of treated ILs—cation or anion. Nowadays, ED is applied for 1-butyl-3-methylimidazolium chloride ([Bmim]Cl), 1-butyl-3-methylimidazolium bromide ([Bmim]Br), 1-butyl-3-methylimidazolium hydrogensulfate ([Bmim]$HSO_4$), 1-allyl-3-methylimidazolium chloride ([Amim]Cl), 1-ethyl-3-methylimidazolium chloride ([Emim]Cl), and triethylammonium hydrogen sulfate [TEA][$HSO_4$]. It was concluded that the IL recovery highly depends on the kind of ILs, the concentration of ILs in solution, and ED parameters. Depending on the treated ILs solution, alkyl chain length, and ILs concentration, the ILs recovery rate ranges from 40 to 95%. Therefore, it is important to check the effectiveness of ILs recovery for the specific ILs [29–35].

The purpose of this work is to investigate the possibility and effectiveness of the recovery of $N,N$-dibutylimidazolium chloride ($[C_4C_4IM]Cl$) using the ED method. The influence of $[C_4C_4IM]Cl$ concentration in diluate solution on the ED efficiency is discussed in detail. The $[C_4C_4IM]Cl$ content in the experimental solutions was selected based on the general ILs content in the wastewater. The recovery and concentration of [C$_4$C$_4$IM]Cl by the ED method has not yet been demonstrated in the available literature, therefore it can be a novel method for [C$_4$C$_4$IM]Cl recovery from wastewater.

## 2. Results and Discussion

The aim of this work is to examine the effectiveness of the *N,N*-dibutylimidazolium chloride ([C$_4$C$_4$IM]Cl) recovery using ED method. The influence of the initial concentration of [C$_4$C$_4$IM]Cl in diluate solution on the ED efficiency was evaluated by the ED effectiveness factors such as recovery ratio, IL concentration rate, IL molar flux across ion-exchange membranes, electric current efficiency, and energy consumption. The IL concentration in feed solution influence on the solution conductivity, electrical resistance, and concentration polarization, as well as simultaneously on the ED efficiency. Thus, four sets of ED experiments with IL concentration in feed solution in the range from 5 to 20 g/L were carried out. The process solutions compositions are presented in Table 1. The experiments were conducted using the method described in Section 3.2.

**Table 1.** The experimental solutions composition.

| Exp. No. | Initial Diluate | Initial Concentrate | Electrode Rinse Solution |
| --- | --- | --- | --- |
| 1. | 300 mL of 5 g/L [C$_4$C$_4$IM]Cl | 100 mL of 5 g/L [C$_4$C$_4$IM]Cl | 250 mL of 0.1 M H$_2$SO$_4$ |
| 2. | 300 mL of 10 g/L [C$_4$C$_4$IM]Cl | 100 mL of 10 g/L [C$_4$C$_4$IM]Cl | 250 mL of 0.1 M H$_2$SO$_4$ |
| 3. | 300 mL of 15 g/L [C$_4$C$_4$IM]Cl | 100 mL of 15 g/L [C$_4$C$_4$IM]Cl | 250 mL of 0.1 M H$_2$SO$_4$ |
| 4. | 300 mL of 20 g/L [C$_4$C$_4$IM]Cl | 100 mL of 20 g/L [C$_4$C$_4$IM]Cl | 250 mL of 0.1 M H$_2$SO$_4$ |

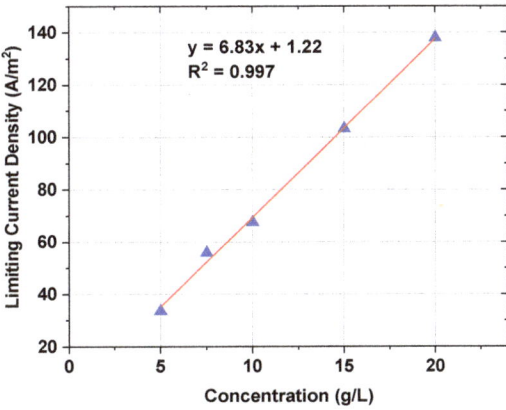

**Figure 1.** The effect of [C$_4$C$_4$IM]Cl concentration in the initial diluate on the LCD.

In the first stage of the work, the LCDs were determined. The LCD is a crucial factor in choosing the operational parameters of the ED [36]. LCD also determines the efficiency of the ED. The ED conducted above LCD is characterized by high electrical resistance in the diluate as a result of depletion of the ions in the laminar boundary layer at the ion-exchange membrane surface. The LCD is highly dependent on the concentration of feed solution [37]. The LCD as a function of the [C$_4$C$_4$IM]Cl concentration in diluate is presented in Figure 1. As was presumed, the LCD increased in a linear manner with increasing [C$_4$C$_4$IM]Cl content in the feed solution. The determined LCDs were in the range of 34 to 138 A/m$^2$. All the ED experiments were conducted below the LCD at constant voltage.

Electrodialytic ILs recovery and concentration degree highly depend on the feed solution concentration. The effect of the initial feed concentration on the [C$_4$C$_4$IM]Cl recovery effectiveness was presented in Figure 2a–e. Figure 2a shows that the [C$_4$C$_4$IM]Cl recovery ratio depended on the initial [C$_4$C$_4$IM]Cl content in the diluate.

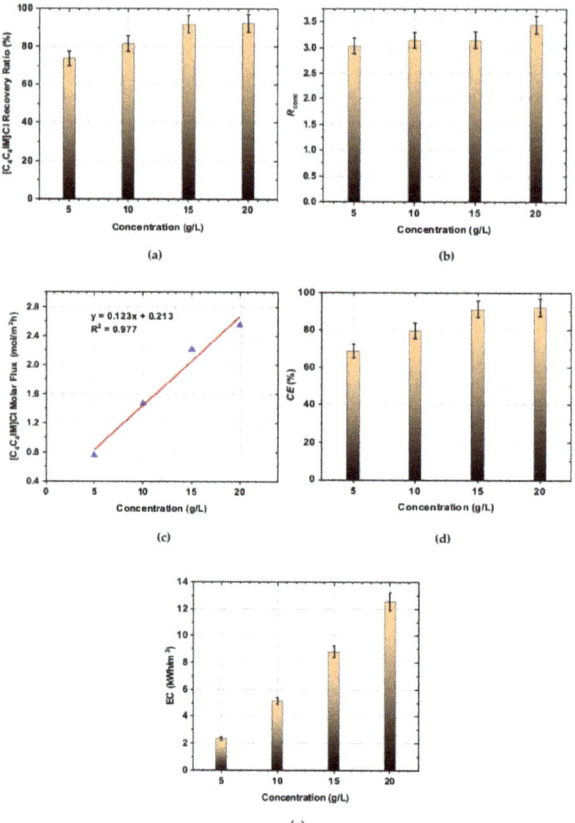

**Figure 2.** The effect of [C$_4$C$_4$IM]Cl concentration in the initial diluate on the: (**a**) recovery ratio, (**b**) concentration rate, (**c**) IL molar flux across ion-exchange membranes, (**d**) electrodialysis current efficiency, (**e**) stack energy consumption.

It was found that the recovery ratio increased with increasing [C$_4$C$_4$IM]Cl concentration in the feed solution. The recovery ratio increased from 73.77% to 92.45% for the feed solution of 5 to 20 g/L of [C$_4$C$_4$IM]Cl, respectively. Thus, it was proved that [C$_4$C$_4$IM]Cl was effectively removed from diluate solution by the ED method. The recovery ratio reached above 90% for solutions with a [C$_4$C$_4$IM]Cl content equal to 15 and 20 g/L. The ED at high [C$_4$C$_4$IM]Cl concentration resulted in a high [C$_4$C$_4$IM]Cl recovery ratio (in the analyzed concentration range). It is in agreement with research on [BMIM]Cl recovery by ED method [28,29]. In the available literature, it is noted that the initial ILs content in diluate solution has an important effect on the ED effectiveness factors. The results presented in the works [28,29,33] confirmed that the ILs recovery by the ED method increases with increasing ILs concentration in the initial diluate. However, the ILs recovery also is highly dependent on the ILs type and chemical character.

It was also noted that the [C$_4$C$_4$IM]Cl content in the all concentrates after ED was three times higher than in the feed solution (Figure 2b). Moreover, the obtained results were correlated with [C$_4$C$_4$IM]Cl molar flux across ion-exchange membranes. It can be clearly seen in Figure 2c that the molar flux of [C$_4$C$_4$IM]Cl increased with increasing [C$_4$C$_4$IM]Cl concentration in the feed solution in an almost linear manner. The [C$_4$C$_4$IM]Cl molar flux across ion-exchange membranes was estimated to be 0.76 mol/m$^2$h at the feed solution concentration of 5 g/L [C$_4$C$_4$IM]Cl and 2.56 mol/m$^2$h at the feed solution concentration of 20 g/L [C$_4$C$_4$IM]Cl.

ED efficiency is also evaluated by electric current efficiency. Current efficiency is defined as the ratio between the current used in the ED stack for effective ion recovery from diluate solution to concentrate and the amount of the total current applied in the ED stack. Current efficiency defines how much of the electric current is effectively used in ion transport across ion-exchange membranes [38,39]. In Figure 2d, the effect of $[C_4C_4IM]Cl$ concentration in the initial diluate on the electrodialysis current efficiency is shown. It was found that current efficiency increased with increasing $[C_4C_4IM]Cl$ concentration in the initial diluate solution. It can be explained by the reduction of electrical resistance of initial diluate with increasing ILs concentration, and the acceleration of ion transport across membranes. Electric current efficiency increased from 68.66% to a maximum value of 92.99% for the feed solutions in the range from 5 to 20 g/L of $[C_4C_4IM]Cl$.

The current efficiency increased linearly for diluates with IL concentrations ranging from 5 to 15 g/L. However, above a concentration of 15 g/L, the current efficiency increases slightly. It was concluded that obtained current efficiencies in the examined range are very satisfactory in comparison to values presented in other works about ILs recovery by ED methods. Current efficiency of the electrodialytic recovery of [BMIM]Cl increased from 37.7% to 70.7% for initial [BMIM]Cl content in feed solutions from 2.24 to 6.90 g/L [30]. In another work [29], current efficiency of [BMIM]Cl recovery was over 70% under the condition of IL concentration equaled 34.94 g/L (0.2 mol/L).

As is shown in Figure 2e, the initial $[C_4C_4IM]Cl$ content in diluate also has an effect on the energy consumption. It was observed that the stack energy consumption highly depended on the initial $[C_4C_4IM]Cl$ content in the diluate. It can be clearly seen that energy consumption increased linearly with increasing of $[C_4C_4IM]Cl$ concentration in the initial diluate solution. When the initial concentration of $[C_4C_4IM]Cl$ increased from 5 to 20 g/L, the energy consumption increased from 2.35 to 12.57 kWh/m$^3$, respectively.

The obtained results confirmed, that the electrodialytic $[C_4C_4IM]Cl$ recovery is influenced by feed solution concentration. The best ED performance was obtained when the $[C_4C_4IM]Cl$ content in the initial diluate was 15 and 20 g/L. When the concentration of $[C_4C_4IM]Cl$ in the initial diluate was 15 g/L, the $[C_4C_4IM]Cl$ recovery ratio, the $[C_4C_4IM]Cl$ concentration rate, the electric current efficiency, as well as energy consumption were 91.87%, 3.15, 91.40%, and 8.77 kWh/m$^3$, respectively. It was also noted that when in the initial diluate the $[C_4C_4IM]Cl$ concentration was 20 g/L, the 3.45-fold concentration degree can be achieved with 92.45% $[C_4C_4IM]Cl$ recovery ratio and current efficiency of 92.99%.

One of the disadvantages of the electrodialytic IL recovery is membrane fouling. Membrane fouling can be described as agglomerations of molecules, inorganic, and organic compounds in the membrane pores or on the membrane surface. Membrane fouling causes reduced membrane separation efficiency. Membrane fouling can be limited by the linear flow velocity of feed solution, ED operational parameters, and the ED module chemical cleaning after process [39,40]. In Figure 3, the SEM micrographs of the tested heterogeneous ion-exchange membranes before and after ED are presented. The AM(H)PP and CM(H)PP membranes were fabricated by a pressing method from polypropylene (as an inert polymer membrane matrix) and ion-exchange resin particles. In Figure 3, the morphology of the pristine AM(H)PP and CM(H)PP membranes are presented. The morphology of pristine membranes are inhomogeneous. The pores, ion-exchange resin grains, and reinforcing net are clearly observed on the pristine membranes' surface (Figure 3). It was also found that the morphology of the anion-exchange membranes' AM(H)PP did not differ from that of the pristine AM(H)PP. On the both AM(H)PP membrane surface micrographs, the ion-exchange grains in the polymer membrane matrix, the reinforcing net, and the pores can be clearly seen. However, in the case of the cation-exchange membranes' (CM(H)PP) SEM micrographs, it can be noted that the fouling occurred. It was found that the morphology of the CM(H)PP differs from that of the pristine CM(H)PP. The layer on the CM(H)PP membrane surface is caused probably by $[C_4C_4IM]^+$. The ionic radius of $[C_4C_4IM]^+$ and the alkyl chain length also influence the fouling ability. As the ion radius increases, the probability of fouling increases. The size of $[C_4C_4IM]^+$ is larger than that of $[C_4MIM]^+$, and in consequence higher fouling of the CM(H)PP membrane was observed in comparison to the ED of [BMIM]Cl [41]. The pores on the surface of the CM(H)PP membrane after ED are clearly smaller (Figure 3). The reinforcing net is also less visible. Fouling of the CM(H)PP membranes can reduce the flux and increase energy consumption [39]. Although some kind of the cation-exchange membranes fouling was observed, the obtained results confirmed that the ED performance was not affected, and the ED can be applied as an efficient $[C_4C_4IM]Cl$ recovery method from wastewater and aqueous solutions. A very important aspect is the cleaning of the membranes after ED. Membranes can be cleaned with distilled water [39], 0.35 wt% HCl [30], as well as 0.4 wt% NaOH [30].

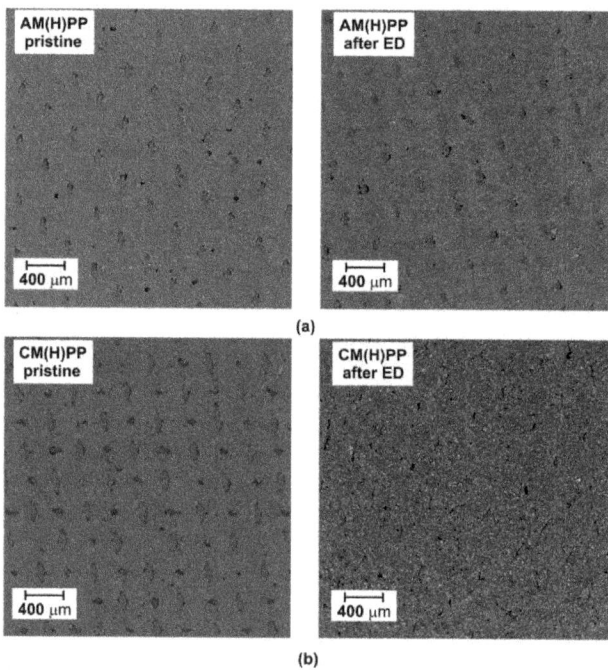

**Figure 3.** SEM micrograph of the tested heterogeneous ion-exchange membranes before and after ED: (**a**) AM(H)PP; (**b**) CM(H)PP membrane.

## 3. Materials and Methods

### 3.1. Experimental Solutions

Experiments were conducted using the model diluate and concentrate solutions containing $[C_4C_4IM]Cl$. The concentration of $[C_4C_4IM]Cl$ in the diluate and concentrate solutions was in the range of 5–20 g/L. The $[C_4C_4IM]Cl$ content in solutions was selected based on the general ILs content in the wastewater. Frequently, the content of IL in the wastewater and post-reaction solution is in the range of 2.24 to 35 g/L [29,30,33].

The $[C_4C_4IM]Cl$ was synthesized as follows: 5 mL of butylimidazole (114 mmol), 14.2 mL of chlorobutane (136.8 mmol), and 30 mL of toluene were put in a round bottom flask, which was then placed in an oil bath at 120 °C and equipped with a reflux condenser. The reaction was carried out for 24 h. After a designated time, the mixture was cooled to room temperature. The upper layer was collected with a syringe. Toluene from the bottom layer was evaporated in a rotary evaporator with a water bath temperature of 100 °C. The ionic liquid was dissolved in methylene chloride, and activated carbon was added to remove possible impurities. The whole solution was passed through a filter and again evaporated in a rotary evaporator with a water bath temperature of 50 °C.

Finally, the ionic liquid was dried to get rid of water and not distilled toluene. $^1$H NMR (400 MHz, CDCl3) δ [ppm]: 0.95–1.00 (t, 6H, >NCH$_2$CH$_2$CH$_2$CH$_3$), 1.35–1.43 (sextet, 4H, >NCH$_2$CH$_2$CH$_2$CH$_3$), 1.88–1.95 (quintet, 4H, >NCH$_2$CH$_2$CH$_2$CH$_3$), 4.35–4.40 (t, 4H, >NCH$_2$CH$_2$CH$_2$CH$_3$), 7.43 (d, 2H, position 4 and 5 in the ring N-CH = CH-N) 10.92 (s, 1H, position 2 in the ring N-CH = N).

The 0.1 M H$_2$SO$_4$ (Avantor Performance Materials, Gliwice, Poland) solution was used as the electrode rinse solution. All experimental solutions were prepared using deionized water (Millipore Elix 10 system, Darmstadt, Germany).

*3.2. Experimental Set-Up*

The electrodialytic [C$_4$C$_4$IM]Cl recovery was carried out at room temperature using the experimental set-up consisting of the EDR-Z/10-0.8 module (MemBrain, Straz pod Ralskem, Czech Republic) with two pairs of the heterogeneous ion-exchange membranes AM(H)PP–CM(H)PP (Mega a.s., Straz pod Ralskem, Czech Republic) in the ED stack. An effective area of the single membrane was 64 cm$^2$. The ED module was connected to a programmable power supply (KORAD KA3010, KORAD Technology Co., Ltd., Dongguan, China). The experiments were performed under constant voltage conditions, which was a maximum value determined by the limiting current density test. The electric current of the electrodialysis system was recorded every 1 min. The electrodialyzer was connected with three tanks named diluate, concentrate, and electrode rinse solution. All process solutions were recirculated using a peristaltic pump (MCP Standard Ismatec, Cole-Parmer, Wertheim, Germany) at a rate corresponding to the linear flow velocity of 2 cm/s. The initial volume of diluate was 300 mL, the initial volume of concentrate was 100 mL, and the volume of the electrode rinse solution was 250 mL. Thus, the diluate-to-concentrate volume ratio was equal to 3. The experiments were conducted until the diluate conductivity dropped to 10% value of initial diluate conductivity, which was monitored using two CX-461 conductivity meters (Elmetron, Zabrze, Poland). Experimental ED set-up for [C$_4$C$_4$IM]Cl recovery is presented in Figure 4. All experiments were replicated three times.

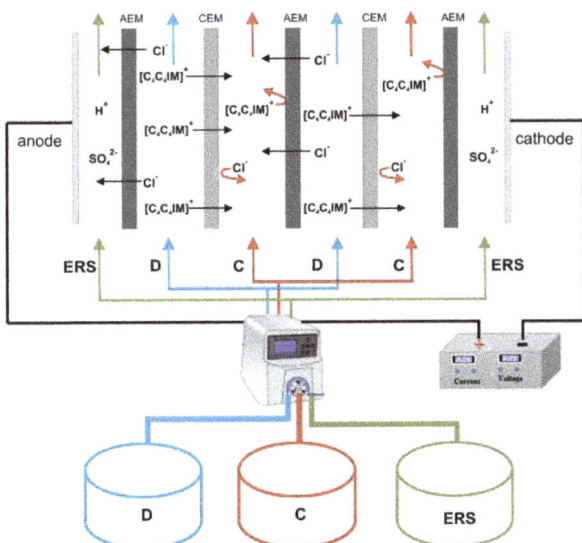

**Figure 4.** Experimental ED set-up for [C$_4$C$_4$IM]Cl recovery. D—diluate, C—concentrate, ERS—electrode rinse solution, AEM—anion-exchange membrane, CEM—cation-exchange membrane.

### 3.3. Membranes

AM(H)PP–CM(H)PP ion-exchange membranes were used in the experiments. The CM(H)PP and AM(H)PP membranes were manufactured by Mega a.s. (Straz pod Ralskem, Czech Republic). The tested membranes comprised an ion exchange resin incorporated within a binder. Before and after the ED experiments, the surface morphology of the tested membranes was investigated using a scanning electron microscope (Hitachi TM3000 table-top TM series, Tokyo, Japan), equipped with a backscattered electron (BSE) detector.

### 3.4. Limiting Current Density (LCD)

The LCDs for solutions with [$C_4C_4IM$]Cl concentrations of 5, 7.5, 10, 15, and 20 g/L were determined by the Cowan–Brown method [36]. During the LCDs determination, the applied voltage was increased stepwise at a speed of 0.5 V/min until the ED cell potential drop reached 20 V. The LCDs were determined from the relationship between the current and the corresponding potential. Therefore, the ED stack resistance–reciprocal current curves for the LCD assessment were drawn.

### 3.5. Analytical Methods

The concentrations of [$C_4C_4IM$]Cl in diluate and concentrate solutions were analyzed using a UV-VIS spectrophotometer (Varian Cary 50 Scan, Agilent, Santa Clara, CA, USA). The maximum absorption wavelength for the [$C_4C_4IM$]$^+$ cation was 211.50 nm. The concentration of [$C_4C_4IM$]Cl was determined based on the standard curve between the concentration and absorbance of [$C_4C_4IM$]Cl. The standard curve between the concentration and absorbance of [$C_4C_4IM$]Cl is presented in Figure S1.

### 3.6. ED Experiments Data Analysis

To estimate the [$C_4C_4IM$]Cl recovery effectiveness, some crucial factors such as the [$C_4C_4IM$]Cl recovery ratio ($R_{[C_4C_4IM]Cl}$), the [$C_4C_4IM$]Cl concentration rate ($R_{conc}$), the [$C_4C_4IM$]Cl molar flux across ion-exchange membranes ($J_{[C_4C_4IM]Cl}$), the electric current efficiency ($CE_{[C_4C_4IM]Cl}$), as well as energy consumption ($EC$) were calculated using the following Equations (1)–(5), respectively:

$$R_{[C_4C_4IM]Cl} = \frac{m_{IL,t}^{conc}}{m_{IL,0}^{dil}} \cdot 100\% \tag{1}$$

where:
- $m_{IL,0}^{dil}$—the initial mass of the [$C_4C_4IM$]Cl in the diluate before ED, [g],
- $m_{IL,t}^{conc}$—the increase in the [$C_4C_4IM$]Cl mass in the concentrate after ED, [g].

$$R_{conc} = \frac{C_{IL,t}^{conc}}{C_{IL,0}^{dil}} \cdot 100\% \tag{2}$$

where:
- $C_{IL,0}^{dil}$—the initial concentration of the [$C_4C_4IM$]Cl in the diluate solution before ED, [g/L],
- $C_{IL,t}^{conc}$—the final concentration of the [$C_4C_4IM$]Cl in the concentrate solution after ED, [g/L].

$$J_{[C_4C_4IM]Cl} = \frac{\left(V_t^{conc} \cdot C_{IL,t}^{conc}\right) - \left(V_0^{conc} \cdot C_{IL,0}^{conc}\right)}{M_{[C_4C_4IM]Cl} \cdot A \cdot t} \tag{3}$$

where:
- $V_t^{conc}$—the volume of the concentrate solution after ED, [L],
- $V_0^{conc}$—the volume of the concentrate solution before ED, [L],

- $C_{IL,t}^{conc}$—the concentration of the [C$_4$C$_4$IM]Cl in the concentrate solution after ED, [g/L],
- $C_{IL,0}^{conc}$—the concentration of the [C$_4$C$_4$IM]Cl in the concentrate solution before ED, [g/L],
- $M_{[C_4C_4IM]Cl}$—the molar mass of [C$_4$C$_4$IM]Cl, [g/mol],
- $A$—the active membrane surface area, [m$^2$],
- $t$—ED time, [h].

$$CE_{[C_4C_4IM]Cl} = \frac{F \cdot z \cdot \frac{C_{IL,t}^{conc}}{M_{[C_4C_4IM]Cl}} \cdot V_t^{conc}}{n \cdot \int_0^t I(t)dt} \cdot 100\% \quad (4)$$

where:
- $F$—the Faraday constant (96,485 C/mol),
- $z$—the charge number of [C$_4$C$_4$IM]$^+$,
- $V_t^{conc}$—the volume of the concentrate solution after ED, [L],
- $C_{IL,t}^{conc}$—the concentration of the [C$_4$C$_4$IM]Cl in the concentrate solution after ED, [g/L],
- $M_{[C_4C_4IM]Cl}$—the molar mass of [C$_4$C$_4$IM]Cl, [g/mol],
- $n$—the number of membrane pairs,
- $I$—the electric current, [A].

$$EC = \frac{U \cdot \int_0^t I(t)dt}{V_0^{dil}} \quad (5)$$

where:
- $EC$—the energy consumption, [kWh/m$^3$],
- $U$—the applied voltage, [V],
- $I$—the electric current, [A],
- $V_0^{dil}$—the initial diluate volume, [L].

## 4. Conclusions

In this work, the possibility and effectiveness of N,N-dibutylimidazolium chloride recovery using the ED method were discussed. It was concluded that ED can be applied as an efficient [C$_4$C$_4$IM]Cl recovery method from wastewater and aqueous solutions. It was proved that the [C$_4$C$_4$IM]Cl content in the feed solution influences the ED performances. The recovery ratio, the [C$_4$C$_4$IM]Cl molar flux, and the electric current efficiency increase with increasing the concentration of [C$_4$C$_4$IM]Cl in the feed solution. Moreover, the energy consumption also highly depends on the initial [C$_4$C$_4$IM]Cl content in diluate. Energy consumption increases linearly with increasing of the [C$_4$C$_4$IM]Cl concentration in the initial diluate solution.

It was also found that in the examined [C$_4$C$_4$IM]Cl concentration range, the best ED performance can be obtained when the [C$_4$C$_4$IM]Cl content in the initial diluate is 15 and 20 g/L. When in the initial diluate the [C$_4$C$_4$IM]Cl concentration is 20 g/L, the 3.45-fold concentration degree can be achieved with 92.45% [C$_4$C$_4$IM]Cl recovery ratio and current efficiency 92.99%.

The ILs concentration in feed solution influences the solution conductivity, electrical resistance, and concentration polarization, as well as simultaneously the ED efficiency. It was found that ED efficiency increased with increasing [C$_4$C$_4$IM]Cl concentration in the initial diluate solution (in the examined range). It was explained by the reduction of electrical resistance of the initial diluate with increasing ILs concentration, and the acceleration of ions transport across membranes.

**Supplementary Materials:** The following supporting information can be downloaded at: https://www.mdpi.com/article/10.3390/ijms23126472/s1.

**Author Contributions:** Conceptualization, D.B.; methodology, D.B. and A.M.; validation, D.B.; formal analysis, D.B.; investigation, D.B. and A.K.-K.; resources, D.B. and A.M.; data curation, D.B.; writing—original draft preparation, D.B.; writing—review and editing, D.B.; visualization, D.B.; supervision, D.B.; project administration, D.B.; funding acquisition, D.B. All authors have read and agreed to the published version of the manuscript.

**Funding:** Publication supported under the Excellence Initiative–Research University Program implemented at the Silesian University of Technology, 2020 (Rector's grant 04/010/SDU/10-21-02), and Rector's Grant in the field of research and development (Silesian University of Technology, Poland, 04/010/RGJ21/1017).

**Institutional Review Board Statement:** Not applicable.

**Informed Consent Statement:** Not applicable.

**Data Availability Statement:** The data presented in this study are available on request from the corresponding author.

**Acknowledgments:** The authors would like to thank Maria Kupczak (Silesian University of Technology) for the technical support in the N,N-dibutylimidazolium chloride synthesis.

**Conflicts of Interest:** The authors declare no conflict of interest.

## References

1. Mai, N.L.; Ahn, K.; Koo, Y.-M. Methods for recovery of ionic liquids—A review. *Process Biochem.* **2014**, *49*, 872–881. [CrossRef]
2. Feng, R.; Zhao, D.; Guo, Y. Revisiting Characteristics of Ionic Liquids: A Review for Further Application Development. *J. Environ. Prot.* **2010**, *1*, 95–104. [CrossRef]
3. Goutham, R.; Rohit, P.; Vigneshwar, S.S.; Swetha, A.; Arun, J.; Gopinath, K.P.; Pugazhendhi, A. Ionic liquids in wastewater treatment: A review on pollutant removal and degradation, recovery of ionic liquids, economics and future perspectives. *J. Mol. Liq.* **2022**, *349*, 118150. [CrossRef]
4. Zhou, J.; Sui, H.; Jia, Z.; Yang, Z.; He, L.; Li, X. Recovery and purification of ionic liquids from solutions: A review. *RSC Adv.* **2018**, *8*, 32832–32864. [CrossRef]
5. Plechkova, N.V.; Seddon, K.R. Applications of ionic liquids in the chemical industry. *Chem. Soc. Rev.* **2008**, *37*, 123–150. [CrossRef]
6. Xu, P.; Liang, S.; Zong, M.-H.; Lou, W.-Y. Ionic liquids for regulating biocatalytic process: Achievements and perspectives. *Biotechnol. Adv.* **2021**, *51*, 107702. [CrossRef]
7. Lee, S.Y.; Khoiroh, I.; Ooi, C.W.; Ling, T.C.; Show, P.L. Recent Advances in Protein Extraction Using Ionic Liquid-based Aqueous Two-phase Systems. *Sep. Purif. Rev.* **2017**, *46*, 291–304. [CrossRef]
8. Adawiyah, N.; Moniruzzaman, M.; Hawatulaila, S.; Goto, M. Ionic liquids as a potential tool for drug delivery systems. *MedChemComm* **2016**, *7*, 1881–1897. [CrossRef]
9. Ferraz, R.; Branco, L.C.; Prudêncio, C.; Noronha, J.P.; Petrovski, Ž. Ionic Liquids as Active Pharmaceutical Ingredients. *ChemMedChem* **2011**, *6*, 975–985. [CrossRef]
10. Magina, S.; Barros-Timmons, A.; Ventura, S.P.M.; Evtuguin, D.V. Evaluating the hazardous impact of ionic liquids—Challenges and opportunities. *J. Hazard. Mater.* **2021**, *412*, 125215. [CrossRef]
11. Palimkar, S.S.; Siddiqui, S.A.; Daniel, T.; Lahoti, R.J.; Srinivasan, K.V. Ionic Liquid-Promoted Regiospecific Friedlander Annulation: Novel Synthesis of Quinolines and Fused Polycyclic Quinolines. *J. Org. Chem.* **2003**, *68*, 9371–9378. [CrossRef]
12. Zimmermann, J.; Ondruschka, B.; Stark, A. Efficient Synthesis of 1,3-Dialkylimidazolium-Based Ionic Liquids: The Modified Continuous Radziszewski Reaction in a Microreactor Setup. *Org. Process Res. Dev.* **2010**, *14*, 1102–1109. [CrossRef]
13. Kubisa, P. Ionic liquids in the synthesis and modification of polymers. *J. Polym. Sci. Part A Polym. Chem.* **2005**, *43*, 4675–4683. [CrossRef]
14. Singha, N.K.; Hong, K.; Mays, J.W. Polymerization in Ionic Liquids. In *Polymerized Ionic Liquids*; Eftekhari, A., Ed.; Royal Society of Chemistry: London, UK, 2018; Chapter 1. [CrossRef]
15. Kudłak, B.; Owczarek, K.; Namiesnik, J. Selected issues related to the toxicity of ionic liquids and deep eutectic solvents—A review. *Environ. Sci. Pollut. Res.* **2015**, *22*, 11975–11992. [CrossRef]
16. Pham, T.P.T.; Cho, C.-W.; Yun, Y.-S. Environmental fate and toxicity of ionic liquids: A review. *Water Res.* **2010**, *44*, 352–372. [CrossRef]
17. Zhang, C.; Zhu, L.; Wang, J.; Wang, J.; Zhou, T.; Xu, Y.; Cheng, C. The acute toxic effects of imidazolium-based ionic liquids with different alkyl-chain lengths and anions on zebrafish (*Danio rerio*). *Ecotoxicol. Environ. Saf.* **2017**, *140*, 235–240. [CrossRef]
18. Flieger, J.; Flieger, M. Ionic Liquids Toxicity—Benefits and Threats. *Int. J. Mol. Sci.* **2020**, *21*, 6267. [CrossRef]

19. Kuzmina, O. Methods of IL Recovery and Destruction. In *Application, Purification, and Recovery of Ionic Liquids*; Kuzmina, O., Hallett, J.P., Eds.; Elsevier: Amsterdam, the Netherlands, 2016; Chapter 5. [CrossRef]
20. Lei, Z.; Chen, B.; Koo, Y.-M.; MacFarlane, D.R. Introduction: Ionic Liquids. *Chem. Rev.* **2017**, *117*, 6633–6635. [CrossRef]
21. Haerens, K.; De Vreese, P.; Matthijs, E.; Pinoy, L.; Binnemans, K.; Van der Bruggen, B. Production of ionic liquids by electrodialysis. *Sep. Purif. Technol.* **2012**, *97*, 90–95. [CrossRef]
22. Al-Amshawee, S.; Yunus, M.Y.B.M.; Azoddein, A.A.M.; Hassell, D.G.; Dakhil, I.H.; Hasan, H.A. Electrodialysis desalination for water and wastewater: A review. *Chem. Eng. J.* **2020**, *380*, 122231. [CrossRef]
23. Xu, T.; Huang, C. Electrodialysis-based separation technologies: A critical review. *AIChE J.* **2008**, *54*, 3147–3159. [CrossRef]
24. Strathmann, H. *Ion-Exchange Membrane Separation Processes, Membrane Science and Technology Series 9*; Elsevier: Amsterdam, the Netherlands, 2004.
25. Siddiqui, M.U.; Generous, M.M.; Qasem, N.A.; Zubair, S.M. Explicit prediction models for brackish water electrodialysis desalination plants: Energy consumption and membrane area. *Energy Convers. Manag.* **2022**, *261*, 115656. [CrossRef]
26. Gurreri, L.; Tamburini, A.; Cipollina, A.; Micale, G. Electrodialysis Applications in Wastewater Treatment for Environmental Protection and Resources Recovery: A Systematic Review on Progress and Perspectives. *Membranes* **2020**, *10*, 146. [CrossRef] [PubMed]
27. Babilas, D.; Dydo, P. Zinc salt recovery from electroplating industry wastes by electrodialysis enhanced with complex formation. *Sep. Sci. Technol.* **2020**, *55*, 2250–2258. [CrossRef]
28. Li, W.; Wang, J.; Nie, Y.; Wang, D.; Xu, H.; Zhang, S. Separation of soluble saccharides from the aqueous solution containing ionic liquids by electrodialysis. *Sep. Purif. Technol.* **2020**, *251*, 117402. [CrossRef]
29. Wang, X.; Nie, Y.; Zhang, X.; Zhang, S.; Li, J. Recovery of ionic liquids from dilute aqueous solutions by electrodialysis. *Desalination* **2012**, *285*, 205–212. [CrossRef]
30. Trinh, L.T.P.; Lee, Y.J.; Lee, J.-W.; Bae, H.-J.; Lee, H.-J. Recovery of an ionic liquid [BMIM]Cl from a hydrolysate of lignocellulosic biomass using electrodialysis. *Sep. Purif. Technol.* **2013**, *120*, 86–91. [CrossRef]
31. Liang, X.; Fu, Y.; Chang, J. Recovery of ionic liquid via a hybrid methodology of electrodialysis with ultrafiltration after biomass pretreatment. *Bioresour. Technol.* **2016**, *220*, 289–296. [CrossRef]
32. Liang, X.; Wang, J.; Liu, H. Quantitative recovery and regeneration of acidic ionic liquid 1-butyl-3-methylimidazolium hydrogen sulphate via industrial strategy for sustainable biomass processing. *Bioresour. Technol.* **2021**, *325*, 124726. [CrossRef]
33. Liang, X.; Fu, Y.; Chang, J. Research on the quick and efficient recovery of 1-allyl-3-methylimidazolium chloride after biomass pretreatment with ionic liquid-aqueous alcohol system. *Bioresour. Technol.* **2017**, *245*, 760–767. [CrossRef]
34. Babilas, D.; Kowalik-Klimczak, A.; Dydo, P. Study on the Effectiveness of Simultaneous Recovery and Concentration of 1-Ethyl-3-methylimidazolium Chloride Ionic Liquid by Electrodialysis with Heterogeneous Ion-Exchange Membranes. *Int. J. Mol. Sci.* **2021**, *22*, 13014. [CrossRef] [PubMed]
35. Liang, X.; Wang, J.; Bao, H.; Liu, H. Accurately-controlled recovery and regeneration of protic ionic liquid after Ionosolv pretreatment via bipolar membrane electrodialysis with ultrafiltration. *Bioresour. Technol.* **2020**, *318*, 124255. [CrossRef] [PubMed]
36. Cowan, D.A.; Brown, J.H. Effect of Turbulence on Limiting Current in Electrodialysis Cells. *Ind. Eng. Chem.* **1959**, *51*, 1445–1448. [CrossRef]
37. Lee, H.-J.; Strathmann, H.; Moon, S.-H. Determination of the limiting current density in electrodialysis desalination as an empirical function of linear velocity. *Desalination* **2006**, *190*, 43–50. [CrossRef]
38. Hyder, A.G.; Morales, B.A.; Cappelle, M.A.; Percival, S.J.; Small, L.J.; Spoerke, E.D.; Rempe, S.B.; Walker, W.S. Evaluation of Electrodialysis Desalination Performance of Novel Bioinspired and Conventional Ion Exchange Membranes with Sodium Chloride Feed Solutions. *Membranes* **2021**, *11*, 217. [CrossRef]
39. Bai, L.; Wang, X.L.; Nie, Y.; Dong, H.F.; Zhang, X.P.; Zhang, S.J. Study on the recovery of ionic liquids from dilute effluent by electrodialysis method and the fouling of cation-exchange membrane. *Sci. China Ser. B Chem.* **2013**, *56*, 1811–1816. [CrossRef]
40. Nady, N.; Franssen, M.C.R.; Zuilhof, H.; Eldin, M.S.M.; Boom, R.; Schroën, K. Modification methods for poly(arylsulfone) membranes: A mini-review focusing on surface modification. *Desalination* **2011**, *275*, 1–9. [CrossRef]
41. Babilas, D.; Kowalik-Klimczak, A.; Dydo, P. The effectiveness of imidazolium ionic liquid concentration by electrodialysis with heterogeneous and homogeneous ion-exchange membranes. *Desalin. Water Treat.* **2021**, *241*, 350–358. [CrossRef]

*International Journal of*
*Molecular Sciences*

Article

# Water-Tree Characteristics and Its Mechanical Mechanism of Crosslinked Polyethylene Grafted with Polar-Group Molecules

Xiao-Xia Zheng [1], You-Cheng Pan [2] and Wei-Feng Sun [3,*]

[1] College of Computer Science and Technology, Heilongjiang Institute of Technology, Harbin 150050, China
[2] Key Laboratory of Cold Region Urban and Rural Human Settlement Environment Science and Technology, Ministry of Industry and Information Technology, School of Architecture, Harbin Institute of Technology, Harbin 150006, China
[3] School of Electrical and Electronic Engineering, Nanyang Technological University, Singapore 639798, Singapore
* Correspondence: weifeng.sun@ntu.edu.sg

**Citation:** Zheng, X.-X.; Pan, Y.-C.; Sun, W.-F. Water-Tree Characteristics and Its Mechanical Mechanism of Crosslinked Polyethylene Grafted with Polar-Group Molecules. *Int. J. Mol. Sci.* **2022**, *23*, 9450. https://doi.org/10.3390/ijms23169450

Academic Editor: Dippong Thomas

Received: 28 July 2022
Accepted: 19 August 2022
Published: 21 August 2022

**Publisher's Note:** MDPI stays neutral with regard to jurisdictional claims in published maps and institutional affiliations.

**Copyright:** © 2022 by the authors. Licensee MDPI, Basel, Switzerland. This article is an open access article distributed under the terms and conditions of the Creative Commons Attribution (CC BY) license (https://creativecommons.org/licenses/by/4.0/).

**Abstract:** In order to restrain electric-stress impacts of water micro-droplets in insulation defects under alternating current (AC) electric fields in crosslinked polyethylene (XLPE) material, the present study represents chemical graft modifications of introducing chloroacetic acid allyl ester (CAAE) and maleic anhydride (MAH) individually as two specific polar-group molecules into XLPE material with peroxide melting approach. The accelerated water-tree aging experiments are implemented by means of a water-blade electrode to measure the improved water resistance and the affording mechanism of the graft-modified XLPE material in reference to benchmark XLPE. Melting–crystallization process, dynamic viscoelasticity and stress-strain characteristics are tested utilizing differential scanning calorimeter (DSC), dynamic thermomechanical analyzer (DMA) and electronic tension machine, respectively. Water-tree morphology is observed for various aging times to evaluate dimension characteristics in water-tree developing processes. Monte Carlo molecular simulations are performed to calculate free-energy, thermodynamic phase diagram, interaction parameter and mixing energy of binary mixing systems consisting of CAAE or MAH and water molecules to evaluate their thermodynamic miscibility. Water-tree experiments indicate that water-tree resistance to XLPE can be significantly improved by grafting CAAE or MAH, as indicated by reducing the characteristic length of water-trees from 120 to 80 μm. Heterogeneous nucleation centers of polyethylene crystallization are rendered by the grafted polar-group molecules to ameliorate crystalline microstructures, as manifested by crystallinity increment from 33.5 to 36.2, which favors improving water-tree resistance and mechanical performances. The highly hydrophilic nature of CAAE can evidently inhibit water molecules from aggregating into water micro-droplets in amorphous regions between crystal lamellae, thus acquiring a significant promotion in water-tree resistance of CAAE-modified XLPE. In contrast, the grafted MAH molecules can enhance van der Waals forces between polyethylene molecular chains in amorphous regions much greater than the grafted CAAE and simultaneously act as more efficient crystallization nucleation centers to ameliorate crystalline microstructures of XLPE, resulting in a greater improvement (relaxation peak magnitude increases by >10%) of mechanical toughness in amorphous phase, which primarily accounts for water-tree resistance promotion.

**Keywords:** crosslinked polyethylene; polar-group molecule; peroxide graft; water-tree resistance

## 1. Introduction

Water-tree aging in polymeric insulation materials has attracted considerable attention due to the inevitable moisture and water-bath environments that cause insulation failures under long-term mechanical fatigues. Nevertheless, there is, so far, no comprehensive explanation and clarified mechanism of water-tree initiation and propagation. Insulation defects and internal electric field distortion in crosslinked polyethylene (XLPE) material under thermal-water environments are dominant triggers leading to water-tree aging of electric insulation [1,2]. The electric field strength at water-tree terminals gradually increases with aging time, resulting in serious local discharge, which forms electric-trees and finally causes dielectric breakdown of XLPE main insulation in power cables. Alternating current (AC) electric field is more likely to render water-trees than DC electric field. The frequency of AC electric fields is one important factor accounting for water-trees, and the electrical, chemical and mechanical characteristics are comprehensively related to water-tree resistant performance of polymeric insulation materials. At present, mechanical damage and electrochemical oxidation are the two most recognized fundamental mechanisms for water-tree initiation and propagation [3].

Electromechanical stress theory is the focus of mechanical damage theory, indicating that the macroscopic molecular-chains of XLPE material are not neatly arranged and thus give rise to substantial holes and frail areas where periodic elastic deformations (local motions of polymer molecular-chains) frequently occur, causing mechanical fatigue under Maxwell's stress derived from AC electric field [4]. Moreover, XLPE molecular-chains will be oriented under the electric field greater than 30 MV/m at water-tree terminal [5]. Meanwhile, the orientation behavior and thermal motions of polymer molecular-chains are two contrary processes in opposite trends, in which the thermal motions of polymers intensify with increasing temperature in disfavor of orientation behaviors. In semi-crystalline polymers, such as XLPE, there are large numbers of molecular branches in amorphous regions, which are in irregular arrangements and will be oriented under AC electric field. The molecular-chains perpendicular to the orientation direction subjected to Maxwell stress is more prone to break under electric-stress fatigue than the unoriented molecular-chains, thus forming micro-cracks along orientation direction in amorphous regions, which accounts for the higher water-tree propagation speed in the electric field direction than the other directions.

Due to the remarkable discrepancy in dielectric permittivity, the water molecules will undergo electrostrictive actions under electric field, rendering local pressures at insulation defects in polyethylene amorphous regions [6]. Accordingly, the amorphous molecular-chains of XLPE suffer continuous impacts of water micro-droplets filled in micro-cracks under AC electric-stress to be further broken, leading to the enlargement of micro-cracks full of water micro-droplets and forming water-fill channels, which will randomly extend along electric field [7]. Compared to amorphous regions, the polyethylene molecular-chains assemble into condensed periodic lattices with much higher stability in crystalline regions, so water-trees are only initiated and propagate in amorphous regions of XLPE material [8]. Water-trees consist of tree-shaped channels of water-filling defects produced inside XLPE insulation under electric field in water-bath environments, which is classified into bow-tie and tube types according to their macroscopic shape [9]. The condition of producing water-trees resides in the micro-droplets forming at structural defects, which should be inhibited for alleviating water-tree aging [10,11]. Three-dimensional crosslinking networks in XLPE amorphous regions can restrict the impacts of micro-droplets on amorphous molecular-chains from forming water-trees when local deformation occurs along electric field direction, which essentially accounts for the high water-tree resistance of XLPE material [12,13]. It has been reported that the larger size and fewer grains of XLPE spherulites conduce to a higher speed of water-tree propagation [14]. Employing elastomer (SEBS and EVA), inorganic nanofiller (nano MgO) or auxiliary crosslinker (HAV2) for XLPE modifications can ameliorate crystallization morphology, increase crystallinity or crosslink-degree so as to improve water-tree resistance and electrical insulation performances of XLPE material [8,15–17].

By grafting polar-group molecules, the hydrophilicity of XLPE molecular-chains can be significantly improved to restrain water molecules in structural defects from aggregating into micro-droplets but resolve them into water membranes, which will inhibit growth of water-trees. Both chloroacetic acid allyl ester (CAAE) and maleic anhydride (MAH) molecules containing multiple polar groups as well as vinyl group, which should be competent for grafting on polyethylene molecular-chains to fulfill molecular-level modifications on XLPE material, are expected to acquire a substantial improvement in water-tree resistance. In the present study, by the melting blend method and thermal chemistry, we make chemical graft modifications on XLPE material with CAAE and MAH, focusing on improving water-tree resistant performance and trying to reveal the correlated modification mechanism, which is expected to provide strategic technical support for developing water-tree resistant XLPE materials for power cables.

## 2. Results and Discussion

### 2.1. Infrared Spectroscopic Analysis

The successful modification of chemical grafting on XLPE can be verified by FTIR spectroscopy, which is characterized by stretching vibration peaks at 907 and 1735 cm$^{-1}$ of vinyl (C=C) and ester (C=O) groups on CAAE and at 912 cm$^{-1}$ of MAH molecules [18,19], as shown in Figure 1. In comparison to both the raw material blends before and after vacuum hot-degassing treatment, the CAAE-grafted XLPE (XLPE-g-CAAE) with 1.0 wt% grafting content gives rise to the characteristic peak of C=O groups. The infrared absorption peak at 907 cm$^{-1}$ of C=C groups in CAAE molecules arises in the mixtures of LDPE, DCP and CAAE without crosslinking and grafting reactions. By contrast, the characteristic peak of C=C disappears in XLPE-g-CAAE, demonstrating that CAAE molecules have been successfully grafted onto XLPE molecular-chains. Moreover, the characteristic absorption peak at 912 cm$^{-1}$ of unsaturated chemical bonds in MAH molecules disappears in MAH-grafted XLPE (XLPE-g-MAH) with 1.0 wt% grafting content, giving rise to a new peak at 1792 cm$^{-1}$ of carbonyl groups after grafting reaction, indicating the successful graft of MAH molecules to XLPE. These FTIR results are also identically presented for the other modified XLPE materials with grafting contents of 0.5 and 1.5 wt%. Therefore, it has been verified by FTIR spectroscopy that chemical modifications of grafting CAAE and MAH to XLPE have been fulfilled.

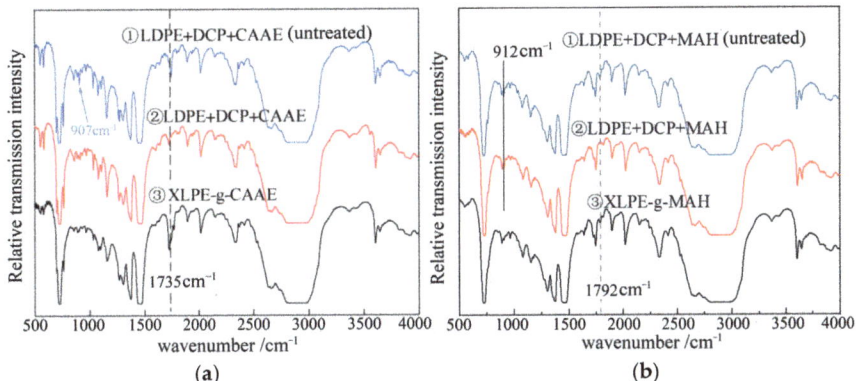

**Figure 1.** Fourier-transform infrared spectra of uncrosslinked raw material blends before and after vacuum hot-degassing (① and ②) and the modified XLPE (③): (**a**) grafting 1.0 wt% CAAE and (**b**) grafting 1.0 wt% MAH.

### 2.2. Water-Tree Morphology and Growth Length

Taking the ungrafted XLPE material as a benchmark, the water-tree morphologies and growth lengths of XLPE-g-CAAE and XLPE-g-MAH with grafted contents of 0.5, 1.0 and 1.5 wt% (XLPE-g-wt%CAAE and XLPE-g-wt%MAH) are observed and evaluated by optical microscopy, as illustrated by photo images of water-trees aging for 192 h and by water-tree length versus aging time profiles in Figure 2. Water-tree resistant performance is characterized by water-tree length at the same aging time, which means the smaller water-tree length represents the higher water-tree resistance. As shown by curve profiles in the top-right two panels of Figure 2, the water-tree lengths of all the tested samples increase with aging time. When the CAAE grafting content reaches 1.0 wt%, the water-tree resistance is improved most significantly, as manifested by water-tree lengths for aging longer than 48 h. In addition, XLPE-g-1.5wt%CAAE are more severe for water-tree aging than XLPE-g-0.5wt%CAAE.

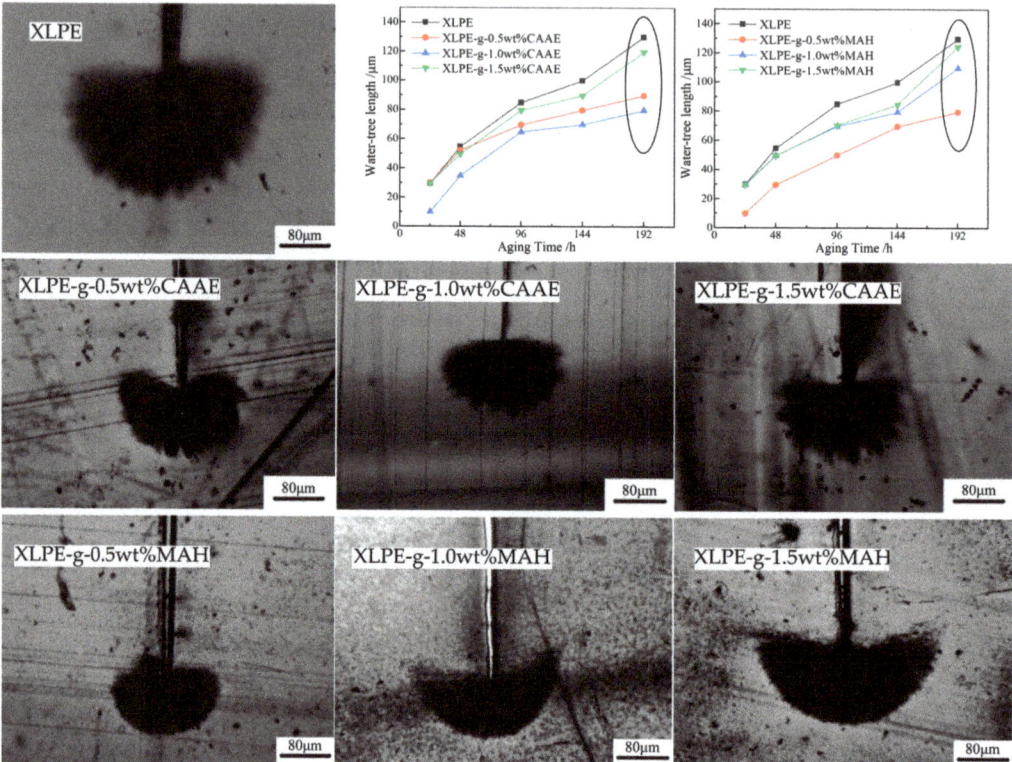

**Figure 2.** Water-tree morphology observed with optical microscopy for XLPE modified by grafting CAAE or MAH (image photos), and water-tree length versus aging time for XLPE-g-CAAE and XLPE-g-MAH compared with benchmark XLPE (top-right curve profiles).

Polar groups on the grafted CAAE and MAH molecules evidently promote the hydrophilicity of XLPE so that the water molecules penetrating into XLPE amorphous regions are mostly adsorbed near these polar groups and disfavor aggregating into micro-droplets at structural defects, which alleviates the impacts of electrostrictive water micro-droplets on polyethylene molecular-chains, as manifested by the appreciable inhibition of water-tree propagation. Meanwhile, the grafted molecules increase molecular-chain density in XLPE amorphous regions, which also accounts for improving water-tree resistance. Further, the excessively high grafting content of CAAE leads to incompatible interfaces in XLPE, which will produce structural defects that initiate water-tree and benefit water-tree propagation, thereby reducing water-tree resistance instead. In contrast, when the grafting content of MAH exceeds 0.5 wt%, the inhibition of water-tree aging fades away, indicating the different underlying mechanisms of grafting CAAE and MAH for restraining water-tree aging that will be further revealed in the following sections.

## 2.3. Melting–Crystallization Characteristics

Melting–crystallization characteristics of XLPE benchmark, XLPE-g-1.0wt%CAAE and XLPE-g-1.0wt%MAH (as the paradigm of graft-modified XLPE) are represented by endothermic and exothermic heat-flows of DSC temperature spectra in reverse processes of heated-melting and cooled-crystallization to evaluate semi-crystalline crystallinity by comparing melting and crystallizing peaks, as shown in Figure 3 and Table 1. In contrast to benchmark XLPE, the modified XLPE grafted with CAAE or MAH acquire substantial elevations in both crystallizing and melting temperatures and in crystallinity, implying that the polar groups on the grafted molecules lead to higher intermolecular forces between polyethylene molecular-chains. Meanwhile, the grafted molecules undergo as the heterogeneous nucleation centers for polyethylene molecular-chain crystallization, which accounts for the smaller size and higher density of spherulites in XLPE microscopic crystallization structure, resulting in the reductions of the spacing between lamellae and the volume of amorphous regions.

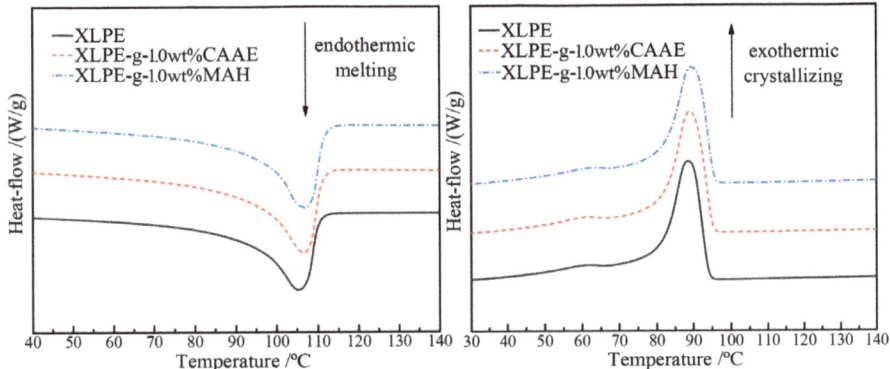

**Figure 3.** DSC temperature spectra of XLPE, XLPE-g-1.0wt%CAAE and XLPE-g-1.0wt%MAH in melting (**left panel**) and crystallizing (**right panel**) processes.

**Table 1.** Melting–crystallization characteristics and the calculated crystallinities.

| Material | Melting Peak $T_m/°C$ | Crystallizing Peak $T_c/°C$ | Melting Enthalpy $\Delta H_m/(J/g)$ | Crystallinity $X_c/\%$ |
|---|---|---|---|---|
| XLPE | 88.76 | 105.13 | 98.35 | 33.5 |
| XLPE-g-1.0wt%CAAE | 89.37 | 106.62 | 105.11 | 35.8 |
| XLPE-g-1.0wt%MAH | 89.61 | 106.98 | 106.29 | 36.2 |

The initial increase slope in crystallizing peak indicates crystallization nucleation rate, whilst the difference between initial temperature and peak temperature characterizes the inverse rate of crystal growth, which are, respectively, higher and lower for the graft-modified XLPE than that for XLPE benchmark. It is thus verified from DSC tests that both the nucleation rate and crystal growth rate of the graft-modified XLPE are higher than those of XLPE benchmark. It is hereby verified that the polar-group molecules (especially for MAH) grafted on polyethylene molecule-chains are capable of enhancing van der Waals interactions between polyethylene molecule-chains and acting as heterogeneous nucleation centers to expedite crystallization nucleation and crystal growth, resulting in higher densities of lamellae and spherulites, as manifested by the higher crystallization enthalpy and crystallinity, respectively.

Optical microscopic images of semi-crystalline morphologies with spherullites and amorphous regions in XLPE, XLPE-g-1.0wt%CAAE and XLPE-g-1.0wt%MAH are shown in Figure 4. Compared with XLPE benchmark, the grafted XLPE materials represent a higher density of spherulites in smaller diameters, resulting in a considerably smaller volume ratio of amorphous phase, for which XLPE-g-MAH is more evident than XLPE-g-CAAE. Based on crystallization kinetics, the polyethylene lamellae are formed by folding regular molecular-chains, while a spherulite is formed by a large number of lamellae arranged radically from a crystallizing nucleus. Therefore, the grafted CAAE or MAH molecules could act as crystallization nuclei to initiate the crystallization of polyethylene lamellae, which will finally develop into spherulites, resulting in a higher crystallinity as manifested by a higher density of smaller-sized spherulites.

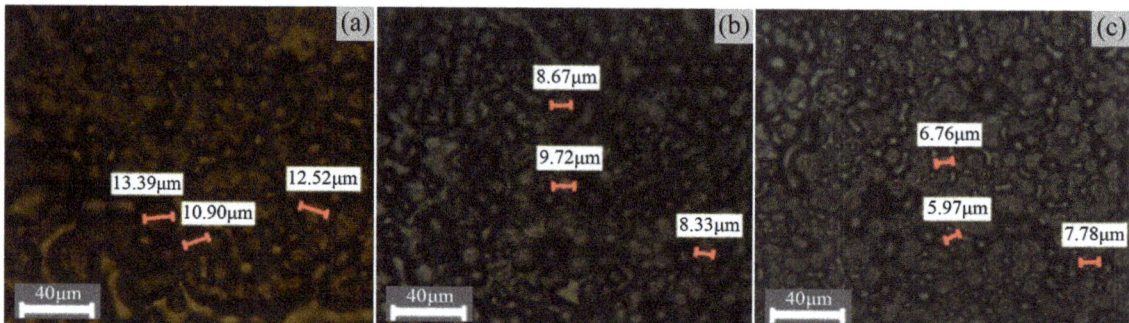

**Figure 4.** Spherulitic semi-crystalline morphologies of (**a**) XLPE, (**b**) XLPE-g-1.0wt%CAAE and (**c**) XLPE-g-1.0wt%MAH observed by PLM.

### 2.4. Viscoelasticity and Stress-Strain Characteristics

In dynamic relaxation temperature spectra of DMA, the storage modulus $E'$ indicates the mechanical stiffness of polymer materials, while loss modulus $E''$ and loss factor $\tan\theta$ characterize the mechanical toughness of polymer materials. DMA temperature spectra and static tensile characteristics of 1.0wt%-graft XLPE in reference to XLPE are shown in Figure 5a–c, with the peak magnitudes and water-tree lengths (aging for 192 h) listed in Table 2. Mechanical toughness of amorphous regions can be measured from the intensities of $E''$ and $\tan\theta$ peaks at glass-transition temperature ($\beta$ relaxation peak at −25 °C) [20,21], which are dominantly derived from the relaxations of amorphous molecular-chains connecting to lamella surfaces. As the density of lamellae in XLPE increases, the $\beta$ relaxation peak shifts toward a lower temperature, and the peak intensity of $E''$ increases, implying that the molecular-chain relaxations on lamella surfaces are intensified. The $\beta$ peak intensities of $E''$ and $\tan\theta$ for XLPE-g-1.0wt%CAAE and XLPE-g-1.0wt%MAH are distinctly higher than that for XLPE benchmark, in which XLPE-g-1.0wt%MAH shows the highest values. Therefore, the mechanical toughness in amorphous regions between the lamellae of XLPE material is enhanced by grafting CAAE and MAH molecules, which improves the resistance to the electric-stress damage from micro-droplets on polyethylene molecular-chains, as manifested consistently by the higher crystallinities.

**Table 2.** DMA $\beta$ relaxation peak magnitude (loss modulus $E''$ and factor $\tan\theta$), crystallinity $X_c$ and water-tree dimension (characteristic length $L_c$).

| Material | $E''$/MPa | $\tan\theta$ | $X_c$/% | $L_c$/μm |
|---|---|---|---|---|
| XLPE | 81 | 0.079 | 33.5 | 129.8 |
| XLPE-g-1.0wt%CAAE | 87 | 0.085 | 35.8 | 80.1 |
| XLPE-g-1.0wt%MAH | 90 | 0.090 | 36.2 | 79.6 |

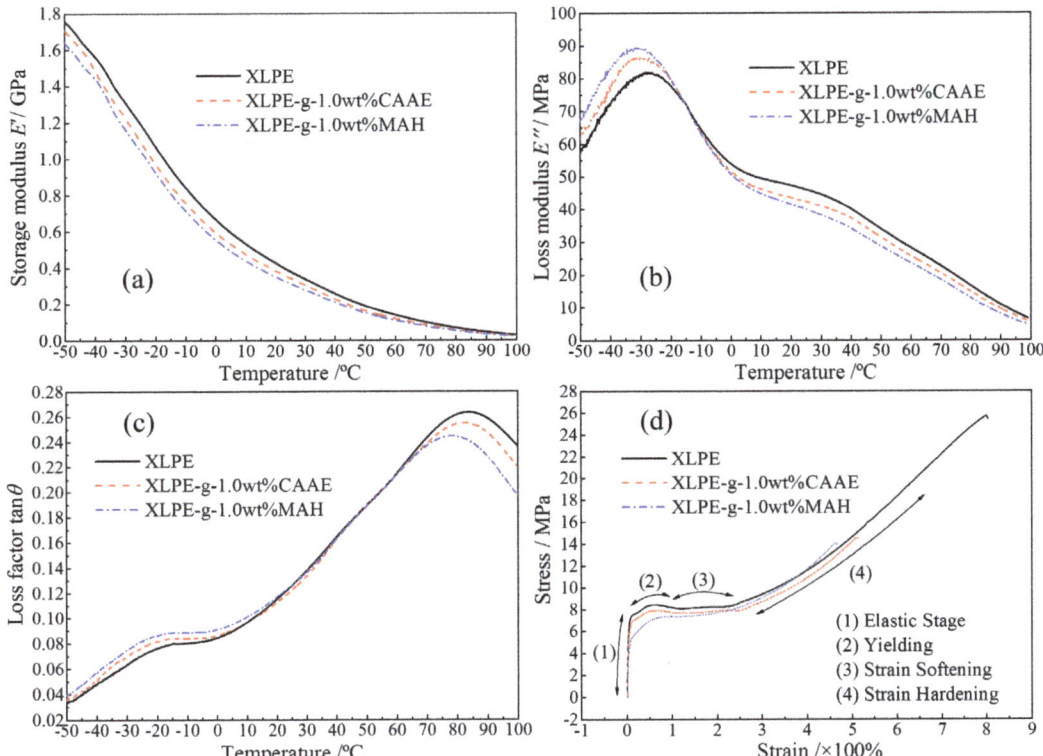

**Figure 5.** DMA temperature spectra of (**a**) storage modulus, (**b**) loss modulus, (**c**) loss factor and (**d**) stress-strain characteristics for XLPE-g-1.0wt%CAAE and XLPE-g-1.0wt%MAH in comparison to XLPE benchmark.

The tan$\theta$ peaks appearing at the higher temperature of 80 °C ($\alpha$ relaxation peak) come from the rotation and slip of lamellae and the relaxation of fold molecular-chains on lamella surfaces [22]. The graft of polar-group molecules ameliorates crystalline characteristics by reducing the volume ratio of XLPE amorphous phase and the size of spherulites, resulting in the effective restriction on $\alpha$ relaxation, which accounts for the lowest $\alpha$ peak of XLPE-g-1.0wt%MAH. Although XLPE-g-1.0wt%MAH shows the highest crystallinity and mechanical toughness, its water-tree resistant performance is similar to XLPE-g-1.0wt%CAAE (as shown in Figure 2), implying a discrepant mechanism for improving water-tree resistance by grafting CAAE.

Stress-strain characteristics include four stages during dynamic mechanical tensile process: elastic, yielding, strain-softening and strain hardening stages, as shown in Figure 5d. Because the grafting reaction causes breakages in parts of polyethylene molecular-chains, the fracture stress and elongation of the graft-modified XLPE materials are distinctly lower than that of benchmark XLPE [23,24]. After grafting modifications, both elastic modulus and yield strength decrease, whilst strain softening and elastic (cold stretching) processes are shortened, and even XLPE-g-1.0wt%MAH exhibits no perceptible strain softening stage. Amorphous regions between lamellae in XLPE material are diminished by chemically introducing polar-group molecules, which restricts the slips between lamellae, as indicated by the shortened strain softening and cold stretching processes.

## 2.5. Miscibility of Grafted Molecules with Water

Employing the modified Flory–Huggins model [25] for the thermodynamics of binary molecular-mixture systems, the free energy, phase diagram and mixing energy of PE/$H_2O$, CAAE/$H_2O$ and MAH/$H_2O$ binary molecular mixtures are calculated to analyze the hydrophilicity of polyethylene molecular-chains and polar-group molecules, as shown in Figure 6. Although the solubility of $H_2O$ in polyethylene (PE) is very low and the critical point temperature reaches 1374K, the metastable phase region located between Spinodal and Binodal boundaries remains largely in phase diagram, implying that ~5 mol% of $H_2O$ molecules will penetrate into polyethylene amorphous regions to form a metastable mixture above 600 K.

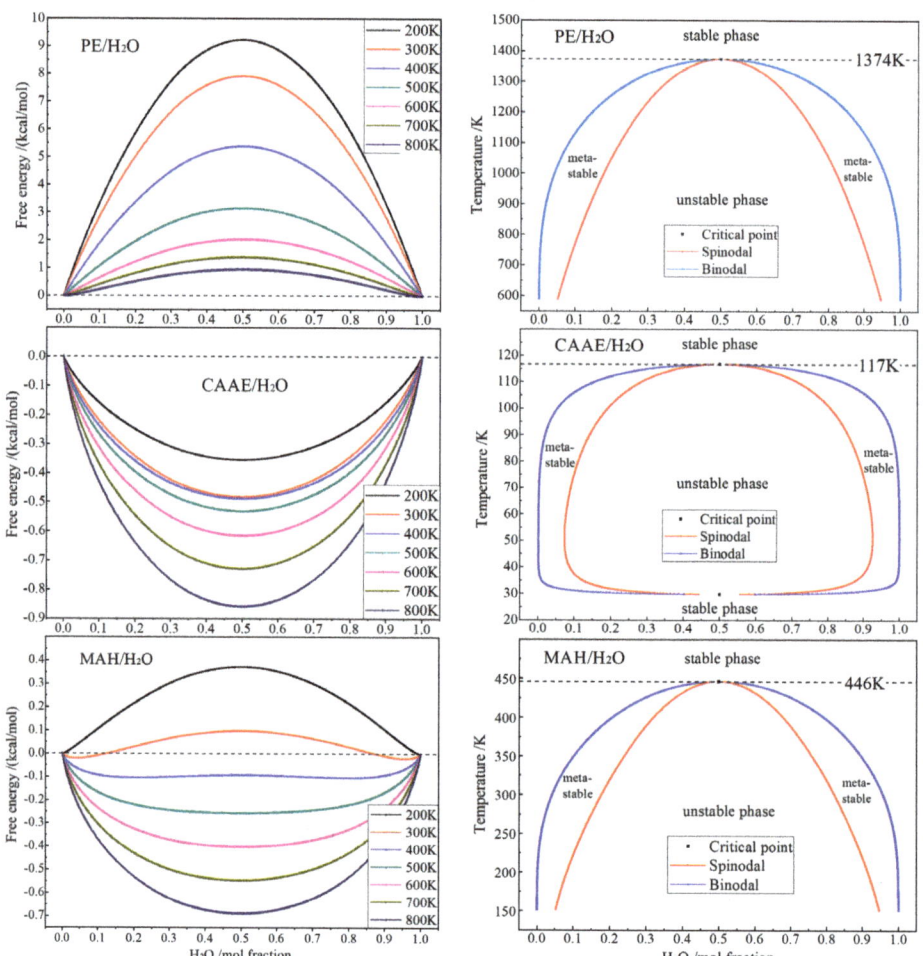

**Figure 6.** Free energy (**left panels**) at 200–800 K and phase diagram (**right panels**) of PE/$H_2O$, CAAE/$H_2O$ and MAH/$H_2O$ binary molecular mixtures.

Due to the existence of polar groups, both CAAE and MAH molecules represent much higher hydrophilic features than PE molecules, with the critical point temperatures approaching 117 and 446 K, respectively. In contrast, the molecular compatibility of CAAE with $H_2O$ is significantly higher than that of MAH, and the critical point temperature is lower than room temperature, implying that CAAE and water can completely dissolve with each other in any proportion. Therefore, it is suggested that water is able to exist as molecules around MAH and CAAE in graft-modified materials, but water molecules cannot disperse uniformly and will aggregate into water droplets to form phase separation around polyethylene molecular chains without polar groups.

A small or negative interaction parameter ($\chi$) indicates that the two molecular components interact strongly at a specific temperature to form a binary mixing system under thermodynamic equilibrium. While for a large positive $\chi$, the two different molecules prefer to cluster with their peers to form a separate two-phase system. Interaction parameters and mixing energies ($E_{mix}$) of PE/H$_2$O, CAAE/H$_2$O and MAH/H$_2$O binary mixing systems as a function of temperature are shown in Figure 7. Binary mixing systems of CAAE/H$_2$O and MAH/H$_2$O present a higher magnitude of $\chi$ and $E_{mix}$ than PE/H$_2$O. The $\chi$ and $E_{mix}$ of CAAE/H$_2$O are negative at temperatures below 350 K, while MAH/H$_2$O represents two positive parameters in temperature range of 200–800 K, approaching maximum at 300 K, which is remarkably higher than that of CAAE/H$_2$O. Since the polar-group molecules in graft-modified XLPE materials are all chemically bonded to polyethylene molecular-chains, the water molecules infiltrated into XLPE-g-CAAE prefer to disperse and dissolve around polar groups rather than assembling into micro-droplets.

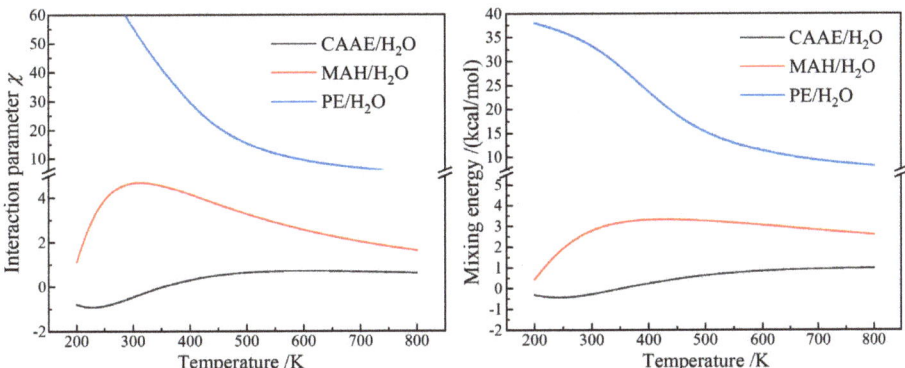

**Figure 7.** Interaction parameter (**left panel**) and mixing energy (**right panel**) of PE/H$_2$O, CAAE/H$_2$O and MAH/H$_2$O binary molecular mixtures.

### 2.6. Mechanism of Water-Tree Resistance

According to the melting–crystallization characteristics and DMA temperature spectra with the derived water-tree resistant mechanisms introduced by grafting polar-group molecules, it is reasonable to predict that besides improving water resistance by ameliorating crystallization structure of XLPE, the imported hydrophilia of polar groups can restrain water molecules from aggregating into micro-droplets, resulting in improved water-tree resistance. Macroscopic molecular-chains of XLPE pass through several crystallization regions and tangle with each other. Hence, polyethylene molecular-chains in amorphous phase, which determine slides between lamellae, undergo deformations and relaxation motions under electric-stress impacts of water micro-droplets. When semi-crystalline XLPE materials bear a mechanical stretching process in strain hardening stage, the molecular-chains connecting lamellae in amorphous regions mainly bear the applied mechanical forces, which means the stronger connecting molecular-chains or the smaller amorphous regions require a higher external force to produce macroscopic deformations.

Although XLPE-g-MAH has a greater strain hardening strength and higher/lower $\beta/\alpha$ relaxation peaks (as shown in Figure 5), the hydrophilicity of CAAE far exceeds MAH, so XLPE-g-CAAE possesses a nearly identical water-tree resistance as XLPE-g-MAH. The grafted hydrophilic CAAE can effectively disperse water micro-droplets gathered in amorphous regions between lamellae, thus alleviating electrical-mechanical damage of water micro-droplets on defect areas under AC electric field, as shown in Figure 8, in which $R'$ and $R''$ respectively denote the grafted MAH and CAAE on polyethylene molecular-chains. In contrast to XLPE-g-MAH, the water micro-droplets in amorphous regions of XLPE-g-CAAE are much smaller, which will not further aggregate in insulation defects to form water-filling holes and thus are difficult to render water-tree channels.

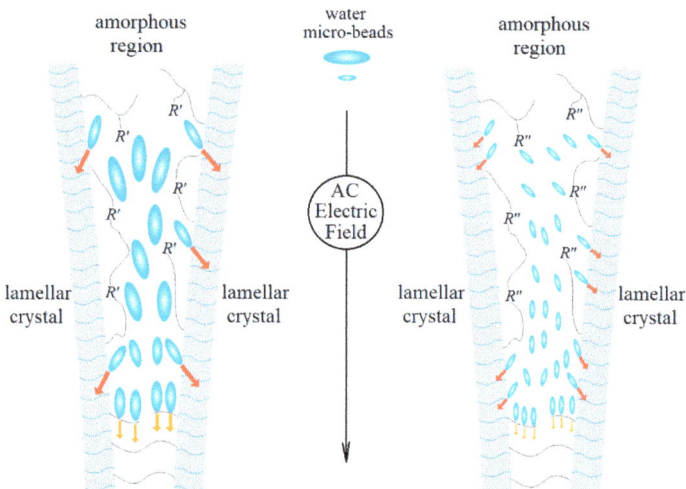

**Figure 8.** Schematic electromechanical mechanism of water micro-droplets breaking through the amorphous regions in XLPE-g-CAAE (**left panel**) and XLPE-g-MAH (**right panel**).

## 3. Methods and Materials

### 3.1. Material Preparation

Raw materials for the melting blend process are comprised of dicumyl peroxide (DCP, Nobel Company Ltd., Aksu, China) as the crosslinking and grafting initiator, low-density polyethylene (LDPE, LD200GH, Sinopec Company Ltd., Beijing, China) as the basis material, and chloroacetic acid allyl ester or maleic anhydride (CAAE or MAH, Ruierfeng Chemical Co. Ltd. Guangzhou, China) as the grafting modification agents. In melting blend process of preparing initial mixture materials before crosslinking and grafting reactions, the pristine LDPE material is heated for melting at 110 °C temperature for 3 min with a rotating speed of 40 rpm in torque rheometer (RM200C, Hapro Co. Ltd., Harbin, China), and then 2 wt% DCP and 0.3 wt% 1010 antioxidant together with 0, 0.5 and 1.0 wt% CAAE or MAH are added into torque rheometer blending for 17 min and cooled down to room temperature, thus obtaining the uniformly mixed material. For crosslinking/grafting chemical reactions, the prepared blend is first heated to 120 °C for melting in plate vulcanizer, and then is further heated to 175 °C at a rate of 5 °C/min with the pressure being raised to 15 MPa by a rate of 1 MPa/min. After 35 min, the crosslinked and grafted polyethylene material is cooled down to room temperature and then compressed into molding films. Eventually, the film samples are hot-degassed under short-circuit in vacuum drying chamber at 80 °C for 48 h so as to clear off the residual reactants and reaction by-products and to relax mechanical stresses in samples.

### 3.2. Water-Tree Experiment

Water-tree experiments are fulfilled by accelerating electric aging process with a water blade electrode being continuously applied by AC high voltage of 4 kV at high frequency of 3 kHz for 24–192 h, as schematically shown in Figure 9. The blade electrode in 0.03 mm thickness and with a blade edge in 0.01 mm curvature radius is vertically inserted into the 4 mm thick film sample whilst keeping 2 mm away from the sample's bottom surface, which causes knife-like defects at blade edge, as shown in Figure 9b. Sodium chloride solution of 1.8 mol/L is used as water medium for water-tree aging experiments. Before water-tree experiment, the entire equipment is degassed in vacuum for 60 min to completely remove residual air in blade defects. After water-tree growth has been finished, the cuboid film sample is sliced (into ~100 μm thickness) through the defects at blade edge along the direction perpendicular to both the blade plane and sample surface, as implemented by the manual rotary microtome (Leica RM2235, Chuangxun Medical Equipment Co., Ltd., Shanghai, China). The sliced samples are then immersed in methylene blue solution persisting at a constant temperature of 90 °C for 4 h, making water-trees legible.

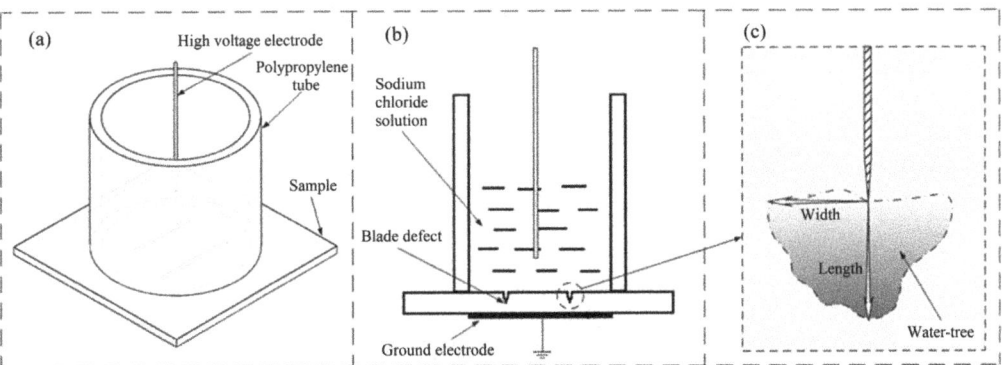

**Figure 9.** Schematics of accelerated water-tree aging experiments with a water blade electrode: (**a**) overview, (**b**) side sectional view, (**c**) generating water-tree.

### 3.3. Characterization and Measurement

Infrared absorption spectra are tested for the raw blends and the crosslinked XLPE materials to determine whether or not the grafting process is successful by comparing the characteristic peaks of specific chemical groups before and after crosslinking/grafting reaction process. The film samples in 0.3 mm thickness are tested by Fourier-transform infrared (FTIR) spectrometer (FT/IR-6100, Jiasco Trading Co., Ltd., Shenyang, China) in wavelength range of 500–4000 cm$^{-1}$ with a resolution of 2 cm$^{-1}$.

Differential scanning calorimetry (DSC) is employed to test the heat flow from and out of samples when being gradually heated/cooled at a rate of 5 °C/min in nitrogen atmosphere, as implemented in differential scanning calorimeter (DSC-3, METTLER TOLEDO, Zurich, Switzerland). Polyethylene crystallinity is calculated from enthalpy change in melting process of DSC test, according to $X_c = \Delta H_m / \Delta H_{100}$, where $\Delta H_{100} = 293.6$ J/g and $\Delta H_m$ denote the melting enthalpies of the 100% crystallized material and the tested semi-crystalline samples, respectively.

Water-tree morphology is observed to evaluate water-tree length with an optical microscope (SteREO DiscoveryV20, Carl Zeiss AG, Berlin, Germany). Semi-crystalline morphologies of the grafted and benchmark XLPE materials in 120 μm thick slice samples are observed by polarizing fluorescence microscope (PLM, DM2500P, Leica Co., Berlin, Germany).

Dynamic thermomechanical analysis (DMA) is performed on $15 \times 6 \times 1$ mm$^3$ cuboid film samples to evaluate viscoelastic characteristics, as described by energy-storage modulus $E'$, loss modulus $E''$ and loss factor $\tan\theta = E'/E''$, in the heating process by a rate of 3 °C/min from −50 to 100 °C under nitrogen atmosphere, as implemented in dynamic thermomechanical analyzer (Q800DMA, TA apparatus Co. Ltd., DE, USA). DMA tests are carried out under tensile stress by specifying target frequency/amplitude and static/dynamic forces of 1 Hz/15 μm and 0.375 N/0.3 N, respectively. Complying with GB/T 1040.2-2006 standard, the stress-strain characteristics of "5A" dumbbell shaped samples with a mark distance of 20 mm in 4 mm width and 2 mm thickness are measured by an elongation speed of 5 mm/min.

*3.4. Miscibility Computation*

Mixing energy, free energy and phase diagram of water ($H_2O$) molecules mixing with CAAE or MAH molecules are calculated by Monte Carlo method combined with the modified Flory–Huggins model, using Blends program of Materials Studio 2020 package (Accelrys Inc., Materials Studio version 2020.08, San Diego, CA, USA). According to thermodynamics theory of miscibility and separation in binary phase systems as described by Flory–Huggins model [26,27], the free energy of a binary mixture system is represented as by:

$$\frac{\Delta G}{RT} = \frac{\phi_b}{n_b} \ln \phi_b + \frac{\phi_s}{n_s} \ln \phi_s + \chi \phi_b \phi_s \quad (1)$$

where $\Delta G$ denotes mixing free energy (per mole), $\phi_i$ and $n_i$ symbolize volume ratio and polymerization degree of component i respectively, $\chi$ represents interaction parameter, $T$ and $R$ indicate absolute temperature and gas constant respectively. The sum of the first two terms are always negative, which is combinatorial entropy in favor of mixed state rather than separating into pure components. The last term of free energy is derived from interaction of mixing different components. Interaction parameter $\chi$ is defined by $\chi = E_{mix}/RT$ ($E_{mix}$ denotes mixing energy), which indicate free energy difference between mixed and separated phase states, in disfavor of mixing when being positive.

## 4. Conclusions

For the first time, we propose and demonstrate how to improve water-tree aging resistance of XLPE material by grafting polar-group molecules, in which the underlying mechanism is elucidated by the hydrophilia of graft molecules, the crystalline characteristics of heterogeneous nucleation, and the mechanical properties derived from polyethylene molecular-chains in amorphous phase. By means of peroxide-initiated thermochemical grafting method, two species of molecules named by CAAE and MAH, which possess polar-groups, are successfully grafted onto XLPE molecular-chains through free radical addition reactions. Infrared spectroscopy, water-tree aging experiments, crystallization characteristics and mechanical properties are conducted to elucidate the mechanism of improving water resistance by chemically grafting CAAE or MAH molecules. Combined with Monte Carlo molecular simulations, it is verified that water-tree resistance can been significantly promoted by grafting polar-group molecules. It is consistently manifested by DSC spectra, DMA peaks and stress-strain characteristics that the grafted polar groups can enhance Van der Waals' force between polyethylene molecules and are available as heterogeneous nucleation centers for polyethylene crystallization, which lead to the increased densities of spherulites with the reduced-volume and increased-tenacity amorphous regions between lamellae, accounting for the considerable improvement in water-tree resistance. It is indicated from Monte Carlo molecular simulations that CAAE/$H_2O$ binary system possesses negative interaction parameters and mixing energies throughout a large temperature range, implying that water molecules will be dispersed without aggregating into water droplets in CAAE grafted XLPE, which alleviates electric-stress impacts on amorphous regions. Inspired by the chemical modifications of XLPE material by grafting polar-group molecules which have been recently verified for improving insulation performances, this paper suggests a feasible strategy of chemical grafts to simultaneously improve water-tree resistances and insulation performances of polyethylene insulation materials in applications of high voltage cable manufactures.

**Author Contributions:** Investigation and writing—original draft preparation, X.-X.Z.; data curation and formal analysis, Y.-C.P.; conceptualization, methodology, writing—review and editing, W.-F.S. All authors have read and agreed to the published version of the manuscript.

**Funding:** This research received no external funding.

**Institutional Review Board Statement:** Not applicable.

**Informed Consent Statement:** Not applicable.

**Data Availability Statement:** The theoretical and experimental methods and results are available from all authors.

**Conflicts of Interest:** The authors declare no conflict of interest.

## References

1. Kurihara, T.; Okamoto, T.; Hozumi, N.; Miyajima, K.; Uchida, K. Evaluation of relationship between residual charge signal and AC breakdown strength of water-tree degraded 22 to 77 kV classes XLPE cables removed from service using pulsed voltages. *IEEE Trans. Dielectr. Electr. Insul.* **2017**, *24*, 656–665. [CrossRef]
2. El-Zein, A.; Mohamed, K.; Talaat, M. Water trees in polyethylene insulated power cables: Approach to water trees initiation mechanism. *Electr. Pow. Syst. Res.* **2020**, *180*, 106158. [CrossRef]
3. Lee, J.B.; Jung, C.K. Technical review on parallel ground continuity conductor of underground cable systems. *J. Int. Coun. Electr. Eng.* **2012**, *2*, 250–256. [CrossRef]
4. Gao, L.Y.; Guo, W.Y.; Tu, D.M. Interfacial and withstand voltage of polyethylene for power cables. *IEEE Trans. Dielectr. Electr. Insul.* **2003**, *10*, 233–238. [CrossRef]
5. Dang, Z.M.; Kang, J.; Tu, D.M. Relationship between the space charge and the property of water absorbing in new cable material of retarding water treeing. *Trans. Chin. Electrotechn. Soc.* **2002**, *17*, 68–71.
6. Wang, Z.; Marcolongo, P.; Lemberg, J.A.; Panganiban, B.; Evans, J.W.; Ritchie, R.O.; Wright, P.K. Mechanical fatigue as a mechanism of water tree propagation in TR-XLPE. *IEEE Trans. Dielectr. Electr. Insul.* **2012**, *19*, 321–330. [CrossRef]
7. Li, K.; Zhou, K.; Zhu, G. Toward understanding the relationship between the microstructure and propagation behavior of water trees. *IEEE Trans. Dielectr. Electr. Insul.* **2019**, *26*, 1116–1124. [CrossRef]
8. Chen, J.Q.; Wang, X.; Sun, W.F.; Zhao, H. Water-tree resistability of UV-XLPE from hydrophilicity of auxiliary crosslinkers. *Molecules* **2020**, *25*, 4147. [CrossRef]
9. Arief, Y.Z.; Shafanizam, M.; Adzis, Z.; Makmud, M.Z.H. Degradation of polymeric power cable due to water treeing under AC and DC stress. In Proceedings of the IEEE International Conference on Power and Energy (PECon), Kota Kinabalu, Malaysia, 2–5 December 2012; pp. 950–955.
10. Teyssedre, G.; Laurent, C. Advances in high-field insulating polymeric materials over the past 50 years. *IEEE Electr. Insul. Mag.* **2018**, *29*, 26–36. [CrossRef]
11. Yang, J.S.; Huang, X.Y.; Wang, G.L.; Liu, F.; Jiang, P.K. Effects of styrene-B-(ethylene-co-butylene)-B-styrene on electrical properties and water tree resistance of cross-linked polyethylene. *High Volt. Eng.* **2010**, *36*, 946–951.
12. Wang, J.F.; Liu, Z.M.; Li, Y.X.; Wu, J.; Hang, X.F.; Zheng, X.Q. Influence of chemical cross-linking on water treeing in polyethylene. *High Volt. Eng.* **2011**, *37*, 2477–2484.
13. Ciuprina, F.; Teissedre, G.; Filippini, J.C.; Smedberg, A.; Campus, A.; Hampton, N. Chemical crosslinking of polyethylene and its effect on water tree initiation and propagation. *IEEE Trans. Dielectr. Electr. Insul.* **2010**, *17*, 709–715. [CrossRef]
14. Ma, Z.S.; Huang, X.Y.; Jiang, P.K.; Wang, G.L. Effect of silane-grafting on water tree resistance of XLPE cable insulation. *J. Appl. Polym. Sci.* **2010**, *115*, 3168–3176. [CrossRef]
15. Chen, J.Q.; Wang, X.; Sun, W.F.; Zhao, H. Improved water-tree resistances of SEBS/PP semi-crystalline composites under crystallization modifications. *Molecules* **2020**, *25*, 3669. [CrossRef] [PubMed]
16. Ma, Z.S.; Huang, X.Y.; Yang, J.S.; Jiang, P.K. Synergetic effects of silane-grafting and EVA on water tree resistance of LDPE. *Chin. J. Polym. Sci.* **2010**, *28*, 1–11. [CrossRef]
17. Zhang, Y.Q.; Wang, X.; Yu, P.L.; Sun, W.F. Water-tree resistant characteristics of crosslinker-modified-$SiO_2$/XLPE nanocomposites. *Materials* **2021**, *14*, 1398. [CrossRef]
18. Zhao, X.D.; Sun, W.F.; Zhao, H. Enhanced insulation performances of crosslinked polyethylene modified by chemically grafting chloroacetic acid allyl ester. *Polymers* **2019**, *11*, 592. [CrossRef]
19. Zhao, H.; Xi, C.; Zhao, X.-D.; Sun, W.-F. Elevated-temperature space charge characteristics and trapping mechanism of cross-linked polyethylene modified by UV-initiated grafting MAH. *Molecules* **2020**, *25*, 3973. [CrossRef]
20. Stadler, F.J. Dynamic-mechanical behavior of polyethylenes and ethene/alpha-olefin-copolymers: Part II. Alpha- and beta-relaxation. *Korean J. Chem. Eng.* **2011**, *28*, 954–963. [CrossRef]
21. Yash, P.K.; Edith, A.T.; Thomas, J.T.; Virgil, V.V.; Richard, F.A. Dynamic mechanical relaxations in polyethylene. *Macromolecules* **1985**, *18*, 1302–1309.
22. Colson, J.P. Alpha relaxation in polyethylene. *J. Appl. Phys.* **1971**, *42*, 5902–5903. [CrossRef]
23. Celina, M.; George, G.A. Characterisation and degradation studies of peroxide and silane crosslinked polyethylene. *Polym. Degrad. Stabil.* **1995**, *48*, 297–312. [CrossRef]
24. Zhang, X.; Yang, H.; Song, Y.; Zheng, Q. Influence of crosslinking on physical properties of low density polyethylene. *Chin. J. Polym. Sci.* **2012**, *30*, 837–844. [CrossRef]

25. Vetere, A. Rules for predicting vapor-liquid equilibria of amorphous polymer solutions using a modified Flory-Huggins equation. *Fluid Phase Equilib.* **1994**, *97*, 43–52. [CrossRef]
26. Bergfeldt, K.; Piculell, L.; Linse, P. Segregation and association in mixed polymer solutions from Flory-Huggins model calculations. *J. Phys. Chem.* **1996**, *100*, 3680–3687. [CrossRef]
27. Costa, G.M.N.; Dias, T.; Cardoso, M.; Guerrieri, Y.; Pessoa, F.L.P.; Vieira de Melo, S.A.B.; Embiruçu, M. Prediction of vapor–liquid and liquid–liquid equilibria for polymer systems: Comparison of activity coefficient models. *Fluid Phase Equilib.* **2008**, *267*, 140–149. [CrossRef]

## Article

# Sensitivity of the Transport of Plastic Nanoparticles to Typical Phosphates Associated with Ionic Strength and Solution pH

Xingyu Liu, Yan Liang *, Yongtao Peng, Tingting Meng, Liling Xu and Pengcheng Dong

School of Resources, Environment and Materials, Guangxi University, Nanning 530004, China
* Correspondence: liangyan@gxu.edu.cn

**Abstract:** The influence of phosphates on the transport of plastic particles in porous media is environmentally relevant due to their ubiquitous coexistence in the subsurface environment. This study investigated the transport of plastic nanoparticles (PNPs) via column experiments, paired with Derjaguin–Landau–Verwey–Overbeek calculations and numerical simulations. The trends of PNP transport vary with increasing concentrations of $NaH_2PO_4$ and $Na_2HPO_4$ due to the coupled effects of increased electrostatic repulsion, the competition for retention sites, and the compression of the double layer. Higher pH tends to increase PNP transport due to the enhanced deprotonation of surfaces. The release of retained PNPs under reduced IS and increased pH is limited because most of the PNPs were irreversibly captured in deep primary minima. The presence of physicochemical heterogeneities on solid surfaces can reduce PNP transport and increase the sensitivity of the transport to IS. Furthermore, variations in the hydrogen bonding when the two phosphates act as proton donors will result in different influences on PNP transport at the same IS. This study highlights the sensitivity of PNP transport to phosphates associated with the solution chemistries (e.g., IS and pH) and is helpful for better understanding the fate of PNPs and other colloidal contaminants in the subsurface environment.

**Keywords:** plastic nanoparticles; phosphates; solution chemistry; retention; release

**Citation:** Liu, X.; Liang, Y.; Peng, Y.; Meng, T.; Xu, L.; Dong, P. Sensitivity of the Transport of Plastic Nanoparticles to Typical Phosphates Associated with Ionic Strength and Solution pH. *Int. J. Mol. Sci.* **2022**, *23*, 9860. https://doi.org/10.3390/ijms23179860

Academic Editor: Dippong Thomas

Received: 13 July 2022
Accepted: 25 August 2022
Published: 30 August 2022

**Publisher's Note:** MDPI stays neutral with regard to jurisdictional claims in published maps and institutional affiliations.

**Copyright:** © 2022 by the authors. Licensee MDPI, Basel, Switzerland. This article is an open access article distributed under the terms and conditions of the Creative Commons Attribution (CC BY) license (https://creativecommons.org/licenses/by/4.0/).

## 1. Introduction

Plastic nanoparticles (PNPs) are commonly defined as plastic debris smaller than 1 μm in diameter across its widest dimension and distinct from the larger microplastics (1–5000 μm) and macroplastics (larger than 5000 μm) [1]. It is reported that more than 300 million tons of plastics are manufactured each year [2,3]. The sources of PNPs in the environment may come from various materials and processes related to our daily life such as synthetic fibers [4], personal care products [5–7], washing [8], and packaging [9,10]. PNPs in the environment can be primary materials or degradation products of large plastic wastes as secondary production [11]. In addition, the wide application of agricultural mulch in farms or greenhouses, irrigation with waters containing plastics, and the use of sewage sludge all potentially bring a significant amount of plastics into soils [12–16]. Previous studies reported that the average amount of plastics in soils in southwestern China was as high as 18,760 particles per kilogram [17]. It is estimated that the annual total amount of plastics in European and North American farmlands can reach 44,000–430,000 tons per year [13].

The toxicity of PNPs to the ecosystem has been studied [18,19]. The presence of PNPs may influence the physical (e.g., hydraulic and pore distribution), chemical (e.g., contaminant adsorption), and biological properties (e.g., microbial communities) of soils [3,20,21]. Previous studies confirm that PNPs in the soil can affect the transportation, reproduction, and metabolism of soil biota [22,23]. In addition, microorganisms can act as carriers that transfer PNPs from soil to plants and eventually to other organisms through food chains [24]. Studies indicated that PNPs can pass important biological barriers (e.g., the intestinal barrier, blood–air barrier, blood–brain barrier, and placental barrier) and potentially produce adverse effects on human beings [3,25]. Furthermore, PNPs can adsorb other pollutants (organic and inorganic) and facilitate their mobility in the aqueous environment or soils thus increase the risk of coupled contaminations in the environment and groundwater [26]. PNPs show colloidal properties and are less affected by gravity due to their light weight and long-term durability [27]. The transport of PNPs in the subsurface environment is expected to be highly affected by a wide range of processes, including sedimentation, aggregation, re-suspension, and entrapment [28,29]. These processes are significantly influenced by the properties of PNPs (i.e., particle size and surface properties), solution chemistries (i.e., ionic strength (IS), cation type, and pH), porous media (i.e., grain size and surface heterogeneity), flow condition, and coexisting pollutants [30–33]. However, the transport behaviors of PNPs and the mechanisms involved are still far from being fully understood.

Phosphates are ubiquitous in agricultural drainage and municipal wastewater [34,35] and may reach high levels in surface water and groundwater, e.g., ranging from 0.0035 to 0.1 mM after a long-term accumulation [36–39]. Furthermore, phosphate is also abundant in soils due to the wide application of phosphate fertilizers and sewage sludge on farms [38]. The presence of phosphates in the environment inevitably alters the composition of solution chemistry and the properties of the natural collector surface (i.e., soil grain surface), thus influencing colloid transport [40]. It has been demonstrated that phosphates can facilitate the transport of graphene oxide, $nTiO_2$, and ZnO-NPs by increased electrostatic repulsion and the competition for retention sites between colloids and phosphates [38,41,42]. However, the influence of phosphates on the interaction of plastic particles and collector surfaces is still rarely studied and poorly understood. In addition, due to the high burdens of both phosphates and PNPs in soils, the influence of phosphates can be a critical issue in the fate of PNPs in the subsurface environment. To the best of our knowledge, the relevant information has not been reported.

Therefore, the objective of this study was to explore the potential coupled effects of typical phosphates ($NaH_2PO_4$ and $Na_2HPO_4$) associated with solution pH, ionic strength, and the presence of NaCl on the transport behaviors of PNPs using column experiments. Batch adsorption experiments, interaction energy calculations based on the classic Derjaguin–Landau–Verwey–Overbeek (DLVO) theory [43,44], and numerical simulations were also performed to better deduce the mechanisms of PNP transport. Findings in this study are helpful for better understanding the fate of PNPs and other colloidal contaminants in the subsurface environment.

## 2. Results and Discussion

### 2.1. Characterization of PNPs and Porous Media

Tables S1–S3 summarize the zeta potentials of PNPs, porous media, and hydrodynamic diameters ($d_p$) of PNPs under all experimental conditions. In general, the $d_p$ values of PNPs in $NaH_2PO_4$ or $Na_2HPO_4$ were similar. In the absence of NaCl, the $d_p$ of PNPs did not increase much when phosphates increased from 0 to 1 mM under pH 7. Under the same phosphate concentration and pH, the $d_p$ of PNPs in the presence of NaCl was slightly larger than that without NaCl; e.g., when pH was 7, $d_p$ ranged from 124 to 150 nm and from 131 to 164 nm in $NaH_2PO_4$ and $NaH_2PO_4$–NaCl systems, respectively (Table S1). The $d_p$ values were in a larger range between 131 and 200 nm in the $Na_2HPO_4$–NaCl system under pH 7 (Table S2). The $d_p$ was also slightly larger under pH 7 compared with pH 10 under the same electrolyte. However, in 0.25 mM $NaH_2PO_4$ with 1 mM NaCl, $d_p$ was stable (128–135 nm) when the pH increased from 5 to 10. Generally, the $d_p$ was in the order of pH 10 < pH 7 with phosphate alone < pH 7 with both phosphate and NaCl. These results indicate that the charge screening under a higher IS and deprotonation of the surface hydroxyl groups under a higher pH [42] play important roles in the charge of the PNP surface.

The trends of PNP zeta potentials in two phosphates showed only minor differences. Although slight fluctuation occurred, the zeta potentials of PNPs were less negative at a higher IS as phosphate concentrations increased (less than 1 mM) with or without 1 mM NaCl and sometimes tended to become more negative under 1 mM phosphates (Tables S1 and S2). Conversely, the zeta potentials of porous media became more negative with increasing IS. This may arise from the adsorption of phosphates that creates charge density due to the deprotonation of the phosphate [45]. Figures S3 and S4 demonstrate the adsorption capacities of sand and PNPs for phosphate increased with the increase in phosphate concentrations, while the adsorption on PNPs is higher than that on sand under comparable phosphate concentrations as in column experiments (Figure S4). The adsorption behaviors can be attributed to the irreversible chemical absorption (hydrogen bonding) on quartz sand [45,46] and the reversible physical absorption on PNPs. The trends of more negative charge with increasing IS in a low-level range were also reported in previous studies for PNPs [47] and porous media [48]. Table S3 shows that the zeta potentials of PNPs and porous media under IS = 1 mM with the mixture of 0.3 mM $Na_2HPO_4$ and 0.1 mM NaCl were more negative than those of 1 mM NaCl. In addition, the zeta potentials of both PNPs and sand were more negative as pH increased due to the deprotonation of the surfaces. Values of $d_p$ and zeta potentials provided in Tables S1–S3 were used to determine the interaction energy of PNPs–sand based on DLVO theory. Figures S5–S8 show the depths of the primary minima ($\Phi_{1min}$), the secondary minima ($\Phi_{2min}$), the energy barrier height ($\Phi_{max}$), and the energy barrier to detachment ($\Delta\Phi_d$). According to DLVO theory, larger $\Phi_{max}$ values indicate stronger repulsions between two surfaces. Figures S5 and S6 indicate that the $\Phi_{max}$ decreases and the depths of $\Phi_{2min}$ are deeper when phosphate concentrations increase. In the presence of phosphate with or without NaCl, the $\Phi_{max}$ values of $Na_2HPO_4$ were higher than those of $NaH_2PO_4$. The $\Phi_{max}$ tends to be larger under a higher pH (Figure S7) or under $Na_2HPO_4$ in the mixture of NaCl and phosphates at an IS of 1 mM (Figure S8). The shallow $\Phi_{2min}$ values indicate a low potential of retention in a secondary minimum. The $\Phi_{1min}$ is deeper and $\Delta\Phi_d$ is increased as phosphate concentration or IS increases or under lower pH, implying the potential of irreversible retention (Table S4).

*2.2. Transport of PNPs in the Presence of Phosphate*

Figure 1 presents the breakthrough curves (BTCs) and releases curves (RCs) of PNPs when concentrations of two phosphates equal 0.00, 0.25, 0.50, and 1.00 mM under pH 7. Table 1 shows the mass recoveries of PNPs from phases 1–3 in the column effluent. As the $NaH_2PO_4$ concentration increased from 0 to 1 mM, PNPs collected in the column effluent in phase 1 ($M_{eff}$) increased from 91% to 98%, and then decreased to 43% and 0 (under detection limit). Figure 1b presents the transport of PNPs in the presence of $Na_2HPO_4$. Different from $NaH_2PO_4$, PNP transport monotonically decreased from 91% to 82%, 32.0%, and 0 as the $Na_2HPO_4$ concentration increased from 0 to 1 mM. In general, at the same concentrations of the two phosphates, PNP mobility under $Na_2HPO_4$ (Figure 1b) was much weaker than under $NaH_2PO_4$ (Figure 1a), mainly due to the higher IS of $Na_2HPO_4$ that resulted in a more pronounced compression of the electrical double layer and a reduction in repulsive force. However, when phosphate concentration was 1 mM, no breakthrough of PNPs occurred under both phosphates. Fitted values of $k_1$ and $S_{max}/C_0$ also indicate the non-monotonic and monotonic (increased) trends in the mass transfer rates and retention capacities as $NaH_2PO_4$ and $Na_2HPO_4$ increase, respectively. However, the calculated energy barriers ($\Phi_{max}$) show fluctuations as phosphates increase. This deviation is attributed to the similar fluctuations of the zeta potentials for PNPs and sand that also display different trends (Tables S1 and S2). The dispersive distribution of PNPs on the sand surface shown in SEM images (Figure S9) is in agreement with $d_p$ measurements that indicate insignificant differences in the particle size under the used IS (Tables S1 and S2). This observation indicates that aggregation is insignificant for enhanced retention and that potential physical straining under a higher IS can be excluded within the tested range of phosphate concentrations. Note that column experiments exhibit good reproducibility (Figure 1) with small standard deviations (less than 5%) for the mass recoveries in the effluent.

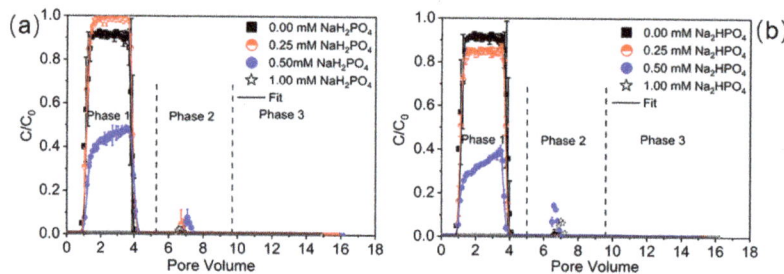

**Figure 1.** Breakthrough curves of PNPs at various NaH$_2$PO$_4$ (0–1 mM) (**a**) or Na$_2$HPO$_4$ (**b**) concentrations in the absence of NaCl under pH 7. The release of PNPs was initiated by eluting with ultrapure water under pH 7 (phase 2) and pH 10 (phase 3). Replicate experiments were performed under all experimental conditions.

**Table 1.** Experimental parameters and the mass recoveries of PNPs under various experimental conditions.

|  | NaH$_2$PO$_4$ mM | Na$_2$HPO$_4$ mM | NaCl mM | pH | IS | Recovery (%) | | |
| --- | --- | --- | --- | --- | --- | --- | --- | --- |
|  |  |  |  |  |  | $M_{eff}$ | $M_2$ | $M_3$ |
| Figure 1a | 0 | 0 | 0 | 7 | 0.01 | 91 | - | - |
|  | 0.25 | 0 | 0 | 7 | 0.25 | 98 | - | 1 |
|  | 0.5 | 0 | 0 | 7 | 0.50 | 43 | 1 | - |
|  | 1 | 0 | 0 | 7 | 1.00 | - | - | - |
| Figure 1b | 0 | 0 | 0 | 7 | 0.01 | 91 | - | - |
|  | 0 | 0.25 | 0 | 7 | 0.75 | 82 | - | - |
|  | 0 | 0.5 | 0 | 7 | 1.50 | 32 | 2 | - |
|  | 0 | 1 | 0 | 7 | 3.00 | - | 2 | - |
| Figure 2a | 0 | 0 | 1 | 7 | 1.00 | 45 | - | 1 |
|  | 0.25 | 0 | 1 | 7 | 1.25 | 10 | - | - |
|  | 0.5 | 0 | 1 | 7 | 1.50 | - | - | - |
|  | 1 | 0 | 1 | 7 | 1.75 | - | 2 | - |
| Figure 2b | 0 | 0 | 1 | 10 | 1.00 | 56 | - | - |
|  | 0.25 | 0 | 1 | 10 | 1.25 | 53 | - | - |
|  | 0.5 | 0 | 1 | 10 | 1.50 | 35 | - | 1 |
|  | 1 | 0 | 1 | 10 | 2.00 | - | - | 1 |
| Figure 2c | 0 | 0 | 1 | 7 | 1.00 | 45 | - | 1 |
|  | 0 | 0.25 | 1 | 7 | 1.75 | - | - | - |
|  | 0 | 0.5 | 1 | 7 | 2.50 | - | - | - |
|  | 0 | 1 | 1 | 7 | 4.00 | - | - | - |
| Figure 2d | 0 | 0 | 1 | 10 | 1.00 | 56 | - | - |
|  | 0 | 0.25 | 1 | 10 | 1.75 | 24 | - | - |
|  | 0 | 0.5 | 1 | 10 | 2.50 | - | - | 2 |
|  | 0 | 1 | 1 | 10 | 4.00 | - | - | 3 |
| Figure 3a | 0.25 | 0 | 1 | 5 | 1.25 | - | - | - |
|  | 0.25 | 0 | 1 | 7 | 1.25 | 10 | - | - |
|  | 0.25 | 0 | 1 | 8.5 | 1.25 | 10 | - | - |
|  | 0.25 | 0 | 1 | 10 | 1.25 | 53 | - | - |
| Figure 3b | 0 | 0.25 | 1 | 5 | 1.75 | - | - | - |
|  | 0 | 0.25 | 1 | 7 | 1.75 | - | - | - |
|  | 0 | 0.25 | 1 | 8.5 | 1.75 | - | - | - |
|  | 0 | 0.25 | 1 | 10 | 1.75 | 24 | - | - |

Table 1. Cont.

|  | NaH$_2$PO$_4$ mM | Na$_2$HPO$_4$ mM | NaCl mM | pH | IS | Recovery (%) | | |
| --- | --- | --- | --- | --- | --- | --- | --- | --- |
|  |  |  |  |  |  | $M_{eff}$ | $M_2$ | $M_3$ |
| Figure 4 | 0 | 0 | 1 | 7 | 1.00 | 45 | - | 1 |
|  | 0.3 | 0 | 0.7 | 7 | 1.00 | 30 | - | - |
|  | 0 | 0.3 | 0.1 | 7 | 1.00 | 86 | 1 | - |

"-" denotes under detection limit; $M_{eff}$ is the mass percentage of PNPs recovered from effluents in the retention (phase 1). $M_2$ and $M_3$ are the mass percentages of PNPs recovered from release phase 2 and phase 3. Note that only one release phase (elution with ultrapure water under pH = 10) was performed when PNPs were retained under pH = 10 in phase 1.

Previous studies demonstrated the enhanced transport of colloids (e.g., graphene oxide, TiO$_2$, and ZnO NPs) in the presence of abundant NaH$_2$PO$_4$ or K$_2$HPO$_4$ under a broad concentration range (e.g., 0.1–10 mM) [38,41,42,49]. In contrast, it was also evident that the transport of TiO$_2$ NPs would be reduced by increasing NaH$_2$PO$_4$ when the phosphate was higher than 1 mM because of the compressed electrical double layer [50]. Different from these trends, as described above, our findings suggest non-monotonic or monotonic decreased trends of PNP transport as phosphate concentrations increase in a narrow range of 0–1 mM and exhibit significant sensitivity. In particular, except for the condition of 0.25 mM NaH$_2$PO$_4$, the presence of phosphates under selected experimental parameters tended to inhibit PNP transport. The zeta potentials of porous media became slightly more negative as phosphate concentration increased at low levels (Tables S1 and S2). The enhanced transport that occurred in the presence of 0.25 mM NaH$_2$PO$_4$ can be explained by the competition for retention sites and the increased electrostatic repulsion attributed to the absorption of the phosphates, which can function as proton donors for hydrogen bonding on the collector surface [38,41,42]. However, the charge screening and the compression of the electrical double layer under a higher IS became more significant, as demonstrated by the less negatively charged PNPs and the greater retention with increasing Na$_2$HPO$_4$ and 1 mM NaH$_2$PO$_4$. Results from adsorption experiments show that the adsorption capacity of phosphate by quartz sand (Figure S3) is only up to 0.01 mg g$^{-1}$. In comparison, higher adsorption of phosphate onto PNPs (Figure S4) reaches a value of 137 mg g$^{-1}$ due to the larger specific surface area of PNPs than sand. RCs in Figure 1 show a minimal release of PNPs. The greater release occurred when a larger number of PNPs were retained in the previous phase (phase 1). In particular, a small portion of PNPs, which were retained under 1 mM phosphates in phase 1, can be released with the elution of ultrapure water in phase 2, whereas no release was observed in both phases when phosphates in phase 1 were less than 0.5 mM. These observations and DLVO calculations (Table S4 and Figure S5) demonstrate that most PNPs were captured in irreversible retention sites and mainly retained in deep primary minima. The negligible release is consistent with the strong energy barriers to detachment ($\Delta\Phi_d$). Therefore, although the presence of NaH$_2$PO$_4$ may facilitate PNP transport under specific concentrations (e.g., 0.25 mM), the compression of the electrical double layer and reduced repulsive force play essential roles, leading to pronounced irreversible retention when IS reaches a threshold. Additionally, certain degrees of micro- and nanoscale surface roughness are demonstrated by SEM images in Figure S9. The presence of phosphates can also increase the surface charge/chemical heterogeneity on the solid–water interfaces. It has been well demonstrated that the surface heterogeneities of colloids and collectors tend to reduce and/or eliminate energy barriers at electrostatically unfavorable locations, thus inhibiting colloid retention [51–54]. Consequently, these surface physicochemical heterogeneities can contribute to the significant sensitivity of PNP transport to phosphate concentration and the deviations of DLVO predictions from BTCs. Therefore, PNP retention in the presence of phosphate was mainly influenced by the coupled effects of increased electrostatic repulsion, competition for retention sites, electrical double layer compression, and increased chemical heterogeneity on the interacting surfaces, depending on the types of phosphates and their concentrations.

## 2.3. Transport of PNPs in the Presence of Phosphate Mixed with NaCl

To further investigate the influence of phosphate and IS, transport experiments were performed under 0–1 mM $NaH_2PO_4$ (Figure 2a,b) or $Na_2HPO_4$ (Figure 2c,d) at pH = 7 or 10 in the presence of 1 mM NaCl (phase 1). The release of retained PNPs was also carried out with the elution of ultrapure water under pH 7 (phase 2) and pH 10 (phase 3). Note that only the elution with water at pH 10 (phase 2) was performed in release experiments when the PNPs were retained under pH 10 in phase 1. Under pH 7 and 1 mM NaCl without phosphate, the recovery of PNPs in column effluent was 45%, whereas, under the phosphate and NaCl mixture, the PNP transport was significantly reduced. For example, at 0.25 mM $NaH_2PO_4$ with 1 mM NaCl, the $M_{eff}$ decreased to 10%, and complete retention occurred when $NaH_2PO_4$ was 0.5 mM or with a higher concentration (Table 1 and Figure 2a). The retention of PNPs was more sensitive to the mixture of $Na_2HPO_4$ and NaCl under pH 7 (Figure 2c). To be specific, the $M_{eff}$ was dramatically decreased to 0 when $Na_2HPO_4$ was 0.25 mM or higher. The more pronounced PNP retention with the presence of $Na_2HPO_4$ may also be attributed to the higher IS, even though the concentrations of the two phosphates were the same (Figure 2a,c). The greater PNP retention under a higher IS further demonstrated the importance of electrical double layer compression. The model was able to describe the BTCs well, and the values of $k_1$ and $S_{max}/C_o$ increased as the IS increased (Table S5), suggesting an increasing tendency for PNP retention. These results can also be explained by the DLVO interaction energy calculations. The repulsive energy barrier ($\Phi_{max}$) declined as phosphate concentration increased, indicating that more PNPs overcame the energy barrier and were retained in the primary minimum in the phosphate–NaCl mixture. The asymmetric shapes of BTCs at higher phosphate concentrations reflect a more pronounced blocking effect due to the gradual filling of retention sites. These influences of the two phosphates will be further discussed below in Section 3.4.

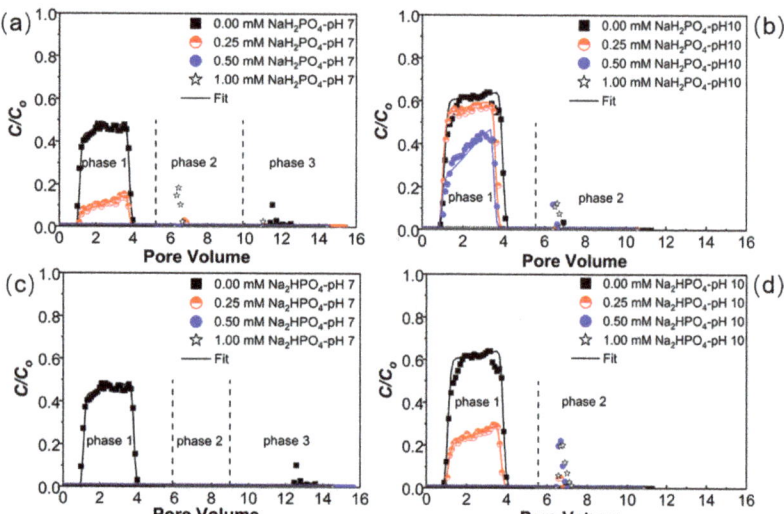

**Figure 2.** Breakthrough curves of PNPs with $NaH_2PO_4$ (0–1 mM) under pH = 7 (**a**) or pH = 10 (**b**); breakthrough curves of PNPs with $Na_2HPO_4$ (0–1 mM) at pH = 7 (**c**) or pH = 10 (**d**). All the experiments were carried out in the presence of 1 mM NaCl. The release of PNPs was initiated by eluting with ultrapure water under pH 7 (phase 2) and pH 10 (phase 3), respectively. Only the elution of water at pH 10 (phase 2) was performed when the PNPs were retained under pH 10 in phase 1.

Figure 2 indicates that the transport of PNPs under pH 10 is considerably higher than that of PNPs at pH 7. Specifically, the $M_{eff}$ values of PNPs dropped from 56% without phosphate to 35% and 24% with 0.5 mM $NaH_2PO_4$ and $Na_2HPO_4$ under pH 10, respectively.

However, the $M_{eff}$ value of PNPs declined from 45% to under the detection limit at 0.5 mM phosphate ($NaH_2PO_4$ and $Na_2HPO_4$) under pH 7. Generally, the zeta potentials of PNPs and porous media (Tables S1 and S2) are more negative at pH 10 (compared with 7) due to deprotonation of the surface, corresponding to stronger energy barriers and shallower primary minima in DLVO calculations (Table S4 and Figure S6). Note that the increase in phosphate concentration also enhances the PNP retention in the alkaline condition, suggesting that PNP transport is sensitive to the presence of phosphate. Thus, the electrical double layer compression is still one of the main factors that influence PNP transport. The influence of solution pH on PNP transport is further discussed in Section 3.4. Similar to the retention in the presence of phosphate without NaCl, the reversible retention in phosphate and NaCl mixtures also accounted for negligible fractions (Figure 2). Only small fractions of PNPs were released in phase 2, when they were retained in the mixture of 1 mM $NaH_2PO_4$ and 1 mM NaCl under pH 7 (phase 1), and in phase 3, when the PNPs were retained without the presence of phosphates (phase 2). At pH 10 (phase 1), the release also occurred when the PNPs were previously retained under 0.25 and 0.5 mM phosphates with 1 mM NaCl. This minor release of retained PNPs further indicated that the interactions of PNPs and quartz sand were strong enough to overcome the forces of diffusion arising from hydrodynamic shear and/or random kinetic energy fluctuations [55].

*2.4. Transport of PNPs under Various Solution pH Levels and Electrolyte Compositions*

Figure 3 and Table 1 present experimental results of PNP transport in the presence of 0.25 mM $NaH_2PO_4$ (a) or $Na_2HPO_4$ (b) with 1 mM NaCl at various levels of solution pH (5–10). As shown in Figure 3, the retention of PNPs decreased as the solution pH increased from 5 to 10, under both conditions of $NaH_2PO_4$ and $Na_2HPO_4$. In the presence of $NaH_2PO_4$–NaCl mixture, no breakthrough occurred under pH 5.0, whereas the $M_{eff}$ values were around 10% under pH 7.0 and 8.5 and dramatically rose to 53% at pH 10. In addition, with the $Na_2HPO_4$–NaCl, the $M_{eff}$ was under the detection limit when the pH was 8.5 or lower but increased to 24% under pH 10. The DLVO prediction certifies the trends of increasing energy barriers/repulsion as pH increases (Table S4 and Figure S7) and is consistent with $M_{eff}$ values.

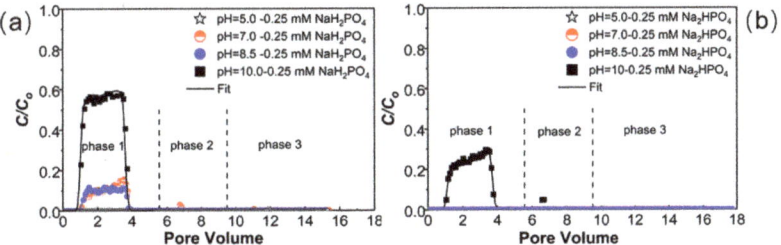

**Figure 3.** Breakthrough curves of PNPs at $NaH_2PO_4$ (0.25 mM) under pH = 5–10 (**a**); breakthrough curves of PNPs at $Na_2HPO_4$ (0.25 mM) under pH = 5–10 (**b**). All the experiments were carried out in the presence of 1 mM NaCl. The release of PNPs was initiated by eluting with ultrapure water under pH 7 (phase 2) and pH 10 (phase 3), respectively.

Previous studies evidenced various effects of solution pH on colloid transport in the presence of phosphates. For example, the transport of ZnO NPs is negligibly influenced by pH when the $K_2HPO_4$ is abundant in the solution [38]. However, the presence of $NaH_2PO_4$ can slightly reduce the transport of $TiO_2$ NPs under a higher pH due to less adsorption of phosphate [49]. In contrast, enhanced transport of graphene oxide NPs occurs as pH increases because of increasing electrostatic repulsion [41]. The forms of phosphates depend on solution pH. In particular, when the solution pH is in a range of 2.2 to 7.2, $H_2PO_4^-$ is the major fraction of phosphates, while in pH 7.2 to 12.3, $HPO_4^{2-}$ and $PO_4^{3-}$ become the main forms, and $PO_4^{3-}$ is more important when the pH is higher than 10 [56,57].

Therefore, the major fractions of phosphates are independent of the initial phosphate forms when the concentrations of $NaH_2PO_4$ and $Na_2HPO_4$ are the same and under the same solution pH. With the same phosphate concentrations, the stronger retention of PNPs under $Na_2HPO_4$ than under $NaH_2PO_4$ is mainly attributed to the higher IS of $Na_2HPO_4$ (higher $Na^+$ concentration), which leads to more pronounced electrical double layer compression, and surface chemical heterogeneity. As pH increases to 10, excessive $OH^-$ on the surfaces of PNPs and porous media enhances the negativity of the surfaces (Tables S1 and S2) due to the enhanced magnitude of deprotonation [58]. This process intensifies the repulsive interactions between the interacting surfaces [59,60]. Consequently, the increase in PNP transport with an increased pH is mainly due to the enhanced deprotonation.

To further deduce the combined influence of phosphates associated with IS and electrolytes on PNP transport, column experiments were conducted under the same phosphate concentration and IS under pH 7; e.g., 0.3 mM $NaH_2PO_4$ (IS = 0.3 mM) was mixed with 0.7 mM NaCl, and 0.3 mM $Na_2HPO_4$ (IS = 0.9) was mixed with 0.1 mM NaCl, to keep a constant total IS of 1 mM. Experimental results showed that $M_{eff}$ accounted for 30% and 86% under 0.3 mM $NaH_2PO_4$–0.7 mM NaCl and 0.3 mM $Na_2HPO_4$–0.1 mM NaCl, respectively, compared with the 45% under 1 mM NaCl (Figure 4 and Table 1). The trend of $M_{eff}$ is consistent with zeta potentials, the DLVO calculations, and fitted $S_{max}/C_o$; e.g., the calculated energy barrier between PNPs and sand is highest (Table S4) and the $S_{max}/C_o$ is smallest at 0.3 mM $Na_2HPO_4$–0.1 mM NaCl, compared to other conditions tested in Figure 4 (Table S5). The larger amount of cations ($Na^+$) in 0.3 mM $NaH_2PO_4$–0.7 mM NaCl mixture could be adsorbed in the diffuse layer by electrostatic force and decreased the surface charges from inner-sphere P-adsorption [50], thus reducing the repulsive force between PNPs and sand, leading to enhanced PNP retention. Additionally, compared to $NaH_2PO_4$ under the same concentration, $Na_2HPO_4$ forms less hydrogen bonding with quartz due to the sole hydrogen bonding donors [61], and its two $P-O^-$ units share more negative charge. Therefore, the interaction of $Na_2HPO_4$ and quartz will increase the electrostatic repulsion because the intermediate will carry more negative charges on oxygen [34]. Consequently, with the presence of 0.3 mM $Na_2HPO_4$–0.1 mM NaCl, the increased repulsion and the weaker charge screening/heterogeneity lead to the greatest mobility, as shown in Figure 4. In contrast, the compression of the double layer, charge screening, and the chemical/charge heterogeneity generated by the adsorption of cations will be more pronounced with an increase in $Na^+$ concentration [62,63], leading to high retention of PNPs under 0.3 mM $NaH_2PO_4$–0.7 mM NaCl. These findings suggest that the association and coupled effects of electrostatic repulsion (attributed to adsorbed phosphates), the compression of the electrical double layer, and the surface chemical heterogeneity (contributed by the adsorption of cations) significantly influence the transport, depending on the phosphate concentrations, IS, and solution pH.

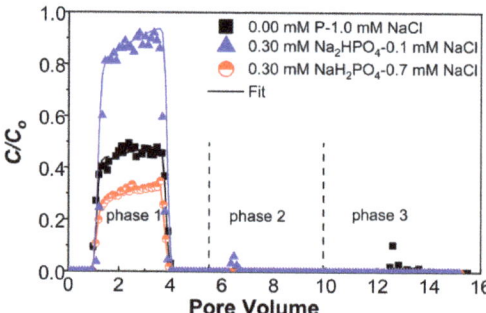

**Figure 4.** Breakthrough curves of PNPs under IS = 1 with different combinations of phosphate and NaCl at pH = 7. The release of PNPs was initiated by eluting with ultrapure water under pH 7 (phase 2) and pH 10 (phase 3), respectively.

## 3. Materials and Methods

### 3.1. Solution Chemistry and Porous Media

Electrolyte solutions were prepared by diluting $NaH_2PO_4$, $Na_2HPO_4$, and/or NaCl in Milli-Q water, and their pH values were adjusted to 5, 7, 8.5, or 10 using HCl or NaOH. Analytically pure quartz sand (Tianjin Guangfu Fine Chemical Research Institute, Tianjing, China) was used as porous media. The sand was purified by washing in tap water, followed by soaking in $HNO_3$ (65%) and $H_2O_2$ (10%) for 24 h [54]. Later, the sand was washed with water again, followed by soaking in 100 mM NaCl and ultrapure water (pH 10) with ultrasonication to remove the potential attached colloidal impurities by cation exchange and expand the electrical double layer. Finally, the quartz sand was sieved within the size range between 250 μm and 380 μm. Zeta potentials of quartz sand were measured using a ZetaSizer (Nano ZS9, Malvern Instruments, Worcestershire, UK). A scanning electron microscope (SEM, ZEISS Sigma 300, Neustadt, Germany) was used to visualize the surface morphology of quartz sand and investigate the interactions of PNPs and sand surfaces. More detailed information is provided in Supplementary Materials.

### 3.2. Plastic Nanoparticles

Polystyrene nanoparticles (purchased from Suzhou Smart Nanotechnology Co., Ltd., Suzhou, China) with regularly spherical shape and a nano size of around 50 nm were used as PNPs in this study. Polystyrene nanoparticles have been frequently employed as model colloids and representative plastics [12,64–67]. The polystyrene was confirmed by an attenuated total reflectance-Fourier transforms infrared spectroscopy (ATR-FTIR) (Nicolet iS50, Thermo Fisher Scientific, Waltham, CA, USA) (Figure S1). The initial/input concentration of PNPs in this study was set as 10 mg $L^{-1}$ by diluting a raw suspension (1 g $L^{-1}$) into selected electrolyte solutions and then sonicating them for 20 min in an ultrasonication bath before use. The zeta potentials and hydrodynamic diameters of PNPs were also measured using the ZetaSizer.

### 3.3. Transport and Release Experiment

Water-saturated column experiments were performed following the processes outlined in previous studies [68]. Columns made of stainless steel with a length of 12 cm and an inner diameter of 3 cm were wet-packed with purified quartz sand. A constant velocity was set as 0.7 cm $min^{-1}$ for all experiments by a peristaltic pump that introduced PNP suspensions and PNP-free electrolyte solutions upward into the vertical columns. The columns were firstly conditioned with background solutions (30 pore volumes) of 0–1 mM $NaH_2PO_4$ or $Na_2HPO_4$ under different pH values with or without the presence of NaCl. Later, the transport of the tracer and PNPs was investigated in each column experiment by injecting a 100 mL pulse of tracer (2–4 times the concentrations of background solution) or PNP suspension, followed by elution with 100 mL background solution. Column effluent samples (4 mL of each) were continuously collected via a fraction collector. A conductivity meter and a fluorescence spectrophotometer were used to determine the concentrations and obtain the corresponding breakthrough curves (BTCs) of the tracer and PNPs, respectively. Experimental conditions are summarized in Table 1. To test the reproducibility, some experimental conditions were repeated in column experiments.

After the completion of transport experiments named phase 1, the release of retained PNPs was conducted to examine the potential detachment from the secondary or primary minimum. The retained PNPs were rinsed with several pore volumes of ultrapure water under the same pH (phase 2) as in phase 1 and then with ultrapure water at pH 10 (phase 3). Release curves (RCs) of PNPs were also determined using a fluorescence spectrophotometer.

### 3.4. Batch Experiments and Theory

Batch adsorption experiments were performed to investigate the adsorption of phosphates on 50 nm PNPs (100 mg $L^{-1}$) and sand ($2 \times 10^5$ mg $L^{-1}$) under different phosphate concentrations (0.25–6 mM) at pH 7. The mixtures were shaken by a water bath oscillator at 25 °C for 24 h. Before the quantification of phosphates, 40 μL $CaCl_2$ (2 mol $L^{-1}$) was added to form larger PNP aggregates to overcome the challenge in the separation of PNPs from the liquid phase [69], followed by centrifugation at 15,000 rpm (23,120× $g$) for 15 min and then filtration through a 0.22 μm membrane. The phosphates were then treated and determined by a colorimetric method at a fixed wavelength of 700 nm using a visible spectrophotometer [70] (Section S2 and Figure S2 in Supporting Information). All adsorption experiments were carried out in triplicate.

Classical DLVO theory was used to calculate the interaction energy between PNPs and quartz sand under various solution chemistries as in column experiments. The total interaction energy includes electrical double layer repulsive and van der Waals attractive forces [71,72]. The transport of PNPs was described by inverse fitting to experimental BTCs to obtain the parameters of $k_1$ and $S_{max}/C_o$ by HYDRUS-1D computer code [73]. The $k_1$ and $S_{max}/C_o$ represent the first-order retention coefficient and the maximum solid-phase concentration of deposited PNPs, respectively. Tracer experiments were used to determine the values of dispersivity and pore water velocity in the simulations for PNP transport. The simulation did not perform when the PNP effluent concentration was under the detection limit. Further information on the DLVO interaction energy calculations and the descriptions of numerical simulations are provided in the Supplementary Information (Sections S4 and S5).

## 4. Conclusions

PNP transport is significantly sensitive to phosphate concentrations at low levels. The transport of PNPs is non-monotonically influenced by $NaH_2PO_4$ due to the increased electrostatic repulsion, the competition for retention sites, and the compression of the double layer varying with IS. The transport is inhibited when $NaH_2PO_4$ concentration reaches a threshold. However, an increase in $Na_2HPO_4$ tends to result in a monotonic decrease in PNP transport. These observations are different from the findings that show an increase in colloid transport even at a much larger range of $NaH_2PO_4$ concentrations in previous studies. Higher pH increases PNP transport due to the deprotonation of surfaces. A minimal fraction for the release of retained PNPs under reduced IS and increased pH indicate that the PNPs are mainly captured in deep primary minima on irreversible retention sites. The presence of $Na_2HPO_4$ tends to result in greater PNP retention than $NaH_2PO_4$ under the same concentration due to the fact of higher IS and cation concentration for $Na_2HPO_4$. Additionally, hydrogen bonding from two phosphates that act as proton donors contributes to variations in the interactions of PNPs and porous media and thus influences PNP transport. These findings further demonstrate that the compression of the electrical double layer tends to be dominant over the electrostatic repulsion arising from the adsorption of phosphates on the interacting surfaces. The adsorption of phosphates can also increase chemical heterogeneity, thus reducing PNP transport and increasing the sensitivity of particle transport to IS, due to the potential reduction/elimination of the energy barrier. Classical DLVO needs an extension to include the influence of physicochemical heterogeneities for a better explanation of experimental results.

This study highlights the sensitivity of PNP transport to phosphates associated with the solution chemistry and indicates enhanced retention of PNPs in the presence of phosphate ($\geq 1$ mM), higher IS, and low pH. The findings are helpful for better understanding the fate of PNPs and colloidal contaminants in the subsurface environment. They also imply that the presence of phosphates will influence PNP transport in engineering processes (e.g., deep bed filtration), agricultural soil, or contaminated subsurface environments. However, polystyrene spheres could not perfectly represent all PNPs in the real soil environment. Further investigations on PNPs from different sources with varying shapes and surface properties are also needed for a better understanding of the environmental fate of colloidal plastics.

**Supplementary Materials:** The supporting information can be downloaded at: https://www.mdpi.com/article/10.3390/ijms23179860/s1. References [74–79] are cited in the supplementary materials.

**Author Contributions:** Conceptualization, methodology, writing—review and editing, Y.L.; writing—original draft preparation, methodology, investigation, X.L.; investigation, Y.P.; investigation, T.M.; investigation, L.X.; investigation, P.D.; All authors have read and agreed to the published version of the manuscript.

**Funding:** This research was funded by the National Natural Science Foundation of China (42167051) and the China Postdoctoral Science Foundation (2021M690763).

**Institutional Review Board Statement:** Not applicable.

**Informed Consent Statement:** Not applicable.

**Data Availability Statement:** The data presented in this study are available in Supplementary Material.

**Conflicts of Interest:** The authors declare no conflict of interest.

# References

1. Gigault, J.; Halle, A.T.; Baudrimont, M.; Pascal, P.Y.; Gauffre, F.; Phi, T.L.; El Hadri, H.; Grassl, B.; Reynaud, S. Current opinion: What is a nanoplastic? *Environ. Pollut.* **2018**, *235*, 1030–1034. [CrossRef] [PubMed]
2. Chen, Y.; Awasthi, A.K.; Wei, F.; Tan, Q.; Li, J. Single-use plastics: Production, usage, disposal, and adverse impacts. *Sci. Total Environ.* **2021**, *752*, 141772. [CrossRef] [PubMed]
3. Allouzi, M.M.A.; Tang, D.Y.Y.; Chew, K.W.; Rinklebe, J.; Bolan, N.; Allouzi, S.M.A.; Show, P.L. Micro (nano) plastic pollution: The ecological influence on soil-plant system and human health. *Sci. Total Environ.* **2021**, *788*, 147815. [CrossRef] [PubMed]
4. Hernandez, E.; Nowack, B.; Mitrano, D.M. Polyester Textiles as a Source of Microplastics from Households: A Mechanistic Study to Understand Microfiber Release During Washing. *Environ. Sci. Technol.* **2017**, *51*, 7036–7046. [CrossRef] [PubMed]
5. Cheung, P.K.; Fok, L. Characterisation of plastic microbeads in facial scrubs and their estimated emissions in Mainland China. *Water Res.* **2017**, *122*, 53–61. [CrossRef]
6. Hernandez, L.M.; Yousefi, N.; Tufenkji, N. Are There Nanoplastics in Your Personal Care Products? *Environ. Sci. Technol. Lett.* **2017**, *4*, 280–285. [CrossRef]
7. Lei, K.; Qiao, F.; Liu, Q.; Wei, Z.; Qi, H.; Cui, S.; Yue, X.; Deng, Y.; An, L. Microplastics releasing from personal care and cosmetic products in China. *Mar. Pollut. Bull.* **2017**, *123*, 122–126. [CrossRef]
8. Yang, T.; Luo, J.; Nowack, B. Characterization of Nanoplastics, Fibrils, and Microplastics Released during Washing and Abrasion of Polyester Textiles. *Environ. Sci. Technol.* **2021**, *55*, 15873–15881. [CrossRef]
9. Ossmann, B.E.; Sarau, G.; Holtmannspotter, H.; Pischetsrieder, M.; Christiansen, S.H.; Dicke, W. Small-sized microplastics and pigmented particles in bottled mineral water. *Water Res.* **2018**, *141*, 307–316. [CrossRef]
10. Schymanski, D.; Goldbeck, C.; Humpf, H.-U.; Fürst, P. Analysis of microplastics in water by micro-Raman spectroscopy: Release of plastic particles from different packaging into mineral water. *Water Res.* **2018**, *129*, 154–162. [CrossRef]
11. Brewer, A.; Dror, I.; Berkowitz, B. The Mobility of Plastic Nanoparticles in Aqueous and Soil Environments: A Critical Review. *ACS EST Water* **2021**, *1*, 48–57. [CrossRef]
12. Dong, Z.; Qiu, Y.; Zhang, W.; Yang, Z.; Wei, L. Size-dependent transport and retention of micron-sized plastic spheres in natural sand saturated with seawater. *Water Res.* **2018**, *143*, 518–526. [CrossRef] [PubMed]
13. Nizzetto, L.; Futter, M.; Langaas, S. Are Agricultural Soils Dumps for Microplastics of Urban Origin? *Environ. Sci. Technol.* **2016**, *50*, 10777–10779. [CrossRef] [PubMed]
14. Nizzetto, L.; Langaas, S.; Futter, M. Pollution: Do microplastics spill on to farm soils? *Nature* **2016**, *537*, 488. [CrossRef]
15. Rillig, M.C. Microplastic in terrestrial ecosystems and the soil? *Environ. Sci. Technol.* **2012**, *46*, 6453–6454. [CrossRef]
16. Scheurer, M.; Bigalke, M. Microplastics in Swiss Floodplain Soils. *Environ. Sci. Technol.* **2018**, *52*, 3591–3598. [CrossRef]
17. Zhang, G.S.; Liu, Y.F. The distribution of microplastics in soil aggregate fractions in southwestern China. *Sci. Total Environ.* **2018**, *642*, 12–20. [CrossRef]
18. Luo, T.; Zhang, Y.; Wang, C.; Wang, X.; Zhou, J.; Shen, M.; Zhao, Y.; Fu, Z.; Jin, Y. Maternal exposure to different sizes of polystyrene microplastics during gestation causes metabolic disorders in their offspring. *Environ. Pollut.* **2019**, *255*, 113122. [CrossRef]
19. Rodrigues, M.O.; Abrantes, N.; Gonçalves, F.J.M.; Nogueira, H.; Marques, J.C.; Gonçalves, A.M.M. Impacts of plastic products used in daily life on the environment and human health: What is known? *Environ. Toxicol. Pharmacol.* **2019**, *72*, 103239. [CrossRef]
20. Yu, Y.; Flury, M. Current understanding of subsurface transport of micro- and nanoplastics in soil. *Vadose Zone J.* **2021**, *20*, e20108. [CrossRef]
21. Iqbal, S.; Xu, J.; Allen, S.D.; Khan, S.; Nadir, S.; Arif, M.S.; Yasmeen, T. Unraveling consequences of soil micro- and nano-plastic pollution on soil-plant system: Implications for nitrogen (N) cycling and soil microbial activity. *Chemosphere.* **2020**, *260*, 127578. [CrossRef] [PubMed]

22. Shen, M.; Zhang, Y.; Zhu, Y.; Song, B.; Zeng, G.; Hu, D.; Wen, X.; Ren, X. Recent advances in toxicological research of nanoplastics in the environment: A review. *Environ. Pollut.* **2019**, *252*, 511–521. [CrossRef]
23. Huang, D.; Chen, H.; Shen, M.; Tao, J.; Chen, S.; Yin, L.; Zhou, W.; Wang, X.; Xiao, R.; Li, R. Recent advances on the transport of microplastics/nanoplastics in abiotic and biotic compartments. *J. Hazard. Mater.* **2022**, *438*, 129515. [CrossRef] [PubMed]
24. Chai, B.; Wei, Q.; She, Y.; Lu, G.; Dang, Z.; Yin, H. Soil microplastic pollution in an e-waste dismantling zone of China. *Waste Manag.* **2020**, *118*, 291–301. [CrossRef] [PubMed]
25. Lai, H.; Liu, X.; Qu, M. Nanoplastics and Human Health: Hazard Identification and Biointerface. *J. Nanomater.* **2022**, *12*, 1298. [CrossRef]
26. Yu, S.; Shen, M.; Li, S.; Fu, Y.; Zhang, D.; Liu, H.; Liu, J. Aggregation kinetics of different surface-modified polystyrene nanoparticles in monovalent and divalent electrolytes. *Environ. Pollut.* **2019**, *255*, 113302. [CrossRef]
27. Pradel, A.; Hadri, H.E.; Desmet, C.; Ponti, J.; Reynaud, S.; Grassl, B.; Gigault, J. Deposition of environmentally relevant nanoplastic models in sand during transport experiments. *Chemosphere* **2020**, *255*, 126912. [CrossRef]
28. Wagner, S.; Reemtsma, T. Things we know and don't know about nanoplastic in the environment. *Nat. Nanotechnol.* **2019**, *14*, 300–301. [CrossRef]
29. Reynaud, S.; Aynard, A.; Grassl, B.; Gigault, J. Nanoplastics: From model materials to colloidal fate. *Curr. Opin. Colloid Interface Sci.* **2022**, *57*, 101528. [CrossRef]
30. Shaniv, D.; Dror, I.; Berkowitz, B. Effects of particle size and surface chemistry on plastic nanoparticle transport in saturated natural porous media. *Chemosphere* **2021**, *262*, 127854. [CrossRef]
31. Wu, X.; Lyu, X.; Li, Z.; Gao, B.; Zeng, X.; Wu, J.; Sun, Y. Transport of polystyrene nanoplastics in natural soils: Effect of soil properties, ionic strength and cation type. *Sci. Total Environ.* **2020**, *707*, 136065. [CrossRef] [PubMed]
32. Jiang, Y.; Yin, X.; Xi, X.; Guan, D.; Sun, H.; Wang, N. Effect of surfactants on the transport of polyethylene and polypropylene microplastics in porous media. *Water Res.* **2021**, *196*, 117016. [CrossRef]
33. Zhou, D.; Cai, Y.; Yang, Z. Key factors controlling transport of micro- and nanoplastic in porous media and its effect on coexisting pollutants. *Environ. Pollut.* **2022**, *293*, 118503. [CrossRef] [PubMed]
34. King, K.; Williams, M.; Macrae, M.; Fausey, N.R.; Frankenberger, J.; Smith, D.; Kleinman, P.; Brown, L. Phosphorus Transport in Agricultural Subsurface Drainage: A Review. *J. Environ. Qual.* **2015**, *44*, 467–485. [CrossRef] [PubMed]
35. Mihelcic, J.R.; Fry, L.M.; Shaw, R. Global potential of phosphorus recovery from human urine and feces. *Chemosphere* **2011**, *84*, 832–839. [CrossRef]
36. Bierman, P.M.; Rosen, C.J.; Bloom, P.R.; Nater, E.A. Soil Solution Chemistry of Sewage-Sludge Incinerator Ash and Phosphate Fertilizer Amended Soil. *J. Environ. Qual.* **1995**, *24*, 279–285. [CrossRef]
37. Wang, L.; Xu, S.; Li, J. Effects of Phosphate on the Transport of Escherichia coli O157:H7 in Saturated Quartz Sand. *Environ. Sci. Technol.* **2011**, *45*, 9566–9573. [CrossRef]
38. Li, L.; Schuster, M. Influence of phosphate and solution pH on the mobility of ZnO nanoparticles in saturated sand. *Sci. Total Environ.* **2014**, *472*, 971–978. [CrossRef]
39. Mu, Y.; Ai, Z.; Zhang, L. Phosphate Shifted Oxygen Reduction Pathway on Fe@Fe$_2$O$_3$ Core–Shell Nanowires for Enhanced Reactive Oxygen Species Generation and Aerobic 4-Chlorophenol Degradation. *Environ. Sci. Technol.* **2017**, *51*, 8101–8109. [CrossRef]
40. Gérard, F. Clay minerals, iron/aluminum oxides, and their contribution to phosphate sorption in soils—A myth revisited. *Geoderma* **2016**, *262*, 213–226. [CrossRef]
41. Chen, J.; Chen, W.; Lu, T.; Song, Y.; Zhang, H.; Wang, M.; Wang, X.; Qi, Z.; Lu, M. Effects of phosphate on the transport of graphene oxide nanoparticles in saturated clean and iron oxide-coated sand columns. *J Env. Sci.* **2021**, *103*, 80–92. [CrossRef]
42. Guo, P.; Xu, N.; Li, D.; Huangfu, X.; Li, Z. Aggregation and transport of rutile titanium dioxide nanoparticles with montmorillonite and diatomite in the presence of phosphate in porous sand. *Chemosphere* **2018**, *204*, 327–334. [CrossRef]
43. Derjaguin, B.; Landau, L. Theory of the stability of strongly charged lyophobic sols and of the adhesion of strongly charged particles in solutions of electrolytes. *Prog. Surf. Sci.* **1993**, *43*, 30–59. [CrossRef]
44. Verwey, E.J.W.; Overbeek, J.T.G. Theory of the Stability of Lyophobic Colloids. *Nature* **1948**, *162*, 315–316. [CrossRef]
45. Morris, J.H.; Perkins, P.G.; Rose, A.E.A.; Smith, W.E. Interaction between aluminium dihydrogen phosphate and quartz. *J. Appl. Chem. Biotechnol.* **1976**, *26*, 385–390. [CrossRef]
46. Li Wang, Z.B.; Rong, R.D.; Huang, Y.Q.; Wen, L.J. The adsorption behavior of phosphorus in microplastics in water and soil. *Nong Ye Huan Jing Ke Xue Xue Bao* **2021**, *40*, 1758–1764. [CrossRef]
47. Lu, S.; Zhu, K.; Song, W.; Song, G.; Chen, D.; Hayat, T.; Alharbi, N.S.; Chen, C.; Sun, Y. Impact of water chemistry on surface charge and aggregation of polystyrene microspheres suspensions. *Sci. Total Environ.* **2018**, *630*, 951–959. [CrossRef]
48. Liang, Y.; Luo, Y.; Lu, Z.; Klumpp, E.; Shen, C.; Bradford, S.A. Evidence on enhanced transport and release of silver nanoparticles by colloids in soil due to modification of grain surface morphology and co-transport. *Environ. Pollut.* **2021**, *276*, 116661. [CrossRef]
49. Chen, M.; Xu, N.; Cao, X.; Zhou, K.; Chen, Z.; Wang, Y.; Liu, C. Facilitated transport of anatase titanium dioxides nanoparticles in the presence of phosphate in saturated sands. *J. Colloid Interface Sci.* **2015**, *451*, 134–143. [CrossRef]
50. Wang, S.; Li, D.; Zhang, M.; Chen, M.; Xu, N.; Yang, L.; Chen, J. Competition between fulvic acid and phosphate-mediated surface properties and transport of titanium dioxide nanoparticles in sand porous media. *J. Soils Sediments* **2020**, *20*, 3681–3687. [CrossRef]

51. Bendersky, M.; Davis, J.M. DLVO interaction of colloidal particles with topographically and chemically heterogeneous surfaces. *J. Colloid Interface Sci.* **2011**, *353*, 87–97. [CrossRef] [PubMed]
52. Bradford, S.A.; Torkzaban, S. Colloid Interaction Energies for Physically and Chemically Heterogeneous Porous Media. *Langmuir* **2013**, *29*, 3668–3676. [CrossRef]
53. Shen, C.; Lazouskaya, V.; Zhang, H.; Li, B.; Jin, Y.; Huang, Y. Influence of surface chemical heterogeneity on attachment and detachment of microparticles. *Colloids Surf. A Physicochem. Eng. Asp.* **2013**, *433*, 14–29. [CrossRef]
54. Liang, Y.; Bradford, S.A.; Šimůnek, J.; Klumpp, E. Mechanisms of graphene oxide aggregation, retention, and release in quartz sand. *Sci. Total Environ.* **2019**, *656*, 70–79. [CrossRef] [PubMed]
55. Torkzaban, S.; Bradford, S.A. Critical role of surface roughness on colloid retention and release in porous media. *Water Res.* **2016**, *88*, 274–284. [CrossRef] [PubMed]
56. Xu, N.; Yin, H.W.; Chen, Z.G.; Chen, M.; Liu, S.Q. Mechanisms of Phosphate Removal by Synthesized Calcite. *Mater. Sci. Forum* **2013**, *743–744*, 597–602. [CrossRef]
57. Tejedor-Tejedor, M.I.; Anderson, M.A. Protonation of phosphate on the surface of goethite as studied by CIR-FTIR and electrophoretic mobility. *Langmuir* **1990**, *6*, 602–611. [CrossRef]
58. Xu, N.; Cheng, X.; Wang, D.; Xu, X.; Huangfu, X.; Li, Z. Effects of Escherichia coli and phosphate on the transport of titanium dioxide nanoparticles in heterogeneous porous media. *Water Res.* **2018**, *146*, 264–274. [CrossRef]
59. Das, J.; Patra, B.S.; Baliarsingh, N.; Parida, K.M. Adsorption of phosphate by layered double hydroxides in aqueous solutions. *Appl. Clay Sci.* **2006**, *32*, 252–260. [CrossRef]
60. Gisbert, R.; García, G.; Koper, M.T.M. Adsorption of phosphate species on poly-oriented Pt and Pt(111) electrodes over a wide range of pH. *Electrochim. Acta* **2010**, *55*, 7961–7968. [CrossRef]
61. Weng, L.; Elliott, G.D. Distinctly Different Glass Transition Behaviors of Trehalose Mixed with $Na_2HPO_4$ or $NaH_2PO_4$: Evidence for its Molecular Origin. *Pharm. Res.* **2015**, *32*, 2217–2228. [CrossRef] [PubMed]
62. Roy, S.B.; Dzombak, D.A. Colloid release and transport processes in natural and model porous media. *Colloids Surf. A Physicochem. Eng. Asp.* **1996**, *107*, 245–262. [CrossRef]
63. Grosberg, A.Y.; Nguyen, T.T.; Shklovskii, B.I. Colloquium: The physics of charge inversion in chemical and biological systems. *Rev. Mod. Phys.* **2002**, *74*, 329–345. [CrossRef]
64. Bradford, S.A.; Kim, H.N.; Haznedaroglu, B.Z.; Torkzaban, S.; Walker, S.L. Coupled Factors Influencing Concentration-Dependent Colloid Transport and Retention in Saturated Porous Media. *Environ. Sci. Technol.* **2009**, *43*, 6996–7002. [CrossRef]
65. Chu, X.; Li, T.; Li, Z.; Yan, A.; Shen, C. Transport of Microplastic Particles in Saturated Porous Media. *Water* **2019**, *11*, 2474. [CrossRef]
66. Li, T.; Shen, C.; Wu, S.; Jin, C.; Bradford, S.A. Synergies of surface roughness and hydration on colloid detachment in saturated porous media: Column and atomic force microscopy studies. *Water Res.* **2020**, *183*, 116068. [CrossRef]
67. Tong, M.; He, L.; Rong, H.; Li, M.; Kim, H. Transport behaviors of plastic particles in saturated quartz sand without and with biochar/$Fe_3O_4$-biochar amendment. *Water Res.* **2020**, *169*, 115284. [CrossRef]
68. Liang, Y.; Zhou, J.; Dong, Y.; Klumpp, E.; Šimůnek, J.; Bradford, S.A. Evidence for the critical role of nanoscale surface roughness on the retention and release of silver nanoparticles in porous media. *Environ. Pollut.* **2020**, *258*, 113803. [CrossRef]
69. Wang, J.; Liu, X.; Liu, G.; Zhang, Z.; Wu, H.; Cui, B.; Bai, J.; Zhang, W. Size effect of polystyrene microplastics on sorption of phenanthrene and nitrobenzene. *Ecotoxicol. Environ. Saf.* **2019**, *173*, 331–338. [CrossRef]
70. Qi, T.; Su, Z.; Jin, Y.; Ge, Y.; Guo, H.; Zhao, H.; Xu, J.; Jin, Q.; Zhao, J. Electrochemical oxidizing digestion using $PbO_2$ electrode for total phosphorus determination in a water sample. *RSC Adv.* **2018**, *8*, 6206–6211. [CrossRef]
71. Hogg, R.; Healy, T.W.; Fuerstenau, D.W. Mutual coagulation of colloidal dispersions. *Phys. Chem. Chem. Phys.* **1966**, *62*, 1638–1651. [CrossRef]
72. Gregory, J. Approximate expressions for retarded van der waals interaction. *J. Colloid Interface Sci.* **1981**, *83*, 138–145. [CrossRef]
73. Šimůnek, J.; van Genuchten, M.T.; Šejna, M. Development and applications of the hydrus and stanmod software packages and related codes. *Vadose Zone J.* **2008**, *7*, 587–600. [CrossRef]
74. Umamaheswari, S.; Margandan, M.M. FTIR Spectroscopic Study of Fungal Degradation of Poly(ethylene terephthalate) and Polystyrene Foam. Available online: https://www.researchgate.net/publication/258316463 (accessed on 1 January 2013).
75. Li, M.; Zhang, X.; Yi, K.; He, L.; Han, P.; Tong, M. Transport and deposition of microplastic particles in saturated porous media: Co-effects of clay particles and natural organic matter. *Environ Pollut.* **2021**, *287*, 117585. [CrossRef] [PubMed]
76. Sze, A.; Erickson, D.; Ren, L.; Li, D. Zeta-potential measurement using the Smoluchowski equation and the slope of the current–time relationship in electroosmotic flow. *J. Colloid Interface Sci.* **2003**, *261*, 402–410. [CrossRef]
77. Deshpande, P.A.; Shonnard, D.R. Modeling the effects of systematic variation in ionic strength on the attachment kinetics of Pseudomonas fluorescens UPER-1 in saturated sand columns. *Water Resour. Res.* **1999**, *35*, 1619–1627. [CrossRef]
78. Israelachvili, J.N. Intermolecular and Surface Forces Academic Pres. In *Intermolecular and Surface Forces*, 3rd ed.; Academic Press: San Diego, CA, USA, 1992; p. iv. [CrossRef]
79. He, L.; Wu, D.; Rong, H.; Li, M.; Tong, M.; Kim, H. Influence of Nano- and Microplastic Particles on the Transport and Deposition Behaviors of Bacteria in Quartz Sand. *Environ. Sci. Technol.* **2018**, *52*, 11555–11563. [CrossRef]

Article

# Impact of Acetate in Reduction of Perchlorate by Mixed Microbial Culture under the Influence of Nitrate and Sulfate

Hosung Yu [1], Kang Hoon Lee [2,*,†] and Jae-Woo Park [1,*,†]

[1] Department of Civil and Environmental Engineering, Hanyang University, 222 Wangsimni-ro, Seongdong-gu, Seoul 04763, Korea
[2] Department of Energy and Environmental Engineering, The Catholic University of Korea, 43 Jibong-ro, Bucheon-si 14662, Korea
* Correspondence: diasyong@catholic.ac.kr (K.H.L.); jaewoopark@hanyang.ac.kr (J.-W.P.)
† These authors contributed equally to this work.

**Abstract:** The biological reduction of slow degradation contaminants such as perchlorate ($ClO_4^-$) is considered to be a promising water treatment technology. The process is based on the ability of a specific mixed microbial culture to use perchlorate as an electron acceptor in the absence of oxygen. In this study, batch experiments were conducted to investigate the effect of nitrate on perchlorate reduction, the kinetic parameters of the Monod equation and the optimal ratio of acetate to perchlorate for the perchlorate reducing bacterial consortium. The results of this study suggest that acclimated microbial cultures can be applied to treat wastewater containing high concentrations of perchlorate. Reactor experiments were carried out with different hydraulic retention times (HRTs) to determine the optimal operating conditions. A fixed optimal HRT and the effect of nitrate on perchlorate reduction were investigated with various concentrations of the electron donor. The results showed that perchlorate reduction occurred after nitrate removal. Moreover, the presence of sulfate in wastewater had no effect on the perchlorate reduction. However, it had little effect on biomass concentration in the presence of nitrate during exposure to a mixed microbial culture, considering the nitrate as the inhibitor of perchlorate reduction by reducing the degradation rate. The batch scale experiment results illustrated that for efficient operation of perchlorate reduction, the optimal acetate to perchlorate ratio of 1.4:1.0 would be enough. Moreover, these experiments found the following results: the kinetic parameters equivalent to $Y = 0.281$ mg biomass/mg perchlorate, $K_s = 37.619$ mg/L and $q_{max} = 0.042$ mg perchlorate/mg biomass/h. In addition, anoxic–aerobic experimental reactor results verify the optimal HRT of 6 h for continuous application. Furthermore, it also illustrated that using 600 mg/L of acetate as a carbon source is responsible for 100% of nitrate reduction with less than 50% of the perchlorate reduction, whereas at 1000 mg/L acetate, approximately 100% reduction was recorded.

**Keywords:** perchlorate; mixed microbial culture; bio reduction; Monod equation; hydraulic retention time (HRT)

## 1. Introduction

Widespread usage of perchlorate ($ClO_4^-$) in propellants, polytechnic devices and explosives for the defense production industry led to its release into the environment [1–5]. In addition, perchlorate has also been used in electropolishing, air bag initiators, paints, perchloric acid production and utilization, and fireworks, which contribute to further pollution while being released into the environment, particularly in water. Moreover, perchlorate also exists as a salt with the combination of various cations, including ammonium, sodium, and potassium perchlorate [6–8]. From these sources, 90% production of perchlorate is in the form of ammonium perchlorate due to low-cost manufacturing and ease of utilization [9,10]. Perchlorate contamination in food and water is long-lasting and toxic to humans because perchlorate requires high activation energy for degradation/reduction [1,6,11,12]. Perchlorate interferes with iodine uptake in the thyroid gland [3,13,14]. Various technologies described in the literature have been used for the reduction of perchlorate from the water and soil, which includes adsorption [15–18], membrane filtration [3,8,19–22], electrochemical reduction [10,14,23,24], biochemical reduction [1,12,25–27], ion exchange [11,13,28–30] and bio reduction [2,6,31–34]. In addition, the utilization of integrating methodologies for perchlorate reduction were also mentioned in the literature and have shown fascinating results in perchlorate reduction, which includes combination of ion exchange with bio reduction [3,11,35], but no method has provided a clear solution for the complete reduction of perchlorate.

The biological removal of perchlorate is assumed to be a promising method for the reduction of perchlorate. In previous studies, several microbial strains were used and developed for the reduction of perchlorate, which is similar to other contaminants such as nitrates, phosphate, chlorate and 1,4-dioxane [2,4,10,31,36–41]. In most of the cases, dissimilar perchlorate reducing bacteria were grown that were capable of nitrate reduction in the environment, and they preferentially utilized nitrate over perchlorate as a terminal electron acceptor, which hindered the reduction of perchlorate [6]. The bacterial strains used as the perchlorate reducing bacteria (PRBs) are *Dechloromonas* and *Dechlorsomna* (efficient for perchlorate and chlorate reduction) [31,34], *Dechloromonas PC1* and *Dechlorosomna* sp. (reduces both perchlorate and chlorate along with a transient accumulation of chlorate for utilization in syntrophic association by PRBs) [32,42], and other strains used for perchlorate reduction are denitrifying bacteria, including *Rhodobacter capsulatus*, *Rhodobacter sphaeroides*, etc. [2,4,39]. Moreover, the PRBs were isolated from a wide variety of habitats, including soils, gold mine drainage sediments, hot springs, etc. [6,31,33,40]. However, the isolation process was not energy efficient and required a high amount of energy, thus to overcome it, various acetate-oxidizing PRBs have been cultivated in which the number of bacterial cells ranged increases from $2.31 \pm 1.33 \times 10^3$ to $2.4 \pm 1.74 \times 10^6$ cells $g^{-1}$ in a sample with same energy consumption and comparatively higher perchlorate reduction efficiency [6,43,44].

Using *Acinetobactor berziniae* Gram-negative strains as PRBs is responsible for increase in reduction to 1.33-fold by using 1.067 mM/L of anthraquinone-1-sulfonate ($\alpha$-AQS) as a mediator [41]. In addition, results from other studies elucidated that washed cells (*Dechloromonas* and *Azospira* species) readily reduced 90 mg/L perchlorate in a bioelectrical reactor in the presence of a mediator (2,6-anthraquinone disulfonate (AQDS)) [10]. The cultivated microbial strains have several environmental limitations, which reduce their real wastewater application. In order to overcome the issues, mixed microbial culture was used to degrade the slow degradation contaminants, including perchlorate [5,45,46]. The mixed microbial culture provides an efficient and stable solution for perchlorate reduction by using different carbon sources as the electron donor during the process [33,45,47]. Moreover, limited studies were available on utilizing acetate as the sole carbon source for mixed microbial culture acclimation and for evaluating the impact of nitrate and sulfate on perchlorate reduction.

The goal of the study was to evaluate the impact of enriched mixed microbial culture grown using sludge from a wastewater treatment plant and using acetate as the carbon source. Moreover, the effect of nitrate on perchlorate reduction and in acclimation of mixed microbial culture was also tested along with acetate, using it as the substrate in batch testing. Moreover, the Monod equation was used to model the kinetic parameters for the perchlorate degradation by varying the acetate concentration. In addition, the variation in biomass concentration was also determined through standard methods. Loss of a carbon source in terms of COD removal was recorded throughout the study, and the impact of nitrate and sulfate as the inhibition role was also elaborated. A lab-scale aerobic reactor was used to optimize the HRT and acetate perchlorate ratio for optimum operation.

## 2. Results and Discussions

### 2.1. Impact of Medium Enrichment by Nitrate on Perchlorate Reduction

Batch testing during enrichment of mixed microbial culture with nitrate as the electron acceptor has significant effects on perchlorate reduction, as elucidated in Figure 1. As Figure 1a illustrated, culture A (nitrate and perchlorate as the electron acceptor) was responsible for the reduction of perchlorate to ≈0 mg/L after 12 h of exposure, but the mixed microbial culture B (perchlorate as the electron acceptor) and culture C (nitrate as the electron acceptor) required 24 h for the complete reduction of the perchlorate from the feed water. In addition, the inhibition impact of the different electron acceptors was also tested for the perchlorate reduction for mixed microbial culture A where perchlorate, nitrate and sulfate were used as the electron acceptor in different combinations, as shown in Figure 1b. The concentration used for the degradation analysis of perchlorate, nitrate and sulfate by culture A was 400 mg/L, 500 mg/L and 500 mg/L, respectively. The results revealed that after 12 h of exposure in batch testing, a 100% reduction in perchlorate was recorded in different combinations. In the initial stages, due to the presence of nitrate and sulfate, the reduction rate of perchlorate was slow as compared to reduction alone. However, after a significant reduction in nitrate during the initial 6 h, there was an increase in the reduction of perchlorate, which clarified that nitrate could degrade faster than perchlorate, and sulfate has little effect on the degradation of perchlorate, as shown in Figure 1c. Thus, the presence of nitrate hindered the mixed microbial culture from degrading the perchlorate due to the higher reduction affinity of nitrate. Moreover, nitrate is a faster degradation compound compared to perchlorate, and for efficient perchlorate reduction by the mixed microbial culture, nitrate concentration would be minimized [6,10,48,49].

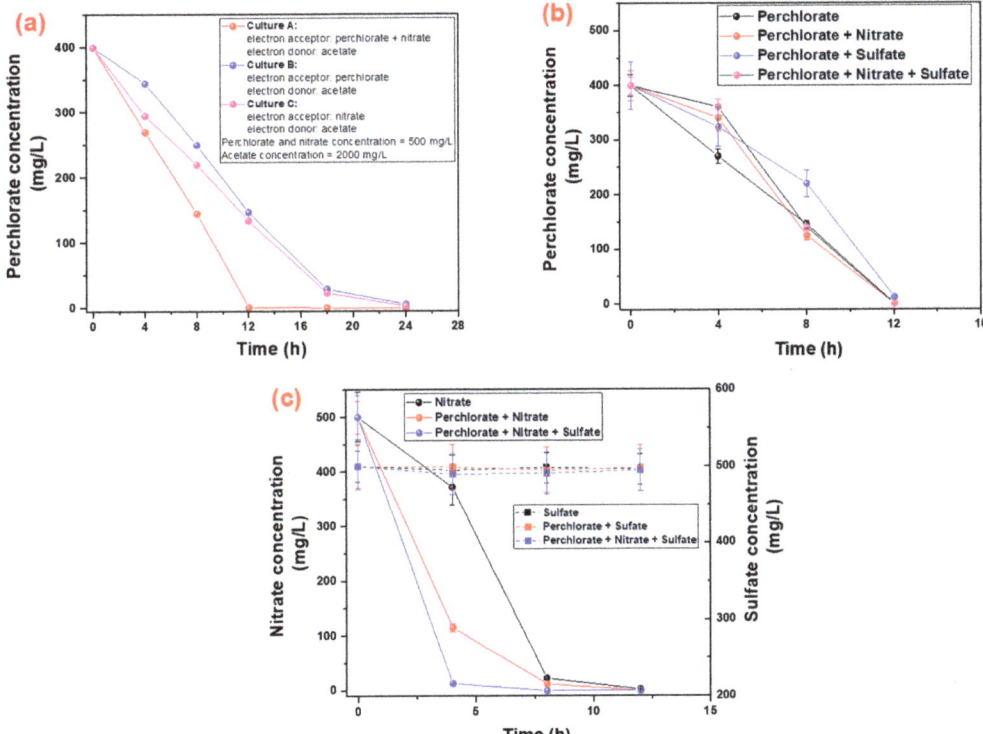

**Figure 1.** (**a**): Perchlorate reduction by mixed microbial culture grown under different combinations of perchlorate and nitrate as the electron acceptor and the acetate as electron donor; (**b**): impact of

different combined electron acceptors in perchlorate reduction by culture A; (**c**): reduction of nitrate and sulfate in exposure to mixed microbial culture A (error bar represents the SD of 3 readings with the maximum SD: 8.6%) (concentration of 400 mg/L, 500 mg/L and 500 mg/L of perchlorate, nitrate and sulfate was used in feed solution as electron acceptor).

Figure 2a illustrates the increase in biomass concentration during the reduction of perchlorate by mixed microbial culture A in the presence of different electron acceptors. The increase in biomass concentration from 1000 mg MLVSS/L to 1250 ± 31.25, 1200 ± 9.6, 1300 ± 19.5, 1150 ± 10.34 and 1450 ± 29 mg MLVSS/L were recorded for the feed solution containing perchlorate, nitrate, perchlorate + nitrate, perchlorate + sulfate and perchlorate +nitrate + sulfate solution, respectively, as the electron acceptor. The experiment was performed with a combination of nitrate + perchlorate, perchlorate + sulfate, nitrate, perchlorate, sulfate, and nitrate + perchlorate + sulfate as the electron acceptor, as mentioned above. In this study, we were unable to perform experiments in combination with nitrate + sulfate as the electron acceptor due to a major concern relating to evaluating the perchlorate reduction. Moreover, the degradation rate of nitrate is faster than perchlorate, and perchlorate degradation starts after the complete degradation of nitrate, so we expected that the presence of sulfate facilitated the nitrate reduction due to biomass production. However, individually it would have no significant effect. Furthermore, faster degradation of perchlorate and nitrate by a mixed microbial culture provides proof of the compatibility of these two chemicals in the microbial community, which was ultimately responsible for the reduction of the feed COD. The acetate concentration was recorded as the COD during the process and elucidated in Figure 2b. The reduction in the COD was from 2000 mg/L to 1400 ± 70, 1500 ± 119, 900 ± 63, 1500 ± 124 and 750 ± 30 mg/L for perchlorate, nitrate, perchlorate + nitrate, perchlorate + sulfate and perchlorate + nitrate + sulfate solution, respectively. The reduction in the COD was particularly high in the feed containing the perchlorate combination with nitrate and nitrate + sulfate. Due to the faster reduction of nitrate as compared to the perchlorate, it was elucidated that after the nitrate reduction, the microbial culture reduction potential for the perchlorate was enhanced due to a rapid increase in the biomass concentration (due to the presence of an active microbial community). However, no significant variation in a reduction in sulfate was observed in a mixed microbial culture where acetate was the primary source of the electron donor; thus, no variation in the COD and biomass concentrations of sulfate as the electron acceptor during 12 h of exposure time were observed.

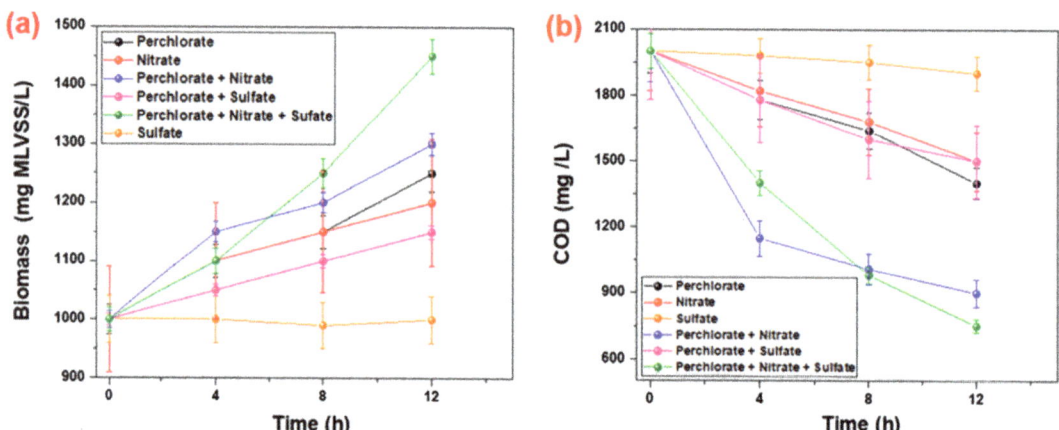

**Figure 2.** Biomass and COD variation under the influence of various electron acceptors. (**a**): Biomass concentration (mg MLVSS/L); (**b**): COD variation (mg/L) (error bar represents the SD with a reading of 3 with maximum SD: 5%).

## 2.2. Optimizing the Electron Donor Dose for Perchlorate Reduction

A batch test reactor was used to evaluate the endogenous decay rate of perchlorate at different concentrations of acetate as an electron donor component. The fixed concentration of 400 mg/L of perchlorate for it and the results are shown in Figure 3a. The endogenous decay of perchlorate by enriched microbial culture under a dose of acetate concentration from 0 to 1500 mg/L for 72 h revealed that the reduction in perchlorate increased with an increase in the concentration of acetate. The results showed that after 72 h, the concentration of 305.17 ± 12.2, 107.24 ± 2.14, 63.7 ± 1.26 and less than 1.5 ± 0.09 mg/L were observed for acetate concentrations of 0, 150, 300 and greater than 600 mg/L, respectively. Moreover, we considered the concentration of 600 mg/L to be optimum, as no significant difference was recorded for a decrease in perchlorate concentration. Figure 3a also shows that 24 h was enough for the retention time of the maximum degradation of perchlorate at 600 mg acetate/L. The figure also defined the optimum acetate to perchlorate ratio for the complete removal of perchlorate, which was 1.4 mg acetate/mg perchlorate at an acetate concentration of 600 mg/L.

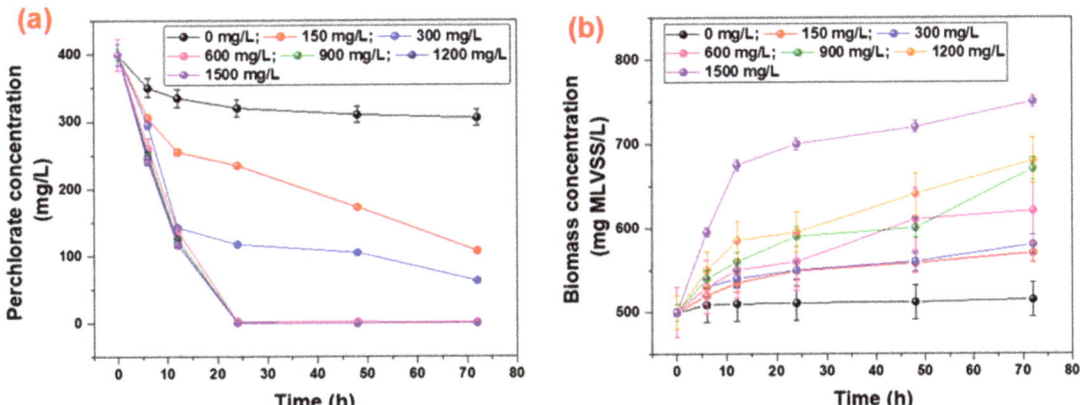

**Figure 3.** Impact of acetate concentration on perchlorate reduction by a mixed microbial culture and biomass concentration (**a**): perchlorate reduction (mg/L); (**b**): biomass concentration (mg MLVSS/L) (error bar represents the SD with a reading of 3 with maximum SD: 5%).

An initial biomass concentration of 500 mg MLVSS/L was used for the batch experiment for endogenous decay of the perchlorate, and the results are shown in Figure 3b. The results show that with an increase in acetate concentration, the production of biomass increases due to the utilization of acetate and perchlorate as food by an enriched mixed microbial culture and significantly decreases the concentration of perchlorate and acetate in the form of COD. The increase in biomass from 500 mg MLVSS/L to 515.65–750.79 mg MLVSS/L was observed for 0–1500 mg/L of acetate at 400 mg/L of perchlorate. Specific substrate utilization was recorded at the 600 mg/L of acetate concentration to evaluate the kinetic coefficients of the degradation of perchlorate for 24 h experiments, and the results are summarized in Table 1. The results showed a good correlation with the previous studies, in which mixed microbial culture was used for the perchlorate reduction with different food sources [5,49]. Logan et al. [45] used the fixed bed bioreactor along with ethanol for the perchlorate reduction and elucidated the optimized kinetic parameters ($K_s$ = 20 mg/L; $q_{max}$ = 1.821 mg/mg·d) for the reduction in perchlorate in the drinking water level. Moreover, the reaction kinetics of perchlorate degradation showed that 190 mg/L of sodium acetate as a carbon source would be enough for complete degradation in the absence of any inhibition chemicals (such as nitrate) [33,45]. Similar results were found in which a single strain of microbial culture was used for the degradation of perchlorate [49]. Yu et al. [49] explained the reduction of perchlorate by using the combined

PRMs and zero-valent ions in batch testing where kinetic parameters of $K_s$ = 8.9 mg/L; $q_{max}$ = 0.65 mg/mg.d were determined and responsible for the increase in perchlorate reduction of up to 4 times compared to a single process. Furthermore, the addition of nitrate and pH delays the reduction in perchlorate, but it is not inhibited completely.

**Table 1.** Summary of kinetic parameter evaluation by a batch test of perchlorate biodegradation.

| Kinetic Parameters | Values |
|---|---|
| $K_s$ (mg Perchlorate/L) | 37.61 |
| $q_{max}$ (mg Perchlorate/mg MLVSS/d) | 0.042 |
| $Y_t$ (mg MLVSS/mg Perchlorate) | 0.281 |

*2.3. Lab Scale Testing*

Lab scale experiments were conducted at different HRT times ranging from 0 to 18 h for five days, each with a 48 h adaption time of the reactor. The results on perchlorate degradation and COD variation are given in Figure 4. The results show that with an increase in the HRT, the degradation of perchlorate increased, as shown in Figure 4a. We considered an HRT of 6 h as optimum because most of the perchlorate was degraded, and the concentration was 2.45 mg/L. In addition to the reduction of perchlorate, the carbon source (acetate) also decreased to a level of 28 mg/L in terms of the COD, and no further reduction was observed for greater than 6 h HRT as seen in Figure 4b. The COD concentration results for the anoxic tank showed that a major reduction took place in the anoxic tank and that recycling of the active biomass played a significant role in the faster degradation of acetate and perchlorate.

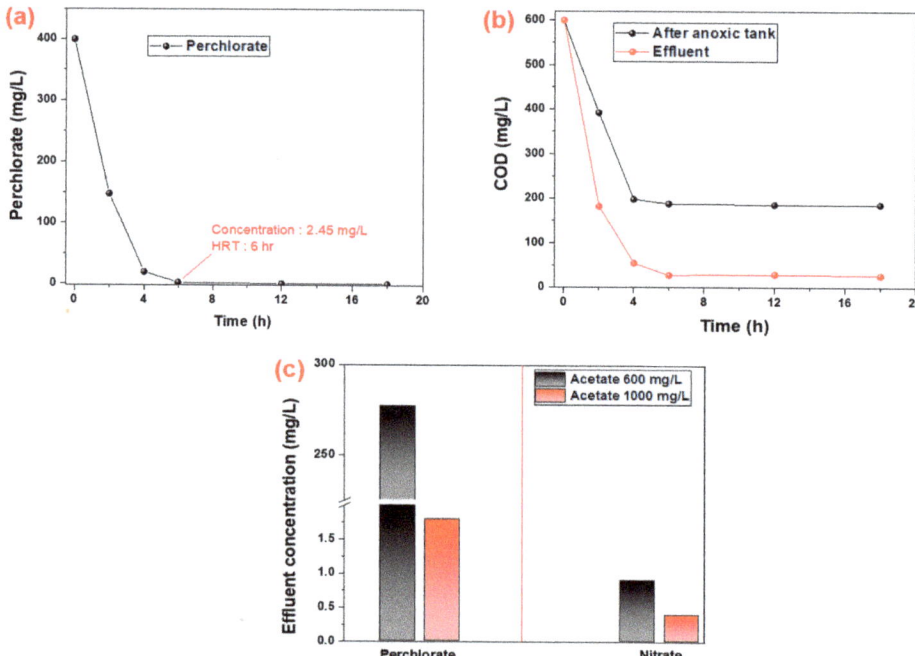

**Figure 4.** Lab scale operation of activated sludge treatment for perchlorate biodegradation by mixed microbial culture. (**a**): HRT optimization at a biomass concentration of 1000 mg MLVSS/L; (**b**): COD variation from the reactor while optimizing the HRT; (**c**): impact of acetate concentration on perchlorate and nitrate reduction at 6 h HRT.

Additional experiments to evaluate the impact of nitrate and the concentration of acetate were also performed, and the results showed that despite the carbon source concentration of the mixed microbial culture, the nitrate experienced a faster degradation rate compared to the perchlorate, as seen in Figure 4c. At a lower concentration of carbon source, there was a 100% nitrate reduction, but perchlorate had less than a 50% reduction, while at a 1000 mg/L acetate dose, both had a ≈100% reduction, which strengthened the observation of the inhibition potential of nitrate toward the perchlorate and was responsible for slowing down its degradation. Thus, for efficient reduction of perchlorate through feed water by a mixed microbial culture, it is highly recommended to remove the nitrate from the feed water or to provide additional carbon, which should be higher than a 1.4 mg acetate/mg perchlorate ratio. Moreover, the pilot scale plant associated with the combination of a plug flow reactor was planned for the real wastewater application for perchlorate degradation through a mixed microbial culture at these design parameters.

## 3. Materials and Methods

### 3.1. Materials

Anaerobic sludge was collected from the digester of a wastewater treatment plant in Suwon, South Korea and used as the seed inoculum for the enrichment of the perchlorate-reducing culture. A mineral salt medium (MSM) was used to grow the mixed culture and enrichment prepared in ultra-pure water according to the methodology mentioned elsewhere [36,38]. The chemicals for MSM preparation were of analytical grade. Concentration and chemicals included in MSM were $NH_4Cl$ (2000 g/L), $KH_2PO_4$ (1000 g/L), $Na_2HPO_4 \cdot 2H_2O$ (1000 g/L), $MgSO_4 \cdot 7H_2O$ (100 g/L), $FeSO_4 \cdot 7H_2O$ (12.6 g/L), $NaMoO_4 \cdot 2H_2O$ (10 g/L), $H_3BO_3$ (0.6 g/L) and $NaSeO_3$ (0.6 g/L). The sodium perchlorate solution was obtained from Sigma-Aldrich Chemicals, Darmstadt, Germany. Figure 5 shows an overview of the experimental study for the bioreduction of $ClO_4^-$.

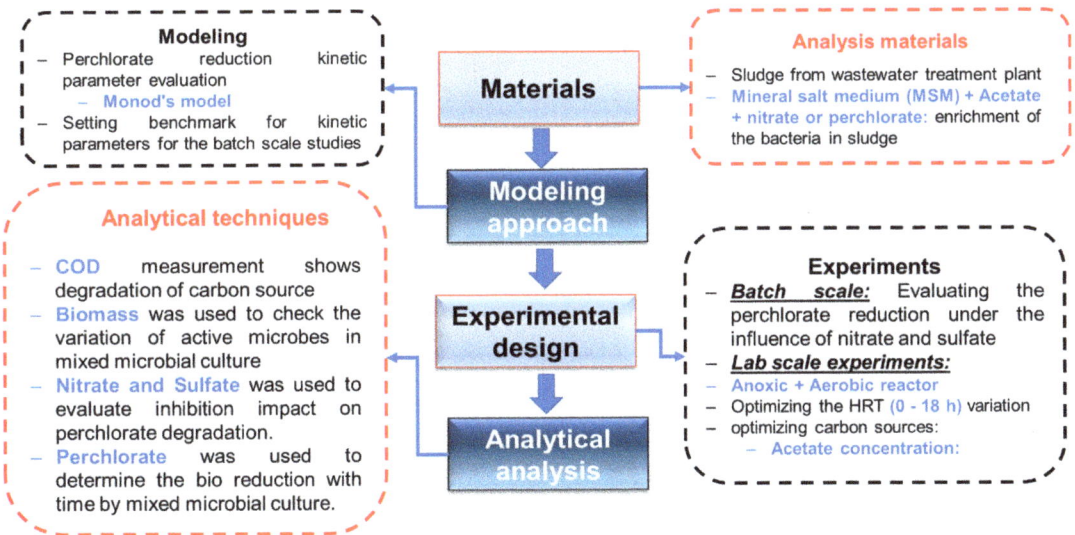

**Figure 5.** Schematic diagram of the experimental study.

### 3.2. Enrichment of Perchlorate Reducing Mixed Bacteria

Enrichments, growth of mixed culture and all experiments were performed in an MSM with the predetermined concentration of perchlorate as the electron acceptor and acetate as the electron donor. They were injected into prepared synthetic wastewater and placed in a bottle for stirring. Synthetic wastewater was prepared by dissolving the predetermined mass of pollutants in DI water, in our case, perchlorate with the combination of nitrate and sulfate. After culturing for 24~48 h, the microorganisms were centrifuged and injected again into newly prepared synthetic wastewater. The anaerobic sludge was centrifuged, all the supernatant was decanted and the settled sludge was washed with MSM two to three times. The sludge obtained after washing was used for the experiment. 1000 mg/L acetate and 500 mg/L perchlorate were added to four 500 mL bottles, each containing equal amounts of biomass and 300 mL of the MSM. All the bottles were capped and shaken at 150 rpm at room temperature (21 ± 2 °C). After 48 h, the biomass was allowed to settle for 1 h. The supernatant was decanted and replaced with a fresh MSM, along with the addition of 1000 mg/L of acetate and 500 mg/L of perchlorate before the start of the next cycle.

### 3.3. Experimental Procedure and Reactor Design

The batch scale experiments were conducted with a fixed biomass concentration of 1000 mg MLVSS/L to evaluate the bioreduction of perchlorate. Moreover, the effect of the medium (enriched mixed culture) acclimation with or without nitrate on perchlorate reduction was also examined. For that, we used 500 mL bottles containing three different acclimated cultures, containing perchlorate (500 mg/L), nitrate (500 mg/L) and acetate (2000 mg/L) in different combinations: *Culture A* (perchlorate and nitrate as the electron acceptor, acetate as the electron donor), *Culture B* (perchlorate/acetate as the electron acceptor/donor) and *Culture C* (nitrate/acetate as the electron acceptor/donor).

We also examined the effect of acetate concentration to determine the optimum relationship between acetate concentration and perchlorate by using the acetate concentration range of 0–1500 mg/L, which was used for medium acclimation to evaluate the reduction of a fixed 400 mg/L perchlorate concentration. In this experiment, 300 mL of mixed microbial culture amended with an acetate concentration of 0–1500 mg/L was shifted to different 500 mL bottles, which contained fixed 400 mg/L perchlorate. The bottles were purged with $N_2$ and then sealed and incubated in a shaker at 150 rpm. All the experiments were conducted at room temperature (21 ± 2), and the samples were collected at regular intervals for analysis. Moreover, these batch scale experiments were conducted to evaluate the decaying parameters by Monod's equation, as mentioned in Section 3.5. Modeling Approach.

A continuous reactor was used to determine the optimized hydraulic retention time (HRT) for the bioreduction of perchlorate under the predetermined optimized acetate and nitrate concentration; the schematic is shown in Figure 6, which shows the anoxic tank and aerobic tank of 3.2 L and 2.7 L. The pump was used to regulate the flow and HRT of the system. The anoxic tank was used to maintain the mixed liquid volatile suspended solids (MLVSS) in suspension by a stirrer. Moreover, a small amount of the mixed microbial culture was injected with 400 mg/L of perchlorate for 48 h in the anoxic tank as an adaptation time. Consideration of 48 h as the adaptation time was based on the batch test analysis. In the batch test, 24 h was used for the degradation of perchlorate in 500 mL bottles by various enriched mixed microbial cultures; thus, we expected that in a larger tank, the bacteria need more time to adapt to the environmental conditions. Therefore, we chose the 48 h adaptation time in an anoxic tank. Air was supplied to the aerobic tank from the bottom at a uniform rate, and sludge was returned from the settling tank to the anoxic tank. The samples were collected after every 24 h period for five days of operation for each HRT duration. The perchlorate concentration, chemical oxidation demand (COD) and nitrate concentration were measured (the average results are given). The biomass concentration of 1000 mg MLVSS/L was maintained during the process.

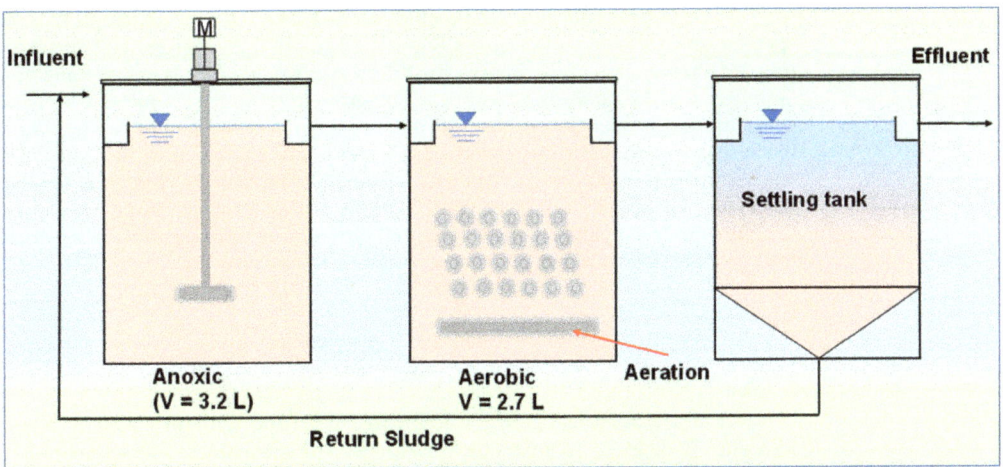

**Figure 6.** Schematic of lab scale reactor setup with a schematic for biodegradation of perchlorate.

*3.4. Analytical Procedure*

Aliquot samples of 2 mL from bottle tests were taken by using a 5 mL sampling syringe, filtered through a 0.2 μm membrane filter and stored in a refrigerator at 4 °C prior to being analyzed. Perchlorate was measured using Liquid Chromatograph/Mass Spectrometry (LC/MS) equipped with an XBrige™ C18 3.5 μm, 2.1 mm × 100 mm column, Waters, Milford, MA, USA and a conductivity detector. The detection limit for perchlorate was 0.05 mg/L. 0.1% formic acid in distilled water, 0.1% formic acid in methanol, 10% methanol and 90% methanol served as the eluent. The flow rate of the eluent was 0.4 mL/min. A 100 mg/L of perchlorate solution was confected to prepare perchlorate calibration standards. Moreover, the COD was measured for acetate concentration measurement and measured by centrifuging the liquid culture sample at 14,000 rpm for 5 min to separate the biomass portion completely. An appropriate volume of supernatant was used for COD measurement in accordance with the standard method.

The biomass concentration was measured by a mass measuring method. An appropriate 50 mL volume of culture was filtered (vacuum filtration by using 0.2 μm PTFE standard filter paper) and weighed after the samples were dried at 105 °C for 4 h, which gave the biomass concentration value. Nitrate was measured using the Chromotropic Acid Method; the high range (0.2 to 30.0 mg/L $NO_3$–N) and sulfate were measured using the SulfaVer 4 Method (2 to 70 mg/L $SO_4^{2-}$) with HACH spectrophotometry at 410 nm and 450 nm wavelength.

*3.5. Modeling Approach*

We investigated the biodegradation of $ClO_4^-$ by various microbial metabolism processes of pure cultures, with a single strain isolated and characterized for $ClO_4^-$ biodegradation. Generally, the perchlorate-reducing bacteria derived from the "16S rDNA sequence" belongs to the subclasses of *Proteobacteria* (α, β, γ and ε), as mentioned earlier [6,33,41]. However, in this study, the mixed culture for the perchlorate degradation was tested by an anoxic–aerobic bioreactor for the lab scale, which was limited in previous studies, as shown in Figure 6. The coefficient of kinetic parameters for these reactors for the rate of perchlorate utilization was determined by applying the Monod model Equation (1).

$$\frac{dS}{dt} = -\frac{q_m S}{K_s + S}\left[X_a^0 + Y_t(S_0 - S)\right] \quad (1)$$

where $K_s$ represents half-saturation coefficients (mg $ClO_4^-$/L), $q_m$ represents the maximum specific substrate utilization ((mg $ClO_4^-$/mg-MLSS/h), $Y_t$ represents the true cell yield (mg-MLSS/mg $ClO_4^-$), $X_a^0$ represents the concentration of the biomass initially, (mg/L) $S_0$ represents the concentration of the substrate initially (mg/L) and $S$ represents the concentration of substrate at a function of time $t$ (mg/L), as mentioned earlier [9,45,49]. The batch scale experiments were conducted to determine the substrate utilization constants and estimate the unknown parameters, where the optimized concentration of acetate was used from a range varying from 0 to 1500 mg/L as the electron donor in Equation (1).

The observed yield strength of the cell growth was estimated by the linearization of the substrate (1,4-Dioxane) utilization, as shown in Equation (2).

$$Y_{ob} = \frac{dX_a/dt}{dS/dt} = \frac{C_{growth}}{Perchlorate_{utilization}} \quad (2)$$

## 4. Conclusions

All mixed microbial cultures acclimated with and without nitrate had the ability to reduce perchlorate in anaerobic conditions. Moreover, negligible reduction in sulfate concentration was observed by mixed culture and had no significant contribution to perchlorate reduction. However, there was an increase in biomass production in the presence of all three chemicals as electron acceptors, which tended to reduce the COD level and increase the perchlorate degradation. Additionally, we confirmed that prior to perchlorate degradation, nitrate removal took place, which showed that an enriched mixed microbial culture acclimated with acetate as a carbon source has a great affinity for nitrate and was an inhibition factor for the perchlorate. With an excess carbon source, the effect of various concentrations of nitrate on perchlorate reduction was similar. In the batch experiment, the optimal acetate to perchlorate ratio of 1.4 mg acetate/mg perchlorate was observed at 600 mg/L of acetate concentration. Along with that the kinetic parameters of the Monod equation were Y = 0.281 mg biomass/ mg perchlorate, $K_s$ = 37.619 mg/L and $q_{max}$ = 0.042 mg perchlorate/ mg biomass/h. The reactor experiment to determine the optimal operation condition had an HRT of 6 h with 600 mg/L of acetate. However, in elucidating the impact of nitrate at optimal conditions, a reduction in perchlorate was observed with a removal efficiency of less than 60% and complete removal of nitrate due to the inhibition impact of nitrate. Furthermore, an increase in the carbon source to 1000 mg/L achieved the 100% removal of nitrate and perchlorate at 6 h HRT time. These results illustrated the significance of mixed microbial culture for the removal of perchlorate in an anaerobic environment and that it is necessary to reduce the nitrate concentration prior to perchlorate reduction or to provide an additional carbon source for simultaneous degradation of nitrate and perchlorate. Furthermore, through a reactor experiment, we also verified that the acetate concentration as a carbon source has a significant impact on the reduction of nitrate and perchlorate. Future experiments on designing real-world applications for wastewater treatment plants are under consideration for a continuous flow reactor with various contaminants of inhibition.

**Author Contributions:** Conceptualization, K.H.L. and H.Y.; methodology, H.Y.; validation, K.H.L. and J.-W.P.; formal analysis, H.Y.; investigation, H.Y.; resources, K.H.L.; data curation, K.H.L.; writing—original draft preparation, H.Y.; writing—review and editing, K.H.L.; visualization, H.Y.; supervision, J.-W.P.; project administration, J.-W.P.; funding acquisition, K.H.L. All authors have read and agreed to the published version of the manuscript.

**Funding:** This research received no external funding.

**Institutional Review Board Statement:** Not applicable.

**Informed Consent Statement:** Not applicable.

**Conflicts of Interest:** The authors declare no conflict of interest.

## References

1. Achenbach, L.A.; Bender, K.S.; Sun, Y.; Coates, J.D. The biochemistry and genetics of microbial perchlorate reduction. In *Perchlorate*; Springer: Berlin/Heidelberg, Germany, 2006; pp. 297–310.
2. Coates, J.D.; Achenbach, L.A. Microbial perchlorate reduction: Rocket-fuelled metabolism. *Nat. Rev. Microbiol.* **2004**, *2*, 569–580. [CrossRef] [PubMed]
3. Li, T.; Ren, Y.; Zhai, S.; Zhang, W.; Zhang, W.; Hua, M.; Lv, L.; Pan, B. Integrating cationic metal-organic frameworks with ultrafiltration membrane for selective removal of perchlorate from Water. *J. Hazard. Mater.* **2020**, *381*, 120961. [CrossRef] [PubMed]
4. Okeke, B.C.; Giblin, T.; Frankenberger, W.T., Jr. Reduction of perchlorate and nitrate by salt tolerant bacteria. *Environ. Pollut.* **2002**, *118*, 357–363. [CrossRef]
5. Wallace, W.; Beshear, S.; Williams, D.; Hospadar, S.; Owens, M. Perchlorate reduction by a mixed culture in an up-flow anaerobic fixed bed reactor. *J. Ind. Microbiol. Biotechnol.* **1998**, *20*, 126–131. [CrossRef]
6. Bardiya, N.; Bae, J.-H. Dissimilatory perchlorate reduction: A review. *Microbiol. Res.* **2011**, *166*, 237–254. [CrossRef]
7. Hu, J.; Xian, Y.; Wu, Y.; Chen, R.; Dong, H.; Hou, X.; Liang, M.; Wang, B.; Wang, L. Perchlorate occurrence in foodstuffs and water: Analytical methods and techniques for removal from water—A review. *Food Chem.* **2021**, *360*, 130146. [CrossRef]
8. Roach, J.D.; Tush, D. Equilibrium dialysis and ultrafiltration investigations of perchlorate removal from aqueous solution using poly (diallyldimethylammonium) chloride. *Water Res.* **2008**, *42*, 1204–1210. [CrossRef]
9. Nozawa-Inoue, M.; Scow, K.M.; Rolston, D.E. Reduction of perchlorate and nitrate by microbial communities in vadose soil. *Appl. Environ. Microbiol.* **2005**, *71*, 3928–3934. [CrossRef]
10. Thrash, J.C.; Van Trump, J.I.; Weber, K.A.; Miller, E.; Achenbach, L.A.; Coates, J.D. Electrochemical stimulation of microbial perchlorate reduction. *Environ. Sci. Technol.* **2007**, *41*, 1740–1746. [CrossRef]
11. Lehman, S.G.; Badruzzaman, M.; Adham, S.; Roberts, D.J.; Clifford, D.A. Perchlorate and nitrate treatment by ion exchange integrated with biological brine treatment. *Water Res.* **2008**, *42*, 969–976. [CrossRef]
12. Liebensteiner, M.G.; Oosterkamp, M.J.; Stams, A.J. Microbial respiration with chlorine oxyanions: Diversity and physiological and biochemical properties of chlorate-and perchlorate-reducing microorganisms. *Ann. N. Y. Acad. Sci.* **2016**, *1365*, 59–72. [CrossRef] [PubMed]
13. Darracq, G.; Baron, J.; Joyeux, M. Kinetic and isotherm studies on perchlorate sorption by ion-exchange resins in drinking water treatment. *J. Water Process Eng.* **2014**, *3*, 123–131. [CrossRef]
14. Wang, D.; Shah, S.I.; Chen, J.; Huang, C. Catalytic reduction of perchlorate by $H_2$ gas in dilute aqueous solutions. *Sep. Purif. Technol.* **2008**, *60*, 14–21. [CrossRef]
15. Hou, P.; Cannon, F.S.; Nieto-Delgado, C.; Brown, N.R.; Gu, X. Effect of preparation protocol on anchoring quaternary ammonium/epoxide-forming compound into granular activated carbon for perchlorate adsorption: Enhancement by Response Surface Methodology. *Chem. Eng. J.* **2013**, *223*, 309–317. [CrossRef]
16. Kim, Y.-N.; Lee, Y.-C.; Choi, M. Complete degradation of perchlorate using Pd/N-doped activated carbon with adsorption/catalysis bifunctional roles. *Carbon* **2013**, *65*, 315–323. [CrossRef]
17. Xu, J.; Gao, N.; Zhao, D.; An, N.; Li, L.; Xiao, J. Bromate reduction and reaction-enhanced perchlorate adsorption by $FeCl_3$-impregnated granular activated carbon. *Water Res.* **2019**, *149*, 149–158. [CrossRef]
18. Xu, J.-h.; Gao, N.-y.; Deng, Y.; Sui, M.-h.; Tang, Y.-l. Perchlorate removal by granular activated carbon coated with cetyltrimethyl ammonium bromide. *J. Colloid Interface Sci.* **2011**, *357*, 474–479. [CrossRef]
19. Han, J.; Kong, C.; Heo, J.; Yoon, Y.; Lee, H.; Her, N. Removal of perchlorate using reverse osmosis and nanofiltration membranes. *Environ. Eng. Res.* **2012**, *17*, 185–190. [CrossRef]
20. Roach, J.D.; Lane, R.F.; Hussain, Y. Comparative study of the uses of poly (4-vinylpyridine) and poly (diallyldimethylammonium) chloride for the removal of perchlorate from aqueous solution by polyelectrolyte-enhanced ultrafiltration. *Water Res.* **2011**, *45*, 1387–1393. [CrossRef]
21. Xie, Y.; Li, S.; Wu, K.; Wang, J.; Liu, G. A hybrid adsorption/ultrafiltration process for perchlorate removal. *J. Membr. Sci.* **2011**, *366*, 237–244. [CrossRef]
22. Yoon, J.; Amy, G.; Chung, J.; Sohn, J.; Yoon, Y. Removal of toxic ions (chromate, arsenate, and perchlorate) using reverse osmosis, nanofiltration, and ultrafiltration membranes. *Chemosphere* **2009**, *77*, 228–235. [CrossRef] [PubMed]
23. Rusanova, M.Y.; Polášková, P.; Muzikař, M.; Fawcett, W.R. Electrochemical reduction of perchlorate ions on platinum-activated nickel. *Electrochim. Acta* **2006**, *51*, 3097–3101. [CrossRef]
24. Wan, D.; Li, Q.; Chen, J.; Niu, Z.; Liu, Y.; Li, H.; Xiao, S. Simultaneous bio-electrochemical reduction of perchlorate and electro-disinfection in a novel Moving-Bed Biofilm Reactor (MBBR) based on proton-exchange membrane electrolysis. *Sci. Total Environ.* **2019**, *679*, 288–297. [CrossRef] [PubMed]
25. Ader, M.; Chaudhuri, S.; Coates, J.D.; Coleman, M. Microbial perchlorate reduction: A precise laboratory determination of the chlorine isotope fractionation and its possible biochemical basis. *Earth Planet. Sci. Lett.* **2008**, *269*, 605–613. [CrossRef]
26. Davila, A.F.; Willson, D.; Coates, J.D.; McKay, C.P. Perchlorate on Mars: A chemical hazard and a resource for humans. *Int. J. Astrobiol.* **2013**, *12*, 321–325. [CrossRef]
27. Gu, B.; Coates, J.D. *Perchlorate: Environmental Occurrence, Interactions and Treatment*; Springer Science & Business Media: Berlin/Heidelberg, Germany, 2006.

28. Faccini, J.; Ebrahimi, S.; Roberts, D.J. Regeneration of a perchlorate-exhausted highly selective ion exchange resin: Kinetics study of adsorption and desorption processes. *Sep. Purif. Technol.* **2016**, *158*, 266–274. [CrossRef]
29. Song, W.; Gao, B.; Wang, H.; Xu, X.; Xue, M.; Zha, M.; Gong, B. The rapid adsorption-microbial reduction of perchlorate from aqueous solution by novel amine-crosslinked magnetic biopolymer resin. *Bioresour. Technol.* **2017**, *240*, 68–76. [CrossRef]
30. Wu, M.; Wang, S.; Gao, N.; Zhu, Y.; Li, L.; Niu, M.; Li, S. Removal of perchlorate from water using a biofilm magnetic ion exchange resin: Feasibility and effects of dissolved oxygen, pH and competing ions. *RSC Adv.* **2016**, *6*, 73365–73372. [CrossRef]
31. Achenbach, L.A.; Michaelidou, U.; Bruce, R.A.; Fryman, J.; Coates, J.D. *Dechloromonas agitata* gen. nov., sp. nov. and *Dechlorosoma suillum* gen. nov., sp. nov., two novel environmentally dominant (per)chlorate-reducing bacteria and their phylogenetic position. *Int. J. Syst. Evol. Microbiol.* **2001**, *51*, 527–533. [CrossRef]
32. Dudley, M.; Salamone, A.; Nerenberg, R. Kinetics of a chlorate-accumulating, perchlorate-reducing bacterium. *Water Res.* **2008**, *42*, 2403–2410. [CrossRef]
33. Logan, B.E.; Wu, J.; Unz, R.F. Biological Perchlorate Reduction in High-Salinity Solutions. *Water Res.* **2001**, *35*, 3034–3038. [CrossRef]
34. Tan, Z.; Reinhold-Hurek, B. *Dechlorosoma suillum* Achenbach et al. 2001 is a later subjective synonym of *Azospira oryzae* Reinhold-Hurek and Hurek 2000. *Int. J. Syst. Evol. Microbiol.* **2003**, *53*, 1139–1142. [CrossRef]
35. Kim, Y.-N.; Choi, M. Synergistic integration of ion-exchange and catalytic reduction for complete decomposition of perchlorate in waste water. *Environ. Sci. Technol.* **2014**, *48*, 7503–7510. [CrossRef] [PubMed]
36. Lee, K.H.; Wie, Y.M.; Jahng, D.; Yeom, I.T. Effects of Additional Carbon Sources in the Biodegradation of 1,4-Dioxane by a Mixed Culture. *Water* **2020**, *12*, 1718. [CrossRef]
37. Malmqvist, Å.; Welander, T.; Moore, E.; Ternström, A.; Molin, G.; Stenström, I.-M. *Ideonella dechloratans* gen. nov., sp. nov., a new bacterium capable of growing anaerobically with chlorate as an electron acceptor. *Syst. Appl. Microbiol.* **1994**, *17*, 58–64. [CrossRef]
38. Parales, R.E.; Adamus, J.E.; White, N.; May, H.D. Degradation of 1,4-dioxane by an actinomycete in pure culture. *Appl. Environ. Microbiol.* **1994**, *60*, 4527–4530. [CrossRef]
39. Roldan, M.; Reyes, F.; Moreno-Vivian, C.; Castillo, F. Chlorate and nitrate reduction in the phototrophic bacteria *Rhodobacter capsulatus* and *Rhodobacter sphaeroides*. *Curr. Microbiol.* **1994**, *29*, 241–245. [CrossRef]
40. Weelink, S.A.; Tan, N.C.; ten Broeke, H.; van den Kieboom, C.; van Doesburg, W.; Langenhoff, A.A.; Gerritse, J.; Junca, H.; Stams, A.J. Isolation and characterization of *Alicycliphilus denitrificans* strain BC, which grows on benzene with chlorate as the electron acceptor. *Appl. Environ. Microbiol.* **2008**, *74*, 6672–6681. [CrossRef]
41. Zhang, Y.; Lu, J.; Guo, J.; Wang, Q.; Lian, J.; Wang, Y.; Zhang, C.; Yang, J. Isolation and characterization of a perchlorate-reducing *Acinetobacter bereziniae* strain GWF. *Biotechnol. Biotechnol. Equip.* **2016**, *30*, 935–941. [CrossRef]
42. Nerenberg, R.; Kawagoshi, Y.; Rittmann, B.E. Kinetics of a hydrogen-oxidizing, perchlorate-reducing bacterium. *Water Res.* **2006**, *40*, 3290–3296. [CrossRef]
43. Coates, J.D.; Michaelidou, U.; Bruce, R.A.; O'Connor, S.M.; Crespi, J.N.; Achenbach, L.A. Ubiquity and diversity of dissimilatory (per) chlorate-reducing bacteria. *Appl. Environ. Microbiol.* **1999**, *65*, 5234–5241. [CrossRef]
44. Wu, J.; Unz, R.F.; Zhang, H.; Logan, B.E. Persistence of perchlorate and the relative numbers of perchlorate-and chlorate-respiring microorganisms in natural waters, soils, and wastewater. *Bioremediation J.* **2001**, *5*, 119–130. [CrossRef]
45. Logan, B.E. Evaluation of Biological Reactors to Degrade Perchlorate to Levels Suitable for Drinking Water. In *Perchlorate in the Environment*; Urbansky, E.T., Ed.; Springer: Boston, MA, USA, 2000; pp. 189–197.
46. Ricardo, A.R.; Carvalho, G.; Velizarov, S.; Crespo, J.G.; Reis, M.A. Kinetics of nitrate and perchlorate removal and biofilm stratification in an ion exchange membrane bioreactor. *Water Res.* **2012**, *46*, 4556–4568. [CrossRef]
47. Kim, K.; Logan, B.E. Microbial reduction of perchlorate in pure and mixed culture packed-bed bioreactors. *Water Res.* **2001**, *35*, 3071–3076. [CrossRef]
48. Lee, K.H.; Khan, I.A.; Inam, M.A.; Khan, R.; Wie, Y.M.; Yeom, I.T. Efficacy of Continuous Flow Reactors for Biological Treatment of 1,4-Dioxane Contaminated Textile Wastewater Using a Mixed Culture. *Fermentation* **2022**, *8*, 143. [CrossRef]
49. Yu, X.; Amrhein, C.; Deshusses, M.A.; Matsumoto, M.R. Perchlorate Reduction by Autotrophic Bacteria in the Presence of Zero-Valent Iron. *Environ. Sci. Technol.* **2006**, *40*, 1328–1334. [CrossRef]

Article

# Effective Removal of Methylene Blue on EuVO$_4$/g-C$_3$N$_4$ Mesoporous Nanosheets via Coupling Adsorption and Photocatalysis

Xia Ran, Li Wang, Bo Xiao, Li Lei, Jinming Zhu, Zuoji Liu, Xiaolan Xi, Guangwei Feng, Rong Li and Jian Feng *

Engineering Research Center for Molecular Medicine, School of Basic Medical Sciences, Guizhou Medical University, Guiyang 550025, China
* Correspondence: jfeng@gmc.edu.cn; Tel.: +86-851-88174017

**Abstract:** In this study, we first manufactured ultrathin g-C$_3$N$_4$ (CN) nanosheets by thermal etching and ultrasonic techniques. Then, EuVO$_4$ (EV) nanoparticles were loaded onto CN nanosheets to form EuVO$_4$/g-C$_3$N$_4$ heterojunctions (EVCs). The ultrathin and porous structure of the EVCs increased the specific surface area and reaction active sites. The formation of the heterostructure extended visible light absorption and accelerated the separation of charge carriers. These two factors were advantageous to promote the synergistic effect of adsorption and photocatalysis, and ultimately enhanced the adsorption capability and photocatalytic removal efficiency of methylene blue (MB). EVC-2 (2 wt% of EV) exhibited the highest adsorption and photocatalytic performance. Almost 100% of MB was eliminated via the adsorption–photocatalysis synergistic process over EVC-2. The MB adsorption capability of EVC-2 was 6.2 times that of CN, and the zero-order reaction rate constant was 5 times that of CN. The MB adsorption on EVC-2 followed the pseudo second-order kinetics model and the adsorption isotherm data complied with the Langmuir isotherm model. The photocatalytic degradation data of MB on EVC-2 obeyed the zero-order kinetics equation in 0–10 min and abided by the first-order kinetics equation for 10–30 min. This study provided a promising EVC heterojunctions with superior synergetic effect of adsorption and photocatalysis for the potential application in wastewater treatment.

**Keywords:** adsorption; photocatalysis; EuVO$_4$/g-C$_3$N$_4$ heterojunction; methylene blue; Langmuir isotherm

## 1. Introduction

With the rapid development of modern industry and the consumption of fossil energy, environmental pollution has become increasingly serious. It is regarded as one of the major hindrances for the sustainable development of human society [1]. The development of renewable and green water purification technology for effective removal of organic contaminants is an extraordinarily problematic undertaking. In recent decades, the strategy coupling adsorption and photocatalytic degradation based on semiconductor catalyst has attracted considerable attention in wastewater treatment due to its high efficiency, low energy consumption, the wide availability of adsorbates, superior recoverability, and less secondary pollution [2,3]. In this strategy, the adsorption process can concentrate organic contaminants from the aqueous solution on the surface of a catalyst. Thereafter, the photocatalytic degradation process can ultimately mineralize organic contaminants to small molecule compounds such as H$_2$O and CO$_2$. Therefore, the active sites on the adsorbent surface are recovered and utilized for the next adsorption and photocatalysis process. The integration of adsorption and photocatalytic degradation is consequently a renewable and green water purification technology for effective removal of organic contaminants.

Graphitic carbon nitride (g-C$_3$N$_4$, CN) has been widely investigated in environmental remediation as a semiconductor with impressive merits, including a proper band gap for visible light harvesting, low toxicity, high stability, low cost, and so on [4]. Researchers have reported the adsorption process on CN for wastewater treatment [5–7]. However, both the adsorption and photocatalytic degradation performance of CN still need to be improved. Generally, the porous structure and large surface area are beneficial for ameliorating the adsorption capability [3]. The ultrathin and layered structure of CN nanosheets has a higher specific surface area and more abundant adsorption sites than bulk g-$C_3N_4$. Therefore, the exfoliation of bulk g-$C_3N_4$ to obtain ultrathin CN nanosheets is considered to be an effective morphology control technique to promote the adsorption capability [6]. For photocatalysis, on the other hand, the diffusion distance of photoinduced charge carriers is shortened in ultrathin CN nanosheets, which can reduce the recombination of photoinduced charge carriers and boost the photocatalytic performance [8]. In addition, semiconductors with suitable band gaps could be employed to design a g-$C_3N_4$-based heterojunction [9–12]. The spatial separation of photoinduced charge carriers at the interface in heterojunction can efficiently suppress its recombination. It is expected to further improve the photocatalytic performance of g-$C_3N_4$.

Europium vanadate ($EuVO_4$, EV) has been studied as a photocatalyst [13,14], laser material [15], and red phosphor [13,16] for its magnetism, thermal stability, and proper energy band gap. Lin et al. prepared EV with different morphologies by the high-temperature electrochemistry method. This prepared adsorbent exhibited good selective adsorption properties and high removal efficiency for U(VI) [17]. Vosoughifar synthesized EV nanoparticles via a precipitation approach; 82% of methyl orange was eliminated in 80 min under ultraviolet light irradiation [18]. In other research, EV nanoparticles were manufactured via a sonochemical method; 64% of methyl orange was removed after 70 min of UV irradiation [14]. Furthermore, a $EuVO_4$-based heterojunction has been synthesized to promote its photodegradation efficiency. He et al. synthesized a $V_2O_5/EuVO_4$ heterojunction from the aqueous solutions of $Eu(NO_3)_3$ and $NH_4VO_3$, which demonstrated high activity for the photodegradation of acetone under both UV and visible light [19]; 99.4% of acetone could be degraded under visible light. It manifested that a small amount of $V_2O_5$ loaded in EV could significantly enhance the photocatalytic efficiency of vanadates [19,20]. $Fe_2O_3/EuVO_4$/g-$C_3N_4$ ternary nanocomposites were designed and used to degrade rhodamine B [21]; 80.06% of rhodamine B was removed using visible source, exhibiting higher photocatalytic activity than the pristine EV nanoparticles. EV nanoparticles were also loaded onto fluorine-doped graphene sheets to synthesize EV/FG24 nanocomposite; 2,4-dinitrophenol and phenol were completely mineralized in 10 h [22].

So far, only a few works, as mentioned above, have involved the construction of EV-based heterojunctions to improve its visible light photocatalytic activity. However, EV and its heterojunctions still exhibit relatively low photodegradation efficiency. Much more efforts should be made to ameliorate the photodegradation activity of EV. Therefore, we constructed EVC heterojunctions by loading EV nanoparticles onto ultrathin CN nanosheets. EVCs presented ultrathin and porous structures, which increased the specific surface area and reaction active sites, shortened the diffusion distance of photoinduced charge carriers. The formation of EVCs extended the visible light absorption of EV and accelerated the separation of charge carriers and ultimately enhanced the adsorption capability and photocatalytic removal efficiency of methylene blue (MB). The effects of ionic strength, initial concentration, temperature, and pH on the adsorption of MB were investigated. The adsorption isotherm and adsorption kinetics were fitted with different models. EVCs exhibited a synergy of adsorption and photocatalysis to remove MB.

## 2. Results and Discussion
### 2.1. Characterization of EVCs

The morphology and textural properties of EVC-2 were determined by TEM and HRTEM. As displayed in Figure 1a,b, the ultrathin, layered nanosheet structure of CN was observed. The ultrathin layered structure significantly increased the specific surface area of EVC-2, which was further confirmed by BET measurements. It should be advantageous to boost the absorption capability of EVC-2 for MB. EV nanoparticles can be clearly observed in Figure 1a,b. The diameters of EV nanoparticles were 20–50 nm. EV nanoparticles were located on the surface of the CN nanosheets. Distinct lattice fringes can be observed in the HRTEM image (Figure 1c). The spacing of the lattice fringe was 0.362 nm, corresponding to the (200) crystal plane of tetragonal EV nanoparticle (PDF#15-0809) [14,17,18]. These results indicate that the intimate contact between EV nanoparticles and CN nanosheets successfully formed EVC heterojunctions [23]. Figure 1d shows the AFM image of EVC-2. The thickness of the CN nanosheets was approximately 4–5 nm, suggesting that EVCs had a higher specific surface area and more abundant adsorption sites than bulk g-$C_3N_4$. This was instrumental in promoting the adsorption capability. Moreover, the diffusion distance of the photoinduced charge carriers was shortened in the CN ultrathin nanosheets. It could reduce the recombination of photoinduced charge carriers and boost the photocatalytic performance [8].

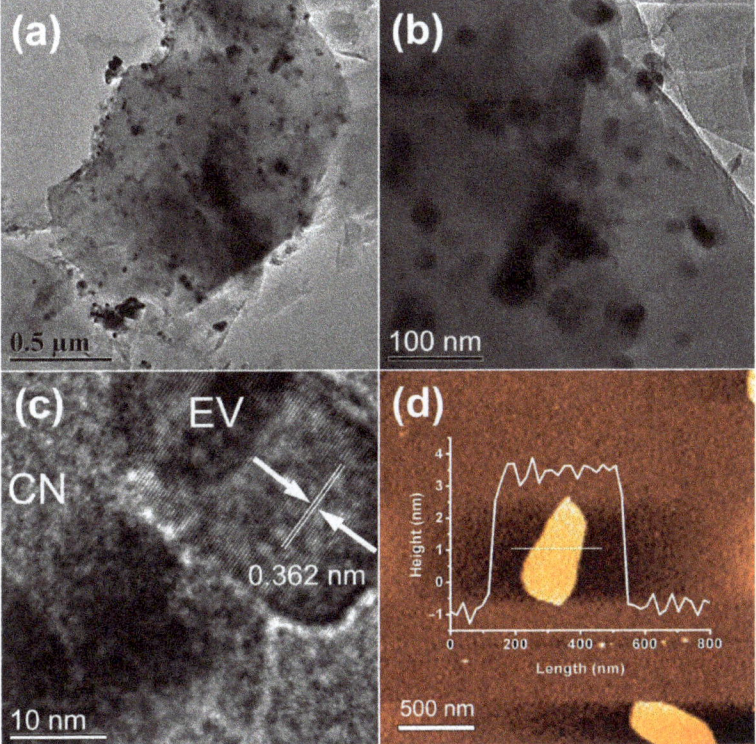

**Figure 1.** (a,b) TEM, (c) HRTEM, and (d) AFM images of EVC-2.

The morphology and chemical composition of EVC-2 were analyzed by SEM, EDS, and element mapping. From the SEM image displayed in Figure 2a, a stacked lamellar morphology emerged. It was the layered structural features of CN. The stacked CN of EVC-2 acquired a three-dimensional porous architecture. Therefore, EVC-2 manifested a mesoporous structure and large pore size distribution, endowing EVC-2 with an exceptional adsorption capability for MB. This has been confirmed by BET measurements. C, N, Eu, V, and O were detected by the energy dispersion spectrum in the selected region (Figure S1). The weight ratios of C, N, Eu, V, and O of EVC-2 are shown in Table S1. The mass percentage of EV in EVC-2 was calculated according to the EDS quantitative results. It was approximately 2.67%, which was slightly higher than the value estimated from the initial $Eu(NO_3)_3$ and $NH_4VO_3$ concentration in the precursor. This might be caused by the mass loss of CN in the reheating process. The uniformity of C, N, Eu, V, and O in EVC-2 was corroborated by elemental mapping, and all appeared in the elemental mapping images (Figure 2b–f). The uniformity of Eu, V, and O demonstrated that the EV nanoparticles were distributed evenly in EVC-2.

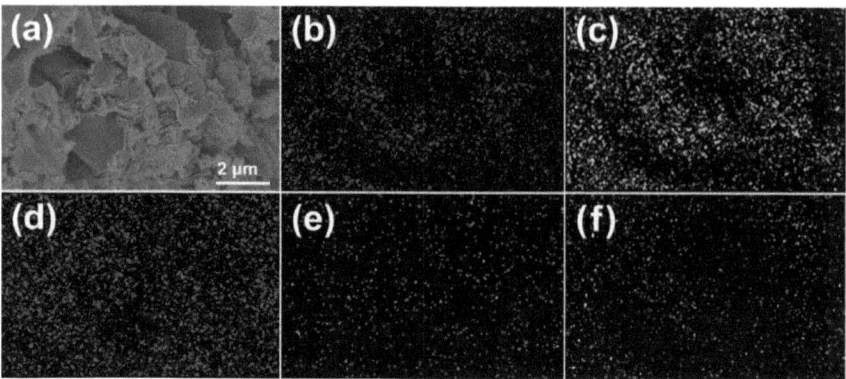

**Figure 2.** (a) SEM and element mapping images of (b) C, (c) N, (d) Eu, (e) V, and (f) O of EVC-2.

The XRD patterns of CN, EVC-2, and EVC-5 are shown in Figure 3a. Two diffraction peaks at 12.8° and 27.7° were observed, corresponding to the (110) and (022) planes of CN (JCPDS#87-1526) [24]. The XRD diffraction peaks of EV nanoparticles were hardly detected in EVC-2 and EVC-5. This could be attributed to the limited dosage of EV in the heterojunctions. The small amount of EF could not be enough to change the chemical skeleton and bulk structure of CN, which is consistent with the results reported in the literature [25,26]. An adsorption band appearing between 2000 and 600 cm$^{-1}$ was displayed in the FTIR spectra of CN, EVC-2, and EVC-5 (Figure 3b). The strong adsorption band located in the range of 1200 to 1700 cm$^{-1}$ was generally originated from the C-N stretching vibrations of CN heterocycles [27]. The sharp band emerging around 805 cm$^{-1}$ can be allocated to the breathing vibration of the s-triazine units of CN. After the adsorption of MB, the variation of the FTIR spectra of EVC-2 was almost undetectable (Figure S2), indicating its excellent structural stability [28]. In addition, the peak at 888 cm$^{-1}$ is characteristic of the =C-H out-of-plane deformation vibration of MB [29].

Figure 3c depicts the XPS survey spectra of CN, EVC-2, and EVC-5. It reveals the existence of C, N, Eu, V, and O in EVC-2 and EVC-5. The O1s peak intensity of CN reduced with the increase of EV dosage, as reported previously, indicating more $H_2O$ being absorbed on CN [25,27]. In the C1s high-resolution XPS spectra of CN, EVC-2, and EVC-5, two peaks at 284.8 and 288.3 eV for CN were observed. The peaks at 284.8 eV originated from the adventitious carbon, and the peaks at 288.3 eV were derived from the sp$^2$-bonded C in CN [27]. The N1s peak of CN at 398.1 eV shifted to 398.8 eV for EVC-2 and EVC-5. These peaks could be attributed to the sp$^2$ nitrogen atoms in C–N=C. The increased N1s binding energy of EVC-2 and EVC-5 demonstrate the reduced electron cloud density of N atoms [30]. This indicates that the coordination of Eu or V ions with pyridinic N occurred where N provided lone-pair electrons and $Eu^{3+}$ or $V^{5+}$ supplied the unoccupied d orbit [31]. The XPS results also suggested that EVCs were successfully formed through the coordination interaction between $Eu^{3+}$ or $V^{5+}$ and N [32]. The construction of EVCs was advantageous to the transfer of photoinduced charge in the photocatalytic reaction, and thus improved the photocatalytic degradation efficiency of MB over EVCs. After the adsorption of MB, the C1s peaks of EVC-2 at 284.8 eV and 288.3 eV shifted to 284.5 eV and 287.6 eV, respectively. The N1s peak of EVC-2 at 398.8 eV shifted to 398.1 eV (Figure S3). This reveals the variation of the electron cloud density of C and N atoms after the adsorption, which could be attributed to the π–π electron donor-acceptor interaction between the CN nanosheets and the MB molecules [33]. The absorption edge of CN and

EV was 457 and 652 nm, respectively, corresponding to 2.71 and 1.90 eV of the band gap (Figure 3f). The UV-vis DRS revealed that EVC-2 and EVC-5 presented stronger visible light absorption than CN, which was instrumental in improving the utilization of visible light in the photocatalytic process. The absorption edge of EVC-2 and EVC-5 lay between the values of CN and EV, further confirming the successful formation of EVC heterojunctions.

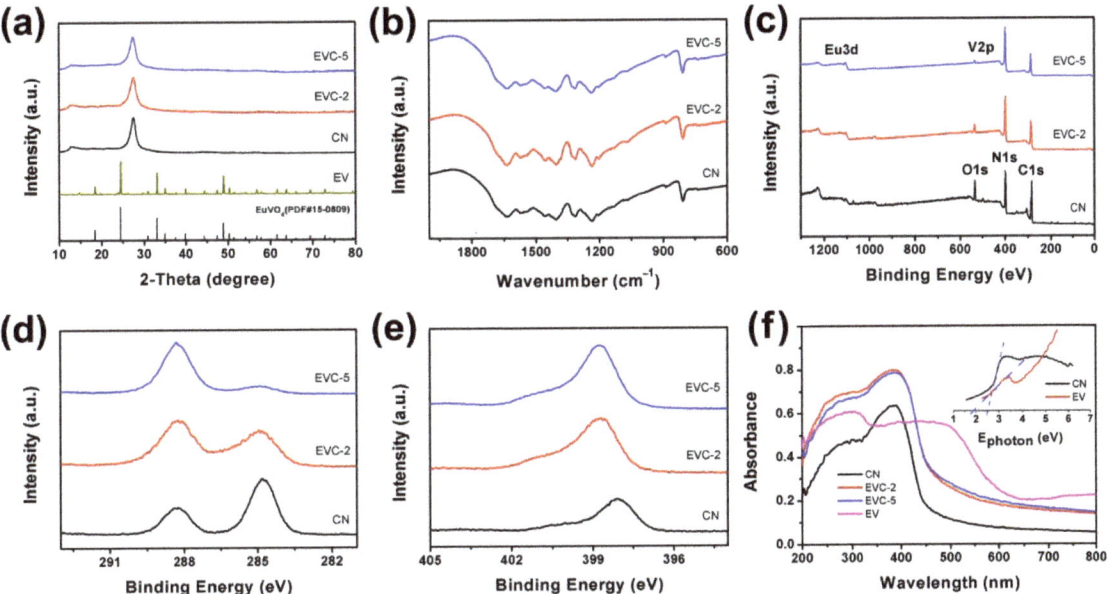

**Figure 3.** (**a**) XRD patterns, (**b**) FTIR spectra, (**c**) XPS survey spectra, high-resolution XPS spectra of (**d**) C1s, (**e**) N1s, and (**f**) UV-vis diffuse reflectance spectrum (Inset: band gap energy of CN and EV) of CN, EVC-2, and EVC-5.

The BET specific surface areas ($S_{BET}$) and mesoporous structures of CN, EVC-2, EVC-5, EVC-10, and EVC-20 were assessed by the $N_2$ adsorption–desorption isotherms and the corresponding BJH pore size distribution curves. All catalysts presented type IV isotherms (Figure 4a), which indicated the mesoporous structures of these catalysts [6,25,34]. The H3 hysteresis loop at high $P/P_0$ from 0.50 to 1.00 suggested that the mesopores of the samples were irregular. The $S_{BET}$ of EVC-2 (80.43 $m^2\ g^{-1}$), EVC-5 (38.14 $m^2\ g^{-1}$), EVC-10 (36.21 $m^2\ g^{-1}$), and EVC-20 (32.53 $m^2\ g^{-1}$) were distinctly higher than that of CN (12.32 $m^2\ g^{-1}$). The pore volumes of EVC-2 and CN were 0.15 and 0.02 $cm^3\ g^{-1}$ (Table S2), respectively. These results reveal that the EVCs possessed higher specific surface areas and more mesopores for the adsorption of contaminants. Meanwhile, this would imply that EVCs had more reaction-active sites than CN for the photocatalytic degradation of MB. The BJH pore size distribution curves of CN, EVC-2, and EVC-5 are depicted in Figure 4b, indicating the remarkable increase of EVC-2 and EVC-5 in the range of 1–50 nm. This might be caused by thermal etching in the reheating process, which is similar to the previous reported results [25]. The ultrathin layered structures increased the specific surface area of EVC-2, which is confirmed by the TEM and AFM results in Figure 1. The stacked CN of EVC-2 acquired a three-dimensional porous architecture, which is corroborated by the SEM results in Figure 2. This further indicates that EVC-2 and EVC-5 possess a higher pore size distribution ranging from 1 to 50 nm. This increased mesoporous structure endows EVC-2 with superior adsorption capability for MB in wastewater [23].

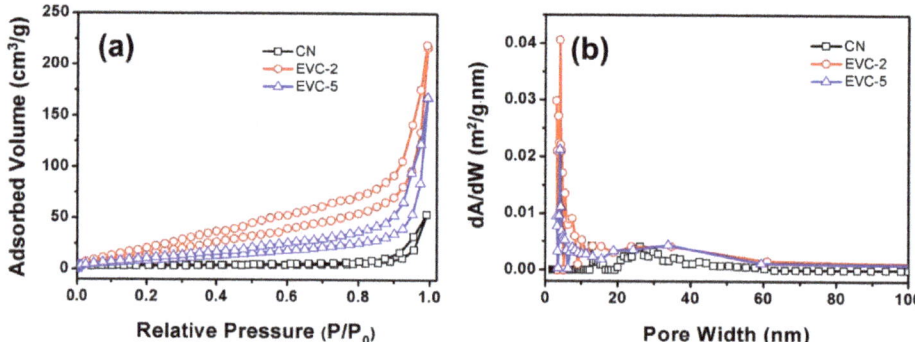

**Figure 4.** (a) The $N_2$ adsorption–desorption isotherms and (b) the corresponding BJH pore size distribution curves of CN, EVC-2, and EVC-5.

## 2.2. Adsorption Kinetics and Isotherm of MB

CN, EV, and EVCs were used to adsorb MB. The effect of different EV mass ratios on the MB adsorption capacity was investigated (Figure 5a). EV content in EVCs significantly affected the adsorption amount of MB. The $q_e$ values of EVCs were higher than that of CN (1.1 mg g$^{-1}$). EVC-2 exhibited a highest MB adsorption capacity of 6.75 mg g$^{-1}$, which was over six times that of CN. This result was well consistent with the BET measurements, as displayed in Figure 5a. It suggested that the reheating synthesis approach was an effective way to ameliorate the adsorption capacity of EVCs. The thermal etching approach could be instrumental in obtaining an ultrathin, layered structure of EVCs and increasing its specific surface area. The adsorption capacity and specific surface areas decreased with the increase of EV content. This might originate from more adsorption active sites being occupied by EV nanoparticles.

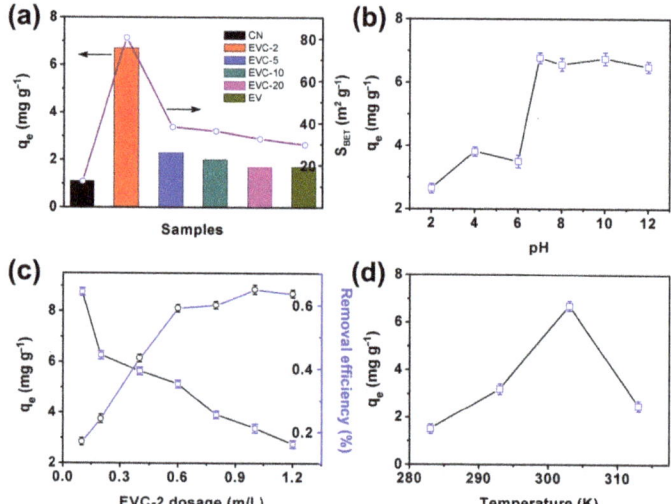

**Figure 5.** (a) Adsorption of MB on CN, EV, and EVCs: the impact of (b) pH, (c) initial dosage, and (d) temperature on the MB adsorption over EVC-2.

MB adsorption capacity with different solutions of pH over EVC-2 at 30 °C was tested and is depicted in Figure 5b. It revealed that the adsorption capacity increased remarkably while the pH was higher than 6. At a pH of less than 6, the adsorption capacity was extremely low. It reached 6.75 mg g$^{-1}$ when pH was 7. The alkaline condition was advantageous to the MB adsorption, which is similar to many reported results [35–37]. These results indicated that the surface charge of EVC-2 was remarkably affected by the solution pH. Electrostatic interaction between EVC-2 and MB contributed to adsorption capacity. MB is a cationic dye and is easily adsorbed on a negatively-charged adsorbent [35]. The point of zero charge pH ($pH_{pzc}$) of EVC-2 was 6.48 (Figure S4). This suggests that the EVC-2 was negatively charged at pH > 6.48 and positively charged at pH < 6.48. The negatively-charged surface of EVC-2 at a higher pH would enhance the electrostatic interaction between EVC-2 and MB, which remarkably increased the adsorption capacity. The electrostatic repulsion between positively-charged EVC-2 and MB resulted in a low adsorption capacity of MB at a solution pH of less than 6.

The effect of initial EVC-2 dosage on MB adsorption at 30 °C was studied and is presented in Figure 5c. The experiment was conducted after 72 h of equilibrium under continuous shaking in the dark. The initial MB concentration was 5 mg $L^{-1}$. It can be seen from Figure 5c that the removal efficiency of MB increased when the dosage of EVC-2 was raised, which is related to the increased active sites of the adsorbent [38]. As the EVC-2 dosage increased from 0.1 g $L^{-1}$ to 1.2 g $L^{-1}$, the adsorption capacity of MB was extremely abated from 8.77 mg $g^{-1}$ to 2.77 mg $g^{-1}$. The reason for this result is mainly that the initial MB concentration remained unchanged in the solution. With the increase of EVC-2 dosage, the adsorption active sites on EVC-2 were redundant and could not be fully utilized [39]. The effect of temperature on MB adsorption capacity at 10, 20, 30, and 40 °C is illustrated in Figure 5d. The adsorption capacity was enhanced when the temperature rose from 10 to 30 °C, implying it was an endothermic process resulting from the π–π interaction between EVC-2 and MB [25]. MB had relative weak molecular thermal motion at low temperature, resulting in fewer molecules being diffused to the EVC-2 surface. The adsorption capacity was consequently less at lower temperatures. With the increase of temperature, MB absorbed the heat and accelerated molecular thermal motion, thus exhibitinga higher adsorption capacity [40,41]. The adsorption capacity decreased at 40 °C. This might be caused by the desorption effect due to the higher velocity of the molecular thermal motion at 40 °C.

The adsorption kinetics curves of MB on CN, EVC-2, and EVC-5 are displayed in Figure 6a. The adsorption equilibrium of MB on CN, EVC-2, and EVC-5 was reached in 60 min. The adsorbed amount of MB was 1.09, 6.75, and 2.27 mg $g^{-1}$, respectively. The adsorption capability was enhanced after the formation of EVCs. The pseudo first-order, pseudo second-order, and intraparticle diffusion kinetics model were adopted to fit the experimental data for the elaboration of the kinetics mechanism of the adsorption process [42]. As can be clearly seen in Figure 6b–d, the adsorption data well met the pseudo second-order kinetics model. The corresponding $R^2$ and RMSE values are listed in Table S3. Therefore, the MB adsorption on CN, EVC-2, and EVC-5 all followed the pseudo second-order kinetics model. This demonstrates that the adsorption rate of MB on the as-prepared adsorbents was controlled by the chemical adsorption, which was the rate-determining step of the adsorption process [43]. The electron transfer or electron sharing between the adsorbent and MB was the primary driving force of the adsorption.

The MB adsorption kinetic parameters from the fitted pseudo second-order equation are displayed in Table S4. The calculated adsorption capacity values were consistent with the experimental values obtained in Figure 6a. EVC-2 possessed the highest adsorption capacity, which was 6.2 times that of CN. This reveals that the formation of EVCs could significantly enhance the adsorption capability. The enhancement of the adsorption capability of EVC-2 and EVC-5 might originate from thermal etching in the reheating process, which would increase the specific surface areas through reducing the thickness of the CN nanosheets and the formation of loose-stacked porous architecture. The SEM and BET results confirmed the significant increase of the pore of EVCs in the whole range of 1–50 nm. The increased specific surface areas could generate more surface reaction sites. As a result, the adsorption and photocatalytic degradation performance would be extremely enhanced.

An adsorption isotherm can be employed to provide information regarding the equilibrium adsorptions. The adsorption capacity increased with the equilibrium concentration of MB (Figure 7a). The experimental equilibrium adsorption data of MB on CN, EVC-2, and EVC-5 were measured and fitted by the Freundlich, Temkin, and Langmuir isotherm models [25,35]. The corresponding $R^2$ and RMSE values are listed in Table S5. As presented in Figure 7b–d, the equilibrium adsorption data conformed to the Langmuir isotherm model according to the correlation coefficients ($R^2$ and RMSE) revealed in Table S5. This result indicates that the adsorption of MB on the surface of CN, EVC-2, and EVC-5 was monolayer adsorption [2]. The adsorption isotherm parameters of MB on CN, EVC-2, and EVC-5 are presented in Table S6. The $q_m$ calculated from the Langmuir isotherm model were 2.24, 20.0, and 4.76 mg g$^{-1}$ for CN, EVC-2, and EVC-5, respectively. This is very similar to the experimental equilibrium adsorption values obtained from Figure 7a. The Langmuir constant ($K_L$) was also calculated and is listed in Table S6; it suggests that EVC-2 had a highest affinity for the adsorption of MB on its surface binding sites.

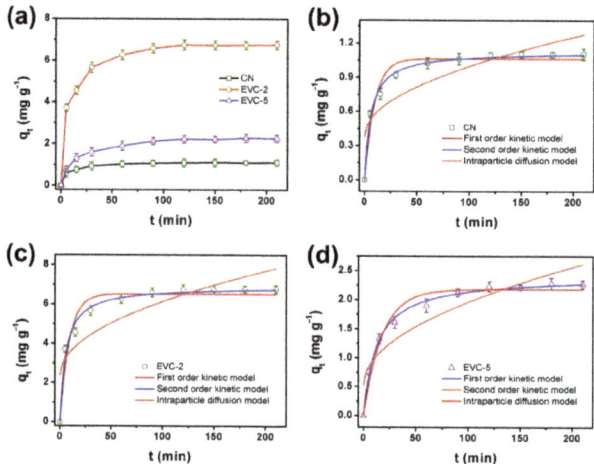

**Figure 6.** (a) The adsorption kinetic curves of MB (5 mg L$^{-1}$) on CN, EVC-2, and EVC-5: the first order, second-order, and intraparticle diffusion kinetics model for adsorption of MB on (b) CN, (c) EVC-2, and (d) EVC-5.

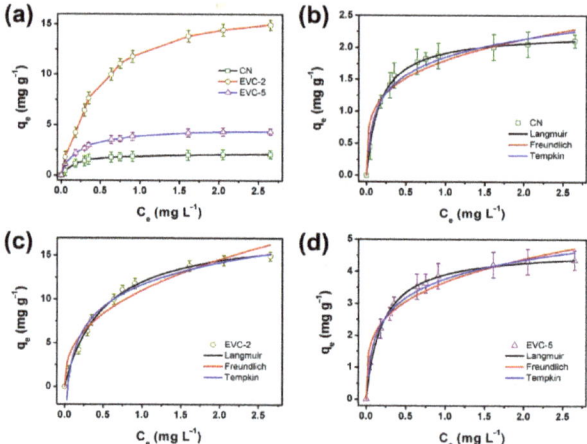

**Figure 7.** (a) Adsorption isotherms of MB (5 mg L$^{-1}$) on CN, EVC-2, and EVC-5: Langmuir, Freundlich, and Tempkin isotherm model for the adsorption of MB on (b) CN, (c) EVC-2, and (d) EVC-5.

## 2.3. Adsorption and Photocatalytic Degradation of MB

The adsorption–photocatalysis synergistic process of MB on the as-prepared catalysts under visible light irradiation was investigated and is illustrated in Figure 8a. As revealed in this figure, the formation of EVCs remarkably improved the removal efficiency of MB. The optimal removal efficiency of MB over EVC-2 ultimately reached 100% after 30 min of degradation. It was 22.7 times higher than that of mere photolysis of MB. All EVC samples presented better MB removal performance than CN, suggesting the higher separation efficiency of photoinduced charge carriers in EVCs. The removal of MB in this degradation process could be considered as the synergistic effect of adsorption and photocatalysis. There was approximately 22–50% of MB eliminated by the adsorption process before light irradiation over EVCs. By comparison, only 8.7% of MB was removed over CN in 60 min of the adsorption process. This confirms that EVCs possess much stronger MB adsorption ability than CN. The same results were also observed in the previous work [25] and are clarified in Figures 6 and 7. As shown in Figure 8b, the adsorption–desorption equilibrium was obtained within 60 min. The removal efficiency of MB reached 49.5% after 60 min of the adsorption process on EVC-2 (Figure 8b), which was 5.7 times that of CN. The photocatalysis process over the as-prepared catalysts was investigated and is exhibited in Figure 8c. Before photocatalytic degradation, the adsorption–desorption equilibrium was achieved first, and thus the influence of the adsorption on the photocatalytic process was ignored. There was approximately 50.5% of MB removed in the photocatalytic degradation process on EVC-2. This was 2.2 times that of CN. These results demonstrate that EVC-2 had optimum adsorption and photocatalytic performance.

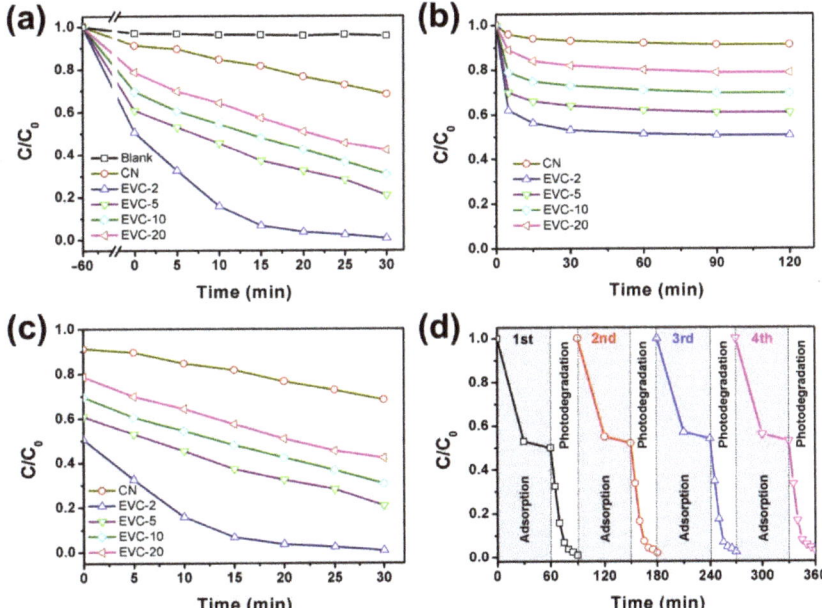

**Figure 8.** (a) The removal of MB over the as-prepared catalysts under visible light irradiation, (b) the adsorption of MB over the catalysts in the dark, (c) the photodegradation of MB over the catalysts under visible light irradiation, (d) the synergistic removal of MB over EVC-2 in the recycle experiment.

The practical application of the catalysts for the removal of contaminants requires the as-prepared sample presented to have stable degradation properties. Therefore, the stability of EVC-2 was implemented by four cycles of the adsorption and photocatalysis experiments. EVC-2 was separated by centrifugation, desorbed in deionized water, and dried at 60 °C after each cyclic experiment. The stability experimental results are displayed in Figure 8d. Only 5% of adsorption and 3% of photocatalysis capability was lost after four cycles of degradation experiments. TEM, FTIR, and XRD (Figures S5 and S6) were measured to investigate the variations of the morphology and the crystal structure between the fresh and used EVC-2. No obvious changes were found between the fresh and used EVC-2. This proves that EVC-2 was highly stable for practical utilization. In addition, TOC analysis was employed to clarify the mineralization efficiency of MB on EVC-2 under visible light irradiation. As shown in Figure S7, the TOC decreased to 42.4% after 30 min of the adsorption and photocatalysis process, corresponding to a mineralization efficiency of 57.6%. In contrast, with 100% photocatalytic degradation efficiency, the lower mineralization efficiency indicated that the mineralization of MB was slower than the decolorization process. It was the degradation intermediates that caused the relatively higher TOC value [44,45]. The mineralization of MB to completely generate $CO_2$ and $H_2O$ needed to react for much longer than 2 h under visible light illumination.

### 2.4. Photocatalytic Kinetics of MB

The photocatalytic degradation of MB on EVC-2 under visible light irradiation is depicted in Figure 9. It obviously manifested two types of kinetics characteristics in the different degradation times. The photocatalytic degradation data obeyed the zero-order kinetics equation in the range of 0–10 min (Figure 9, left). In this stage, the photocatalytic degradation reaction rate was almost extraneous to MB concentration. This might originate from the excessive MB in the initial degradation solution. In the meantime, the photocatalytic degradation reaction rate only related to the amount of surface active sites of the photocatalyst [24]. The zero-order degradation reaction rate constant ($k_0$) was 0.035 mg L$^{-1}$ min$^{-1}$. With the gradual decrease of MB concentration after 10 min of degradation, the MB diffusion rate from the solution to the surface of the photocatalyst was reduced. The influence of the MB concentration could not be disregarded at this stage. The photocatalytic degradation data consequently followed the first-order kinetics equation for 10–30 min (Figure 9, right). The first-order degradation reaction rate constant ($k_1$) was 0.150 min$^{-1}$. As shown in Figure 8a, the adsorption and photocatalytic performances of MB on CN, EVC-5, EVC-10, and EVC-20 were relatively low. All photocatalytic degradation data seemed to comply with the zero-order kinetics equation during the whole degradation period (Figure 8c). This might be caused by the comparatively low photocatalytic degradation reaction rate of MB on these samples, resulting in the excessive MB concentration over 30 min. The $k_0$ values of MB on CN, EVC-5, EVC-10, and EVC-20 were 0.007, 0.013, 0.013, and 0.012 mg L$^{-1}$ min$^{-1}$, respectively (Figure S8). The $k_0$ of EVC-2 was five times that of CN.

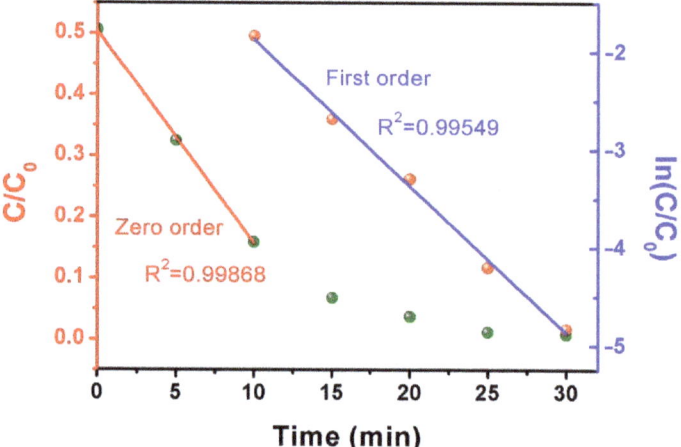

**Figure 9.** The photocatalytic degradation of MB on EVC-2 under 40 W LED irradiation: the corresponding zero-order kinetics curve (left) and the first-order kinetics curve (right).

*2.5. Active Oxidation Species and Possible Mechanism*

Radical scavenger experiments were conducted to identify the major reactive species in the MB photocatalytic degradation on EVC-2 (Figure 10a). NaNO$_3$, AO, IPA, and PBQ were adopted as the scavengers to trap $e^-$, $h^+$, •OH, and •O$_2^-$, respectively [46]. The degradation efficiencies of MB on EVC-2 were obviously decreased with the presence of NaNO$_3$, AO, IPA, and PBQ, indicating that $e^-$, $h^+$, •OH, and •O$_2^-$ were all involved in the photocatalytic degradation process. In particular, when PBQ was added to the degradation solution, the degradation efficiency was enormously reduced, indicating that •O$_2^-$ was absolutely the major active oxidation species in MB degradation. As can be seen in the radical scavenger experimental results shown in Figure 10a, the influence of the reactive species followed the order of •O$_2^-$ > $e^-$ > $h^+$ > •OH, and the degradation efficiencies of MB on EVC-2 were decreased to 45%, 77%, 85%, and 91%, respectively. The degradation efficiency declined slightly in the presence of NaNO$_3$, AO, and IPA, suggesting that $e^-$, $h^+$, and •OH played only a relatively minor role in the MB degradation process. The ESR spectra of DMPO-•OH and DMPO-•O$_2^-$ were employed to further confirm the generation of active species •OH and •O$_2^-$ radicals in the MB photocatalytic degradation process on EVC-2 and EVC-5 (Figure 10b,c). The ESR signals of DMPO-•OH and DMPO-•O$_2^-$ adducts could be clearly detected after 15 min of visible light irradiation, but were not detectable in the dark. This signifies that the •OH and •O$_2^-$ radicals were only produced in the photocatalytic degradation process. In addition, The DMPO-•OH and DMPO-•O$_2^-$ signal intensities of EVC-2 were much stronger than that of EVC-5, revealing that EVC-2 had higher •O$_2^-$ generation capability than EVC-5. This is consistent with the photodegradation results presented in Figure 8c, which indicate that EVC-2 possessed higher carrier separation efficiency. Based on the above-mentioned experimental results, the possible removal mechanism of MB on EVC-2 can be deduced. MB molecules were first adsorbed on the surface of EVC-2. There followed the visible light absorption and the photoinduced electron and hole generation. Then, the photoinduced electron and hole separated and transferred to the surface of EVC-2 and produced •OH and •O$_2^-$. Next, MB was ultimately degraded into small molecules by the complex reaction with •O$_2^-$, $e^-$, $h^+$, and •OH. The formation of ultrathin nanosheet structures and EVC heterojunctions could provide the driving force for the separation and transfer of photoinduced electron-hole pairs.

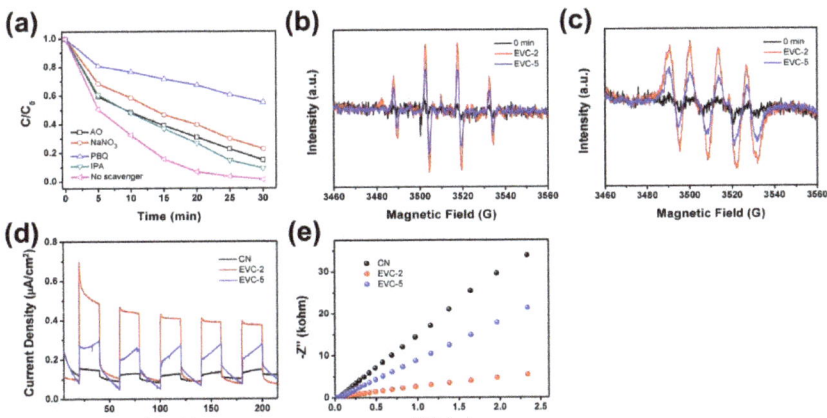

**Figure 10.** (a) the effect of radical scavengers on MB photocatalytic degradation on EVC-2 under 40 W LED irradiation: the ESR spectra of (b) DMPO-•OH in water; (c) DMPO-•O$_2^-$ in methanol with EVC-2 and EVC-5 after 15 min of visible light irradiation; (d) TPC spectra; (e) EIS Nyquist plots of CN, EVC-2, and EVC-5.

Transient photocurrent response (TPC) and electrochemical impedance spectroscopy (EIS) was further conducted to confirm the photoinduced charge separation efficiency of

CN, EVC-2, and EVC-5. As shown in Figure 10d, EVC-2 and EVC-5 had higher photocurrent density than CN, indicating that the formation of EVCs promoted charge separation and transfer efficiency. EIS Nyquist plots of EVC-2 and EVC-5 possessed a smaller arc radius than CN, confirming the same conclusion as the TPC result (Figure 10e). In addition, EVC-2 exhibited a higher photocurrent density and a smaller arc radius than EVC-5, clarifying that EVC-2 had higher charge separation efficiency and superior photocatalytic degradation efficiency than EVC-5. This is consistent with the photocatalytic degradation result of MB over as-prepared catalysts.

In this study, the superior photocatalytic degradation activity of EVC-2 was primarily attributed to the formation of an ultrathin EVCs heterojunction structure and the synergistic effect of adsorption and photocatalysis. We firstly manufactured CN nanosheets with an ultrathin structure by thermal etching and ultrasonic techniques. This thus increased its specific surface area and reaction active sites and shortened the diffusion distance of photoinduced charge carriers. Further, EV nanoparticles were loaded onto CN nanosheets to form an EVCs heterojunction. This consequently accelerated the separation of charge carriers and reduced its recombination. These factors boosted the generation of $e^-$, $h^+$, $\bullet OH$, and $\bullet O_2^-$, which remarkably ameliorated the visible light-driven photocatalytic degradation efficiency. Based on these advantages, EVC-2 emerged with a much higher adsorption capacity and photocatalytic activity than that of pristine CN.

## 3. Materials and Methods

### 3.1. Synthesis of EVCs

Bulk g-$C_3N_4$ was synthesized by the polycondensation of dicyandiamide and $NH_4Cl$, with a 1:1 mass ratio at 550 °C for 4 h according to the previous literature, with certain modifications [47]. The light-yellow powder was collected and ground. An amount of 100 mg of g-$C_3N_4$ was mixed with the 100 mL deionized water and underwent ultrasonic treatment for 4 h [48]. The mixture was centrifuged at 4000 rpm for 20 min. The g-$C_3N_4$ nanosheets in the supernatant were separated by vacuum freeze-drying.

An amount of 200 mg of g-$C_3N_4$ nanosheets and a certain amount of $Eu(NO_3)$, and $NH_4VO_3$ were mixed in 100 mL of deionized water for 2 h. The water was then evaporated and the obtained solid product was ground. EVCs powder was produced after heating the powder for 2 h at 550 °C. The as-prepared samples were named as EVC-2, EVC-5, EVC-10, and EVC-20, respectively, while 2, 5, 10, and 20 wt% of EV was contained in the EVCs. EV was generated under the same reaction conditions except for the absence of g-$C_3N_4$ nanosheets.

### 3.2. Catalyst Characterization

Transmission electron microscopy (TEM, JEOL, Tokyo, Japan) and high-resolution TEM (HRTEM) images were taken on a JEOL-2100F transmission electron microscope. Scanning electron microscopy (SEM, JEOL, Tokyo, Japan) images, energy dispersion spectrum (EDS), and elemental mapping images of the samples were taken on a JSM-4800F scanning electron microscope. Atomic force microscopy (AFM, Bruker, Billerica, MA, USA) images were measured on a Bruker Multimode 8 AFM system. X-ray diffraction (XRD, Rigaku Corporation, Tokyo, Japan) patterns of g-$C_3N_4$, EF, and as-prepared EFC heterojunctions were taken on a Rigaku Smartlab diffractometer. Fourier-transform infrared spectra (FTIR, ThermoFisher, Waltham Mass, MA, USA) were recorded on a Nicolet NEXUS 470 spectrometer in the range of 4000–400 $cm^{-1}$. XPS spectra (ThermoFisher, Waltham, MA, USA) were detected on a Thermo ESCALAB 250XI X-ray photoelectron spectroscopy spectrometer equipped with anAlK$\alpha$X-ray source. The $N_2$ adsorption desorption isotherms and BET specific surface area was measured on a Micromeritics ASAP 2460 analyzer at 77 K (Micrometrics, Londonderry, NH, USA).

### 3.3. Adsorption of MB

The adsorption experiment of MB was carried in the dark. Briefly, a 50 mg catalyst was mixed with 100 mL of different concentrations of MB. The pH value of the MB solution was adjusted to 7 using 0.1 M HCl or NaOH. The mixture solution was agitated in the dark at 150 rpm. An amount of 5 mL adsorption solution was added at certain intervals and removed the catalyst through 0.22 μm PTFE filter membrane. The MB concentration was determined on a UV-vis spectrophotometer at the absorption wavelength of 664 nm. The adsorption capacity ($q_t$) at a given adsorption time was calculated. The impact of pH on MB adsorption over the catalyst was studied from 2 to 11 using 0.1 M HCl or NaOH to adjust the pH value. The effect of temperature on MB adsorption over the catalyst was investigated at 20, 30, and 40 °C, respectively. The adsorption kinetic data were fitted by a pseudo first-order and pseudo second-order adsorption kinetics model and intraparticle diffusion (Weber–Morris) model. Langmuir, Freundlich, and Tempkin models were selected to fit the adsorption isotherm data.

*3.4. Photocatalytic Degradation of MB*

The visible light driven photocatalytic degradation of MB was conducted under the irradiation of a 40 W LED. Typically, a 100 mg catalyst was mixed with 200 mL 5 mg L$^{-1}$ MB solution. The mixture was stirred in the dark for 1 h for the adsorption–desorption equilibrium of MB over the catalysts. An amount of 5 mL degradation suspension was taken out at certain intervals and removed the catalyst through 0.22 μm PTFE filter membrane. The MB concentration was determined on a UV-vis spectrophotometer by absorption at 664 nm. The photocatalytic degradation kinetics curve was obtained by plotting degradation efficiency against degradation time. The photocatalytic stability of the catalyst was performed by four cycles of degradation experiments. After each cycle, the catalyst was separated by centrifugation and washed thoroughly with ethanol and deionized water to eliminate the MB absorbed on the catalyst. The catalyst was then freeze-dried and collected for the next cycle of degradation experiments. To confirm the active species, NaNO$_3$, AO (ammonium oxalate), IPA (isopropyl alcohol), and PBQ (p-benzoquinone) were adopted as the scavengers to trap e$^-$, h$^+$, •OH, and •O$_2^-$, respectively.

## 4. Conclusions

In summary, ultrathin CN nanosheets with a thickness of 4–5 nm were first manufactured by thermal etching and ultrasonic techniques. EV nanoparticles were then loaded onto the CN nanosheets to form EVCs. The as-prepared EVC-2 possessed the optimal adsorption and photocatalytic removal performance. The superior photocatalytic degradation activity of EVC-2 was primarily attributed to the formation of the ultrathin EVCs heterostructure and the synergistic effect of adsorption and photocatalysis. The MB adsorption capability of EVC-2 was 6.2 times that of CN, and the zero-order degradation reaction rate constant ($k_0$) was 5 times that of CN. The MB adsorption on EVC-2 followed the pseudo second-order kinetics model, and the adsorption isotherm data complied with the Langmuir isotherm model. The photocatalytic degradation data of MB on EVC-2 obeyed the zero-order kinetics equation in the range of 0–10 min and abided by the first-order kinetics equation for 10–30 min. By comparison, the photocatalytic degradation data observed the zero-order kinetics equation during the whole degradation period. The radical scavenger experiments demonstrated that •O$_2^-$, e$^-$, h$^+$, and •OH were involved in the photocatalytic degradation process.

**Supplementary Materials:** The supporting information can be downloaded at: https://www.mdpi.com/article/10.3390/ijms231710003/s1.

**Author Contributions:** Methodology, writing—original draft, visualization, X.R.; investigation, data curation, visualization, L.W.; project administration, B.X.; funding acquisition, L.L.; investigation, data curation, J.Z.; investigation, Z.L.; supervision, X.X.; resources, funding acquisition, G.F. and R.L.; conceptualization, writing—review and editing, resources, funding acquisition, J.F. All authors have read and agreed to the published version of the manuscript.

**Funding:** This research was funded by the National Natural Science Foundations of China, grant number 21865006, the Key Project of Basic Research Program of Guizhou Province, China, grant number ZK[2021]022 and the Science and Technology Project of Guizhou Province, China, grant number [2018]5779-18 and ZK[2021]067.

**Institutional Review Board Statement:** Not applicable.

**Informed Consent Statement:** Not applicable.

**Data Availability Statement:** The original data are available from the corresponding author upon reasonable request.

**Conflicts of Interest:** The authors declare no conflict of interest.

## References

1. He, Y.Q.; Zhang, F.; Ma, B.; Xu, N.; Junior, L.B.; Yao, B.; Yang, Q.; Liu, D.; Ma, Z. Remarkably enhanced visible-light photocatalytic hydrogen evolution and antibiotic degradation over g-$C_3N_4$ nanosheets decorated by using nickel phosphide and gold nanoparticles as cocatalysts. *Appl. Surf. Sci.* **2020**, *517*, 146187. [CrossRef]
2. Ashrafi, H.; Akhond, M.; Absalan, G. Adsorption and photocatalytic degradation of aqueous methylene blue using nanoporous carbon nitride. *J. Photochem. Photobiol. A* **2020**, *396*, 112533. [CrossRef]
3. Zou, W.; Gao, B.; Ok, Y.S.; Dong, L. Integrated adsorption and photocatalytic degradation of volatile organic compounds (VOCs) using carbon-based nanocomposites: A critical review. *Chemosphere* **2019**, *218*, 845–859. [CrossRef]
4. Fronczak, M. Adsorption performance of graphitic carbon nitride-based materials: Current state of the art. *J. Environ. Chem. Eng.* **2020**, *8*, 104411. [CrossRef]
5. Yan, L.; Gao, H.; Chen, Y. Na-Doped Graphitic Carbon Nitride for Removal of Aqueous Contaminants via Adsorption and Photodegradation. *ACS Appl. Nano Mater.* **2021**, *4*, 7746–7757. [CrossRef]
6. Shi, J.; Chen, T.; Guo, C.; Liu, Z.; Feng, S.; Li, Y.; Hu, J. The bifunctional composites of AC restrain the stack of g-$C_3N_4$ with the excellent adsorption-photocatalytic performance for the removal of RhB. *Colloids Surf. A* **2019**, *580*, 123701. [CrossRef]
7. Kim, J.-G.; Kim, H.-B.; Choi, J.-H.; Baek, K. Bifunctional iron-modified graphitic carbon nitride (g-$C_3N_4$) for simultaneous oxidation and adsorption of arsenic. *Environ. Res.* **2020**, *188*, 109832. [CrossRef]
8. Xia, P.; Zhu, B.; Yu, J.; Cao, S.; Jaroniec, M. Ultra-thin nanosheet assemblies of graphitic carbon nitride for enhanced photocatalytic $CO_2$ reduction. *J. Mater. Chem. A* **2017**, *5*, 3230–3238. [CrossRef]
9. Chen, M.; Li, M.; Lee, S.; Zhao, X.; Lin, S. Constructing novel graphitic carbon nitride-based nanocomposites—From the perspective of material dimensions and interfacial characteristics. *Chemosphere* **2022**, *302*, 134889. [CrossRef]
10. Fazal, T.; Iqbal, S.; Shah, M.; Mahmood, Q.; Ismail, B.; Alsaab, N.S.; Ibrahim, H.A.; Elkaeed, E.B. Optoelectronic, structural and morphological analysis of $Cu_3BiS_3$ sulfosalt thin films. *Results Phys.* **2022**, *36*, 105453. [CrossRef]
11. Fazal, T.; Iqbal, S.; Shah, M.; Bahadur, A.; Ismail, B.; Abd-Rabboh, H.S.M.; Hameed, R.; Mahmood, Q.; Ibrar, A.; Nasar, M.S.; et al. Deposition of bismuth sulfide and aluminum doped bismuth sulfide thin films for photovoltaic applications. *J. Mater. Sci. Mater. Electron.* **2022**, *33*, 42–53. [CrossRef]
12. Fazal, T.; Iqbal, S.; Shah, M.; Mahmood, Q.; Ismail, B.; Alzhrani, R.M.; Awwad, N.S.; Ibrahium, H.A.; Alam, S.; Yasir, M.; et al. Optoelectronic Analysis of Bismuth Sulfide and Copper-Doped Bismuth Sulfide Thin Films. *JOM* **2022**, *74*, 2809–2816. [CrossRef]
13. Reitz, C.; Smarsly, B.; Brezesinski, T. General Synthesis of Ordered Mesoporous Rare-Earth Orthovanadate Thin Films and Their Use as Photocatalysts and Phosphors for Lighting Applications. *ACS Appl. Nano Mater.* **2019**, *2*, 1063–1071. [CrossRef]
14. Hosseini, S.A. Nanocrystalline $EuVO_4$: Synthesis, characterization, optical and photocatalytic properties. *J. Mater. Sci.-Mater. Electron.* **2016**, *27*, 10775–10779. [CrossRef]
15. Chen, J.; Fang, L.; Li, J.; Tang, Y.; Cheng, K.; Cao, Y. Packing fraction, bond valence and crystal structure of $AVO_4$ (A = Eu, Y) microwave dielectric ceramics with low permittivity. *J. Mater. Sci.-Mater. Electron.* **2020**, *31*, 19180–19187. [CrossRef]
16. Kim, K.Y.; Yoon, S.J.; Parkn, K. Synthesis and photoluminescence properties of $EuVO_4$ red phosphors. *Ceram. Int.* **2014**, *40*, 9457–9461. [CrossRef]
17. Lin, Y.; Liu, Y.; Zhang, S.; Xie, Z.; Wang, Y.; Liu, Y.; Dai, Y.; Wang, Y.; Zhang, Z.; Liu, Y.; et al. Electrochemical synthesis of $EuVO_4$ for the adsorption of U(VI): Performance and mechanism. *Chemosphere* **2021**, *273*, 128569. [CrossRef] [PubMed]
18. Vosoughifar, M. Investigation of the morphologies, optical and magnetic properties of europium vanadate nanoparticles synthesized via a simple method. *J. Mater. Sci.-Mater. Electron.* **2017**, *28*, 2227–2232. [CrossRef]
19. He, Y.; Wu, Y.; Sheng, T.; Wu, X. Photodegradation of acetone by visible light-responsive $V_2O_5/EuVO_4$ composite. *Catal. Today* **2010**, *158*, 209–214. [CrossRef]
20. He, Y.; Wu, Y.; Guo, H.; Sheng, T.; Wu, X. Visible light photodegradation of organics over VYO composite catalyst. *J. Hazard. Mater.* **2009**, *169*, 855–860. [CrossRef]
21. Monsef, R.; Ghiyasiyan-Arani, M.; Salavati-Niasari, M. Design of Magnetically Recyclable Ternary $Fe_2O_3/EuVO_4$/g-$C_3N_4$ Nanocomposites for Photocatalytic and Electrochemical Hydrogen Storage. *ACS Appl. Energy Mater.* **2021**, *4*, 680–695. [CrossRef]
22. Shandilya, P.; Mittal, D.; Soni, M.; Raizada, P.; Lim, J.-H.; Jeong, D.Y.; Dewedi, R.P.; Saini, A.K.; Singh, P. Islanding of $EuVO_4$ on high-dispersed fluorine doped few layered graphene sheets for efficient photocatalytic mineralization of phenolic compounds and bacterial disinfection. *J. Taiwan Inst. Chem. Eng.* **2018**, *93*, 528–542. [CrossRef]

23. Zhang, J.; Mei, J.; Yi, S.; Guan, X. Constructing of Z-scheme 3D g-$C_3N_4$-ZnO@graphene aerogel heterojunctions for high-efficient adsorption and photodegradation of organic pollutants. *Appl. Surf. Sci.* **2019**, *492*, 808–817. [CrossRef]
24. Tian, C.; Zhao, H.; Sun, H.; Xiao, K.; Wong, P.K. Enhanced adsorption and photocatalytic activities of ultrathin graphitic carbon nitride nanosheets: Kinetics and mechanism. *Chem. Eng. J.* **2020**, *381*, 122760. [CrossRef]
25. Xu, K.; Yang, X.; Ruan, L.; Qi, S.; Liu, J.; Liu, K.; Pan, S.; Feng, G.; Dai, Z.; Yang, X.; et al. Superior Adsorption and Photocatalytic Degradation Capability of Mesoporous $LaFeO_3$/g-$C_3N_4$ for Removal of Oxytetracycline. *Catalysts* **2020**, *10*, 301. [CrossRef]
26. Zhao, H.; Sun, S.; Jiang, P.; Xu, Z. Graphitic $C_3N_4$ modified by $Ni_2P$ cocatalyst: An efficient, robust and low cost photocatalyst for visible-light-driven $H_2$ evolution from water. *Chem. Eng. J.* **2017**, *315*, 296–303. [CrossRef]
27. Xu, K.; Feng, J. Superior photocatalytic performance of $LaFeO_3$/g-$C_3N_4$ heterojunction nanocomposites under visible light irradiation. *RSC Adv.* **2017**, *7*, 45369–45376. [CrossRef]
28. Chu, Z.; Li, J.; Lan, Y.-P.; Chen, C.; Yang, J.; Ning, D.; Xia, X.; Mao, X. KCl-LiCl molten salt synthesis of $LaOCl/CeO_2$-g-$C_3N_4$ with excellent photocatalytic-adsorbed removal performance for organic dye pollutant. *Ceram. Int.* **2022**, *48*, 15439–15450. [CrossRef]
29. Chen, Z.; Pan, Y.; Cai, P. Sugarcane cellulose-based composite hydrogel enhanced by g-$C_3N_4$ nanosheet for selective removal of organic dyes from water. *Int. J. Biol. Macromol.* **2022**, *205*, 37–48. [CrossRef]
30. Wang, X.; Chen, X.; Thomas, A.; Fu, X.; Antonietti, M. Metal-containing carbon nitride compounds: A new functional organic-metal hybrid material. *Adv. Mater.* **2009**, *21*, 1609–1612. [CrossRef]
31. Wang, T.; Huang, M.; Liu, X.; Zhang, Z.; Liu, Y.; Tang, W.; Bao, S.; Fang, T. Facile one-step hydrothermal synthesis of α-$Fe_2O_3$/g-$C_3N_4$ composites for the synergistic adsorption and photodegradation of dyes. *RSC Adv.* **2019**, *9*, 29109–29119. [CrossRef] [PubMed]
32. Wang, X.; Lu, W.; Zhao, Z.; Zhong, H.; Zhu, Z.; Chen, W. In situ stable growth of β-FeOOH on g-$C_3N_4$ for deep oxidation of emerging contaminants by photocatalytic activation of peroxymonosulfate under solar irradiation. *Chem. Eng. J.* **2020**, *400*, 125872. [CrossRef]
33. Zhu, T.; Song, Y.; Ji, H.; Xu, Y.; Song, Y.; Xia, J.; Yin, S.; Li, Y.; Xu, H.; Zhang, Q.; et al. Synthesis of g-$C_3N_4$/$Ag_3VO_4$ composites with enhanced photocatalytic activity under visible light irradiation. *Chem. Eng. J.* **2015**, *271*, 96–105. [CrossRef]
34. Liu, X.; Jin, A.; Jia, Y.; Xia, T.; Deng, C.; Zhu, M.; Chen, C.; Chen, X. Synergy of adsorption and visible-light photocatalytic degradation of methylene blue by a bifunctional Z-scheme heterojunction of $WO_3$/g-$C_3N_4$. *Appl. Surf. Sci.* **2017**, *405*, 359–371. [CrossRef]
35. Li, H.; Budarin, V.L.; Clark, J.H.; North, M.; Wu, X. Rapid and efficient adsorption of methylene blue dye from aqueous solution by hierarchically porous, activated starbons®: Mechanism and porosity dependence. *J. Hazard. Mater.* **2022**, *436*, 129174. [CrossRef]
36. Ghereghlou, M.; Esmaeili, A.A.; Darroudi, M. Adsorptive Removal of Methylene Blue from Aqueous Solutions Using Magnetic $Fe_3O_4$@C-dots: Removal and kinetic studies. *Sep. Sci. Technol.* **2022**, *57*, 2005–2023. [CrossRef]
37. Azeez, L.; Adebisi, S.A.; Adejumo, A.L.; Busari, H.K.; Aremu, H.K.; Olabode, O.A.; Awolola, O. Adsorptive properties of rod-shaped silver nanoparticles-functionalized biogenic hydroxyapatite for remediating methylene blue and congo red. *Inorg. Chem. Commun.* **2022**, *142*, 109655. [CrossRef]
38. Fan, H.; Li, F.; Huang, H.; Yang, J.; Zeng, D.; Liu, J.; Mou, H. pH graded lignin obtained from the by-product of extraction xylan as an adsorbent. *Ind. Crops Prod.* **2022**, *184*, 114967. [CrossRef]
39. Liang, C.; Shi, Q.; Feng, J.; Yao, J.; Huang, H.; Xie, X. Adsorption Behaviors of Cationic Methylene Blue and Anionic Reactive Blue 19 Dyes onto Nano-Carbon Adsorbent Carbonized from Small Precursors. *Nanomaterials* **2022**, *12*, 1814. [CrossRef]
40. Salazar-Rabago, J.J.; Leyva-Ramos, R.; Rivera-Utrilla, J.; Ocampo-Perez, R.; Cerino-Cordova, F.J. Biosorption mechanism of Methylene Blue from aqueous solution onto White Pine (*Pinus durangensis*) sawdust: Effect of operating conditions. *Sustain. Environ. Res.* **2017**, *27*, 32–40. [CrossRef]
41. Ehsan, M.F.; Fazal, A.; Hamid, S.; Arfan, M.; Khan, I.; Usman, M.; Shafiee, A.; Ashiq, M.N. $CoFe_2O_4$ decorated g-$C_3N_4$ nanosheets: New insights into superoxide anion mediated photomineralization of methylene blue. *J. Environ. Chem. Eng.* **2020**, *8*, 104556. [CrossRef]
42. Ismail, A.A.; Faisal, M.; Harraz, F.A.; Al-Hajry, A.; Al-Sehemi, A.G. Synthesis of mesoporous sulfur-doped $Ta_2O_5$ nanocomposites and their photocatalytic activities. *J. Colloid Interface Sci.* **2016**, *471*, 145–154. [CrossRef] [PubMed]
43. Heo, J.W.; An, L.L.; Chen, J.S.; Bae, J.H.; Kim, Y.S. Preparation of aminefunctionalizedlignins for the selective adsorption of Methylene blue and Congo red. *Chemosphere* **2022**, *295*, 133815. [CrossRef]
44. Venkatesha, T.C.; Viswanatha, R.; Arthoba Nayaka, Y.; Chethana, B.K. Kinetics and thermodynamics of reactive and vat dyes adsorption on MgO nanoparticles. *Chem. Eng. J.* **2012**, *198–199*, 1–10. [CrossRef]
45. Xiao, T.; Tang, Z.; Yang, Y.; Tang, L.; Zhou, Y.; Zou, Z. In situ construction of hierarchical $WO_3$/g-$C_3N_4$ composite hollow microspheres as a Z-scheme photocatalyst for the degradation of antibiotics. *Appl. Catal. B-Environ.* **2018**, *220*, 417–428. [CrossRef]
46. Sun, L.; Li, J.; Li, X.; Liu, C.; Wang, H.; Huo, P.; Yan, Y. Molecularly imprinted Ag/$Ag_3VO_4$/g-$C_3N_4$ Z-scheme photocatalysts for enhanced preferential removal of tetracycline. *J. Colloid Interface Sci.* **2019**, *552*, 271–286. [CrossRef] [PubMed]
47. Niu, P.; Zhang, L.; Liu, G.; Cheng, H.-M. Graphene-Like Carbon Nitride Nanosheets for Improved Photocatalytic Activities. *Adv. Funct. Mater.* **2012**, *22*, 4763–4770. [CrossRef]
48. Zhang, X.; Xie, X.; Wang, H.; Zhang, J.; Pan, B.; Xie, Y. Enhanced Photoresponsive Ultrathin Graphitic-Phase $C_3N_4$ Nanosheets for Bioimaging. *J. Am. Chem. Soc.* **2013**, *135*, 18–21. [CrossRef]

*Article*

# Effects of Acidic Environments on Dental Structures after Bracket Debonding

Cristina Iosif [1], Stanca Cuc [2,*], Doina Prodan [2], Marioara Moldovan [2,*], Ioan Petean [3], Mîndra Eugenia Badea [4], Sorina Sava [1], Andrada Tonea [1] and Radu Chifor [4]

1. Department of Prosthetic Dentistry and Dental Materials, "Iuliu Hatieganu" University of Medicine and Pharmacy, 32 Clinicilor Street, 400006 Cluj-Napoca, Romania
2. Department of Polymer Composites, Institute of Chemistry "Raluca Ripan", University Babes-Bolyai, 30 Fantanele Street, 400294 Cluj-Napoca, Romania
3. Faculty of Chemistry and Chemical Engineering, University Babes-Bolyai, 11 Arany János Street, 400028 Cluj-Napoca, Romania
4. Department of Preventive Dental Medicine, "Iuliu Hatieganu" University of Medicine and Pharmacy, Avram Iancu 31, 400083 Cluj-Napoca, Romania
* Correspondence: stanca.boboia@ubbcluj.ro (S.C.); marioara.moldovan@ubbcluj.ro (M.M.)

**Abstract:** Brackets are metallic dental devices that are very often associated with acidic soft drinks such as cola and energy drinks. Acid erosion may affect the bonding between brackets and the enamel surface. The purpose of this study was to investigate the characteristics of brackets' adhesion, in the presence of two different commercially available drinks. Sixty human teeth were divided into six groups and bonded with either resin-modified glass ionomer (RMGIC) or resin composite (CR). A shared bond test (SBS) was evaluated by comparing two control groups with four other categories, in which teeth were immersed in either Coca-Cola$^{TM}$ or Red Bull$^{TM}$ energy drink. The debonding between the bracket and enamel was evaluated by SEM. The morphological aspect correlated with SBS results showed the best results for the samples exposed to artificial saliva. The best adhesion resistance to the acid erosion environment was observed in the group of teeth immersed in Red Bull$^{TM}$ and with brackets bonded with RMGIC. The debonded structures were also exposed to Coca-Cola$^{TM}$ and Red Bull$^{TM}$ to assess, by atomic force microscopy investigation (AFM), the erosive effect on the enamel surface after debonding and after polishing restoration. The results showed a significant increase in surface roughness due to acid erosion. Polishing restoration of the enamel surface significantly reduced the surface roughness that resulted after debonding, and inhibited acid erosion. The roughness values obtained from polished samples after exposure to Coca-Cola$^{TM}$ and Red Bull$^{TM}$ were significantly lower in that case than for the debonded structures. Statistical results evaluating roughness showed that Red Bull$^{TM}$ has a more erosive effect than Coca-Cola$^{TM}$. This result is supported by the large contact surface that resulted after debonding. In conclusion, the prolonged exposure of the brackets to acidic drinks affected the bonding strength due to erosion propagation into both the enamel–adhesive interface and the bonding layer. The best resistance to acid erosion was obtained by RMGIC.

**Keywords:** dental brackets; enamel surface; acid environment; shared bond test; surface roughness

## 1. Introduction

Various factors can affect the bonding of brackets during orthodontic treatment, such as enamel structure, bracket surface properties, poor operator technique, the patient's behavior, and masticator forces [1]. The failure of bonding results in a longer period of time for treatment, which is why the cooperation of patients is necessary [2]. One essential factor is the patient's oral hygiene, which can determine the accumulation of dental plaque, whether white spot lesions will occur, and whether the orthodontic therapy will be compromised [2].

The ideal cement used for bracket bonding should have good retention to the surface of the tooth, enough strength to resist and transmit the orthodontic forces during the course of therapy, and should be able to be removed without resulting in iatrogenic damage to the enamel [3]. A bonding strength of 5.9–7.8 MPa is recommended for orthodontic treatment [1].

Composite resins are frequently used for bonding brackets due to their good aesthetic and bond strength [4]. However, to bond with composite resins we need a completely dry surface, and the patient's compliance is essential [5]. Due to the fixed appliance, and depending on the behavior of the patient regarding oral hygiene, the prevalence of plaque accumulation tends to increase during orthodontic treatment [6]. This is why the risk of decalcification is larger during the therapy and tends to occur around the bracket, in particular in the gingival area, which is more difficult to clean [7].

Demineralization can occur as early as one month after the beginning of treatment [6] and can result in white spot lesions or enamel cavities in severe cases. Enamel demineralization after orthodontic treatment can appear in up to 50% of patients [2]. It has been demonstrated that there is a possibility for the initial damage to be repaired with the use of fluoride, which has the capacity to re-mineralize teeth, resulting in an anti-caries effect [7].

In other words, when the brackets are bonded with cement resins it can result in white spot lesions and the risk of enamel damage after bracket removal, which can reduce the success of the treatment [4]. As a result, the use of other materials, such as glass ionomers, has been suggested. The advantages of glass ionomers are that they can be used for bonding in cases with moisture contamination, such as when we cannot realize the proper isolation of the teeth [2], and the fact that they are fluoride-releasing, with the potential to decrease the demineralization of the surface of the enamel around the bracket [6].

However, glass ionomers have a lower bond strength (2.37–5.5 MPa), which will frequently result in bond failure [8], and therefore limits their clinic use.

Hybrid glass ionomer cements were developed to combine the adequate bond strength of cement resins with the fluoride-releasing properties of glass ionomers [6]. Ghubaryi, et al. [9] demonstrated that the bond strength is initially lower, but will increase after 24 h to 5.39–18.9 MPa; such hybrid cements have been successfully used in orthodontic therapy.

Enamel demineralization is a severe complication of orthodontic treatment and is the result of a decrease in salivary pH [5]. If its pH is lower than 5, saliva is sub-saturated in fluoroapatite and hydroxyapatite, and this creates a medium for demineralization. Behavioral factors, such as the pH of drinks consumed, and buffering of saliva can result in enamel demineralization and dental erosion [5];in such cases, the bond between bracket and enamel will be affected and the success of the treatment will be compromised.

More people, especially children, and teenagers, are frequently consuming soft drinks [10]. This kind of beverage includes all drinks except alcohol, mineral water, fruit juices, tea, coffee, or milk-based drinks, and may or may not be carbonated [5]. The carbonated beverages are more aggressive, because of their carbonic acid content [11].

Soft drinks have a negative impact on the structure of enamel because of their content of sugar and citric acid, which are very erosive and will degrade the surface of the tooth around the brackets [1]. The consumption of soft drinks results in a pH lower than 5.5, which can then lead to dental erosion. Dental erosion is a defect on the surface of enamel resulting from exposure to acids and will compromise the bond between the bracket and the tooth, because healthy enamel is essential for the retention of the bracket.

Coca-Cola is the most popular soft drink, and one of the three best-known beverages [12]. Red Bull is the most frequently consumed energy drink. One of the best-known slogans in the United States is "Red Bull gives you wings", because of the effects of the drink on mental and athletic performance due to its caffeine content. They also might influence the dental material color [13] and behavior due to the acid action at the body temperature [14]. The dental plaque proliferation on the bonding layer also might be influenced by acid components of mentioned soft drinks [15].

The purpose of this study is the determination of these two soft drinks' effects on the shear bond strength and microstructural aspects regarding orthodontic metal brackets bonded with two different materials: cement resin (Transbond Color Change) and RMGIC (Fuji Ortho LC).

## 2. Results

### 2.1. Shear Bond Strength

The adhesion was evaluated by recording the shear strength according to the standard procedures [16,17], which is generally used to evaluate adhesive systems in the laboratory. Generally, the comparison between materials or bracket application methods is justified, but the extrapolation of these results in the clinical situation is not easy. The results obtained by us for CR and RMGIC are presented in Table 1. The highest value for SBS was recorded for the RMGIC sample exposed to Saliva (7.65 MPa), followed by the CR control sample, exposed only to artificial saliva (5.20 MPa), and the lowest value for bond strength was recorded for the CR sample exposed to Cola (2.70 MPa). The other sample groups all registered SBS of approximately 3 MPa.

Table 1. Shear bond test results.

| Mechanical Properties | Saliva | | Coca-Cola™ | | Red Bull™ | |
|---|---|---|---|---|---|---|
| | Mean | SD | Mean | SD | Mean | SD |
| CR | | | | | | |
| Bond Strength, MPa | 5.20774 | 1.96879 | 2.72076 | 0.77658 | 2.90572 | 1.49381 |
| Maximum Load, N | 67.5786 | 23.91672 | 32.96726 | 22.66505 | 35.92676 | 18.51022 |
| Breaking Load, N | 60.34059 | 31.61737 | 30.044 | 18.14057 | 31.888897 | 15.86262 |
| RMGIC | | | | | | |
| Bond Strength, MPa | 7.65976 | 2.3793 | 3.53586 | 1.10649 | 3.62436 | 2.38791 |
| Maximum Load, N | 86.92681 | 24.5682 | 41.42943 | 14.38154 | 53.40967 | 36.83688 |
| Breaking Load, N | 82.48911 | 26.11797 | 31.19834 | 10.23796 | 50.90918 | 38.59967 |

Statistical Analysis: The results of the bracket adhesion test for all six investigated teeth groups showed different statistical significance, with a $p$-value = $2.77933 \times 10^{-4}$. Following the Tukey test, it was evident that different statistical significances are only between RMGIC immersed in artificial saliva and the other groups tested (RMGIC in COLA, CR in COLA, CR in HELL, CR in artificial saliva).

The Anova test obtained a value of $p = 0.00906$, regarding the breaking strength of the bracket with little different statistics between groups: RMGIC immersed in artificial saliva and CR in COLA; RMGIC in artificial saliva and RMGIC in COLA. The same differences between groups are also obtained in the case of the maximum strength accepted, with a $p = 0.01026$.

### 2.2. Scanning Electron Microscopy SEM

SEM images were taken on low magnification to observe all morphological details regarding the shearing of the adhesion layer between brackets and enamel. Its failure under shearing stress generated important morphological details that are necessary to explain the variation related to the data in Table 1. The images resulting from CR samples are presented in Figure 1 and the ones for RMGIC samples are presented in Figure 2.

CR control sample exposed to the artificial saliva reveals a complex morphology adhesive layer shearing. The bracket base, Figure 1(Aa), is almost covered with an adhesive having a relatively uniform surface (from a macroscopic point of view). It covers about 60% of the bracket base. The shearing stress during failure caused adhesive layer fragmentation with complete detachment from the other 40% of the bracket base which is clearly observed in the direction of the yellow arrow in Figure 1(Aa). On the opposite, the enamel surface after debonding is macroscopically smooth for about 60%, and the other 40% presents adhesive fragments as indicated with the arrow in Figure 1(Ab). The polishing effect removes the adhesive debris and small unevenness which conducts to a plain surface of good macroscopic quality, Figure 1(Ac). There appear some small microscopic irregularities and scratches which need a detailed microscopic investigation using the atomic force microscope (AFM).

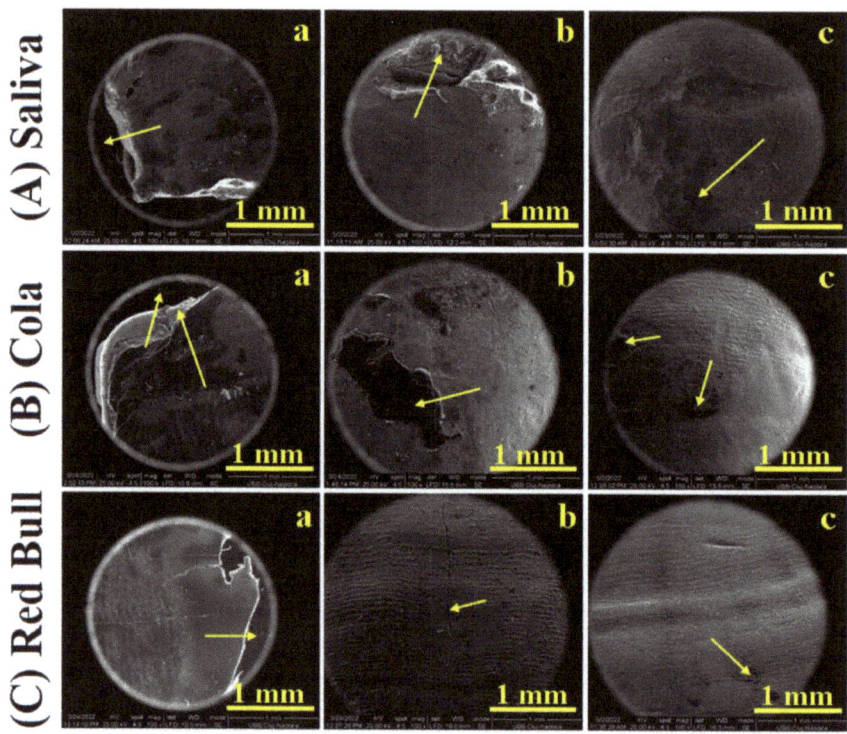

**Figure 1.** SEM images of CR samples: (**a**) bracket after debonding, (**b**) enamel after debonding, and (**c**) debonded enamel after polishing; stored in Saliva (**A**), exposed to Cola (**B**) and exposed to Red Bull™ (**C**).

Figure 1(Ba) presents the bracket base of the CR sample exposed to Coca-Cola™. Adhesive failure after debonding leads to a complex microstructure formed by: a large part of the surface covered with an adhesive of about 70% being relatively uniform and 30% of the surface being irregular due to the adhesive fragmentation. It occurs in two steps: a thinner rest of adhesive is observed on about 15% of the surface due to its internal failure under debonding and about 15% of the free bracket base, see the arrow in Figure 1(Bb). The polishing treatment removes the adhesive debris from the enamel surface in a good manner but some residual adhesive traces are still present, Figure 1(Bc).

Red Bull™ exposure of the CR sample certainly influences the bracket debonding features, Figure 1C. The bracket failure occurred at the enamel-adhesive interface, most of the adhesive being attached to the bracket base on 85% of the surface, the rest being fragmented and completely detached from the debonded parts; see the arrow in Figure 1(Ca). The enamel surface is completely free of adhesive at the macroscopical view, and some microscopic traces may occur. A significant crack in the enamel is observed and indicated with the arrow in Figure 1(Cb). The polishing procedure conducts to a smoother surface free of cracks and only very small traces of adhesive are observed, Figure 1(Cc).

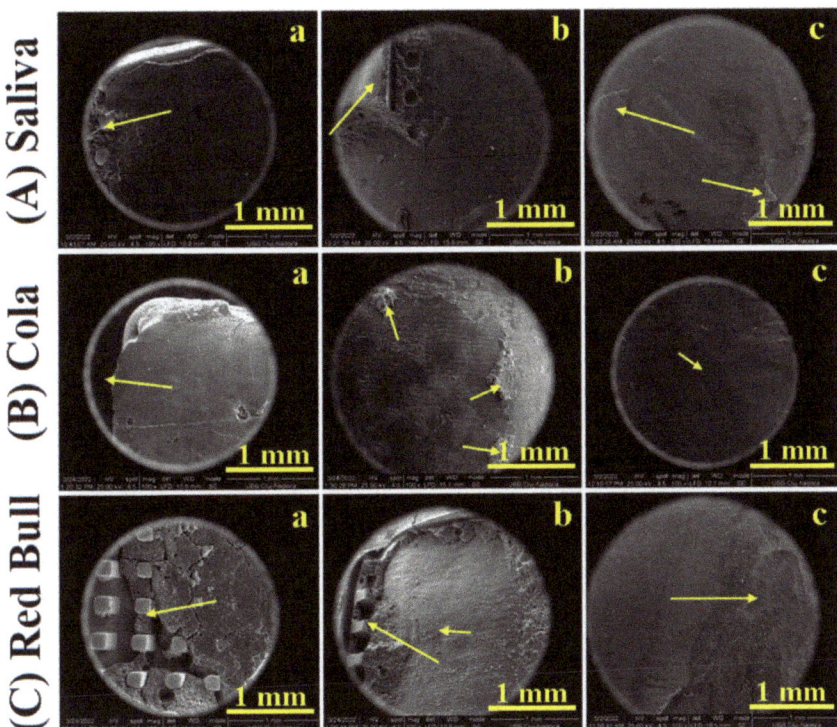

**Figure 2.** SEM images of RMGIC samples: (**a**) bracket after debonding, (**b**) enamel after debonding, and (**c**) debonded enamel after polishing; stored in Saliva (**A**), exposed to Cola (**B**) and exposed to Red Bull[TM] (**C**).

The morphology of the RMGIC sample kept in the artificial saliva after debonding is presented in Figure 2A. Bracket failure during debonding occurred at the enamel adhesive interface. Therefore, the base of the bracket is almost completely covered by the adhesive of about 88% of the surface only 12% of the bracket base being visible on the side indicated with the yellow arrow, Figure 2(Aa). The enamel surface after debonding, Figure 2(Aa), reveals a macroscopic uniform surface which is covered with adhesive only on about 12% of its surface as indicated with an arrow in Figure 2(Ab). Some footprints from the bracket base are visible, imprinted in the adhesive on the observed area. The polishing procedure is required to remove this debonding debris on the enamel surface. It was an improvement during the restoration attempt as observed in Figure 2(Ac). Some irregularities on the enamel surface are observed, due to the adhesive that was not polished enough.

Coca-Cola[TM] exposure to the RMGIC sample influences the debonding parts morphological aspect as observed in Figure 2B. The adhesive failure occurs mainly on the enamel interface assuring proper preservation of the parts. Most of the adhesive layer remains attached to the bracket, Figure 2(Ba), covering about 83% of its base. The other 17% corresponds to the free view of the bracket base as shown with the yellow arrow. The adhesive in this area was broken during debonding and fell away. The enamel surface, Figure 2(Bb), presents a uniform surface after debonding, only small areas with adhesive traces were observed, Figure 2(Bb). The debonding structures and erosion marks were very well removed by the polishing treatment, only few scratches being visible in Figure 2(Bc).

RMGIC samples exposed to Red Bull[TM] after debonding presents a very interesting morphology. Almost 50% of the bracket base surface is covered with an adhesive having internal fissures and the other part is visible, Figure 2(Ca). This internal fragmentation of also influences the morphology of the enamel surface, Figure 2(Cb). There are significant adhesive deposits on 15% of the surface and several small traces as indicated with arrows. A polishing procedure is absolutely necessary. Figure 2(Cc) evidenced on the left side of the observation field a very well-restored area of the enamel (about 55% of the surface) and on the right side appears several remaining defects such as traces of adhesive and erosion features.

*2.3. Atomic Force Microscopy AFM*

The fine microstructure of debonded CR sample exposed to saliva is presented in Figure 3(Aa). The surface topography evidenced several HAP nanostructural units pulled out during the debonding process forming some depressions with irregular margins and the diameter ranging from 300 to 1000 nm. There are also some adhesive debris traces situated on the highest zones of the observation field. The Coca-Cola$^{TM}$ exposure causes a pronounced decaying of the enamel fine microstructure due to the progressive enlarging of the depressions and significant weight loss associated with HAP nanoparticles loss, Figure 3(Ba). It causes a significant increase in surface roughness, Figure 4a,b. The exposure to Red Bull$^{TM}$ energy booster juice leads to an advanced decaying of the enamel surface by pronounced enlarging of the depressions, their aspects becoming similar to the desert dunes, Figure 3(Ca). These fine microstructural aspects are in good agreement with SEM observation in Figure 1(Ab,Ac). The surface roughness value increases significantly because of the observed decayed features.

The polishing treatment leads to a proper CR enamel surface smoothness which is considerably improved, Figure 3(Ab). The adhesive debris traces from the enamel surface were completely removed by polishing and the topographic depressions (caused by HAP units pulling out during debonding) are considerably attenuated. Therefore, the roughness value significantly decreases in Figure 4.

The enamel surface compactness resulting after the polishing treatment determines a better resistance to acid erosion due to the removal of the depressions which act as decaying promoters. Thus, Coca-Cola and Red Bull$^{TM}$ found a compact surface without faults to be penetrated and enlarged. It means that the erosive elements within the mentioned juices must generate the erosion faults prior to being enlarged. Therefore, the roughness values after exposure are lower than the ones observed for the unpolished samples, Figure 4. The Coca-Cola$^{TM}$ exposure generates a surface topography having a washout aspect but with low penetration in-depth, Figure 3(Bb). The surface topography is more rugged after Red Bull$^{TM}$ exposure as observed in Figure 3(Cb). Thus, the second aspect of the null hypothesis is rejected because the roughness resulting after Red Bull is significantly greater than the one caused by Cola, Figure 4.

The nanostructure of the CR enamel sample resulting after debonding is presented in Figure 3(Ac). The observed topography is quite rugged because of the depressions formed by HAP nanostructural units pulling out during debonding. In addition to these certainly affected areas, there are remarked compact portions of healthy enamel featuring nanostructural units of about 40 nm diameter well bonded one to another by the protein binder. The exposure to Coca-Cola$^{TM}$ causes a widening of the above-mentioned depressions along with an advanced erosion of the areas initially unaffected, Figure 3(Bb). Therefore, the diameter of the HAP nano-structural units increases in the range of 60–90 nm, Figure 4c. The exposure to Red Bull$^{TM}$ leads to a more advanced decaying of the nanostructure, Figure 1(Cc). We remark on the increase in the depression's depth along with advanced demineralization signs such as washout surfaces and increased diameter of the nanostructural units around 100 nm. The roughness values are significantly increased as observed in Figure 4a,b.

The nanostructure resulting from the CR enamel after the polishing procedure presents quite a uniform topography, Figure 3(Ad), which corresponds to a compact surface with HAP nanostructural units very well attached one to another. Their diameter is situated at about 40 nm in good agreement with the values for the healthy enamel [18,19]. The surface roughness significantly decreases as observed in Figure 4a,b, also being in good agreement with the data in the literature for healthy enamel [18,19].

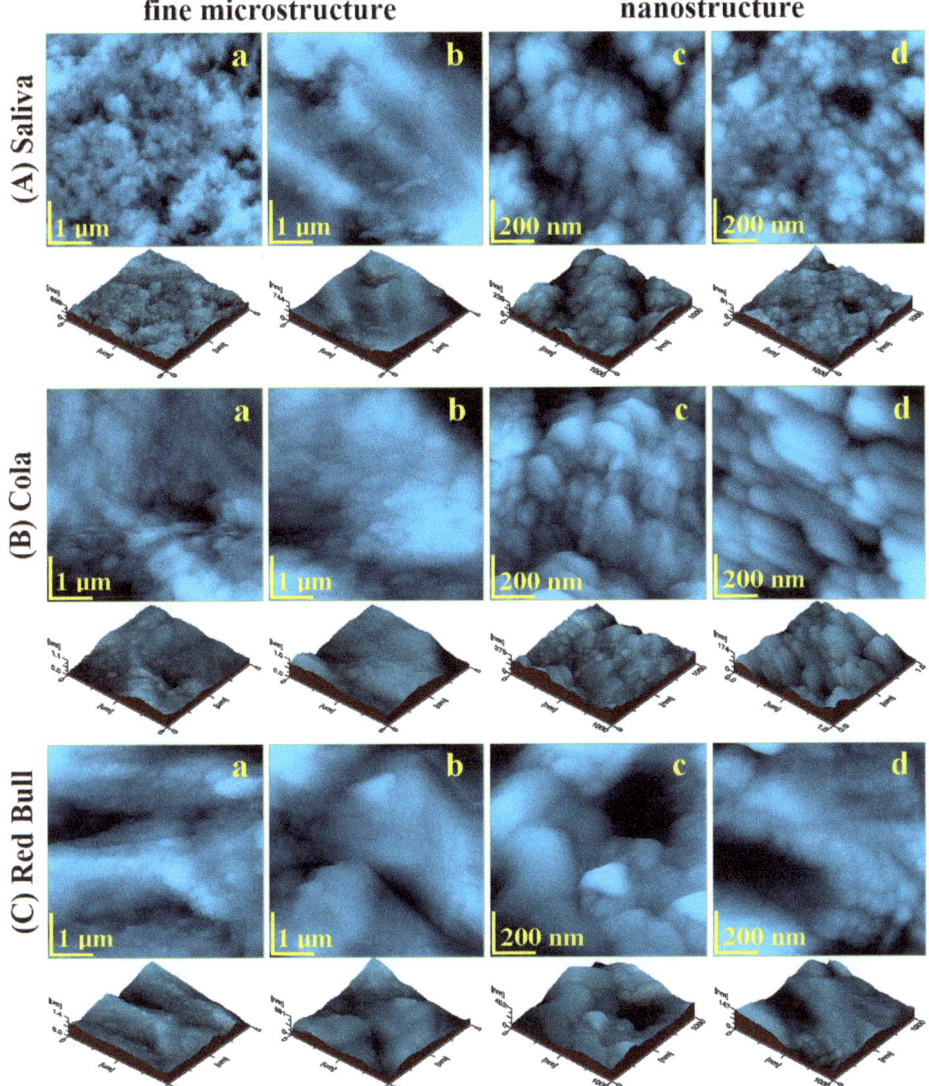

**Figure 3.** AFM imaging of enamel after CR debonding: fine microstructure before polishing (**a**) and after polishing (**b**); nanostructure before polishing (**c**) and after polishing (**d**); stored in Saliva (**A**), exposed to Cola (**B**), and exposed to Red Bull™ (**C**).

Coca-Cola™ exposure causes an advanced erosion of the HAP nanostructural units from the CR enamel surface but without generating the decaying depressions, Figure 3(Bd). The fact determines the severe increase in the nanostructural unit diameter from 40 nm to over 120 nm. The exposure to Red Bull on the CR enamel surface leads to more advanced decay as observed in Figure 3(Cd). We observe an undulated topography associated with HAP crystallites with deep erosion marks and traces of weight loss. This fact leads to increased values of roughness compared to the Cola exposure. Therefore, the null hypothesis was rejected as a nanostructural point of view.

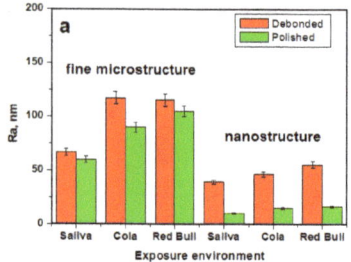 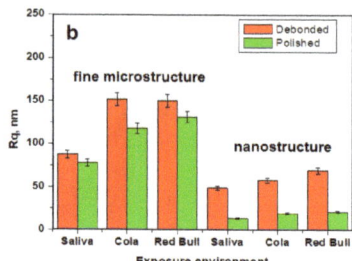 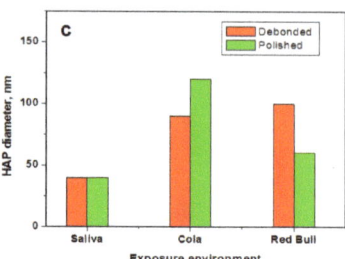

**Figure 4.** The evolution of the parameters measured with AFM for CR samples: (**a**) Ra, (**b**) Rq, and (**c**) nano HAP diameter.

Bracket debonding from RMGIC adhesion on the enamel stored in the artificial saliva leads to a very rugged surface. It presents a mixed characteristic resulting from the topographic combination of the submicron adhesive traces and depressions formed by HAP nanostructural units pulling out during debonding, Figure 5(Aa). These topographic features have the nature to generate a relatively high surface roughness as observed in Figure 6a,b.

Advanced acid erosion on the RMGIC debonded sample exposed to Coca-Cola[TM], is remarked in Figure 5(Ba). The adhesive traces present a better resistance to the erosive effect of the phosphoric acid, meanwhile, HAP nanostructural units are deeply eroded with evidence of significant weight loss. The roughness values proved to be greater than the ones observed for the control sample. Red Bull[TM] proves to be more aggressive than Cola by the significant deepening of the acid erosion depressions, Figure 5(Ca). Such an erosive effect leads to the greater roughness values reached in the present research, Figure 6a,b.

RMGIC enamel sample polishing is an absolutely necessary step for the adhesive submicron traces removal as well as for the depressions (formed by HAP nanostructures unit pulling out during debonding) attenuation. The topography of the sample stored in artificial saliva after polishing treatment is smoother and more compact as observed in Figure 5(Ab). The complete removal of submicron adhesive traces is observed along with a very compact disposal of the Hap nanostructural units similar to the typical structures in the healthy enamel according to the data in the literature [19,20]. The mentioned depressions are well attenuated and their marks are almost unobservable. A direct consequence of these facts is the significant decrease in roughness, as observed in Figure 6a,b.

The exposure to Coca-Cola[TM] of RMGIC polished enamel presents a significant tendency of the acid etching of the HAP pulling out depressions to form acid erosion depressions, Figure 1(Bb). It is a matter of the topography disturbance that increases the roughness value. The exposure to Red Bull[TM] causes slightly enhanced erosion features than those observed for Cola, Figure 1(Cb) and to the formation of several deep valleys such as the one observed in the upper right side of the topographical image in Figure 1(Cb). These additional features lead to an additional increase in roughness, Figure 6.

Considering the AFM evidence, we remark that the polishing treatment on the RMGIC enamel samples is beneficial to prevent the propagation of the acid erosion features into the profoundness of the samples and consequently to prevent the excessive increase in roughness. The resistance of RMGIC adhesive traces to the acid erosion especially to the Red Bull is observed.

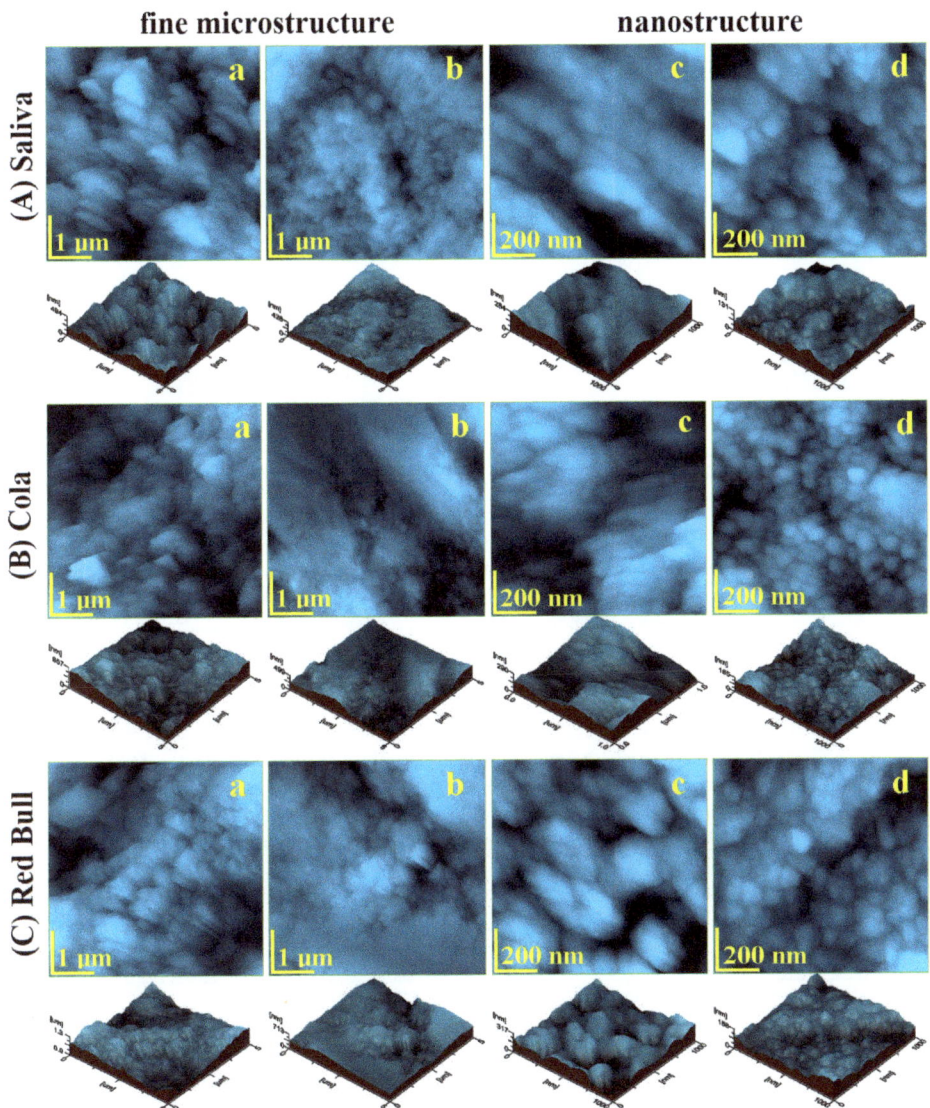

**Figure 5.** AFM imaging of enamel after RMGIC debonding: fine microstructure before polishing (**a**) and after polishing (**b**); nanostructure before polishing (**c**) and after polishing (**d**); stored in Saliva (**A**), exposed to Cola (**B**), and exposed to Red Bull™ (**C**).

The RMGIC enamel sample nanostructure after debonding and storage in artificial saliva presents some small areas of pulled-out nano HAP units but also some submicron adhesive traces which remain from the adhesion interface, Figure 1(Ac). Therefore, the HAP nanostructural unit diameter is about 40 nm, Figure 6c. The exposure of this surface to Coca-Cola™ leads to the enlarging of the depressions by the acid erosion of exposed enamel and due to the tendency to penetrate depth. The nano HAP units present on the surface are significantly affected by the acid causing their diameter to increase to about 55 nm as observed in Figure 5(Bc). Thus, significant roughness increasing occurs. The exposure to Red Bull™ generates a more affected nanostructure by the of rugged topography with an aspect of eroded sand dunes, Figure 5(Cc). The nano HAP unit diameter is situated in the range of 60–90 nm; it indicates that they are close to their material disintegration which is able to cause weight loss. It has a strong influence on the roughness increasing. Therefore, the null hypothesis is totally rejected by the RMGIC enamel samples nanostructure after debonding.

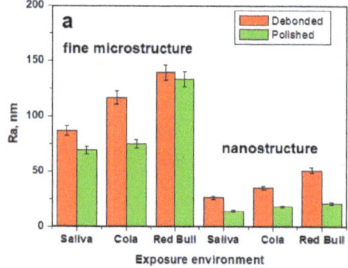 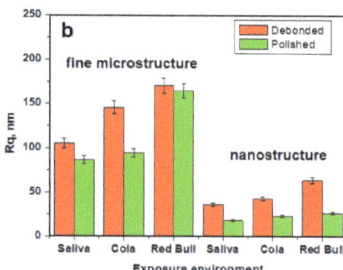 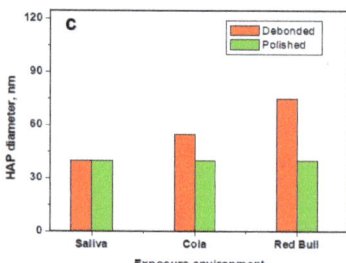

**Figure 6.** The evolution of surface parameters measured with AFM for RMGIC samples: (**a**) $R_a$, (**b**) $R_q$, and (**c**) nano HAP diameter.

The RMGIC polished enamel stored in artificial saliva nanostructure is considerably improved, Figure 5(Ad). The surface topography is more uniform than before treatment due to the adhesive trace removal and attenuation of the nano HAP pulling out depressions. The enamel morphology revealed a nano Hap unit with a diameter of about 40 nm, in good agreement with the data in the literature for healthy enamel [20,21], which are well welded one to another. In fact, this sample is so good that presets the lowest roughness among all samples in the current study, Figure 6a,b. The Coca-Cola$^{TM}$ and Red Bull$^{TM}$ exposure of the polished RMGIC samples leads to very interesting topographies, Figure 5(Bd,Cd). The microstructure general aspect is well preserved with a nano HAP unit of about 40 nm and compactly organized. Only a small erosion of their boundaries is observed which is significantly pronounced for Red Bull compared to Cola. A small roughness increase was observed. In consequence, the null hypothesis is also rejected completely for the RMGIC enamel polished samples.

Roughness statistical analysis for the investigated samples is centralized in Tables 2 and 3.

**Table 2.** Surface roughness statistical analysis for the samples after debonding.

|  | CR + Saliva | CR + Cola | CR + Red Bull | RMGIC + Saliva | RMGIC + Cola | RMGIC + Red Bull |
|---|---|---|---|---|---|---|
| | | | Enamel fine microstructure | | | |
| Ra, nm | 66.90 | 117.66 | 115.56 | 86.90 | 116.83 | 133.30 |
| P1 | | 0.08 | 0.08 | | 0.46 | 0.46 |
| Rq, nm | 87.63 | 151.66 | 150 | 105.50 | 145.66 | 170.30 |
| P2 | | 0.06 | 0.06 | | 0.36 | 0.36 |
| | | | Enamel nanostructure | | | |
| Ra, nm | 39.43 | 46.6 | 55.7 | 25.56 | 35.26 | 51.26 |
| P1 | | 0.24 | 0.24 | | 0.21 | 0.21 |
| Rg, nm | 48.63 | 57.83 | 69 | 36 | 42.53 | 63.83 |
| P2 | | 0.24 | 0.24 | | 0.25 | 0.25 |

The AFM statistical results of the enamel fine microstructure after debonding, but before polishing, present no significant statistical differences after the exposure to Cola or Red Bull in both cases, when the brackets were bonded with composite resin or glass ionomer modified with resins. For the enamel fine nanostructure, statistical differences do not exist between the values after exposure to Cola or Red Bull$^{TM}$ before polishing when was used composite resin or glass ionomer modified for bonding.

In this case as well, the AFM statistical results of the enamel fine microstructure and nanostructure after debonding and polishing present no significant statistical differences after exposure to Cola or Red Bull™, for composite resin and glass ionomer modified.

**Table 3.** Surface roughness statistical analysis for the samples after polishing.

| | CR + Saliva | CR + Cola | CR + Red Bull | RMGIC + Saliva | RMGIC + Cola | RMGIC + Red Bull |
|---|---|---|---|---|---|---|
| | | | Enamel fine microstructure | | | |
| Ra, nm | 66 | 90.03 | 105.13 | 69.53 | 75.2 | 133.43 |
| P1 | | 0.40 | 0.40 | | 0.24 | 0.24 |
| Rq, nm | 77.56 | 117.63 | 131.33 | 86.43 | 94.7 | 164.66 |
| P2 | | 0.41 | 0.41 | | 0.20 | 0.20 |
| | | | Enamel nanostructure | | | |
| Ra, nm | 10.4 | 15.16 | 16.70 | 14.26 | 18.23 | 21.23 |
| P1 | | 0.22 | 0.22 | | 0.29 | 0.29 |
| Rq, nm | 13.28 | 19.10 | 21.23 | 18.03 | 23.10 | 26.26 |
| P2 | | 0.22 | 0.22 | | 0.28 | 0.28 |

There are no significant statistical differences when comparing the enamel fine microstructure both before and after polishing for Ra and Rq in all three sample groups (the control samples immersed in artificial saliva, the samples exposed to Cola, and those exposed to Red Bull™), as all $p$ values are higher than 0.05. Ra for: control sample—$p = 0.33$, Cola exposure—$p = 0.40$; Red Bull™ exposure $p = 0.78$; Rq for control sample—$p = 0.24$, and Cola exposure—$p = 0.40$; Red Bull™ exposure $p = 0.68$.

There are significant statistical differences when comparing the enamel nanostructure both before and after polishing for UM Ra and Rq in all three sample groups (the control samples immersed in artificial saliva, the samples exposed to Cola, and those exposed to Red Bull™): Ra for control sample—$p = 0.04$, Cola exposure—$p = 9.02897 \times 10^{-4}$; Red Bull™ exposure—$p = 7.23641 \times 10^{-4}$; Rq for control sample—$p = 0.04$, Cola exposure—$p = 5.9025 \times 10^{-4}$; Red Bull™ exposure $p = 0.00156$.

## 3. Discussion

Oral hygiene, orthodontic bonding technique, and nutrition are the factors determining the development of dental erosion during orthodontic treatment [11,22,23]. It is a matter of complex interaction between the enamel surface and bracket towards the adhesive layer. Poor oral hygiene causes food residue deposits around the brackets and bonds which generate erosive compounds during the putrefaction process. Poor bonding technique will generate weaker adhesion of the bracket to the enamel and may generate local erosion [24]. Nutrition is very often associated with acidic food and drink ingestion. The acid compounds within the masticated and ingested aliments will decrease the retention of the brackets and will affect the results of the therapy.

The percentage of teenagers who are consuming soft drinks has increased in the last few years. This is the reason we chose to use Coca-Cola™, the most famous of carbonated drinks, and Red Bull™, the most well-known energy booster drink. Red Bull™ contains citric acid and Coca-Cola™ phosphoric acid, both of which are used in acid etching for bonding orthodontic brackets [10]. These acids are also reported in the literature to be the main cause of enamel surface demineralization [25,26]. It is demonstrated that even short periods of consuming soft drinks diminish the enamel microhardness and increase the roughness. It is crucial to limit exposure to this type of beverage to prevent erosive lesions [1,26]. Acidic drinks reduce oral cavity pH lower than 5.5 which will favors the dissolution of hydroxyapatite and fluor-apatite within the teeth enamel [5] and limits the re-mineralizing effect of the saliva. The tooth enamel is a non-remodeling tissue, for this reason, any changes that occur on its surface will be permanent [27].

The aim of this study is to observe how much shear bond strength can be affected by the consumption of soft drinks and for what type of material (composite resin or glass ionomer modified) it can be obtained a better adhesion between bracket and enamel.

Composite resins are most frequently used to bond brackets during orthodontic treatment. It is necessary to use phosphoric acid for etching, and to obtain the micropores on the enamel for mechanical bonding with the cement adhesive to obtain a satisfactory bond between the enamel and the tooth [19]. These cements have micro-mechanical mechanisms for adhesion [3], but they have enamel demineralization as a side effect. For this reason, fluoride-releasing cements were developed to inhibit the loss of minerals.

The glass ionomers can release fluoride and prevent the demineralization of the enamel, but they have lower bond strength. RMGIC were developed to improve the adhesion between brackets and the teeth and presents the advantage of a higher SBS than composite resins and fluoride-releasing due to the presence of the glass ionomers. A polyacrilic acid of 10% is used with RMGIC, because phosphoric acid is too aggressive and demineralization will be deeper. After mixing powder with liquid, the curing-mechanism is chemical, with an acid-base reaction. When light-curing begins, polymerization is initiated [28]. Previous research demonstrated that the bond strength of RMGIC is higher after 24 h, but not immediately. One of the advantages of using RMGIC for bonding brackets is that the surface of enamel is almost intact after debonding [2].

Considering all materials involved in the orthodontic devices treatment in current research, resulting in three possible ways to promote erosion [29,30]:

- the first is within the bonding structure, in our case CR and RMGIC adhesive;
- the second is on the enamel-adhesive interface;
- and the third is on the bracket-adhesive interface.

SEM microscopy evidences that the bracket-adhesive interface is very cohesive due to the bracket base equipped with adhesion pillars. The adhesive after debonding is found predominantly on the bracket base with a surface coverage between 70–85%. On the opposite, the enamel-adhesive interface after debonding has less adhesive after debonding with coverage ranging from 15 to 30%. Our SEM evidence shows that the adhesive layer presents some internal fractures during debonding, especially for CR samples exposed to Coca-Cola$^{TM}$ and Red Bull$^{TM}$. The RMGIC adhesive layer proves to be more cohesive without significant internal fracture after debonding. This observation shows that RMGIC is more resistant to acid erosion attempts than CR, a fact in good agreement with the data in the literature [31,32].

Debonding the orthodontic brackets from the enamel of premolars causes damage to their surfaces. The force used to mechanically debond the metallic bracket must be great enough to exceed the bond strength of the adhesive compound. Consequently, the bonding interface is sheared from the enamel, leading to the violent extraction of structural components from the enamel surface, as well as areas where the adhesive remains attached. These aspects must be investigated by a detailed microstructural investigation.

Tooth enamel is a complex and hierarchic structure based on the hydroxyapatite (HAP) nanoparticles stuck together with a protein binder. HAP nanoparticles are organized in nanostructural units with rounded shapes and diameters of 40 nm which forms the enamel nanostructure. The nanostructural units are further bonded together into the HAP prisms having a diameter of about 5 μm which represents the fine microstructure of the enamel.

SEM evidence agrees that the results of SBS testing and proves that the enamel-adhesive interface is the most affected by the shearing failure. CR adhesive presents evidence of internal fracturing under shearing forces. Both aspects are influenced by acid erosion. RMGIC prove to be more resistant than the others samples. The best situation observed by SEM and SBS is resulted from control samples stored in artificial saliva. The best results after acid environment exposure is obtained for RMGC exposed to Red Bull and the worst results were obtained for CR samples exposed to Coca-Cola. These aspects require a more detailed microscopic and topographic investigation with the AFM.

It is expected that cola is more erosive against RMGIC bonding layer than Red Bull because of the pH difference between them. However, data in the literature shows that the polymer coating over mineral particles assures good insulation against acid erosion [33]. SEM images prove that our RMGIC sample has very good insulation of the mineral filler particle. The acid liquid has to penetrate the insulation within the bonding layer to cause the weakening of the shear bond strength. Therefore, besides the pH value, the liquid viscosity is very important because it directly influences the liquid absorption within RMGIC bulk. Red Bull$^{TM}$ is more viscous than the artificial saliva and Coca-Cola$^{TM}$ due to a large number of dissolved vitamins [34,35], respectively it has a lower penetration potential and in consequence, it is an explanation why it assures the best SBS value of the RMGIC sample. The fact requires more investigation on the sample liquid absorption and solubility which is the subject of a further article.

However, AFM studies in the literature have shown that citric acid is more erosive than phosphoric acid [1,11]. Data in the literature also shows that Red Bull$^{TM}$ is also hazardous for aquatic life forms [36]. This is how it can be explained why SBS was higher for the brackets bonded with RMGIC and immersed in Red Bull$^{TM}$, which contains citric acid, and determine adverse effects in the structure of the teeth but the adhesion interface and enamel area behind it is protected by the good insulation of mineral filler [37].Coca-Cola contains phosphoric acid that can produce enamel surface modification, which appeared as irregularities in enamel morphology and will determine the decrease in SBS with a negative effect on bracket retention and on the success of orthodontic treatment. The results of SBS determinations show significant differences and demonstrate that the two acids had produced a very deep demineralization. Considering our AFM topographic and morphologic observation, we can conclude that the erosive effect of Cola exposure is in accordance with data in the literature [27] and is caused by phosphoric acid etching. The Red Bull$^{TM}$ energy drink contains a multitude of potentially erosive ingredients, which translates into a greater aggressive upon exposure to sample surfaces when compared to Cola-exposure. The more advanced enamel demineralization, in this case, is in accordance with data in the literature [18].

Bond failure may occur within the bracket, at the bracket-adhesive interface, within the adhesive, or in the tooth-adhesive interface [27]. The presence of micro-leakage at the enamel-adhesive interface determines the appearance of white spot lesions, while microleakage at the adhesive-bracket interface is responsible for bracket failure [1]. Bracket failure should occur at the enamel-adhesive interface, which will make adhesive removal and polishing much easier. Bracket debonding and mechanical adhesive removal can determine iatrogenic effects including rough surface and enamel cracking or fracture. The low residual adhesive on the surface of the enamel reduces the time for polishing, is less harmful to the structural integrity of the tooth. AFM images show that the debonded samples stored in the artificial saliva preserve the best enamel fine microstructure and nanostructure. Exposure to Coca-Cola and Red Bull induces acid erosion by enlarging the unevenness that may occur on the enamel surface forming erosion depressions and alteration that generated advanced decaying with possibly weight loss. The most erosive agent for CR was Coca-Cola$^{TM}$ and for RMGIC was Red Bull$^{TM}$ as observed in the roughness variations presented in Figures 4 and 6.

The enamel was polished after the brackets debonded, aiming for the removal of the remained adhesive and to obtain a smoother and less eroded surface. Data in the literature shows that a rough surface will determine the retention of bio-film and decrease light reflection [38]. RMGIC removal can be accomplished easily, as compared to composite resin [18]. AFM results show that the lowest roughness in each group was obtained after the polishing treatment.

The polished surface exposure to the acid environment is affected by this fact is sustained by the lower values of the roughness of these samples compared to the unpolished ones exposed to the acid environment. Even after the polishing, the AFM results shows that Coca-Cola$^{TM}$ is more erosive for CR samples and Red Bull is most erosive for RMGIC, fact in good agreement with the data in the literature [11,14,28]. Overall, we can conclude that polishing the enamel causes a delay in acid etching compared to the unpolished samples. The correlation of all obtained results rejects both aspects of the null hypothesis.

## 4. Materials and Methods

### 4.1. Samples Preparation

Sixty premolars extracted for orthodontic reasons were used for this study. They had no caries, no surface cracks, or chemical treatment before the extraction. The teeth were cleaned with a low-speed rotary instrument and a prophylactic brush and a fluoride-free paste (Cleanic, Kerr, Kloten, Switzerland).

Roth brackets of 0.022-inch stainless steel were used. They were bonded in the third middle of the buccal surface of each tooth with the long axis parallel to the axis of the tooth. The specimens were randomly divided into two groups. Brackets in each of these groups were bonded according to the standard procedures required by the adhesives manufacturer detailed below:

Group 1: Normal metallic brackets were bonded with Transbond Plus Color Change (3M Unitek, St. Paul, MN, USA) on the vestibular surface of thirty teeth:

(1) 37% phosphoric acid was applied for 30 s, the acid was rinsed off and the enamel was dried.
(2) The primer with the adhesive was applied with a sponge for 20 s, dry easily and light-cured for 20 s.
(3) The cement resin was applied on the base of the bracket and the bracket was placed and pressed on the vestibular surface of the teeth. Removing the excess around the bracket could be difficult due to the fact that the color of tooth enamel and the cement are alike. That is why a colored orthodontic cements system has been created, which changes the color during polymerization as observed by Turkkahraman et al., 2010 [13]. After the excess was removed, it was light cured for 20 s, with 5 s light incidence over each marginal side of it.

Group 2: The brackets were bonded with Fuji Ortho LC (GC Company, Tokyo, Japan) capsules on the surface of thirty teeth. Bracket bonding takes about 1–2 min (depending on the environment temperature) after the RMGIC was mixed, or the adhesive will be rough [14].

(1) Thirty teeth were etched using a 10% acid polyacrilic enamel conditioner. The conditioner was applied for 20 s and the teeth were rinsed and dried. We did not use phosphoric acid with RMGIC, because this type of acid will demineralize too profoundly.
(2) The capsules containing the RMGIC (Fuji Ortho LC) were activated and triturated at 4000 rpm for 8 s. Capsules were placed in the GC Capsule Applier to place the adhesive on the base of the bracket
(3) RMGIC was applied and the bracket was placed on buccal enamel and pressed firmly into place. The excess of adhesive material may increase the retention of dental plaque and will determine the increase in the incidence of gingivitis and caries according to Naranjo et al., 2006 [15]. The excess was removed with a sharp scaler and the bracket was light-cured for 20 s.

The acid exposure protocol was identical for all tested samples as follows. All samples were kept in artificial saliva at 37 Celsius for 24 h to allow the complete polymerization. We used their own prepared artificial saliva produced at the Department of Polymer Composites, Institute of Chemistry "Raluca Ripan", Babes-Bolyai University. It contains $Na_2HPO_4$, $NaHCO_3$, $CaCl_2$, and HCl in an aqueous solution with pH = 7. Each group was divided into three other groups, with ten teeth in every group.

Group A: was kept in artificial saliva, pH = 7.

Group B: was immersed in Coca-Cola$^{TM}$ (pH = 2.5) for 15 min, three times a day, with equal intervals. This procedure was repeated for 20 days.

Group C: the teeth were immersed in Red Bull™ (pH = 3.3), three times a day for 15 min over 20 days. Teeth were kept in artificial saliva for the rest of the time. Artificial saliva was refreshed daily. After 20 days, all teeth were mounted vertically in acrylic blocks up to the clinical crown level, so the force could be applied to the bracket-tooth interface parallel to the buccal tooth surface in an occlusion apical direction. Duracryl Plus, from Spofa Dental (Kerr, Kloten, Switzerland), was used.

*4.2. Shear Bond Strength Test*

The shear bond strength (SBS) test was effectuated with the ASTM D638 test method according to the data in the literature [16,17] using the Lloyd LR5k Plus dual-column mechanical testing machine (Ametek/Lloyd Instruments, Germany), force capacity 5 kN. The universal materials testing machine features an electronic system to test and measure compression. The used load used was 0.5 N with a crosshead speed of 1 mm/min in the occluso-gingival direction. The data collected were processed using NEXYGEN Plus 3.0 software.

The value of SBS for each group is the mean of 10 mechanical tests (n = 10). Further ANOVA and Tukey testing was run on the datasets for post hoc comparison between the groups, and the significance level was set at $\alpha = 0.05$, using the Origin2019b Graphing and Analysis software, Microcal Co., Northampton, MA, USA.

*4.3. Scanning Electron Microscopy SEM*

Tooth enamel was studied using scanning electron microscopy (SEM Inspect™ S, FEI, Hillsboro, OR, USA) after the shear bond test, bracket debonding, and polishing, the surface morphology of the. This electron microscope generates the image of a sample surface scanning with a focused electron beam. The electrons interact with the enamel atoms and thus produce various signals that provide information regarding the surface topography and chemical composition of the tooth. The specimens were removed from the artificial saliva they were stored in, patted dry with filter paper and analyzed in a low vacuum, 4.5 spot and 25.00 kV, 100× magnifications.

*4.4. Atomic Force Microscopy AFM*

This study analyzes the tooth enamel surface at the precise place where orthodontic brackets had been previously attached. The samples are divided into two groups: "CR-Enamel" and "RMGIC-Enamel". The specimens were sectioned parallel to the enamel surface to ensure optimal positioning for the AFM investigation. We prepared several sets of specimens to accommodate the required testing. As such, the enamel samples include debonded surface sections, prior to polishing, and uniform surface sections, after polishing.

A specimen of both before and after polishing samples were kept as control samples, while the rest were divided into two groups that were exposed to an acid etching medium: one group was exposed to a Coca-Cola™ beverage (acidifier-phosphoric acid) and another group was exposed to the Red Bull™ energy drink (acidifier-citric acid) and a vitamin mixture. The acid exposure for each group follows the protocol described in Section 4.1.

All samples were then cleansed in bi-distilled water and stored in the artificial saliva in individual containers at room temperature. Each sample was extracted from their container, cleansed with bi-distilled water, dried with filter paper, and mounted onto the required AFM support. After the AFM investigation, each sample was reinserted into itsartificial saliva container.

We aim to observe the evolution of the morphology and topography of the sample surfaces in response to the applied treatments. During the first stage, we evaluate if there are any improvements in the surface quality of the enamel samples after polishing. During the second stage, we evaluate the effect of acid etching on the samples, before and after any polishing is applied.

The null hypothesis of this study has two conjectures: the first is that the effect of acid etching does not depend on the polishing treatment, and the second is that there is no difference between the degree of degradation caused by Coca-Cola™ and Red Bull™.

The atomic force microscopy (AFM) analysis was effectuated using a JEOL microscope (JEOL JSPM 4210, JEOL, Tokyo, Japan). All samples were evaluated in tapping mode using NSC 15 cantilevers produced by MikroMasch, Bulgaria Headquarters, Sofia. The cantilever resonant frequency was 330 kHz, and the spring constant was 48 N/m. In accordance with data in the literature [20,39–42], the topographic images scanned and recorded were as follows: a 5 μm × 5 μm area for the fine microstructure of enamel and a 1 μm × 1 μm area for the enamel nanostructure. Three separate macroscopic areas were scanned for each sample. The images were processed using the WinSPM 2.0 JEOL software (WinSPM 2.0 JEOL, Tokyo, Japan) in accordance with standard procedures, presenting 2D topographic images, 3D images and the Ra and Rq surface roughness parameters were measured. Ra represents the arithmetic average of the profile height and is described by Equation (1) and Rq root mean square of the profile height and is described by Equation (2):

$$Ra = \frac{1}{l_r} \int_0^{l_r} |z(x)| dx, \qquad (1)$$

and,

$$Rq = \sqrt{\frac{1}{l_r} \int_0^{l_r} |z(x)^2| dx}. \qquad (2)$$

where: $l$ is the profile length and $z$ is the height at $x$ point. Both $Ra$ and $Rq$ are important for various research applications.

The 3D images are also called 3D profiles; they are a graphic representation of the depth profile of the enamel surface, closely related to the measured values of surface roughness.

For each test group, the Ra and Rq values represent the mean for three measurement areas (n = 3). The data were then run through the ANOVA One-Way test to compare the effect of acid etching both before and after polishing, and the significance level was set at α = 0.05, using the Origin 2019b Graphing and Analysis software, Microcal Co., Northampton, MA, USA.

## 5. Conclusions

The acid environment influences the shear bond strength of the adhesion layer of CR and RMGIC samples because of the pH values associated with the micro-structural aspects (e.g., pores and fissures) allowing the erosive liquid penetration into the bonding material. The proper polymeric insulation of the mineral filler particle assures the proper resistance to the erosive agent penetration on the inside of the bonding layer and preserves the bracket retention. RMGIC samples prove to be more resistant in certain conditions than CR. The enamel surface after CR debonding is more affected by Coca-Cola™ due to the phosphoric acid content and the enamel after RMGIC debonding was more affected by Red Bull™ due to the citric acid content. Polishing treatment of the debonded enamel areas further assures a good resistance against the acid erosive effect of both Coca-Cola™ and Red Bull™.

**Author Contributions:** Conceptualization, M.M. and M.E.B.; methodology, C.I.; software, D.P.; validation, R.C.; formal analysis, S.S.; investigation, I.P. and S.C.; resources, A.T.; data curation, R.C.; writing—original draft preparation, C.I.; writing—review and editing, I.P. and M.M.; visualization, S.C. and I.P.; supervision, M.E.B. All authors have read and agreed to the published version of the manuscript.

**Funding:** This research received no external funding.

**Institutional Review Board Statement:** The study was conducted according to the guidelines of the Declaration of Helsinki, and approved by the Ethics Committee of "Iuliu Hatieganu" University of Medicine and Pharmacy from Cluj Napoca No. DEP213/27.06.2022.

**Informed Consent Statement:** Not applicable because the humans were not directly involved in the present research.

**Data Availability Statement:** The data presented in this study are available on request from the corresponding author.

**Acknowledgments:** The authors acknowledge the Research Centre in Physical Chemistry "CECHIF" of Babes Bolyai University for AFM assistance.

**Conflicts of Interest:** The authors declare no conflict of interest.

## References

1. Vorachart, W.; Sombuntham, N.; Parakonthun, K. The effect of beer and milk tea on the shear bond strength of adhesive precoated brackets: An in vitro comparative study. *Heliyon* **2020**, *8*, 10260. [CrossRef] [PubMed]
2. Mocuta, D.-E.; Grad, O.; Mateas, M.; Luca, R.; Carmen Todea, D. Comparative Evaluation of Influence of Nd:YAG Laser (1064 nm) and 980 nm Diode Laser on Enamel around Orthodontic Brackets: An In Vitro Study. *Medicina* **2022**, *58*, 633.
3. Bucur, S.M.; Bud, A.; Gligor, A.; Vlasa, A.; Cocoș, D.I.; Bud, E.S. Observational Study Regarding Two Bonding Systems and the Challenges of Their Use in Orthodontics: An In Vitro Evaluation. *Appl. Sci.* **2021**, *11*, 7091. [CrossRef]
4. Chee, S.; Mangum, J.; Teeramongkolgul, T.; Tan, S.; Schneider, P. Clinician preferences for orthodontic bracket bonding materials: A quantitative analysis. *Australas. Orthod. J.* **2022**, *38*, 173–182. [CrossRef]
5. Santos, C.N.; Souza Matos, F.; Mello Rode, S.; Cesar, P.F.; Nahsan, F.P.S.; Paranhos, L.R. Effect of two erosive protocols using acidic beverages on the shear bond strength of orthodontic brackets to bovine enamel. *Dent. Press J. Orthod.* **2018**, *23*, 1–5. [CrossRef]
6. Simunovic Anicic, M.; Goracci, C.; Juloski, J.; Miletic, I.; Mestrovic, S. The Influence of Resin Infiltration Pretreatment on Orthodontic Bonding to Demineralized Human Enamel. *Appl. Sci.* **2020**, *10*, 3619. [CrossRef]
7. Nica, I.; Stoleriu, S.; Iovan, A.; Tărăboanță, I.; Pancu, G.; Tofan, N.; Brânzan, R.; Andrian, S. Conventional and Resin-Modified Glass Ionomer Cement Surface Characteristics after Acidic Challenges. *Biomedicines* **2022**, *10*, 1755. [CrossRef]
8. Padmaja, S.; Ashma, V.; Ankit, A.; Sachin, A. A comparative evaluation of metallic brackets bonded with resin-modified glass ionomer cement under different enamel preparations: A pilot study. *Contemp. Clin. Dent.* **2013**, *4*, 1–7.
9. Ghubaryi, A.A.; Ingle, N.; Basser, M.A. Surface treatment of RMGIC to composite resin using different photosensitizers and lasers: A bond assessment of closed Sandwich restoration. *Photodiagnosis Photodyn. Ther.* **2020**, *32*, 101965. [CrossRef]
10. Hammad, S.M.; Enan, E.T. In vivo effects of two acidic soft drinks on shear bond strength of metal orthodontic brackets with and without resin infiltration treatment. *Angle Orthod.* **2013**, *83*, 648–652. [CrossRef]
11. Nam, H.-J.; Kim, Y.-M.; Kwon, Y.H.; Yoo, K.-H.; Yoon, S.-Y.; Kim, I.-R.; Park, B.-S.; Son, W.-S.; Lee, S.-M.; Kim, Y.-I. Fluorinated Bioactive Glass Nanoparticles: Enamel Demineralization Prevention and Antibacterial Effect of Orthodontic Bonding Resin. *Materials* **2019**, *12*, 1813. [CrossRef] [PubMed]
12. Kessler, P.; Turp, J.P. Influence of Coca-Cola on orthodontic material. *Swiss Dent. J. SSO* **2020**, *130*, 777–780.
13. Türkkahraman, H.; Adanir, N.; Gungor, A.Y.; Alkis, H. In vitro evaluation of shear bond strengths of colour change adhesives. *Eur. J. Orthod.* **2010**, *32*, 571–574. [CrossRef] [PubMed]
14. Elnafar, A.A.S.; Alam, M.K.; Hasan, R. The impact of surface preparation on shear bond strength of metallic orthodontic brackets bonded with a resin-modified glass ionomer cement. *J. Orthod.* **2014**, *41*, 201–207. [CrossRef]
15. Condò, R.; Mampieri, G.; Pasquantonio, G.; Giancotti, A.; Pirelli, P.; Cataldi, M.E.; La Rocca, S.; Leggeri, A.; Notargiacomo, A.; Maiolo, L.; et al. In Vitro Evaluation of Structural Factors Favouring Bacterial Adhesion on Orthodontic Adhesive Resins. *Materials* **2021**, *14*, 2485. [CrossRef] [PubMed]
16. Eslamian, L.; Borzabadi-Farahani, A.; Karimi, S.; Saadat, S.; Badiee, M.R. Evaluation of the Shear Bond Strength and Antibacterial Activity of Orthodontic Adhesive Containing Silver Nanoparticle, an In-Vitro Study. *Nanomaterials* **2020**, *10*, 1466. [CrossRef] [PubMed]
17. Ju, G.-Y.; Oh, S.; Lim, B.-S.; Lee, H.-S.; Chung, S.H. Effect of Simplified Bonding on Shear Bond Strength between Ceramic Brackets and Dental Zirconia. *Materials* **2019**, *12*, 1640. [CrossRef]
18. Panpan, L.; Chungik, O.; Hongjun, K.; Chen-Glasser, M.; Park, G.; Jetybayeva, A.; Yeom, J.; Kim, H.; Ryu, J.; Hong, S. Nanoscaleeffects of beverages on enamel surface of human teeth: An atomic force microscopy study. *J. Mech. Behav. Biomed. Mater.* **2020**, *110*, 103930.
19. Torres-Gallegos, I.; Zavala-Alonso, V.; Patino-Marin, N.; Martinez-Castanon, G.A.; Anusavice, K.; Loyola-Rodriguez, J.P. Enamel roughness and depth profile after phosphoric acid etching of healthy and fluorotic enamel. *Aust. Dent. J.* **2012**, *57*, 1–6. [CrossRef]
20. Lei, L.; Zheng, L.; Xiao, H.; Zheng, J.; Zhou, Z. Wear mechanism of human tooth enamel: The role of interfacial protein bonding between HA crystals. *J. Mech. Behav. Biomed. Mater.* **2020**, *110*, 103845. [CrossRef]
21. Agrawal, N.; Shashikiran, N.D.; Singla, S.; Ravi, K.S.; Kulkarni, V.K. Atomic force microscopic comparison of remineralization with casein-phosphopeptide amorphous calcium phosphate paste, acidulated phosphate fluoride gel and iron supplement in primary and permanent teeth: An in-vitro study. *Contemp. Clin. Dent.* **2014**, *5*, 75–80. [CrossRef]
22. Urichianu, M.; Makowka, S.; Covell, D., Jr.; Warunek, S.; Al-Jewair, T. Shear Bond Strength and Bracket Base Morphology of New and Rebonded Orthodontic Ceramic Brackets. *Materials* **2022**, *15*, 1865. [CrossRef] [PubMed]

23. Sfondrini, M.F.; Pascadopoli, M.; Gallo, S.; Ricaldone, F.; Kramp, D.D.; Valla, M.; Gandini, P.; Scribante, A. Effect of Enamel Pretreatment with Pastes Presenting Different Relative Dentin Abrasivity (RDA) Values on Orthodontic Bracket Bonding Efficacy of Microfilled Composite Resin: In Vitro Investigation and Randomized Clinical Trial. *Materials* **2022**, *15*, 531. [CrossRef] [PubMed]
24. Vartolomei, A.-C.; Serbanoiu, D.-C.; Ghiga, D.-V.; Moldovan, M.; Cuc, S.; Pollmann, M.C.F.; Pacurar, M. Comparative Evaluation of Two Bracket Systems' Kinetic Friction: Conventional and Self-Ligating. *Materials* **2022**, *15*, 4304. [CrossRef]
25. Ga, Y.; Okamoto, Y.; Matsuya, S. The effects of treated time of acidulated phosphate fluoridesolutions on enamel erosion. *Pediatr. Dent. J.* **2012**, *22*, 1–7. [CrossRef]
26. Cerci, B.B.; Roman, L.S.; Guariza-Filho, O.; Camargo, E.S.; Tanaka, O.M. Dental enamel roughness with different acid etching times: Atomic force microscopy study. *Eur. J. Gen. Dent.* **2012**, *1*, 187–191. [CrossRef]
27. Pulgaonkar, R.; Chitra, P. Stereomicroscopic analysis of microleakage, evaluation of shear bond strengths and adhesive remnants beneath orthodontic brackets under cyclic exposure to commonly consumed commercial "soft" drinks. *Indian J. Dent. Res.* **2021**, *32*, 98–103.
28. Zheng, B.-W.; Cao, S.; Ali-Somairi, M.A.; He, J.; Liu, Y. Effect of enamel-surface modifications on shear bond strength using different adhesive materials. *BMC Oral Health* **2022**, *22*, 224. [CrossRef]
29. Ghoubril, V.; Ghoubril, J.; Khoury, E. A comparison between RMGIC and composite with acid-etch preparation or hypochlorite on the adhesion of a premolar metal bracket by testing SBS and ARI: In vitro study. *Int. Orthod.* **2020**, *18*, 127–136. [CrossRef]
30. Farshidfar, N.; Agharokh, M.; Ferooz, M.; Bagheri, R. Microtensile bond strength of glass ionomer cements to a resin composite using universal bonding agents with and without acid etching. *Heliyon* **2022**, *8*, 08858. [CrossRef]
31. Wiegand, A.; Lechte, C.; Kanzow, P. Adhesion to eroded enamel and dentin: Systematic review and meta-analysis. *Dent. Mater.* **2021**, *37*, 1845–1853. [CrossRef]
32. Yang, H.; Hong, D.-W.; Attin, T.; Cheng, T.; Yu, H. Erosion of CAD/CAM restorative materials and human enamel: An in situ/in vivo study. *J. Mech. Behav. Biomed. Mater.* **2020**, *110*, 103903. [CrossRef]
33. Sabău, E.; Bere, P.; Moldovan, M.; Petean, I.; Miron-Borzan, C. Evaluation of Novel Ornamental Cladding Resistance, Comprised of GFRP Waste and Polyester Binder, within an Acid Environment. *Polymers* **2021**, *13*, 448. [CrossRef] [PubMed]
34. Zielińska, A.; Mazurek, A.; Siudem, P.; Kowalska, V.; Paradowska, K. Qualitative and quantitative analysis of energy drinks using 1H NMR and HPLC methods. *J. Pharm. Biomed. Anal.* **2022**, *213*, 114682. [CrossRef] [PubMed]
35. Frias, F.J.L. The "big red bull" in the esports room: Anti-doping, esports, and energy drinks. *Perform. Enhanc. Health* **2022**, *10*, 100205. [CrossRef]
36. Mokkarala, P.; Shekarabi, A.; Wiah, S.; Rawls, M. Energy drink produces aversive effects in planarians. *Physiol. Behav.* **2022**, *225*, 113933. [CrossRef] [PubMed]
37. Saini, S.; Meena, A. A comparative study of the effect of fillers and monomer on dental restorative material. *Mater. Today Proc.* **2021**, *44*, 5023–5027. [CrossRef]
38. Caixeta, R.V.; Berger, S.B.; Lopes, M.B.; Paloco, E.A.C.; Faria-Junior, E.M.; Contreras, E.F.R.; Gonini-Junior, A.; Guiraldo, R.D. Evaluation of enamel roughness after the removal of brackets bonded with different materials: In vivo study. *Braz. Dent. J.* **2021**, *32*, 30–40. [CrossRef]
39. Tisler, C.E.; Moldovan, M.; Petean, I.; Buduru, S.D.; Prodan, D.; Sarosi, C.; Leucuța, D.-C.; Chifor, R.; Badea, M.E.; Ene, R. Human Enamel Fluorination Enhancement by Photodynamic Laser Treatment. *Polymers* **2022**, *14*, 2969. [CrossRef]
40. Voina, C.; Delean, A.; Muresan, A.; Valeanu, M.; Mazilu Moldovan, A.; Popescu, V.; Petean, I.; Ene, R.; Moldovan, M.; Pandrea, S. Antimicrobial activity and the effect of green tea experimental gels on teeth surfaces. *Coatings* **2020**, *10*, 537. [CrossRef]
41. Chisnoiu, A.M.; Moldovan, M.; Sarosi, C.; Chisnoiu, R.M.; Rotaru, D.I.; Delean, A.G.; Pastrav, O.; Muntean, A.; Petean, I.; Tudoran, L.B.; et al. Marginal adaptation assessment for two composite layering techniques using dye penetration, AFM, SEM and FTIR: An In-Vitro comparative study. *Appl. Sci.* **2021**, *11*, 5657.
42. Chisnoiu, R.M.; Moldovan, M.; Prodan, D.; Chisnoiu, A.M.; Hrab, D.; Delean, A.G.; Muntean, A.; Rotaru, D.I.; Pastrav, O.; Pastrav, M. In-Vitro comparative adhesion evaluation of bioceramic and dual-cure resin endodontic sealers using SEM, AFM, Push-Out and FTIR. *Appl. Sci.* **2021**, *11*, 4454. [CrossRef]

MDPI
St. Alban-Anlage 66
4052 Basel
Switzerland
Tel. +41 61 683 77 34
Fax +41 61 302 89 18
www.mdpi.com

*International Journal of Molecular Sciences* Editorial Office
E-mail: ijms@mdpi.com
www.mdpi.com/journal/ijms

www.ingramcontent.com/pod-product-compliance
Lightning Source LLC
LaVergne TN
LVHW070239100526
838202LV00015B/2155